Antigen Presenting Cells

Edited by
Harald Kropshofer and
Anne B. Vogt

Related Titles

R. A. Meyers (Ed.)

Encyclopedia of Molecular Cell Biology and Molecular Medicine, 2nd Edition

2005
ISBN 3-527-30542-4

M. Lutz, N. Romani, A. Steinkasserer

Handbook of Dendritic Cells – Biology, Diseases, and Therapy

2006
ISBN 3-527-31109-2

J. R. Kalden, M. Herrmann

Apoptosis and Autoimmunity – From Mechanisms to Treatments

2003
ISBN 3-527-30442-8

S. H. E. Kaufmann

Novel Vaccination Strategies

2004
ISBN 3-527-30523-8

A. Hamann, B. Engelhardt

Leukocyte Trafficking – Mechanisms, Therapeutic Targets, and Methods

2005
ISBN 3-527-31228-5

Antigen Presenting Cells

From Mechanisms to Drug Development

Edited by
Harald Kropshofer and Anne B. Vogt

WILEY-VCH Verlag GmbH & Co. KGaA

Editors

PD Dr. Harald Kropshofer
Roche Centre for Medical Genomics
Building 86/805
F. Hoffmann-La Roche AG
Grenzacherstr. 124
4070 Basel
Switzerland

Dr. Anne B. Vogt
Roche Centre for Medical Genomics
Building 86/804
F. Hoffmann-La Roche AG
Grenzacherstr. 124
4070 Basel
Switzerland

■ All books published by Wiley-VCH are carefully produced. Nevertheless, authors, editors, and publisher do not warrant the information contained in these books, including this book, to be free of errors. Readers are advised to keep in mind that statements, data, illustrations, procedural details or other items may inadvertently be inaccurate.

Library of Congress Card No.: applied for

British Library Cataloguing-in-Publication Data
A catalogue record for this book is available from the British Library.

Bibliographic information published by Die Deutsche Bibliothek
Die Deutsche Bibliothek lists this publication in the Deutsche Nationalbibliografie; detailed bibliographic data is available in the Internet at <http://dnb.ddb.de>.

© 2005 WILEY-VCH Verlag GmbH & Co. KGaA, Weinheim

All rights reserved (including those of translation into other languages).
No part of this book may be reproduced in any form – nor transmitted or translated into machine language without written permission from the publishers. Registered names, trademarks, etc. used in this book, even when not specifically marked as such, are not to be considered unprotected by law.

Printed in the Federal Republic of Germany.
Printed on acid-free paper.

Typesetting Kühn & Weyh, Satz und Medien, Freiburg
Printing betz-druck GmbH, Darmstadt
Bookbinding J. Schäffer GmbH, Grünstadt

ISBN-13: 978-3-527-31108-8
ISBN-10: 3-527-31108-4

Contents

Preface *XV*

List of Contributors *XIX*

List of Abbreviations *XXIII*

Color Plates *XXVII*

Part I	**Antigen Presentation in the Immune System** *1*	
1	**Some Old and Some New Findings on Antigen Processing and Presentation** *3* *Emil R. Unanue*	
1.1	Introduction *3*	
1.2	HEL Processing *4*	
1.3	Selection of Peptide Segments of HEL *9*	
1.4	HEL: Conformational Isomers *11*	
1.4.1	Biology of Type B T Cells *15*	
1.5	Negative Selection and Peripheral Activation to HEL Peptides *16*	
1.6	Response to HEL Immunization in the Draining Lymph Node *17*	
Part II	**Molecular Mechanisms of Antigen Processing** *25*	
2	**Antigen Entry Routes – Where Foreign Invaders Meet Antigen Presenting Cells** *27* *Percy A. Knolle*	
2.1	Introduction *27*	
2.2	Antigen Entry via the Gastrointestinal Tract *28*	
2.2.1	Peyer's Patches *29*	
2.2.2	Mesenteric Lymph Node *30*	
2.2.3	Dendritic Cells of the Lamina Propria *31*	
2.2.4	Pathogens Target Intestinal Antigen Presenting Cells *33*

Antigen Presenting Cells: From Mechanisms to Drug Development. Edited by H. Kropshofer and A. B. Vogt
Copyright © 2005 WILEY-VCH Verlag GmbH & Co. KGaA, Weinheim
ISBN: 3-527-31108-4

2.3	Antigen Entry via the Skin	35
2.4	Systemic Dissemination of Antigens/Infectious Microorganisms	38
2.5	Antigen Presenting Cells in the Liver	39
2.5.1	Dendritic Cells	39
2.5.2	Kupffer Cells	41
2.5.3	Liver Sinusoidal Endothelial Cells	42
2.6	Conclusion	44

3 Antigen Processing in the Context of MHC Class I Molecules   51
Frank Momburg   51

3.1	Tracing the Needle in the Haystack: The Efficiency of Antigen Processing and Presentation by MHC Class I Molecules	51
3.2	The "Classical" Route: Loading of MHC Class I Molecules With Peptides Generated in the Cytoplasm	53
3.2.1	Cytosolic Peptide Processing by Proteasomes and other Proteases	53
3.2.1.1	Structure and Function of the Proteasomal Core and Interferon-induced Subunits	56
3.2.1.2	Targeting Proteins for ATP-dependent Degradation by 26S Proteasomes	56
3.2.1.3	Cleavage Properties of (Immuno)Proteasomes	57
3.2.1.4	Peptide Processing by Nonproteasomal Cytosolic Peptidases	59
3.3	Crossing the Border – Peptide Translocation into the ER by TAP	60
3.3.1	Structure and Function of TAP	60
3.3.2	Substrate Specificity of TAP	62
3.3.3	TAP-independent Peptide Entry into the ER	63
3.4	Fitting in the Best: TAP-associated Peptide Loading Complex Optimizes MHC-I Peptide Binding	63
3.4.1	Structure of MHC-I Molecules	64
3.4.2	Early Steps in the Maturation of MHC-I Molecules	64
3.4.3	Structure and Molecular Interactions of Tapasin	66
3.4.4	Optimization of Peptide Loading in the TAP-associated Loading Complex	67
3.5	On the Way Out: MHC-I Antigen Processing along the Secretory Route	70
3.6	Closing the Circle – Cross-presentation of Endocytosed Antigens by MHC-I Molecules	73
3.6.1	Phagosome-to-cytosol Pathway of MHC-I Peptide Loading	73
3.6.2	Endolysosomal Pathway of MHC-I Peptide Loading	76

4 Antigen Processing for MHC Class II   89
Anne B. Vogt, Corinne Ploix and Harald Kropshofer

4.1	Introduction	89
4.2	Types of Antigen Presenting Cells	90

4.2.1	Macrophages, B Lymphocytes and DCs	90
4.2.2	Tissue-resident APCs	91
4.2.3	Maturation State of APCs	92
4.2.3.1	Immature APCs	92
4.2.3.2	Mature APCs	92
4.3	Antigen Uptake by APCs	93
4.3.1	Macropinocytosis	93
4.3.2	Phagocytosis	94
4.3.3	Receptors for Endocytosis	95
4.4	Generation of Antigenic Peptides	97
4.4.1	Reduction of Disulfide Bonds: GILT	97
4.4.2	Regulation of the Proteolytic Milieu	98
4.4.3	Protease/MHC Interplay in Antigen Processing	99
4.5	Assembly of MHC II Molecules	102
4.5.1	Structural Requirements of MHC II	102
4.5.2	Biosynthesis of MHC II	103
4.5.3	Chaperones for Peptide Loading	104
4.5.3.1	HLA-DM/H2-DM	104
4.5.3.2	HLA-DO/H2-DO	107
4.6	Export of MHC II and Organization on the Cell Surface	109
4.6.1	Membrane Microdomains	109
4.6.2	Tubular Transport	112
4.7	Viral and Bacterial Interference	114
4.8	Concluding Remarks	116
5	**Antigen Processing and Presentation by CD1 Family Proteins**	**129**
	Steven A. Porcelli and D. Branch Moody	
5.1	Introduction	129
5.2	CD1 Genes and Classification of CD1 Proteins	129
5.3	Structure and Biosynthesis of CD1 Proteins	130
5.3.1	Three-dimensional (3D) Structures of CD1 Proteins	132
5.3.2	Molecular Features of CD1–Lipid Complexes	133
5.3.3	CD1 Pockets and Portals	135
5.4	Foreign Lipid Antigens Presented by Group 1 CD1	136
5.5	Self Lipid Antigens Presented by CD1	137
5.6	Group 2 CD1-restricted T Cells	138
5.6.1	Antigens Recognized by Group 2 CD1-restricted T Cells	139
5.7	Tissue Distribution of CD1 Proteins	140
5.8	Subcellular Distribution and Intracellular Trafficking of CD1	140
5.8.1	Trafficking and Localization of CD1a	141
5.8.2	Trafficking and Localization of CD1b	141
5.8.3	Trafficking and Localization of CD1c	143
5.8.4	Trafficking and Localization of CD1d	144
5.8.5	Trafficking and Localization of CD1e	145

5.9	Antigen Uptake, Processing and Loading in the CD1 Pathway	146
5.9.1	Cellular Uptake of CD1-presented Antigens	146
5.9.2	Endosomal Processing of CD1-presented Antigens	147
5.9.3	Accessory Molecules for Endosomal Lipid Loading of CD1	148
5.9.4	Non-endosomal Loading of Lipids onto CD1 Molecules	149
5.10	Conclusions	150

Part III Antigen Presenting Cells' Ligands Recognized by T- and Toll-like Receptors 157

6 Naturally Processed Self-peptides of MHC Molecules 159
Harald Kropshofer and Sebastian Spindeldreher

6.1	Introduction	159
6.2	Milestone Events	160
6.2.1	Nomenclature	160
6.2.1.1	Autologous Peptides	160
6.2.1.2	Endogenous Peptides	161
6.2.1.3	Natural Peptides Ex Vivo and In Vitro	161
6.2.2	Extra Electron Density Associated to MHC Molecules	162
6.2.3	Acidic Peptide Elution Approach	163
6.2.4	First Natural Foreign Peptides on MHC Class II	165
6.2.5	First Natural Viral Epitopes on MHC Class I	165
6.2.6	Self-peptide Sequencing on MHC Class I: the First Anchor Motifs	166
6.2.7	First Murine MHC Class II-associated Self-peptides: Nested Sets	167
6.2.8	First Human MHC Class II-bound Self-peptides: Hydrophobic Motifs	169
6.3	Progress in Sequence Analysis of Natural Peptides	172
6.3.1	Edman Microsequencing	172
6.3.2	Electrospray Ionization Tandem Mass Spectrometry	173
6.3.3	Automated Tandem Mass Spectrometry	175
6.3.4	MAPPs: MHC-associated Peptide Proteomics	176
6.4	Natural Class II MHC-associated Peptides from Different Tissues and Cell-types	177
6.4.1	Peripheral Blood Mononuclear Cells	177
6.4.2	Myeloid Dendritic Cells	178
6.4.3	Medullary Thymic Epithelial Cells	179
6.4.4	Splenic APCs	181
6.4.5	Tumor Cells	181
6.4.6	Autoimmunity-related Epithelial Cells	182
6.5	The CLIP Story	183
6.5.1	CLIP in APCs Lacking HLA-DM	184
6.5.2	Flanking Residues and Self-release of CLIP	184
6.5.3	CLIP in Tetraspan Microdomains	185

6.5.4	CLIP as an Antagonist of T_H1 Cells *188*	
6.6	Outlook: Natural Peptides as Diagnostic or Therapeutic Tools *189*	

7	**Target Cell Contributions to Cytotoxic T Cell Sensitivity** *199*	
	Tatiana Lebdeva, Michael L. Dustin and Yuri Sykulev *199*	
7.1	Introduction *199*	
7.2	Intercellular Adhesion Molecule 1 (ICAM-1) *200*	
7.2.1	Adhesion Molecules on the Surface of APC and Target Cells *200*	
7.2.2	ICAM-1 Structure and Topology on the Cell Surface *200*	
7.2.3	ICAM-1 as Co-stimulatory Ligand and Receptor *201*	
7.2.4	ICAM-1-mediated Signaling *203*	
7.2.5	Role of ICAM-1 in Endothelial Response to Leukocytes *206*	
7.2.6	ICAM-1 Association with Lipid Rafts *206*	
7.3	Major Histocompatability Complex (MHC) *208*	
7.3.1	MHC Molecules *208*	
7.3.2	Molecular Associations of MHC-I Molecules *208*	
7.3.3	Association of MHC-I and ICAM-1 *211*	
7.3.4	Could APC and Target Cells Play an Active Role in Ag Presentation? *212*	
7.3.5	Identical pMHCs are Clustered in the Same Microdomain *212*	
7.3.6	Identical pMHC can be Recruited to the Same Microdomain During Target Cell–T Cell Interaction *213*	
7.3.7	Co-clustering of MHC and Accessory Molecules *213*	
7.3.8	Role of Cytoskeleton *214*	
7.4	Conclusion *215*	

8	**Stimulation of Antigen Presenting Cells: from Classical Adjuvants to Toll-like Receptor (TLR) Ligands** *221*	
	Martin F. Bachmann and Annette Oxenius	
8.1	Synopsis *221*	
8.2	Pathogen-associated Features that Drive Efficient Immune Responses *221*	
8.3	Composition and Function of Adjuvants *222*	
8.4	TLR Protein Family in Mammals *224*	
8.4.1	TLR4 *226*	
8.4.2	TLR2 *227*	
8.4.3	TLR5 *227*	
8.4.4	TLR11 *228*	
8.4.5	TLR12 and TLR13 *228*	
8.4.6	Nucleic Acids as PAMPs *228*	
8.4.6.1	TLR3 *228*	
8.4.6.2	TLR7 and TLR8 *229*	
8.4.6.3	TLR9 *229*	

8.4.7	Compartmentalization of Sensing Renders the Nucleic Acid PAMPs *229*	
8.5	TLR Signaling *230*	
8.5.1	Signal Transduction Across the Membrane *231*	
8.5.2	MyD88-dependent Pathways *231*	
8.5.3	MyD88-independent Pathways *232*	
8.6	TLR-independent Recognition of PAMPs: Nods, PKR and Dectin-1 *233*	
8.6.1	Nods *233*	
8.6.2	PKR (IFN-inducible dsRNA-dependent Protein Kinase) *234*	
8.6.3	Dectin-1 *234*	
8.7	Therapeutic Potential of TLRs and their Ligands *235*	
8.8	Conclusion *237*	

Part IV The Repertoire of Antigen Presenting Cells *245*

9 Evolution and Diversity of Macrophages *247*
Nicholas S. Stoy

9.1	Evolution of Macrophages: Immunity without Antigen Presentation *247*
9.1.1	Introduction *247*
9.1.2	*Drosophila*: a Window into Innate Immunity *247*
9.1.3	Evolution of Adaptive Immunity: Macrophages in a New Context *255*
9.2	Diversity of Macrophages in Mammalian Tissues *257*
9.2.1	Classifying Heterogeneity *257*
9.2.2	Phenotypic Manipulations and Transdifferentiations: Routes to and from Macrophages *258*
9.2.3	Function-related 'Markers' in Macrophages and DCs *262*
9.2.4	Macrophage Phenotypic Diversity in Response to Microbial Challenge *266*
9.2.5	Interactions between Tissue Microenvironments and Macrophages Generate Diversity *283*
9.2.6	Sequential and Regulatory Changes in Macrophage Phenotypes: Limiting Pro- and Antiinflammatory Responses *292*
9.2.6.1	Pre-TLR and TLR Regulation of Immune Responses *293*
9.2.6.2	Signal Transduction in the Regulation of Immune Responses *294*
9.2.6.3	Regulation of Immune Responses by Cytokines and other Bioactive Molecules *299*
9.2.6.4	Regulation of Immune Responses by Decoys *300*
9.2.6.5	Regulation of Immune Responses by the Adaptive Immune System *300*
9.2.6.6	Regulation of Immune Responses by Apoptosis *301*

9.2.6.7	Interaction of Regulatory Mechanisms during Immune Responses *301*	
9.2.7	Macrophage Diversity: an Overview *302*	

10 Macrophages – Balancing Tolerance and Immunity *331*
Nicholas S. Stoy 331

10.1	Balancing Tolerance and Immunity *331*
10.1.1	Introduction *331*
10.1.2	Macrophage Phenotypes: Effects on Immunity and Tolerance *332*
10.1.3	Concept of Innate (Peripheral) Tolerance *334*
10.1.4	Concept of Adaptive Tolerance *335*
10.1.5	Innate Tolerance: Receptors, Responses and Mechanisms *342*
10.1.6	Incorporating NK and NT Cells into the Innate Tolerance/Innate Immunity Paradigm *349*
10.1.7	Definitions and Terminology *354*
10.2	Ramifications of the Paradigm: Asthma *356*
10.3	Ramifications of the Paradigm: Autoimmunity *362*
10.4	Summary and Conclusions: Towards Immune System Modeling and Therapeutics *378*

11 Polymorphonuclear Neutrophils as Antigen-presenting Cells *415*
Amit R. Ashtekar and Bhaskar Saha

11.1	Introduction *415*
11.2	PMN as Antigen-presenting Cells *417*
11.2.1	Basic Criteria of an APC for T Cells *417*
11.2.2	Acquisition of Antigens *418*
11.2.3	Antigen Processing *420*
11.2.4	Expression of MHC Class I/II and Co-stimulatory Molecules *424*
11.2.5	Delivery of Second Signal *427*
11.2.6	Alteration in Cytokine Milieu *430*
11.3	Evolution of Newer Thoughts as PMN March to a Newer Horizon *434*

12 Microglia – The Professional Antigen-presenting Cells of the CNS? *441*
Monica J. Carson

12.1	Introduction: Microglia and CNS Immune Privilege *441*
12.1.1	What are Microglia? *441*
12.1.2	Is Immune Privilege Equivalent to Immune Isolation? *442*
12.2	Do Microglia Differ from Other Macrophage Populations? *444*
12.2.1	Microglia are Likely of Mesodermal Origin *444*
12.2.2	Parenchymal Microglia are not the only Myeloid Cells in the CNS *444*
12.2.3	In Contrast to other Macrophages, Parenchymal Microglia are not Readily Replaced by Bone Marrow Stem Cells *444*

12.2.4	Microglia Display Stable Differences in Gene Expression that Distinguish them from Other Macrophage Populations *446*
12.2.5	Morphology is not a Reliable Parameter to Differentiate Microglia from Other Macrophage Populations *447*
12.3	To What Extent is Microglial Phenotype Determined by the CNS Microenvironment? *448*
12.4	Microglia versus Macrophages/Dendritic Cells as Professional Antigen-presenting Cells *449*
12.4.1	In vitro and Ex Vivo Assays of Antigen-presentation *449*
12.4.2	Culture Conditions can have Profound Effects on Microglia Effector Functions as Assayed In Vitro *450*
12.4.3	In Vivo Assays of Antigen-presentation *451*
12.4.4	Antigen-presentation by Microglia is Necessary to Evoke or Sustain Neuroprotective T Cell Effector Function *451*
12.4.5	Why were Microglia Unable to Initiate Protective T Cell Responses? *453*
12.5	TREM-2 Positive Microglia may Represent Subsets Predisposed to Differentiate into Effective Antigen-presenting Cells *454*
12.6	Are Microglia the "Professional Antigen-presenting Cell of the CNS?" *456*

13   Contribution of B Cells to Autoimmune Pathogenesis *461*
Thomas Dörner and Peter E. Lipsky

13.1	Introduction *461*
13.2	Autoimmunity and Immune Deficiency *463*
13.2.1	Basic Mechanisms Providing Diversity to the B Cell Receptor *463*
13.2.2	Ig V Gene Usage by B Cells of Healthy Individuals *465*
13.2.3	Potential Abnormalities in Molecular Mechanisms Underlying IgV Gene Usage in Systemic Autoimmune Diseases *465*
13.2.4	Lack of Molecular Differences in V(D)J Recombination in Patients with Systemic Autoimmune Diseases *466*
13.2.5	Receptor Editing/Revision and Autoimmunity *467*
13.2.6	Selective Influences Shaping the Ig V Gene Repertoire in Autoimmune Diseases *469*
13.2.6.1	IgV Gene Usage by Autoantibodies *469*
13.2.7	Role of Somatic Hypermutation in Generating Autoantibodies *470*
13.3	Disturbed Homeostasis of Peripheral B Cells in Autoimmune Diseases *472*
13.4	Signal Transduction Pathways in B Cells *473*
13.4.1	B Cell Function Results from Balanced Agonistic and Antagonistic Signals *474*
13.4.1.1	Altered B Cell Longevity can Lead to Autoimmunity *474*
13.4.1.2	Altered B Cell Activation can Lead to Autoimmunity *476*
13.4.1.3	Inhibitory Receptors of B Cells *477*

13.4.1.4	Inhibitory Receptor Pathways and Autoimmunity *480*	
13.5	B Cell Abnormalities Leading to Rheumatoid Arthritis *482*	
13.5.1	Activated B Cells may Bridge the Innate and Adaptive Immune System *483*	
13.5.2	"Humoral Imprinting" in Rheumatoid Arthritis *484*	
13.5.3	Indications of Enhanced B Cell Activity in RA *485*	
13.5.4	T Cell Independent B Cell Activation *486*	
13.6	Depleting anti-B Cell Therapy as a Novel Therapeutic Strategy *487*	

14 **Dendritic Cells (DCs) in Immunity and Maintenance of Tolerance** *503*
Magali de Heusch, Guillaume Oldenhove and Muriel Moser

14.1	Introduction *503*
14.2	Dendritic Family *503*
14.3	DCs at Various Stages of Maturation *504*
14.4	Immature DCs *505*
14.5	Homing of DCs into Secondary Lymphoid Organs *505*
14.6	DCs as Adjuvants *507*
14.7	DC Subsets *508*
14.7.1	Classical DCs *508*
14.7.2	Plasmacytoid DCs *508*
14.8	DCs in T Cell Polarization *509*
14.9	Tolerogenic DC *510*
14.10	Mechanisms of Tolerance *512*
14.10.1	Lack of Co-stimulation *512*
14.10.2	Peripheral Deletion of Autoreactive T Cells *512*
14.10.3	Dynamics of Cellular Contacts *512*
14.10.4	Induction of Regulatory T Cells *513*
14.11	CD28-B7 Bidirectional Signaling *514*
14.12	Crosspriming *515*
14.13	Cross-presentation and Cross-tolerization *515*
14.14	DC as Regulators of T Cell Recirculation *516*
14.15	DC-based Immunotherapy of Cancer *517*
14.16	Conclusion *517*

15 **Thymic Dendritic Cells** *523*
Kenneth Shortman and Li Wu

15.1	Thymic Dendritic Cells *523*
15.2	Localisation and Isolation of Thymic DC *523*
15.3	Pickup of Antigens by Thymic DC *524*
15.4	Subtypes of Thymic DC *525*
15.5	Major Thymic cDC Population *525*
15.6	Minor Thymic cDC Population *526*
15.7	Thymic pDC *527*
15.8	Maturation State and Antigen Processing Capacity of Thymic DC *527*

15.9	Cytokine Production by Thymic DC	528
15.10	DC of the Human Thymus	529
15.11	Turnover Rate and Lifespan of the Thymic DC	530
15.12	Endogenous versus Exogenous Sources of Thymic DC	530
15.13	Lineage Relationship and Differentiation Pathways of Thymic cDC	531
15.14	Lineage Relationships and Developmental Pathways of Thymic pDC	532
15.15	Thymic cDC do not Mediate Positive Selection	533
15.16	Thymic cDC and Negative Selection	533
15.17	Role of pDC in the Thymus	535

Part V Antigen Presenting Cell-based Drug Development 539

16 Antigen Presenting Cells as Drug Targets 541
Siquan Sun, Robin Thurmond and Lars Karlsson

16.1	Introduction	541
16.2	Roles of DC in disease	542
16.2.1	Transplantation	542
16.2.2	Autoimmune Diseases	542
16.2.3	Allergy/Asthma	543
16.2.4	Cancer	543
16.3	Marketed Drugs Affecting APC function	544
16.4	New Potential APC Drug Targets	547
16.4.1	APC Activation	547
16.4.2	Antigen Presentation	550
16.4.3	Co-stimulation	553
16.4.4	Cell Adhesion	555
16.4.5	APC Chemotaxis	557
16.4.6	APC Survival	558
16.4.7	Intracellular Signaling	559
16.4.8	APC Depletion	560
16.5	APC per se as Drugs – DC-based Immunotherapy Therapy	561
16.5.1	DC-based Cancer Vaccines	561
16.5.2	Targeting and Activating DC In Vivo	562
16.5.3	DC-based Immunotherapy for Transplantation and Autoimmune Diseases	563
16.6	Conclusion	564

Glossary 585

Index 599

Preface

The history of Antigen Presenting Cells (APC) started in the second half of the 19th century when novel staining techniques entered the field of histopathology: the German pathologist Paul Langerhans was the first to describe irregularly shaped cells in the epidermis of the skin. These cells displayed long dendritic processes, thereby forming an almost continuous meshwork – an ideal feature to capture pathogens invading the body via the skin. The cells were termed "Langerhans cells"; however, at the time of their discovery nobody could be aware that Langerhans had stained the archetype of an APC that, subsequently, turned out to be key in triggering an immune response. Almost a century later, in the late 1960s, when it became immunology textbook knowledge that Langerhans cells contain racket-like Birbeck granules and can move from the epidermis to regional lymph nodes as so-called "veiled cells", APC became a frequently reoccurring term in immunological journals. In the 1970s and 1980s, pioneering work with macrophages taught us what almost all APC have in common: they share the capacity to engulf exogenous antigens, decompose them into proteolytic fragments and present these peptide fragments on their cell surfaces bound to molecules of the major histocompatibility complex (MHC). The last 15 years of research on APC has been governed, on the one hand, by the elucidation of the molecular mechanisms of antigen processing, e.g., which molecules decide which antigenic peptides become immunodominant in triggering T lymphocytes, and, on the other hand, by exploring which role APC play in tolerance induction against self versus autoimmunity.

In *Antigen Presenting Cells*, the principal investigators in the field present the development of central principles in the regulation of immunity versus tolerance, as accomplished through the activity of APC. E. Unanue, one of the pioneers in antigen processing, describes how APC may contribute to the diversity of T cell responses against a given T cell epitope (Chapter 1). P. Knolle focuses on liver sinusoidal epithelial cells and their capacity to induce tolerance against antigens entering the body via the oral route (Chapter 2). Contributions by F. Momburg (Chapter 3) and A. Vogt (Chapter 4) give updates on MHC class I and class II processing pathways, respectively; in particular, they discuss accessory molecules explored during the last decade. S. Porcelli and D. B. Moody provide us with state-of-the-art knowledge on presentation of lipid antigens by non-MHC molecules of the CD1 family (Chapter 5). H. Kropshofer and S. Spindeldreher give an overview of progress in the identification of naturally processed self-peptide antigens that are essential for the maintenance of a self-tolerant T cell repertoire (Chapter 6). In Chapter 7, M. Dustin and colleagues collected evidence on how the spatial organisation of key surface receptors on APC contribute to cytotoxic T cell activation. Chapter 8 by M. Bachmann and A. Oxenius highlight the more recent seminal discovery of Toll-like receptors that link innate and adaptive immune system functions, in particular through the expression on dendritic cells and macrophages. Two chapters by N. Stoy discuss the multiple roles that macrophages play in the maintenance of health or the induction of diseases (Chapters 9 and 10). Three other types of APC frequently implicated in diseases mediated by the immune system are polymorphonuclear neutrophils, microglial cells and B cells. Their significance in functioning as professional APC and their contribution to disease pathogenesis are outlined in Chapters 11–13 by B. Saha and A. R. Ashtekar, M. Carson and T. Dörner and P. E. Lipsky, respectively. M. Moser and colleagues and K. Shortman and L. Wu summarize recent progress in our understanding of how peripheral or thymic dendritic cells are principal inducers of self-tolerance (Chapters 14 and 15). Finally, L. Karlsson and colleagues discuss, from a pharmaceutical viewpoint, current options to exploit APC as drug targets in attempts to treat autoimmunity or cancer (Chapter 16).

Processing and presentation of antigens are central to the initiation of adaptive immune responses. Future therapeutics to control immune responses will come from a deeper understanding of the molecular details involved in antigen processing. In autoimmunity, a major goal can be to interfere with the proteolytic machinery that generates autoepitopes in dendritic cells or B cells. Targeting the B cell surface marker CD20 with monoclonal antibodies in rheumatoid arthritis is another example underlining the potential importance of APC in the design of new therapeutic approaches to control autoimmune diseases. New vaccine strategies for the treatment of infectious diseases and tumors will also rely on our combined knowledge about the functioning of APC. Breakage of tolerance imposed by dendritic cells, identification of appropriate tumor peptides presented on APC or the activation of dendritic cells by novel adjuvants through Toll-like receptors, thereby enhancing the presentation of tumor-associated antigens, could lead to successful immunotherapy of tumors. Achieving these goals in clinical trials should become reality. The fundamental work of our colleagues presented in this book will help to pave the way towards novel therapeutics and individualized health care.

Basel, April 2005

H. Kropshofer
A.B. Vogt
Pharmaceutical Research
F. Hoffmann La Roche Ltd

Acknowledgments

This compilation of meaningful aspects of antigen presenting cells has been accomplished with the help of a selected group of authors, each of whom has performed an admirable job in reviewing critical and cutting-edge findings in their field of research. We are grateful for their valuable contributions. Moreover, we thank Andreas Sendtko and his team from Wiley-VCH for working together on this book and for their continuous and inspiring support during its production. In addition, we would like to thank Günter Hämmerling, German Cancer Research Center, Heidelberg, for being our mentor in immunology. Finally, we extend our gratitude to Jonathan Knowles, Klaus Lindpaintner and Theodor Guentert, F. Hoffmann La Roche AG, Basel, for their encouragement to keep on deepening the understanding of the immune system.

List of Contributors

Amit R. Ashtekar
National Centre for Cell Science
Ganeshkhind
Pune 411007
India

Martin F. Bachmann
Cytos Biotechnology AG
Wagistr. 25
8952 Zürich-Schlieren
Switzerland

Monica J. Carson
University of California Riverside
Division of Biomedical Science
1274 Webber Hall
Riverside, CA 92521
USA

Magali de Heusch
Université Libre de Bruxelles
Laboratoire de Physiologie Animale,
IBMM
Rue des Pr. Jeener et Brachet 12
6041 Gosselies
Belgium

Thomas Dörner
University Medicine Berlin
Campus Mitte, Coagulation Unit
Schumannstr. 20/21
10098 Berlin
Germany

Michael L. Dustin
Skirball Institute of Biomolecular
Medicine
New York University School of
Medicine
504 First Avenue
New York, NY 10016
USA

Lars Karlsson
Johnson & Johnson
Department of Immunology
Pharmaceutical Research and
Development, L.L.C.
3210 Merryfield Row
San Diego, CA 92121
USA

Percy A. Knolle
Rheinische Friedrich-Wilhelms-
University Bonn
Institute of Molecular Medicine
and Experimental Immunology
Sigmund-Freud-Str. 25
53105 Bonn
Germany

Harald Kropshofer
F. Hoffmann La Roche AG
Non-clinical Immunology
Grenzacherstr. 124
4070 Basel
Switzerland

Antigen Presenting Cells: From Mechanisms to Drug Development. Edited by H. Kropshofer and A. B. Vogt
Copyright © 2005 WILEY-VCH Verlag GmbH & Co. KGaA, Weinheim
ISBN: 3-527-31108-4

Tatiana Lebdeva
Department of Microbiology and
Immunology and Kimmel Cancer
Institute
Thomas Jefferson University
Philadelphia, PA 19107
USA

Peter E. Lipsky
National Institutes of Health
NIAMS, Autoimmunity Branch
9000 Rockville Pike
Bethesda, MD 20892-1560
USA

Frank Momburg
German Cancer Research Center
Division of Molecular Immunology
(DKFZ/D050)
Im Neuenheimer Feld 280
69120 Heidelberg
Germany

D. Branch Moody
Brigham and Women's Hospital
Division of Rheumatology,
Immunology & Allergy
Room 514 Smith Building
1 Jimmy Fund Way
Boston, MA 02115
USA

Muriel Moser
Université Libre de Bruxelles
Laboratoire de Physiologie Animale,
IBMM
Rue des Pr. Jeener et Brachet 12
6041 Gosselies
Belgium

Guillaume Oldenhove
Université Libre de Bruxelles
Laboratoire de Physiologie Animale,
IBMM
Rue des Pr. Jeener et Brachet 12
6041 Gosselies
Belgium

Annette Oxenius
Swiss Federal Institute of Technology
Institute for Microbiology
Wolfgang-Pauli-Str.
8093 Zürich
Switzerland

Corinne Ploix
F. Hoffmann La Roche AG
Roche Center for Medical Genomics
Grenzacherstr. 124
4070 Basel
Switzerland

Steven A. Porcelli
Albert Einstein College of Medicine
Department of Microbiology and
Immunology and Department of
Medicine
Room 416 Forchheimer Building
1300 Morris Park Avenue
Bronx, NY 10461
USA

Bhaskar Saha
National Centre for Cell Science
Ganeshkhind
Pune 411007
India

Kenneth Shortman
The Walter and Eliza Hall Institute
1G Royal Parade
3050 Parkville
Australia

Sebastian Spindeldreher
F. Hoffmann La Roche AG
Roche Center for Medical Genomics
Grenzacherstr. 124
4070 Basel
Switzerland

Nicholas S. Stoy
Royal Hospital for Neurodisability
Huntington's Disease Unit
West Hill
Putney
London, SW15 3SW
UK

Siquan Sun
Johnson & Johnson Pharmaceutical
Department of Immunology
Research and Development, L.L.C.
3210 Merryfield Row
San Diego, CA 92121
USA

Yuri Sykulev
Department of Microbiology and
Immunology and Kimmel Cancer
Institute
Thomas Jefferson University
Philadelphia, PA 19107
USA

Robin Thurmond
Johnson & Johnson Pharmaceutical
Department of Immunology
Research and Development, L.L.C.
3210 Merryfield Row
San Diego, CA 92121
USA

Emil Unanue
Washington University School of
Medicine
Department of Pathology and
Immunology
660 South Euclid Ave.
St Louis, MO 63110
USA

Anne B. Vogt
F. Hoffmann La Roche AG
Roche Center for Medical Genomics
Grenzacherstr. 124
4070 Basel
Switzerland

Li Wu
The Walter and Eliza Hall Institute
1G Royal Parade
3050 Parkville
Australia

List of Abbreviations

AC	Adenylate cyclase
ADC	Arginine decarboxylase
AI	Arginase I
AL	Argininosuccinate lyase
AM	Alveolar macrophage
AMP	Anti-microbial peptide
AP-1	Activator protein-1
AS	Argininosuccinate synthase
BMP	Bone morphogenetic protein
Bsk	Basket (*Drosophila*)
C/EPB	CCAAT/enhancer binding protein
cAMP	Adenosine 3′,5′-cyclic monophosphate
CAT	Cationic amino acid transporter
CBP	CREB-binding protein
CK2	Casein kinase 2
COX	Cyclooxygenase
CREB	cAMP response element binding protein
CRP	C-reactive protein
Dach1	Dachshund protein (currently known to suppressor Smads in some cells)
DAP12	Adaptor protein of 12kDA (adaptor for TREM2)
DC-LAMP	DC-lysosomal-associated membrane protein
DC-SIGN	DC-specific ICAM-3-grabbing nonintegrin
DEC-205	DC receptor comprising decalectin with 10 contiguous, C-type lectin domains; homologous to the macrophage mannose receptor (MR)
DIF	Dorsal-related immune factor
Dome	Domeless (*Drosophila*)
DREDD	A death domain-containing Drosophila homolog of caspase 8
EAE	Experimental autoimmune (allergic) encephalomyelitis
EIF2α	Eukaryotic translation initiation factor 2α
ELAM-1	Endothelial leukocyte adhesion molecule-1

EP	E-prostanoid (PGE$_2$) receptor
EPAC1	Exchange protein activated by cAMP 1
ERK	Extracellular signal-regulated kinase
ET-1	Endothelin-1
FADD	Fas-associated protein with death domain
FAN	Factor associated with neutral sphingomyelinase activation
FIZZ	A chitinase marker of alternative macrophage activation
GAS	γ-activated sequences
GCN2K	General control non-derepressible-2 kinase
G-CSF	Granulocyte colony-stimulating factor
GM-CSF	Granulocyte-macrophage colony-stimulating factor
HDAC	Histone deacetylase
HMGB1	High-mobility group box 1 protein
Hop	Hopscotch (*Drosophila*)
HSC	Hepatic stellate cell
HSP	Heat shock protein
ICE	IL-1β-converting enzyme (caspase-1)
ICER	Inducible cAMP early repressor
IDO	Indoleamine 2,3-dioxygenase
IFRL	Innate functional response linkage
IKKγ	IκB kinase γ
IL-18BP	IL-18-binding protein
IL-1Ra	IL-1 receptor antagonist
iNOS	Inducible nitric oxide synthase
IRF	Interferon regulatory factor
ISG	Interferon-stimulated gene
ISGF3	A complex comprising STAT1, STAT2 and IRF-9
ISRE	IFN-α/β-stimulated response element
Jak	Janus kinase
JNK	c-Jun N-terminal kinase
KC	Kupffer cell
LAM	Lipoarabinomannan
LBP	Lipopolysaccharide binding protein
LC	Langerhans cell
LLO	Listeriolysin O
LPS	Lpopolysaccharide
Mal/TIRAP	MyD88 adaptor-like/TIR domain-containing adaptor protein
MAM	Microbial-associated molecule
ManLAM	Mannosylated (mannose-capped) lipoarabinomannan
MAPK	Mitogen-activated protein kinase
MBL	Mannose-binding lectin
MCP	Monocyte chemoattractant protein
M-CSF	Macrophage colony-stimulating factor
MD-2	Myeloid differential protein-2
MDL-1	Myeloid DAP12-associating lectin-1

MEK	Mitogen-activated protein kinase/extracellular signal-regulated kinase kinase
MEKK3	Mitogen-activated protein kinase/extracellular signal-regulated kinase kinase kinase 3
MGL	Macrophage galactose-type C-type lectin
MIG	Monokine induced by IFN-gamma
MIP	Macrophage Inflammatory Protein
MKP-1	Mitogen-activated protein kinase phosphatase-1
MMP	Matrix metalloproteinase
MMRR	Microbial molecular recognition receptor
MR	Mannose receptor
MSK1	Mitogen and stress-activated protein kinase 1
MTB	Mycobacterium tuberculosis
MyD88	Myeloid differentiation primary-response protein 88
NEMO	Nuclear factor-κB essential modulator (IKKγ)
NF-κB	Nuclear factor-κB
NO	Nitric oxide
NOD mouse	Non-obese diabetic mouse
NOD1/2	NOD-LRRs 1/2: nucleotide-binding oligomerization domain-leucine-rich repeats 1 and 2
NOHA	N^G-hydroxy-L-arginine
OAT	Ornithine acetyltransferase
ODC	Ornithine decarboxylase
OVA	Ovalbumin
P5C	Pyrroline-5-carboxylate
PACAP	Pituitary adenylate cyclase-activating polypeptide
PAMP	Pathogen-associated molecular pattern
PAMP	Pathogen-associated molecular pattern (see also MMAR)
PDE	Phosphodiesterase
PG	Peptidoglycan
PGRP	Peptidoglycan recognition protein
PI3K	Phosphoinositide 3-kinase
PKA	Protein kinase A
PRR	Pattern recognition receptor
Puc	Puckered (*Drosophila*)
RAGE	Receptor for advanced glycation end-products
RANK(L)	Receptor activator of NF-κB
RANTES	Regulated on activation, normal T cell-expressed and secreted protein (CCL5)
RIP	Receptor-interacting protein
S2	Steiner 2 cells (*Drosophila*)
SCID	Severe combined immunodeficiency
SHP2	Src homology 2 domain-containing protein-tyrosine phosphatase-2 (PTPN11)
SOCS	Suppressors of cytokine signaling

STAT	Signal transducer and activator of transcription
TAK1	Transforming growth factor-β-activated kinase
TAM	Tumor-associated macrophage
TGF-β	Transforming growth factor-β
TICAM1	(also called TRIF)
TICAM2	TICAM1 bridging adaptor
TIMP	Tissue inhibitor of matrix metalloproteinase
TIMP	Tissue inhibitor of metalloproteinase
TIR	Toll-IL-1 receptor
Tollip	Toll-interacting protein
TORC1	Target of rapamycin complex 1
TotA	Turandot A (*Drosophila*)
TRADD	TNF-receptor-associated death domain
TRAF	TNF receptor-associated factor
TRAP	Tartrate-resistant acid phosphatase
TREM	Triggering receptor expressed on myeloid cells
TRIF	Toll-IL-1 receptor homology domain-containing adaptor inducing interferon-β (also called TICAM-1)
Upd	Unpaired
VCAM-1	Vascular cell adhesion molecule-1
VEGF	Vascular endothelial growth factor
VIP	Vasoactive intestinal peptide
Ym1	A chitinase marker of alternative macrophage activation

Color Plates

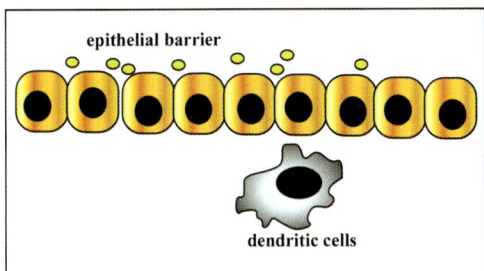
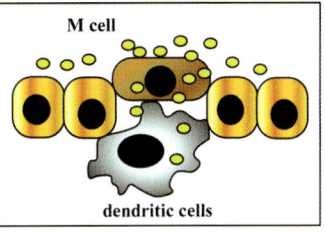
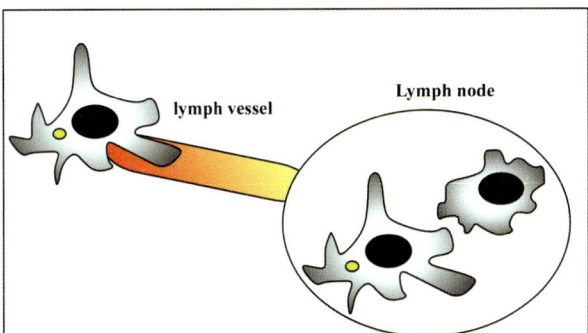

Figure 2.1 Outline of the conventional view that antigens are in peripheral tissue by dendritic cells and are transported by migratory dendritic cells to the lymph node where interaction with naive T lymphocytes occurs. Initial contact of antigens or pathogens in these peripheral microenvironments is believed to determine the functional outcome of ensuing immune responses. (This figure also appears on page 28.)

Antigen Presenting Cells: From Mechanisms to Drug Development. Edited by H. Kropshofer and A. B. Vogt
Copyright © 2005 WILEY-VCH Verlag GmbH & Co. KGaA, Weinheim
ISBN: 3-527-31108-4

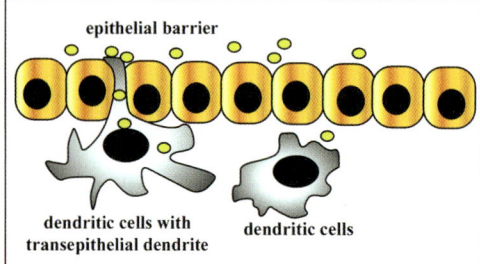
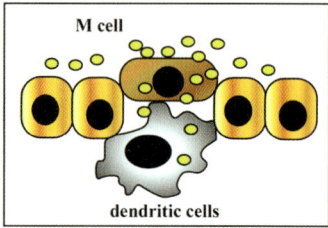
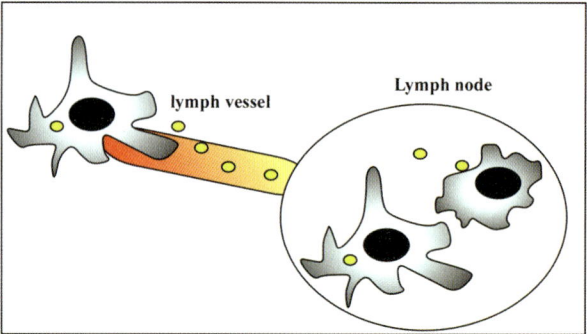

Figure 2.2 Recent work has provided evidence that dendritic cells reach via dendritic extensions, e.g. transepithelial dendritis in the gut, the outside world and directly sample antigens as well as microorganisms. Moreover, antigens that have reached peripheral tissue may directly gain access to lymph nodes, via defined anatomic structures, and in the lymph node interact with a, presumably, resident dendritic cell population. Bypassing peripheral antigen sampling migratory dendritic cells it remains an open question how specific information about the need for induction of immune responses is transferred to the lymph node, or whether this pathway is uniquely used for induction of immune tolerance towards tissue-autoantigens.
(This figure also appears on page 32.)

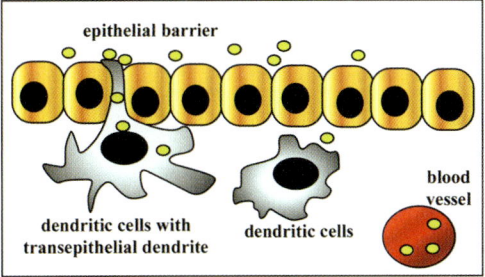
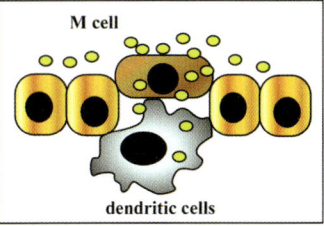
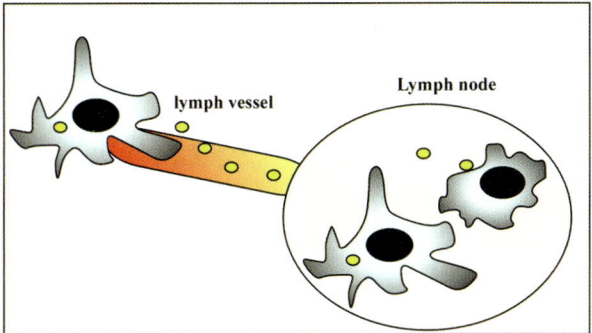
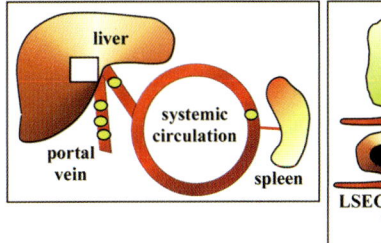
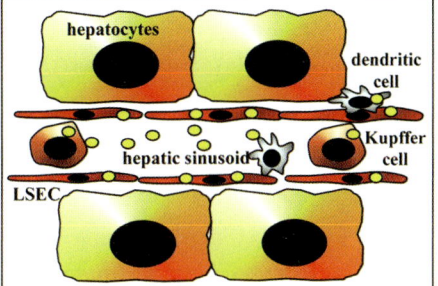

Figure 2.3 Antigens and pathogens may breach physical barriers and escape local immune control by migratory antigen presenting cells and disseminate in the entire organisms via the blood stream. Once systemically distributed, antigens and pathogens come in contact with antigen presenting cells in the spleen and liver. Within the liver a unique population of organ-resident antigen presenting cells, the liver sinusoidal endothelial cells, shape together with Kupffer cells and immature hepatic dendritic cells the immune response towards blood-borne antigens.
(This figure also appears on page 40.)

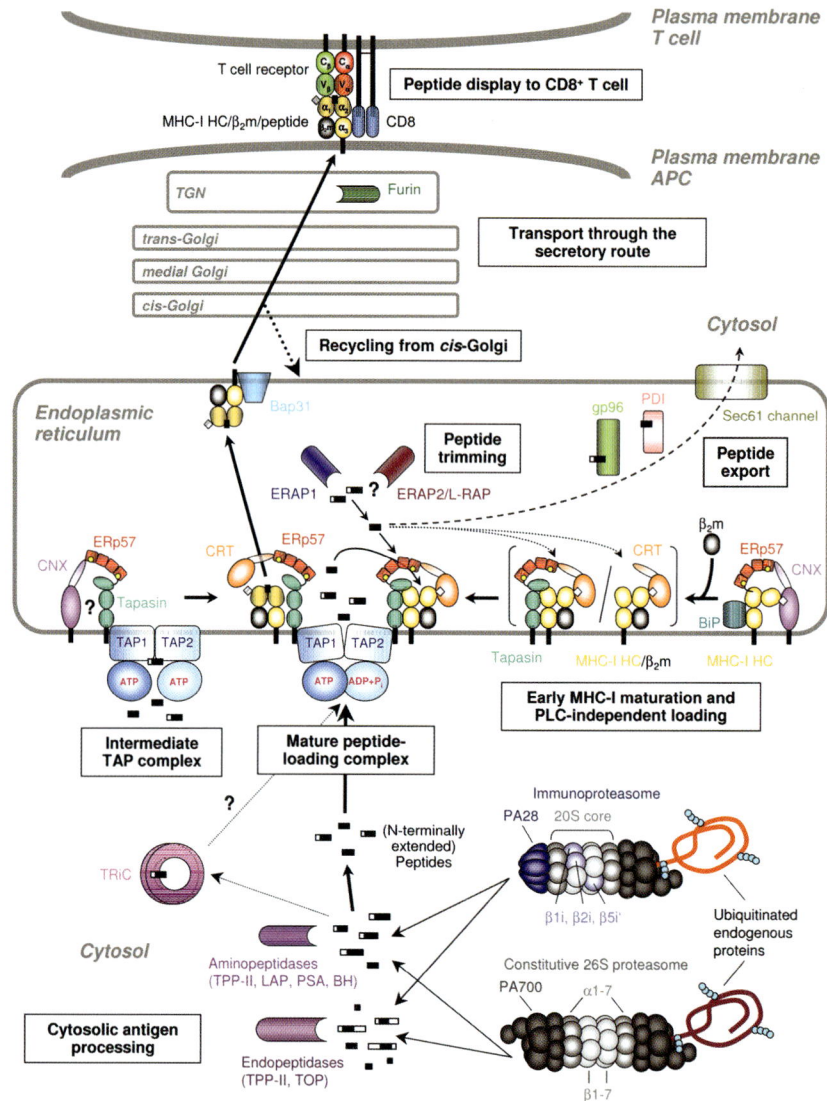

Figure 3.1 Classical pathway of MHC-I antigen processing. 26S proteasomes degrade cytosolic proteins tagged by multiubiquitin chains into peptides of 3–25 amino acids. The C termini of final MHC-I ligands are mostly generated by the proteasomal cut. In the presence of interferon-γ, the proteolytically active β subunits of the constitutively expressed 20S core complex are replaced by the respective immunoproteasome subunits β1i, β2i and β5i, which entail an altered proteasomal cleavage specificity. Also, the constitutively expressed ATP-dependent PA700 regulator complex can be replaced by the interferon-induced PA28 regulator complex. Through the endopeptidase activity of cytosolic tripeptidyl peptidase II (TPP-II) larger proteasomal products (>16 residues) can be processed into MHC-I-binding peptides, while thimet oligopeptidase (TOP) produces fragments that are too small for subsequent TAP transport and MHC-I binding. Cytosolic aminopeptidases, e.g., TPP-II, leucine aminopeptidase (LAP), puromycin-sensitive aminopeptidase (PSA), and bleomycin hydrolase (BH), can trim N-terminal amino acids before ER translocation by TAP. Cytosolic chaperones such as TRiC may protect peptides from rapid hydrolysis. The transporter associated with antigen processing (TAP) transports peptides of 8–16 amino acids in an ATP-dependent fashion. Subunits TAP1 and TAP2 undergo conformational changes upon peptide binding, upon ATP binding to the nucleotide binding domains, and upon ATP hydrolysis that is linked to substrate release into the ER lumen. N-terminally extended variants of final MHC-I-binding ligands are subjected to trimming by the ER-resident aminopeptidase ERAP1, and potentially also by ERAP2/L-RAP. Peptides can bind to ER-resident chaperones such as gp96 or PDI and are removed from the ER through the Sec61 channel. Peptides of 8–10 amino acids are loaded onto MHC class I heavy chain (HC)/β_2-microglobulin (β_2m) dimers that are tethered to TAP by means of the dedicated chaperone tapasin. Tapasin also recruits the oxidoreductase ERp57, which might isomerize the MHC-I α_2 domain disulfide bridge during peptide loading. Together with the lectin-like chaperone calreticulin (CRT), these components form the peptide-loading complex (PLC). Tapasin seems to retain MHC-I in a peptide-receptive conformation and its editing function results in an optimized affinity of MHC-I ligands and improved stability of cell surface HC/β_2m/peptide complexes. Tapasin also stabilizes TAP heterodimers by virtue of its transmembrane domain. Intermediate TAP complexes contain tapasin, ERp57 and calnexin but interaction modalities have not been defined. After biosynthesis MHC-I HC assemble first with the chaperones BiP and calnexin (CNX). ERp57 is also present in early HC complexes with calnexin. Following binding of β_2m to HC, CNX is exchanged by CRT that binds to the α_1 domain N-glycan of MHC-I HC in the monoglucosylated form. Intermediate MHC-I complexes, shown in brackets, are also able to bind peptides. Tapasin-independent allelic MHC-I molecules are not, or are only poorly, detectable in the PLC, but are regularly loaded with peptides. Successfully loaded MHC-I molecules are released from the PLC and exit the ER in cargo vesicles. The cargo receptor Bap31 is involved ER egress of peptide-loaded MHC-I. Tapasin-mediated recycling of MHC-I molecules from the cis-Golgi has been reported. HC/β_2m/peptide complexes traffick through the Golgi apparatus, where N-glycans are modified, and the trans-Golgi network (TGN) to the plasma membrane. In the TGN, the endopeptidase furin can contribute to antigen processing. HC/β_2m/peptide complexes displayed on the cell surface of antigen presenting cells (APC) are recognized by antigen-specific T cell receptors together with CD8.
(This figure also appears on page 54.)

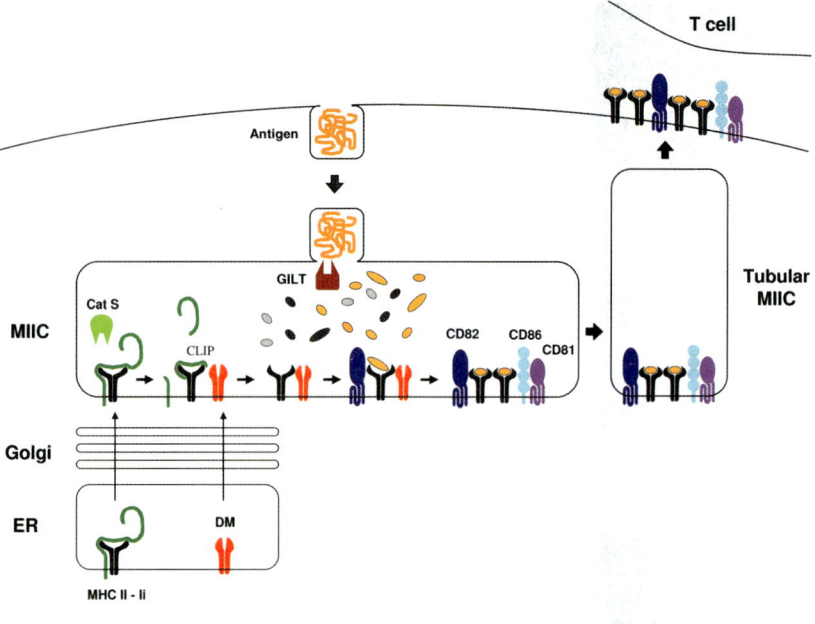

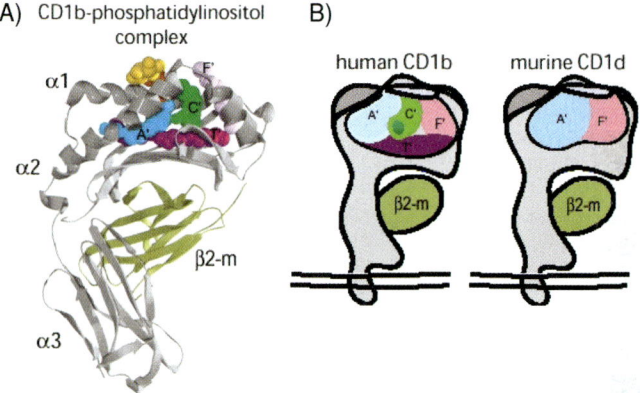

Figure 5.1 Structures of CD1 antigen binding grooves. Antigen binding grooves of CD1 proteins vary in size, architecture and ligand specificity. (A) The crystal structure of human CD1b bound to phosphatidylinositol shows that the two alkyl chains of the antigen are inserted into the A′ and C′ pockets, and the inositol moiety protrudes from the groove so that it can contact T cell antigen receptors [30]. (B) The grooves of CD1 proteins are composed of two or more discrete pockets. A schematic of the CD1b groove shows how it is divided into four pockets (named A′, C′, F′ and T′), and illustrates the two portals located at the top of the F′ pocket and the bottom of the C′ pocket. The smaller mCD1d groove has two pockets and only one portal.
(This figure also appears on page 133.)

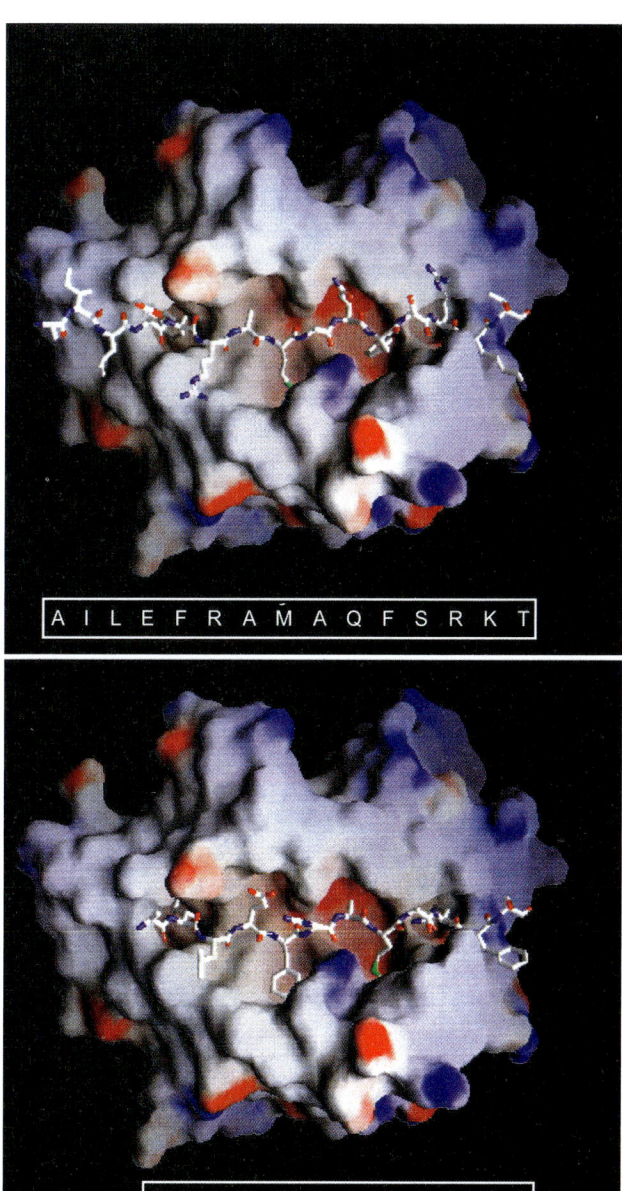

Figure 6.3 Computer modeling of the HLA-DR1:SP3 self-peptide complex. According to the X-ray crystal structure of the HLA-DR1 molecule, the self-peptide SP3 can bind into the peptide binding cleft in two different registers: either F-5 binds into the P1 pocket (upper panel) or I-2 (lower panel). The latter option foresees the hydrophobic residues L-3 and F-11 pointing towards the hydrophilic outer milieu, which is energetically disfavored. (Kindly provided by L. Mozyak, Harvard University, Boston, USA.) (This figure also appears on page 171.)

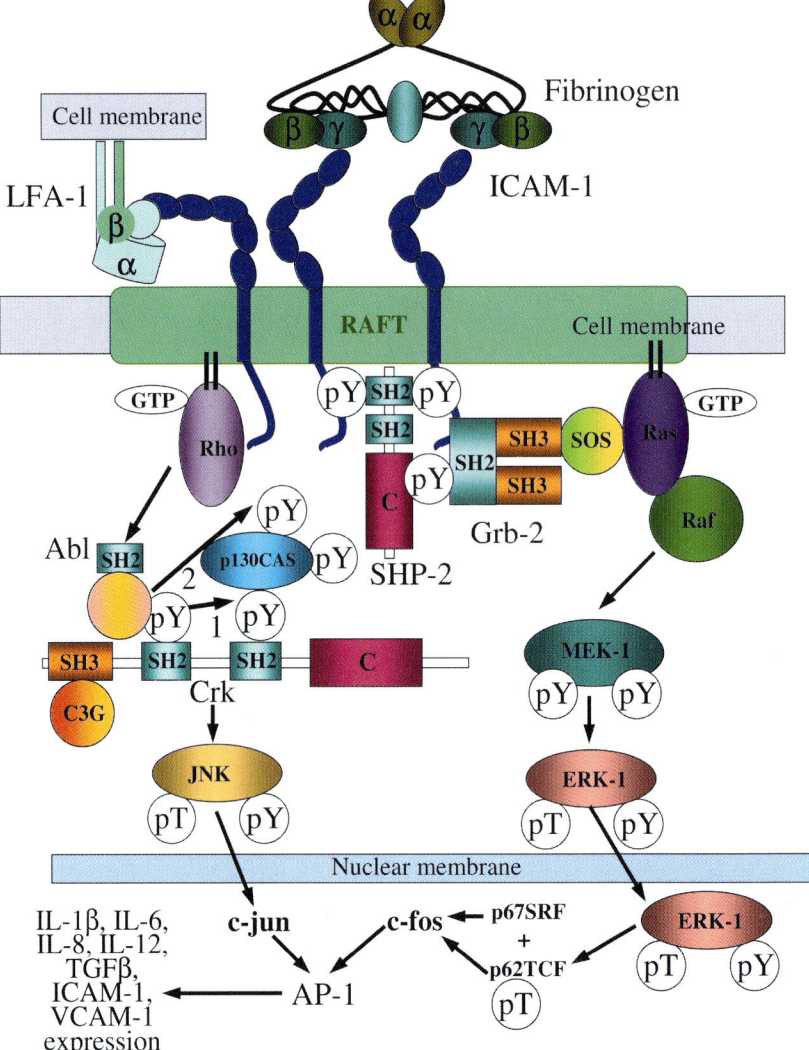

Figure 7.1 Productive engagement of ICAM-1 activates several signaling pathways. ICAM-1 cross-linking with multivalent ligands (such as fibrinogen) results in ICAM-1 recruitment to lipid rafts and phosphorylation of both ICAM-1 (at Y485) and SHP-2, Src homology 2 domain (SH2)-containing tyrosine phosphatase (Pluskota et al., 2000). SHP-2 recruitment to rafts is usually mediated by high levels of membrane cholesterol and interaction with annexin II, Ca^{2+}-dependent phospholipid-binding protein (Burkart et al., 2003). SHP-2 contains two N-terminal SH2 domains, catalytic domain (C) and a short C-terminal tail. Even though pY485 containing amino acid sequence does not resemble ITIM (immune receptor tyrosine-based inhibitory motif) consensus motif, association of SHP-2 and phosphorylated ICAM-1 occurs via SH2 domain of SHP-2. ICAM-1 is associated with SHP-2 on endothelial cells, Raji B cell line and human kidney fibroblast line 293 (Pluskota et al., 2000). Along with its

enzymatic activity, SHP-2 serves as adaptor molecule; being phosphorylated, it interacts with Grb2 (growth factor receptor binding protein 2) and via its connection with SOS (guanidine exchange factor for Ras) can activate Ras, small G-protein. Both signaling through the phosphatase catalytic domain of SHP-2 and signaling through Ras are necessary for activation of mitogen-activated protein kinase (MAPK). The activated Ras associates with the serine/threonine kinase Raf. Ras is localized to the plasma membrane due to its prenylation, while Raf is recruited to membrane by association with Ras. Its localization at the membrane results in activation and subsequent phosphorylation of the dual specific MAPK–ERK kinase (MEK) that, in turn, phosphorylates ERK (extracellular signal regulated kinase) on both tyrosine (pY) and threonine (pT) residues. Phosphorylated ERK dimerizes and exposes a signal peptide that mediates its transfer to the nucleus. Inside the nucleus ERK phosphorylates p62TCF (ternary complex factor) that then associates with p67SRF (serum response factor) to form active transcription factor complex to promote transcription of c-fos gene. Initially it was found that ICAM-1-SHP-2 interactions are necessary for cellular survival, but they are implicated in regulation of cytokines and growth factors expression during cell proliferation and differentiation as well as in upregulation of some membrane proteins, including immune receptors. Another signal transduction route activated upon ICAM-1 engagement is associated with activation of RhoA-family G-proteins (Etienne et al., 1998; Thompson et al., 2002) that activates downstream Abl tyrosine kinase capable of autophosphorylation. Abl initially phosphorylates Crk-accosiated substrate, CAS, at a single tyrosine residue (1) and at the same time binds to SH2 domain of adaptor protein Crk (CT10 regulator of kinase), contributing to additional phosphorylation of tyrosine residues of p130CAS (2). The Crk-based scaffold also provides a binding site for C3G, guanidine exchange factor protein. Assembly of Crk, p130CAS and C3G (Etienne et al., 1998) leads to activation of JNK (c-jun N-terminal kinase) that, similar to ERK, can translocate to nucleus and induce transcription of c-jun gene. C-Fos and c-Jun proteins together form activator protein complex-1 (AP-1), which controls expression of cytokines and genes encoding other proteins, ICAM-1 and VCAM-1, in particular (Koyama et al., 1996; Lawson et al., 1999; Poudrier and Owens, 1994; Sano et al., 1998). ICAM-1 molecules lacking the intracellular domain can not activate Rho proteins upon cross-linking (Greenwood et al., 2003). ICAM-1-associated signaling is not limited to the above pathways. Engagement of ICAM-1 also induces activation of Src-family kinases, in particular, p53Lyn upstream PLCγ. This cascade is responsible for Ca^{2+} influx and phosphorylation of cortactin p85 and other actin-associated proteins involved in cytoskeletal rearrangement (Etienne-Manneville et al., 2000). Src-family kinase pathway is also implicated in activation of p38 MAPK via ezrin phosphorylation and is responsible for generation of reactive oxygen species and nitric oxide (Wang and Doerschuk, 2001; Wang et al., 2003), suggesting an ICAM-1 role in mediating inflammation. The ability of ICAM-1 to induce various cell responses could depend on the nature of its ligand, cell type, ICAM-1 expression level and molecular distribution on the cell surface, suggesting that ICAM-1 mediated signaling is involved in modulation of immune response by a wide spectrum of various effects.

(This figure also appears on page 204.)

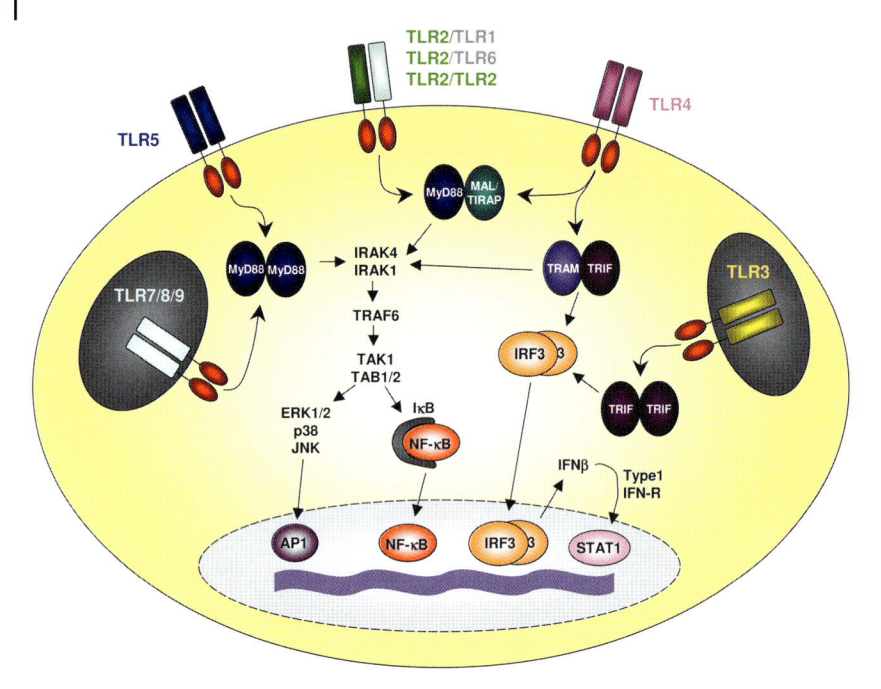

Figure 8.2 TLR signaling cascades. (This figure also appears on page 231.)

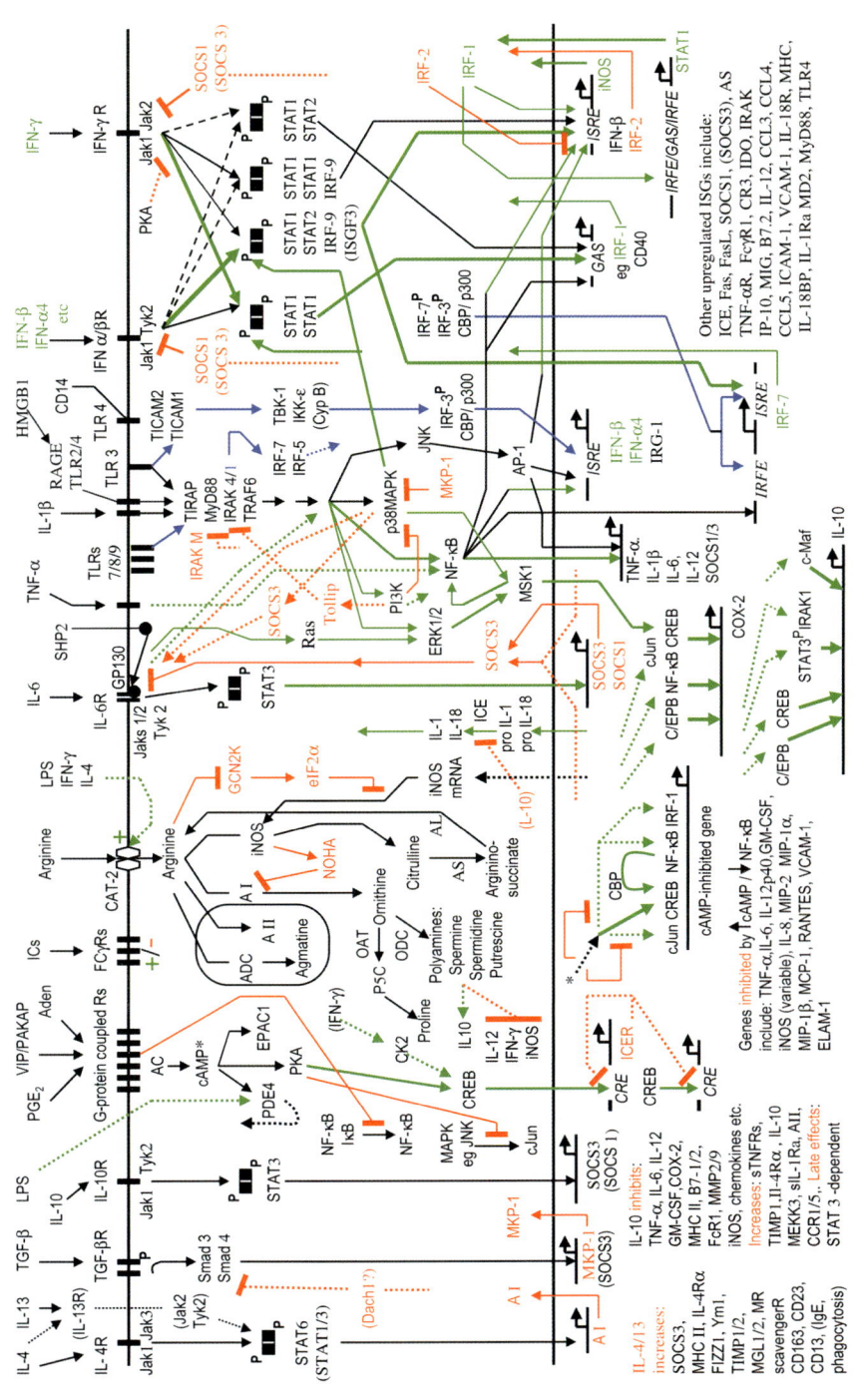

Figure 9.2 Balancing proinflammatory and antiinflammatory responses in macrophages. Some of the better-known intracellular transduction/signalling pathways and their interactions are shown. These control upregulation, downregulation, and the limits and sequencing of innate and adaptive immune responses, initiated by a variety of external stimuli. Most are discussed in the text, but this figure is also intended as a 'reference' diagram, of possible help in the interpretation of future research findings. (Green lines: facilitatory/proinflammatory; red lines: inhibitory/antiinflammatory; blue lines: pathways involving IRFs; dotted lines: detailed pathways omitted for clarity or because unknown.) (This figure also appears on page 251.)

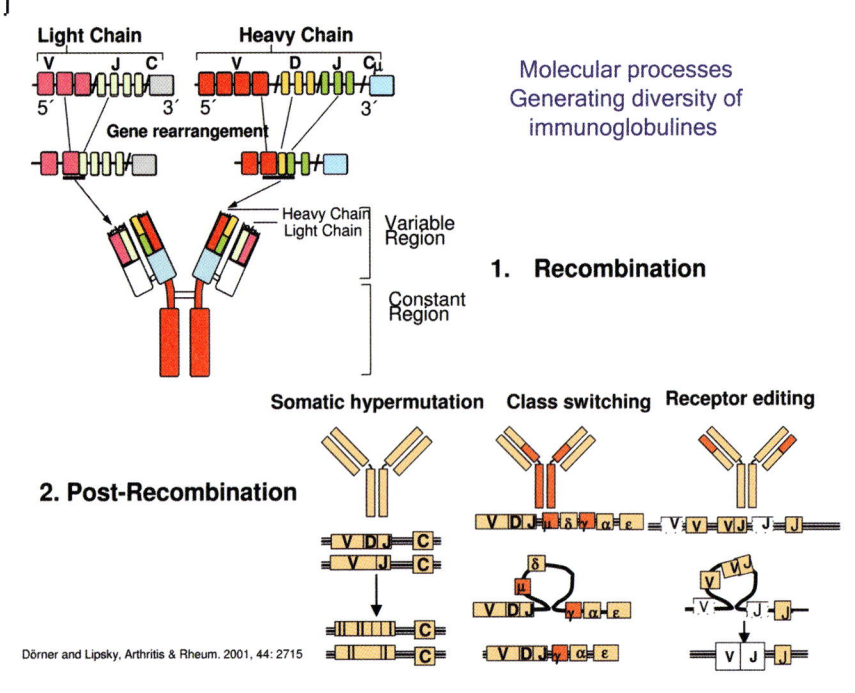

Figure 13.3 Molecular mechanisms involved in the generation of IgV variability leading to the diversity of the BCR as well as secreted Ig. (This figure also appears on page 464.)

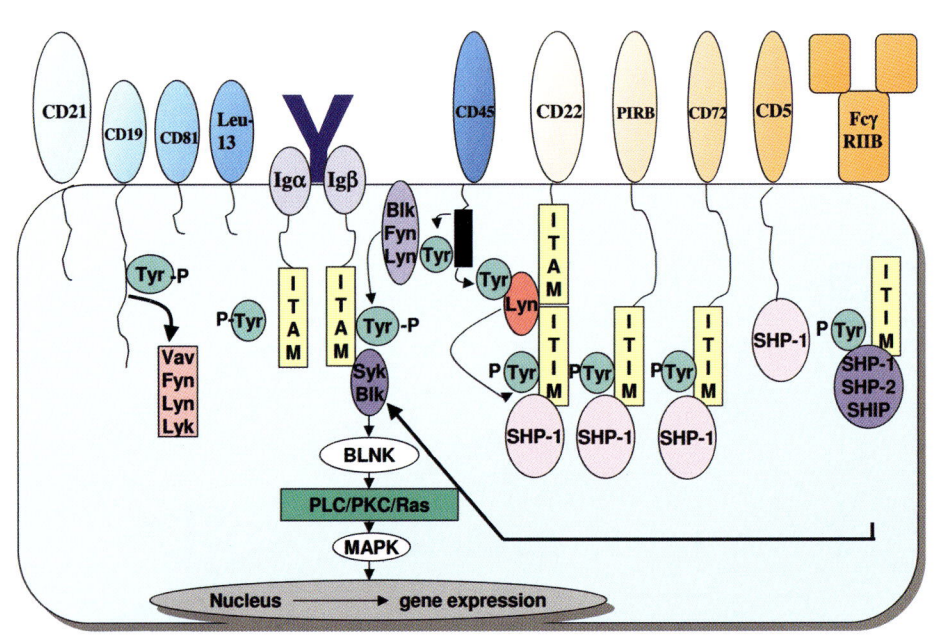

Figure 13.6 Activation and inhibitory markers on B cells and their signaling pathways involved (for further details see text). (This figure also appears on page 473.)

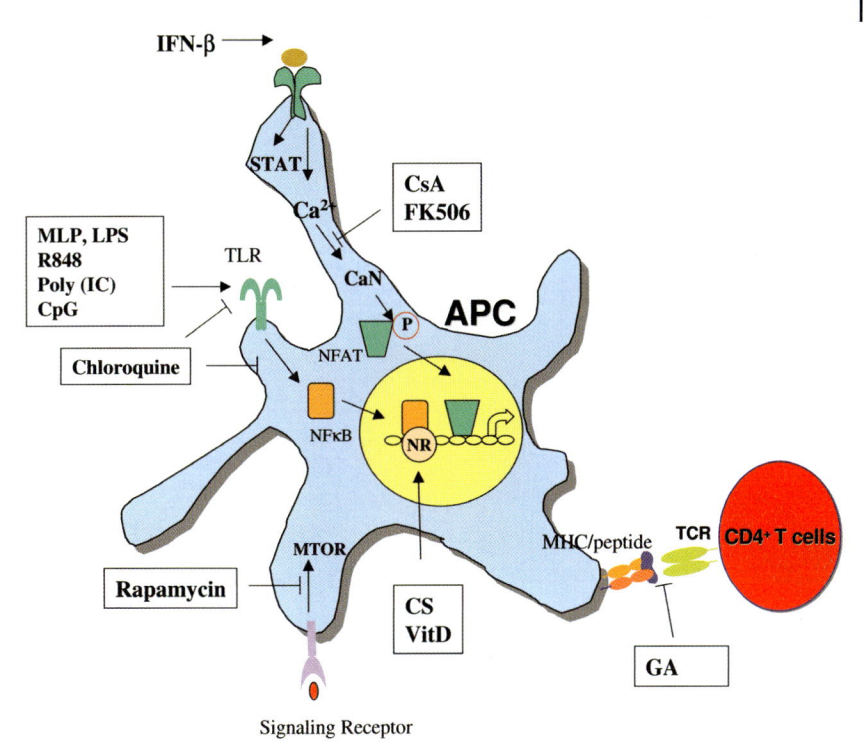

Figure 16.1 Mechanisms of action of marketed drugs affecting functions of APC. (This figure also appears on page 544.)

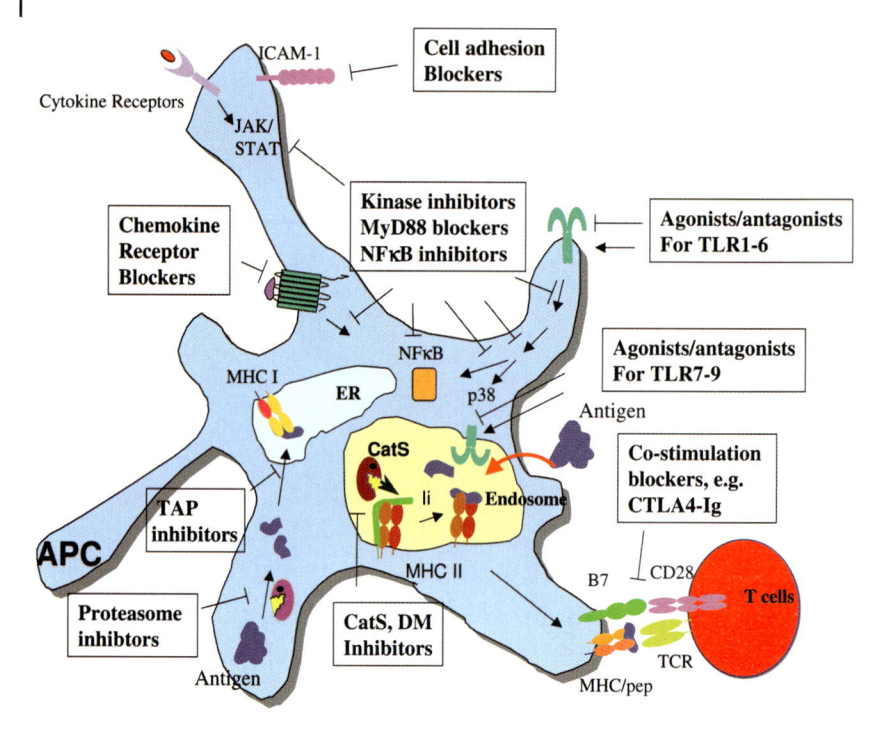

Figure 16.2 New potential APC drug targets. (This figure also appears on page 548.)

Part I
Antigen Presentation in the Immune System

1
Some Old and Some New Findings on Antigen Processing and Presentation

Emil R. Unanue

1.1
Introduction

This chapter summarizes our investigations on antigen processing and presentation using the model protein antigen hen-egg white lysozyme (HEL). It covers mostly the work from our laboratory and does not contain a comprehensive analysis of the work of others, who are credited in the original papers and reviews from the laboratory. This chapter is based on recent ones on the same topic [1, 2].

Our research for many years has centered on the symbiotic relationship between the phagocyte or antigen presenting cell (APC) system and the T cells. Using HEL has several advantages for studying antigen processing by the class II-MHC molecules, which has been our central focus: first, HEL is one of the best studied proteins at a biochemical level; second it is a strong antigen for many strains of mice (we use the strains bearing the k haplotype, and have examined, particularly, I-A^k molecules, referred to here as Ak); and, third, it has been extensively analyzed immunologically [3–5].

Our initial studies on antigen processing analyzed for the first time how phagocytes handled an antigen that was recognized by CD4 T cells. We first examined the bacterium *Listeria monocytogenes* in short-term interactions between T cells and peritoneal macrophages containing *Listeria*. It was established that antigen recognition by T cells required an APC, which needed to internalize the antigen (*Listeria*) into acidic vesicles, after which an epitope was presented on its cell surface [6, 7]. This peptide associated with an MHC molecule allowed T cells to bind to the macrophage.

Biochemical analysis of processing was extended using HEL, leading to several fundamental observations. Early studies [5] and those of Grey's laboratory using ovalbumin [8] were the two systems that defined the basic parameters of antigen processing using purified proteins. These early studies went against the long-standing paradigm that protein antigens were recognized before catabolism by the host (which is what happens with the B cell, as we now understand, but not with the T cell system, reviewed in Ref. 1). Moreover, the finding that antigen was immunogenic only when associated with, presented by, phagocytic cells was unexpected [9].

Antigen Presenting Cells: From Mechanisms to Drug Development. Edited by H. Kropshofer and A. B. Vogt
Copyright © 2005 WILEY-VCH Verlag GmbH & Co. KGaA, Weinheim
ISBN: 3-527-31108-4

1.2
HEL Processing

HEL is processed by APC, and suffers denaturation and unfolding. Peptides of HEL or unfolded HEL can bind to the class II MHC molecules, in our case to Ak molecules. In fact, the interaction of Ak molecules with the chemically dominant peptide of HEL, that found in the trypsin fragment from residues 46-61, was the first direct documentation that MHC molecules were peptide-binding molecules [10, 11]. It was also demonstrated that the MHC molecules bind to autologous peptides, and do not make the self–non-self discrimination [11]. Progress made in understanding the biochemistry and biology of antigen processing led into a parallel examination of how autologous proteins were presented in autoimmunity.

In APC, HEL needs to be taken into a deep vesicular compartment to be unfolded and processed [12]. With HEL the evidence points to its reduction followed by the selection and assembly of the various peptide segments by the Ak molecules, with the catalytic help of the H-2 DM molecules (referred as DM) in such a deep vesicular compartment. To test this scenario, HEL was encapsulated in liposomes of different chemical compositions. These liposomes would release their content in either deep, highly acidic, vesicles, or in recycling, lightly acidic, vesicles: very strong presentation took place when the peptides were encapsulated in liposomes made of phosphatidylcholine and phosphatidylserine that were disrupted, releasing their cargo, only in a late, highly acidic vesicle – see the next section [13–15]. References 16 and 17 discuss the importance of protein reduction. In toto, our findings, both in macrophages and dendritic cells (DC), and those of other investigators, indicate that the assembly site of the peptide-MHC complex is an endocytic vesicle that receives newly synthesized Ak molecules bound to the invariant chain, and that also contains the auxiliary molecules DM.

As a result of processing, several peptides are selected and displayed on Ak molecules; all the Ak-bound HEL peptides were identified by combinations of cellular, immunological and biochemical analysis. Table 1.1 summarizes the characteristics of HEL peptides presented by the class II MHC molecules [18–22]. As can be noted, the selection of peptides by Ak is dominated by one family centered on the 52-60 segment and which can occupy up to 10% of the bound peptides. As discussed below, the 52-60 segment is flanked, usually by four residues on the amino side and two on the carboxy-end.

Ak bound peptides were examined by electrospray tandem mass spectrometry (MS), including sequence analysis of the peptide mixture (Figure 1.1). Although peptides bound to MHC were first isolated and their general features established by bulk sequencing [23], MS is now the best approach for chemically isolating and characterizing MHC-bound peptides – an approach strongly emphasized by Donald Hunt, see Refs. 24–26. (Our MS operation is run by our colleague Michael Gross and his staff in our Department of Chemistry.) For chemical analysis we mostly used the C3F.6 B lymphoma line, a variant of the M12 line transfected with the alpha and beta chains of Ak molecules. Chemical analysis was facilitated by examining lines transfected with a membrane form of HEL. These

1.2 HEL Processing

Table 1.1 Major peptides of HEL presented by A^k molecules.[a]

			Binding IC-50 (μM)	Amounts (10^6pmol)
48-63	=	D G S T **D** Y G I L **Q** I **N** S R W W	0.04	404
31-47	=	A A K F E **S** N F **N** T **Q** A T **N** R N T	0.9	7
20-35	=	Y R G Y **S** L G N **W** V **C** A A K F E	1.4	2
114-129	=	R C K G T **D** V Q **A** W **I** R **G** C R L	5.0	<1

a) Indicated are the four major families of peptides selected by A^k of APC. Core sequences are underlined. Flanking residues most frequently identified are indicated. MHC contact residues are in bold. Binding was done using purified A^k molecules obtained from recombinant baculovirus. Amounts were estimated by ELISA assays using monoclonal antibodies. The main anchor residue of 48-63 is Asp 52. In 31-47 and 20-35, it is the asparagines at P4. Peptide 114-129 has the aspartic acid at P1 as the main anchor, but it contains hindering residues at P7 and P9.

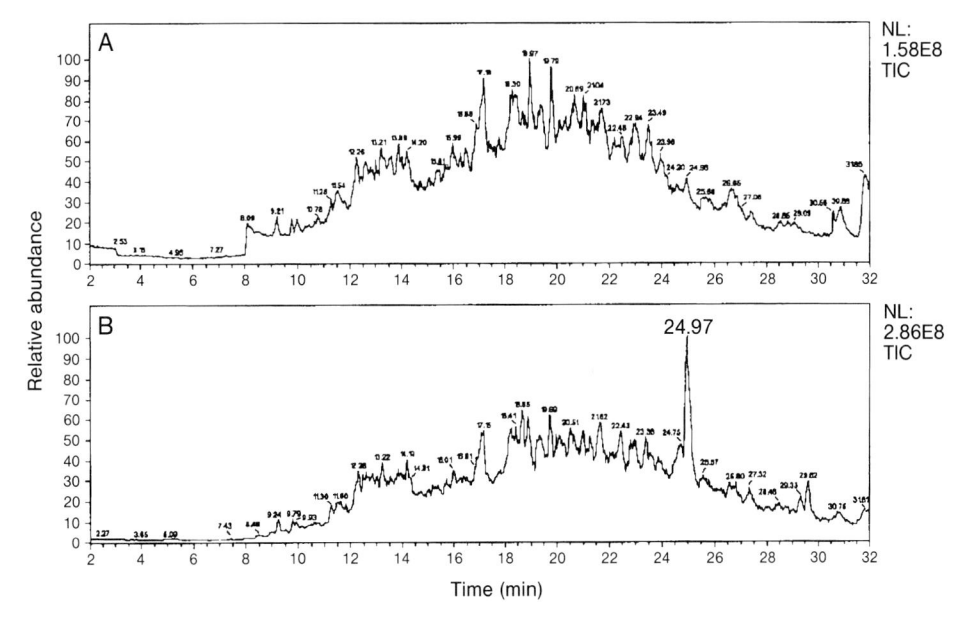

Figure 1.1 Peptides were isolated from Ak molecules of the C3.F6 APC line. (A) cultured with HEL; (B) cultured with HEL. Note the many peaks of autologous peptides in both parts; and the prominent peak of the chemically dominant 48-62 peptide in (B). (From Ref. 21, with permission.)

lines handled HEL identically, as if the soluble protein had been offered as an exogenous protein. The lines have the advantage that the mHEL gene can be mutated, allowing different residues in peptide selection and presentation to be evaluated. Peptides were also analyzed from the standard APC of HEL transgenic mice with identical results.

Our studies were facilitated by using monoclonal antibodies to peptides or to peptide–MHC complexes as capture reagents. We used monoclonal antibodies against the main core segment of the MHC-bound peptides of HEL [20–22], or to specific peptide-MHC complexes [27]. An important first use of these reagents was to quantitate the amounts of peptide bound to Ak using ELISA techniques [28]: bound peptides were released from the isolated Ak molecules from APC and their amounts were estimated by ELISA, by inhibition assays using standards of peptides. Figure 1.2 shows an example of quantitating one of the minor peptides from HEL. This allowed us for the first time to estimate the exact amounts displayed by APC of the various peptides. A second use was as an antigen capture device for the identification of peptides. Peptides released from the class II molecules were bound to the antibodies attached to Sepharose particles and then released into the MS [op. cit.].

Most class II MHC-bound peptides are presented as families having a core nine amino acid segment plus flanking residues that vary both at the amino and carboxy termini. The peptides vary from 14 up to 20+ residues long [1, 20–22, 24–26]. The variation in length of flanking residues determines the number of peptides in a family. The "core" segment is responsible for the specificity of the interaction: usually amino acid side chains of the peptide stretched in the binding groove establish contact with four or five pocket sites, an issue well established by the X-ray crystal structure of several peptide-class II MHC complexes [29–37]. Depending on the particular MHC genotype, peptides are selected that contain favorable motifs for interaction. Flanking residues contribute to binding energy by adding to the conservative interactions between the main chain of the peptide and the surrounding hydrogen network (reviewed in Ref. 29).

The peptide segment selected by, and bound to Ak molecules, is protected from catabolism, an issue first shown by Paul Allen [38; see also 39]. Indeed the 48-62-Ak complex was highly resistant to proteolytic enzymes, and was also resistant to SDS denaturation (in SDS-PAGE examinations) [40] (Figure 1.3). But the MHC-bound peptide can be subjected to amino and carboxypeptidases that trim the overhanging portions close to the class II combining site. Notably, the naturally selected peptides containing amino terminal prolines are the longest bound to MHC molecules [41]. Prolines, because of the structure of their peptide bonds formed by their amino group, block exopeptidases. We examined this issue directly by making mutations in the HEL molecules at the sites flanking the 48-62 segment [42]. Most of the 52-60 family have four flanking residues on the amino terminus starting at amino acid 48 (Table 1.1); by placing a proline at 48, which is P-4, the resulting peptide bound to Ak was extended by one or two residues. The nature of the aminopeptidase has not been determined.

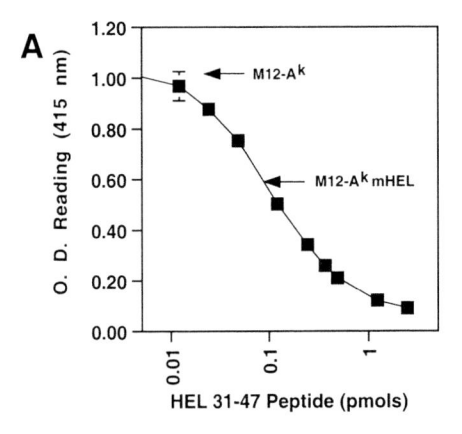

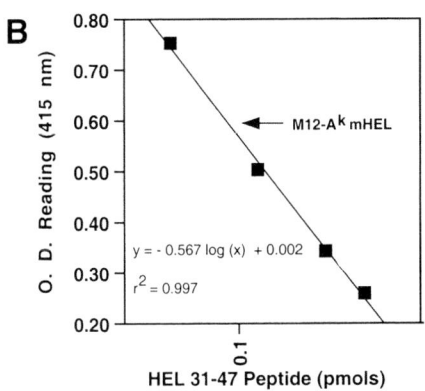

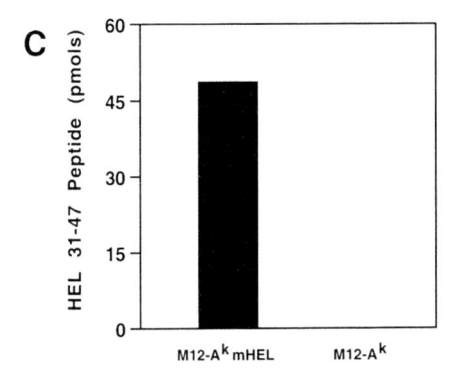

Figure 1.2 Quantitation of the 31-47 family of HEL peptide using specific anti-peptide monoclonal antibodies. (A) Actual results with the anti-31-47 antibody binding to the peptide attached to the plate in the presence of increasing amounts of soluble peptide as a competitor. (B) Calibration isotherm from which the amounts of peptide in the test samples (arrow) can be determined (C). (From Ref. 20, with permission.)

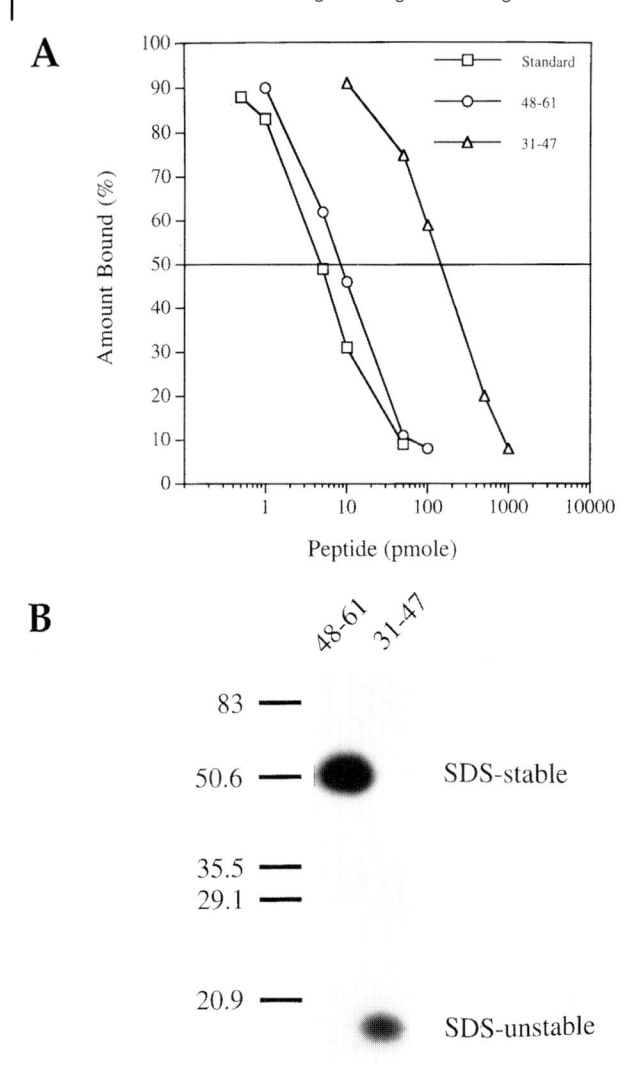

Figure 1.3 Binding features of the chemically dominant peptide 48-61 (i.e., (–O–)) and the minor peptide 31-47. (A) The inhibitory curve when purified Ak molecules are bound to a standard radioactive peptide. Each curve indicates the amounts bound in the presence of unlabeled peptide; a standard unlabeled, and the two HEL peptides. 48-61 is a strong binding peptide as evidenced by the amounts required to inhibit by 50%, which are much less than those required from 31-47. (B) Indication that the complex of 48-61 with Ak is SDS-resistant, whereas that with 31-47 readily dissociates. (From Ref. 20, with permission.)

1.3
Selection of Peptide Segments of HEL

The chemically dominant HEL peptide epitope is that having the 52-60 residues as a core segment (Table 1.1): DYGILQINS. The family usually encompasses peptide starting at residue 48 and ending with one or two tryptophans at residues 62 or 63 (in about equal amounts). This family is the chemically dominant family selected during processing (Figure 1.1). The main interaction responsible for binding involves the aspartic acid at P1, residue 52, which establishes an ion pair with residue arginine 62 in the alpha chain that forms the base of the pocket 1 [19, 34]. An indication of the importance of the P1 interaction is that a peptide made only of alanines, except for an acidic residue for P1, will bind as strongly as 48-62 to Ak molecules [19]. Similar results were obtained by Jardetzky, Wiley and colleagues: a peptide made of alanines, except for a tyrosine at P1 was responsible for binding to HLA DR2 molecules [43]. For the HEL 48-62 family, the other MHC-contact amino acids, at P4, P6, P7 and P9, did not contribute much binding energy, provided the strong driving interaction was at P1.

Importantly, changes in the Arg 62 to alanine of the alpha-k molecules, or the HEL 52 residue also to alanine, abolished the extensive selection of this peptide family during processing. Thus, the high level of selection is dominated by a single amino acid of the peptide in which its side chain has a highly favorable interaction with one of the allelic sites in the Ak binding groove.

Another major consideration both in the binding and in the selection of HEL peptides was the presence of negative or hindering residues at the auxiliary positions, which weakened the interaction with the P1 favorable site [44]. Although other groups had given evidence of hindering residues [45, 46], their strong influence became very evident in our analysis of the selection of various HEL segments, particularly of the 52-60 family (Figures 1.4 and 1.5). For example, only four peptides were presented of 21 potential peptides having an aspartic acid to interact at the P1 site. Not only did we find single residues that were strongly hindering, but we found combinations of them that together affected the binding. Glycines were particularly strong negative residues, probably by causing entropic

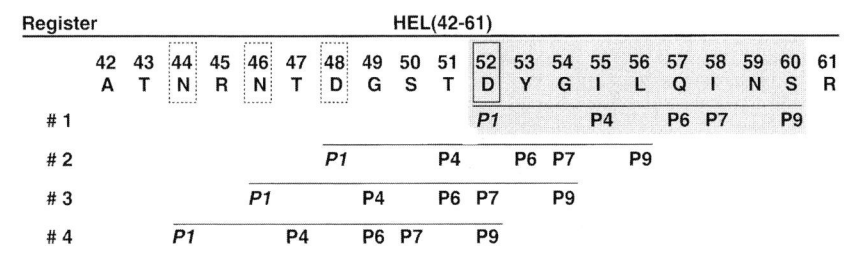

Register									HEL(42-61)										
	42	43	44	45	46	47	48	49	50	51	52	53	54	55	56	57	58	59	60	61
	A	T	N	R	N	T	D	G	S	T	D	Y	G	I	L	Q	I	N	S	R
# 1											P1			P4		P6	P7		P9	
# 2							P1			P4		P6	P7		P9					
# 3				P1			P4		P6	P7		P9								
# 4		P1		P4		P6	P7		P9											

Figure 1.4 Segment of 52-60 is selected from a long stretch of the HEL molecule. Four possible registers are indicated here but only register # 1 is used. MHC contacts of P4, P5, P7 and P9, of register # 2, # 3, or # 4 were substituted instead of those in register # 1 in the sequence of 52-60: no binding was found (Figure 1.5B). (From Ref. 44, with permission.)

1 Some Old and Some New Findings on Antigen Processing and Presentation

A.

	48	49	50	51	P1 52	53	54	P4 55	56	P6 57	P7 58	59	P9 60	61	RIC⁻¹
HEL(48-61)	D	G	S	T	D	Y	G	I	L	Q	I	N	S	R	1.3
Transferrin-R								L		Q	L		G		2.5
Aβk								V		E	F		A		3.0
Cathepsin H								I		G	E		S		5.8
DNA B								N		E	N		T		28

B.

Register	48	49	50	51	P1 52	53	54	P4 55	56	P6 57	P7 58	59	P9 60	61	RIC⁻¹
#1	D	G	S	T	D	Y	G	I	L	Q	I	N	S	R	1.3
#2								T		Y	G		L		NB
#3								G		T	D		G		NB
#4								T		G	S		D		NB

Figure 1.5 Importance of the "auxiliary" binding sites for the 52-60 segment binding. Peptide 48-61 binds well to Ak; note the binding results on the last columns (the lower the number, the better the binding). (A) Secondary residues in the indicated peptides were substituted in 48-61. All allow for the binding, which was similar to that of the peptide, all of which have aspartic acid at P1. (B) Putative registers 2–4 shown in Figure 1.4 inhibit the binding (NB). (From Ref. 44.)

disorganization at the binding sites. Figures 1.4 and 1.5 show the results of experiments in which the residues corresponding to P4, P6, P7 and P9 of different peptides were placed instead of those of the 52-60 sequence. For example, Figure 1.4 shows the sequence of HEL, from residue 42 to 61. Of the four possible binding registers, only one was used, and this was as a result of hindering amino acids.

Finally, to note in Table 1.1, the other two peptide families 20-35 and 31-45 did not use an acidic residue at P1: their main MHC anchor was the asparagine at P4. Both these sets of peptides bound with lower affinities to Ak and were selected to a much lesser degree. Peptide 114-129 showed a weak interaction, despite the favorable aspartic acid at P1, due to two unfavorable residues, the arginine at P7 and the cysteine at P9: changing them to alanines restored high-affinity binding (our unpublished data with Ravi Veraswamy).

Thus, high-affinity interactions result from favorable residues in the core segments of the peptide, together with the absence of negative or hindering amino acids, plus the contributions of the peptide backbone with conserved residues along the binding groove [reviewed in Ref. 29]. The findings with HEL peptide selection are mimicked by the analysis of natural peptides bound to Ak molecules in which a very high number contain acidic residues at P1.

In most bound peptides, three amino acids are potential contacts with the T cell receptor (TCR); for the 52-60 segment Tyr 53, Leu 56 and Asn 59 are the TCR contacts; substitution of any of them results in a loss of T cell responses [47, 48].

No evidence was found of intramolecular competition among different segments of the HEL molecules (Figure 1.2). For example, take the processing of the HEL segments 31-47 of low binding strength, and 48-62, the highest affinity segment [21]. In the normal selection the family of 31-47 ends at residue 47 while the family of 52-60 starts at 48: there is no overlap between most family members. If proline is placed at residue 48, then the peptides identified by MS have substantial overlap in their sequence. However, despite this overlap there was no effect in the *amounts* of each segment selected chemically by the APC. These results are supported by an examination of HEL in which the 52-60 segment is not selected because of introduction of hindering residues. Such an HEL did not affect the amounts of minor HEL peptides of lower affinities, i.e., the dominant 48-62/3 segment was not competing out the minor, and each were selected from different molecules [21].

1.4
HEL: Conformational Isomers

Chemical analysis of naturally presented peptides from HEL led to the identification of conformational isomers of a peptide-MHC complex. Our initial observations challenged the concept of peptide determinants termed "cryptic" or hidden in the native protein, i.e., determinants that were not processed.

Several groups had indicated that CD4 T cells directed to HEL peptides did not recognize HEL presented by APC, leading to the interpretation that these peptides were not generated during processing [49]. However, our biochemical studies proved directly that such peptides from the processing of HEL were selected and bound to MHC molecules [50]. To reconcile these conflicting findings, we focused on the chemically dominant 48-62 epitope of HEL described above and found two sets of CD4 T cells to that epitope [51]. One set recognized both the peptide derived from HEL processing as well as the peptide 48-62 peptide given exogenously to the APC. These are the conventional T cells that we termed type A. In contrast, the type B T cells recognized only the exogenous peptide but not the peptide derived from the processing of HEL! Table 1.2 summarizes the various experimental manipulations that led to the characterization of type A and B peptide-MHC complexes.

Table 1.2 Type A and B complexes.[a]

APC Status	Antigen	Type A	Type B
Live	Native HEL	+	−
Live	Synthetic 48-61 peptide	+	+
Live	Tryptic peptide 48-61	+	+
Live	Extracted peptide from APC	+	+
Live	Covalent peptide-MHC complex	+	−
Live	Denatured HEL	+	+
Fixed[b]	Native HEL	−	−
Fixed[b]	48-62	+	+
Fixed[b]	Denatured HEL	+	+
Chloroquine-treated live[c]	Native HEL	−	−
Chloroquine-treated live[c]	48-62	+	+
DM-deficient[d]	Native protein	−	−
DM-deficient[d]	Peptide	+	+
Soluble I-A^{k} [e]	Covalent peptide-MHC complex	+	−
Soluble I-A^{k} [e]	Peptide-exchanged complex	+	+

a Indicated are the various experimental manipulations. Live APC included peritoneal macrophages, spleen DC, or B cells.
b Fixation in paraformaldehyde, which inhibits processing but allows for peptide exchange at the cell surface.
c APC also treated with chloroquine to inhibit intracellular processing.
d Refers to APC from mice with genetic ablation of the DM gene.
e Soluble A^{k} refers to baculovirus purified molecules.

The differences were neither explained by contaminants in the preparations of synthetic peptide nor by post-translational changes [52, 53]. Moreover, the peptide could be extracted from the class II molecules of APC, after HEL processing, and offered to the type B T cells, which then recognized it. In addition, recombinant complexes in which the 48-62 peptide was covalently linked to Ak molecules only stimulated the type A sets, while complexes formed by peptide exchange stimulated both subsets [53]. Presentation of the peptide by the type B complex was resistant to chloroquine, and did not require any processing, since fixed APC were able to present it, indicating that the peptide did not require further processing in acid vesicles [48]. Thus, the two T cells recognized the *identical* linear sequence. The mode of processing and assembly of the peptide was responsible for the difference in recognition, as will be explained below.

Zheng Pu and Javier Carrero went on to identify the T cell receptor gene segment usage of multiple type A and B T cell hybridomas, and found no skewing in either subset. However, type B T cells displayed unique recognition of the TCR-contact residues. As mentioned above, the three TCR contact residues for most if not all type A T cells are at P2 (tyrosine), P5 (leucine), and P9 (asparagine). Substitution of any of them by alanines resulted in complete loss of the response. Type B T cells segregated into two subsets: *type B-long* required contact with the P5 and P8 side chains and were indifferent to substitutions at P2. While t*ype B-short* T cells required contact with the P2 and P5 side chains and were indifferent to substitutions at P8. Type B-short T cells responded to a peptide consisting only of the P1–P7 residues [48].

Two sets of data led to the conclusion that the T cells differed in their unique recognition of conformational isomers of the same peptide-MHC complex in which the DM molecule played the pivotal role. Zheng Pu in our group developed an assay system in which the 48-62-Ak complex was formed in vitro either in the presence or absence of DM molecules. Both purified molecules were isolated from cultures of insect cells infected with recombinant baculoviruses (Figure 1.6). Peptide 48-62 was incubated with Ak molecules, and the complex was isolated and cultured with T cells. Both type A and B T cells responded equally well to the complex. Addition of DM increased the amount of peptide bound to Ak. But the addition of DM to the peptide-Ak complex had no effect on the response of type A T cells. However, type B response was completely eliminated. In a different manipulation, complexes were formed first and then incubated with DM briefly.

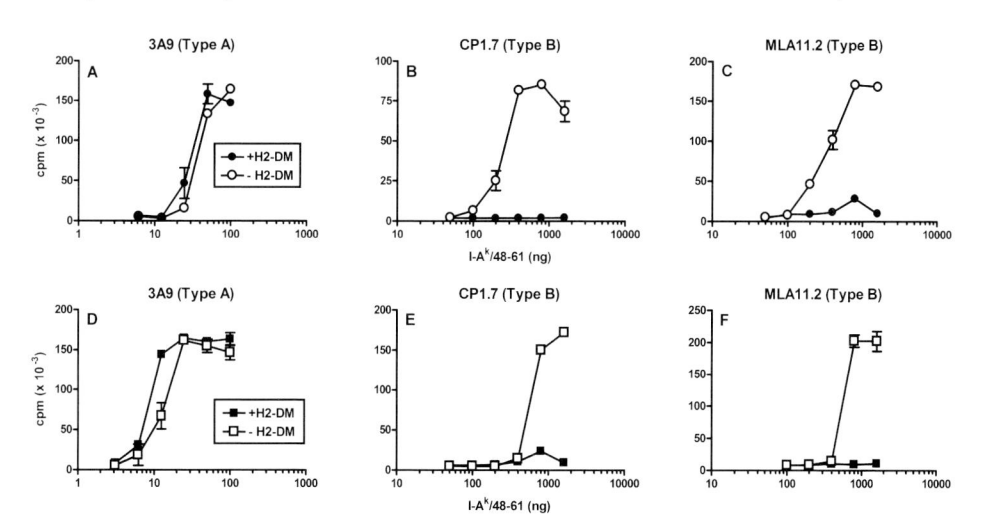

Figure 1.6 Formation of type A and B complexes in vitro, in the presence or absence of DM molecules. (A)–(C) Ak molecules were incubated with 48-62 in the presence or absence of DM. The complex was then isolated and incubated with one type A T cell (3A9) or two type B T cells (Cp1.7 and MLA11.2). (D)–(F) Ak molecules were first incubated with 48-61, after which Dm molecules were added. See the text for an explanation. (From Ref. 15.)

Such addition of DM also eliminated the type B conformer. This result indicated that the editing function of DM was exerted on previously formed complexes and not during assembly.

Scott Lovitch set out to evaluate the involvement of DM in the APC by using the system of liposome-mediated antigen delivery. Peptides were encapsulated in liposomes of different chemical composition to be delivered to either late endocytic vesicles, the sites of loading of peptides generated through processing of the protein, or to early endocytic vesicles, where exogenous peptide loading occurs.

Type A T cells responded to peptide delivered to both compartments, although targeting of peptide to late vesicles resulted in a markedly enhanced response; in contrast, type B T cells did not respond to peptide delivered to late vesicles, and responded only to peptide delivered to early endosomes [48]. DM molecules eliminated the type B complex in late vesicles. However, in DM-deficient APC, type B T cells responded to peptide delivered to late vesicles, in contrast to the results with wild-type APC.

We concluded that DM molecules are active in late vesicles, but essentially inactive in early endosomes and at the cell surface; furthermore, DM edits the repertoire of class II MHC-bound peptides, favoring high-affinity epitopes [54–58]. We reasoned that DM, likewise, was eliminating flexible conformers and favoring stable ones.

We envisage the following scenario: a globular protein such as HEL requires reduction and denaturation for processing; as a result, peptide-MHC loading only occurs in late endocytic vesicles. Here, protein antigens are unfolded and degraded, and peptides complex with nascent MHC class II molecules; DM then edits the conformation of the complex so that only the most stable conformer, type A, emerges and is exported to the cell surface (Figure 1.7A). In contrast, loading of exogenous peptides occurs in early endocytic vesicles, or at the cell surface, by exchange with weakly-bound peptides. Due to the absence of functional DM in these compartments, the flexible type B conformer forms in addition to the type A conformer (Figure 1.7B).

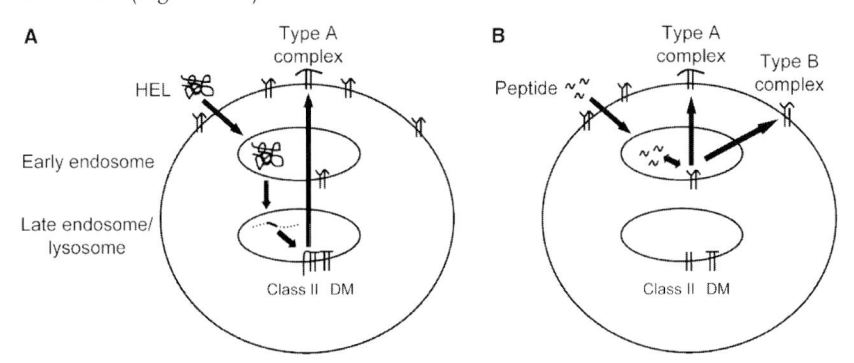

Figure 1.7 Model for the formation of distinct conformers of a peptide-MHC complex. (A) Processing of an intact protein affords a peptide-MHC complex in late endosomal vesicles under the influence of DM molecules, giving only type A complexes. (B) However, peptides can load by exchange in recycling vesicles without the editing role of DM and can generate various isomers. (From Ref. 15.)

1.4.1
Biology of Type B T Cells

The frequency of type B T cells in mice immunized with 46-61 in complete Freund's adjuvant varied from 30 to 50% of the responding T cells [53]. This was measured by limiting dilution analysis of cells in the draining lymph node. Thus the type B T cells did not represent a rare or aberrant T cell, but a significant component of the T cell repertoire. Very importantly, transgenic mice expressing HEL as a membrane-linked protein in their APC showed complete deletion of type A T cells. However, the type B T cells were not deleted, indicating that they had escaped negative selection. The HEL transgenic mice did not respond to immunization with native HEL (which induces mostly type A T cells), but responded to immunization with the peptide. Instead of representing the 30–50% of the anti-48-62 repertoire, type B now represented 100%.

To determine whether such self-reactive type B T cells were present in the normal T cell repertoire of non-transgenic mice, Scott Lovitch immunized B10.BR mice with an abundant autologous peptide derived from the beta chain of I-A^k; this particular peptide was shown by Kappler and Marrack and by us to be present on ~10% of MHC molecules in the spleen and abundantly present in the thymus [18, 19, 59]. Immunization with this peptide did, indeed, result in priming of type B T cells. That the naturally processed peptide was bound to Ak but the T cells failed to recognize it was proven by isolating it from the Ak molecules of the APC and then offering it as an exogenous peptide. Under these circumstances, the peptide was recognized [60].

Thus, naive type B T cells reactive to some self-peptides are found – but does their activation result in autoimmune pathology? Teleologically speaking, type B T cells are unlikely to have evolved to induce autoimmunity. Because of their abundance, we suggest that they represent a specific mode of recognition of exogenous peptides with biological importance, perhaps in antimicrobial immunity. Peptides from processed microbes (or dead and dying infected cells) may reach the extracellular milieu and in this way amplify the repertoire of the anti-microbial T cells. In this context, type B T cells that preferentially recognize short peptides suggest a preference for peptides generated under conditions of high proteolytic activity at inflammatory sites. Recently, Scott Lovitch generated mice with transgenic type B T cell receptors, which should prove extremely useful in investigating the mechanisms whereby peptides are released from proteins to become the type B epitopes.

Other laboratories have reported on T cell reactivity that could be explained by conformational differences. Janeway's group studied the presentation of the dominant peptide derived from Eβ chain presented by I-A^b. They identified T cells that appeared to respond to distinct forms of the complex and that varied in their susceptibility to negative selection in strains of mice that express Eβ [61–65]. Ward's group identified different specificities in the T cell response to the acetylated 1-11 peptide of myelin basic protein (MBP) presented by I-A^u explained by conformational differences [64]. Similar observations were made with respect to the MHC class I allele HLA-B27 [65]. Finally, McConnell's group presented biochemical evidence for conformational isomers of cytochrome peptides [66].

1.5
Negative Selection and Peripheral Activation to HEL Peptides

The relationship between antigen presentation and an ensuing T cell response was addressed in the framework of the biochemical analysis on MHC-bound peptides. The biology of the T cell response to HEL was studied in two biological contexts. The first was central thymic tolerance: we examined how HEL as a self-protein affected T cell development and the response, particularly in situations where TCR transgenic anti-HEL T cells were present. The second issue was the T cell response after immunization of mice with HEL in adjuvant, focusing on their clonal distribution following immunization with HEL in Complete Freund's Adjuvant (CFA). The main issue was to understand how T cell clonal distribution correlated with 300-fold differences in peptide display [28].

To understand the biology of self-antigen presentation, Dan Peterson and Rich DiPaolo first measured the in vitro response of transgenic T cell receptor 3A9 cells that recognizes the complex of the 48-62-Ak. As few as 2–3 complexes per APC [67] deleted double positive thymocytes while 100-fold more complexes were required to activate single positive mature T cells [see also Ref. 68]. The highly efficient process of deletion was confirmed in vivo when transgenic mice expressing membrane HEL driven by the class II promoter were immunized with HEL and the clonal response to its various peptides examined. HEL expression in the thymic APC negatively selected T cells against all HEL displayed peptides [53], even against those with display levels 300-fold less than the chemically dominant 48-63 epitope.

We postulated a "biochemical margin of safety" for self-peptides: thymocytes required fewer peptide-MHC complexes to be deleted than mature T cells needed for their activation. This ensures that a mature T cell that escapes negative selection must see high levels of peptide-MHC complex in the periphery to become spontaneously activated. However, conditions of inflammation may circumvent the need for high-level peptide-MHC to attain peripheral activation [69].

We are evaluating the "biochemical margin of safety", in models of autoimmune diabetes in which HEL is expressed on pancreatic beta cells that express HEL as a membrane protein under the rat insulin promoter [70]. The first transgenic mouse, developed by Goodnow, called ILK-3, had about two million mHEL molecules per beta cells. A cross of the 3A9 TCR transgenic to the RIP-mHEL mice resulted in the development of diabetes [70, 71]. The 3A9 x ILK-3 double transgenic mice developed diabetes, despite extensive negative selection and limited number of T cells in the lymph nodes [71]. Diabetes was likely caused by the high level of HEL in the beta cells (2×10^6 molecules per cell), and its effective cross presentation by APCs in the draining peri-pancreatic lymph node These studies, first by my colleague Rich DiPaolo and most recently by Craig Byersdorfer, suggested that 48-63-Ak complexes presented in a localized fashion exceeded the peripheral "margin of safety" and activated the T cells, beginning the autoimmune process.

1.6
Response to HEL Immunization in the Draining Lymph Node

We studied the clonal distribution of T cells following immunization with HEL in adjuvants by using a sensitive limiting dilution assay for measuring T cell responses [72]. The number of HEL reactive T cells was directly proportional to the amounts of HEL used for immunization. However, a strikingly poor correlation was found between the amount of each epitope displayed by Ak molecules of APC and the *relative* distribution of the reactive T cells following immunization [73]. For example, while the 52-60 peptide family was displayed in very high numbers the T cell clones to it were about the same as for the 20-35 and 114-129 peptide families, which are displayed at considerably lower amounts. Even when immunizing with 100-fold lower levels of HEL as immunogen, the relative reactivity to the four characterized HEL epitopes remained consistent.

Immunization experiments were performed in mice lacking CD40 [74], B7-1 & B7-2 concurrently [75], or in mice treated with an antibody to block CTLA-4 [76], to investigate the role of co-stimulatory molecules. The total anti-HEL response was lowered to varying degrees in the gene knockout mice, but the relative distribution (i.e., percentage of clones found against each epitope) did not change [77]. No one epitope was more or less sensitive to the effects of either co-stimulation or lowered antigen levels. Neither was there an influence of CIITA (Table 1.3).

Table 1.3 Distribution of T cells in lymph nodes of mice immunized to HEL.

Mouse	Dose (nmol)	Frequency	% Specificity			
			48-63	31-47	18-33	115-129
B10.BR	10	1/5000	30	9	17	25
B10.BR	1	1/9500	38	9	15	17
B10.BR	0.1	1/50 000	37	3	17	11
B10.BR	10	1/5000	30	9	17	25
$CD40^{-/-}$	10	1/12 600	45	7	21	27
$B7-1/B7-2^{-/-}$	10	1/64 500	37	16	18	8
Anti-CTLA-4	1	1/32 000	25	5	11	20
Control IgG	1	1/27 500	32	5	17	25

This table summarizes experiments published in Refs. 73 and 77.

The APC responsible for presentation of HEL in vivo was examined by tracing radiolabeled HEL into the draining lymph node [78]. The major subsets of cells bearing HEL at 24 h were CD11c$^+$ or CD11b$^+$/CD11c$^-$ (i.e., DCs and macrophages). At the peak times of presentation, 24–96 h post immunization, the HEL was highly concentrated in a small percentage of DC and macrophages (ca. 1%). This highly efficient concentration indicated that each APC had enough HEL to present all the different peptide-MHC, albeit at widely different densities. If the total HEL had been distributed among most of the APC of the node, rather than concentrated in a few, it would have been mathematically and biochemically impossible to present the minor peptides.

Following high-dose immunization, B cells were found to be positive for 48-62-Ak complexes and functionally capable of presenting HEL peptides, but only following high-dose immunization. The role of B cells is likely to be as an accessory APC, amplifying the response initiated by DCs.

The question as to why immunodominance (the distribution of the reactive T cells) did not correlate with chemical dominance (hierarchy of epitope presentation) remains unresolved. The leveling of the clonal distribution could be the result of (i) a complex interplay between negative and positive co-stimulatory molecules and/or cells not yet investigated; (ii) cooperativity among activated T cell clones, shown in one example in the CD4 response [79], but argued against with respect to CD8 T cells [80]; or (iii) a difference in the primary repertoire of reactive T cells that is then expanded during immunization. All three areas are actively being pursued in the laboratory.

Acknowledgments

We acknowledge the very competent and experienced technical staff of our laboratory and our pre- and post-doctoral fellows who have made the contributions referred to in this chapter. Our colleagues in the Immunology Program at Washington University have provided us with much encouragement and help. The National Institutes of Health, and the National Institute for Allergy and Infectious Disease, supported these investigations.

References

1 Unanue ER. Perspective on antigen processing and presentation. *Immunol. Rev.* 2002, 185, 86–102.

2 Unanue E, Byersdorfer C, Carrero J, Levisetti M, Lovitch S, Pu Z, Suri A. Antigen presentation. Autoimmune diabetes and Listeria – What do they have in common? *Immunol. Res.* 2005, in press.

3 Gammon G, Shastri N, Cogswell J, Wilbur S, Sadegh-Nasseri S, Krzych U, Miller A, Sercarz E. The choice of T-cell epitopes utilized on a protein antigen depends on multiple factors distant from, as well as at the determinant site. *Immunol. Rev.* 1987, 98, 53–73.

4 Moudgil KD, Sekiguchi D, Kim SY, Sercarz EE. Immunodominance is independent of structural constraints: Each region within hen egg white lysozyme is potentially available upon processing of native antigen. *J. Immunol.* 1997, 159, 2574–2579.

5 Allen PM, Strydom DJ, Unanue ER. Processing of lysozyme by macrophages: identification of the determinant recognized by two T cell hybridomas. *Proc. Natl. Acad. Sci. U.S.A.* 1984, 81, 2489–2495.

6 Ziegler K, Unanue ER. Identification of a macrophage antigen-processing event required for I-region-restricted antigen presentation to lymphocytes. *J. Immunol.* 1981, 127, 1869–1877.

7 Ziegler K, Unanue ER. Decrease in macrophage antigen catabolism by ammonia and chloroquine is associated with inhibition of antigen presentation to T cells. *Proc. Natl. Acad. Sci. U.S.A.* 1982, 78, 175–180.

8 Shimonkevitz R, Kappler J, Marrack P, Grey H. Antigen recognition by H-2 restricted T cells. Cell-free antigen processing. *J. Exp. Med.* 1983, 158, 303–316.

9 Unanue ER, Askonas BA. The immune response of mice to antigen in macrophages. *Immunology* 1968, 15, 287.

10 Babbitt BP, Allen PM, Matsueda G, Haber E, Unanue ER. Binding of immunogenic peptides to Ia histocompatibility molecules. *Nature* 1985, 317, 359–362.

11 Babbitt BP, Matsueda G, Haber E, Unanue ER, Allen PM. Antigenic competition at the level of peptide - Ia binding. *Proc. Natl. Acad. Sci. U.S.A.* 1986, 83, 4509–4513.

12 Lindner R, Unanue ER. Distinct antigen MHC class II complexes generated by separate processing pathways. *EMBO J.* 1996, 15, 6910–6920.

13 Harding CV, Collins DS, Slot JW, Geuze HJ, Unanue ER. Liposome-encapsulated antigens are processed in lysosomes, recycled, and presented to T cells. *Cell* 1991, 64, 393–401.

14 Collins DS, Unanue ER, Harding CV. Reduction of disulphide bonds within lysosomes is a key step in antigen processing. *J. Immunol.* 1991, 147, 4054–4059.

15 Pu, Z, Lovitch, S.B., Bikoff, E.K., and E. R. Unanue. 2004. T cells distinguish MHC-peptide complexes formed in separate vesicles and edited by H2-DM. *Immunity* 20, 467–476.

16 Jensen PE. Reduction of disulfide bonds during antigen processing. Evidence from a thiol-dependent insulin determinant. *J. Exp. Med.* 1991, 174, 1121–1130.

17 Arunachalam B, Phan UT, Dong C, Garrett WS, Cannon KS, Alonso C, Karlsson L, Flavell RA, Cresswell P. Defective antigen processing in GILT-free mice. *Science* 2001, 294, 1361–1365.

18 Nelson CA, Roof RW, McCourt DW, Unanue ER. Identification of the naturally processed form of hen egg white lysozyme bound to the murine major histocompatibility complex class II molecule I-A^k. *Proc. Natl. Acad. Sci. U.S.A.* 1992, 89, 7380–7383.

19 Nelson CA, Viner NJ, Young SP, Petzold SJ, Unanue ER. A negatively charged anchor residue promotes high-affinity binding to the MHC molecule I-A^k. *J. Immunol.* 1996, 157, 755–762.

20 Gugasyan R, Vidavsky I, Nelson CA, Gross ML, Unanue ER. Isolation and quantitation of a minor determinant of hen egg white lysozyme bound to I-A^k

20. by using peptide-specific immunoaffinity. *J. Immunol.* 1998, 161, 6074–6083.
21. Gugasyan R, Velazquez C, Vidavsky I, Deck BM, van der Drift K, Gross ML, Unanue ER. Independent selection by I-A^k molecules of two epitopes found in tandem in an extended polypeptide. *J. Immunol.* 2000, 165, 3206–3213.
22. Velazquez C, Vidavsky I, van der Drift, K, Gross ML, Unanue ER. Chemical identification of a low abundant lysozyme peptide family bound to I-A^k histocompatibility molecules. *J. Biol. Chem.* 2002, 277, 42 514–42 522.
23. Falk K, Ratzschke O, Stevanovic S, Jung G, Rammensee HG. Allele-specific motifs revealed by sequencing of self-peptides eluted from MHC molecules. *Nature* 1991, 351, 290–296.
24. Chicz RM, Urban RG, Lane WS, Gorga JC, Stern LJ, Vignali DA, Strominger JL. Predominant naturally processed peptides bound to HLA-DR1 are derived from MHC-related molecules and are heterogeneous in size. *Nature* 1992, 358, 764–768.
25. Rudensky AY, Preston-Hurlburt P, Al-Ramadi BK, Rothbard J, Janeway Jr CA. Predominant naturally processed peptides bound to HLA-DR1 are derived from MHC-related molecules and are heterogeneous in size. *Nature* 1992, 429–431.
26. Hunt DF, Michel H, Dickinson TA, Shabanowitz J, Cox AL, Sakaguchi K, Appella E, Grey HM, Sette A. Peptides presented to the immune system by the murine class II major histocompatibility complex molecule I-A^d. *Science* 1992, 256, 1817–1820.
27. Dadaglio G, Nelson CA, Deck MB, Petzold SJ, Unanue ER. Characterization and quantitation of peptide-MHC complexes produced from hen egg lysozyme using a monoclonal antibody. *Immunity* 1997, 6, 727–738.
28. Velazquez C, DiPaolo R, Unanue ER. Quantitation of lysozyme peptides bound to class II MHC molecules indicates very large differences in levels of presentation. *J. Immunol.* 2001, 166, 5488–5494.
29. Nelson CA, Fremont DH. Structural prinicples of MHC class II antigen presentation. *Rev. Immunogen.* 1999, 1, 47–59.
30. Brown JH, Jardetzky TS, Gorga JC, Stern LJ, Urban RG, Strominger JL, Wiley DC. Three-dimensional structure of the human class II histocompatibility antigen HLA-DR1. *Nature* 1993, 364, 33–39.
31. Jardetzky TS, Brown JH, Gorga JC, Stern LJ, Urban RG, Chi YI, Stauffacher C, Strominger JL, Wiley DC. Three-dimensional structure of a human class II histocompatibility molecule complexed with superantigen. *Nature* 1994, 368, 711–718.
32. Dessen A, Lawrence CM, Cupo S, Zaller DM, Wiley DC. X-ray crystal structure of HLA-DR4 (DRA*0101, DRB1*0401) complexed with a peptide from human collagen II. *Immunity* 1997, 7, 473–481.
33. Fremont DH, Hendrickson WA, Marrack P, Kappler J. Structures of an MHC class II molecule with covalently bound single peptides. *Science* 1996, 272, 1001–1004.
34. Fremont DH, Monnaie D, Nelson CA, Hendrickson WA, Unanue ER. Crystal structure of I-A^k in complex with a dominant epitope of lysozyme. *Immunity* 1998, 8, 305–317.
35. Corper AL, Stratmann T, Apostolopoulos V, Scott CA, Garcia KC, Kang AS, Wilson IA, Teyton L. A structural framework for deciphering the link between I-A^{g7} and autoimmune diabetes. *Science* 2000, 288, 505–511.
36. Scott CA, Peterson PA, Teyton L, Wilson IA. Crystal structures of two I-Ad-peptide complexes reveal that high affinity can be achieved without large anchor residues. *Immunity* 1998, 319–329.
37. Latek RR, Suri A, Petzold SJ, Nelson CA, Kanagawa O, Unanue ER, Fremont DH. Structural basis of peptide binding and presentation by the type 1 diabetes-associated MHC class II molecule of NOD mice. *Immunity* 2000, 12, 699–710.
38. Donermeyer DL, Allen PM. Binding to Ia protects an immunogenic peptide from proteolytic degradation. *J. Immunol.* 1989, 142, 1063–1068.

39 Werdelin O, Mouritsen S, Peterson BL, Sette A, Buus S. Facts on the fragmentation of antigens in presenting cells, on the association of antigen fragments with MHC molecules in cell-free systems, and speculation on the cell biology of antigen processing. *Immunol. Rev.* 1988, 106, 181–193.

40 Carrasco-Marin E, Petzold S, Unanue ER. Two structural states of peptide-class II MHC complexes revealed by photoaffinity labeled peptides. *J. Biol. Chem.* 1999, 274, 31 333–31 340.

41 Falk K, Rotzschke O, Stevanovic S, Jung G, Rammensee HG. Pool sequencing of natural HLA-DR, DQ, and DP ligands reveals detailed peptide motifs, constraints of processing, and general rules. *Immunogenetics* 1994, 39, 230–242.

42 Nelson CA, Vidavsky I, Viner NJ, Gross ML, Unanue ER. Amino-terminal trimming of peptides for presentation on major histocompatibility complex class II molecules. *Proc. Natl. Acad. Sci. U.S.A.* 1997, 94, 628–633.

43 Jardetzky TS, Gorga JC, Busch R, Rothbard J, Strominger JL, Wiley DC. Peptide binding to HLA-DR1: a peptide with most residues substituted to alanine retains MHC binding. *EMBO J.* 1990, 9, 1797–1803.

44 Latek RR, Petzold S, Unanue ER. Hindering auxiliary anchors are potent modulators of peptide binding and selection by I-A^k class II molecules. *Proc. Natl. Acad. Sci. U.S.A.* 2000, 97, 11 460–11 465.

45 Boehncke WH, Takeshita T, Pendleton CD, Houghten RA, Sadeghi-Nasseri S, Racioppi L, Berzofsky JA, Germain RN. The importance of dominant negative effects of amino acid side chain substitution in peptide-MHC molecule interactions and T cell recognition. *J. Immunol.* 1993, 150, 331–341.

46 Sette A, Sidney J, Oseroff C, del Guercio MF, Southwood S, Arrhenius T, Powell MF, Colon SM, Gaeta FC, Grey HM. HLA DR4w4-binding motifs illustrate the biochemical basis of degeneracy and specificity in peptide-DR interactions. *J. Immunol.* 1993, 151, 3163–3170.

47 Allen PM, Matsueda GR, Evans RJ, Dunbar JB Jr, Marshall G, Unanue, ER. Identification of the T-cell and Ia contact residues of a T-cell antigenic epitope. *Nature* 1987, 327, 713–715.

48 Pu, Z., Carrero, J. A., and E. R. Unanue. Distinct recognition by two subsets of T cells of an MHC class II-peptide complex. *Proc. Natl. Acad. Sci. U.S.A.* 2002, 99, 8844–8849.

49 Sercarz EE, Lehmann PV, Ametani A, Benichou G, Miller A, Mougdil K. Dominance and crypticity of T cell antigenic determinants. *Annu. Rev. Immunol.* 1993, 11, 729–766.

50 Viner NJ, Nelson CA, Unanue ER. Identification of a major I–E^k-restricted determinant of hen egg lysozyme: Limitations of lymph node proliferation studies in defining immunodominance and crypticity. *Proc. Natl. Acad. Sci. U.S.A.* 1995, 92, 2214–2218.

51 Viner NJ, Nelson CA, Deck B, Unanue ER. Complexes generated by the binding of free peptides to class II MHC molecules are antigenically diverse compared with those generated by intracellular processing. *J. Immunol.* 1996, 156, 2365–2368.

52 Cirrito TP, Pu Z, Deck MB, Unanue ER. Deamidation of asparagine in a major histocompatibility complex-bound peptide affects T cell recognition but does not explain type B reactivity. *J. Exp. Med.* 2001, 194, 1165–1170.

53 Peterson DA, DiPaolo RJ, Kanagawa O, Unanue ER. Quantitative analysis of the T cell repertoire that escapes negative selection. *Immunity* 1999, 11, 453–62.

54 Sloan VS, Cameron P, Porter G, Gammon M, Amaya M, Mellins E, Zaller DM. Mediation by HLA-DM of dissociation of peptides from HLA-DR. *Nature* 1995, 375, 802–806.

55 Kropshofer H, Vogt AB, Moldenhauer G, Hammer J, Blum JS, Hammerling GJ. Editing of the HLA-DR-peptide repertoire by HLA-DM. *EMBO J.* 1996, 15, 6144–6154.

56 van Ham SM, Gruneberg U, Malcherek G, Broker I, Melms A, Trowsdale J. Human histocompatibility leukocyte antigen (HLA)-DM edits peptides pre-

sented by HLA-DR according to their ligand binding motifs. *J. Exp. Med.* 1996, 184, 2019–2024.

57 Lovitch SB, Petzold SJ, Unanue ER. Cutting edge: H-2DM is responsible for the large differences in presentation among peptides selected by I-A^k during antigen processing. *J. Immunol.* 2003, 171, 2183–2186.

58 Doebele RC, Busch R, Scott HM, Pashine A, Mellins ED. Determination of the HLA-DM interaction site on HLA-DR molecules. *Immunity* 2000, 13, 517–527.

59 Marrack P, Ignatowicz L, Kappler JW, Boymel J, Freed JH. 1993. Comparison of peptides bound to spleen and thymus class II. *J. Exp. Med.* 1993, 178, 2173–2183.

60 Lovitch SB, Walters JJ, Gross ML, Unanue ER. APCs present A^k-derived peptides that are autoantigenic to type B T cells. *J. Immunol.* 2003, 170, 4155–4160.

61 Barlow AK, He X, Janeway CA Jr. Exogenously provided peptides of a self-antigen can be processed into forms that are recognized by self-T cells. *J. Exp. Med.* 1998, 187, 1403–1415.

62 Viret C, He X, Janeway CA Jr. Paradoxical intrathymic positive selection in mice with only a covalently presented agonist peptide. *Proc. Natl. Acad. Sci. U.S.A.* 2001, 98, 9243–9248.

63 Viret C, He X, Janeway CA Jr. Altered positive selection due to corecognition of floppy peptide/MHC II conformers supports an integrative model of thymic selection. *Proc. Natl. Acad. Sci. U.S.A.* 2003, 100, 5354–5359.

64 Huang JC, Han M, Minguela A, Pastor S, Qadri A, Ward ES. T cell recognition of distinct peptide: I-A^u conformers in murine experimental autoimmune encephalomyelitis. *J. Immunol.* 2003, 171, 2467–2477.

65 Hulsmeyer M, Fiorillo M, Fiorillo, MT, Bettosini F, Sorrentino R, Saenger W, Ziegler A, Uchanska-Ziegler B. Dual, HLA-B27 Subtype-dependent conformation of a self-peptide. *J. Exp. Med.* 2004, 199, 271–281.

66 Beeson C, Anderson TG, Lee C, McConnell HM. Isomeric complexes of peptides with class II proteins of the major histocompatibility complex. *J. Am. Chem. Soc.* 1995, 117, 10 429–10 433.

67 Peterson DA, DiPaolo RJ, Kanagawa O, Unanue ER. Cutting edge: negative selection of immature thymocytes by a few peptide-MHC complexes: differential sensitivity of immature and mature T cells. *J. Immunol.* 1999, 162, 3117–3120.

68 Reay PA, Matsui K, Haase K, Wulfing C, Chien YH, Davis MM. Determination of the relationship between T cell responsiveness and the number of MHC-peptide complexes using specific monoclonal antibodies. *J. Immunol.* 2000, 164, 5626–5634.

69 Pape KA, Khoruts A, Mondino A, Jenkins MK. Inflammatory cytokines enhance the *in vivo* clonal expansion and differentiation of antigen-activated CD4+ T cells. *J. Immunol.* 1997, 159, 591.

70 Akkaraju S, Ho WY, Leong D, Canaan K, Davis MM, Goodnow CC. A range of CD4 T cell tolerance: partial inactivation to organ-specific antigen allows nondestructive thyroiditis or insulitis. *Immunity* 1997, 7, 255–271.

71 DiPaolo RJ, Unanue ER. The level of peptide-MHC complex determines the susceptibility to autoimmune diabetes: studies in HEL transgenic mice. *Eur. J. Immunol.* 2001, 31, 3453–3459.

72 Kanagawa O, Martin SM, Vaupel BA, Carrasco-Marin E, Unanue ER. Autoreactivity of T cells from nonobese diabetic mice: an I-Ag7-dependent reaction. *Proc. Natl. Acad. Sci. U.S.A.* 1998, 95, 1721–1724.

73 DiPaolo RJ, Unanue ER. Cutting edge: chemical dominance does not relate to immunodominance: studies of the CD4+ T cell response to a model antigen. *J. Immunol.* 2002, 169, 1–4.

74 Castigli E, Alt FW, Davidson L, Bottaro A, Mizoguchi E, Bhan AK, Reha RS. CD40-deficient mice generated by recombination-activating gene-2-deficient blastocyst complementation. *Proc. Natl. Acad. Sci. U.S.A.* 1994, 91, 12 135–12 141.

75 Borriello F, Sethna MP, Boyd SD, Schweitzr AN, Tivol EA, Jacoby D, Strom TB, Simpson EM, Freeman GJ, Sharpe AH. B7-1 and B7-2 have overlapping, critical roles in immunoglobulin class switching and germinal center formation. *Immunity* 1997, 6, 303–311.

76 Luhder, Hoglund P, Allison JP, Benoist C, Mathis D. Cytotoxic T lymphocyte-associated antigen 4 (CTLA-4) regulates the unfolding of autoimmune diabetes. *J. Exp. Med.* 1998, 187, 427–432.

77 DiPaolo RJ, Unanue ER. Cutting edge: the relative distribution of T cells responding to chemically dominant or minor epitopes of lysozyme is not affected by CD40-CD40 ligand and B7-CD28-CTLA-4 costimulatory pathways. *J. Immunol.* 2002, 169, 2832–2836.

78 Byersdorfer C, DiPaolo R, Petzold S, Unanue E. Following immunization antigen becomes concentrated in a limited number of antigen presenting cells including B cells. *J. Immunol.* 2004, 173, 6627–6634.

79 Creusot RJ, Thomsen LL, Tite JP, Chain BM. Local cooperation dominates over competition between CD4+ T cells of different antigen/MHC specificity. *J. Immunol.* 2003, 171, 240–246.

80 Kedl RM, Rees WA, Hildeman DA, Schaefer B, Mitchell T, Kappler J, Marrack P. T cells compete for access to antigen-bearing antigen-presenting cells. *J. Exp. Med.* 2000, 192, 1105–1113.

Part II
Molecular Mechanisms of Antigen Processing

2
Antigen Entry Routes – Where Foreign Invaders Meet Antigen Presenting Cells

Percy A. Knolle

2.1
Introduction

The immune system is designed to fight infection and thereby preserve the functional integrity of the host. The different components of the immune system, i.e., innate and adaptive immunity, operate in a coordinate fashion to assure continuous immune surveillance at those anatomical sites most likely to be place of invasion by infectious microorganisms. Although infection is normally localized and control of infection by the immune system requires local immune activity, induction of efficient antigen-specific immunity requires multiple steps that occur in different anatomic compartments. This chapter will attempt to shed light on the mechanisms of local immune control by antigen presenting cells, which are located either at external body surfaces (to prevent local infection) or in the spleen and liver to fight infectious microorganisms once they have disseminated via the blood stream.

Conceptually, the immune system is strongly compartmentalized. Antigen presenting cells reside in non-lymphatic tissue where they sample antigens that have crossed the physical barriers preventing direct access of infectious microorganisms into tissue. Exit of antigen presenting cells from these sites occurs either at a constant rate or may be enhanced by signals derived from invading microorganisms. Leaving peripheral tissues via the lymphatics, antigen presenting cells enter lymphatic tissue, i.e. draining lymph nodes, where they encounter in a specialized microenvironment naive T or B lymphocytes in distinct anatomic locations [1]. If antigen presenting cells have been stimulated in an adequate fashion to achieve a fully activated state, the antigen-loaded cells can mount a strong antigen-specific immune response mediating immunity [1, 2]. The molecular mechanisms leading to full activation or "licensing" of antigen presenting cells are still under intensive investigations, but it has become evident that binding of certain highly conserved microbial structures to Toll-like receptors expressed on antigen presenting cells is instrumental in this crucial activation process [3]. Following cognate interaction with activated antigen presenting cells, T and B lymphocytes undergo clonal expansion and large numbers of antigen-specific lymphocytes are released from

Antigen Presenting Cells: From Mechanisms to Drug Development. Edited by H. Kropshofer and A. B. Vogt
Copyright © 2005 WILEY-VCH Verlag GmbH & Co. KGaA, Weinheim
ISBN: 3-527-31108-4

lymphatic tissue to execute their effector function at peripheral sites of infection (Figure 2.1). Recently, evidence has accumulated that immune surveillance in peripheral tissues by antigen presenting cells is a highly elaborate process and that the initial encounter of infectious microorganisms with different types of antigen presenting cells in these peripheral sites largely determines the subsequent development of immune responses. Moreover, organ anatomy is shaped by function, and local immune surveillance has to adapt to the need to maintain organ function.

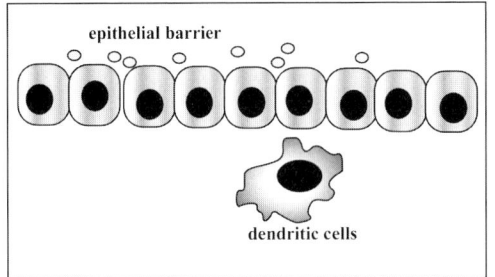

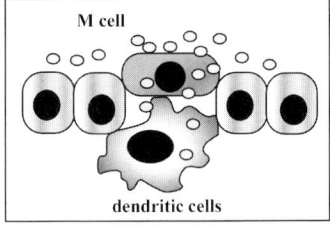

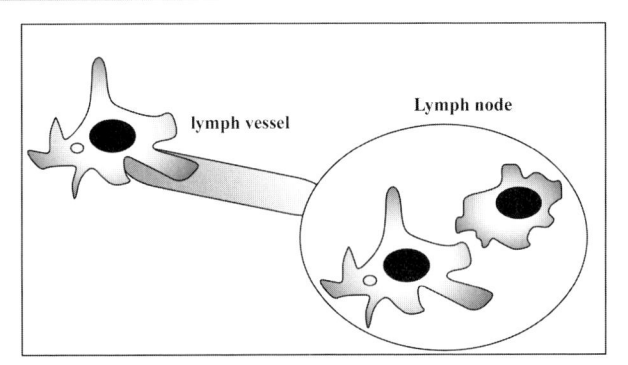

Figure 2.1 Outline of the conventional view that antigens are in peripheral tissue by dendritic cells and are transported by migratory dendritic cells to the lymph node where interaction with naive T lymphocytes occurs. Initial contact of antigens or pathogens in these peripheral microenvironments is believed to determine the functional outcome of ensuing immune responses. (This figure also appears with the color plates.)

2.2
Antigen Entry via the Gastrointestinal Tract

The mucosa of the intestinal tract is exposed to myriads of microorganisms that colonize the gut, but only a few of these can enter the body and cause infection and disease. Microorganisms are limited in their ability to enter the body partly because gastrointestinal epithelial cells are connected by tight junctions and form a physical barrier that prevents paracellular movement of microorganisms, metabolites or degradation products [4]. Interestingly, only a single epithelial cell layer separates intestinal tissue from antigens and microorganisms contained in the gut lumen. For further protection, the surface of epithelial cells is covered by secretory products such as mucins, defensins and secretory antibodies [5]. Effective immune surveillance requires that intact antigens and microorganisms are transported to immune cells located on the other side of the epithelial cell layer. This transport occurs at specialized sites that contain organized mucosal lymphoid follicles and where a close collaboration of specialized epithelial cells with antigen presenting cells has developed [6]. The distribution of these sites in the body reflects the local abundance of foreign antigens and microorganisms.

2.2.1
Peyer's Patches

The best studied examples of these organized mucosa associated lymphoid tissue (MALT) are Peyer's Patches (PP), which are found in the distal ileum and are characterized by the appearance of numerous lymphoid follicles in patches. In close resemblance to lymph nodes, PP are organized into an assembly of B lymphocytes, which is surrounded by follicular dendritic cells and is flanked by T lymphocyte rich areas. Within the T lymphocyte area high endothelial venules allow entry and exit of migrating cells. The follicle is separated from the epithelial cell layer by a so-called "dome" region that is rich in antigen presenting cells and T lymphocytes [7].

Epithelial cells overlying this "dome"-structure bear a unique phenotype. In contrast to normal villous epithelium that is specialized in digestion and absorption of nutrients, the follicle-associated epithelium contains specialized (microfold) M cells, which are operative in transepithelial vesicular transport from the lumen directly to intra-epithelial antigen presenting cells [8]. These cells differentiate from enterocytes under the influence of membrane bound $LT\alpha_1\beta_2$, which is expressed by local lymphoid cells, especially B cells [9]. M cells do not express polymeric Ig receptors and are therefore unable to transport protective IgA from the interstitium to their luminal surface. Interaction with microorganisms in the gut lumen is further facilitated by different expression of glycosyltransferases, resulting in a unique glycosylation pattern [10]. M cells lack a brush border but possess numerous microvilli interspersed with large plasma membrane subdomains that participate in clathrin-mediated endocytosis of antigens and microorganisms [11]. Furthermore, M cells execute fluid-phase endocytosis, phagocytosis and macropinocytosis [12]. Numerous microorganisms are taken up M cells, such as poliovirus and shigella [13]. Transcytotic transport by polarized M cells is the main pathway for endocytosed material in M cells and little material is delivered into the lysosomal pathway for destruction [8]. A large network of dendritic cells is found in direct vicinity to M cells [14]. Dendritic cells found at this site are either $CD11c^+CD11b^{int}$ or are $CD11c^+CD8\alpha^{neg}$, $CD11c^+CD11b^{neg}$ and appear to be in an immature state as they do not express markers such as DEC-205, which are associated with dendritic cell maturation. Conceivably, the close proximity of M cells, which sample antigens and microorganisms from the intestinal lumen, and dendritic cells allows for efficient uptake of the transcytosed materials. Indeed uptake of pathogenic microorganisms such as salmonella or listeria has been observed into dendritic cells in the "dome" region [15].

Dendritic cells in the "dome" region bear further unique functional characteristics. They express the cytokine IL-10 after ligation of the co-stimulatory molecule RANK, which in splenic dendritic cells induces strong expression of IL-12. Along the same line, dendritic cells in this location polarize antigen-specific T cells to production of IL-10 and Th2-cytokines [16, 17].

Instrumental in the recruitment of dendritic cells into the subepithelial "dome" region is the expression of the chemokine receptor CCR6, whose ligand CCL20 is expressed in the intestinal tract only by M cells [7]. Mice deficient for CCR6 lack dendritic cells in the "dome" region and are unable to mount an immune response although size of the PP and distribution of other lymphoid cells appear normal [18]. These experiments demonstrated that the presence of dendritic cells in this location is not accidental but rather is an essential component in local immune surveillance of the intestinal tract. Interestingly, contact of M cells with microorganisms or microbial structures leads to increased CCL20 expression and might thus enhance recruitment of dendritic cells from the subepithelial area to the "dome" region [19].

The anatomical situation in PP is unique as the site of initial antigen entry and uptake into dendritic cells is in close to organized lymphoid tissue containing T and B lymphocytes. Instead of time-consuming migration along lymphatic vessels, dendritic cells loaded with antigen previously transported by M cells from the intestinal lumen migrate to adjacent interfollicular T cell zones, where they undergo maturation and present luminal antigens [7]. However, migration of dendritic cells only occurs in the presence of appropriate microbial "signals"; immunologically inert cargo is transported by M cells to dendritic cells and is rapidly endocytosed but dendritic cells remain in the subepithelial "dome" region for long periods (Figure 2.1).

These data assigned an important role of antigen presenting cells within PP to immune surveillance of the intestinal tract. The immune system needs to discriminate pathogens from innocuous food antigens and commensal bacteria that are equally present in the intestinal tract. Local immune-regulatory mechanisms operative in the unique and specialized microenvironment of PP seemed to be sufficient to determine the outcome of immune responses towards local intestinal antigens/microorganisms [20]. However, experimental results obtained from mice deficient for PP gave conflicting data, challenging the view that PP were the only site important for intestinal immune surveillance.

2.2.2
Mesenteric Lymph Node

There is unequivocal evidence that mesenteric lymph nodes are absolutely required for T cell priming towards intestinal antigens to occur. Different experimental approaches, using adoptive transfer of TCR-transgenic T cells and LTα-deficient animals (lacking mesenteric lymph nodes), all pointed to this important function of the mesenteric lymph node. This role of mesenteric lymph nodes in intestinal immune surveillance is not surprising, as lymph nodes in general are required for initiation of immune responses towards antigen present in peripheral tissues. Antigen sampling by M cells and loading of dendritic cells with luminal antigens may be followed by migration of dendritic cells into the mesenteric lymph node where then T cell priming occurs. However, undisturbed priming of T cells towards intestinal antigens in B cell-deficient mice, which lack almost all M cells, argues that delivery of intestinal antigens by M cells is not essential [21]. Probably, dendritic cells loaded with luminal antigen in PP migrate to mesenteric lymph nodes. Alternatively, antigen may be delivered in a different way to the regional lymph nodes. Free antigen may reach mesenteric lymph nodes, especially if large amounts are present in the intestinal lumen. Antigens may further be distributed as exosomes secreted by intestinal epithelial cells [22]. In both cases antigen not only reaches regional lymph nodes but is systemically distributed via the blood stream, a process occurring within minutes to hours after oral ingestion of antigen [23, 24]. The consequences of antigen uptake by antigen presenting cell populations that survey the blood compartment will be dealt with later.

2.2.3
Dendritic Cells of the Lamina Propria

Distribution of dendritic cells in the gastrointestinal tract is not restricted to PP. Dendritic cells with typical features of immature cells have also been found in the lamina propria of the gut inserting into the epithelial cell layer [25]. In this subepithelial location, antigens or microorganisms contact dendritic cells only after traversing the epithelial cell monolayer. Involvement of lamina propria dendritic cells in the immune surveillance of the gastrointestinal tract is indicated by the observation that penetration of M cells by bacteria requires expression of bacterial invasion proteins. Nevertheless, *Salmonella typhimurium*, which is deficient in invasion genes, still manages to infect dendritic cells and disseminate to the spleen [26], which suggests an M-cell independent pathway. As CD18-deficient mice were protected from infection by *Salmonella* [27], it was initially postulated that $CD18^+$ phagocytic antigen presenting cells were involved in bacterial invasion. This conceptual contradiction was solved by the finding that dendritic cells open up the tight junctions between epithelial cells and send dendrites outside the epithelium [28]. Dendritic cells express tight junction proteins such as occluding, claudin-1, E-cadherin and β-catenin and thus preserve the integrity of the epithelial barrier function [28]. The molecular mechanism mediating formation of transepithelial dendrites is expression of the chemokine CX_3CL1 (fractalkine) by intestinal epithelial cells and expression of CX_3CR1 by dendritic cells. No formation of transepithelial dendrites is observed if lamina propria dendritic cells lack CX_3CR1 expression. Transepithelial dendrite formation is restricted to certain areas of the gastrointestinal tract and was only observed in the villi of the terminal ileum. Such dendrites enable lamina propria dendritic cells to directly sample luminal antigens and microorganisms [29]. Interestingly, uptake of bacteria occurs both directly through transepithelial dendrites into lamina propria lymphocytes and indirectly via M cells into dendritic cells present in the "dome" region of PP. Using knockout animals for CX_3CR1 it became clear that formation of transepithelial dendrites is absolutely required for uptake of non-invasive bacteria by lamina propria dendritic cells, whereas uptake of non-invasive bacteria in dendritic cells located in PP was not affected [29]. However, only lamina propria dendritic cells loaded with bacteria are found in mesenteric lymph nodes, whereas dendritic cells loaded with antigen in PP remain stationary. Dendritic cells present in the lamina propria are phenotypically different from those in the "dome" region of PP, because they lack CCR6 expression [29]. These studies underline the importance of lamina propria dendritic cells in sampling bacteria from the gastrointestinal tract (Figure 2.2).

Interestingly, invasive bacteria are detected mainly in the CX_3CR1 dendritic cell population within the lamina propria, which implies a further role for these unique dendritic cells in control of bacterial infection. In CX_3CR1 knockout animals, invasive bacteria are taken up by another phagocytosing cell population.

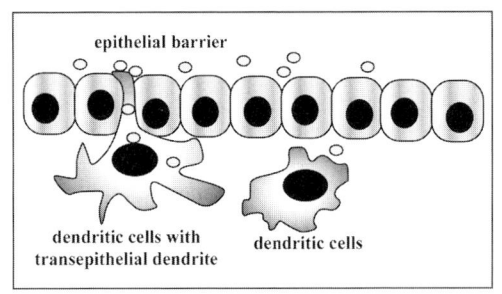

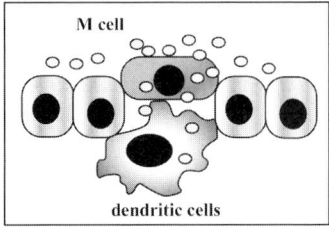

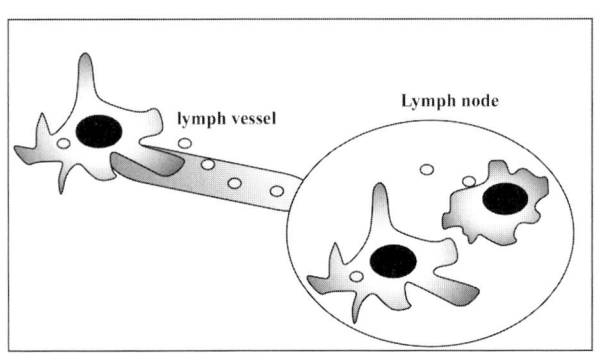

Figure 2.2 Recent work has provided evidence that dendritic cells reach via dendritic extensions, e.g. transepithelial dendritis in the gut, the outside world and directly sample antigens as well as microorganisms. Moreover, antigens that have reached peripheral tissue may directly gain access to lymph nodes, via defined anatomic structures, and in the lymph node interact with a, presumably, resident dendritic cell population. Bypassing peripheral antigen sampling migratory dendritic cells it remains an open question how specific information about the need for induction of immune responses is transferred to the lymph node, or whether this pathway is uniquely used for induction of immune tolerance towards tissue-autoantigens. (This figure also appears with the color plates.)

More importantly, CX_3CR1 knockout animals succumb to infection with invasive bacteria, whereas heterozygous knockout animals or wild-type littermates showed resistance [29]. Infection in CX_3CR1 knockout animals was characterized by significantly higher bacterial loads. Delayed uptake of bacteria by dendritic cells in the lamina propria from the gut lumen may thus constitute a mechanistic explanation for the impairment of local immune responses in the gut in these knockout animals. As dendritic cell populations present in the lamina propria or in PP show so fundamentally different functions it seems likely that they belong to different dendritic cell subsets that await more detailed analysis.

The interaction of dendritic cells in the lamina propria with bacteria is rather dynamic and far from short-lived. In contrast to $CD11b^+$ macrophages, which rapidly destroy bacteria after phagocytosis, dendritic cells are relatively inefficient in killing phagocytosed bacteria [30]. Bacterial degradation products enhance expression of CX_3CL1 and thus increase formation of transepithelial dendrites. Uptake of commensal non-invasive bacteria occurs again by both lamina propria dendritic cells and dendritic cells within PP. Uptake and survival of commensal bacteria in intestinal dendritic cells leads to local stimulation of IgA^+ B cells. As a consequence increased expression of IgA is induced locally, providing protective effects for intestinal tissue [31]. In contrast to the paradigmatic view that immune responses must undergo a phase of systemic stimulation, dendritic cells loaded with commensal bacteria locally induce protective B cell dependent mechanisms to prevent tissue infection. Immune responses towards commensal bacteria are kept strictly local. Intestinal dendritic cells migrate to mesenteric lymph nodes after sampling of commensal bacteria but do not enter the systemic circulation. Removal of the mesenteric lymph nodes leads to rapid distribution of commensal bacteria into the systemic circulation and development of a widespread inflammatory reaction [31]. Local dendritic cell populations are operative in maintaining a finely tuned balance between host and commensal bacteria. However, the exact molecular mechanisms that allow the intestinal immune system to raise fast and efficient immunity towards entero-invasive pathogens remain undefined.

2.2.4
Pathogens Target Intestinal Antigen Presenting Cells

Two salient features of antigen presenting cells are their ability to take up antigens or pathogens and their migratory capacity. As discussed above, uptake of invasive and non-invasive bacteria as well as migration of intestinal antigen presenting cells to mesenteric lymph nodes is instrumental for immune surveillance of the gastrointestinal tract. Exactly these functions of antigen presenting cells are exploited by pathogens to enter the body and establish infection. However, pathogens or their components can activate antigen presenting cells directly [3] and thus induce maturation of antigen presenting cells, which is the prerequisite to mount strong pathogen-specific immunity. Modulation of antigen-presenting cell function by the pathogen is therefore necessary to achieve successful infection from a pathogen's point of view.

The central role of CD18$^+$ antigen presenting cells in infection by *Salmonella* has been elegantly shown in CD18 knockout mice, which are resistant to infection with *Salmonella* [27]. Invasive and non-invasive *Salmonella* are phagocytosed by macrophages and dendritic cells that are present in PP or in lamina propria forming transepithelial dendrites [28, 29]. Breaching of the physical barrier of the intestinal epithelial cells by either invasion of M cells or exploitation of the phagocytic capacity of "snorqueling" dendritic cells constitutes the first step in *Salmonella* infection. Within minutes of oral administration of non-invasive bacteria, CD18$^+$ cells harboring *Salmonella* are found in the blood [27], clearly demonstrating the efficiency of this process. Infection of antigen presenting cells, however, allows direct presentation of pathogen-derived antigens to CD4 and CD8 T lymphocytes [32, 33]. To escape degradation in the phagolysosomal compartment and subsequent induction of antigen specific immune responses *Salmonella* mediate apoptosis of infected macrophages and dendritic cells. Using a "molecular syringe" (type III secretion system) *Salmonella* inject invasins that induce apoptosis through binding to caspases [34]. But targeted elimination of infected antigen-presenting cells does not abrogate pathogen-specific immune responses. Apoptotic cellular material containing *Salmonella* are efficiently taken up by neighboring dendritic cells. These dendritic cells are then able to process and present bacterial antigens on MHC II molecules to CD4 T cells. Moreover, dendritic cells can cross-present these exogenous antigens on MHC I molecules to CD8 T cells [35]. The mechanism of cross-presentation is of utmost importance to mount pathogen-specific CD8 T cell immunity in those cases where pathogens do not directly infect antigen presenting cells [36]. However, cross-presentation is equally important to maintain immune tolerance towards tissue autoantigens [37]. Appropriate signals are therefore required to shift the balance from default tolerance to immunity [3]. In general, to fight *Salmonella* infection different antigen presenting cell populations in different anatomic locations act cooperatively to contain initial infection and to prevent further dissemination, even in the presence of bacterial immune evasion mechanisms.

Another example of exploitation of antigen presenting cells is infection with the human immunodeficiency virus (HIV). Migration of pathogen-loaded dendritic cells from peripheral sites to lymph nodes not only serves to activate pathogen-specific T cell responses but may serve to contribute to dissemination of infection within the host. HIV uses a novel mechanism to exploit dendritic cell migration and subsequent interaction with T cells in lymphatic tissue for its replication and spread. HIV can bind to a dendritic cell specific C-type II lectin, DC-SIGN, which is strongly expressed on dendritic cells in mucosal tissues. Upon binding to DC-SIGN, HIV is endocytosed by dendritic cells but endosomal maturation and subsequent endosomal-lysosomal fusion with subsequent virus degradation does not occur [38]. Uptake of HIV does not result in infection of dendritic cells. Rather, HIV now survives for long time periods (≥5 days) within dendritic cells, having escaped the rather hostile extracellular environment. Migration of HIV-bearing dendritic cells to peripheral lymph nodes now enables HIV to find its final target cell, CD4 T lymphocytes. Infection of CD4 T lymphocytes occurs in "trans" when

HIV bound to DC-SIGN has re-emerged on the dendritic cell surface [38]. Immune surveillance function by dendritic cells, i.e., continuous sampling of external body surfaces, thus provides a mean for HIV to overcome the physical barrier of the mucosal epithelium. Indeed, atraumatic application of SIV to mucosal sites leads to infection [39]. Meanwhile it has become apparent that other pathogenic microorganisms equally use DC-SIGN as a receptor. But DC-SIGN not only serves as an entry receptor for pathogens. Binding of pathogens to DC-SIGN is now recognized to lead to functional alteration of dendritic cells rather supporting development of immune escape [40].

Encounter of antigen presenting cells with pathogens has fundamental consequences if pathogens have evolved strategies to escape the ensuing immune response. Conversely, immune surveillance at mucosal sites is of enormous importance to prevent unnecessary inflammation in response to commensal microorganisms. Consequently, maintenance of this balance between immune response and commensal microorganisms may be achieved either by keeping immune reactivity strictly local or even by inducing immune tolerance. Central to this to decision as to mount immunity or immune tolerance are antigen presenting cells. Their activation status and local environmental signals provide the complex mixture of determinants that influence the outcome of immune responses.

2.3
Antigen Entry via the Skin

The predominant dendritic cell population of the skin is the Langerhans cells. Much of our knowledge on the migration of dendritic cells derives from studies conducted with Langerhans cells (LC). These cells are typically localized in the basal and suprabasal layer of the epidermis and represent the principal barrier of hematopoietic cells to the external environment. LC are well equipped to take up antigens or pathogens that have breached the physical skin barrier. In contrast to dendritic cell populations in other non-lymphoid organs, LC are rather tissue-resident and remain stationary for months [41]. Replacement of LC occurs from bone-marrow derived precursors in a CCR2-dependent fashion [41]. Numerous studies have demonstrated that LC take up antigens in the skin and migrate to the draining lymph node if appropriate inflammatory signals were locally present in the skin. Considerable numbers of LC migrated to draining lymph nodes (5000 antigen-bearing LC per draining lymph node) if they were stimulated by gene gune immunization [42]. Migration of LC occurred over three days and, more importantly, LC remained in draining lymph nodes for more than two weeks [42]. Interestingly, transport via LC is not the only means for antigen to reach skin-draining lymph nodes. Unprocessed antigen is detected in lymph nodes already several hours after subcutaneous inoculation of antigen and does not require cell-mediated transport [43]. Remarkably, dendritic cells residing in the lymph node were able to take up and process the antigen and subsequently present it efficiently to naive antigen-specific T lymphocytes [43]. At later time points (≥ 24 h) skin-derived dendritic cells arrive in the lymph node that express MHC-restricted high levels of peptide derived from the antigen originally applied to the skin. Despite induction of T lymphocyte priming and proliferation after arrival of soluble antigen in the lymph node, complete CD4 T lymphocyte differentiation requires further stimuli only provided by skin-derived dendritic cells [43]. These results already challenged the view that antigen uptake in the periphery, migration and maturation followed by antigen-presentation to T lymphocytes is a linear process carried out by the same dendritic cell. Rather, specific anatomic structures have evolved to guarantee transport of free antigen from peripheral sites into the lymph node [44]. These corridors or conduits are separate from the pathways used by migratory dendritic cells and end in specific anatomic compartments within the lymph node. A presumably resident dendritic cell populations surveys this corridor and samples antigens delivered directly in free form to the lymph node [44]. So far, conflicting results exist with respect to the identity of this dendritic cell subset [45]. Future studies will be required to clarify the functional relevance of these different modes of antigen delivery to different antigen presenting cell populations in the periphery or directly in lymph nodes with respect to the influence on developing immune responses.

An even more divergent picture evolves if immune responses to pathogens are studied. Following epidermal infection with Herpes Simplex Virus (HSV) dermal dendritic cells loaded with viral antigens rapidly migrate to the lymph node, resulting in activation and proliferation of T lymphocytes within hours of infection. A closer look at the time kinetics of the different steps revealed that uptake of de novo synthesized viral proteins from infected epithelial cells by dermal dendritic cells, migration to the draining lymph node and induction of T cell activation as determined by increased expression levels of CD69 all occurred within 6 h. There was significant proliferation of virus-specific T lymphocytes after 24 h [46]. Clearly, free virus or virus-containing dendritic cells were not detected in draining lymph nodes. This underlines the importance of antigen-sampling of dendritic cells in the spleen and cross-presentation of these antigens derived from pathogens to CD8 T lymphocytes. These observations indicate an extremely fast induction of protective immune responses following antigen-sampling by dendritic cells in the skin after infection with a cytopathic virus.

Moreover, several different experimental assay systems revealed that dendritic cells sampling antigens from pathogens in the skin are not identical to those dendritic cells presenting antigen to T lymphocytes in the draining lymph node. Although LC with a mature phenotype ($CD80^+$ $CD86^+$ MHC II^+ DEC 205^+) are found within the T cell zones of skin-draining lymph nodes [47], and present antigen to T lymphocytes following contact sensitization of the skin [48], there is no evidence for a role of LC in presentation of antigens derived from skin-infecting microorganisms. Following epidermal infection with HSV, dendritic cells cross-presenting viral antigens to CD8 T lymphocytes in the draining lymph node bear CD8 molecules, which are absent from LC. A more detailed analysis revealed that $CD8^+$ dendritic cells were not prototypical plasmacytoid dendritic cells, because they lacked CD45RA expression typical of such cells [49]. Unequivocal evidence for the lack of antigen presentation by LC comes from experiments employing bone marrow chimaeric mice. Dendritic cells in the skin were almost entirely of recipient origin, which is compatible with the low turn-over and long half-life of LC in the skin. The composition of dendritic cells found in draining lymph nodes was different. Recipient-derived LC still were the prominent population, but a significant number of donor-derivedCD8$^+$ DEC205$^+$ dendritic cells was detected [50]. After sorting of these cell populations and in vitro testing for their ability to present viral antigens to specific T lymphocytes, it became apparent that only donor-derived CD8$^+$ dendritic cells can cross-present viral antigens to T lymphocytes [50]. These experiments suggest that antigen presenting dendritic cells in lymph nodes do not directly acquire their antigen within the site of infection but rather rely on migrating dendritic cells for transfer of antigen. Definitive proof for this hypothesis came from bone-marrow chimaera experiments where donor derived CD8$^+$ dendritic cells (in the lymph node) were unable to present antigen in an MHC I restricted fashion due to a point mutation in the binding groove of MHC I. In these mice, no presentation of viral antigens from peripheral tissue is observed in draining lymph nodes, although LC were still capable of MHC I restricted immune stimulation [50].

CD8$^+$ dendritic cells obviously do not migrate from peripheral tissue via the lymphatics into draining lymph nodes. They appear to be largely absent from non-lymphatic tissue [51, 52], although CD8$^+$ dendritic cells are capable of presenting MHC I and MHC II restricted antigens derived from these sites. Thus, CD8$^+$ dendritic cells appear to represent a predominantly non-migratory population, which are derived from blood borne precursors and are "resident" to lymphatic tissue potentially specialized in antigen presentation rather than antigen-sampling in peripheral tissues.

Similar observations were made for infection with other pathogenic microorganisms at other anatomical sites. Following intravaginal infection with HSV-2 a rapid recruitment of submucosal dendritic cells to infected epithelial cells was observed. At later time points, CD11c$^+$ dendritic cells were seen in draining lymph nodes presenting viral antigens in the context of MHC II molecules to CD4 T lymphocytes. Closer analysis of the dendritic cell population involved in antigen presentation demonstrated that it was not LC or CD8$^+$ dendritic cells but CD11c$^+$ CD11b$^+$ CD8$^-$ dendritic cells that were responsible for efficient virus-specific T cell priming [53]. During experimental influenza infection of the lung, both CD11c$^+$ CD11b$^-$ CD8$^-$ dendritic cells migrating from airways and lymph-node resident CD8$^+$ dendritic cells contribute to MHC I restricted presentation of viral antigens to specific CD8 T lymphocytes [54].

Collectively, the continuously accumulating data suggest that not only the population of dendritic cells patrolling peripheral tissue but also lymph-node resident CD8$^+$ dendritic cells are operative in shaping the immune response towards infecting microorganisms. While these findings nicely illustrate a division of labor between these functionally different dendritic cell populations it remains ill-defined how exactly "local information" is conveyed by migratory dendritic cells to antigen presenting CD8$^+$ dendritic cells in draining lymph nodes. The maturation state of dendritic cells is critical as to the decision of whether antigen presentation to T lymphocytes leads to development of immunity or immune tolerance. Migratory, antigen-sampling dendritic cells in the periphery are confronted with antigen in the local context of additional stimuli that may lead to activation of the dendritic cell. At present it is unclear how such information can be conveyed from a migratory dendritic cell population to lymph-node resident dendritic cells. Priming of naive T lymphocytes by $CD8^+$ dendritic cells may lead either to immunity or immune tolerance [50, 54–56], supporting the notion that lymph node resident $CD8^+$ dendritic cells display the same functional plasticity described for other dendritic cell populations. However, activation of T lymphocytes by antigen presenting dendritic cells is not the sole factor determining whether immunity arises in peripheral tissues [57]. Clearly, for T cell mediated immunity to rise, additional local activation steps are required, which are triggered by local activation of parenchymal cell population, for instance through TLR ligands [58].

2.4
Systemic Dissemination of Antigens/Infectious Microorganisms

Although considerable effort is undertaken by the immune system to contain infectious microorganisms locally at the site of infection, systemic blood borne dissemination of these microorganisms occurs. Furthermore, transmission of infectious microorganisms by athropod vectors, e.g., *Plasmodium* spp., circumvents physical and immune barriers of the skin and directly results in delivery of microorganisms into the blood. But not only microorganisms gain access to the blood. Even simple protein antigens present at body surfaces rapidly enter the blood stream. For instance, antigens ingested orally can be detected within minutes circulating with the blood stream [23]. Already 6 h after antigen ingestion, oral antigens have reached the spleen and provoked T lymphocyte activation, cytokine expression and proliferation [24]. The rapid transition of antigens from the intestinal lumen into the blood is not a unique peculiarity of the gastrointestinal tract. Within 7 min after application to the skin, antigen has already reached the liver and provoked expression of effector cytokines from local cell populations following local antigen presentation [59, 60] (see below). Thus, we have to consider that the interaction of the immune system with foreign antigen and microorganisms is not at all restricted to peripheral tissues, which form the external barrier of the organism.

The two organs mainly involved in elimination of blood borne macromolecules and microorganisms are the spleen and the liver. Many studies have described the contribution of the spleen to induction of antigen-specific immunity. I will rather concentrate here on the role of the liver in antigen/pathogen elimination from the blood, because the liver not only serves as a blood filter but at the same time controls development of antigen-specific immune responses at the local and systemic level. About 2500 L of blood pass daily through the liver, which drains not only the venous blood derived from the gastrointestinal tract but also receives arterial blood supply. Within the liver, arterial and venous blood merge in the hepatic sinusoids, giving rise to a mixed arterio-venous perfusion. About 20% of cardiac output passes through the liver, underlining its importance as metabolic organ [61].

2.5 Antigen Presenting Cells in the Liver

The liver bears several different antigen presenting cell populations. The macrophage population of the liver, the Kupffer cells, represents the largest macrophage population of the body. Hepatic dendritic cells are present in relatively low numbers and bear an immature phenotype. In addition to macrophages and dendritic cells, the liver contains a third antigen presenting cell population, i.e., the liver sinusoidal endothelial cells (LSEC). Hepatic antigen presenting cells are ideally positioned to execute their function to filter the blood. After numerous ramifications the portal vein drains into the microvessels, which are called sinusoids. To achieve maximal filtration and to optimize the metabolic function of hepatocytes, the liver is organized like a sponge. Every hepatocyte is surrounded by LSEC and thus has access to sinusoidal blood. Hepatic sinusoids are lined by LSEC and Kupffer cells as well as dendritic cells, which interact either on the luminal or the basolateral side with LSEC (Figure 2.3).

2.5.1 Dendritic Cells

Dendritic cells in the liver are composed of at least four different $CD11c^+$ subpopulations based on the expression of CD8 and CD11b [62, 63]. $B220^+$ plasmacytoid dendritic cells are rather abundant in the liver, accounting for ca. 20% of dendritic cells. $CD8^+$ dendritic cells in the liver are phenotypically similar to $CD8^+$ dendritic cells in lymph nodes as they show little or no DEC205 expression [64]. In contrast to dendritic cells in spleen or draining lymph nodes, hepatic dendritic cells are rather poor stimulators of T cell activation. Although, especially, $CD8^+$ hepatic dendritic cells have a somewhat mature phenotype their capacity to prime naive T lymphocytes is diminished compared to dendritic cells derived from spleen or lymph node [63]. Interestingly, human hepatic dendritic cells expressed IL-10 and showed reduced ability to stimulate T lymphocyte proliferation [65]. Moreover, T lymphocytes primed by hepatic dendritic cells released immune-regulatory mediators, such as IL-4 and IL-10, upon re-stimulation, suggesting an immune control function of hepatic dendritic cells [65]. The extent to which hepatic dendritic cells are involved in triggering immune responses towards antigens/pathogens from blood passing through the liver awaits more detailed investigation. They can present bacterial antigens following blood borne dissemination of *Salmonella* and prime naive CD4 and CD8 T lymphocytes in vitro [66].

Hepatic dendritic cells appear to migrate via lymphatic vessels into the hepatic lymph node [67, 68]. However, the composition of dendritic cell subpopulations in

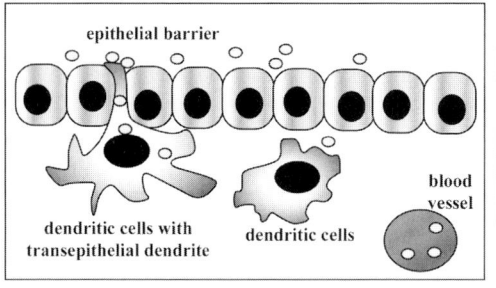

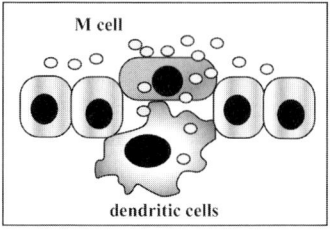

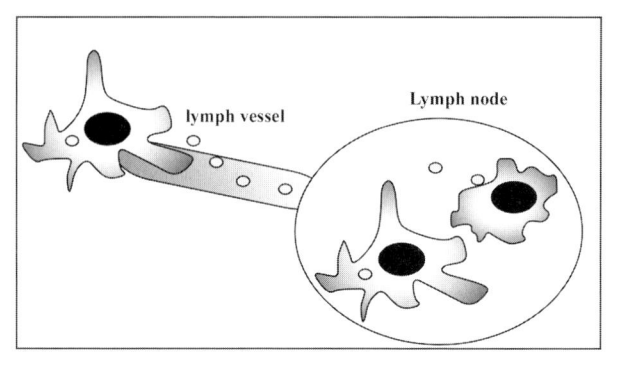

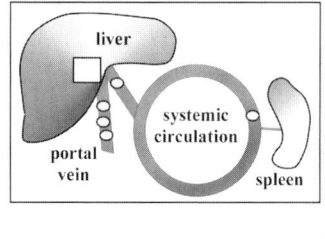

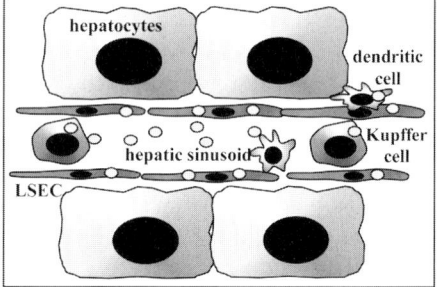

Figure 2.3 Antigens and pathogens may breach physical barriers and escape local immune control by migratory antigen presenting cells and disseminate in the entire organisms via the blood stream. Once systemically distributed, antigens and pathogens come in contact with antigen presenting cells in the spleen and liver. Within the liver a unique population of organ-resident antigen presenting cells, the liver sinusoidal endothelial cells, shape together with Kupffer cells and immature hepatic dendritic cells the immune response towards blood-borne antigens. (This figure also appears with the color plates.)

the liver-draining lymph node does not reflect the composition found in the liver [69]. It is therefore tempting to speculate that certain populations of hepatic dendritic cells may be rather organ-resident. If this were the case, hepatic dendritic cells would be exposed to the unique hepatic microenvironment, which is especially rich in tolerogenic mediators such as IL-10 and PGE_2 (derived from Kupffer cells) and TGF_b (derived from hepatic stellate cells). As exposure of dendritic cells to these substances induces a tolerogenic phenotype [70], extended sojourn of dendritic cells in the liver may render them tolerogenic. Indeed, hepatic dendritic cells have reduced capacity to induce T lymphocyte immunity in vitro [63, 65, 71]. If these dendritic cells do not migrate to the draining lymph node, they will exert their immune-regulatory function locally in the liver. However, of course not all hepatic dendritic cells remain resident to the liver. The liver even represents a stage for transition of blood borne dendritic cells to the lymphatic compartment. The functional relevance of migrating versus tissue-resident dendritic cells in the liver for the induction of antigen-specific immune responses has not been addressed yet.

2.5.2
Kupffer Cells

Kupffer cells are the macrophage population of the liver that is repopulated from the bone marrow. The tissue half-life of Kupffer cells is rather variable but can be up to several months. These cells are specialized in elimination of particulate material and pathogens by phagocytosis. Kupffer cells contribute to the unique hepatic microenvironment by release of anti-inflammatory mediators such as IL-10 and prostanoids [72]. Although they are capable of antigen presentation to CD4 and CD8 T lymphocytes [73–75], their contribution to T lymphocyte stimulation relative to other local antigen presenting cell populations is rather low. Nevertheless, Kupffer cells take up antigens from portal venous blood and are involved in the induction of tolerance towards antigens applied to the portal vein [76].

With respect to their function as phagocytosing cells they contribute to elimination of bacteria from the blood stream. To achieve the most efficient elimination of bacteria Kupffer cells use a rather unique mechanism. Secretory IgA prevents attachment and invasion of mucous membranes by adhering to microorganisms in the intestinal lumen [6]. In this way, secretory IgA contributes to local immune defense to prevent infection. Serum IgA, however, enhances phagocytosis of coated microorganisms by FcaRI-expression phagocytes and triggers activatory responses [77]. Pro-inflammatory mediators strongly increase expression levels of FcaRI on Kupffer cells. Bacteria coated with serum IgA are most efficiently eliminated by FcaRI-expression Kupffer cells in the liver [78]. Kupffer cells can, therefore, be considered to form a "second line of defense" against gut-derived bacteria that reach the liver with portal venous blood. Elimination of bacteria in portal venous blood prevents further systemic dissemination of bacteria and subsequent development of systemic inflammation.

Other pathogens, however, may exploit Kupffer cells to infect the liver. Malaria is caused by *Plasmodium* spp., which are transmitted via the bite of infected mosquitoes and transmission of sporozoites into the dermis. Within minutes sporozoites reach the liver, where they infect hepatocytes and undergo differentiation as well as massive proliferation. Targeting of hepatic tissue is specific and efficient as infection with one single sporozoite is sufficient to cause infection. Kupffer cells are targeted by sporozoites and are invaded. Targeting of Kupffer cells allows sporozoites to escape from the blood circulation and now enter hepatocytes for further completion of their life cycle [79]. These examples show again that antigen presenting cell populations are, on one hand, operative in immune surveillance but, on the other hand, may be exploited by microorganisms to establish infection.

2.5.3
Liver Sinusoidal Endothelial Cells

Liver sinusoidal endothelial cells form a part of the so-called reticulo-endothelial system that is functional in clearance of particulate matter and macromolecules from the blood circulation. As already mentioned, LSEC are strategically positioned in the hepatic sinusoid. Because hepatic sinusoids form a dense meshwork of vessels with a large overall surface and because the low perfusion pressure leads to slow blood flow in the sinusoids, LSEC have perfect conditions to eliminate macromolecules from the blood circulation. Although LSEC bear the capacity to perform phagocytosis of small particles, up 200 nm [80], their main form of antigen uptake is receptor-mediated endocytosis. They express a number of pattern-recognition receptors, such as the mannose receptor and several scavenger receptors [81].

The phenotype of LSEC is rather unique compared to other endothelial cell populations. LSEC constitutively express MHC I and MHC II molecules as well as low levels of the co-stimulatory molecules CD80, CD86 and CD40 [82, 83]. They further express the CD4 molecule and L-SIGN, which is a homologue of DC-SIGN [84]. Together with their constitutive high expression levels of CD54 and CD106 they express all molecules necessary to initiate contact and engage in cognate interaction with T lymphocytes. The phenotype of LSEC resembles immature dendritic cells and it has been suggested that these cells have antigen-presenting function.

Early experiments revealed that sinusoidal lining cells were MHC II positive and could present antigen to T lymphocytes [74]. With the development of techniques to isolate pure populations of LSEC [85], it was possible to investigate the antigen presenting function of these cells in well-defined in vitro cultures. Using primary cultures of LSEC isolated from murine liver, it was possible to demonstrate that LSEC could take up and process antigen for MHC II-restricted presentation to CD4 T lymphocytes. In contrast to other endothelial cells, LSEC did not require additional pro-inflammatory stimuli such as $TNF\alpha$ or $IFN\gamma$ to engage in cognate interaction with CD4 T lymphocytes. Interestingly, bacterial degradation products like endotoxin affected antigen-uptake and endolysosomal acidification, thereby decreasing MHC II restricted antigen presentation in LSEC [86]. Low levels of endotoxin are present in portal venous blood, the LSEC scavenge endotoxin and were responsive to these low endotoxin concentrations [87]. These observations reveal that the liver not only functions as a filter for gut-derived molecules and pathogens but also bears antigen presenting cells with a sentinel function for microenvironmental changes.

LSEC even further resemble dendritic cells, because they are able to prime naive CD4 T lymphocytes in an antigen-specific fashion. No maturation is required for LSEC to engage in cognate interaction with naive CD4 T lymphocytes and to induce initial cytokine expression and proliferation [83]. In contrast to dendritic cells, naive CD4 T lymphocytes primed by antigen presenting LSEC failed to differentiate into Th1 cells. Upon restimulation via the T cell receptor, CD4 T lymphocytes primed by LSEC released significant amounts of IL-10 and IL-4 whereas T lymphocytes primed by dendritic cells released mainly IFNγ [83]. These experiments suggested that antigen-specific immune responses can be initiated outside lymphatic tissue.

LSEC share another feature with dendritic cells, i.e. the capacity to present exogenous antigens on MHC I molecules to CD8 T lymphocytes (cross presentation). Cross presentation in LSEC is most efficient, requiring only low antigen concentrations and a short time period after receptor-mediated endocytosis [75]. LSEC not only cross present antigen to armed effector CD8$^+$ T cells but have the capacity to stimulate naive CD8$^+$ T lymphocytes. Following encounter with cross presenting LSEC, naive CD8$^+$ T lymphocytes release cytokines and start proliferation in vitro. However, antigen-specific restimulation of these T cells revealed that they lost the ability to express effector cytokines such as IL-2 and IFNγ and that they lost their cytotoxic activity [75]. In vivo, LSEC cross present antigen to naive CD8$^+$ T lymphocytes outside the lymphatic system. So far, stimulation of naive T cells was believed to occur exclusively in the highly specialized lymphatic microenvironment. Following stimulation by cross presenting LSEC, naive CD8$^+$ T lymphocytes start to proliferate locally in the liver. However, the outcome of cross presentation by LSEC in vivo is the induction of systemic immune tolerance. Similar to CD8$^+$ T cells stimulated by cross presenting LSEC in vitro, CD8$^+$ T cells in vivo lose the capacity to express effector cytokines and to exert cytotoxic activity against their specific target antigens once stimulated by cross presenting LSEC [75]. Deletion of antigen specific CD8$^+$ T cells occurs to some extent but is not the main mechanism of immune tolerance induced by LSEC [75]. Mice rendered tolerant by LSEC cross presenting a model antigen fail to develop an immune response against a tumor carrying this model antigen, which constitutes the prime target of the immune response in non-tolerant littermates, leading to immunity and tumor rejection in control animals [75].

Collectively, these experiments reveal the presence of a hepatic antigen presenting cell population with unique characteristics: organ-resident, most efficient in scavenging blood-borne antigens from the circulation, not requiring maturation for induction of immune function, induction of antigen-specific T lymphocyte tolerance by default. Encounter of antigens with this cell type in various situations, e.g., after systemic dissemination of gut-derived antigens, is likely to lead to induction of immune tolerance. As mentioned above, other hepatic antigen presenting cell populations may equally induce immune tolerance. A recent study has demonstrated that exclusive expression and presentation of an antigen in the liver leads to induction of immune tolerance towards this antigen, whereas expression and presentation in lymphatic tissue gives rise to strong immunity [88]. Distribution of antigen and the type of antigen presenting cell mediating the initial contact with T lymphocytes are important determining factors for the quality of the ensuing immune response.

Similar to antigen presenting cells in various other tissues, LSEC are exploited by microorganisms to achieve infection. Certain viruses target the liver with high specificity, like HepaDNA viruses or Hepatitis C Virus (HCV). In an animal model of Hepatitis B Virus infection, the Duck Hepatitis B Virus (DHBV), it was shown that DHBV following intravenous inoculation rapidly located in the liver in LSEC. As very few viruses are sufficient to establish successful infection in hepatocytes, it was assumed that DHBV initially taken up by LSEC was further available for hepatocellular infection after transcytotic delivery to the basolateral side of LSEC, which is in close physical contact with hepatocytes [89]. More mechanistic insights came from experiments studying the mechanisms mediating infection of the liver with HCV. L-SIGN was described as a receptor for HCV that allows HCV to escape lysosomal degradation [90] and mediates transinfection of hepatocytes [91], similar to the mechanisms observed for DC-SIGN mediated trans-infection of CD4 T lymphocytes [38]. LSEC are the only cell population of the liver expressing L-SIGN and, therefore, are a good candidate for mediating HCV targeting of the liver [84].

2.6
Conclusion

This chapter has illustrated the cellular and molecular mechanisms of the encounter between antigen presenting cells and microorganisms in the context of the anatomic compartment and organ-specific microenvironment. The distinction made by antigen presenting cells between innocuous commensal microorganisms and harmless antigens on one side and pathogenic microorganisms on the other side is most important for survival of the organism. Immune surveillance has to discriminate between these and at the same time preserve functional organ integrity in the presence of infection. Antigen presenting cells may either function solely to eliminate pathogens without alerting other parts of the immune system or may take up microorganisms for MHC restricted antigen presentation, resulting in immunity or tolerance. In stark contrast to the previous belief that antigen uptake, migration and antigen presentation all occur in the same dendritic cell, new experimental evidence demonstrates that, for skin-derived antigens, antigen uptake and antigen presentation are carried out by different dendritic cell populations.

Antigens and pathogens gain access to the systemic circulation and therefore additional immune defense lines are operative to prevent damage to the host. The liver with its different populations of antigen presenting cells, together with the spleen, plays an important role to eliminate blood borne pathogens and to contain antigen-specific immune responses once antigens are systemically distributed via the blood stream. In the liver, another unique population of organ-resident antigen presenting cells, the liver sinusoidal endothelial cells, exerts all the salient functions of antigen presenting cells at the same time, i.e., antigen uptake, processing and presentation. These organ-resident antigen presenting cells induce immune tolerance in T lymphocytes. It can be concluded that immune surveillance is not only operative at external surfaces of the organism to prevent or contain infection locally but further entails control of systemic pathogen spread and immune-regulatory function towards blood borne antigens.

Acknowledgments

I thank Andreas Limmer, Christian Kurts and Linda Diehl for critical discussion. P.K. is supported by grants from the Deutsche Forschungsgemeinschaft, the European Union and the Volkswagenstiftung.

References

1 Banchereau, J., and R. M. Steinman. 1998. Dendritic cells and the control of immunity. *Nature* 392, 245.

2 Mellman, I. and R. M. Steinman. 2001. Dendritic cells: specialized and regulated antigen processing machines. *Cell* 106: 255.

3 Hoebe, K., E. Janssen, and B. Beutler. 2004. The interface between innate and adaptive immunity. *Nat. Immunol.* 5, 971.

4 Madara, J. L., S. Nash, R. Moore, and K. Atisook. 1990. Structure and function of the intestinal epithelial barrier in health and disease. *Monogr. Pathol.*, 306.

5 Lamm, M. E. 1997. Interaction of antigens and antibodies at mucosal surfaces. *Annu. Rev. Microbiol.* 51, 311.

6 Brandtzaeg, P., E. S. Baekkevold, I. N. Farstad, F. L. Jahnsen, F. E. Johansen, E. M. Nilsen, and T. Yamanaka. 1999. Regional specialization in the mucosal immune system: what happens in the microcompartments? *Immunol. Today* 20, 141.

7 Iwasaki, A., and B. L. Kelsall. 2000. Localization of distinct Peyer's patch dendritic cell subsets and their recruitment by chemokines macrophage inflammatory protein (MIP)-3alpha, MIP-3beta, and secondary lymphoid organ chemokine. *J. Exp. Med.* 191, 1381.

8 Kraehenbuhl, J. P., and M. R. Neutra. 2000. Epithelial M cells: differentiation and function. *Annu. Rev. Cell Dev. Biol.* 16, 301.

9 Golovkina, T. V., M. Shlomchik, L. Hannum, and A. Chervonsky. 1999. Organogenic role of B lymphocytes in mucosal immunity. *Science* 286, 1965.

10 Sharma, R., E. J. van Damme, W. J. Peumans, P. Sarsfield, and U. Schumacher. 1996. Lectin binding reveals divergent carbohydrate expression in human and mouse Peyer's patches. *Histochem. Cell Biol.* 105, 459.

11 Neutra, M. R., T. L. Phillips, E. L. Mayer, and D. J. Fishkind. 1987. Transport of membrane-bound macromolecules by M cells in follicle-associated epithelium of rabbit Peyer's patch. *Cell Tissue Res.* 247, 537.

12 Neutra, M. R., N. J. Mantis, and J. P. Kraehenbuhl. 2001. Collaboration of epithelial cells with organized mucosal lymphoid tissues. *Nat. Immunol.* 2, 1004.

13 Neutra, M. R., A. Frey, and J. P. Kraehenbuhl. 1996. Epithelial M cells: gateways for mucosal infection and immunization. *Cell* 86, 345.

14 Kelsall, B. L. and W. Strober. 1996. Distinct populations of dendritic cells are present in the subepithelial dome and T cell regions of the murine Peyer's patch. *J. Exp. Med.* 183, 237.

15 Hopkins, S. A., F. Niedergang, I. E. Corthesy-Theulaz, and J. P. Kraehenbuhl. 2000. A recombinant Salmonella typhimurium vaccine strain is taken up and survives within murine Peyer's patch dendritic cells. *Cell Microbiol.* 2, 59.

16 Iwasaki, A. and B. L. Kelsall. 2001. Unique functions of CD11b+, CD8 alpha+, and double-negative Peyer's patch dendritic cells. *J. Immunol.* 166, 4884.

17 Williamson, E., J. M. Bilsborough, and J. L. Viney. 2002. Regulation of mucosal dendritic cell function by receptor activator of NF-kappa B (RANK)/RANK ligand interactions: impact on tolerance induction. *J. Immunol.* 169, 3606.

18 Cook, D. N., D. M. Prosser, R. Forster, J. Zhang, N. A. Kuklin, S. J. Abbondanzo, X. D. Niu, S. C. Chen, D. J. Manfra, M. T. Wiekowski, L. M. Sullivan, S. R. Smith, H. B. Greenberg, S. K. Narula, M. Lipp, and S. A. Lira. 2000. CCR6 mediates dendritic cell localization, lymphocyte homeostasis, and immune responses in mucosal tissue. *Immunity* 12, 495.

19 Izadpanah, A., M. B. Dwinell, L. Eckmann, N. M. Varki, and M. F. Kagnoff. 2001. Regulated MIP-3alpha/CCL20 production by human intestinal epithelium: mechanism for modulating mucosal immunity. *Am. J. Physiol. Gastrointest. Liver Physiol.* 280, G710.

20 Mowat, A. M. 2003. Anatomical basis of tolerance and immunity to intestinal antigens. *Nat. Rev. Immunol.* 3, 331.

21 Alpan, O., G. Rudomen, and P. Matzinger. 2001. The role of dendritic cells, B cells, and M cells in gut-oriented immune responses. *J. Immunol.* 166, 4843.

22 Karlsson, M., S. Lundin, U. Dahlgren, H. Kahu, I. Pettersson, and E. Telemo. 2001. "Tolerosomes" are produced by intestinal epithelial cells. *Eur. J. Immunol.* 31, 2892.

23 Smith, K. M., J. M. Davidson, and P. Garside. 2002. T-cell activation occurs simultaneously in local and peripheral lymphoid tissue following oral administration of a range of doses of immunogenic or tolerogenic antigen although tolerized T cells display a defect in cell division. *Immunology* 106, 144.

24 Gutgemann, I., A. M. Fahrer, J. D. Altman, M. M. Davis, and Y. H. Chien. 1998. Induction of rapid T cell activation and tolerance by systemic presentation of an orally administered antigen. *Immunity* 8, 667.

25 Maric, I., P. G. Holt, M. H. Perdue, and J. Bienenstock. 1996. Class II MHC antigen (Ia)-bearing dendritic cells in the epithelium of the rat intestine. *J. Immunol.* 156, 1408.

26 Galan, J. E., and R. Curtiss, 3rd. 1989. Cloning and molecular characterization of genes whose products allow *Salmonella typhimurium* to penetrate tissue culture cells. *Proc. Natl. Acad. Sci. U.S.A.* 86, 6383.

27 Vazquez-Torres, A., J. Jones-Carson, A. J. Baumler, S. Falkow, R. Valdivia, W. Brown, M. Le, R. Berggren, W. T. Parks, and F. C. Fang. 1999. Extraintestinal dissemination of Salmonella by CD18-expressing phagocytes. *Nature* 401, 804.

28 Rescigno, M., M. Urbano, B. Valzasina, M. Francolini, G. Rotta, R. Bonasio, F. Granucci, J. P. Kraehenbuhl, and P. Ricciardi-Castagnoli. 2001. Dendritic cells express tight junction proteins and penetrate gut epithelial monolayers to sample bacteria. *Nat. Immunol.* 2, 361.

29 Niess, J. H., S. Brand, X. Gu, L. Landsman, S. Jung, B. A. McCormick, J. M. Vyas, M. Boes, H. L. Ploegh, J. G. Fox, D. R. Littman, and H. C. Reinecker. 2005. CX3CR1-mediated dendritic cell access to the intestinal lumen and bacterial clearance. *Science* 307, 254.

30 Nagl, M., L. Kacani, B. Mullauer, E. M. Lemberger, H. Stoiber, G. M. Sprinzl, H. Schennach, and M. P. Dierich. 2002. Phagocytosis and killing of bacteria by professional phagocytes and dendritic cells. *Clin. Diagn. Lab. Immunol.* 9, 1165.

31 Macpherson, A. J., and T. Uhr. 2004. Induction of protective IgA by intestinal dendritic cells carrying commensal bacteria. *Science* 303, 1662.
32 Svensson, M., B. Stockinger, and M. J. Wick. 1997. Bone marrow-derived dendritic cells can process bacteria for MHC-I and MHC-II presentation to T cells. *J. Immunol.* 158, 4229.
33 Svensson, M., and M. J. Wick. 1999. Classical MHC class I peptide presentation of a bacterial fusion protein by bone marrow-derived dendritic cells. *Eur. J. Immunol.* 29, 180.
34 Hersh, D., D. M. Monack, M. R. Smith, N. Ghori, S. Falkow, and A. Zychlinsky. 1999. The Salmonella invasin SipB induces macrophage apoptosis by binding to caspase-1. *Proc. Natl. Acad. Sci. U.S.A.* 96, 2396.
35 Yrlid, U., and M. J. Wick. 2000. Salmonella-induced apoptosis of infected macrophages results in presentation of a bacteria-encoded antigen after uptake by bystander dendritic cells. *J. Exp. Med.* 191, 613.
36 Sigal, L. J., S. Crotty, R. Andino, and K. L. Rock. 1999. Cytotoxic T-cell immunity to virus-infected non-haematopoietic cells requires presentation of exogenous antigen. *Nature* 398, 77.
37 Kurts, C., H. Kosaka, F. R. Carbone, J. F. Miller, and W. R. Heath. 1997. Class I-restricted cross-presentation of exogenous self-antigens leads to deletion of autoreactive CD8(+) T cells. *J. Exp. Med.* 186, 239.
38 Geijtenbeek, T. B., D. S. Kwon, R. Torensma, S. J. van Vliet, G. C. van Duijnhoven, J. Middel, I. L. Cornelissen, H. S. Nottet, V. N. KewalRamani, D. R. Littman, C. G. Figdor, and Y. van Kooyk. 2000. DC-SIGN, a dendritic cell-specific HIV-1-binding protein that enhances trans-infection of T cells. *Cell* 100, 587.
39 Stahl-Hennig, C., R. M. Steinman, K. Tenner-Racz, M. Pope, N. Stolte, K. Matz-Rensing, G. Grobschupff, B. Raschdorff, G. Hunsmann, and P. Racz. 1999. Rapid infection of oral mucosal-associated lymphoid tissue with simian immunodeficiency virus. *Science* 285, 1261.
40 van Kooyk, Y., and T. B. Geijtenbeek. 2003. DC-SIGN: escape mechanism for pathogens. *Nat. Rev. Immunol.* 3, 697.
41 Merad, M., M. G. Manz, H. Karsunky, A. Wagers, W. Peters, I. Charo, I. L. Weissman, J. G. Cyster, and E. G. Engleman. 2002. Langerhans cells renew in the skin throughout life under steady-state conditions. *Nat. Immunol.* 3, 1135.
42 Garg, S., A. Oran, J. Wajchman, S. Sasaki, C. H. Maris, J. A. Kapp, and J. Jacob. 2003. Genetic tagging shows increased frequency and longevity of antigen-presenting, skin-derived dendritic cells in vivo. *Nat. Immunol.* 4, 907.
43 Itano, A. A., S. J. McSorley, R. L. Reinhardt, B. D. Ehst, E. Ingulli, A. Y. Rudensky, and M. K. Jenkins. 2003. Distinct dendritic cell populations sequentially present antigen to CD4 T cells and stimulate different aspects of cell-mediated immunity. *Immunity* 19, 47.
44 Sixt, M., N. Kanazawa, M. Selg, T. Samson, G. Roos, D. P. Reinhardt, R. Pabst, M. B. Lutz, and L. Sorokin. 2005. The conduit system transports soluble antigens from the afferent lymph to resident dendritic cells in the T cell area of the lymph node. *Immunity* 22, 19.
45 Anderson, A. O., and S. Shaw. 2005. Conduit for privileged communications in the lymph node. *Immunity* 22, 3.
46 Mueller, S. N., C. M. Jones, C. M. Smith, W. R. Heath, and F. R. Carbone. 2002. Rapid cytotoxic T lymphocyte activation occurs in the draining lymph nodes after cutaneous herpes simplex virus infection as a result of early antigen presentation and not the presence of virus. *J. Exp. Med.* 195, 651.
47 Hoefsmit, E. C., A. M. Duijvestijn, and E. W. Kamperdijk. 1982. Relation between langerhans cells, veiled cells, and interdigitating cells. *Immunobiology* 161, 255.
48 Macatonia, S. E., S. C. Knight, A. J. Edwards, S. Griffiths, and P. Fryer. 1987. Localization of antigen on lymph node dendritic cells after exposure to the contact sensitizer fluorescein iso-

thiocyanate. Functional and morphological studies. *J. Exp. Med.* 166, 1654.

49 Nakano, H., M. Yanagita, and M. D. Gunn. 2001. CD11c(+)B220(+)Gr-1(+) cells in mouse lymph nodes and spleen display characteristics of plasmacytoid dendritic cells. *J. Exp. Med.* 194, 1171.

50 Allan, R. S., C. M. Smith, G. T. Belz, A. L. van Lint, L. M. Wakim, W. R. Heath, and F. R. Carbone. 2003. Epidermal viral immunity induced by CD8alpha+ dendritic cells but not by Langerhans cells. *Science* 301, 1925.

51 Scheinecker, C., R. McHugh, E. M. Shevach, and R. N. Germain. 2002. Constitutive presentation of a natural tissue autoantigen exclusively by dendritic cells in the draining lymph node. *J. Exp. Med.* 196, 1079.

52 Vermaelen, K. Y., I. Carro-Muino, B. N. Lambrecht, and R. A. Pauwels. 2001. Specific migratory dendritic cells rapidly transport antigen from the airways to the thoracic lymph nodes. *J. Exp. Med.* 193, 51.

53 Zhao, X., E. Deak, K. Soderberg, M. Linehan, D. Spezzano, J. Zhu, D. M. Knipe, and A. Iwasaki. 2003. Vaginal submucosal dendritic cells, but not Langerhans cells, induce protective Th1 responses to herpes simplex virus-2. *J. Exp. Med.* 197, 153.

54 Belz, G. T., C. M. Smith, L. Kleinert, P. Reading, A. Brooks, K. Shortman, F. R. Carbone, and W. R. Heath. 2004. Distinct migrating and nonmigrating dendritic cell populations are involved in MHC class I-restricted antigen presentation after lung infection with virus. *Proc. Natl. Acad. Sci. U.S.A.* 101, 8670.

55 Belz, G. T., G. M. Behrens, C. M. Smith, J. F. Miller, C. Jones, K. Lejon, C. G. Fathman, S. N. Mueller, K. Shortman, F. R. Carbone, and W. R. Heath. 2002. The CD8alpha(+) dendritic cell is responsible for inducing peripheral self-tolerance to tissue-associated antigens. *J. Exp. Med.* 196, 1099.

56 Belz, G. T., C. M. Smith, D. Eichner, K. Shortman, G. Karupiah, F. R. Carbone, and W. R. Heath. 2004. Cutting edge: conventional CD8 alpha+ dendritic cells are generally involved in priming CTL immunity to viruses. *J. Immunol.* 172, 1996.

57 Hamilton-Williams, E. E., A. Lang, D. Benke, G. M. Davey, K. H. Wiesmuller, and C. Kurts. 2005. Cutting edge: TLR ligands are not sufficient to break cross-tolerance to self-antigens. *J. Immunol.* 174, 1159.

58 Sacher, T., P. Knolle, T. Nichterlein, B. Arnold, G. J. Hammerling, and A. Limmer. 2002. CpG-ODN-induced inflammation is sufficient to cause T-cell-mediated autoaggression against hepatocytes. *Eur. J. Immunol.* 32, 3628.

59 Campos, R. A., M. Szczepanik, A. Itakura, M. Akahira-Azuma, S. Sidobre, M. Kronenberg, and P. W. Askenase. 2003. Cutaneous immunization rapidly activates liver invariant Valpha14 NKT cells stimulating B-1 B cells to initiate T cell recruitment for elicitation of contact sensitivity. *J. Exp. Med.* 198, 1785.

60 Askenase, P. W., M. Szczepanik, A. Itakura, C. Kiener, and R. A. Campos. 2004. Extravascular T-cell recruitment requires initiation begun by Valpha14+ NKT cells and B-1 B cells. *Trends Immunol.* 25, 441.

61 Knolle, P. A., and G. Gerken. 2000. Local control of the immune response in the liver. *Immunol. Rev.* 174, 21.

62 Jomantaite, I., N. Dikopoulos, A. Kroger, F. Leithauser, H. Hauser, R. Schirmbeck, and J. Reimann. 2004. Hepatic dendritic cell subsets in the mouse. *Eur. J. Immunol.* 34, 355.

63 Pillarisetty, V. G., A. B. Shah, G. Miller, J. I. Bleier, and R. P. DeMatteo. 2004. Liver dendritic cells are less immunogenic than spleen dendritic cells because of differences in subtype composition. *J. Immunol.* 172, 1009.

64 Henri, S., D. Vremec, A. Kamath, J. Waithman, S. Williams, C. Benoist, K. Burnham, S. Saeland, E. Handman, and K. Shortman. 2001. The dendritic cell populations of mouse lymph nodes. *J. Immunol.* 167, 741.

65 Goddard, S., J. Youster, E. Morgan, and D. H. Adams. 2004. Interleukin-10 secretion differentiates dendritic cells from human liver and skin. *Am. J. Pathol.* 164, 511.

66. Johansson, C., and M. J. Wick. 2004. Liver dendritic cells present bacterial antigens and produce cytokines upon Salmonella encounter. *J. Immunol.* 172, 2496.
67. Kudo, S., K. Matsuno, T. Ezaki, and M. Ogawa. 1997. A novel migration pathway for rat dendritic cells from the blood: hepatic sinusoids-lymph translocation. *J. Exp. Med.* 185, 777.
68. Matsuno, K., T. Ezaki, S. Kudo, and Y. Uehara. 1996. A life stage of particle-laden rat dendritic cells in vivo: their terminal division, active phagocytosis, and translocation from the liver to the draining lymph [see comments]. *J. Exp. Med.* 183, 1865.
69. Tanis, W., S. Mancham, R. Binda, H. L. Janssen, G. Bezemer, I. Jzermans, H. W. Tilanus, J. D. Laman, H. de Wit, H. A. Drexhage, S. W. Schalm, and J. Kwekkeboom. 2004. Human hepatic lymph nodes contain normal numbers of mature myeloid dendritic cells but few plasmacytoid dendritic cells. *Clin. Immunol.* 110, 81.
70. Steinbrink, K., M. Wolfl, H. Jonuleit, J. Knop, and A. H. Enk. 1997. Induction of tolerance by IL-10-treated dendritic cells. *J. Immunol.* 159, 4772.
71. Lu, L., C. A. Bonham, X. Liang, Z. Chen, W. Li, L. Wang, S. C. Watkins, M. A. Nalesnik, M. S. Schlissel, A. J. Demestris, J. J. Fung, and S. Qian. 2001. Liver-derived DEC205+B220+CD19- dendritic cells regulate T cell responses. *J. Immunol.* 166, 7042.
72. Knolle, P., J. Schlaak, A. Uhrig, P. Kempf, K. H. Meyer zum Buschenfelde, and G. Gerken. 1995. Human Kupffer cells secrete IL-10 in response to lipopolysaccharide (LPS) challenge. *J. Hepatol.* 22, 226.
73. Richman, L. K., R. J. Klingenstein, J. A. Richman, W. Strober, and J. A. Berzofsky. 1979. The murine Kupffer cell. I. Characterization of the cell serving accessory function in antigen-specific T cell proliferation. *J. Immunol.* 123, 2602.
74. Rubinstein, D., A. K. Roska, and P. E. Lipsky. 1987. Antigen presentation by liver sinusoidal lining cells after antigen exposure in vivo. *J. Immunol.* 138, 1377.
75. Limmer, A., J. Ohl, C. Kurts, H. G. Ljunggren, Y. Reiss, M. Groettrup, F. Momburg, B. Arnold, and P. A. Knolle. 2000. Efficient presentation of exogenous antigen by liver endothelial cells to CD8+ T cells results in antigen-specific T-cell tolerance. *Nat. Med.* 6, 1348.
76. Roland, C. R., M. J. Mangino, B. F. Duffy, and M. W. Flye. 1993. Lymphocyte suppression by Kupffer cells prevents portal venous tolerance induction: a study of macrophage function after intravenous gadolinium. *Transplantation* 55, 1151.
77. Monteiro, R. C., H. Kubagawa, and M. D. Cooper. 1990. Cellular distribution, regulation, and biochemical nature of an Fc alpha receptor in humans. *J. Exp. Med.* 171, 597.
78. van Egmond, M., E. van Garderen, A. B. van Spriel, C. A. Damen, E. S. van Amersfoort, G. van Zandbergen, J. van Hattum, J. Kuiper, and J. G. van de Winkel. 2000. FcalphaRI-positive liver Kupffer cells: reappraisal of the function of immunoglobulin A in immunity. *Nat. Med.* 6, 680.
79. Pradel, G. and U. Frevert. 2001. Malaria sporozoites actively enter and pass through rat Kupffer cells prior to hepatocyte invasion. *Hepatology* 33, 1154.
80. Steffan, A. M., J. L. Gendrault, R. S. McCuskey, P. A. McCuskey, and A. Kirn. 1986. Phagocytosis, an unrecognized property of murine endothelial liver cells. *Hepatology* 6, 830.
81. Smedsrod, B. 2004. Clearance function of scavenger endothelial cells. *Comp. Hepatol.* 3(Suppl 1), S22.
82. Lohse, A. W., P. A. Knolle, K. Bilo, A. Uhrig, C. Waldmann, M. Ibe, E. Schmitt, G. Gerken, and K. H. Meyer Zum Buschenfelde. 1996. Antigen-presenting function and B7 expression of murine sinusoidal endothelial cells and Kupffer cells. *Gastroenterology* 110, 1175.
83. Knolle, P. A., E. Schmitt, S. Jin, T. Germann, R. Duchmann, S. Hegenbarth, G. Gerken, and A. W. Lohse. 1999.

Induction of cytokine production in naive CD4(+) T cells by antigen-presenting murine liver sinusoidal endothelial cells but failure to induce differentiation toward Th1 cells. *Gastroenterology* 116, 1428.

84 Bashirova, A. A., T. B. Geijtenbeek, G. C. van Duijnhoven, S. J. van Vliet, J. B. Eilering, M. P. Martin, L. Wu, T. D. Martin, N. Viebig, P. A. Knolle, V. N. KewalRamani, Y. van Kooyk, and M. Carrington. 2001. A dendritic cell-specific intercellular adhesion molecule 3-grabbing nonintegrin (DC-SIGN)-related protein is highly expressed on human liver sinusoidal endothelial cells and promotes HIV-1 infection. *J. Exp. Med.* 193, 671.

85 Knook, D. L. and E. C. Sleyster. 1976. Separation of Kupffer and endothelial cells of the rat liver by centrifugal elutriation. *Exp. Cell Res.* 99, 444.

86 Knolle, P. A., T. Germann, U. Treichel, A. Uhrig, E. Schmitt, S. Hegenbarth, A. W. Lohse, and G. Gerken. 1999. Endotoxin down-regulates T cell activation by antigen-presenting liver sinusoidal endothelial cells. *J. Immunol.* 162, 1401.

87 Knolle, P. A., E. Loser, U. Protzer, R. Duchmann, E. Schmitt, K. H. zum Buschenfelde, S. Rose-John, and G. Gerken. 1997. Regulation of endotoxin-induced IL-6 production in liver sinusoidal endothelial cells and Kupffer cells by IL-10. *Clin. Exp. Immunol.* 107, 555.

88 Bowen, D. G., M. Zen, L. Holz, T. Davis, G. W. McCaughan, and P. Bertolino. 2004. The site of primary T cell activation is a determinant of the balance between intrahepatic tolerance and immunity. *J. Clin. Invest.* 114, 701.

89 Breiner, K., H. Schaller, and P. Knolle. 2001. Endothelial cell-mediated uptake of a hepatitis B virus – a new concept of liver-targeting of hepatotropic microorganisms. *Hepatology* 34, 803.

90 Ludwig, I. S., A. N. Lekkerkerker, E. Depla, F. Bosman, R. J. Musters, S. Depraetere, Y. van Kooyk, and T. B. Geijtenbeek. 2004. Hepatitis C virus targets DC-SIGN and L-SIGN to escape lysosomal degradation. *J. Virol.* 78, 8322.

91 Cormier, E. G., R. J. Durso, F. Tsamis, L. Boussemart, C. Manix, W. C. Olson, J. P. Gardner, and T. Dragic. 2004. L-SIGN (CD209L) and DC-SIGN (CD209) mediate transinfection of liver cells by hepatitis C virus. *Proc. Natl. Acad. Sci. U.S.A.* 101, 14067.

3
Antigen Processing in the Context of MHC Class I Molecules

Frank Momburg

3.1
Tracing the Needle in the Haystack: The Efficiency of Antigen Processing and Presentation by MHC Class I Molecules

Major histocompatibility complex class I (MHC-I) molecules belong to the key elements of the adaptive immune system as they alert cytotoxic T lymphocytes to the presence of intracellular pathogens such as viruses or certain types of bacteria or indicate mutations of self antigens occurring in tumor cells. They do this by displaying huge arrays of short peptides of 8–11 amino acids representing degradation products of the foreign or altered proteins. From mass spectrometric analysis of peptides acid-eluted from different MHC-I isoforms it was estimated that over 10 000 peptide species were presented at the level of >1 fmol per 10^8 cells. The complexity of peptide spectra was similar in B lymphoblastoid and melanoma cells and it was estimated that ca. 90% of these peptides were derived from commonly expressed proteins (reviewed in Engelhard et al., 2002). It was calculated that B-lymphoblastoid cells with 8×10^5 MHC-I molecules on the cell surface display individual peptide/MHC-I complexes at varying frequencies of 1 to 4000 per cell. The peptide repertoire could thus represent degradation products of a substantial fraction of the proteome (~30 000 proteins). Moreover, several cryptic translation products from non-coding regions, alternative reading frames or introns have been identified as MHC-I peptides, which extend the great diversity of peptide/MHC-I complexes even more (reviewed in Shastri et al., 2000). In addition, MHC-I presented peptides can harbor various post-translational modifications such as $N\alpha$-acetylation, cysteinylated cysteines, dimethylated arginines, phosphorylations on serine, threonine or tyrosine residues, and *O*- or *N*-glycosylations (reviewed in Engelhard et al., 2002).

To ensure rapid responses during acute virus infections the immune system cannot afford to wait until metabolically stable or compartmentalized viral proteins eventually undergo age-associated damage and turnover by the cellular degradation machinery. There is evidence that CD8$^+$ T cells recognize long-lived viral proteins rapidly after infection, thereby linking the period of active viral protein biosynthesis with antigen presentation (Khan et al., 2001). Errors in mRNA synthesis and splicing, incorrect or abortive mRNA translation on ribosomes, post-translational misfolding, mistargeting to organelles, and defective assembly into multisubunit complexes can lead to faulty protein variants that never attain a native structure. Recently, these were collectively termed "defective ribosomal products" (DRiPs) (Yewdell, 2001). Using proteasome inhibitors, DRiPs can make up 25–40% of newly synthesized proteins in different tumor and normal cells (Schubert et al., 2000; Princiotta et al., 2003). DRiPs are mostly polyubiquitinated (Schubert et al., 2000) and hence represent substrates for 26S proteasomes (Section 3.2). Because of their large quantities DRiPs appear to be an important source for MHC-I binding peptides. In a model system employing short- or long-lived epitope-containing proteins, 440–3000 substrates had to be degraded for each K^b/ovalbumin peptide complex (Princiotta et al., 2003). On average, each of the 8×10^5 proteasomes in one L cell degraded 2.5 substrates per minute, resulting in $\sim 2 \times 10^6$ peptides s^{-1}, of which, however, only ~150 peptides s^{-1} (~0.1%) formed complexes with MHC-I molecules (Princiotta et al., 2003; Yewdell et al., 2003).

A comparably low efficiency of epitope formation (1/3900) was found using vaccinia virus-expressed β-galactosidase as model antigen (Montoya and Del Val, 1999). Likewise, less than 1% of peptides cleaved by proteasomes out of various ubiquitin-tagged fusion proteins were recruited for MHC-I presentation (Fruci et al., 2003). Such a wasteful antigen processing seems to be unharmful because viral proteins are often produced in large amounts and the numbers of MHC-I/peptide complexes necessary for triggering of specific CD8$^+$ T cells can vary from several thousands to as few as one (reviewed in Engelhard et al., 2002). By contrast, proteins secreted by *Listeria monocytogenes* bacteria into the cytosol of infected cells were reported to be processed into MHC-I ligands with much higher efficiencies. For instance, only ~35 murein hydrolase molecules were required to yield one K^d-bound epitope (Villanueva et al., 1994). The reasons for these discrepancies in processing efficiencies are probably multifaceted and presently unresolved.

By monitoring the decay of fluorescent peptides injected into the cytosol of living cells it was noted that the half-life of 9-mer peptides was of the order of <10 s owing to abundant aminopeptidase activities (Reits et al., 2003). It was concluded that 99% of newly produced peptides that freely diffuse in the cytoplasm and nucleus are destroyed before reaching the peptide transporter in the ER membrane (Reits et al., 2003; Yewdell et al., 2003). Transient binding to heat shock proteins may protect antigenic peptides from destruction by cytosolic peptidases (reviewed in Srivastava, 2002). Following siRNA-mediated silencing of the cytosolic chaperonin TRiC, the amounts of total cell surface MHC-I molecules and of particular MHC-I/peptide complexes were reduced (Kunisawa and Shastri, 2003). This suggests that TRiC protects a substantial subset of N-terminally extended peptide intermediates, but it was not excluded whether TRiC has other functions in the MHC-I pathway. The chances of freely diffusing peptides escaping degradation and entering the ER would probably be increased if proteasomes and trimming aminopeptidases were close to the ER membrane. Interestingly, in this regard, immunoproteasomes, but not house-keeping proteasomes (Section 3.2), were found significantly enriched at the cytosolic face of the ER membrane (Brooks et al., 2000).

Resulting from conformational changes upon ATP and peptide binding, green fluorescence protein (GFP)-tagged TAP heterodimers attain a slower lateral mobility in the ER membrane. TAP mobility has been used as a read-out system for the presence of cytosolic peptides in living cells (Reits et al., 2000). Normally, only one third of all TAP molecules actively transport peptides. However, TAP molecules became fully engaged during an acute influenza virus infection. When protein translation was inhibited, the peptide-dependent, low-mobility conformation of TAP subsided within 15–30 min, in accordance with the short half-life of cytosolic peptides. These findings imply that the turnover products from long-lived proteins represent only a minor quantitative source of peptides and that newly synthesized proteins chiefly contribute to the MHC-I bound peptide pool. Furthermore, the total degradative capacity of proteasomes is engaged to only ~20% in unstressed cells (Dantuma et al., 2000). In conclusion, the cellular antigen processing machinery has considerable idle capacity that can be rapidly recruited during an ongoing production of viral proteins.

3.2
The "Classical" Route: Loading of MHC Class I Molecules With Peptides Generated in the Cytoplasm

3.2.1
Cytosolic Peptide Processing by Proteasomes and other Proteases

Large multisubunit protease complexes, the proteasomes, account for the main proteolytic activity that digests proteins into oligopeptides (reviewed in Kloetzel, 2001, 2004; Rock et al., 2002; Rock and Goldberg, 1999) (see Figure 3.1). Evidence for the crucial involvement of proteasomes in antigen processing was obtained in studies using proteasome-inhibiting peptide aldehydes as well as highly proteasome-specific inhibitors such as lactacystin (reviewed in Rock and Goldberg, 1999; Rock et al., 2002; Grommé and Neefjes, 2002; Kessler et al., 2002). Proteasome inhibitors caused a significant reduction of MHC-I surface expression and diminished the presentation of CTL epitopes. The generation of some MHC-I ligands was, however, either not negatively affected or their formation was even enhanced in the presence of proteasome inhibitors (reviewed in Rock et al., 2002; Grommé and Neefjes, 2002). This suggests that alternative degradation pathways exist in which proteasomes may not be rate limiting and that, in certain cases, proteasomes may destroy antigenic peptides. The surface expression of allelic HLA-A, B, C molecules requiring peptides with different anchor residues (Section 3.4.1) was unequally affected by proteasome inhibitors (Benham et al., 1998; Luckey et al., 2001), suggesting that non-proteasomal and proteasomal cleavage specificities may complement each other during peptide processing.

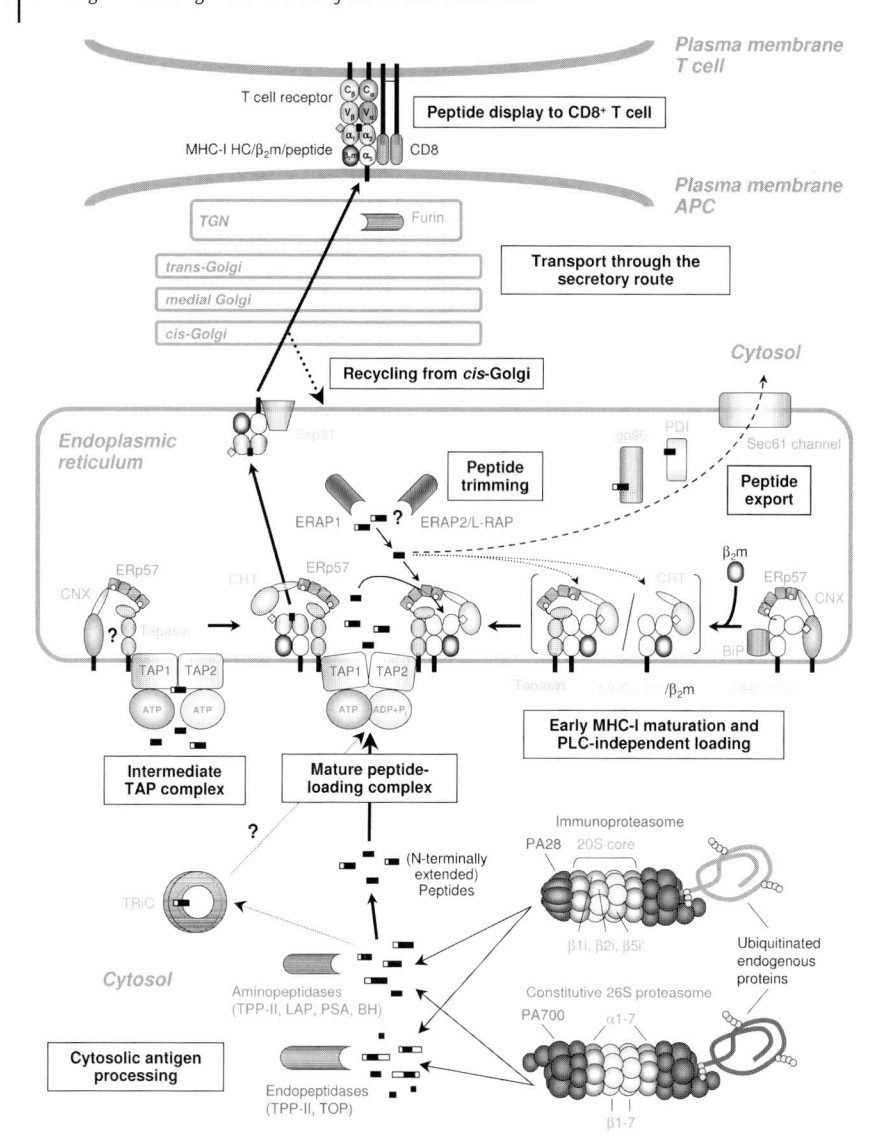

3 Antigen Processing in the Context of MHC Class I Molecules

Figure 3.1 Classical pathway of MHC-I antigen processing. 26S proteasomes degrade cytosolic proteins tagged by multiubiquitin chains into peptides of 3–25 amino acids. The C termini of final MHC-I ligands are mostly generated by the proteasomal cut. In the presence of interferon-γ, the proteolytically active β subunits of the constitutively expressed 20S core complex are replaced by the respective immunoproteasome subunits β1i, β2i and β5i, which entail an altered proteasomal cleavage specificity. Also, the constitutively expressed ATP-dependent PA700 regulator complex can be replaced by the interferon-induced PA28 regulator complex. Through the endopeptidase activity of cytosolic tripeptidyl peptidase II (TPP-II) larger proteasomal products (>16 residues) can be processed into MHC-I-binding peptides, while thimet oligopeptidase (TOP) produces fragments that are too small for subsequent TAP transport and MHC-I binding. Cytosolic aminopeptidases, e.g., TPP-II, leucine aminopeptidase (LAP), puromycin-sensitive aminopeptidase (PSA), and bleomycin hydrolase (BH), can trim N-terminal amino acids before ER translocation by TAP. Cytosolic chaperones such as TRiC may protect peptides from rapid hydrolysis. The transporter associated with antigen processing (TAP) transports peptides of 8–16 amino acids in an ATP-dependent fashion. Subunits TAP1 and TAP2 undergo conformational changes upon peptide binding, upon ATP binding to the nucleotide binding domains, and upon ATP hydrolysis that is linked to substrate release into the ER lumen. N-terminally extended variants of final MHC-I-binding ligands are subjected to trimming by the ER-resident aminopeptidase ERAP1, and potentially also by ERAP2/L-RAP. Peptides can bind to ER-resident chaperones such as gp96 or PDI and are removed from the ER through the Sec61 channel. Peptides of 8–10 amino acids are loaded onto MHC class I heavy chain (HC)/β_2-microglobulin (β_2m) dimers that are tethered to TAP by means of the dedicated chaperone tapasin. Tapasin also recruits the oxidoreductase ERp57, which might isomerize the MHC-I α_2 domain disulfide bridge during peptide loading. Together with the lectin-like chaperone calreticulin (CRT), these components form the peptide-loading complex (PLC). Tapasin seems to retain MHC-I in a peptide-receptive conformation and its editing function results in an optimized affinity of MHC-I ligands and improved stability of cell surface HC/β_2m/peptide complexes. Tapasin also stabilizes TAP heterodimers by virtue of its transmembrane domain. Intermediate TAP complexes contain tapasin, ERp57 and calnexin but interaction modalities have not been defined. After biosynthesis MHC-I HC assemble first with the chaperones BiP and calnexin (CNX). ERp57 is also present in early HC complexes with calnexin. Following binding of β_2m to HC, CNX is exchanged by CRT that binds to the α_1 domain N-glycan of MHC-I HC in the monoglucosylated form. Intermediate MHC-I complexes, shown in brackets, are also able to bind peptides. Tapasin-independent allelic MHC-I molecules are not, or are only poorly, detectable in the PLC, but are regularly loaded with peptides. Successfully loaded MHC-I molecules are released from the PLC and exit the ER in cargo vesicles. The cargo receptor Bap31 is involved ER egress of peptide-loaded MHC-I. Tapasin-mediated recycling of MHC-I molecules from the cis-Golgi has been reported. HC/β_2m/peptide complexes traffick through the Golgi apparatus, where N-glycans are modified, and the trans-Golgi network (TGN) to the plasma membrane. In the TGN, the endopeptidase furin can contribute to antigen processing. HC/β_2m/peptide complexes displayed on the cell surface of antigen presenting cells (APC) are recognized by antigen-specific T cell receptors together with CD8. (This figure also appears with the color plates.)

3.2.1.1 Structure and Function of the Proteasomal Core and Interferon-induced Subunits

The 26S proteasome consists of the barrel-shaped 20S proteasomal core and two ATP-dependent 19S regulatory complexes bound at either end (reviewed in Baumeister et al., 1998; Voges et al., 1999). The 20S core is a cylindrical stack of four seven-membered rings with a central cavity and narrow entrances at the top and the bottom. In eukaryotic cells, the two outer rings each contain the non-identical but related subunits $\alpha 1$ to $\alpha 7$, which possess no protease activity (reviewed in Baumeister et al., 1998; Kloetzel, 2001). Each of the two inner rings consists of the structurally related subunits $\beta 1$ to $\beta 7$. Confined to the surface of the inner cavity, the proteolytic activity resides in subunits $\beta 1$, $\beta 2$ and $\beta 5$ containing N-terminal threonine residues as active nucleophiles, whereas the other β subunits are enzymatically inactive. During immune responses, the constitutively expressed $\beta 1$, $\beta 2$ and $\beta 5$ subunits are replaced by the IFN-γ-inducible "immunoproteasome" subunits $\beta 1i$ (LMP2), $\beta 2i$ (MECL-1) and $\beta 5i$ (LMP7) (reviewed in Baumeister et al., 1998; Rock et al., 2002; Kloetzel et al., 2001, 2004). LMP2 and LMP7 are encoded in the class II region of the major histocompatibility complex (reviewed in Momburg and Hämmerling, 1998; Rock and Goldberg, 1999). The incorporation of $\beta 1i$, $\beta 2i$ and $\beta 5i$ modulates the proteasomal cleavage specificity as discussed below.

In addition to the immuno-subunits of the 20S core complex, IFN-γ induces the expression of PA28 (11S) regulator complex (reviewed in Baumeister et al., 1998; Rock et al., 2002; Kloetzel, 2001, 2004). PA28 is a hexa- or heptameric complex composed of PA28α and PA28β subunits that bind to the outer α rings of the 20S core. The PA28 regulator opens the narrow gates of the 20S particle and may stimulate the ATP-independent entry or exit of substrates (Whitby et al., 2000; reviewed in Kloetzel, 2001; 2004; Rock et al., 2002). In vitro, the PA28 regulator complex markedly increased protein hydrolysis and quantitatively altered proteasomal cleavage site preferences (reviewed in Rock et al., 2002). Mice deficient for PA28α and PA28β have been analyzed (Preckel et al., 1999; Murata et al., 2001), but it is controversial whether CTL responses are generally or selectively impaired in the absence of PA28. Overexpression of PA28 in cells improved the generation of a subset MHC-I ligands (Groettrup et al., 1996; van Hall et al., 2000; Sijts et al., 2002). Hybrid proteasomes consisting of the 20S core assembled with the 19S and the PA28 complex at opposite ends generated unique peptides in vitro that were not produced by 26S proteasomes (Cascio et al., 2002). In addition, regulator-free 20S (immuno)proteasomal core complexes have been suggested to function downstream of 26S and PA28/20S proteasomes (Kloetzel, 2004).

3.2.1.2 Targeting Proteins for ATP-dependent Degradation by 26S Proteasomes

Proteins are usually marked for proteasomal processing by isopeptidic conjugation of lysine residues with a chain of four or more moieties of ubiquitin, a small protein of 76 amino acids (reviewed in Hershko and Ciechanover, 1998). This requires a cascade of enzymatic reactions, including the ATP-dependent ubiquitin-activating enzyme E1, several ubiquitin-conjugating (E2) enzymes, and multiple substrate-specific ubiquitin ligases (E3) (reviewed in Hershko and Ciechanover, 1998; Weissman, 2001). Deubiquitinating enzymes regulate the multiubiquitin chain length or disassemble multiubiquitin trees for recycling (reviewed in Wilkinson et al., 2000). Although most studies underpinned an essential role of ubiquitin-targeted proteolysis for antigen processing, there are also examples for ubiquitin-independent protein degradation in cells (reviewed in Rock et al., 2002). In vitro, 26S and 20S proteasomes degraded various non-ubiquitinated proteins, some of which, however, required extensive denaturation (Benaroudj et al., 2001). The ubiquitin-independent ornithine decarboxylase–antizyme complex employs a special mechanism for proteasomal targeting (reviewed in Coffino, 2001).

The major function of the 19S regulatory complex is to bind ubiquitinated substrates and to thread them into inner cavity of the 20S core. The 19S complex is composed of a "base" and a "lid" subcomplex with 8 and 9–10 different components, respectively (reviewed in Voges et al., 1999; Kloetzel, 2001). The base, which is attached to the two α rings, contains besides non-ATPase subunits a ring of 6 AAA-type ATPases that facilitate the unfolding and channeling of substrates through the axial gates of the 20S core (Navon and Goldberg, 2001; reviewed in Rock et al., 2002). The ATP-dependent function of the 19S regulator is thus linked to the earlier observation that ATP hydrolysis is strictly required for non-lysosomal protein degradation. Multiubiquitin trees need to be removed before an unfolded protein can pass the narrow axial entry of the core complex. Consistently, deubiquitinating enzymes have been identified as subunits of the 19S regulatory complex (Yao and Cohen, 2002).

3.2.1.3 Cleavage Properties of (Immuno)Proteasomes

The three active proteasome subunits $\beta 1$, $\beta 2$ and $\beta 5$ show differential proteolytic specificities as they preferentially cleave after acidic, basic and hydrophobic residues in short fluorogenic substrates, respectively (reviewed in Rock and Goldberg, 1999). In natural protein substrates these cleavage specificities are less well defined as cuts can occur after almost every amino acid, thus providing a high degree of flexibility to proteasomes (Niedermann et al., 1996; Nussbaum et al., 1998; Toes et al., 2001). Protein hydrolysis is, however, by no means random as only 10–15% of all peptide bonds are used in model protein substrates (Kisselev et al., 1999). Proteasomes generally prefer to cut after large hydrophobic and aromatic amino acids (Niedermann et al., 1996; Nussbaum et al., 1998; Toes et al., 2001). The sequential context within the degraded protein can strongly influence the yield of particular MHC-I ligands (Nussbaum et al., 1998; reviewed in Rock and Goldberg, 1999; Rock et al., 2002). Whether finally processed ligands or N-terminally extended precursors are generated or preferentially destroyed by proteasomes in vitro appears to correlate with the immunodominance of these epitopes in vivo and can be used to predict CTL epitopes (Niedermann et al., 1995; Kessler et al., 2001).

As compared with "house-keeping" 20S proteasomes, 20S immunoproteasomes cleaved more often after hydrophobic and aromatic residues in the protein substrate enolase 1, whereas cuts after acidic residues occurred less frequently (Toes et al., 2001). Thus, the cleavage specificity of immunoproteasomes appears to be altered in favor of C-terminal peptide residues that predominate in natural MHC-I ligands. Although the IFN-γ inducible 20S subunits β1i (LMP2) and β5i (LMP7) are no general prerequisite for antigen presentation, the study of IFN-γ treated or immuno-subunit transfected cells has provided ample evidence for the modulation of proteasomal cleavage specificities by immuno-subunits (reviewed in Momburg and Hämmerling, 1998; Rock and Goldberg, 1999; Kloetzel, 2001, 2004; Rock et al., 2002; Grommé and Neefjes, 2002). Slightly reduced levels of MHC-I surface expression and $CD8^+$ T cells as well as a reduced capacity to process and present certain MHC-I peptides have been noted for LMP2- and LMP7-deficient mice (Van Kaer et al., 1994; Fehling et al., 1994). Together, these studies demonstrated that immunoproteasomes are required for efficient generation of a subset of MHC-I epitopes. Hence, the degradation machinery can rapidly adapt to the greater demands of antigen presentation during infections. In contrast to the augmented presentation of virus-derived epitopes in the presence of immunoproteasomes, two tumor-associated self peptides were produced only by constitutive but not by immunoproteasomes expressed in professional antigen-presenting cells (Morel et al., 2000). This is of importance for immunization strategies using dendritic cells for the presentation of tumor-derived proteins.

Although proteasomes generate a subset of peptides in vitro that need no additional processing for MHC-I binding (Dick et al., 1998; Lucchiari-Hartz et al., 2000), many proteasomal products are intermediate products that carry the exact C terminus of the definitive MHC-I-ligand, but contain N-terminal (Nt) extensions of various lengths (Niedermann et al., 1996; Cascio et al., 2001). The proteasomal cleavage products produced in vitro were between 3 and >25 amino acids long, with most products being <10 amino acids (Niedermann et al., 1996; Nussbaum et al., 1998; Kisselev et al., 1999). Interestingly, both 26S and 20S immunoproteasomes processed significantly more N-terminally extended versions of the K^b-binding ovalbumin peptide SIINFEKL than constitutive proteasomes (Cascio et al., 2001). It was noted that about 15% of proteasomal products fall within the size range (8–9 residues) of typical MHC-I ligands, another 15–20% are larger and thus require additional trimming, but the remaining two-thirds were too short for TAP transport and MHC-I binding. Proteasomes are ancient enzyme complexes that evolved long before the development of the adaptive immune system in vertebrates (reviewed in Baumeister et al., 1998) and whose apparent purpose is amino acid recycling and the maintenance of homeostasis in the presence of continuous protein neosynthesis. Unsurprisingly, therefore, most proteasomal products have no apparent value for the immune system. Notably, however, the mentioned size range of proteasomal products in vitro does not necessarily reflect the situation in living cells. 26S proteasomes can degrade or activate proteins by a single endoproteolytic cut, resulting in fragments of more than 100 amino acids (reviewed in Rape and Jentsch, 2002). Limited proteasomal proteolysis may thus occur more frequently in vivo than predicted from the degradation of proteins and oligopeptides in vitro.

3.2.1.4 Peptide Processing by Nonproteasomal Cytosolic Peptidases

Proteasome-derived intermediates of MHC-I ligands can undergo further processing in the cytosol. Various cytosolic amino- and endopeptidases have been identified with the capacity to either degrade proteasomal products into short fragments or to exert limited trimming at the aminoterminal end of the peptide (reviewed in Rock et al., 2002, 2004; Saveanu et al. 2002; Del Val and López, 2002). Proteasome inhibitors completely blocked the generation of mature epitopes from C-terminally extended epitope precursors while, conversely, these inhibitors did not affect the presentation of precursors with N-terminal extensions or of finally processed peptide epitopes (Craiu et al., 1997; Kunisawa and Shastri, 2003; reviewed in Rock et al., 2002, 2004). Thus, with a reported few exceptions, proteasome appear to be generally responsible for the generation of the final C-terminal residues of MHC-I ligands.

The ubiquitous cytosolic metallo-endopeptidase, thimet oligopeptidase (TOP), usually destroys MHC-I ligands. TOP hydrolyzed most antigenic peptides in cell extracts to small fragments (Saric et al., 2001). Its overexpression in cells reduced MHC-I expression by decreasing the pool of available MHC-I ligands while the silencing of TOP by siRNA had the opposite effect (York et al., 2003). Leucine aminopeptidase (LAP) was shown to remove N-terminal extensions from ovalbumin-derived precursor peptides (Beninga et al., 1998). This cytoplasmic zinc aminopeptidase is present in many tissues and is inducible by IFN-γ (Beninga et al., 1998). Bleomycin hydrolase and puromycin-sensitive aminopeptidase (PSA) trimmed a viral precursor peptide in cytosolic extracts (Stoltze et al., 2000). Tripeptidyl peptidase II (TPP-II) has been implicated in the processing of some antigenic peptides (Lévy et al., 2002; Seifert et al., 2003; Kloetzel, 2004). The generation of an HLA-A3-restricted epitope from the HIV Nef protein, which could not be generated by proteasomes, critically depended on the activity of TPP-II (Seifert et al., 2003). TPP-II is a huge multimeric complex with a rod-like structure (Geier et al., 1999). TPP-II removes three residues at a time from the N terminus of oligopeptides, but also displays an as yet incompletely characterized endopeptidase activity in vitro (Geier et al., 1999; Seifert et al., 2003; Reits et al., 2004). TPP-II was reported to act sequentially with PSA (Lévy et al., 2002). Since TPP-II, but no other aminopeptidase activity, attacks peptides with 16 or more residues (Reits et al., 2004), TPP-II may have a specific function in processing longer proteasome-derived oligopeptides (16–25 residues). The down-regulation of surface MHC-I molecules in the presence of the TPP-II inhibitor butabindide suggests that the production of final MHC-I ligands partially depends on TPP-II-mediated cleavage of proteasomal products (Reits et al., 2004). It is unclear whether the different cytosolic amino- and endopeptidases exert redundant or unique functions in the MHC-I presentation pathway. Analysis of knock-out mice deficient for the respective peptidases will help to resolve this question.

3.3
Crossing the Border – Peptide Translocation into the ER by TAP

3.3.1
Structure and Function of TAP

The transporter associated with antigen processing (TAP) represents the principal ER conduit for those peptides that survive the degradative environment of the cytosol (reviewed in Elliott, 1997a; Momburg and Hämmerling, 1998) (see Figure 3.1). Because MHC-I molecules are unstable in the absence of peptides, TAP1-deficient mice show severely reduced levels of MHC-I surface expression and <1% of wild-type CD8$^+$ T cell counts (Van Kaer et al., 1992). Nevertheless, TAP1$^{-/-}$ mice are able to develop a diverse CD8$^+$ T cell repertoire and mount peptide-specific T cell responses (Aldrich et al., 1994), suggesting that TAP-independent pathways warrant a minor supply with MHC-I-binding peptides (Sections 3.5 and 3.6).

TAP belongs to the large family of ABC (ATP-binding cassette) transporters that translocate a vast variety of substrates across membranes (reviewed in Higgins, 1992). TAP consists of the two non-covalently linked subunits TAP1 and TAP2, each possessing an N-terminal membrane-spanning portion and a C-terminal, cytosol-exposed nucleotide-binding domain (NBD) (reviewed in Elliott, 1997a; Momburg and Hämmerling, 1998). Despite their common architecture, TAP1 and TAP2 proteins share only about 36% of their amino acids (reviewed in Momburg and Hämmerling, 1998; Elliott, 1997a). Together with the genes coding for the immunoproteasome subunits LMP2 and LMP7, TAP1 and TAP2 genes are located in the class II region of the MHC and their expression is likewise inducible by IFN-γ (reviewed in Momburg and Hämmerling, 1998; Elliott, 1997a). TAP expression is confined to membranes of the ER and the *cis*-Golgi (Kleijmeer et al. 1992; Vos et al. 2000).

The complementation with the lacking subunit is required to rescue peptide transport and MHC-I antigen presentation by TAP1- and TAP2-deficient cells. These findings indicated that both subunits are required to form a functional peptide transporter, although a few studies indicated that TAP1 homodimers may possess a residual transport function (reviewed in Momburg and Hämmerling, 1998; Elliott, 1997a). Analysis of the 3D structure of TAP1/TAP2 single particles by electron microscopy confirmed that TAP1 and TAP2 form heterodimers (Velarde et al., 2001). This analysis revealed a central pore of 3 nm diameter on the predicted ER luminal side, a compact globular structure and cytoplasmatic lobes that likely represent the dimerized NBDs of TAP1 and TAP2. The X-ray crystal structure of the NBD of TAP1 shows significant similarities with NBDs of other ABC transporters (Gaudet and Wiley, 2001). Owing to the lack of a high-resolution crystal structure of the complete TAP heterodimer, the complex organisation of the membrane-spanning hydrophobic domains of TAP1 and TAP2 remains a controversial issue. Ten transmembrane segments (TMS) in TAP1 and 9 TMS in TAP2 have been predicted on the basis of hydrophobicity algorithms and comparisons with other ABC transporters (Elliott, 1997a; Abele and Tampé,

1999), implicating that the N terminus of TAP1 is orientated towards the cytoplasm and the N terminus of TAP2 is located in the ER lumen. By determining the ER-lumenal vs. cytoplasmic orientation of C-terminally truncated variants of TAP1 and TAP2 proteins a topology model was developed that implicates a head-head/tail-tail arrangement of the heterodimer and a hypothetical peptide translocation pore consisting of TMS 1–6 in TAP1 and TMS 1–5 in TAP2 (Vos et al., 1999, 2000). By analogy with the structure of other ABC transporters, another group proposed that the translocation pore of TAP be formed by the six symmetrically arranged TMS forming the C-terminal part of the membrane-spanning domain of both TAP1 (TMS 4–9) and TAP2 (TMS 5–10) (Abele and Tampé, 1999). The functionality of such a 6+6 membrane-spanning core region in terms of heterodimeric assembly, peptide binding and transport has recently been demonstrated in insect cells (Koch et al., 2004). Truncation of the hydrophobic N-terminal TAP1 and TAP2 sequences outside the putative 6+6 core resulted in the loss of tapasin association with TAP (Koch et al., 2004; see Section 3.4). Further studies are required to elucidate the exact membrane topology of TAP1/TAP2 and interaction sites with additional molecules.

Binding of peptide substrates induces conformational changes in the TAP heterodimer and stimulates the hydrolysis of ATP in the NBDs of both TAP1 and TAP2 (reviewed in Abele and Tampé, 2004). It is unclear, however, how the binding and hydrolysis of ATP is coupled to the actual membrane translocation of peptide (reviewed in van Endert et al., 2002; Abele and Tampé, 2004). In permeabilized cells or microsomal membranes, peptide translocation by TAP strictly depends on ATP hydrolysis (Neefjes et al., 1993a; Shepherd et al., 1993), whereas peptide binding to TAP can occur in the absence of ATP (van Endert et al., 1994; Androlewicz and Cresswell, 1994). Given the high cytoplasmic ATP concentration (1–10 mM), peptide binding to nucleotide-free TAP appears, however, to be unlikely in living cells. Several studies indicate that TAP1 and TAP2 function in an asymmetrical manner during the translocation cycle. From mutational studies the picture emerges that ATP binding and hydrolysis by the TAP2 NBD is critical for substrate binding and translocation by human TAP, whereas hydrolysis-defective mutant TAP1 subunits still allowed for low levels of peptide transport (Karttunen et al., 2001; Lapinski et al., 2001; Saveanu et al., 2001). Conflicting data were obtained in another study, showing the importance of ATP hydrolysis by rat TAP1 for peptide translocation and peptide-dependent release of TAP-associated MHC-I molecules (Alberts et al., 2001). In a recently proposed hypothetical model, the sequential hydrolysis of ATP at NBD2 and NBD1 would only be required to reset the transporter for the next cycle, while a substrate-induced dimerization of the two ATP-loaded NBDs would elicit conformational rearrangements that move the peptide through the translocation channel (Abele and Tampé, 2004).

Cross-linking studies with photoactivatable peptides and mutational deletions provided evidence for both TAP1 and TAP2 contributing to a composite binding site mapping to sequences that flank the most C-terminal putative pair of TMS (Androlewicz and Cresswell, 1994; Nijenhuis et al., 1996; Nijenhuis and Hämmerling, 1996; Ritz et al., 2001). The use of chimeras of allelic forms rat TAP2 allowed for the identification of polymorphic residues that determine the differential transport of peptides with C-terminal hydrophobic or basic side chains, respectively (see below). Consistent with the cross-linking studies, these residues map to the membrane boundaries of putative cytoplasmic loops within the core membrane-spanning domain of rat $TAP2^a$ and $TAP2^u$ (Momburg et al., 1996; Deverson et al., 1998). Peptides bind to TAP with relatively low affinities (0.2–1.6 μM for selected peptides; Uebel et al., 1995, 1997; Koopmann et al., 1996) and similar peptide concentrations are required for half-maximal transport rates (Bangia et al., 1999; Momburg and Tan, 2002). Considering cytoplasmic concentrations of free peptides far below 1 μM, transport rates should increase proportionally to the effective peptide concentrations. Enhanced protein expression during viral infections should therefore imply an enhanced TAP-mediated peptide translocation.

3.3.2
Substrate Specificity of TAP

The substrate specificity of TAP has been investigated by using radiolabeled peptides containing an ER-dependent glycosylation motif or by peptide binding assays. TAP preferably transports peptides 8–16 amino acids long, but longer peptides up to 40 residues can be translocated at low rates (Momburg et al., 1994a; van Endert et al., 1994; Koopmann et al., 1996). Thus, a subset of TAP substrates requires additional processing in the ER to obtain the usual MHC-I ligands of 8–10 residues (Section 3.5). Despite its remarkable promiscuity in terms of peptide lengths and sequences, TAP binds and transports individual peptide sequences with greatly varying efficiencies (reviewed in Momburg and Hämmerling, 1998; Elliott, 1997a). Using combinatorial peptide libraries, sequence selectivity has been noted for positions 1–3 and, most significantly, for position 9 in ninemer peptides, while internal residues contribute to the binding affinity to a much lesser extent (van Endert et al., 1995; Uebel et al., 1997). Proline at position 2 or 3 is strongly disfavored in TAP substrates (Neisig et al., 1995; van Endert et al., 1995; Uebel et al., 1997; reviewed in Grommé and Neefjes, 2002). The available data collectively suggest that peptides contact the binding site predominantly with the carboxy-terminal and with amino-terminal residues involving free terminal carboxylate and ammonium groups, while long peptides might be accommodated by bulging out in the middle (reviewed in Momburg and Hämmerling, 1998; Grommé and Neefjes, 2002).

Allelic forms of rat TAP2 exhibit different specificities for the C-terminal peptide residue. Restrictive transporters of the TAP-B group exclusively translocate peptides with C-terminal aliphatic or aromatic residues, whereas permissive TAP-A transporters also accept substrates with charged and polar C-terminal residues (Heemels et al., 1993; Momburg et al., 1994b). The functional polymorphism of rat TAP can cause significant sequence variations in the peptide pool available for MHC-I binding in the ER (Powis et al., 1996). In rat inbred strains, permissive TAP-A transporters are co-expressed with MHC-I allomorphs that can bind peptides with basic C-terminal residues, suggesting remarkable co-evolution of the substrate specificities of TAP and MHC-I (Joly et al., 1998). The specificity of mouse TAP is similar to the restrictive rat TAP-B (Momburg et al., 1994b; Schumacher et al., 1994). This concords with the exclusive presence of hydrophobic C-terminal residues in murine MHC-I ligands (Section 3.4). By contrast, human TAP prefers C-terminal basic, aromatic and aliphatic but disfavors C-terminal acidic residues (van Endert et al., 1995; Momburg et al., 1994b; Uebel et al., 1997). This in turn matches the peptide requirements of human MHC-I molecules.

3.3.3
TAP-independent Peptide Entry into the ER

Not all peptides require the peptide transporter to enter the ER. ER signal sequences are co-translationally cleaved by signal peptidase, giving rise to hydrophobic peptides released into the ER lumen. Such peptides have been found in association with HLA-A2 molecules (Wei and Cresswell, 1992). Hydrophobic peptides that are processed out of the multiple membrane-spanning Epstein-Barr virus protein LMP2 in a proteasome-dependent fashion are presented by TAP-deficient cells (reviewed in Lautscham et al., 2003). Peptides that have the capacity to traverse the ER membrane independently of TAP can also be expressed as cytosolic minigene products (Zweerink et al., 1993; Lautscham et al., 2002). It is unclear how such hydrophobic peptides travel in the cytosol and how they are released from the ER membrane to associate with the peptide binding groove of MHC-I molecules. Peptide-binding ER chaperones such calreticulin, gp96, PDI, or ERp72 might transiently bind such peptides and transfer them onto MHC-I (Lammert et al., 1997; Spee and Neefjes, 1997). Curiously, peptides can reach the ER of TAP-deficient cells also when fused to the arginine/lysine-rich membrane-translocating sequence of the HIV Tat protein (Lu et al., 2001). Such "Trojan" peptides can undergo further processing within the secretory pathway and may represent a therapeutic strategy to render TAP-deficient cells susceptible to CTL lysis.

3.4
Fitting in the Best: TAP-associated Peptide Loading Complex Optimizes MHC-I Peptide Binding

After ER translocation by TAP, peptides meet their MHC-I receptors in an assembly of auxiliary proteins that assist during the loading and the optimization the bound peptide spectrum. The "peptide loading complex" (PLC) is composed of MHC-I heavy chain/β_2-microglobulin heterodimers associated with the chaperones tapasin, calreticulin/calnexin and ERp57, all of which assemble around the peptide transporter and whose interactions will be described in this chapter (see Figure 3.1).

3.4.1
Structure of MHC-I Molecules

MHC class I molecules are type I glycoproteins of 45 kDa that associate non-covalently with the soluble 12 kDa β_2-microglobulin (β_2m) and a short antigenic peptide to form a stable heterotrimeric complex. X-ray crystal structures of MHC-I molecules show that the membrane-proximal α_3 domain of the MHC-I heavy chain (HC) and β_2m adopt immunoglobulin-like folds that support a peptide binding platform built by the HC α_1 and α_2 domains (reviewed in Madden, 1995; Bouvier, 2003). Intramolecular disulfide bonds in the α_2 and α_3 domains and in β_2m stabilize the structure. The peptide binding platform contains a structure of 8 β strands lying underneath the α_1 and α_2 helices that form the side walls of the binding cleft. The binding groove is closed at both ends, thereby limiting the lengths of peptides to 8–10 residues (reviewed in Madden, 1995; Engelhard, 1994). The free ammonium and carboxylate groups of the peptide are held in a network of hydrogen bonds and thus contribute to the binding energy of the peptide (reviewed in Madden, 1995). The peptide backbone is bound in an extended conformation that can slightly bulge out in the central part in case of unusually long peptides. Some 73–83% of the peptide ligands are buried in the binding groove and are thus invisible to the T cell receptor (reviewed in Madden, 1995).

The highly polymorphic classical MHC class I HC is encoded by the MHC HLA-A, -B, and -C loci in humans. The co-dominant expression of alleles results in up to six different HC molecules expressed per individual; 250 HLA-A alleles, 490 HLA-B alleles and 119 HLA-C alleles have been identified (Marsh et al., 2002). In the mouse the polymorphic H-2K, -D and -L locus products correspond to human HLA-A, -B and -C molecules. The vast majority of polymorphic residues are at the bottom and in the walls of the peptide binding groove (Bjorkman et al., 1987; Zhang et al., 1998). These polymorphic residues determine the physico-chemical properties of structural pockets in the binding site that make tight contact with peptide side chains and thus enable the allele-specific binding of subsets of antigenic peptides with suited sequences. The C-terminal peptide residue usually serves as the C-terminal pocket-binding anchor, while residue 2, or sometimes residue 3 or 5, serves as N-terminal anchor. N-terminal anchor residues show a diverse chemical nature (reviewed in Rammensee et al., 1993; Engelhard et al., 1994). The C-terminal residues are either aliphatic, aromatic or basic in human MHC-I binding peptides, whereas in the mouse only aliphatic and aromatic C-terminal anchor residues have been described (reviewed in Rammensee et al., 1993; Engelhard et al., 1994).

3.4.2
Early Steps in the Maturation of MHC-I Molecules

After their biosynthesis into the ER membrane and co-translational N-glycosylation, MHC-I HCs assemble with the membrane-bound, ER-resident chaperone calnexin, which assists during its folding and promotes the assembly with the

3.4 Fitting in the Best: TAP-associated Peptide Loading Complex Optimizes MHC-I Peptide Binding

β_2m (reviewed in Momburg and Tan, 2002; Paulsson and Wang, 2003; Bouvier, 2003; Wright et al., 2004). In addition, the chaperone BiP was reported to interact with β_2m-free human HC. Calnexin assembles with nascent MHC-I HC through lectin-like binding of MHC-I-linked glycans and by recognition of the polypeptide chain (Zhang et al., 1995). HC folding, assembly with β_2m and subsequent peptide loading can, however, also occur in the absence of calnexin, or if the glycan-dependent assembly with calnexin is disturbed (reviewed in Momburg and Tan, 2002; Paulsson and Wang, 2003; Bouvier, 2003; Wright et al., 2004). In β_2m-expressing cells the folding of HC and assembly with β_2m precedes the binding of peptides (Neefjes et al., 1993b).

There is evidence, that the ER-resident, thiol-dependent oxidoreductase ERp57(ER60) binds to incompletely oxidized, β_2m-free HC during the early calnexin-associated stage (Farmery et al., 2000; reviewed in Momburg and Tan, 2002; Paulsson and Wang, 2003; Wright et al., 2004). Disulfide-bonded conjugates of HC and ERp57 have been detected (Lindquist et al., 2001; Antoniou et al., 2002b). ERp57 is thought to facilitate the formation of the α_2 domain disulfide bond. In calnexin-deficient cells another member of the thiol-reductase family, ERp72, could be co-precipitated with HC, showing that functional redundancy may also apply to HC-oxidizing enzymes (Lindquist et al., 2001). As partner of calreticulin, ERp57 is also found as a constituent of the PLC (see below).

Following β_2m binding, human MHC-I molecules exchange calnexin for the soluble, lectin-like chaperone calreticulin that binds to the N-glycan attached to residue 86 that is at the C-terminal end of the α_1 helix (Sadasivan et al., 1996; reviewed in Momburg and Tan, 2002; Bouvier, 2003; Wright et al., 2004). Some, but apparently not all, allelic mouse MHC-I molecules also assemble with calreticulin, but do not release calnexin after HC-β_2m dimerization until late stages of MHC-I maturation in the ER (reviewed in Momburg and Tan, 2002; Paulsson and Wang, 2003). The difference between the characteristics of chaperone association of H-2 and HLA-A, -B and -C molecules is likely due to the presence of an additional N-glycan at residue 176 in the α_2 domain of mouse MHC-I recognized by calnexin. Both calnexin and calreticulin bind to monoglucosylated, mannose-rich N-glycans. An important ER quality control mechanism involves the constitutive deglucosylation by ER glucosidases I and II and the conformation-dependent reglucosylation by UDP-glucose:protein glucosyltransferase (UGGT). UGGT preferentially glucosylates glycans in unfolded, flexible parts of proteins. Reglucosylation allows to calnexin to rebind and start another folding cycle (reviewed in Ellgard and Helenius, 2003). If the folding fails, ER mannosidase I and the mannosidase-like protein EDEM eventually target misfolded HC to the retrotranslocation and cytoplasmic degradation pathway (Wilson et al., 2000; reviewed in Ellgard and Helenius, 2003; Tsai et al., 2002).

3.4.3
Structure and Molecular Interactions of Tapasin

The 48 kDa ER-resident glycoprotein tapasin is the cornerstone in the PLC because it forms the crucial bridge between calreticulin-associated MHC-I/β_2m dimers on the one hand and TAP on the other (Ortmann et al., 1997; Sadasivan et al., 1996; reviewed in Momburg and Tan, 2002; Paulsson and Wang, 2003; Bouvier, 2003; Wright et al., 2004). During immune responses, the expression of the tapasin gene, which is located at the centromeric end of the MHC class II region, is upregulated by IFN-γ (Herberg et al., 1998). Tapasin is composed of a membrane-proximal immunoglobulin-like domain containing an intradomain disulfide bond, an additional Ig-like fold without disulfide bond (except chicken tapasin), and an N-terminal domain without significant homology to other proteins (Ortmann et al., 1997; reviewed in Momburg and Tan, 2002). The N-terminal domain harbors a second conserved pair of cysteines that probably form an intramolecular disulfide bridge (Dick et al., 2002). The cytoplasmic domain of tapasin is poorly conserved in different species but always ends with a KKXX ER retrieval motif. Consistently, tapasin was reported to target improperly loaded MHC-I molecules to COP-I vesicles that recycle from the *cis*-Golgi to the ER (Paulsson et al., 2002). Another study, however, found no evidence for recycling as tapasin was excluded from ER exit sites (Pentcheva et al., 2002). The anterograde transport of peptide-loaded MHC-I by COP-II transport vesicles from the ER seems to be regulated by the Bap31 cargo receptor (Spiliotis et al., 2000; Paquet et al., 2004). Bap31 can, independently of MHC-I, also associate with tapasin and TAP and thus appears to check for potential cargo already in the PLC (Paquet et al., 2004).

N-terminally truncated forms of tapasin lose their interaction with MHC-I molecules (Bangia et al., 1999; Momburg and Tan, 2002), suggesting that structural elements of N-terminal domain, e.g., the putative cysteine-linked loop, are critical for this interaction. In addition, surface-exposed residues in the membrane-proximal domain of tapasin are involved in MHC-I binding (Turnquist et al., 2001). These may contact the acidic residues D227 and E229 in a protruding loop within the α_3 domain of MHC-I. Mutation of these α_3 residues resulted in the loss of tapasin-mediated TAP association (Carreno et al., 1995; reviewed in Momburg and Tan, 2002; Wright et al., 2004). A solvent-exposed loop of the MHC-I α_2 domain (residues 128–137) is another functionally important contact site with tapasin. Various point mutations in this loop, e.g., Thr134 to Lys, abolished the assembly of MHC-I with tapasin and indirectly also with TAP and calreticulin (Yu et al., 1999; reviewed in Momburg and Tan, 2002; Bouvier, 2003). Two additional residues in the MHC-I α_2 (115 and 122) domain, identified through mutational analysis, influence the affinity of the interaction between HC and tapasin (Beißbarth et al., 2000). These residues point downwards from the bottom of the peptide binding platform and contact β_2m. Taken together, these studies delineate a contact area of tapasin on the surface of HC/β_2m dimers that is similar to the contact area of CD8.

Binding of the side chain of the C-terminal peptide amino acid is largely influenced by the properties of the polymorphic MHC-I residues 114 and 116 that point upwards from the bottom of the groove toward the peptide (Bjorkman et al., 1987). Strikingly, these polymorphic residues determine whether a given MHC-I allomorph shows a strong, an intermediate or no tapasin dependency with regard to surface expression and antigen presentation (Peh et al., 1998; Williams et al., 2002; Park et al., 2003). A His residue at position 114 correlates with tapasin-dependent peptide loading (Park et al., 2003). In addition, aromatic amino acids at position 116 reportedly correlate with an increased PLC association of some HLA allomorphs (Turnquist et al., 2000; reviewed in Momburg and Tan, 2002). Although in the absence of tapasin various HLA allomorphs can be effectively loaded with peptides, they nevertheless benefit to a certain extent from tapasin as the bound peptide spectrum is altered and MHC-I maturation and stability is improved in its presence (Peh et al., 1998; Purcell et al., 2001; Williams et al., 2002; Zernich et al., 2004). The biological function of tapasin-independent alleles has recently been illustrated for HLA-B*4402 and B*4405 molecules that differ only in amino acid 116. Surface expression of B*4405 (Tyr116), which was not detectably incorporated into the PLC, was much less affected than B*4402 (Asp116) by the herpesviral TAP inhibitor ICP47 (Zernich et al., 2004). Thus, the development of tapasin-independent HC alleles due to micropolymorphism may represent an evolutionary response to viruses that interfere with the proper formation of the PLC.

3.4.4
Optimization of Peptide Loading in the TAP-associated Loading Complex

A substantial fraction of ER-resident MHC-I molecules is contained in the PLC (Suh et al., 1994; reviewed in Momburg and Tan, 2002). Likewise, most calreticulin-associated HC has been found in complexes with TAP (Diedrich et al., 2001). This correlation can be explained by experiments showing that the binding of tapasin, which tethers MHC-I to TAP (see below), and the binding of calreticulin to MHC-I are highly cooperative processes. Point mutations in MHC-I disrupting the interaction with tapasin and TAP simultaneously abolished the association with calreticulin, with the notable exception of certain HLA-B27 and H-2D^b mutants that still bound calreticulin (Lewis and Elliott, 1998; Yu et al., 1999; Beißbarth et al., 2000; reviewed in Momburg and Tan, 2002; Paulsson and Wang, 2003). Moreover, several studies showed that in tapasin-deficient cells, or in the presence of N- or C-terminally truncated forms of tapasin, the association of MHC-I with calreticulin was reduced or undetectable (Lewis and Elliott, 1998; Harris et al., 2001; Bangia et al., 1999; Tan et al., 2002). Conversely, tapasin/TAP did not efficiently assemble with MHC-I molecules when the binding of calreticulin to the α_1 domain glycan was prevented, indicating that calreticulin also promoted the recruitment of tapasin (Sadasivan et al., 1996; Yu et al., 1999; Harris et al., 2001), although it still occurred in calreticulin knock-out cells (Gao et al., 2002). Evidence for a specific function of calreticulin during peptide loading, which could not be compensated for by overexpression of soluble calnexin, was provided by a study using cells from calreticulin-deficient mice. The surface expression of MHC-I molecules was reduced to 25–30% of wild-type levels and the presentation of the majority of analyzed T cell epitopes was impaired (Gao et al., 2002). In a concerted action with tapasin, calreticulin thus appears to plays an important role in the ER retention of suboptimally loaded HC/β_2m dimers (Lewis and Elliott, 1998; Gao et al., 2002).

MHC-I HC seem to enter the PLC with fully oxidized intramolecular disulfide bonds (Harris et al., 2001; reviewed in Momburg and Tan, 2002). The presence of the oxidoreductase ERp57 in the PLC, however, suggests that disulfide bond isomerizations may occur during the editing of peptide cargo entailing structural optimization of MHC-I molecules. ERp57 is recruited to the PLC by means of a labile intermolecular disulfide bridge formed between the unpaired Cys95 of tapasin and Cys57 in the thioredoxin domain 1 of ERp57 (Dick et al., 2002). In addition, the activity of the thioredoxin domain 2 is required for efficient trapping of ERp57 (Dick et al., 2002). In the PLC, ERp57 is probably in direct contact with calreticulin, which binds to ERp57 with the tip of its arm-like P-domain (reviewed in Bouvier, 2003). In the presence of a tapasin mutant (Cys95 to Ala) that is unable to trap ERp57, PLC-associated B4402 HC were incompletely oxidized, the surface expression of B4402 molecules was slightly reduced and their turnover was enhanced (Dick et al., 2002). These findings suggest that, subsequent to reduction of the α_2 domain disulfide bond, ERp57 may re-oxidize HC in the presence of wild-type tapasin. The reducing activity mediating the initial reduction of the HC α_2 domain disulfide bond is presently unknown. Notably, ERp57 itself has only a minimal reducing activity for folded HC (Antoniou et al., 2002b).

The TAP1/TAP2 heterodimer acts as a scaffold for the PLC. Tapasin binds to TAP by virtue of its transmembrane segment with a possible contribution of the cytoplasmic domain (Lehner et al., 1998; Tan et al., 2002). The TMS of tapasin contains a conspicuous Lys residue (Ortmann et al., 1997), which may interact with negatively charged residues present in different TMS of TAP1 or TAP2. Tapasin has evolved a twofold chaperone function as it not only improves MHC-I peptide loading but also increases steady-state expression levels of TAP and peptide transport rates without altering the affinity of TAP for peptide substrates (Lehner et al., 1998; Bangia et al., 1999; Momburg and Tan, 2002). This stabilizing effect may be cell type-dependent as the tapasin-induced increase in TAP levels was reported to be ~3-fold after re-transfection of tapasin in the tapasin-deficient, human B-lymphoblastoid cell line .220 (Lehner et al., 1998), but more than hundred times reduced in spleen cells from tapasin knockout mice (Garbi et al., 2003). Tapasin can bridge HC to individually expressed TAP1 and, with lower efficiency, TAP2 subunits (Antoniou et al., 2002a). On average, four tapasin, four HC and 3–4 calreticulin molecules per TAP1/TAP2 heterodimer have been detected in macromolecular complexes (Ortmann et al., 1997). Analysis of the lateral membrane diffusion of fluorescence-tagged TAP1 molecules suggested that TAP exists in very large arrays of perhaps hundreds of molecules (Marguet et al., 1999). It is unclear whether tapasin, which is highly clustered in the ER membrane (Pentcheva et al., 2002), only assists during the initial folding of TAP1 and TAP2 polypeptides or also prevents TAP heterodimers from denaturation through continuous association.

Tapasin–ERp57–TAP–calnexin precursor complexes have been reported to precede the recruitment of HC/β_2m and calreticulin into mature PLC that no longer contain calnexin (Diedrich et al., 2001). It is possible that, in addition to tapasin, the chaperone calnexin contributes to the formation of TAP heterodimers (Diedrich et al., 2001). TAP itself appears to possess a chaperone-like function for the assembly of the other components of the PLC. In TAP-deficient cells the interaction of tapasin, HC/β_2m, ERp57 and calreticulin with each other was significantly diminished (Sadasivan et al., 1996; Momburg and Tan, 2002). Speculatively, TAP might induce subtle conformational changes in tapasin that, in turn, support the recruitment of the other PLC components.

Suitable peptide is required to release MHC-I from the tapasin–TAP complex (Suh et al., 1994; Carreno et al., 1995). A prolonged association of HC with the PLC is observed when the supply with suitable peptides is limiting, e.g., for highly peptide-selective HLA-Cw4 molecules (Neisig et al., 1998), or for rat RT1A^a molecules expressed in the presence of the restrictive TAP^u transporter (Knittler et al., 1998). The nucleotide-dependent transport cycle of TAP is coupled to productive MHC-I peptide loading and subsequent release from the PLC (Neisig et al., 1998; Knittler et al., 1999; Alberts et al., 2001).

Several lines of evidence indicate that tapasin and the other PLC components serve to improve the peptide loading of MHC-I molecules in a way that the latter acquire a higher structural stability, resulting in an increased half-life on the cell surface. Impaired peptide loading in the absence of tapasin interaction can lead to either prolonged ER retention and degradation in the early secretory or to the premature ER egress of peptide-receptive, relatively unstable MHC-I molecules (reviewed in Momburg and Tan, 2002). Some studies suggested that tapasin actively mediates ER retention of peptide-receptive MHC-I, whereas other investigators could not confirm this (reviewed in Momburg and Tan, 2002). Tapasin deficiency results in a higher proportion of peptide-receptive forms on the cell surface and a shorter half-life of surface MHC-I molecules (Yu et al., 1999; Garbi et al., 2000; Tan et al., 2002; Zarling et al., 2003), and a diminished resistance of HC/β_2m complexes to heat denaturation in lysates (Garbi et al., 2000; Tan et al., 2002; Williams et al., 2002). The half-life of highly tapasin-dependent B4402 molecules increased significantly, from ~2 h in the absence of tapasin to ~20 h in its presence (Tan et al., 2002). The reduction of surface-expressed MHC-I molecules in tapasin-deficient .220 cells varied, however, significantly with the analyzed allelic MHC-I molecules (Peh et al., 1998; Park et al., 2003; reviewed in Momburg and Tan, 2003). Splenocytes from tapasin-deficient mice displayed only 10–15% of wild-type K^b/D^b cell surface levels (Garbi et al., 2000; Grandea et al., 2000). Soluble tapasin, which is unable to tether MHC-I molecules to TAP, conferred half-lives and thermostabilities of MHC-I molecules that were intermediate compared with wild-type tapasin and the absence of tapasin (Tan et al., 2002; Williams et al., 2002; Zarling et al., 2003). It was concluded that for tapasin-dependent MHC-I alleles, tapasin-mediated bridging to TAP is required to attain an optimized stability.

Depending on the tapasin dependency of the MHC-I allomorph, the presentation of some but not all studied T cell epitopes were compromised in the absence of tapasin (Ortmann et al., 1997; Peh et al., 1998; Garbi et al., 2000; Grandea et al., 2000). Analysis of peptide spectra eluted from MHC-I molecules demonstrated qualitative differences in terms of unique peptides expressed in the presence or absence of tapasin, respectively, but also quantitative differences were obvious as the total recovery of peptides from tapasin-deficient cells was significantly reduced (Purcell et al., 2001; Zarling et al., 2003). No striking bias toward higher affinity ligands was noted, however, for individual peptides isolated from tapasin-sufficient as compared with tapasin-deficient deficient cells (Purcell et al., 2001; Zarling et al., 2003). Using minigene-encoded variants of the K^b ligand SIINFEKL it was shown that in the presence, but not the absence of tapasin, peptides were preferentially loaded in the ER which conveyed a longer half-life to surface MHC-I/peptide complexes (Howarth et al., 2004). This recent finding is the first evidence that tapasin may indeed optimize the MHC-I peptide cargo in a way that is comparable to the peptide editing function that HLA-DM has for polymorphic HLA-DR molecules (Brocke et al., 2002); however, further studies are required.

The molecular mechanism by which tapasin influences the spectrum of MHC-I bound peptides is speculative. Using a conformation-sensitive antibody, there is a predominance of immature, peptide-receptive MHC-I conformations in the PLC (Carreno et al., 1995; Yu et al., 1999). Tapasin binding may require a particular conformation displayed by the protruding HC α_2 domain loop 128–137 (and the adjacent HC-β_2m interface), which may be characteristic for an immature, not fully closed conformation induced by suboptimal peptides or for empty MHC-I conformers. Following the interaction of tapasin with HC, only peptides with low off-rates might induce a mature conformation triggering the release of MHC-I/β_2m from the tapasin/TAP complex. In contrast to the more rigid N-terminal portion of the peptide binding groove, the part of the α_2 helix contacting the peptide C terminus is thermodynamically quite flexible and its orientation is influenced by the C-terminal peptide residue contacting the F pocket in the binding cleft (Wright et al., 2004). By tightly interacting with the exposed α_2 loop, tapasin might cause an outward rolling movement of the α_2 helix positioned above this loop and the loss of hydrogen bonds fixing the peptide C terminus, which, in turn, might facilitate the exchange of peptide (Elliott, 1997b; Wright et al., 2004).

3.5
On the Way Out: MHC-I Antigen Processing along the Secretory Route

Owing to its relaxed substrate specificity, TAP facilitates the entry of peptides of 8–16 residues or even longer ones into the ER (Section 3.3). Some N-terminally extended precursors have an even higher affinity for TAP than the finally trimmed antigenic peptides (Neisig et al., 1995; Lauvau et al., 1999). In particular, this is the case for peptides with proline residues at position 2, which are inefficiently transported by TAP (Section 3.3) but are nevertheless often found as ligands for human and murine MHC-I allelic isoforms (reviewed in Rammensee et al., 1993). The transport of such precursor peptides establishes a clear need for aminopeptidase trimming in the ER (see Figure 3.1).

Efficient N-terminal trimming has been demonstrated using minigene-encoded peptide precursors targeted into the ER of TAP-deficient cells by leader sequences, whereas significant carboxypeptidase activity was not detectable in the secretory pathway (Powis et al., 1996; Craiu et al., 1997; Serwold et al., 2001; reviewed in Del Val and López, 2002; Saveanu et al., 2002; Grommé and Neefjes, 2002). Biochemical studies using synthetic peptides imported into TAP-containing microsomes documented the existence of trimming aminopeptidases in the ER (Brouwenstijn et al., 2001; Komlosh et al., 2001; Fruci et al., 2001; Serwold et al., 2001). The trimming enzymes displayed a broad cleavage specificity but spared peptides with a proline residue at position 2 (Serwold et al., 2001).

The zinc-dependent aminopeptidase ERAP1/ERAAP has been identified as the major peptide trimming enzyme of the ER (Serwold et al., 2002; Saric et al., 2002). The expression of this IFN-γ inducible enzyme shows a good correlation with the broad MHC-I tissue distribution (Serwold et al., 2002). In vitro, ERAP1 removes N-terminal amino acids in a promiscuous fashion but cleaves X-Pro bonds only poorly (Serwold et al., 2002). Being a unique property among aminopeptidases, ERAP1 prefers peptide substrates of 9–16 residues and peptides possessing hydrophobic C termini (York et al., 2002; Rock et al., 2004). These characteristics of ERAP1 strikingly resemble the substrate length and sequence preferences of restrictive TAP alleles (Section 3.3). Importantly, the capacity to trim peptides of ≥ 9 amino acids can lead to the destruction of potential MHC-I ligands (York et al., 2002). Blockade of ERAP1 function resulted in the decreased presentation of many though not all tested T cell epitopes (Serwold et al., 2001, 2002; York et al., 2002). The effect of ERAP1 inactivation on MHC-I expression varies significantly between alloforms, e.g., K^k and L^d were decreased by ERAP1 blockade, K^b expression remained unchanged and HLA-A, -B, -C expression was even increased (Serwold et al., 2002; York et al., 2002). Following treatment with IFN-γ, however, ERAP1 silencing resulted in decreased K^b and HLA-A, -B, -C surface expression (York et al., 2002). This appears to be in keeping with the increased generation of N-terminally extended epitope precursors by immunoproteasomes (Cascio et al., 2001; see Section 3.2). Whether ERAP1 generally plays a relevant role during MHC-I antigen presentation needs to be confirmed in ERAP1-deficient mice.

A second, less-well characterized ER-resident aminopeptidase, L-RAP, may also be involved in peptide processing (Tanioka et al., 2003). L-RAP/ERAP2 is also induced by IFN-γ but has a more limited tissue expression than ERAP1 (Tanioka et al., 2003). ERAP2 has a strong preference for arginine and lysine dipeptides but can also remove other residues from oligopeptides (Tanioka et al., 2003; Rock et al., 2004). In vitro, ERAP2 does not stop peptides trimming when reaching a substrate length of 8–9 residues, and in living cells ERAP2 seems to generally decrease MHC-I peptide presentation (Rock et al., 2004).

Observations by Rammensee and colleagues that the composition of the total cellular peptide repertoire is strongly influenced by the presence of particular MHC-I molecules led to the hypothesis that MHC-I molecules play an instructive role during peptide processing (reviewed in Rammensee et al., 1993). The critical relevance of peptide-binding MHC-I molecules for the detectability of these peptides was confirmed in other studies (Paz et al., 1999; Komlosh et al., 2001). It is, however, difficult to resolve whether MHC-I molecules simply bind properly sized ligands and rescue them from further degradation in the ER and from ER export through the Sec61 channel (Koopmann et al., 2000) or alternatively, whether MHC-I molecules play an active role by first binding precursor peptides and then recruiting a trimming aminopeptidase that is released when the appropriate length for binding is reached. The latter scenario was suggested by the demonstration of MHC-I association of N-terminally extended epitope precursors and their stepwise conversion into the definitive epitope by a soluble aminopeptidase activity (Brouwenstijn et al., 2001). This conversion required the presence of $K^b/\beta_2 m$ dimers in the ER. The view of MHC-guided peptide trimming was, however, challenged by the finding that certain TAP-imported precursor peptides could be trimmed in the absence of the restricting MHC-I molecule (Fruci et al., 2001). Furthermore, purified ERAP1 was able to degrade peptides in the absence of MHC-I molecules and spared 8-mer and many 9-mer peptides from further degradation (York et al., 2002; Serwold et al., 2002). ERAP1 apparently does not detectably associate with the PLC (Yewdell et al., 2003), but it remains to be investigated whether MHC-I molecules otherwise cooperate with ERAP1 or ERAP2 during peptide trimming.

Several studies have demonstrated that selected antigenic peptides can also be processed from soluble or membrane-bound proteins targeted to the ER of TAP-deficient cells (Snyder et al., 1998; reviewed in Del Val and López, 2002; Saveanu et al., 2002; Grommé and Neefjes, 2002). Since these MHC-I ligands were liberated from internal sequences or from the C-termini of proteins these findings strongly suggest the involvement of extracytoplasmic endopeptidases in antigen processing. The *trans*-Golgi network endoprotease furin, a subtilisin-like preprotein convertase, which cleaves after polybasic motifs, is the only involved endoprotease in the secretory pathway identified so far. In TAP-deficient cells, furin releases a fragment from a chimeric hepatitis B virus secretory core (HBe) protein harboring a MHC-I ligand (Gil-Torregrosa et al., 2000; reviewed in Del Val and López, 2002). This HBe fragment is subsequently processed by as yet unknown proteases to the final antigenic peptide.

Proteins expressed in the secretory pathway may also undergo processing following retrotranslocation to the cytosol for proteasomal degradation (reviewed in Tsai et al., 2002). Deglycosylation and asparagine deamidation of peptides by cytosolic *N*-glycanase showed that retrotranslocated ER glycoproteins provide a source for TAP-dependent MHC-I ligands (Bacik et al., 1997).

3.6
Closing the Circle – Cross-presentation of Endocytosed Antigens by MHC-I Molecules

3.6.1
Phagosome-to-cytosol Pathway of MHC-I Peptide Loading

MHC-I molecules not only present peptides derived from endogenously synthesized protein but also from antigens that enter the cell through the endocytic pathway in a process termed "cross-presentation". Soluble and immobilized proteins, chaperoned peptides, inactivated viruses, virus-like particles, bacteria, and various antigens from apoptotic or necrotic cells can be cross-presented (reviewed in Jondal et al., 1996; Rock, 1996; Yewdell et al., 1999). Although cross-presentation can be mediated by different cell types with endocytic activity, including macrophages, B cells and some non-professional antigen presenting cells (Rock, 1996), the most efficient cross-presenting cells for CTL priming in vivo appear to be $CD8^+$ dendritic cells (DC) (reviewed in Ackerman and Cresswell, 2004). Different forms of endocytosis, including phagocytosis, receptor-mediated endocytosis, macropinocytosis and fluid phase endocytosis, have been implicated in cross-presentation (reviewed in Rock, 1996; Yewdell et al., 1999; Ackerman and Cresswell, 2004).

Two major mechanisms of cross-presentation can be distinguished, which either involve the cytoplasmic proteolytic machinery, the TAP peptide transporter and peptide binding to newly synthesized MHC-I molecules or, alternatively, the endolysosomal system as processing compartment and peptide loading of recycling MHC-I molecules (see Figure 3.2). The existence of an endosome-to-cytosol pathway was indicated by the abrogation of cross-presentation in the presence of proteasome inhibitors such as lactacystin, and in cells that do not express functional peptide transporters (Fonteneau et al., 2003; reviewed in Yewdell et al., 1999; Ackerman and Cresswell, 2004). In this pathway, cross-presentation is also inhibited by the fungal metabolite brefeldin A that interferes with the egress of MHC-I molecules from the ER (reviewed in Yewdell et al., 1999). The involvement of proteasomes and TAP indicate that internalized antigens can be shuttled from endocytic vesicles to the cytosol for degradation and subsequent ER translocation of resulting peptides. Indeed, nondegradable, fluorochrome-conjugated dextrans up to 40 kDa were observed to enter the cytosol of dendritic cells, whereas dextran molecules of 500 kDa and larger were mostly retained inside phagosomes (Rodriguez et al., 1999). In keeping with these size limits, virus-like particles and phagocytosed cells require partial digestion by endolysosomal hydrolases such as cathepsin D before cytosolic transfer (Fonteneau et al., 2003; Morón et al., 2003). Several studies have demonstrated efficient cross-presentation of peptides bound to heat shock proteins (reviewed in Srivastava, 2002). Recent studies suggested, however, that proteasomal protein substrates rather than chaperoned peptides are transferred from internalized dead cells for cross-presentation (Shen and Rock, 2004; Norbury et al., 2004). The TAP-dependent pathway of cross-presentation by hematopoietic APC is crucially involved in the in vivo priming and generation of

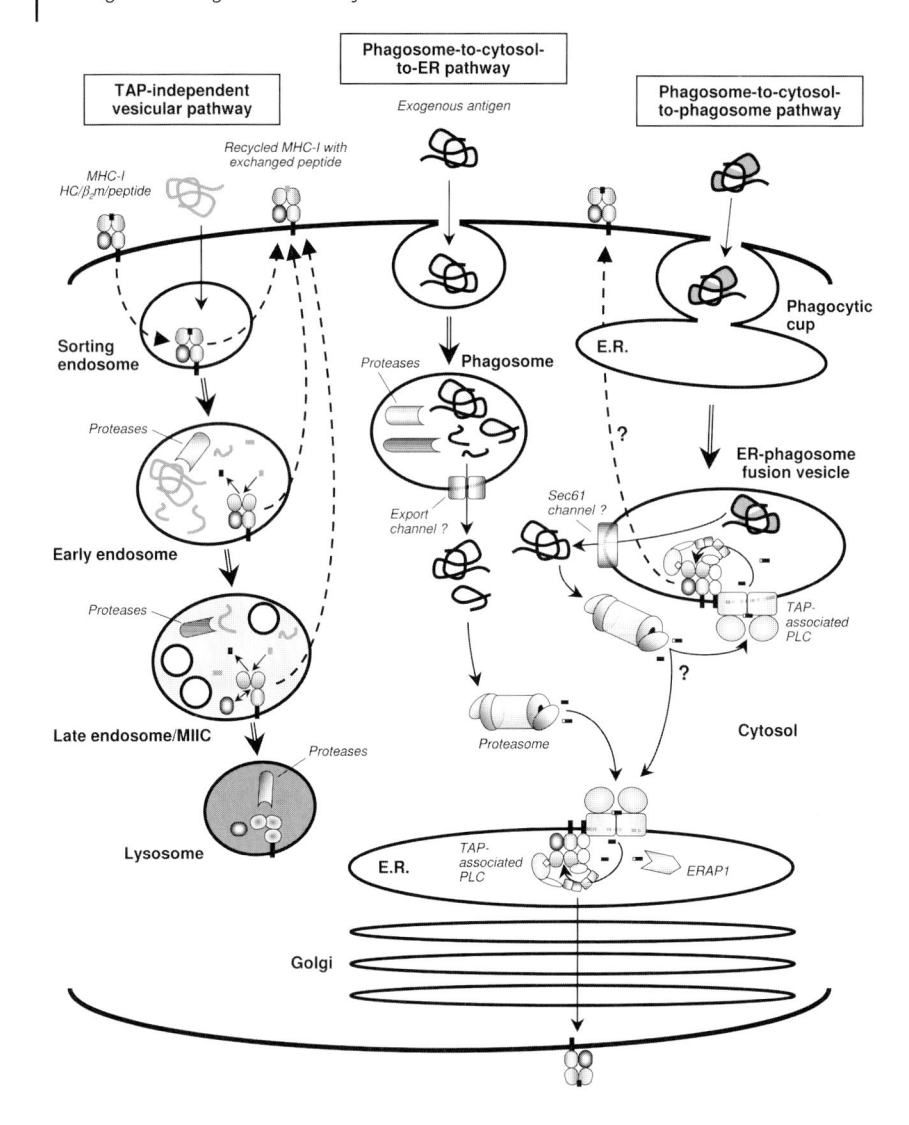

3 Antigen Processing in the Context of MHC Class I Molecules

Figure 3.2 Alternative pathways of MHC-I antigen processing. (1) TAP-independent vesicular pathway: Surface MHC-I HC/β_2m/peptide complexes are constitutively internalized into recycling early endosomes that have a sorting function. From there, MHC-I can be targeted to slightly acidic early endosomes as well as to more acidic late endosomes and the MHC class II-rich compartment (MIIC) in antigen presenting cells. In endosomes exogenous antigens are degraded into peptides by acidic hydrolases (various endopeptidases and exopeptidases of the cathepsin family). MHC-I-bound peptides can be exchanged by a poorly understood mechanism that may involve pH-dependent destabilization of the heterotrimer and re-binding of β_2m and peptide. Through carrier vesicles, MHC-I molecules traffic back to the cell surface. In lysosomes MHC-I molecules are denatured and degraded. (2) Phagosome-to-cytosol-to-ER pathway of cross-presentation: Phagocytosed particulate antigens and proteins, internalized by macropinocytosis or receptor-mediated endocytosis in dendritic cells and macrophages, are translocated across the vesicular membrane through an unknown export channel either in intact form or after partial degradation by acidic hydrolases. Upon arrival in the cytosol, the polypeptides are degraded by proteasomes. Resulting oligopeptides are translocated by TAP and loaded onto MHC-I molecules following the classical pathway. (3) Phagosome-to-cytosol-to-phagosome pathway of cross-presentation: In dendritic cells and macrophages, ER domains can fuse with phagocytic cups to form vesicles that contain all ER-typical components of the TAP-associated PLC together with phagocytosed protein antigens. Proteins are transferred to the cytosol, which might involve the ER-resident Sec61 (retro)translocation channel. Phagosome-attached ubiquitinating enzymes and proteasomes may process exported proteins into peptides that may be able to re-enter ER-phagosome fusion vesicles through TAP for loading onto MHC-I. It is unclear whether ER-phagosomal MHC-I molecules recycle to the cell surface using a vesicular route, or whether a transient continuity with the ER proper facilitates their access to the ordinary ER-to-Golgi-to-surface route.

protective immunity against tumor-associated antigens and antigens supplied by virally infected cells (Huang et al., 1996; Sigal et al., 1999). This is of particular relevance for immune responses against viruses that do not infect APC themselves (Sigal et al., 1999) or viruses that would either kill the infected APC or interfere with the MHC-I antigen presentation pathway (Fonteneau et al., 2003; reviewed in Ackerman and Cresswell, 2004).

The conduit used by proteins to leave endosomal vesicles is still unclear but circumstantial evidence suggests that it may be the Sec61 channel, which is also implicated in the dislocation of misfolded proteins from ER membranes to the cytosol (reviewed in Tsai et al., 2002). The cell-biological observation that nascent phagosomes can acquire ER membrane components led to the discovery of ER-phagosome fusion vesicles that contain, in addition to the phagocytosed particulate antigen, MHC-I molecules and the other components of the TAP-associated PLC (Guermonprez et al., 2003; Houde et al., 2003; Ackerman et al., 2003). The presence of the Sec61 channel in these ER-derived phagosomes and the colocalization of ubiquitinated proteins and proteasomes at the cytoplasmic face of these vesicles suggest that they may represent self-sufficient processing compartments. The access of external proteins into ER-endosome fusion vesicles has not only been observed following phagocytosis, but also following the macropinocytosis of antibodies (Ackermann et al., 2003). It is unclear whether MHC-I molecules loaded in ER-derived phagosomes recycle back to the plasma membrane through early endosomes or whether they take the conventional route via the ER and Golgi to the surface (reviewed Ackermann and Cresswell, 2004).

3.6.2
Endolysosomal Pathway of MHC-I Peptide Loading

In the alternative, TAP-independent route of cross-presentation MHC-I molecules are subjected to the rules governing the well-known pathways of MHC-II antigen processing and presentation involving acidic endolysosomal vesicles (reviewed in Grommé and Neefjes, 2002; Yewdell et al., 1999). At first glance it seems unlikely that acidic hydrolases can produce the correct 8–10-mer MHC-I ligands that are usually generated by the peptide processing machinery in the cytosol and ER. The study of TAP-deficient APC in vitro provided, however, evidence for the existence of an endolysosomal MHC-I cross-presentation pathway. In some cell types, MHC-I molecules undergo constitutive internalization from the plasma membrane and recycling through early endosomes back to the surface (reviewed in Grommé and Neefjes, 2002). In B cells, MHC-I molecules have been detected in MHC class II-rich late endosomal vesicles from where they recycle back to the cell surface (Grommé et al., 1999). For the targeting into late endosomes of DC, a conserved Tyr residue in the cytoplasmic tail of K^b molecules is crucial since cross-presentation of exogenous antigens by K^b was abolished after mutation of this Tyr (Lizée et al., 2003). The HC/β_2m/peptide heterotrimer tends to dissociate in the acidic environment of late endosomes (Grommé et al., 1999). A sufficient supply with exogenous β_2m seems to facilitate the re-assembly of HC with endosome-derived peptides (reviewed in Schirmbeck and Reimann, 2002).

The TAP-independent and brefeldin A/lactacystin-resistant generation of MHC-I ligands from model antigens such as bacterial-derived recombinant ovalbumin (Pfeifer et al., 1993; Campbell et al., 2000) and some viral proteins has been reported (reviewed in Grommé and Neefjes, 2002). In some cases peptides generated in endosomes are secreted and can bind to surface MHC-I molecules on neighboring cells in a process termed "peptide regurgitation" (Pfeifer et al., 1993). Consistent with the requirements of endolysosomal vesicular trafficking and antigen processing by proteases that are active in acidic conditions, the vacuolar MHC-I pathway was shown to be sensitive to proton-depleting lysosomotropic amines such as chloroquine or ammonium chloride (Schirmbeck and Reimann, 1994; Grommé et al., 1999). Cathepsin S and possibly other cysteine proteases seem to play a major role in this pathway (Shen et al., 2004; Campbell et al., 2000), which is also relevant for CTL priming in vivo (Shen et al., 2004). Dendritic cells from cathepsin S-deficient mice lack the TAP-independent pathway. These mice show a strongly reduced cross-priming of immune responses against cell-associated antigens and influenza virus (Shen et al., 2004).

A growing body of evidence indicates that cross-priming plays an important role during the induction of immunity to viruses and possibly also to tumors (reviewed in Heath and Carbone, 2001). Cross-presentation offers the general advantage that the antigen-presenting cell need not express the tumor antigen itself and need not be infected by viruses that might severely compromise its functions.

References

Abele, R., Tampé, R. (1999): Function of the transport complex TAP in cellular immune recognition. *Biochim. Biophys. Acta* 1461: 405–419.

Abele, R., Tampé, R. (2004): The ABCs of immunology: structure and function of TAP, the transporter associated with antigen processing. *Physiology* 19: 216–224.

Ackerman, A.L., Kyritsis, C., Tampé, R., Cresswell, P. (2003): Early phagosomes in dendritic cells form a cellular compartment sufficient for cross presentation of exogenous antigens. *Proc. Natl. Acad. Sci. U.S.A.* 100: 12889–12894.

Ackerman, A.L., Cresswell, P. (2004): Cellular mechanisms governing cross-presentation of exogenous antigens. *Nat. Immunol.* 5: 678–684.

Alberts, P., Daumke, O., Deverson, E.V., Howard, J.C., Knittler, M.R. (2001): Distinct functional properties of the TAP subunits coordinate the nucleotide-dependent transport cycle. *Curr. Biol.* 11: 242–251.

Aldrich, C.J., Ljunggren, H.G., Van Kaer, L., Ashton-Rickardt, P.G., Tonegawa, S., Forman, J. (1994): Positive selection of self- and alloreactive CD8$^+$ T cells in *Tap-1* mutant mice. *Proc. Natl. Acad. Sci. U.S.A.* 91: 6525–6528.

Androlewicz, M.J, Cresswell, P. (1994): Human transporters associated with antigen processing possess a promiscuous peptide-binding site. *Immunity* 1: 7–14.

Antoniou, A.N., Ford, S., Pilley, E.S., Blake, N., Powis, S.J. (2002a): Interactions formed by individually expressed TAP1 and TAP2 polypeptide subunits. *Immunology* 106: 182–189.

Antoniou, A.N., Ford, S., Alphey, M., Osborne, A., Elliott, T., Powis, S.J. (2002b): The oxidoreductase ERp57 efficiently reduces partially folded in preference to fully folded MHC class I molecules. *EMBO J.* 21: 2655–2663.

Bacik, I., Snyder, H.L., Antón, L.C., Russ, G., Chen, W., Bennink, J.R., Urge, L., Otvos, L., Dudkowska, B., Eisenlohr, L., Yewdell, J.W. (1997): Introduction of a glycosylation site into a secreted protein provides evidence for an alternative antigen processing pathway: transport of precursors of major histocompatibility complex class I-restricted peptides from the endoplasmic reticulum to the cytosol. *J. Exp. Med.* 186: 479–487.

Bangia, N., Lehner, P.J., Hughes, E.A., Surman, M., Cresswell, P. (1999): The N-terminal region of tapasin is required to stabilize the MHC class I loading complex. *Eur. J. Immunol.* 29: 1858–1870.

Baumeister, W., Walz, J., Zühl, F., Seemüller, E. (1998): The proteasome: paradigm of a self-compartmentalizing protease. *Cell* 92: 367–380.

Beißbarth, T., Sun, J., Kavathas, P.B., Ortmann, B. (2000) Increased efficiency of folding and peptide loading of mutant MHC class I molecules. *Eur. J. Immunol.* 30: 1203–1213.

Benaroudj, N., Tarcsa, E., Cascio, P., Goldberg, A.L. (2001): The unfolding of substrates and ubiquitin-independent protein degradation by proteasomes. *Biochimie* 83: 311–318.

Benham, A.M., Grommé, M., Neefjes, J. (1998): Allelic differences in the relationship between proteasome activity and MHC class I peptide loading. *J. Immunol.* 161: 83–89.

Beninga, J., Rock, K.L., Goldberg, A.L. (1998): Interferon-γ can stimulate postproteasomal trimming of the N terminus of an antigenic peptide by inducing leucine aminopeptidase. *J. Biol. Chem.* 273: 18 734–18 742.

Bjorkman, P.J., Saper, M.A., Samraoui, B., Bennett, W.S., Strominger, J.L., Wiley, D.C. (1987): The foreign antigen binding site and T cell recognition regions of class I histocompatibility antigens. *Nature* 329: 512–518.

Bouvier M. (2003): Accessory proteins and the assembly of human class I MHC molecules: a molecular and structural perspective. *Mol. Immunol.* 39: 697–706.

Brocke, P., Garbi, N., Momburg, F., Hämmerling, G.J. (2002): HLA-DM, HLA-DO and tapasin: functional similarities and differences. *Curr. Opin. Immunol.* 14: 22–29.

Brooks, P., Murray, R.Z., Mason, G.G., Hendil, K.B., Rivett, A.J. (2000): Association of immunoproteasomes with the endoplasmic reticulum. *Biochem. J.* 352: 611–615.

Brouwenstijn, N., Serwold, T., Shastri, N. (2001): MHC class I molecules can direct proteolytic cleavage of antigenic precursors in the endoplasmic reticulum. *Immunity* 15: 95–104.

Campbell, D.J., Serwold, T., Shastri, N. (2000): Bacterial proteins can be processed by macrophages in a transporter associated with antigen processing-independent, cysteine protease-dependent manner for presentation by MHC class I molecules. *J. Immunol.* 164: 168–175.

Carreno, B.M., Solheim, J.C., Harris, M., Stroynowski, I., Connolly, J.M., Hansen, T.H. (1995): TAP associates with a unique class I conformation, whereas calnexin associates with multiple class I forms in mouse and man. *J. Immunol.* 155: 4726–4733.

Cascio, P., Hilton, C., Kisselev, A.F., Rock, K.L., Goldberg, A.L. (2001): 26S proteasomes and immunoproteasomes produce mainly N-extended versions of an antigenic peptide. *EMBO J.* 20: 2357–2366.

Cascio, P., Call, M., Petre, B.M., Walz, T., Goldberg, A.L. (2002): Properties of the hybrid form of the 26S proteasome containing both 19S and PA28 complexes. *EMBO J.* 21: 2636–2645.

Coffino, P. (2001): Regulation of cellular polyamines by antizyme. *Nat. Rev. Mol. Cell Biol.* 2: 188–194.

Craiu, A., Akopian, T., Goldberg, A., Rock, K.L. (1997): Two distinct proteolytic processes in the generation of a major histocompatibility complex class I-presented peptide. *Proc. Natl. Acad. Sci. U.S.A.* 94: 10850–10855.

Dantuma, N.P., Heessen, S., Lindsten, K., Jellne, M., Masucci, M.G. (2000): Inhibition of proteasomal degradation by the Gly-Ala repeat of Epstein-Barr virus is influenced by the length of the repeat and the strength of the degradation signal. *Proc. Natl. Acad. Sci. U.S.A.* 97: 8381–8385.

Del Val, M., López, D. (2002): Multiple proteases process viral antigens for presentation by MHC class I molecules to $CD8^+$ T lymphocytes. *Mol. Immunol.* 39: 235–247.

Deverson, E.V., Leong, L., Seelig, A., Coadwell, W.J., Tredgett, E.M., Butcher, G.W., Howard, J.C. (1998): Functional analysis by site-directed mutagenesis of the complex polymorphism in rat transporter associated with antigen processing. *J. Immunol.* 160: 2767–2779.

Dick, T.P., Stevanovic, S., Keilholz, W., Ruppert, T., Koszinowski, U., Schild, H., Rammensee, H.G. (1998): The making of the dominant MHC class I ligand SYFPEITHI. *Eur. J. Immunol.* 28: 2478–22486.

Dick, T.P., Bangia, N., Peaper, D.R., Cresswell, P., 2002. Disulfide bond isomerization and the assembly of MHC class I-peptide complexes. *Immunity* 16, 87–98.

Diedrich, G., Bangia, N., Pan, M., Cresswell, P. (2001): A role for calnexin in the assembly of the MHC class I loading complex in the endoplasmic reticulum. *J. Immunol.* 166: 1703–1709.

Ellgard, L., Helenius, A. (2003): Quality control in the endoplasmic reticulum. *Nat. Rev. Mol. Cell Biol.* 4: 181–191.

Elliott, T. (1997a): Transporter associated with antigen processing. *Adv. Immunol.* 65: 47–109.

Elliott, T. (1997b): How does TAP associate with MHC class I molecules? *Immunol. Today* 18: 375–379.

Engelhard, V.H. (1994): Structure of peptides associated with MHC class I molecules. *Curr. Opin. Immunol.* 6: 13–23.

Engelhard, V.H., Brickner, A.G., Zarling, A.L. (2002): Insights into antigen processing gained by direct analysis of the naturally processed class I MHC associated peptide repertoire. *Mol. Immunol.* 39: 127–137.

Farmery, M.R., Allen, S., Allen, A.J., Bulleid, N.J. (2000): The role of ERp57 in disulfide bond formation during the assembly of major histocompatibility complex class I in a synchronized semipermeabilized cell translation system. *J. Biol. Chem.* 275: 14933–14938.

Fehling, H.J., Swat, W., Laplace, C., Kühn, R., Rajewsky, K., Müller, U., von Boehmer, H. (1994): MHC class I expression in mice lacking the proteasome subunit LMP-7. *Science* 265: 1234–1237.

Fonteneau, J.F., Kavanagh, D.G., Lirvall, M., Sanders, C., Cover, T.L., Bhardwaj, N., Larsson, M. (2003): Characterization of the MHC class I cross-presentation pathway for cell-associated antigens by human dendritic cells. *Blood* 102: 4448–4455.

Fruci, D., Lauvau, G., Saveanu, L., Amicosante, M., Butler, R.H., Polack, A., Ginhoux, F., Lemonnier, F., Firat, H., van Endert, P.M. (2003): Quantifying recruitment of cytosolic peptides for HLA class I presentation: impact of TAP transport. *J. Immunol.* 170: 2977–2984.

Fruci, D., Niedermann, G., Butler, R.H., van Endert, P.M. (2001): Efficient MHC class I-independent amino-terminal trimming of epitope precursor peptides in the endoplasmic reticulum. *Immunity* 15: 467–476.

Gao, B., Adhikari, R., Howarth, M., Nakamura, K., Gold, M.C., Hill, A.B., Knee, R., Michalak, M., Elliott, T. (2002): Assembly and antigen-presenting function of MHC class I molecules in cells lacking the ER chaperone calreticulin. *Immunity* 16: 99–109.

Garbi, N., Tan, P., Diehl, A.D., Chambers, B.J., Ljunggren, H.G., Momburg, F., Hämmerling, G.J. (2000): Impaired immune responses and altered peptide repertoire in tapasin-deficient mice. *Nat. Immunol.* 1: 234–238.

Garbi, N., Tiwari, N., Momburg, F., Hämmerling, G.J. (2003): A major role for tapasin as a stabilizer of the TAP peptide transporter and consequences for MHC class I expression. *Eur. J. Immunol.* 33: 264–273.

Gaudet, R., Wiley, D.C. (2001): Structure of the ABC ATPase domain of human TAP1, the transporter associated with antigen processing. *EMBO J.* 20: 4964–4972.

Geier, E., Pfeifer, G., Wilm, M., Lucchiari-Hartz, M., Baumeister, W., Eichmann, K., Niedermann, G. (1999): A giant protease with potential to substitute for some functions of the proteasome. *Science* 283: 978–981.

Gil-Torregrosa, B.C., Castaño, A.R., López, D., Del Val, M. (2000): Generation of MHC class I peptide antigens by protein processing in the secretory route by furin. *Traffic* 1: 641–651.

Grandea, A.G., 3rd, Golovina, T.N., Hamilton, S.E., Sriram, V., Spies, T., Brutkiewicz, R.R., Harty, J.T., Eisenlohr, L.C., Van Kaer, L. (2000): Impaired assembly yet normal trafficking of MHC class I molecules in Tapasin mutant mice. *Immunity* 13: 213–222.

Groettrup, M., Soza, A., Eggers, M., Kuehn, L., Dick, T.P., Schild, H., Rammensee, H.G., Koszinowski, U.H., Kloetzel, P.M. (1996): A role for the proteasome regulator PA28α in antigen presentation. *Nature* 381: 166–168.

Grommé, M., Uytdehaag, F.G., Janssen, H., Calafat, J., van Binnendijk, R.S., Kenter, M.J., Tulp, A., Verwoerd, D., Neefjes, J. (1999): Recycling MHC class I molecules and endosomal peptide loading. *Proc. Natl. Acad. Sci. U.S.A.* 96: 10326–10331.

Grommé, M., Neefjes, J. (2002): Antigen degradation or presentation by MHC class I molecules via classical and non-classical pathways. *Mol. Immunol.* 39: 181–202.

Guermonprez, P., Saveanu, L., Kleijmeer, M., Davoust, J., Van Endert, P., Amigorena, S. (2003): ER-phagosome fusion defines an MHC class I cross-presentation compartment in dendritic cells. *Nature* 425: 397–402.

Harris, M.R., Lybarger, L., Yu, Y.Y., Myers, N.B., Hansen, T.H. (2001): Association of ERp57 with mouse MHC class I molecules is tapasin dependent and mimics that of calreticulin and not calnexin. *J. Immunol.* 166: 6686–6692.

Heath, W.R., Carbone, F.R. (2001): Cross-presentation in viral immunity and self-tolerance. *Nat. Rev. Immunol.* 1: 126–134.

Heemels, M.T., Schumacher, T.N., Wonigeit, K., Ploegh, H.L. (1993): Peptide translocation by variants of the transporter associated with antigen processing. *Science* 262: 2059–2063.

Herberg, J.A., Sgouros, J., Jones, T., Copeman, J., Humphray, S.J., Sheer, D., Cresswell, P., Beck, S., Trowsdale, J. (1998): Genomic analysis of the *Tapasin* gene, located close to the *TAP* loci in the MHC. *Eur. J. Immunol.* 28: 459–467.

Hershko, A., Ciechanover, A. (1998): The ubiquitin system. *Annu. Rev. Biochem.* 67: 425–479.

Higgins, C.F. (1992): ABC transporters: from microorganisms to man. *Annu. Rev. Cell Biol.* 8: 67–113.

Houde, M., Bertholet, S., Gagnon, E., Brunet, S., Goyette, G., Laplante, A., Princiotta, M.F., Thibault, P., Sacks, D., Desjardins, M. (2003): Phagosomes are competent organelles for antigen cross-presentation. *Nature* 425: 402–406.

Howarth, M., Williams, A., Tolstrup, A.B., Elliott, T. (2004): Tapasin enhances MHC class I peptide presentation according to peptide half-life. *Proc. Natl. Acad. Sci. U.S.A.* 101: 11 737–11 742.

Huang, A.Y., Bruce, A.T., Pardoll, D.M., Levitsky, H.I. (1996): In vivo cross-priming of MHC class I-restricted antigens requires the TAP transporter. *Immunity* 4: 349–355.

Joly, E., Le Rolle, A.F., González, A.L., Mehling, B., Stevens, J., Coadwell, W.J., Hünig, T., Howard, J.C., Butcher, G.W. (1998): Co-evolution of rat TAP transporters and MHC class I RT1-A molecules. *Curr. Biol.* 8: 169–172.

Jondal, M., Schirmbeck, R., Reimann, J. (1996): MHC class I-restricted CTL responses to exogenous antigens. *Immunity* 5: 295–302.

Karttunen, J.T., Lehner, P.J., Gupta, S.S., Hewitt, E.W., Cresswell, P. (2001): Distinct functions and cooperative interaction of the subunits of the transporter associated with antigen processing (TAP). *Proc. Natl. Acad. Sci. U.S.A.* 98: 7431–7436.

Kessler, B.M., Glas, R., Ploegh, H.L. (2002): MHC class I antigen processing regulated by cytosolic proteolysis – short cuts that alter peptide generation. *Mol. Immunol.* 39: 171–179.

Kessler, J.H., Beekman, N.J., Bres-Vloemans, S.A., Verdijk, P., van Veelen, P.A., Kloosterman-Joosten, A.M., Vissers, D.C., ten Bosch, G.J., Kester, M.G., Sijts, A., Drijfhout, J.W., Ossendorp, F., Offringa, R., Melief, C.J. (2001): Efficient identification of novel HLA-A*0201-presented cytotoxic T lymphocyte epitopes in the widely expressed tumor antigen PRAME by proteasome-mediated digestion analysis. *J. Exp. Med.* 193: 73–88.

Khan, S., de Giuli, R., Schmidtke, G., Bruns, M., Buchmeier, M., van den Broek, M., Groettrup, M. (2001): Cutting edge: neo-synthesis is required for the presentation of a T cell epitope from a long-lived viral protein. *J. Immunol.* 167: 4801–4804.

Kisselev, A.F., Akopian, T.N., Woo, K.M., Goldberg, A.L. (1999): The sizes of peptides generated from protein by mammalian 26 and 20 S proteasomes. Implications for understanding the degradative mechanism and antigen presentation. *J. Biol. Chem.* 274: 3363–3371.

Kleijmeer, M.J., Kelly, A., Geuze, H.J., Slot, J.W., Townsend, A., Trowsdale, J. (1992): Location of MHC-encoded transporters in the endoplasmic reticulum and *cis*-Golgi. *Nature* 357: 342–344.

Kloetzel, P.M. (2001): Antigen processing by the proteasome. *Nat. Rev. Mol. Cell Biol.* 2: 179–187.

Kloetzel, P.M. (2004): Generation of major histocompatibility complex class I antigens: functional interplay between proteasomes and TPPII. *Nat. Immunol.* 5: 661–669.

Knittler, M.R., Gülow, K., Seelig, A., Howard, J.C. (1998): MHC class I molecules compete in the endoplasmic reticulum for access to transporter associated with antigen processing. *J. Immunol.* 161: 5967–5977.

Koch, J., Guntrum, R., Heintke, S., Kyritsis, C., Tampé, R. (2004): Functional dissection of the transmembrane domains of the transporter associated with antigen processing (TAP). *J. Biol. Chem.* 279: 10 142–10 147.

Komlosh, A., Momburg, F., Weinschenk, T., Emmerich, N., Schild, H., Nadav, E., Shaked, I., Reiss, Y. (2001): A role for a novel luminal endoplasmic reticulum aminopeptidase in final trimming of 26 S proteasome-generated major histocompatability complex class I antigenic peptides. *J. Biol. Chem.* 276: 30 050–30 056.

Koopmann, J.O., Post, M., Neefjes, J.J., Hämmerling, G.J., Momburg, F. (1996): Translocation of long peptides by transporters associated with antigen processing (TAP). *Eur. J. Immunol.* 26: 1720–1728.

Koopmann, J.O., Albring, J., Hüter, E., Bulbuc, N., Spee, P., Neefjes, J., Hämmerling, G.J., Momburg F. (2000): Export of antigenic peptides from the endoplasmic reticulum intersects with retrograde protein translocation through the Sec61p channel. *Immunity* 13: 117–127.

Kunisawa, J., Shastri, N. (2003): The group II chaperonin TRiC protects proteolytic intermediates from degradation in the MHC class I antigen processing pathway. *Mol. Cell* 12: 565–576.

Lammert, E., Stevanovic, S., Brunner, J., Rammensee, H.G., Schild, H. (1997): Protein disulfide isomerase is the dominant acceptor for peptides translocated into the endoplasmic reticulum. *Eur. J. Immunol.* 27: 1685–1690.

Lapinski, P.E., Neubig, R.R., Raghavan, M. (2001): Walker A lysine mutations of TAP1 and TAP2 interfere with peptide translocation but not peptide binding. *J. Biol. Chem.* 276: 7526–7533.

Lautscham, G., Rickinson, A., Blake, N. (2003): TAP-independent antigen presentation on MHC class I molecules: lessons from Epstein-Barr virus. *Microbes Infect.* 5: 291–299.

Lauvau, G., Kakimi, K., Niedermann, G., Ostankovitch, M., Yotnda, P., Firat, H., Chisari, F.V., van Endert, P.M. (1999): Human transporters associated with antigen processing (TAPs) select epitope precursor peptides for processing in the endoplasmic reticulum and presentation to T cells. *J. Exp. Med.* 190: 1227–1240.

Lizée, G., Basha, G., Tiong, J., Julien, J.P., Tian, M., Biron, K.E., Jefferies, W.A. (2003): Control of dendritic cell cross-presentation by the major histocompatibility complex class I cytoplasmic domain. *Nat. Immunol.* 4: 1065–1073.

Lehner, P.J., Surman, M.J., Cresswell, P. (1998): Soluble tapasin restores MHC class I expression and function in the tapasin-negative cell line .220. *Immunity* 8: 221–231.

Lévy, F., Burri, L., Morel, S., Peitrequin, A.L., Lévy, N., Bachi, A., Hellman, U., Van den Eynde, B.J., Servis, C. (2002): The final N-terminal trimming of a subaminoterminal proline-containing HLA class I-restricted antigenic peptide in the cytosol is mediated by two peptidases. *J. Immunol.* 169: 4161–4171.

Lewis, J.W., Elliott, T. (1998): Evidence for successive peptide binding and quality control stages during MHC class I assembly. *Curr. Biol.* 8: 717–720.

Lindquist, J.A., Hämmerling, G.J., Trowsdale, J. (2001): ER60/ERp57 forms disulfide-bonded intermediates with MHC class I heavy chain. *FASEB J.* 15: 1448–1450.

Lucchiari-Hartz, M., van Endert, P.M., Lauvau, G., Maier, R., Meyerhans, A., Mann, D., Eichmann, K., Niedermann, G. (2000): Cytotoxic T lymphocyte epitopes of HIV-1 Nef: Generation of multiple definitive major histocompatibility complex class I ligands by proteasomes. *J. Exp. Med.* 191: 239–252.

Luckey, C.J., Marto, J.A., Partridge, M., Hall, E., White, F.M., Lippolis, J.D., Shabanowitz, J., Hunt, D.F., Engelhard, V.H. (2001): Differences in the expression of human class I MHC alleles and their associated peptides in the presence of proteasome inhibitors. *J. Immunol.* 167: 1212–1221.

Lu, J., Wettstein, P.J., Higashimoto, Y., Appella, E., Celis, E. (2001): TAP-independent presentation of CTL epitopes by Trojan antigens. *J. Immunol.* 166: 7063–7071.

Madden, D.R. (1995): The three-dimensional structure of peptide-MHC complexes. *Annu. Rev. Immunol.* 13: 587–622.

Marguet, D., Spiliotis, E.T., Pentcheva, T., Lebowitz, M., Schneck, J., Edidin, M. (1999): Lateral diffusion of GFP-tagged H2L^d molecules and of GFP-TAP1 reports on the assembly and retention of these molecules in the endoplasmic reticulum. *Immunity* 11: 231–240.

Marsh, S.G., Albert, E.D., Bodmer, W.F., Bontrop, R.E., Dupont, B., Erlich, H.A., Geraghty, D.E., Hansen, J.A., Mach, B., Mayr, W.R., Parham, P., Petersdorf, E.W., Sasazuki, T., Schreuder, G.M., Strominger, J.L., Svejgaard, A., Terasaki, P.I.

(2002): Nomenclature for factors of the HLA system. *Hum. Immunol.* 63: 1213–1268.

Momburg, F., Roelse, J., Hämmerling, G.J., Neefjes, J.J. (1994a): Peptide size selection by the major histocompatibility complex-encoded peptide transporter. *J. Exp. Med.* 179:1613–1623.

Momburg, F., Roelse, J., Howard, J.C., Butcher, G.W., Hämmerling, G.J., Neefjes, J.J. (1994b): Selectivity of MHC-encoded peptide transporters from human, mouse and rat. *Nature* 367: 648–651.

Momburg, F., Armandola, E.A., Post, M., Hämmerling, G.J. (1996): Residues in TAP2 peptide transporters controlling substrate specificity. *J. Immunol.* 156: 1756–1763.

Momburg, F., Hämmerling, G.J. (1998). Generation and TAP-mediated transport of peptides for major histocompatibility complex class I molecules. *Adv. Immunol.* 68, 191–256.

Momburg, F., Tan, P. (2002): Tapasin – the keystone of the loading complex optimizing peptide binding by MHC class I molecules in the endoplasmic reticulum. *Mol. Immunol.* 39: 217–233.

Montoya, M., Del Val, M. (1999): Intracellular rate-limiting steps in MHC class I antigen processing. *J. Immunol.* 163: 1914–1922.

Morel, S., Lévy, F., Burlet-Schiltz, O., Brasseur, F., Probst-Kepper, M., Peitrequin, A.L., Monsarrat, B., Van Velthoven, R., Cerottini, J.C., Boon, T., Gairin, J.E., Van den Eynde, B.J. (2000): Processing of some antigens by the standard proteasome but not by the immunoproteasome results in poor presentation by dendritic cells. *Immunity* 12: 107–117.

Morón, V.G., Rueda, P., Sedlik, C., Leclerc, C. (2003): In vivo, dendritic cells can cross-present virus-like particles using an endosome-to-cytosol pathway. *J. Immunol.* 171: 2242–2250.

Murata, S., Udono, H., Tanahashi, N., Hamada, N., Watanabe, K., Adachi, K., Yamano, T., Yui, K., Kobayashi, N., Kasahara, M., Tanaka, K., Chiba, T. (2001): Immunoproteasome assembly and antigen presentation in mice lacking both PA28α and PA28β. *EMBO J.* 20: 5898–5907.

Navon, A., Goldberg, A.L. (2001): Proteins are unfolded on the surface of the ATPase ring before transport into the proteasome. *Mol. Cell* 8: 1339–1349.

Neefjes, J.J., Momburg, F., Hämmerling, G.J. (1993a): Selective and ATP-dependent translocation of peptides by the MHC-encoded transporter. *Science* 261: 769–771.

Neefjes, J.J., Hämmerling, G.J., Momburg, F. (1993b): Folding and assembly of major histocompatibility complex class I heterodimers in the endoplasmic reticulum of intact cells precedes the binding of peptide. *J. Exp. Med.* 178: 1971–1980.

Neisig, A., Roelse, J., Sijts, A.J., Ossendorp, F., Feltkamp, M.C., Kast, W.M., Melief, C.J., Neefjes, J.J. (1995): Major differences in transporter associated with antigen presentation (TAP)-dependent translocation of MHC class I-presentable peptides and the effect of flanking sequences. *J. Immunol.* 154: 1273–1279.

Neisig, A., Melief, C.J., Neefjes, J. (1998): Reduced cell surface expression of HLA-C molecules correlates with restricted peptide binding and stable TAP interaction. *J. Immunol.* 160: 171–179.

Niedermann, G., Butz, S., Ihlenfeldt, H.G., Grimm, R., Lucchiari, M., Hoschutzky, H., Jung, G., Maier, B., Eichmann, K. (1995): Contribution of proteasome-mediated proteolysis to the hierarchy of epitopes presented by major histocompatibility complex class I molecules. *Immunity* 2: 289–299.

Niedermann, G., King, G., Butz, S., Birsner, U., Grimm, R., Shabanowitz, J., Hunt, D.F., Eichmann, K. (1996): The proteolytic fragments generated by vertebrate proteasomes: structural relationships to major histocompatibility complex class I binding peptides. *Proc. Natl. Acad. Sci. U.S.A.* 93: 8572–8577.

Nijenhuis, M., Schmitt, S., Armandola, E.A., Obst, R., Brunner, J., Hämmerling, G.J. (1996): Identification of a contact region for peptide on the TAP1 chain of the transporter associated with antigen processing. *J. Immunol.* 156: 2186–2195.

Nijenhuis, M., Hämmerling, G.J. (1996): Multiple regions of the transporter associated with antigen processing (TAP) contribute to its peptide binding site. *J. Immunol.* 157: 5467–5477.

Norbury, C.C., Basta, S., Donohue, K.B., Tscharke, D.C., Princiotta, M.F., Berglund, P., Gibbs, J., Bennink, J.R., Yewdell, J.W. (2004): CD8+ T cell cross-priming via transfer of proteasome substrates. *Science* 304: 1318–1321.

Nussbaum, A.K., Dick, T.P., Keilholz, W., Schirle, M., Stevanovic, S., Dietz, K., Heinemeyer, W., Groll, M., Wolf, D.H., Huber, R., Rammensee, H.G., Schild, H. (1998): Cleavage motifs of the yeast 20S proteasome beta subunits deduced from digests of enolase 1. *Proc. Natl. Acad. Sci. U.S.A.* 95: 12 504–12 509.

Ortmann, B., Copeman, J., Lehner, P.J., Sadasivan, B., Herberg, J.A., Grandea, A.G., Riddell, S.R., Tampé, R., Spies, T., Trowsdale, J., Cresswell, P. (1997): A critical role for tapasin in the assembly and function of multimeric MHC class I–TAP complexes. *Science* 277: 1306–1309.

Paquet, M.E., Cohen-Doyle, M., Shore, G.C., Williams, D.B. (2004): Bap29/31 influences the intracellular traffic of MHC class I molecules. *J. Immunol.* 172: 7548–7555.

Park, B., Lee, S., Kim, E., Ahn, K. (2003): A single polymorphic residue within the peptide-binding cleft of MHC class I molecules determines spectrum of tapasin dependence. *J. Immunol.* 170: 961–968.

Paulsson, K.M., Kleijmeer, M.J., Griffith, J., Jevon, M., Chen, S., Anderson, P.O., Sjögren, H.O., Li, S., Wang, P. (2002): Association of tapasin and COPI provides a mechanism for the retrograde transport of major histocompatibility complex (MHC) class I molecules from the Golgi complex to the endoplasmic reticulum. *J. Biol. Chem.* 277: 18 266–18 271.

Paulsson K, Wang P. (2003): Chaperones and folding of MHC class I molecules in the endoplasmic reticulum. *Biochim. Biophys. Acta* 1641: 1–12.

Paz, P., Brouwenstijn, N., Perry, R., Shastri, N. (1999): Discrete proteolytic intermediates in the MHC class I antigen processing pathway and MHC I-dependent peptide trimming in the ER. *Immunity* 11: 241–251.

Peh, C.A., Burrows, S.R., Barnden, M., Khanna, R., Cresswell, P., Moss, D.J., McCluskey, J. (1998): HLA-B27-restricted antigen presentation in the absence of tapasin reveals polymorphism in mechanisms of HLA class I peptide loading. *Immunity* 8: 531–542.

Pentcheva, T., Spiliotis, E.T., Edidin, M. (2002): Cutting edge: tapasin is retained in the endoplasmic reticulum by dynamic clustering and exclusion from endoplasmic reticulum exit sites. *J. Immunol.* 168: 1538–1541.

Pfeifer, J.D., Wick, M.J., Roberts, R.L., Findlay, K., Normark, S.J., Harding, C.V. (1993): Phagocytic processing of bacterial antigens for class I MHC presentation to T cells. *Nature* 361: 359–362.

Powis, S.J., Young, L.L., Joly, E., Barker, P.J., Richardson, L., Brandt, R.P., Melief, C.J., Howard, J.C., Butcher, G.W. (1996): The rat cim effect: TAP allele-dependent changes in a class I MHC anchor motif and evidence against C-terminal trimming of peptides in the ER. *Immunity* 4: 159–165.

Preckel, T., Fung-Leung, W.P., Cai, Z., Vitiello, A., Salter-Cid, L., Winqvist, O., Wolfe, T.G., Von Herrath, M., Angulo, A., Ghazal, P., Lee, J.D., Fourie, A.M., Wu, Y., Pang, J., Ngo, K., Peterson, P.A., Früh, K., Yang, Y. (1999): Impaired immunoproteasome assembly and immune responses in PA28-/- mice. *Science* 286: 2162–2165.

Princiotta, M.F., Finzi, D., Qian, S.B., Gibbs, J., Schuchmann, S., Buttgereit, F., Bennink, J.R., Yewdell, J.W. (2003): Quantitating protein synthesis, degradation, and endogenous antigen processing. *Immunity* 18: 343–354.

Purcell, A.W., Gorman, J.J., Garcia-Peydro, M., Paradela, A., Burrows, S.R., Talbo, G.H., Laham, N., Peh, C.A., Reynolds, E.C., López de Castro, J.A., McCluskey, J. (2001): Quantitative and qualitative influences of tapasin on the class I peptide repertoire. *J. Immunol.* 166: 1016–1027.

Rammensee, H.G., Falk, K., Rötzschke, O. (1993): Peptides naturally presented by MHC class I molecules. *Annu. Rev. Immunol.* 11: 213–244.

Rape, M., Jentsch, S. (2002): Taking a bite: proteasomal protein processing. *Nat. Cell Biol.* 4: E113–6.

Reits, E.A., Vos, J.C., Grommé, M., Neefjes, J., 2000. The major substrates for TAP in vivo are derived from newly synthesized proteins. *Nature* 404, 774–778.

Reits, E., Griekspoor, A., Neijssen, J., Groothuis, T., Jalink, K., van Veelen, P., Janssen, H., Calafat, J., Drijfhout, J.W., Neefjes, J. (2003): Peptide diffusion, protection, and degradation in nuclear and cytoplasmic compartments before antigen presentation by MHC class I. *Immunity* 18: 97–108.

Reits, E., Neijssen, J., Herberts, C., Benckhuijsen, W., Janssen, L., Drijfhout, J.W., Neefjes, J. (2004): A major role for TPPII in trimming proteasomal degradation products for MHC class I antigen presentation. *Immunity* 20: 495–506.

Ritz, U., Momburg, F., Pircher, H.P., Strand, D., Huber, C., Seliger, B. (2001): Identification of sequences in the human peptide transporter subunit TAP1 required for transporter associated with antigen processing (TAP) function. *Int. Immunol.* 13: 31–41.

Rock, K.L. (1996): A new foreign policy: MHC class I molecules monitor the outside world. *Immunol. Today* 17: 131–137.

Rock, K.L., Goldberg, A.L. (1999): Degradation of cell proteins and the generation of MHC class I-presented peptides. *Annu. Rev. Immunol.* 17: 739–779.

Rock, K.L., York, I.A., Saric, T., Goldberg, A.L. (2002): Protein degradation and the generation of MHC class I-presented peptides. *Adv. Immunol.* 80: 1–70.

Rock, K.L., York, I.A., Goldberg, A.L. (2004): Post-proteasomal antigen processing for major histocompatibility complex class I presentation. *Nat. Immunol.* 5: 670–677.

Rodriguez, A., Regnault, A., Kleijmeer, M., Ricciardi-Castagnoli, P., Amigorena, S. (1999): Selective transport of internalized antigens to the cytosol for MHC class I presentation in dendritic cells. *Nat. Cell Biol.* 1: 362–368.

Sadasivan, B., Lehner,. P.J., Ortmann, B., Spies, T., Cresswell, P. (1996): Roles for calreticulin and a novel glycoprotein, tapasin, in the interaction of MHC class I molecules with TAP. *Immunity* 5: 103–114.

Saric, T., Beninga, J., Graef, C.I., Akopian, T.N., Rock, K.L., Goldberg, A.L. (2001): Major histocompatibility complex class I-presented antigenic peptides are degraded in cytosolic extracts primarily by thimet oligopeptidase. *J. Biol. Chem.* 276: 36 474–36 481.

Saric, T., Chang, S.C., Hattori, A., York, I.A., Markant, S., Rock, K.L., Tsujimoto, M., Goldberg, A.L. (2002): An IFN-γ-induced aminopeptidase in the ER, ERAP1, trims precursors to MHC class I-presented peptides. *Nat. Immunol.* 3: 1169–1176.

Saveanu, L., Daniel, S., van Endert, P.M. (2001): Distinct functions of the ATP binding cassettes of transporters associated with antigen processing: a mutational analysis of Walker A and B sequences. *J. Biol. Chem.* 276: 22 107–22 113.

Saveanu, L., Fruci, D., van Endert, P.M. (2002): Beyond the proteasome: trimming, degradation and generation of MHC class I ligands by auxiliary proteases. *Mol. Immunol.* 39: 203–215.

Schirmbeck, R., Reimann, J. (1994): Peptide transporter-independent, stress protein-mediated endosomal processing of endogenous protein antigens for major histocompatibility complex class I presentation. *Eur. J. Immunol.* 24: 1478–1486.

Schirmbeck, R., Reimann, J. (2002): Alternative processing of endogenous or exogenous antigens extends the immunogenic, H-2 class I-restricted peptide repertoire. *Mol. Immunol.* 39: 249–259.

Schubert, U., Anton, L.C., Gibbs, J., Norbury, C.C., Yewdell, J.W., Bennink, J.R. (2000): Rapid degradation of a large fraction of newly synthesized proteins by proteasomes. *Nature* 404: 770–774.

Schumacher, T.N., Kantesaria, D.V., Heemels, M.T., Ashton-Rickardt, P.G., Shepherd, J.C., Früh, K., Yang, Y., Peterson, P.A., Tonegawa, S., Ploegh, H.L. (1994): Peptide length and sequence specificity of the mouse TAP1/TAP2 translocator. *J. Exp. Med.* 179: 533–540.

Seifert, U., Marañón, C., Shmueli, A., Desoutter, J.F., Wesoloski, L., Janek, K., Henklein, P., Diescher, S., Andrieu, M., de la Salle, H., Weinschenk, T., Schild, H., Laderach, D., Galy, A., Haas, G., Kloetzel, P.M., Reiss, Y., Hosmalin, A. (2003): An essential role for tripeptidyl peptidase in the generation of an MHC class I epitope. *Nat. Immunol.* 4: 375–379.

Serwold, T., Gaw, S., Shastri, N. (2001): ER aminopeptidases generate a unique pool of peptides for MHC class I molecules. *Nat. Immunol.* 2: 644–651.

Serwold, T., Gonzalez, F., Kim, J., Jacob, R., Shastri N. (2002): ERAAP customizes peptides for MHC class I molecules in the endoplasmic reticulum. *Nature* 419: 480–483.

Shastri, N. Schwab, S., Serwold, T. (2000): Producing nature's gene chips – The generation of peptides for display by MHC class I molecules. *Annu. Rev. Immunol.* 20: 463–493.

Shen, L., Rock, K.L. (2004): Cellular protein is the source of cross-priming antigen in vivo. *Proc. Natl. Acad. Sci. U.S.A.* 101: 3035–3040.

Shen, L., Sigal, L.J., Boes, M., Rock, K.L. (2004): Important role of cathepsin S in generating peptides for TAP-independent MHC class I crosspresentation in vivo. *Immunity* 21: 155–165.

Shepherd, J.C., Schumacher, T.N., Ashton-Rickardt, P.G., Imaeda, S., Ploegh, H.L., Janeway, C.A., Jr., Tonegawa, S. (1993): TAP1-dependent peptide translocation in vitro is ATP dependent and peptide selective. *Cell* 74: 577–584.

Sigal, L.J., Crotty, S., Andino, R., Rock, K.L. (1999): Cytotoxic T-cell immunity to virus-infected non-haematopoietic cells requires presentation of exogenous antigen. *Nature* 398: 77–80.

Sijts, A., Sun, Y., Janek, K., Kral, S., Paschen, A., Schadendorf, D., Kloetzel, P.M. (2002): The role of the proteasome activator PA28 in MHC class I antigen processing. *Mol. Immunol.* 39: 165–169.

Snyder, H.L., Bacik, I., Yewdell, J.W., Behrens, T.W., Bennink, J.R. (1998): Promiscuous liberation of MHC-class I-binding peptides from the C termini of membrane and soluble proteins in the secretory pathway. *Eur. J. Immunol.* 28: 1339–1346.

Spee, P., Neefjes, J. (1997): TAP-translocated peptides specifically bind proteins in the endoplasmic reticulum, including gp96, protein disulfide isomerase and calreticulin. *Eur. J. Immunol.* 27: 2441–2449.

Spiliotis, E.T., Manley, H., Osorio, M., Zúñiga, M.C., Edidin, M. (2000): Selective export of MHC class I molecules from the ER after their dissociation from TAP. *Immunity* 13: 841–851.

Srivastava, P. (2002): Interaction of heat shock proteins with peptides and antigen presenting cells: chaperoning of the innate and adaptive immune responses. *Annu. Rev. Immunol.* 20: 395–425.

Stoltze, L., Schirle, M., Schwarz, G., Schröter, C., Thompson, M.W., Hersh, L.B., Kalbacher, H., Stevanovic, S., Rammensee, H.G., Schild, H. (2000): Two new proteases in the MHC class I processing pathway. *Nat. Immunol.* 1: 413–418.

Suh, W.K., Cohen-Doyle, M.F., Früh, K., Wang, K., Peterson, P.A., Williams, D.B. (1994): Interaction of MHC class I molecules with the transporter associated with antigen processing. *Science* 264, 1322–1326.

Tan, P., Kropshofer, H., Mandelboim, O., Bulbuc, N., Hämmerling, G.J., Momburg, F. (2002): Recruitment of MHC class I molecules by tapasin into the transporter associated with antigen processing-associated complex is essential for optimal peptide loading. *J. Immunol.* 168: 1950–1960.

Tanioka, T., Hattori, A., Masuda, S., Nomura, Y., Nakayama, H., Mizutani, S., Tsujimoto, M. (2003): Human leukocyte-derived arginine aminopeptidase. The third member of the oxytocinase subfamily of aminopeptidases. *J. Biol. Chem.* 278: 32 275–32 283.

Toes, R.E., Nussbaum, A.K., Degermann, S., Schirle, M., Emmerich, N.P., Kraft, M., Laplace, C., Zwinderman, A., Dick, T.P., Müller, J., Schönfisch, B., Schmid, C., Fehling, H.J., Stevanovic, S., Rammensee, H.G., Schild, H. (2001): Discrete cleavage motifs of constitutive and immunoproteasomes revealed by quantitative analysis of cleavage products. *J. Exp. Med.* 194: 1–12.

Tsai, B., Ye, Y., Rapoport, T.A. (2002): Retrotranslocation of proteins from the endoplasmic reticulum into the cytosol. *Nat. Rev. Mol. Cell Biol.* 3: 246–255.

Turnquist, H.R., Vargas, S.E., Reber, A.J., McIlhaney, M.M., Li, S., Wang, P., Sanderson, S.D., Gubler, B., van Endert, P., Solheim, J.C. (2001): A region of tapasin that affects Ld binding and assembly. *J. Immunol.* 167: 4443–4449.

Turnquist, H.R., Thomas, H.J., Prilliman, K.R., Lutz, C.T., Hildebrand, W.H., Solheim, J.C. (2000): HLA-B polymorphism affects interactions with multiple endoplasmic reticulum proteins. *Eur. J. Immunol.* 30: 3021–3028.

Uebel, S., Meyer, T.H., Kraas, W., Kienle, S., Jung, G., Wiesmüller, K.H., Tampé, R. (1995): Requirements for peptide binding to the human transporter associated with antigen processing revealed by peptide scans and complex peptide libraries. *J. Biol. Chem.* 270: 18512–18516.

Uebel, S., Kraas, W., Kienle, S., Wiesmüller, K.H., Jung, G., Tampé, R. (1997): Recognition principle of the TAP transporter disclosed by combinatorial peptide libraries. *Proc. Natl. Acad. Sci. U.S.A.* 94: 8976–8981.

van Endert, P.M., Tampé, R., Meyer, T.H., Tisch, R., Bach, J.F., McDevitt, H.O. (1994): A sequential model for peptide binding and transport by the transporters associated with antigen processing. *Immunity* 1: 491–500.

van Endert, P.M., Riganelli, D., Greco, G., Fleischhauer, K., Sidney, J., Sette, A., Bach, J.F. (1995): The peptide-binding motif for the human transporter associated with antigen processing. *J. Exp. Med.* 182: 1883–1895.

van Endert, P.M., Saveanu, L., Hewitt, E.W., Lehner, P. (2002): Powering the peptide pump: TAP crosstalk with energetic nucleotides. *Trends Biochem. Sci.* 27: 454–461.

van Hall, T., Sijts, A., Camps, M., Offringa, R., Melief, C., Kloetzel, P.M., Ossendorp, F. (2000): Differential influence on cytotoxic T lymphocyte epitope presentation by controlled expression of either proteasome immunosubunits or PA28. *J. Exp. Med.* 192: 483–494.

Van Kaer, L., Ashton-Rickardt, P.G., Ploegh, H.L., Tonegawa, S. (1992): TAP1 mutant mice are deficient in antigen presentation, surface class I molecules, and $CD4^-8^+$ T cells. *Cell* 71: 1205–1214.

Van Kaer, L., Ashton-Rickardt, P.G., Eichelberger, M., Gaczynska, M., Nagashima, K., Rock, K.L., Goldberg, A.L., Doherty, P.C., Tonegawa, S. (1994): Altered peptidase and viral-specific T cell response in LMP2 mutant mice. *Immunity* 1: 533–541.

Velarde, G., Ford, R.C., Rosenberg, M.F., Powis, S.J. (2001): Three-dimensional structure of transporter associated with antigen processing (TAP) obtained by single Particle image analysis. *J. Biol. Chem.* 276: 46054–46063.

Villanueva, M.S., Fischer, P., Feen, K., Pamer, E.G. (1994): Efficiency of MHC class I antigen processing: a quantitative analysis. *Immunity* 1: 479–489.

Voges, D., Zwickl, P., Baumeister, W. (1999): The 26S proteasome: a molecular machine designed for controlled proteolysis. *Annu. Rev. Biochem.* 68: 1015–1068.

Vos, J.C., Spee, P., Momburg, F., Neefjes, J. (1999): Membrane topology and dimerization of the two subunits of the transporter associated with antigen processing reveal a three-domain structure. *J. Immunol.* 163: 6679–6685.

Vos, J.C., Reits, E.A., Wojcik-Jacobs, E., Neefjes, J. (2000): Head-head/tail-tail relative orientation of the pore-forming domains of the heterodimeric ABC transporter TAP. *Curr. Biol.* 10: 1–7.

Wei, M.L., Cresswell, P. (1992): HLA-A2 molecules in an antigen-processing mutant cell contain signal sequence-derived peptides. *Nature* 356: 443–446.

Weissman, A.M. (2001): Themes and variations on ubiquitylation. *Nat. Rev. Mol. Cell Biol.* 2: 169–178.

Whitby, F.G., Masters, E.I., Kramer, L., Knowlton, J.R., Yao, Y., Wang, C.C., Hill, C.P. (2000): Structural basis for the activation of 20S proteasomes by 11S regulators. *Nature* 408: 115–120.

Wilkinson, K.D. (2000): Ubiquitination and deubiquitination: Targeting of proteins for degradation by the proteasome. *Sem. Cell Dev. Biol.* 11: 141–148.

Williams, A.P., Peh, C.A., Purcell, A.W., McCluskey, J., Elliott, T. (2002): Optimization of the MHC class I peptide cargo is dependent on tapasin. *Immunity* 16: 509–520.

Wilson, C.M., Farmery, M.R., Bulleid, N.J. (2000): Pivotal role of calnexin and mannose trimming in regulating the endoplasmic reticulum-associated degradation of major histocompatibility complex class I heavy chain. *J. Biol. Chem.* 275: 21224–21232.

Wright, C.A., Kozik, P., Zacharias, M., Springer, S. (2004): Tapasin and other chaperones: models of the MHC class I loading complex. *Biol. Chem.* 385: 763–778.

Yao, T., Cohen, R.E. (2002): A cryptic protease couples deubiquitination and degradation by the proteasome. *Nature* 419: 403–407.

Yewdell, J.W., Norbury, C.C., Bennink, J.R. (1999): Mechanisms of exogenous antigen presentation by MHC class I molecules in vitro and in vivo: implications for generating $CD8^+$ T cell responses to infectious agents, tumors, transplants, and vaccines. *Adv. Immunol.* 73: 1–77.

Yewdell, J.W. (2001): Not such a dismal science: the economics of protein synthesis, folding, degradation and antigen processing. *Trends Cell Biol.* 11: 294–297.

Yewdell, J.W., Reits, E., Neefjes, J. (2003): Making sense of mass destruction: quantitating MHC class I antigen presentation. *Nat. Rev. Immunol.* 3: 952–961.

York, I.A., Chang, S.C., Saric, T., Keys, J.A., Favreau, J.M., Goldberg, A.L., Rock, K.L. (2002): The ER aminopeptidase ERAP1 enhances or limits antigen presentation by trimming epitopes to 8–9 residues. *Nat. Immunol.* 3: 1177–1184.

York, I.A., Mo, A.X., Lemerise, K., Zeng, W., Shen, Y., Abraham, C.R., Saric, T., Goldberg, A.L., Rock, K.L. (2003): The cytosolic endopeptidase, thimet oligopeptidase, destroys antigenic peptides and limits the extent of MHC class I antigen presentation. *Immunity* 18: 429–440.

Yu, Y.Y., Turnquist, H.R., Myers, N.B., Balendiran, G.K., Hansen, T.H., Solheim, J.C. (1999): An extensive region of an MHC class I $\alpha 2$ domain loop influences interaction with the assembly complex. *J. Immunol.* 163: 4427–4433.

Zarling, A.L., Luckey, C.J., Marto, J.A., White, F.M., Brame, C.J., Evans, A.M., Lehner, P.J., Cresswell, P., Shabanowitz, J., Hunt, D.F., Engelhard, V.H. (2003): Tapasin is a facilitator, not an editor, of class I MHC peptide binding. *J. Immunol.* 171: 5287–5295.

Zernich, D., Purcell, A.W., Macdonald, W.A., Kjer-Nielsen, L., Ely, L.K., Laham, N., Crockford, T., Mifsud, N.A., Bharadwaj, M., Chang, L., Tait, B.D., Holdsworth, R., Brooks, A.G., Bottomley, S.P., Beddoe, T., Peh, C.A., Rossjohn, J., McCluskey, J. (2004): Natural HLA class I polymorphism controls the pathway of antigen presentation and susceptibility to viral evasion. *J. Exp. Med.* 200: 13–24.

Zhang, Q., Tector, M., Salter, R.D. (1995): Calnexin recognizes carbohydrate and protein determinants of class I major histocompatibility complex molecules. *J. Biol. Chem.* 270: 3944–3948.

Zhang, C., Anderson, A., DeLisi, C. (1998): Structural principles that govern the peptide-binding motifs of class I MHC molecules. *J. Mol. Biol.* 281: 929–947.

Zweerink, H.J., Gammon, M.C., Utz, U., Sauma, S.Y., Harrer, T., Hawkins, J.C., Johnson, R.P., Sirotina, A., Hermes, J.D., Walker, B.D., Biddison, W.E. (1993): Presentation of endogenous peptides to MHC class I-restricted cytotoxic T lymphocytes in transport deletion mutant T2 cells. *J. Immunol.* 150: 1763–1771.

4
Antigen Processing for MHC Class II

Anne B. Vogt, Corinne Ploix and Harald Kropshofer

4.1
Introduction

Modern technologies in the field of proteomics aim to analyze and compare highly complex mixtures of proteins obtained from living organisms under various conditions, e.g., the serum proteom from tumor patients versus healthy individuals. The goal is to identify marker proteins that are significantly altered, qualitatively and/or quantitatively, and thus can be used as biomarkers for diagnostic purposes, e.g., as an index of disease severity or to monitor treatment efficacy. Ideally, proteomics technologies can identify patient-specific target proteins that can be used to customize individualized treatment, e.g., to design a specific therapeutic monoclonal antibody.

Unsurprisingly, nature has been already using a similar concept since the invention of adaptive immunity in vertebrates and is actively applying this principle in fighting foreign invaders and pathogens in order to limit their propagation. Instead of using tryptic digests and mass spectrometry, the vertebrate immune system employs specialized immune cells, so-called professional antigen presenting cells (APCs) that are able to take up proteins from their environment, proteolytically process them within intracellular compartments and present a representative repertoire of the protein fragments via specialized immune receptors, the major histocompatibility complex class II (MHC II) molecules. MHC II molecules or their equivalents in humans, human leukocyte antigen class II (HLA II) molecules, are membrane-bound molecules, displaying a peptide binding groove where they can accommodate a linear peptide of at least 12–15 amino acid length. Such a peptide is loaded onto MHC II molecules inside APCs and then transported to the cell surface for presentation to T cells and recognition by a T cell receptor with the corresponding specificity.

Thus, APCs are equipped to give a snapshot of the protein environment they have encountered during a certain time window and present it via specialized receptors, the MHC II molecules, to other immune cells, the T cells. When some T cells recognize in this image what they have been trained to consider as a foreign structure they will react by proliferating and producing various cytokines,

Antigen Presenting Cells: From Mechanisms to Drug Development. Edited by H. Kropshofer and A. B. Vogt
Copyright © 2005 WILEY-VCH Verlag GmbH & Co. KGaA, Weinheim
ISBN: 3-527-31108-4

thereby eliciting a specific immune response against a defined foreign protein. Finally, a specific antibody production may be induced against the same protein from which a peptide fragment has been presented via MHC II. Thus, a fully folded protein needs to be transformed first into a panel of linear peptide sequences presentable via MHC II to finally elicit the generation of antibodies capable of recognizing 3-dimensional complex epitopes of the fully folded protein.

The present chapter focuses on the first part of this process, namely the sequence of mechanisms involved in the generation of MHC II molecules presenting linear peptides sampled from proteins that originate from the external and internal environment. This process is called "antigen processing" and has been the subject of intense research during the past 26 years. Broad, detailed insight has been gained during the last 10 years, especially concerning the involvement of accessory molecules, the understanding of the cell biology of the most professional APCs, the dendritic cells, the interplay of receptors of the innate immune system with APCs and with the emergence of novel microscopy techniques that have helped to visualize processes within an APC.

4.2
Types of Antigen Presenting Cells

Whereas all nucleated cells express class I MHC, only a limited group of cells express class II MHC. Constitutive expression of MHC II molecules is mainly found on phagocytic cells and confers them with the ability to present foreign antigens they generate after engulfment and proteolysis of exogenous pathogens. These "professional" type of APCs include dendritic cells (DCs), macrophages and B cells. Although all MHC II positive cells are broadly called "APCs", IFNγ-inducible expression of MHC II molecules has also been reported on the cell surface of non-phagocytic cells. Due to their lack of the internal machinery for protein processing, the relevance of their ability to present antigen is still controversial. Such "non-professional" APCs include fibroblasts, glial cells, pancreatic beta cells, thymic epithelial cells, thyroid epithelial cells, keratinocytes and vascular endothelial cells [1–4].

4.2.1
Macrophages, B Lymphocytes and DCs

Mononuclear phagocytes, i.e., macrophages and DCs, are crucial mediators of host defense against pathogens, linking innate and adaptive immune responses. As a consequence, they are particularly abundant in tissues in close contact with the environment. Additionally, during infections, macrophages can be recruited to peripheral tissues to phagocyte foreign antigens while initiating T cell immunity by the release of inflammatory signals. IFNγ and GM-CSF, often present at infected sites, contribute to activate macrophages, thus enhancing their capacity to function as APCs.

In contrast, antigen presentation by B cells occurs primarily in secondary lymphoid organs, where the cognate B-T cell interaction allows the development of T cell-dependent humoral immunity. Although B lymphocytes are not actively phagocytic, unlike macrophages and DCs, they can capture and internalize antigens via their B cell receptor (BCR), which consists of membrane-bound immunoglobulin.

Little was known about DCs, apart from their CD11c expression, until the recent development of in vitro DC culture methods led to the differentiation of multiple DC subsets and various functional stages. Hence, large numbers of immature DCs were generated from bone marrow progenitors or from blood monocytes and pulsed with proteins for in vitro functional assays [5]. Although the sequence of events leading to DC activation is not fully elucidated, it is becoming increasingly clear that in vivo human plasmacytoid DCs and monocyte-derived DCs use a complex ensemble of pattern-recognition receptors (PRRs), cytokine and membrane receptors to sense alterations of their environment and phagocyte incoming pathogens [6]. Located close to the mucosal surfaces and acquiring motility as they get activated, DCs play a crucial role in conveying innate information to lymphocytes and in orchestrating adaptive immune responses. An increasing body of evidence suggests that different stimuli may trigger qualitatively different types of mature DCs, allowing for the development of immunogenic vs. tolerogenic DCs or for the polarization of T cell responses towards the Th1 vs. Th2 phenotype.

4.2.2
Tissue-resident APCs

While some DCs access the secondary lymphoid organs directly from the blood, others migrate from peripheral tissues. Langerhans cells located in the epidermal layer of the skin are the principal model of tissue DC [7]. Generated from $CD1a^+$ precursors, they are characterized by the expression of Langerin, E-cadherin and by the presence of Birbeck granules. They have limited capacity for self-renewal but can persist in the epidermis for extended periods. After antigen exposure they are induced to migrate out of the skin, which they do in part by downregulating E-cadherin, thus losing the ability to interact with the surrounding keratinocytes [8]. Additional tissue resident DCs can be found in peripheral tissues, such as interstitial DCs in the dermis, the gut and in pulmonary epitheliums and Kupffer cells in the liver [9].

Microglial cells are a CNS-resident immune cell population sharing phenotypic characteristics as well as lineage properties with bone marrow-derived monocytes/macrophages. The smallest of the glial cells, they represent 5% of total brain cells and produce neurotrophins and growth factors promoting neuronal and oligodendrocyte survival. As true phagocytes, microglial cells clean up CNS debris and, under pathological conditions, they have been shown to process antigens, secrete interleukins and contribute to inflammation within the CNS. To date, the lack of specific phenotypical marker(s) has hampered their characterization and the appreciation of their plasticity and heterogeneity in vivo. Therefore, most of our knowledge about the nature, activation status and function of microglial cells comes from in vitro mixed glial cell cultures. Much like immature APCs, microglia have a weak MHC II expression and, although bearing ICAM-1 and the co-stimulatory molecules CD86 and CD40 on their surface, they fail to present antigens [10]. Upregulation of MHC II expression triggered by various stress, inflammatory and environmental signals allows microglial cells to interact with T cells in an antigen-specific manner. But unlike for DCs, there is currently no data supporting the hypothesis that upon maturation microglial cells become capable of homing to the lymph nodes in the vicinity of the CNS. Therefore, antigen presentation by microglia probably occurs locally during their interaction with infiltrating lymphocytes. The full scope of immune responses subsequent to this interaction and the role of CNS environment on the regulation of these responses are yet to be clarified.

4.2.3
Maturation State of APCs

Upon activation, macrophages and DCs change their antigen processing program and upregulate their surface expression levels of MHC II and co-stimulatory molecules, thereby becoming mature and potent APCs. The distinction between immature and mature stages of DCs is now well established.

4.2.3.1 Immature APCs

Immature DCs are true sentinels of the body in that they are specialized in taking up, internalizing and processing various particulate and soluble antigenic molecules derived from foreign invaders. Representing a first line of defense against infectious agents, DCs are readily found in skin, lung, gut, blood [7, 11, 12] as well as in liver and spleen.

As immature DCs retain MHC II molecules within endocytic compartments formed by late endosomes and lysosomes [12], they do not efficiently present antigens. In the absence of antigenic stimulation, these MHC II molecules are the object of intense recycling with an estimated half-life of 12 h, as opposed to 40–100 h in mature DCs [13, 14].

4.2.3.2 Mature APCs

Upon encountering inflammatory signal mediators, such as LPS, TNFα, IL-1β, bacterial lipids or lipopeptides, CpG DNA or viral ds RNA, during a microbial attack, immature DCs differentiate into mature DCs with high T cell stimulatory capacity [15]. This functional change is associated with a profound alteration in the architecture of endocytic processing compartments, an enormous increase in the cell surface area and a decrease in their capacity to endocytose [16]. In the initial phase of DC maturation, protease activity and lysosomal acidification are also enhanced, promoting the degradation of antigenic proteins and peptide loading to MHC II molecules. Within the first hours of stimulation, MHC II molecules relocate from perinuclear lysosomal compartments to peripheral vesicles. After 12–24 h, they are again redistributed and massively access the cell surface [13]. Abundant expression of co-stimulatory molecules (e.g., CD86, CD80, CD40) and ligands of T cell receptors [17] add to transform these APCs into powerful inducers of T cell activation. Meanwhile, the acquisition of motility allows DCs to migrate while morphological rearrangements are ongoing. Hence, during the initial 12–24 h of their maturation, DCs re-localize from sites of antigen capture in the periphery to areas of T cell priming within lymphoid organs.

4.3
Antigen Uptake by APCs

Internalization of whole pathogens or parts thereof into the endocytic pathway of an APC is a critical step as, in most cases, the particular mode of internalization is the limiting step in antigen processing and presentation. Macrophages, microglia or DCs use phagocytosis, pinocytosis and, in particular, receptor-mediated uptake mechanisms, whereas B cells widely rely on B cell receptor-mediated endocytosis to internalize antigen.

4.3.1
Macropinocytosis

Macropinocytosis is a cytoskeleton-dependent type of fluid-phase endocytosis that is limited to few cell types, such as macrophages, DCs and epithelial cells [18]. In macrophages, cytokines and growth factors can induce macropinocytosis, whereas it is a constitutive phenomenon in immature DCs [19]. Through macropinocytosis, immature DCs constantly sample large amounts of extracellular fluid, attaining uptake rates per hour that roughly equal its own cell volume. This is the promiscuous way of how DCs can screen their environment for microbial products. Apart from that, sampling of "self" by non-activated DCs is emerging as an important means for the induction and maintenance of tolerance under steady-state conditions [20]. In accordance, antigens captured by macropinocytosis have access to endocytic loading compartments where MHC and CD1 molecules reside. Most recently, evidence has emerged that ligands of Toll-like receptors (TLRs), shortly after receptor engagement, stimulate macropinocytosis of DCs, leading to enhanced presentation on class I and class II MHC molecules [21]. Macropinocytosis was apparently enhanced through recruitment of actin from podosomes, mediated by signal-regulated activation of the p38 kinase pathway. Thus, DCs can mobilize their actin cytoskeleton in response to innate immune stimuli in order to boost macropinocytosis-dependent antigen capture, most likely at the transient expense of their migratory capacity.

4.3.2
Phagocytosis

Phagocytosis is essential in both host defense against microbial pathogens and also in clearance of apoptotic or necrotic cells [22]. Phagocytosis is initiated by the engagement of specific surface receptors, as discussed below. DCs and macrophages can phagocytose whole bacteria, yeast cells and hyphae, as well as debris from cells that have undergone cell death by apoptosis or necrosis. Human DCs take up apoptotic bodies derived from B or T lymphocytes, virus-infected apoptotic monocytes or tumor cells. Phagocytosis of apoptotic bodies mainly relies on a surface complex composed of the scavenger receptor CD36 and the integrins $a_v\beta_5$ or $a_v\beta_3$ [23].

Both microbial and apoptotic cells are delivered on a common route from the cell surface of macrophages or DCs into phagosomes and finally lysosomes for degradation (Figure 4.1). Recently, it was demonstrated that activation of the TLR2/4 signaling pathway by bacteria, but not apoptotic cells, regulates phagocytosis [24]: Strikingly, phagolysosomal fusion in macrophages could be significantly accelerated through a signal emanating from the TLR of the same phagosome by engaging its bacterial cargo. Hence, the critical TLR triggering event for phagosome maturation is spatially confined and relies on phagocytosed cargo molecules bearing innate stimulatory capacity. In conclusion, activation of the phagosome signaling pathway has a critical impact on processing of bacterial antigen long before the conventional TLR-mediated transcription program associated with inflammation is initiated.

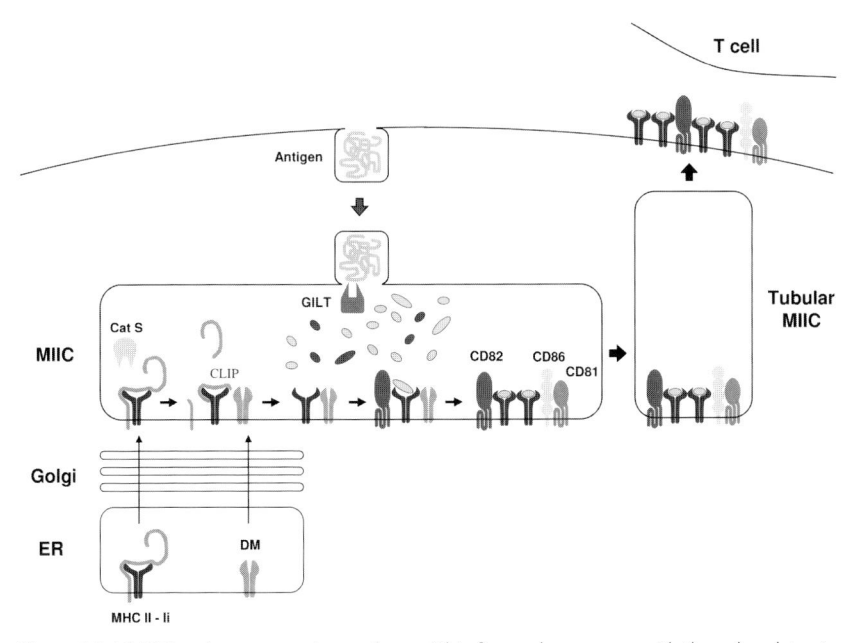

Figure 4.1 MHC II antigen processing pathway. (This figure also appears with the color plates.)

4.3.3
Receptors for Endocytosis

The principle behind receptor-mediated endocytosis is reminiscent of the mode of action described for some hormones, such as insulin: binding to surface receptors leads to receptor dimerization, initiation of kinase-driven signaling events that lead to specific effector functions. On APCs, recognition of antigen by antigen receptors leads to internalization, in most instances being driven by the polyvalency of antigenic structures, leading to receptor oligomerization that triggers G-protein mediated membrane invaginations. A most thoroughly studied example is provided by B cells that express the BCR: BCR mediates endocytosis of BCR-antigen complexes via clathrin-coated pits into endo-lysosomal compartments where MHC II molecules capture antigenic peptides [25].

A related principle relies on endocytosis of opsonized microbes: soluble antibodies decorating the surface of viruses or bacteria in a polyvalent fashion are recognized by the receptors for the F_c portion of antibodies. These F_c receptors belong to the prominent receptors on the surface of most APCs [26]. Human monocyte-derived DCs express mainly FcγRII and FcαR, Langerhans cells of the skin express FcγRI and FcϵRI, whereas blood DCs stain positive for FcγRII and FcγRI (Table 4.1). As indicated by the Greek letter in the names of the various Fc receptor classes, PCs differ in their Fc receptor repertoires and, hence, in their specificity and capacity to endocytose antigens decorated by immunoglobulins of different isotypes.

Another important receptor for antigen uptake is the mannose receptor (MR), CD206. It is a C-type lectin, mainly expressed on macrophages and immature DCs, that contains multiple carbohydrate-recognition domains [27]. MR binds glycoligands containing exposed mannose, N-acetylglucosamine or fucose residues. It allows APCs to endocytose a large variety of bacterial and yeast antigens, e.g., mannan-containing proteins or glycolipids, and also desialyted immunoglobulins. Agalactosyl IgG (G0IgG), increased levels of which are found in several autoimmune diseases, including rheumatoid arthritis, can be efficiently internalized by MR on DCs and macrophages [28]. Due to the ability of the MR to recycle, antigen uptake by APCs is essentially continuous, allowing accumulation of large quantities of antigen in internal loading compartments, where MR co-localizes with MHC II molecules and lysosomal hydrolases [29]. DEC-205 (CD205), another C-type lectin, is another mannose receptor, described in mice, that is expressed by DCs and thymic APCs and plays a similar role as the MR in antigen uptake and processing [30].

DC-SIGN (DC-specific ICAM-3-Grabbing Nonintegrin), CD209, is a mannose-binding lectin that is selectively expressed on immature and mature DCs. Although it is a known adhesion receptor, enabling transient DC-T cell interactions through binding to ICAM-3 on naive T cells, an ICAM-2-dependent rolling receptor and an HIV-1 trans-receptor, binding to the HIV-1 coat protein gp120, DC-SIGN also internalizes antigen into late endosomes for presentation to CD4$^+$ T cells [31].

Table 4.1 C-type lectin and Fc receptors.

	DC subsets	C-type lectin receptor	Fc receptor type
Blood	Plasmacytoid DC	BDCA2	
	Myeloid DC	(Blood DC antigen)	
		Dectin-1	FcγRI
		DC-SIGN (CD209)	FcγRII
		DEC-205 (CD205)	
Tissue	Langerhans cells	Langerin	FcγRI
			FcεRI (CD23)
	Monocyte-derived DC	Mannose receptor (CD206)	
		DEC-205 (CD205)	
		DC-SIGN (CD209)	
		MGL-1	FcαR (CD89)
		(Mannose galactose, N-acetyl-galactosamine specific lectin)	FcγRII (CD32)
		C-LEC1 (C-type lectin rec.1)	
		DCIR (DC immunoreceptor)	
		DCAL (DC-associated lectin)	

Langerhans cells (LCs) do not express the MR but, besides CD205, the LC-specific receptor Langerin (CD207), another C-type lectin with mannose specificity. Unlike the MR, CD207 induces the formation of Birbeck granules [32], endosomal organelles of so far unknown function. Langerin-deficient mice, which are devoid of Birbeck granules, were not impaired in their capacity to take up antigen and in their pathogen susceptibility [33]. There are indications that the C-type lectin receptors on APCs may crosstalk with TLRs that recognize characteristic molecular patterns present in microbial lipids, lipoproteins, lipopolysaccharides, bacterial DNA, viral RNA, as well as factors secreted upon tissue damage, such as heat shock proteins (Hsp70) [34] (cf. Chapter 8). Interestingly, a recent report demonstrated that TLR2 itself can shuttle antigen to endosomal compartments for presentation to CD4+ T cells [35].

According to current concepts, binding of antigen by C-type lectins alone may favor immune suppression, whereas recognition in a situation of danger, where both TLRs and C-type lectins carry a ligand, may induce immune activation. In other words, phagocytosis mediated by C-type lectins without involvement of TLRs may contribute to maintenance of tolerance, whereas co-incident binding of ligands to TLRs may trigger breakage of tolerance [36].

4.4
Generation of Antigenic Peptides

Almost two decades ago, when immunologists realized that MHC-mediated antigen presentation relies on peptide fragments derived from antigenic proteins, the hypothesis emerged that APCs may internalize antigenic proteins into specialized compartments where, on the one hand, proteolytic generation of appropriate peptide epitopes occurs, and, on the other hand, where these peptides are loaded onto newly synthesized MHC II molecules. Later on, when it became clear that the multicatalytic protease complex, the proteasome, is mainly responsible for the generation of epitopes to be presented by MHC class I molecules [37], an equivalent set of defined proteases was expected to be responsible for antigen processing in the class II MHC pathway.

Meanwhile, it became increasingly obvious that APCs utilize the whole endocytic pathway, including lysosomes, for the generation of peptide epitopes [38] and that this task is accomplished by numerous types of proteases – several with overlapping specificities – as this is the most efficient way that CD4+ T cells will detect the presence of at least a single foreign peptide [39]. Apart from offering several types of compartments and a complex set of hydrolytic enzymes, the endocytic route from early endosomes to lysosomes provides an increasingly acidic milieu and conditions favoring disulfide-bond reduction. Both aspects favor protein unfolding so that proteases and MHC molecules gain access to previously buried parts of antigenic proteins.

4.4.1
Reduction of Disulfide Bonds: GILT

The reduction of inter- and intra-chain disulfide bonds within endosomal/lysosomal compartments is a key step in antigen processing, as shown with several model antigens [40]. Disulfide bonds can impact the hierarchy of epitope generation in at least two ways: either when a T cell epitope contains one or more cysteine residues that are derived from disulfide bonds in the native antigen or when the accessibility of a T cell epitope is controlled by a proximal or distant disulfide bond.

Reduction of disulfide bonds requires an appropriate redox potential. The cytoplasm and the ER are highly reducing compartments with millimolar concentrations of reduced glutathione. The redox state of endocytic vesicles, however, is still a matter of debate, as strong acidic pH renders disulfide reduction and exchange reactions less likely. Therefore, the possibility that an enzyme may catalyze antigen disulfide reduction was considered early on [40]. An enzyme such as the ER-resident protein disulfide isomerase (PDI) was proposed to serve as a candidate reductase; however, PDI loses its catalytic activity rapidly under acidic conditions [40]. This problem was resolved by the recent discovery that γ-interferon-inducible lysosomal thiol reductase (GILT), originally termed IP-30, is constitutively present in late endocytic compartments of APCs [41]. GILT actually reduces disulfide-bonded or cysteinylated antigens very efficiently under acidic pH (Figure 4.1). Accordingly, GILT-deficient mice were severely impaired in their T cell response against 2 of the 4 immunodominant epitopes of hen egg lysozyme (HEL) that contains four intrachain disulfide bonds [42]. Likewise, the lack of GILT in melanoma cells reduced the presentation of an immunodominant IgG epitope. Transfection of melanoma cells with GILT restored presentation of the IgG epitope and of an endogenous epitope derived from the melanoma antigen tyrosinase [43].

4.4.2
Regulation of the Proteolytic Milieu

Apart from the redox potential, the local pH is one of the most critical parameters that govern the efficacy of antigen processing in endosomal/lysosomal compartments of APCs. At least four aspects can be listed that underscore the importance of the pH.

1. Antigenic proteins enter the endocytic pathway at neutral pH, which favors their native folding. With increasing acidity, most proteins lose their quaternary and tertiary structure, thereby exposing potential T cell epitopes, which may be liberated by endoprotease cleavage or be directly captured by MHC II molecules.
2. The drop in pH in endosomes promotes the removal of the N-terminal propeptides of endosomal/lysosomal cathepsins, thereby converting inactive zymogens into active proteases [44].
3. Treatment with drugs, such as chloroquine, monensin or bafilomycin that interfere with endosome acidification, led to the accumulation of invariant chain processing intermediates, such as p22 or p10, that remain in a trimeric complex with MHC II $\alpha\beta$ dimers and thereby prevent antigenic peptide loading [45].
4. The peptide loading catalyst HLA-DM displays optimal activity around pH 5.0, at higher pH the removal of the invariant chain fragment CLIP (see Section 5.2) becomes rate-limiting for antigenic peptide loading [46].

DCs have been intensively studies with regard to how antigen proteolysis is regulated and what consequences this regulatory means may have on the capacity of immature versus mature DCs to present antigen to CD4$^+$ T cells. Murine immature DCs have been reported to readily internalize exogenous, foreign antigenic proteins, such as HEL, and bring it into late endosomal and lysosomal compartments, but fall short in loading immunogenic HEL peptides onto MHC II molecules, unless DCs are triggered by a maturation stimulus [47]. Accordingly, cystatin C has been described as downregulating cathepsin S activity in murine immature DCs, thereby preventing the conversion of $\alpha\beta$-invariant chain complexes into peptide-receptive MHC II molecules [48]. An even more profound explanation was provided in a recent report demonstrating that maturation signals induce the formation of higher numbers of functional V-ATPases in the lysosomal membrane of immature DCs, thereby increasing the acidification capacity of lysosomes

[49]. Consequently, the intralysosomal pH of ~5.4 in immature DCs was decreased to ~4.5 in mature DCs, a pH also obtained for macrophages or fibroblasts [49]. This shift in lysosomal acidification is supposed to augment antigen processing by favoring the activation of the reductase GILT and lysosomal proteases, such as cathepsin L or legumain that display activity optima around pH 4.5, and the catalytic activity of HLA-DM [41, 50].

Other DC subsets, however, behave differently. Immature DCs isolated ex vivo from lymphoid organs were able to constitutively present self-antigens [51]. Likewise, the HLA-DR-associated self-peptide repertoire of immature and mature DCs derived from human blood monocytes was widely similar, suggesting that endosomes and lysosomes of immature human DCs do not restrict processing of self-antigens [52]. The finding that DCs can process and present antigens across a wide range of maturation stages, irrespective of whether endogenous or exogenous antigens were investigated [53], underlines the plasticity of the DC family with regard to the regulation of antigen processing.

Most recently, an unexpected association of MHC II molecules with endocytically processed zwitterionic polysaccharides (ZPS) from pathogenic bacteria, such as *Streptococcus pneumoniae* or *Bacteroides fragilis*, was shown [54]. It turned out that APCs partially depolymerize ZPS inside endocytic compartments before ZPS fragments form antigenic complexes with MHC II molecules. ZPS processing relies on a NO-mediated mechanism, as mice lacking the NO synthetase iNOS were unable to activate ZPS-specific T cells [54]. These novel findings extend our view on how APCs perform antigen processing, as not only hydrolytic but also radical mechanisms apparently contribute to the generation of antigenic ligands.

4.4.3
Protease/MHC Interplay in Antigen Processing

Formation of MHC II–peptide complexes occurs under conditions that are rather unfavorable for the survival of epitopes over 13 amino acids long, as late endosomes and lysosomes normally degrade proteins to dipeptides and amino acids [55] (Table 4.2). Therefore, APCs need to balance destructive and creative steps in their antigen processing compartments such that not only endo- and exopeptidases control the selection of antigenic determinants but MHC II molecules strongly contribute to shaping immunodominance of epitopes [56]. The MHC II peptide-binding groove being open at both ends and the broad peptide specificity may be viewed as critically important for competing out proteases displaying sequence-dependent cleavage specificity in the search for appropriate polypeptide ligands [56]. Indeed, several observations support the concept that MHC II molecules engage precursor polypeptides of much greater length than those eluted as naturally processed peptides from the groove, and that antigen processing may occur in an MHC-guided fashion [57]. In cases where this "bind first, trim later" model reflects the reality, the type, specificity and cleavage order of proteases is less important, as the MHC groove determines which epitopes become dominant or cryptic. Endo- and exoproteases would, then, stochastically remove the areas not buried in the peptide binding groove [56].

Table 4.2 Endosomal/lysosomal proteases.

Protease family	Protease	Cleavage specificity	Expression	Processing involvement	
				Ii	Ag
Cysteine	Legumain	Endo/Asn	Ubiquitous	+	+
	Cat B	Endo	Ubiquitous	–	+/–
	Cat C	Amino	Ubiquitous	+	?
	Cat F	Endo	Ubiquitous	–	?
	Cat H	Endo/amino	Ubiquitous	–	?
	Cat Z	Endo	Ubiquitous	–	?
	Cat O	Endo	Ubiquitous	+	–
	Cat L	Endo	Ubiquitous	–	+
	Cat L2	Endo	Thymus, testis	+	+
	Cat K	Endo	Osteoclasts	–	–
	Cat S	Endo	APCs	–	+
	Cat W	Endo	T cells/NK cells	–	–
	Cat J	Endo	Placenta	–	–
	Cat P	Endo	placenta	–	–
Aspartic	Cat D	Endo	Ubiquitous	–	+/–
	Cat E	Endo	Ubiquitous	–	+/–
	Napsin A	Endo	Lung, kidney	–	–
	Napsin B	?	?	–	?

MHC-guided processing may proceed according to the kinetics of unfolding of native antigenic proteins. Hence, those sequences may have the primary change to give rise to immunodominant epitopes that are the first ones to be accessible for MHC binding during the unfolding process. In particular, in lysosomes that harbor high numbers of very active proteases that may otherwise destroy candidate epitopes, MHC-guided processing would be most beneficial in rescuing epitopes from destructive proteolysis. The situation may be different in vesicles of less acidic pH where proteases are supposed to be less aggressive and fewer peptide-receptive MHC II molecules are present and where, without an initial "unlocking" by endoprotease cleavage, unfolding alone would be too inefficient. Here, the particular set of proteases being expressed may become more relevant for epitope selection so that one or more particular proteases may decide whether an epitope is being generated. In the following, some of the most thoroughly investigated proteases that reside in MHC II loading compartments, are discussed (Figure 4.1).

In contrast to earlier assumptions, it is not necessarily highly abundant proteases, such as the cysteine protease cathepsin B (Cat B) or the aspartic protease cathepsin D (Cat D) that belong to the major players in constructive antigen proteolysis, as revealed in the respective knock-out mice [58]. Although, in one study, IgG F(ab)$_2$ fragments as part of immune complexes were degraded poorly in the absence of Cat B [59], other proteases could obviously replace Cat B in Cat B-deficient mice [58].

Apart form its role in processing of Ii in cortical thymic epithelial cells (see Section 5.2), Cat L proved to be directly involved in the selection of CD4+ T cells: it contributes to shaping the repertoire of positively selecting self-peptide ligands [60]. Another study on peripheral APCs confirmed the impact of Cat L on the antigenic peptide repertoire, albeit in an inverted sense [61]: presentation by murine I-A^b of an epitope from IgM was diminished by Cat L expression as compared to APCs of Cat L$^{-/-}$ mice. The same result was obtained when Cat S$^{-/-}$ and wild-type mice were compared. Conversely, Cat S, which is primarily responsible for the generation of CLIP-MHC II out of Iip10-MHC II complexes in peripheral APCs (see Section 5.2), was also found to be required for presentation of two HEL epitopes: HEL(30-44) and HEL(46-59) are separated by a Cat S cleavage site at position HEL-45 [62].

Unlike the cathepsins, the asparaginyl endopeptidase AEP, also denoted as legumain, has strict specificity for cleaving at asparagine and, in rare cases, aspartic acid residues [63]. AEP bears the dominant processing activity for a C-terminal fragment of tetanus toxin [63]. With the multiple sclerosis (MS)-associated autoepitope MBP(85-99), derived from myelin basic protein (MBP), AEP plays a destructive role: cleavage of MBP at Asn-94 by AEP destroys this epitope [64]. As AEP is expressed in medullary thymic epithelial cells where autoreactive T cells are deleted, central tolerance against MBP(85-99) may fail to be established. This may be a rationale for the pathology of the autoimmune disease MS: in contrast to AEP-deficient APCs of the thymus, APCs in the brain of MS patients generate the MBP(85-99) epitope which is thought to attract autoreactive CD4+ T cells [65]. Thus, differential expression of proteases in different tissues may lead to differential processing of the same antigenic protein.

In contrast to Cat D, the homologous aspartic protease Cat E is less abundant and more restricted in its expression to the gastrointestinal tract, lymphoid tissue and microglia [66]. Cat E is the key protease in generating the immunodominant ovalbumin epitope OVA(266-281) in murine B cells and microglia [66]. Surprisingly, according to most recent observations, Cat E expression is negatively regulated by the class II MHC transactivator CIITA [67]. The fact that MHC II molecules and Cat E share a common transcriptional regulator supports the view that Cat E plays an active role in antigen processing of the MHC II pathway and that its proteolytic activity requires tight regulation.

4.5
Assembly of MHC II Molecules

MHC II molecules expressed on the surface of APCs are $\alpha\beta$ heterodimers complexed with a peptide ligand [68]. Only a small percentage of surface MHC II molecules are thought to be "empty", which means they are devoid of a ligand (Figure 4.1). T cell receptors, expressed on the surface of T cells, are selected to interact with both self-structures of the rim of the MHC class II peptide binding groove and amino acid side chains of the peptide ligand. Determinants on the surface of the antigenic peptide that specifically bind to the T cell receptor are called "T cell epitopes". Thus, in contrast to "B cell epitopes", which are linear or conformational structures on the surface of an antigen to be recognized by soluble or membrane-bound immunoglobulins of B cells, T cell epitopes are generated by proteolytic cleavage of antigenic proteins and presented as linear structures in the context of MHC II. The process of assembly of MHC II $\alpha\beta$ dimers in the ER and the acquisition of the third subunit, the peptide ligand, in endosomal/lysosomal compartments is a coordinated process involving several accessory molecules.

4.5.1
Structural Requirements of MHC II

MHC II molecules are membrane-bound heterodimers consisting of one α and one β chain. The amino-terminal $\alpha1$ and $\beta1$ domains constitute the peptide binding groove – consisting of two flanking α-helices on a bottom of β-sheets – and are followed by Ig-like domains ($\alpha2$ and $\beta2$), transmembrane regions and short cytoplasmic tails [68]. To achieve stability, MHC II $\alpha\beta$ dimers have to accommodate a third ligand, either the Ii or an appropriate peptide ligand at least 12 amino acids long. The Ii interacts with $\alpha\beta$ dimers through contacts within and outside the peptide binding groove of $\alpha\beta$ dimers [69, 70]. In contrast, an appropriate peptide ligand engages nearly exclusively with residues inside the binding groove: about half of the binding enthalpy arises from interactions of the peptide backbone with side chains of the MHC II binding groove, while the other half is due to interactions of amino acid side chains of the peptide with residues of the so-called specificity pockets of the MHC II peptide binding groove [68]. In contrast to MHC I, the peptide groove of MHC II is open at both ends [71], thus allowing peptide ligands of variable lengths to bind. Thus, capturing of precursors of peptide ligands that are bound first and then trimmed by proteases afterwards is observed for MHC II [56, 57] (cf. Chapter 6). The openness of the MHC II binding groove may be the consequence of the co-evolution of MHC II molecules and Ii: as Ii binding to MHC II $\alpha\beta$ heterodimers via the CLIP region is fundamental to its chaperone role [72], the accommodation of CLIP as a part of the Ii polypeptide in the groove was a prerequisite. Hence, fixation in the groove of N- and C-termini of a peptide of a defined length, as with MHC class I molecules, could no longer have been an option.

CLIP may have been acting as a powerful factor shaping the polymorphism, which is restricted to residues localizing to the MHC II binding groove [73]. On the one hand, the increase in polymorphism increases the capacity to cope with the high degree of variability of foreign invaders while, on the other hand, mutations in hypervariable regions of MHC II alleles that prevent CLIP from binding or inhibit CLIP removal are necessarily subject to negative selection during evolution. Due to the high polymorphism, especially of the human HLA-DR (short: DR) system with more than 400 allelic products, many combinations of anchor residues for peptide ligands are possible. Considerable work has been invested to achieve knowledge about the requirements of different HLA alleles to bind their cognate peptide ligands. While DR molecules, especially those encoded on the DRB1 locus, are well characterized [74, 75], anchor motifs for peptide binding to DRB3, DRB4, DRB5, DQ and DP alleles are less well understood and are the subject of current research.

Superantigens of bacterial or mammalian origin represent an exceptional case of antigen that is dependent on MHC II [76]. Superantigens bind outside the binding groove of MHC II, thereby crosslinking MHC II with the TCR [77]. Regarding the TCR, different superantigens display preference for certain $V\beta$ domains. Superantigens can bridge MHC II molecules and TCRs in the absence and presence of self-peptides occupying the binding groove of MHC II [78]. Thus, superantigens circumvent the classical rules of MHC restriction in activation of T cells.

4.5.2
Biosynthesis of MHC II

MHC II molecules assemble in the endoplasmic reticulum (ER), involving the concerted action of the chaperones calnexin and Ii (Figure 4.1). While calnexin serves several types of antigen receptors [79], such as immunoglobulins, TCR and MHC I, Ii is dedicated to MHC II molecules and can rapidly associate with all allelic variants of MHC II [80]. Sequential binding of calnexin and Ii to newly synthesized individual α and β chains facilitates the assembly of $\alpha\beta$Ii heterotrimers followed (after dissociation of calnexin) by the formation of a nonameric $(\alpha\beta$Ii$)_3$ complex. Ii protects $\alpha\beta$ dimers from aggregation in a chaperone-like fashion. Moreover, Ii prevents premature binding of peptides and polypeptides in the ER [81].

The cytosolic tail of Ii harbors two sorting signals that target the nonameric $(\alpha\beta$Ii$)_3$ complex from the trans-Golgi network to endocytic compartments (Figure 4.1). In the transit from endosomes to lysosomes, stepwise processing of Ii takes place; hence, Ii processing and antigen processing mainly co-localize. By using specific protease inhibitors and genetic knockouts of lysosomal proteases, Ii processing was shown to be a staged process, proceeding from the lumenal C-terminus to the membrane-proximal N-terminus [38]. Inhibition of cysteine proteases by leupeptin results in accumulation of a 22 kD Ii fragment (p22) in human cells and in accumulation of a 10 kD fragment (p10) in murine cells [82]. Under these conditions, the leupeptin-insensitive cysteine protease AEP is initiating Ii (p31) processing by targeting two Asn residues close to the C-terminal trimerization region, resulting in generation of the p22 fragment of Ii [83]. Further processing is observed with murine Ii because of additional Asn residues that are closer to the N-terminus. However, proteases different from AEP can also start Ii processing. In human APCs at least two subsequent cleavage steps by Cat S give, finally, rise to the residual fragment CLIP [84]. CLIP keeps on occupying the peptide binding groove, thereby preventing $\alpha\beta$ disassembly. These cleavages, at the same time, release $\alpha\beta$ dimers from their trimeric oligomerization state previously maintained by Ii trimers [39].

In cells devoid of Cat S, such as murine thymic epithelial cells, the related enzyme Cat L processes Ii [85]. Cat S k.o. mice are healthy and normal in most respects, but they show a decrease in Ii degradation in B cells and DCs [86]. The moderate effects seen in macrophages of Cat S-deficient may be due to Cat F taking over Ii processing functions [87]. The critical role of Cat S in Ii processing of APCs translates also into impaired antigen presentation in APCs lacking Cat S activity. Cat S k.o. mice show a reduced susceptibility to collagen-induced arthritis [86]. Likewise, treatment with Cat S inhibitors led to decreased inflammation in both the rat adjuvant-induced arthritis model [88] and the collagen-induced arthritis mouse model [89]. In a mouse model of pulmonary hypersensitivity, reduced IgE titers and eosinophil infiltration were reported upon administration of Cat S inhibitors [90]. Most recently, a non-covalent Cat S inhibitor has been shown to qualify as a new candidate drug for immunosuppressive therapy of allergies and autoimmunity [91].

4.5.3
Chaperones for Peptide Loading

The Ii remnant CLIP needs to be removed from the MHC II peptide binding groove before cognate self-peptide or foreign peptides can have access. Although, in the context of particular MC II allelic products, a CLIP self-release mechanism is operative [92], efficient removal of CLIP and peptide loading require an accessory molecule, designated HLA-DM in humans and H2-DM in mice. Only in B cells is peptide loading modulated by HLA-DO (H2-DO in mice).

4.5.3.1 HLA-DM/H2-DM
HLA-DM (short: DM) was first described in 1991 as a non-classical MHC II molecule [93, 94]. The expression of DM α and β subunits is widely co-regulated with classical MHC II molecules, but DM displays only very limited polymorphism and cannot bind peptide ligands [95]. In further contrast to conventional MHC II molecules, only a small subset of DM is found on the cell surface, in particular on B cells and immature DCs [96], whereas most DM accumulates in endosomal/lysosomal loading compartments [97]. In these compartments, DM catalyses one of the rate-limiting steps in MHC II antigen processing, i.e., the removal of CLIP

[98]. Mutant APCs devoid of DM, therefore, have a defect in presentation of protein antigens, originally leading to the discovery of DM and the awareness that accessory molecules may be critical in antigen processing [99].

DM-mediated exchange of CLIP for cognate peptide follows Michaelis–Menten kinetics derived from enzyme catalysis, demonstrating that DM acts as a true catalyst [100]. Consistent with a catalytic pH optimum around 4.5 in vitro, DM:DR ratios of about 1:5 and acidic conditions characterized by pH 4.5–5.5 were found in typical loading compartments where DM resides [101, 102]. After the release of CLIP, the binding cleft is "empty", which renders $\alpha\beta$ dimers susceptible to denaturation and aggregation, in particular at the low pH of endosomal/lysosomal compartments (Figure 4.1). DM binds to peptide-receptive empty MHC II molecules and prevents them from unfolding [101]. This property is regarded as the chaperone function of DM [101, 103]. In B cells, about 20–25% of DR molecules were shown to engage in DR-DM complexes, thereby constituting a pool of antigen-receptive DR molecules that may respond promptly when confronted with an antigenic challenge [101]. The structure of MHC II-DM complexes has not yet been solved, as no co-crystal has been reported.

In vitro analysis suggests that the MHC II/DM interaction depends strongly on anchoring of both molecules in a shared phospholipid bilayer or detergent micelle [104]. In vivo, this may be accomplished by segregation of MHC II and DM molecules into membrane microdomains [105], as discussed below (cf. Section 6.1). A mapping study revealed a lateral, DM-interacting surface on DR molecules that includes acidic and hydrophobic DR residues around the region where the N-terminus of an antigenic peptide is located [106]. Accordingly, an acidic face that may be protonated in acidic compartments serves, most likely, as the MHC II-interacting surface on DM [107]. Protonated carboxy groups may be primary candidates to explain the chaperone function of DM in biochemical terms, as small phenols, such as parachlorphenol, carrying an H-bond donor group, can induce a peptide-receptive state in DR molecules, which is reminiscent of DM [108]. Thus, an intermolecular hydrogen bonding network between the chaperone DM and MHC II molecules may stabilize the peptide-receptive conformer of the MHC II binding groove.

Recent findings suggest that conformational changes on MHC II molecules imposed by engagement with DM are maintained after DM has dissociated. Initial evidence came from the DR3-specific mAb 16.23 that recognizes DR3-DM complexes and DR3 loaded with cognate peptide in B cells expressing DM [97]. In DM-deficient B cells, however, DR3-peptide complexes remained 16.23-negative [109]. The same study suggested that conformational changes imposed by DM are relevant to T cell recognition, as T cells could discriminate between 16.23-positive and 16.23-negative isoforms of DR3-peptide complexes [109]. There are at least four other examples where defined immunodominant peptides and their respective MHC II restriction element form more than one isoform: in one example, a peptide binds to murine E^k in a single registry, but in two distinct conformations, as determined by NMR spectroscopy [110]. In a second case, myelin basic protein peptide and A^u form two isomeric complexes, where each isomer is recognized by a distinct T cell clone [111]. Recognition of the "aberrant" complex has been suggested to be a critical pathogenic mechanism in the murine experimental autoimmune encephalomyelitis model [112]. Moreover, Janeway and colleagues described conformers of an Eα(52-68)-A^b complex that varied in their ability to be negatively selected in mice that express Eα [113]. The most thoroughly studied example of conformational variability is hen egg lysozyme peptide HEL(48-61) presented by A^k (see Chapter 1). Type A T cells specifically recognize the HEL(48-61)-A^k conformer generated from HEL protein in late endosomal/lysosomal vesicles in the presence of DM, whereas type B T cells recognize complexes generated from exogenous HEL(48-61) in early endosomes or on the cell surface of APCs in the absence of DM [114]. The diversity in conformers of HEL(48-61)-A^k complexes is mediated through the action of DM (Figure 4.2): DM acts directly on the complex by binding to it and abolishing the type B conformer [114]. These observations strongly favor the concept that DM is not merely an endosomal/lysosomal chaperone of APCs but also behaves as a conformational editor of MHC II–peptide complexes [50].

The conformational editor activity of DM, in mechanistic terms, is most likely tightly linked to its earlier discovered potency to act as a peptide editor [115]. DM not only exchanges CLIP for cognate peptide but also exchanges low-stability peptides for peptides that display high-stability binding [116, 117]. Peptides that are DM-resistant display low intrinsic off-rates whereas peptides with moderate or high off-rates are more rapidly released by DM (similar to CLIP) [118]. The presence of appropriate anchor residues at the correct position within the peptide and a length of >12 residues are features that favor DM resistance [116], whereas glycine and proline residues at anchor and non-anchor positions increase susceptibility to DM-mediated release [119]. The ability of DM to catalyze the exchange of a wide variety of peptides suggests that the catalytic mechanism involves features of the MHC-peptide interaction common to all peptides, such as the network of hydrogen bonds between the MHC II and the backbone of a bound peptide [118]. The network is composed of 12–15 hydrogen bonds [71]. Disruption of hydrogen

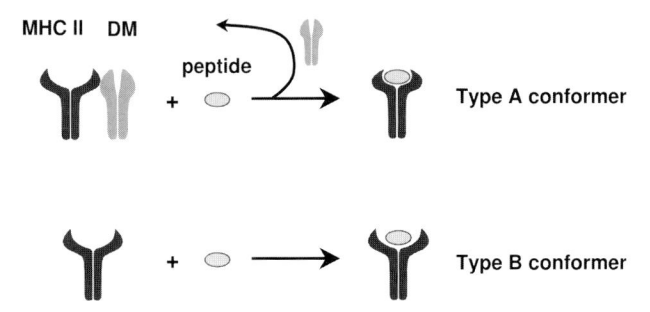

Figure 4.2 Conformational editing by HLA-DM. Peptide loading in the presence of bound HLA-DM produces type A conformers of MHC II–peptide complexes, whereas in the absence of HLA-DM the same MHC molecule and the same peptide give rise to type B conformers.

bonds, involving DR α chain residues 51 and 53, by N-methylation or truncation of the peptide led to accelerated peptide release [120]. These results provide a structural rationale for the earlier finding that peptides shortened at their N-termini are rapidly edited by DM [116].

There is also a large body of evidence that DM alters sequence-dependent MHC II–peptide interactions and, thereby, contributes to the generation of immunodominant epitopes. A recent example is the finding that it is DM that influences the selection of HEL epitopes presented on A^k by favoring high-stability peptides [121]. Further support came from kinetic analysis of the autoantigen human cartilage glycoprotein gp39, demonstrating that kinetic stability of MHC II–peptide complexes is a key determinant of immunogenicity in vivo [122]. DM editing is also critical in the context of another autoimmune disease: presentation of the immunodominant insulin-dependent diabetes autoantigen glutamate decarboxylase GAD(273-285) was significantly diminished with increasing expression of DM [123].

We conclude that DM acting as a peptide editor favors the formation and presentation of long-lived MHC II–peptide complexes on the surface of APCs. The great benefit is that $CD4^+$ T cells in T cell-rich areas of lymph nodes have sufficient time to screen the surface of APCs for appropriate MHC II–peptide complexes – a process that may last for 1–3 days. Conversely, the composition of the self- and foreign peptide repertoire appears to depend on the MHC II:DM ratio, the DM expression level and the transit time of MHC-peptide complexes in the respective loading compartment. This knowledge opens the possibility that future studies may deepen our understanding on the potential impact of aberrant regulation of DM expression or DM activity in autoimmunity or allergy.

4.5.3.2 HLA-DO/H2-DO

HLA-DO (H2-DO in mice) is expressed mainly in B cells and in subsets of thymic epithelial cells [99]. HLA-DO (short: DO) belongs to the non-classical MHC II molecules and, similar to HLA-DM, is unable to bind peptides [124]. DO requires association with DM to leave the ER [125]. DO reportedly remains engaged in DO:DM complexes during intracellular transport and recycles between peptide loading compartments and the cell surface [126]. In peripheral blood B cells, DO is quantitatively associated to DM, whereas roughly 50% of DM is bound to DO [127]. Moreover, in lymphoblastoid B cell lines, DR-DM-DO complexes have been reported to be constituents of tetraspan microdomains in acidic peptide loading compartments [128, 129]. The DOβ cytoplasmic tail redistributes MHC II and DM-DO complexes from internal to the limiting membrane of multivesicular compartments, thereby probably favoring the lateral interaction between MHC II, DM and DO molecules and tetraspanins within membrane microdomains [126, 128]. These findings imply that DO may play a role in modulating the function of DM.

There is coherent evidence that DO affects the repertoire of self- and foreign peptides presented by MHC II molecules [128, 130], although it is still open whether this is due to differential compartments or changes in the composition of microdomains where loading takes place or modulation of DM editing.

The overall impact of DO on the efficacy and selectivity of peptide presentation by B cells remains unresolved. Overexpression of DO (DO:DM $\gg$ 0.5) in various transfectants has led to accumulation of MHC II-CLIP complexes at the cell surface in some studies [131, 132]. This finding could not be confirmed by other studies involving DO-deficient mice [125, 133], and a decline in DR-CLIP levels upon transfection with moderate amounts of DO was shown [128].

A clue to these apparently contradictory observations may be that DO downregulates DM activity in endosomal compartments whereas it favors DM activity in lysosomal compartments where antigen taken up via the BCR is being processed. This concept is supported by most findings in this field. The down-modulatory activity of DO on antigen presentation is observed only when the model antigen is taken up via fluid-phase endocytosis, where loading can occur in any compartment along the endocytic pathway [125, 134]. In contrast, when antigen is taken up via the BCR, presentation of several epitopes is enhanced by DO [99, 125, 135]. The rationale for the latter aspect is that BCRs are likely to transport antigen into late endosomal/lysosomal compartments, where DO-DM complexes are superior to DM alone with regard to facilitating peptide loading. This conclusion is supported by experiments with purified proteins, demonstrating that DO favors peptide loading at pH 4.5–5.0 [125, 128, 130] and appears to reduce loading at pH 6.0–6.5 [125, 130]. Further biochemical studies showed that DO functions as a co-chaperone at low pH in at least two ways: (i) DM–DO complexes prevent inactivation of empty MHC II molecules more efficiently than DM alone [128], (ii) DO stabilizes DM molecules themselves by delaying the pH-induced inactivation of DM [128]. In support of the latter result, studies with murine B cells and DCs ectopically expressing DO show that DO prolongs the half-life of H2-DM [136].

Recently, it was demonstrated that DO is not a static component of the antigen processing pathway in B cells, but instead is regulated during B cell development [127, 137]. Relative to naive and memory B cells, DO was down-modulated in B cells upon entry into germinal centers (GC). The same GC B cells displayed reduced levels of surface MHC II-CLIP complexes. It remains open whether the reduction of DO expression in centrocytes and centroblasts occurs before or after the initial T-B cell interaction that leads to the commitment of the B cell to the GC reaction [127]. Assuming that DO promotes BCR-mediated antigen processing, as described above, it is more likely that downregulation of DO in GC B cells follows the initial T-B cell engagement.

In summary, the current model foresees that DO selectively supports loading of antigen that has entered the B cell via a high-affinity BCR, as low-affinity BCRs might lose their antigen cargo in early compartments where DO down-modulates DM-mediated loading. Thus, DO most likely skews the humoral immune response towards B cells expressing high-affinity Ig.

4.6
Export of MHC II and Organization on the Cell Surface

A decade ago, when immunologists started thinking about how APCs may deliver MHC II–peptide complexes to the surface for inspection by T cells, P. Cresswell wrote in a review article: "it remains possible that class II $\alpha\beta$ dimers follow a direct, as yet uncharacterized route from a late endosome or even a lysosome, directly to the cell surface" [45]. Recent progress, in particular in the field of developing DCs, provided evidence that late-endosomal and lysosomal compartments of APCs, indeed, are equipped in such a way that they can mature and transfer their MHC II cargo directly to the cell surface via tubular processes [138]. Beyond that, the maturation program of APCs also includes the reorganization of MHC II–peptide complexes within the membrane: MHC II–peptide complexes cluster in endosomal/lysosomal microdomains that are maintained further to facilitate peptide presentation to CD4+ T cells on the cell surface [139].

4.6.1
Membrane Microdomains

In an attempt to find accessory molecules that aid classical and non-classical MHC II molecules in endosomal/lysosomal loading compartments, the tetraspan family members CD63 and CD82 were identified [129]. Both tetraspanins engage in complexes with DR, DM and DO molecules, but, interestingly, did not co-precipitate together with $\alpha\beta$Ii complexes [129]. CD63 is a lysosomal marker that co-localizes with MHC II molecules in peptide loading compartments [140]. Regarding their ultrastructural localization inside organelles, CD63 and CD82 have been found enriched on internal vesicles of multivesicular endosomes, as originally described for B cells [141]. This is precisely where most intracellular MHC II molecules reside in naive B cells or immature DCs [138]. Parallel studies revealed that CD82 and CD63 form supramolecular assemblies together with other tetraspanins and MHC II molecules on the surface of APCs [142, 143]. These initial observations gave rise to the hypothesis that tetraspanins expressed in APCs may be involved in peptide loading and in antigen presentation to helper T cells (Figure 4.1).

Tetraspanins, first discovered in the early 1990s, are a large superfamily of more than 150 members [144]. They are characterized by four transmembrane domains and short cytoplasmic regions. They form homo- or heterodimers through their stalk subdomain, while their head subdomain appears to serve as an interface to contact other membrane proteins, such as integrins, signaling molecules or MHC molecules [139, 145]. Tetraspanins are expressed in a wide variety of cell types and contribute to diverse physiological processes, such as cellular adhesion, motility, activation and tumor invasion. APCs express almost 20 different tetraspanins, with the core family members denoted as CD9, CD37, CD53, CD63, CD81 and CD82 [144]. Together with their homologues, they are thought to constitute a novel type of membrane microdomain, also named "tetraspan web", which may act as a kind of membrane skeleton that forms the glue between molecules destined to form a functional unit (Table 4.3).

Table 4.3 Tetraspanins expressed on APCs.

Tetraspanin	Expression				Function		
	DC	Monocyte/ macrophage	B cell	T cell	Ag process/ presentation	Integrin binding	B–T cell interaction
CD9	+	+	+	+	+	+	+
CD37	+	+	+	+	–	–	+
CD53	+	+	+	+	–	+	–
CD63	+	+	+	+	+	+	–
CD81	+	+	+	+	+	+	+
CD82	+	+	+	+	+	+	–
CD151	–	+	+	+	–	–	–
NAG-2	+	?	+	+	?	?	?
Tspan-3	+	+	+	+	?	?	?

In APCs, endocytic tetraspanins, such as CD63 or CD82, apparently keep together multimolecular complexes that are to be conserved for subsequent T cell activation on the cell surface: HLA class II molecules loaded with cognate peptide together with the co-stimulatory molecule CD86 [105]. The rationale is that cognate antigenic peptide is generated in high copy number in a particular loading compartment where antigenic parent protein has accumulated before. Thus, the high local density of MHC II molecules loaded with cognate peptide in the particular loading compartment, a prerequisite for highly efficient T cell stimulation, is worth preserving by preventing dissipation during export to and on the cell surface. This may be best achieved by engagement in tetraspan microdomains [139].

Another critical characteristic of tetraspan microdomains, found with human B cells and DCs, further improves the capacity to trigger T cells: tetraspan microdomains carry a strongly selected set of peptides rather than the complex repertoire representative for whole APCs [139, 146]. This is most likely accomplished by recruitment of the peptide editor DM, which associates to CD82 even in the absence of classical MHC II molecules [105, 129]. The presence of DM obviously leads to accumulation of peptides that bind with moderate stability to MHC II molecules, whereas high-stability MHC II–peptide complexes are excluded [105, 139]. Together, tetraspan microdomains preferentially form around MHC II–peptide complexes of moderate stability, thereby compensating by a gain in avidity for the usually lower copy number of those complexes (compared with those attaining higher stability).

Consequently, dissipation of tetraspan microdomains by treatment with saponin impaired the capacity of APCs to activate T cells [105]. Accordingly, a B cell line deficient for tetraspan microdomains was unable to stimulate a T cell line specific for a peptide that localized to tetraspan microdomains in wild-type B cells [105]. The importance of tetraspan microdomains in the interaction between APCs and T helper cells could also be demonstrated with DCs. Human monocyte-derived DCs upregulate their surface pool of tetraspan microdomains 15–25-fold during maturation, whereas the surface MHC class II subset rises by only 3–5-fold [105].

Strikingly, a single self-peptide, the Ii-derived CLIP, occupied about 50–80% of the MHC II binding sites in these tetraspan microdomains on the surface of mature DCs (Ref. 52). As 9–12% of the DR molecules on mature DCs reside in tetraspan microdomains [105], CLIP covers 5–10% of the total population of DR molecules, which equals ca. $4–8 \times 10^5$ molecules [52]. This is roughly the total number of MHC II molecules expressed by human EBV-transformed B cell clones. A similar enrichment of MHC II-CLIP complexes was found ex vivo in tetraspan microdomains on APCs of human tonsils [52]. Confocal microscopy studies revealed that DR-CLIP complexes can localize to both the central and peripheral molecular activation cluster, apparently depending on the exogenous antigen to be recognized by the T cell receptor.

This is consistent with the observation that tetraspan microdomains segregate to those areas where APCs get in contact with T cells [52] and that the tetraspanin CD81 accumulates at the central zone of the immune synapse formed by B and T lymphocytes [147]. The conclusion is that a network of tetraspan molecules populates the APC side of the immune synapse, thereby stabilizing clusters of MHC molecules loaded with cognate peptide and CLIP (Figure 4.1). Attempts to define the physiological relevance of elevated MHC II-CLIP complexes embedded in tetraspan microdomains on the surface of mature DCs revealed that CLIP antagonizes T helper type 1 polarization [52]. In support of this finding, Ii-deficient mice contain CD4+ T cells, preferentially displaying the T_H1 phenotype [148]; immunization of wild-type mice with CLIP along with cognate antigen resulted in a T_H2-like response [149] and splenocytes from H2-DM k.o. mice, displaying the CLIP(high) phenotype, induced the T_H2 type of helper T cell responses [52].

Monoclonal antibodies (mAb), such as the mAb FN1, that have been instrumental in identifying tetraspan microdomains, recognize superdimers of HLA class II molecules, embedded in a network of tetraspanins [139, 150]. Hence CLIP, as a resident of tetraspan microdomains on the surface of APCs appears to be the first example of an endogenous self-peptide that may aid in the dimerization of T cell receptors (TCRs), as suggested in the "pseudodimer" model pioneered by M. Davis and colleagues [151, 152]. Pseudodimers have been defined as heterologous superdimers of MHC II molecules carrying a self-peptide and an agonistic peptide. In particular, in the initial phase of TCR docking, pseudodimers localizing to the synapse may be essential in overcoming limitations in TCR triggering caused by too low local density of agonistic MHC II–peptide complexes [151–153]. This may apply especially to those TCRs that have been positively selected through MHC II-CLIP complexes in the thymic cortex, as those TCRs are thought to recognize CLIP with moderate affinity. CLIP is one of the most abundant self-peptides on APCs of the thymic cortex (cf. Chapter 6).

Later on, at the stage when the TCR begins to transmit signals, CLIP is supposed to reduce the local density of agonistic ligands in the synapse, as compared to the situation when only agonistic ligands engage the TCR. This would mirror the situation in which very low doses of agonist or low-affinity agonists give rise to low-avidity T cell interactions, thereby favoring T_H2 polarization [154].

Apart from tetraspan microdomains, MHC II molecules in B cell and DCs have been detected in cholesterol- and sphingolipid-rich membrane domains, which are also termed "lipid rafts" [155, 156]. In B cells >90% of MHC II molecules were described as associated to lipid rafts in a constitutive manner [155], whereas other reports could find less than 5% [105]. Lipid rafts are thought to consist mainly of tightly packed sphingolipids, cholesterol, and a subset of peripheral and integral membrane proteins. The tight packing of sphingolipids and cholesterol in lipid rafts is deemed responsible for their stability in 1% Triton X-100 at 4°C (conditions used to isolate raft structures). MHC II–peptide complexes localizing to lipid rafts have been suggested to be critical for T cell stimulation, as they are visible in immune synapses [156] and as lipid rafts concentrate relevant MHC II–peptide complexes, thereby facilitating T cell activation at low antigen dose [155]. In contrast to tetraspan microdomains, raft microdomains prepared in Triton X-100 or CHAPS do not enrich particular subsets of MHC II–peptide complexes [105, 139]. As detergent usage leads to artificial fusion of lipid rafts by removal of non-packed lipids it is still questionable whether the aforementioned observations with raft domains reflect the situation in vivo [157, 158].

4.6.2
Tubular Transport

Upon receiving a maturation stimulus, DCs are transformed into potent APCs characterized by increased surface expression of MHC I, MHC II and multiple co-stimulatory molecules, such as CD86, CD80 and CD40 [159]. The increase in surface expression of MHC II is accomplished mainly by the redistribution of a pre-synthesized pool from intracellular storage sites to the plasma membrane [13]. Similar to those in other APCs, the intracellular storage sites of DCs have the characteristics of late endocytic multivesicular bodies (MVBs) or lysosomes. Collectively, these compartments are referred to as MHC II enriched compartments (MIIC) [160].

In murine immature D1 DCs, most MHC II molecules residing in MVBs were stored in internal vesicles of MVBs [161]. In contrast, the peptide loading chaperone DM was located predominantly at the limiting membrane of MVBs [161]. In melanoma cells transfectants, a sorting signal in the cytoplasmic tail of DO has been suggested to be responsible for the fact that DM/DO complexes are also preferentially found at the limiting membrane of MVBs [162]. As DCs do not express DO, it is still open as to how DM and MHC II molecules are differentially sorted.

4.6 Export of MHC II and Organization on the Cell Surface

Upon receiving a maturation stimulus, internal vesicles carrying stored MHC II molecules are transferred to the limiting membrane of MVBs, thereby allowing DM and MHC II molecules to contact each other laterally in the same membrane domain, and so facilitating DM mediated peptide loading and editing [161]. Protein transport from inner to outer membranes cannot occur laterally in the plane of the membrane, but required fusion between the exoplasmic leaflets of both membrane domains [163]. In the next step, MVBs transform into long tubular organelles that extend into the periphery of the cells towards the plasma membrane [161]. Vesicles formed at the tips of the tubules are supposed to mediate transport of MHC II–peptide complexes to the cell surface. These transport vesicles are reminiscent of the class II MHC vesicles, named "CIIV", originally described in B cells [164] and in bone-marrow derived murine DCs [165].

In contrast to the tubule-derived vesicles of murine DCs, the CIIVs described earlier were devoid of DM [161, 165]. The CIIV-type of transport vesicles in bone-marrow-derived DCs co-expressed clusters composed of MHC I and II and CD86 molecules that persisted after their arrival at the cell surface [165]. Most likely, this is because MHC and the co-stimulator CD86 in transport vesicles are not free in their lateral movement but embedded in tetraspan microdomains, as suggested above [105].

The retrograde transport pathway from lysosomes to the cell surface has been visualized in live DCs by video microscopy most recently: using cells that express MHC II tagged with green fluorescent protein it was shown that developing DCs generate tubules from lysosomal compartments that finally fuse directly with the plasma membrane [166]. In-depth analysis of the direction these tubules take disclosed that antigen-specific CD4+ T cells interacting with antigen-loaded DCs induce the formation of long tubular MIICs that polarize towards the T cell [167]. T cells with TCRs of irrelevant specificity failed to do so. LPS-evoked tubules that form in the absence of T cells are rather short (~ 5 µm) and not necessarily microtubule-driven, whereas those formed from polarized tubules in the presence of maturation stimuli and appropriate T cells are rather long (15–20 µm), strictly antigen-dependent and require an intact microtubular apparatus [167]. Importantly, in myeloid differentiation factor 88 (MyD88)-deficient DCs, T cell polarized endosomal tubules failed to form even in the presence of antigen-specific T cells. Likewise, in the absence of LPS or other pro-inflammatory stimuli no tubulation was observed [168]. The necessity of the adaptor MyD88, involved in signaling through TLRs, clearly shows that a microbe-derived TLR ligand in addition to engagement of the appropriate T cell is critical for the formation of long polarized tubules in immature DCs.

Apart from TLR and TCR, other molecules constituting the immune synapse obviously play a role in polarized tubulation. The adhesion molecules LFA-1 and CD2, known to localize to the peripheral supramolecular activation cluster of the synapse, are involved in both the stabilization of the DC–T cell interaction and the initiation of intracellular signaling [169]. Antibodies against both LFA-1 and CD2 could partially inhibit tubulation [170]. Furthermore, the activation marker CD40L, which was brought to the T cell surface after about 2 h of co-incubation with antigen-pulsed developing DCs, promoted tubulation of MIIC compartments [170]. Most strikingly, antigenic peptide added to DCs did allow T cell stimulation but failed to evoke T cell polarized tubulation – quite in contrast to DCs pulsed with intact antigenic protein [170]. The most likely explanation for the differential signal emanating from MHC-peptide complexes being formed at the cell surface upon binding exogenous peptide versus MHC-peptide complexes originating from MIICs is their organization in the membrane. As discussed above, MHC II–peptide complexes formed in MIICs tend to be incorporated into tetraspan microdomains [52, 105]. This may not apply to peptide-receptive MHC II molecules on the cell surface (Figure 4.1).

In summary, in the pre-tubulation phase, the TCR needs to be activated by microdomains within the immune synapse filled with densely packed cognate MHC II–peptide complexes, co-stimulatory and adhesion molecules. This initial contact, employing a limited number of MHC–peptide complexes, generates a signal that mobilizes further cognate MHC II–peptide complexes from internal MIIC stores, thereby triggering the tubulation phase. Making use of microtubule tracks, tubules emanating from deep endocytic MIICs then deliver a steady supply of class II–peptide molecules to the immune synapse for further engagement of TCRs.

4.7
Viral and Bacterial Interference

During the past decade, several mechanisms have been explored as to how pathogens may evade antigen processing and presentation. Although more work has focused on the MHC class I pathway, knowledge is accumulating on how microbes interfere with the MHC II processing pathway.

A strategy pursued by both bacteria and virus is the suppression of MHC class II expression. The most thoroughly studied example is *Mycobacterium tuberculosis* (MTB) infection. Macrophages are critical to the control of MTB infection, because they harbor the bacteria in early endocytic compartments and stimulate CD4+ T cells [171]. Activated MTB-reactive CD4+ T cells secrete IFN-γ to activate infected macrophages and induce microbicidal function [172]. However, despite vigorous immune responses, MTB persists inside macrophages. Chronic stimulation (>16 h) of TLR2 by MTB 19 kDa lipoprotein inhibits IFN-γ-dependent induction of the class II transactivator CIITA and thereby downregulates MHC II and MHC II-dependent antigen processing [173]. Likewise, other TLR ligands, such as CpG DNA [174], LPS [175] and the MTB 24 kDa lipoprotein LprG [176], inhibit antigen processing in macrophages in a TLR/MyD88 dependent fashion. Recently, the mycolylarabinogalactan peptidoglycan was shown to inhibit macrophage responses to IFNγ independently of TLR2, TLR4 and MyD88 [177]. Together, MTB exploits the fact that, in contrast to B cells or DCs, MHC II expression in macrophages relies on IFN-γ. Moreover, quite different from DCs that respond to prolonged presence of TLR ligands by upregulation of MHC II expression, MTB-infected macrophages downregulate their MHC II-dependent antigen processing machinery.

Although alveolar macrophages are the main targets of MTB infection, in the airway mucosa DCs expressing DC-SIGN have been identified that are infected by MTB as well. DC-SIGN is the main receptor for mycobacteria in DCs [178]; although mycobacteria end up in lysosomes inside DCs, DCs are impaired by MTB in their capacity to mature: increased secretion of the cell wall component lipoarabinomannan (ManLAM) by infected macrophages or DCs targets DC-SIGN and results in inhibitory signals that interfere with TLR-activating stimuli, normally triggering DC maturation [178]. The ManLAM/DC-SIGN interaction also results in secretion of the immunosuppressive cytokine IL-10, thereby interfering with cellular immune responses against MTB. Hence, some pathogens seem able to manipulate the balance between TLR stimulation and C-type lectin occupation, resulting in immune suppression (cf. Section 3.3).

Similar to mycobacteria, other pathogens establish residence in host cells within a membrane-bound vacuole, but have evolved different strategies for inhibiting MHC II expression: *Chlamydia trachomatis* secrete proteases that degrade transcription factors necessary for MHC II biosynthesis [179], whereas *Leishmania amazonensis* amastigotes directly degrade MHC II molecules [180].

Viruses infecting APCs have evolved their own strategies that impair surface expression of MHC class II and co-stimulatory molecules. Immature DCs infected with human cytomegalovirus (HCMV) display reduced expression of HLA class II, CD80 and CD86 molecules [181, 182]. Likewise, in mature DCs, HCMV reduces the expression of MHC I, MHC II and CD83 molecules and the secretion of IL-12 [181]. These effects may, at least in part, be mediated by the HCMV glycoprotein US3: US3 binds to newly synthesized MHC II molecules, thereby preventing invariant chain from binding. The consequence is an impairment in sorting of MHC II to peptide loading compartments and in presentation of endogenous HCMV antigens to CD4+ T cells [183]. Independent from US3, US2 is the crucial protein involved in an additional mechanism that may allow HCMV-infected APCs to remain "invisible" to CD4+ T cells. In macrophages, US2 causes premature degradation of the loading chaperone DM and of HLA class II molecules itself [184]. This is accomplished by inducing selective proteolysis of DMα and DRα chains.

Infection with Herpes simplex virus (HSV) of DCs causes phenotypic changes reminiscent of HCMV: HSV interferes with LPS-mediated upregulation of HLA class I, class II and CD86 molecules [185]. In HSV-infected B cells Ii expression is strongly reduced and the HSV envelope glycoprotein gB binds to both DR and DM, thereby interfering efficiently with antigenic peptide loading [186].

Human immunodeficiency virus-1 (HIV-1) also impacts on the MHC II:Ii ratio, albeit in a distinct way: HIV-1 Nef protein promotes the surface expression of MHC II-Ii at the expense of MHC II–peptide complexes, thereby reducing the capacity of HIV-1 infected APCs to trigger CD4+ T cells [187].

Epstein-Barr virus (EBV) persists lifelong in B cells of infected hosts, indicating that EBV may have also evolved effective ways to evade immune recognition. In the lytic phase, the EBV protein pg42 is critical in hampering T cell recognition. Gp42 binds to HLA class II-peptide complexes, thereby creating a complex that is not compatible with the engagement of TCRs [188]. This mechanism provides a previously undescribed strategy for viral immune escape.

Helicobacter pylori colonizes the human gastric mucosa. This type of infection can persist for decades and in most cases remains asymptomatic. A major virulence factor of *H. pylori* is the protein toxin VacA [189]. VacA interferes with proteolytic processing in the MHC II pathway. Similar to lysosomotropic agents, such as monensin or chloroquin, VacA depresses the number of T cell epitopes generated in MIICs of APCs. This mode of action may contribute to the persistence of *H. pylori* in the gastric mucosa.

4.8
Concluding Remarks

The efficacy and quality of loading of MHC II molecules with antigenic peptides, being the final goal of antigen processing, is influenced by several factors, including the uptake mechanism of antigenic protein, the structure of antigen, its resistance to proteolysis and the accessibility of MHC II molecules in loading compartments. Availability of MHC II is strongly determined by the activation state of APCs, regulating the number of MHC II molecules and accessory molecules, such as DM and DO. In addition, the rate of endocytosis and expression of appropriate receptor molecules for antigen-uptake is controlled by the activation state of APCs and the export of MHC II to the cell surface is subject to regulation by the degree and type of activation. As well as the quality and quantity of peptide loading itself, other phenomena strongly contribute, such as the expression of co-stimulatory molecules or secretion of cytokines that shape the environment of $CD4^+$ T cells recognizing MHC II–peptide complexes.

Both self-peptides and foreign antigenic peptides are taken up, processed and presented via MHC II in the same way. Thus, one might argue that APCs are unable to discriminate between self and foreign. At first glance, this argument is true. The uptake and processing machinery of APCs can not discriminate between an antigen of foreign or self origin. However, receptors, such as TLRs, that can perceive the signal "foreign" may modulate the antigen processing pathway, thus introducing changes to the quality and quantity of antigenic peptides to be presented. In addition, the spacing and timing of the deposition of MHC II–peptide complexes on the surface of an APC may contribute to discriminate the usual "self" situation from an alerted "foreign" situation. Organization of MHC II in certain microdomains might be one way. Continuous presentation of self-peptides derived from proteins taken up by fluid phase or derived from the APC's own membrane proteins may constitute the normal background of MHC II–peptide complexes. As soon as a bolus of foreign antigen, derived from a pathogen or aggregated via immunoglobulin, moves through the endosomal system, the APC is flushed by a distinct set of antigens, high in copy number but low in diversity, giving rise to a few immunodominant antigenic peptide complexes presented in high abundance in a narrow window of time.

Future studies will have to deepen our understanding of how temporal, spatial and systemic organization of antigen processing inside and outside APCs govern the capacity to trigger CD4$^+$ T cell responses.

References

1 Girvin, A. M., Gordon, K. B., Welsh, C. J., Clipstone, N. A., and Miller, S. D. (2002). Differential abilities of central nervous system resident endothelial cells and astrocytes to serve as inducible antigen-presenting cells. *Blood* 99, 3692–3701.

2 Leiter, E. H., Christianson, G. J., Serreze, D. V., Ting, A. T., and Worthen, S. M. (1989). MHC antigen induction by interferon gamma on cultured mouse pancreatic beta cells and macrophages. Genetic analysis of strain differences and discovery of an "occult" class I-like antigen in NOD/Lt mice. *J. Exp. Med.* 170, 1243–1262.

3 Skoskiewicz, M. J., Colvin, R. B., Schneeberger, E. E., and Russell, P. S. (1985). Widespread and selective induction of major histocompatibility complex-determined antigens in vivo by gamma interferon. *J. Exp. Med.* 162, 1645–1664.

4 Steimle, V., Siegrist, C. A., Mottet, A., Lisowska-Grospierre, B., and Mach, B. (1994). Regulation of MHC class II expression by interferon-gamma mediated by the transactivator gene CIITA. *Science* 265, 106–109.

5 Sallusto, F., and Lanzavecchia, A. (1994). Efficient presentation of soluble antigen by cultured human dendritic cells is maintained by granulocyte/macrophage colony-stimulating factor plus interleukin 4 and downregulated by tumor necrosis factor alpha. *J. Exp. Med.* 179, 1109–1118.

6 Jarrossay, D., Napolitani, G., Colonna, M., Sallusto, F., and Lanzavecchia, A. (2001). Specialization and complementarity in microbial molecule recognition by human myeloid and plasmacytoid dendritic cells. *Eur. J. Immunol.* 31, 3388–3393.

7 Schuler, G., Romani, N., and Steinman, R.M. (1985). A comparison of murine epidermal Langerhans cells with spleen dendritic cells. *J. Invest. Dermatol.* 85, 99s–106s.

8 Tang, A., Amagai, M., Granger, L. G., Stanley, J. R., and Udey, M. C. (1993). Adhesion of epidermal Langerhans cells to keratinocytes mediated by E-cadherin. *Nature* 361, 82–85.

9 Matsuno, K., Ezaki, T., Kudo, S., and Uehara, Y. (1996). A life stage of particle-laden rat dendritic cells in vivo: their terminal division, active phagocytosis, and translocation from the liver to the draining lymph. *J. Exp. Med.* 183, 1865–1878.

10 Carson, M. J., Reilly, C. R., Sutcliffe, J. G., and Lo, D. (1998). Mature microglia resemble immature antigen-presenting cells. *Glia* 22, 72–85.

11 Stumbles, P. A., Thomas, J. A., Pimm, C. L., Lee, P. T., Venaille, T. J., Proksch, S., and Holt, P. G. (1998). Resting respiratory tract dendritic cells preferentially stimulate T helper cell type 2 (Th2) responses and require obligatory cytokine signals for induction of Th1 immunity. *J. Exp. Med.* 188, 2019–2031.

12 Nijman, H. W., Kleijmeer, M. J., Ossevoort, M. A., Oorschot, V. M., Vierboom, M. P., van de Keur, M., Kenemans, P., Kast, W. M., Geuze, H. J., and Melief, C. J. (1995). Antigen capture and major histocompatibility class II compartments of freshly isolated and cultured human blood dendritic cells. *J. Exp. Med.* 182, 163–174.

13 Pierre, P., Turley, S. J., Gatti, E., Hull, M., Meltzer, J., Mirza, A., Inaba, K.,

13 Steinman, R. M., and Mellman, I. (1997). Developmental regulation of MHC class II transport in mouse dendritic cells. *Nature* 388, 787–792.

14 Cella, M., Sallusto, F., and Lanzavecchia, A. (1997). Origin, maturation and antigen presenting function of dendritic cells. *Curr. Opin. Immunol.* 9, 10–16.

15 Winzler, C., Rovere, P., Rescigno, M., Granucci, F., Penna, G., Adorini, L., Zimmermann, V. S., Davoust, J., and Ricciardi-Castagnoli, P. (1997). Maturation stages of mouse dendritic cells in growth factor-dependent long-term cultures. *J. Exp. Med.* 185, 317–328.

16 Stossel, H., Koch, F., Kampgen, E., Stoger, P., Lenz, A., Heufler, C., Romani, N., and Schuler, G. (1990). Disappearance of certain acidic organelles (endosomes and Langerhans cell granules) accompanies loss of antigen processing capacity upon culture of epidermal Langerhans cells. *J. Exp. Med.* 172, 1471–1482.

17 Zhou, L. J., and Tedder, T. F. (1995). Human blood dendritic cells selectively express CD83, a member of the immunoglobulin superfamily. *J. Immunol.* 154, 3821–3835.

18 Steinman, R. M. and Cohn, Z. A. (1974). Identification of a novel cell type in peripheral lymphoid organs of mice. II. Functional properties in vitro. *J. Exp. Med.* 139, 380–397.

19 Sallusto, F., Cella, M., Danieli, C. and Lanzavecchia, A. (1995). Dendritic cells use macropinocytosis and the mannose receptor to concentrate macromolecules in the major histocompatibility complex class II compartment: downregulation by cytokines and bacterial products. *J. Exp. Med.* 182, 389–400.

20 Steinman, R. M., Hawiger, D. and Nussenzweig, M. C. (2003). Tolerogenic dendritic cells. *Annu. Rev. Immunol.* 21, 685–711.

21 West, M. A., Wallin, R. P., Matthews, S. P., Svensson, H. G., Zaru, R., Ljunggren, H. G., Prescott, A. R. and Watts, C. (2004). Enhanced dendritic cell antigen capture via toll-like receptor-induced actin remodeling. *Science* 305, 1153–1157.

22 Underhill, D. M. and Ozinsky, A. (2002). Phagocytosis of microbes: complexity in action. *Annu. Rev. Immunol.* 20, 825–852.

23 Rubartelli, A., Poggi, A. and Zocchi, M. R. (1997). The selective engulfment of apoptotic bodies by dendritic cells is mediated by the alpha(v)beta3 integrin and requires intracellular and extracellular calcium. *Eur. J. Immunol.* 27, 1893–1900.

24 Blander, J. M. and Medzhitov, R. (2004). Regulation of phagosome maturation by signals from toll-like receptors. *Science* 304, 1014–1018.

25 John, B., Herrin, B. R., Raman, C., Wang, Y. N., Bobbitt. K. R., Brody. B. A. and Justement, L. B. (2003). The B cell coreceptor CD22 associates with AP50, a clathrin-coated pit adapter protein, via tyrosine-dependent interaction. *J. Immunol.* 170, 3534–3543.

26 Fanger, N. A., Wardwell, K., Shen, L., Tedder, T. F. and Guyre, P. M. (1996). Type I (CD64) and type II (CD32) Fc gamma receptor-mediated phagocytosis by human blood dendritic cells. *J. Immunol.* 157, 541–548.

27 Stahl, P. D. and Ezekowitz, R. A. (1998). The mannose receptor is a pattern recognition receptor involved in host defense. *Curr. Opin. Immunol.* 10, 50–55

28 Dong, X., Storkus, W. J. and Salter, R. D. (1999). Binding and uptake of agalactosyl IgG by mannose receptor on macrophages and dendritic cells. *J. Immunol.* 163, 5427–5434.

29 Prigozy, T. I., Sieling, P. A., Clemens, D., Stewart, P. L., Behar, S. M., Porcelli, S. A., Brenner, M. B., Modlin, R. L. and Kronenberg, M. (1997). The mannose receptor delivers lipoglycan antigens to endosomes for presentation to T cells by CD1b molecules. *Immunity* 6, 187–197.

30 Jiang, W., Swiggard, W. J., Heufler, C., Peng, M., Mirza, A., Steinman, R. M. and Nussenzweig, M. C. (1995). The receptor DEC-205 expressed by dendritic cells and thymic epithelial cells is involved in antigen processing. *Nature* 375, 151–155.

31 Engering, A., Geijtenbeek, T. B., van Vliet, S. J., Wijers, M., van Liempt, E.,

Demaurex, N., Lanzavecchia, A., Fransen, J., Figdor, C. G., Piguet, V. and van Kooyk Y. (2002). The dendritic cell-specific adhesion receptor DC-SIGN internalizes antigen for presentation to T cells. *J. Immunol.* 168, 2118–2126.

32. Valladeau, J., Ravel, O., Dezutter-Dambuyant, C., Moore, K., Kleijmeer, M., Liu, Y., Duvert-Frances, V., Vincent, C., Schmitt, D., Davoust, J., Caux, C., Lebecque, S. and Saeland, S. (2000). Langerin, a novel C-type lectin specific to Langerhans cells, is an endocytic receptor that induces the formation of Birbeck granules. *Immunity* 12, 71–81.

33. Kissenpfennig, A., Ait-Yahia, S., Clair-Moninot, V., Stossel, H., Badell, E., Bordat, Y., Pooley, J. L., Lang, T., Prina, E., Coste, I., Gresser, O., Renno, T., Winter, N., Milon, G., Shortman, K., Romani, N., Lebecque, S., Malissen, B., Saeland, S. and Douillard P. (2005). Disruption of the langerin/CD207 gene abolishes Birbeck granules without a marked loss of Langerhans cell function. *Mol. Cell Biol.* 25, 88–99.

34. Gantner, B. N., Simmons, R. M., Canavera, S. J., Akira, S. and Underhill, D. M. (2003). Collaborative induction of inflammatory responses by dectin-1 and Toll-like receptor 2. *J. Exp. Med.* 197, 1107–1117.

35. Schjetne, K. W., Thompson, K. M., Nilsen, N., Flo, T. H., Fleckenstein, B., Iversen, J. G., Espevik, T. and Bogen, B. (2003). Toll-like receptor 2 internalizes antigen for presentation to CD4+ T cells and could be an efficient vaccine target. *J. Immunol.* 171, 32–36.

36. Geijtenbeek, T. B., van Vliet, S. J., Engering, A, 't Hart, B. A. and van Kooyk, Y. (2004). Self- and nonself-recognition by C-type lectins on dendritic cells. *Annu. Rev. Immunol.* 22, 33–54.

37. Kloetzel, P. M. and Ossendorp, F. (2004). Proteasome and peptidase function in MHC-class-I-mediated antigen presentation. *Curr. Opin. Immunol.* 16, 76–81.

38. Watts, C. (2004). The exogenous pathway for antigen presentation on major histocompatibility complex class II and CD1 molecules. *Nat. Immunol.* 5, 685–692.

39. Villadangos, J. A. and Ploegh, H. L. (2000). Proteolysis in MHC class II antigen presentation: who's in charge? *Immunity* 12, 233–239.

40. Jensen, P. E. (1995). Antigen unfolding and disulfide reduction in antigen presenting cells. *Semin. Immunol.* 6, 347–353.

41. Arunachalam, B., Phan, U. T., Geuze, H. J. and Cresswell, P. (2000). Enzymatic reduction of disulfide bonds in lysosomes: characterization of a gamma-interferon-inducible lysosomal thiol reductase (GILT). *Proc. Natl. Acad. Sci. U.S.A.* 97, 745–750.

42. Maric, M., Arunachalam, B., Phan, U. T., Dong, C., Garrett, W. S., Cannon, K. S., Alfonso, C., Karlsson, L., Flavell, R. A. and Cresswell, P. (2001). Defective antigen processing in GILT-free mice. *Science* 294, 1361–1365

43. Haque, M. A., Li, P., Jackson, S. K., Zarour, H. M., Hawes, J. W., Phan, U. T., Maric M., Cresswell, P. and Blum, J. S. (2002). Absence of gamma-interferon-inducible lysosomal thiol reductase in melanomas disrupts T cell recognition of select immunodominant epitopes. *J. Exp. Med.* 195, 1267–1277.

44. Lennon-Dumenil, A. M., Bakker, A. H., Wolf-Bryant, P., Ploegh, H. L. and Lagaudriere-Gesbert, C. (2002). A closer look at proteolysis and MHC-class-II-restricted antigen presentation. *Curr. Opin. Immunol.* 14, 15–21.

45. Cresswell, P. (1994). Assembly, transport, and function of MHC class II molecules. *Annu. Rev. Immunol.* 12, 259–293.

46. Sherman, M. A., Weber, D. A. and Jensen, P. (1995). DM enhances peptide binding to class II MHC by release of invariant chain-derived peptides. *Immunity* 3, 197–205.

47. Inaba, K., Turley, S., Iyoda, T., Yamaide, F., Shimoyama, S., Reis e Sousa, C., Germain, R. N., Mellman, I. and Steinman, R. M. (2000). The formation of immunogenic major histocompatibility complex class II-peptide ligands in lysosomal compartments of dendritic cells

is regulated by inflammatory stimuli. *J. Exp. Med.* 191, 927–936.
48 Pierre P. and Mellman I. (1998). Developmental regulation of invariant chain proteolysis controls MHC class II trafficking in mouse dendritic cells. *Cell* 93, 1135–1145.
49 Trombetta, E. S., Ebersold, M., Garrett, W., Pypaert, M. and Mellman, I. (2003). Activation of lysosomal function during dendritic cell maturation. *Science* 299, 1400–1403.
50 Vogt, A. B. and Kropshofer, H. (1999). HLA-DM - an endosomal/lysosomal chaperone for the immune system. *Trends Biochem. Sci.* 24, 150–154.
51 Wilson, N. S., El-Sukkari, D. and Villadangos, J. A. (2004). Dendritic cells constitutively present self antigens in their immature state in vivo and regulate antigen presentation by controlling the rates of MHC class II synthesis and endocytosis. *Blood* 103, 2187–2195.
52 Rohn, T. A., Boes, M., Wolters, D., Spindeldreher, S., Muller, B., Langen, H., Ploegh, H., Vogt, A. B. and Kropshofer, H. (2004). Upregulation of the CLIP self peptide on mature dendritic cells antagonizes T helper type 1 polarization. *Nat. Immunol.* 5, 909–918.
53 Veeraswamy, R. K., Cella, M., Colonna, M. and Unanue, E.R. (2003). Dendritic cells process and present antigens across a range of maturation states. *J. Immunol.* 170, 5367–5372.
54 Cobb, B. A., Wang, Q., Tzianabos, A. O., Kasper, D. L. (2004). Polysaccharide processing and presentation by the MHCII pathway. *Cell* 117, 677–687.
55 Kornfeld, S. and Mellman, I. (1989). The biogenesis of lysosomes. *Annu. Rev. Cell Biol.* 5, 483–525.
56 Sercarz, E. E. and Maverakis, E. (2003). MHC-guided processing: binding of large antigen fragments. *Nat. Rev. Immunol.* 3, 621–629.
57 Castellino, F., Zappacosta, F., Coligan, J. E. and Germain, R. N. (1998). Large protein fragments as substrates for endocytic antigen capture by MHC class II molecules. *J. Immunol.* 161, 4048–4057.
58 Deussing, J., Roth, W., Saftig, P., Peters, C., Ploegh, H. L. and Villadangos, J. A. (1998). Cathepsins B and D are dispensable for major histocompatibility complex class II-mediated antigen presentation. *Proc. Natl. Acad. Sci. U.S.A.* 95, 4516–4521
59 Driessen, C., Lennon-Dumenil, A. M. and Ploegh H. L. (2001). Individual cathepsins degrade immune complexes internalized by antigen-presenting cells via Fcgamma receptors. *Eur. J. Immunol.* 31, 1592–1601.
60 Honey, K., Nakagawa, T., Peters, C. and Rudensky, A. (2002). Cathepsin L regulates CD4+ T cell selection independently of its effect on invariant chain: a role in the generation of positively selecting peptide ligands. *J. Exp. Med.* 195, 1349–1358.
61 Hsieh, C. S., deRoos, P., Honey, K., Beers and C., Rudensky, A. Y. (2002). A role for cathepsin L and cathepsin S in peptide generation for MHC class II presentation. *J. Immunol.* 168, 2618–2625.
62 Pluger, E. B., Boes, M., Alfonso, C., Schroter, C. J., Kalbacher, H., Ploegh, H.L. and Driessen, C. (2002). Specific role for cathepsin S in the generation of antigenic peptides in vivo. *Eur. J. Immunol.* 32, 467–476.
63 Manoury, B., Hewitt, E. W., Morrice, N., Dando, P. M., Barrett, A.J. and Watts, C. (1998). An asparaginyl endopeptidase processes a microbial antigen for class II MHC presentation. *Nature* 396, 695–699.
64 Manoury, B., Mazzeo, D., Fugger, L., Viner, N., Ponsford, M., Streeter, H., Mazza, G., Wraith, D. C. and Watts, C. (2002). Destructive processing by asparagine endopeptidase limits presentation of a dominant T cell epitope in MBP. *Nat. Immunol.* 3, 169–174.
65 Krogsgaard, M., Wucherpfennig, K. W., Cannella, B., Hansen, B. E., Svejgaard, A., Pyrdol, J., Ditzel, H., Raine, C., Engberg, J. and Fugger, L. (2000). Visualization of myelin basic protein (MBP) T cell epitopes in multiple sclerosis lesions using a monoclonal antibody specific for the human histocompatibility leuko-

cyte antigen (HLA)-DR2-MBP 85-99 complex. *J. Exp. Med.* 191, 1395–1412.
66. Nishioku, T., Hashimoto, K., Yamashita, K., Liou, S. Y., Kagamiishi, Y., Maegawa, H., Katsube, N., Peters, C., von Figura, K., Saftig, P., Katunuma, N., Yamamoto, K. and Nakanishi, H. (2002). Involvement of cathepsin E in exogenous antigen processing in primary cultured murine microglia. *J. Biol. Chem.* 277, 4816–4822.
67. Yee, C.S., Yao, Y., Li, P., Klemsz, M. J., Blum, J. S. and Chang, C. H. (2004). Cathepsin E: a novel target for regulation by class II transactivator. *J. Immunol.* 172, 5528–5534.
68. Stern, L. J., Brown, J. H., Jardetzky, T. S., Gorga, J. C., Urban, R. G., Strominger, J. L. and Wiley, D. C. (1994). Crystal structure of the human class II MHC protein HLA-DR1 complexed with an influenza virus peptide. *Nature* 368, 215–221.
69. Vogt, A. B., Stern, L. J., Amshoff, C., Dobberstein, B., Hammerling, G. J. and Kropshofer, H. (1995). Interference of distinct invariant chain regions with superantigen contact area and antigenic peptide binding groove of HLA-DR. *J. Immunol.* 155, 4757–4765.
70. Stumptner, P. and Benaroch, P. (1997). Interaction of MHC class II molecules with the invariant chain: role of the invariant chain (81-90) region. *EMBO J.* 16, 5807–5818.
71. Brown, J. H., Jardetzky, T. S., Gorga, J. C., Stern, L. J., Urban, R. G., Strominger. J. L. and Wiley, D. C. (1993). Three-dimensional structure of the human class II histocompatibility antigen HLA-DR1. *Nature* 364, 33–39.
72. Zhong, G., Castellino, F., Romagnoli, P. and Germain, R. N. (1996). Evidence that binding site occupancy is necessary and sufficient for effective major histocompatibility complex (MHC) class II transport through the secretory pathway redefines the primary function of class II-associated invariant chain peptides (CLIP). *J. Exp. Med.* 184, 2061–2066.
73. Doebele, R. C., Pashine, A., Liu, W., Zaller, D. M., Belmares, M., Busch, R. and Mellins, E. (2003). Point mutations in or near the antigen-binding groove of HLA-DR3 implicate class II-associated invariant chain peptide affinity as a constraint on MHC class II polymorphism. *J. Immunol.* 170, 4683–4692.
74. Hammer, J., Valsasnini, P., Tolba, K., Bolin, D., Higelin, J., Takacs, B. and Sinigaglia, F. (1993). Promiscuous and allele-specific anchors in HLA-DR-binding peptides. *Cell* 74, 197–203.
75. Sturniolo, T., Bono, E., Ding, J., Raddrizzani, L., Tuereci, O., Sahin, U., Braxenthaler, M., Gallazzi, F., Protti, M. P., Sinigaglia, F. and Hammer, J. (1999). Generation of tissue-specific and promiscuous HLA ligand databases using DNA microarrays and virtual HLA class II matrices. *Nat. Biotechnol.* 17, 555–561.
76. Herman, A., Kappler, J. W., Marrack, P. and Pullen, A. M. (1991). Superantigens: mechanism of T-cell stimulation and role in immune responses. *Annu. Rev. Immunol.* 9, 745–772.
77. Li, H., Llera, A., Malchiodi, E. L. and Mariuzza, R. A. (1999). The structural basis of T cell activation by superantigens. *Annu. Rev. Immunol.* 17, 435–466.
78. Hogan, R. J., Van Beek, J., Broussard, D. R., Surman, S. L. and Woodland, D. L. (2001). Identification of MHC class II-associated peptides that promote the presentation of toxic shock syndrome toxin-1 to T cells. *J. Immunol.* 166, 6514–6522.
79. Melnick, J. and Argon, Y. (1995). Molecular chaperones and the biosynthesis of antigen receptors. *Immunol. Today* 16, 243–250.
80. Cresswell, P. (1996). Invariant chain structure and MHC class II function. *Cell* 84, 505–507
81. Busch, R., Cloutier, I., Sékaly, R.-P. and Hämmerling, G. J. (1996). Invariant chain protects class II histocompatibility antigens from binding intact polypeptides in the endoplasmic reticulum. *EMBO J.* 15, 418–428.
82. Blum, J. S. and Cresswell, P. (1988). Role for intracellular proteases in the processing and transport of class II HLA antigens. *Proc. Natl. Acad. Sci. U.S.A.* 85, 3975–3979.

83 Manoury, B., Mazzeo, D., Li, D. N., Billson, J., Loak, K., Benaroch. P. and Watts, C. (2003). Asparagine endopeptidase can initiate the removal of the MHC class II invariant chain chaperone. *Immunity* 18, 489–498

84 Riese, R. J., Wolf, P. R., Bromme, D., Natkin, L. R., Villadangos, J. A., Ploegh, H.L and Chapman, H. A. (1996). Essential role for cathepsin S in MHC class II-associated invariant chain processing and peptide loading. *Immunity* 4, 357–366.

85 Nakagawa, T., Roth, W., Wong, P., Nelson, A., Farr, A., Deussing, J., Villadangos, J. A., Ploegh, H., Peters, C. and Rudensky, A. Y. (1998). Cathepsin L: critical role in Ii degradation and CD4 T cell selection in the thymus. *Science* 280, 450–453.

86 Nakagawa, T. Y., Rudensky, A. Y. (1999). The role of lysosomal proteinases in MHC class II-mediated antigen processing and presentation. *Immunol. Rev.* 172, 121–129.

87 Shi, G. P., Bryant, R. A., Riese, R., Verhelst, S., Driessen, C., Li, Z., Bromme, D., Ploegh, H. L. and Chapman, H. A. (2000). Role for cathepsin F in invariant chain processing and major histocompatibility complex class II peptide loading by macrophages. *J. Exp. Med.* 191, 1177–1186.

88 Biroc, S. L., Gay, S., Hummel, K., Magill, C., Palmer, J. T., Spencer, D. R., Sa, S., Klaus, J. L., Michel, B. A., Rasnick, D. and Gay, R. E. (2001). Cysteine protease activity is up-regulated in inflamed ankle joints of rats with adjuvant-induced arthritis and decreases with in vivo administration of a vinyl sulfone cysteine protease inhibitor. *Arthritis Rheum.* 44, 703–711.

89 Podolin, P. L., Gapper, E. A., Bolognese, B. J., Ghalupowicz, D. G., Dong, X., Fox, .J. H., Gao, E. N., Johanson R. A., Katchur, S. and Marshall, L. A. (2001). Inhibition of cathepsin S blocks invariant chain processing and antigen-induced proliferation *in vitro* and reduces the severity of collagen-induced arthritis *in vivo*. *Inflamm. Res.* 50, S159.

90 Riese, R. J., Mitchell, R. N., Villadangos, J. A., Shi, G. P., Palmer, J. T., Karp, E. R., De Sanctis, G. T., Ploegh, H. L. and Chapman, H. A. (1998). Cathepsin S activity regulates antigen presentation and immunity. *J. Clin. Invest.* 101, 2351–2363.

91 Thurmond, R. L., Sun, S., Sehon, C. A., Baker, S. M., Cai, H., Gu, Y., Jiang, W., Riley, J. P., Williams, K. N., Edwards, J. P. and Karlsson, L. (2004). Identification of a potent and selective noncovalent cathepsin S inhibitor. *J. Pharmacol. Exp. Ther.* 308, 268–276.

92 Kropshofer, H., Vogt, A. B., Stern, L. J. and Hämmerling, G. J. (1995). Self-release of CLIP in peptide loading of HLA-DR molecules. *Science* 270, 1357–1359.

93 Kelly, A. P., Monaco, J. J., Cho, S. and Trowsdale, J. (1991). A new human HLA class II-related locus. *Nature* 353, 571–573.

94 Cho, S., Attaya, M. and Monaco, J. J. (1991). New class II-like genes in the murine MHC. *Nature* 353, 573–576.

95 Mosyak, L., Zaller, D. M. and Wiley, D.C. (1998). The structure of HLA-DM, the peptide exchange catalyst that loads antigen onto class II MHC molecules during antigen presentation. *Immunity* 9, 377–383.

96 Arndt, S.O., Vogt, A. B., Markovic-Plese, S., Martin, R., Moldenhauer, G., Wölpl, A., Sun, Y., Schadendorf, D., Hämmerling, G. J. and Kropshofer, H. (2000). Functional HLA-DM on the surface of B cells and immature dendritic cells. *EMBO J.* 19, 1241–1251.

97 Sanderson, F., Kleijmeer, M., Kelly, A., Verwoerd, D., Tulp, A., Neefjes, J., Geuze, H. J. and Trowsdale, J. (1994). Accumulation of HLA-DM, a regulator of antigen presentation, in MHC class II compartments. *Science* 266, 1566–1569.

98 Sloan V. S., Cameron, P., Porter, G., Gammon, M., Amaya, M., Mellins, E. and Zaller, D. M. (1995). Mediation by HLA-DM of dissociation of peptides from HLA-DR. *Nature* 375, 802–806.

99 Alfonso, C. and Karlsson, L. (2000). Nonclassical MHC class II molecules. *Annu. Rev. Immunol.* 18, 113–142.

100 Vogt, A. B., Kropshofer, H., Moldenhauer, G. and Hämmerling, G. J. (1996). Kinetic analysis of peptide loading onto HLA-DR molecules mediated by HLA-DM. *Proc. Natl. Acad. Sci. U.S.A.* 93, 9724–9729.

101 Kropshofer, H., Arndt, S. O., Moldenhauer, G., Hämmerling, G. J. and Vogt, A. B. (1997). HLA-DM acts as a molecular chaperone and rescues empty HLA-DR molecules at lysosomal pH. *Immunity* 6, 293–302.

102 Schafer, P. H., Green, J. M., Malapati, S., Gu, L. and Pierce, S. K. (1996). HLA-DM is present in one-fifth the amount of HLA.DR in the class II peptide loading compartment where it associates with leupeptin-induced peptide (LIP)-HLA-DR complexes. *J. Immunol.* 157, 5487–5495.

103 Denzin, L. K., Hammond, C. and Cresswell, P. (1996). HLA-DM interactions with intermediates in HLA-DR maturation and a role for HLA-DM in stabilizing empty HLA-DR molecules. *J. Exp. Med.* 184, 2153–2165.

104 Weber, D. A., Dao, C. T., Jun, J., Wigal, J. L. and Jensen, P. E. (2001). Transmembrane domain-mediated colocalization of HLA-DM and HLA-DR is required for optimal HLA-DM catalytic activity. *J. Immunol.* 167, 5167–5174.

105 Kropshofer, H., Spindeldreher, S., Rohn, T. A., Platania, N., Grygar, C., Daniel, N., Wolpl, A., Langen, H., Horejsi, V. and Vogt, A. B. (2002). Tetraspan microdomains distinct from lipid rafts enrich select peptide-MHC class II complexes. *Nat. Immunol.* 3, 61–68.

106 Doebele, R. C., Busch, R., Scott, H. M., Pashine, A. and Mellins, E. D. (2000). Determination of the HLA-DM interaction site on HLA-DR molecules. *Immunity* 13, 517–527.

107 Pashine, A., Busch, R., Belmares, M. P., Munning, J. N., Doebele, R. C., Buckingham, M., Nolan, G. P. and Mellins, E. D. (2003). Interaction of HLA-DR with an acidic face of HLA-DM disrupts sequence-dependent interactions with peptides. *Immunity* 19, 183–192.

108 Marin-Esteban, V., Falk, K. and Rotzschke, O. (2004). "Chemical analogues" of HLA-DM can induce a peptide-receptive state in HLA-DR molecules. *J. Biol. Chem.* 279, 50 684–50 690.

109 Verreck, F. A., Fargeas, C. A. and Hammerling, G. J. (2001). Conformational alterations during biosynthesis of HLA-DR3 molecules controlled by invariant chain and HLA-DM. *Eur. J. Immunol.* 31, 1029–1036.

110 Schmitt, L., Boniface, J. J., Davis, M. M. and McConnell, H. M. (1998). Kinetic isomers of a class II MHC-peptide complex. *Biochemistry* 37, 17 371–17 380.

111 Rabinowitz, J. D., Tate, K., Lee, C., Beeson, C. and McConnell, H. M. (1997). Specific T cell recognition of kinetic isomers in the binding of peptide to class II major histocompatibility complex. *Proc. Natl. Acad. Sci. U.S.A.* 94, 8702–8707.

112 He, X. L., Radu, C., Sidney, J., Sette, A., Ward, E. S. and Garcia, K. C. (2002). Structural snapshot of aberrant antigen presentation linked to autoimmunity: the immunodominant epitope of MBP complexed with I-Au. *Immunity* 17, 83–94.

113 Viret, C., He, X. and Janeway, C. A. Jr. (2003). Altered positive selection due to corecognition of floppy peptide/MHC II conformers supports an integrative model of thymic selection. *Proc. Natl. Acad. Sci. U.S.A.* 100, 5354–5359.

114 Pu, Z., Lovitch, S. B., Bikoff, E. K. and Unanue, E. R (2004). T cells distinguish MHC-peptide complexes formed in separate vesicles and edited by H2-DM. *Immunity* 20, 467–476.

115 Kropshofer, H., Hammerling, G. J. and Vogt, A. B. (1997). How HLA-DM edits the MHC class II peptide repertoire: survival of the fittest? *Immunol. Today* 18, 77–82.

116 Kropshofer, H., Vogt, A. B., Moldenhauer, G., Hammer, J., Blum, J. S. and Hammerling, G. J. (1996). Editing of the HLA-DR-peptide repertoire by HLA-DM. *EMBO J.* 15, 6144–6154.

117 van Ham, S. M., Grüneberg, U., Malcharek, G., Bröker, I., Melms, A. and Trowsdale, J. (1996). Human histocompatibility leukocyte antigen (HLA)-DM edits peptides presented by HLA-DR according to their ligand binding motifs. *J. Exp. Med.* 184, 2019–2024.

118 Weber, D. A., Evavold, B. D. and Jensen, P. E. (1996). Enhanced dissociation of HLA-DR-bound peptides in the presence of HLA-DM. *Science* 274, 618–620.

119 Raddrizzani, L., Bono, E., Vogt, A. B., Kropshofer, H., Gallazzi, F., Sturniolo, T., Hämmerling, G. J., Sinigaglia, F. and Hammer, J. (1999). Identification of destabilizing residues in HLA class II-selected bacteriophage display libraries edited by HLA-DM. *Eur. J. Immunol.* 29, 660–668.

120 Stratikos, E., Wiley, D. C. and Stern, L. J. (2004). Enhanced catalytic action of HLA-DM on the exchange of peptides lacking backbone hydrogen bonds between their N-terminal region and the MHC class II alpha-chain. *J. Immunol.* 172, 1109–1117.

121 Lovitch, S. B., Petzold, S. J. and Unanue, E. R. (2003). Cutting edge: H-2DM is responsible for the large differences in presentation among peptides selected by I-Ak during antigen processing. *J. Immunol.* 171, 2183–2186.

122 Hall, F. C., Rabinowitz, J. D., Busch, R., Visconti, K. C., Belmares, M., Patil, N. S., Cope, A. P., Patel, S., McConnell, H. M., Mellins, E. D. and Sonderstrup, G. (2002). Relationship between kinetic stability and immunogenicity of HLA-DR4/peptide complexes. *Eur. J. Immunol.* 32, 662–670.

123 Lich, J. D., Jayne, J. A., Zhou, D., Elliott, J. F. and Blum, J. S. (2003). Editing of an immunodominant epitope of glutamate decarboxylase by HLA-DM. *J. Immunol.* 171, 853–859.

124 Kropshofer, H., Hämmerling, G. J. and Vogt, A.B. (1999).The impact of the nonclassical MHC proteins HLA-DM and HLA-DO on loading of MHC class II molecules. *Immunol. Rev.* 172, 267–278.

125 Liljedahl, M., Kuwana, T., Fung-Leung, W.-P., Jackson, M. R., Peterson, P. A. and Karlsson, L. (1996). HLA-DO is a lysosomal resident which requires association with HLA-DM for efficient intracellular transport. *EMBO J.* 15, 4817–4824.

126 van Lith, M., van Ham, M., Griekspoor, A., Tjin, E., Verwoerd, D., Calafat, J., Janssen, H., Reits, E., Pastoors, L. and Neefjes, J. (2001). Regulation of MHC class II antigen presentation by sorting of recycling HLA-DM/DO and class II within the multivesicular body. *J. Immunol.* 167, 884–892.

127 Chen, X., Laur, O., Kambayashi, T., Li, S., Bray, R. A., Weber, D. A., Karlsson, L. and Jensen, P. E. (2002). Regulated expression of human histocompatibility leukocyte antigen (HLA)-DO during antigen-dependent and antigen-independent phases of B cell development. *J. Exp. Med.* 195, 1053–1062.

128 Kropshofer, H., Vogt, A. B., Thery, C., Armandola, E. A., Li, B.-C., Moldenhauer, G., Amigorena, S. and Hämmerling, G. J. (1998). A role for HLA-DO as a co-chaperone of HLA-DM in peptide loading of MHC class II molecules. *EMBO J.* 17, 2971–2981.

129 Hammond, C., Denzin, L. K., Pan, M., Griffith, J. M., Geuze, H. J. and Cresswell, P. (1998). The tetraspan protein CD82 is a resident of MHC class II compartments where it associates with HLA-DR, -DM, and -DO molecules. *J. Immunol.* 161, 3282–3291.

130 van Ham, M., van Lith, M., Lillemeier, B., Tjin, E., Gruneberg, U., Rahman, D., Pastoors, L., van Meijgaarden, K., Roucard, C., Trowsdale, J., Ottenhoff, T., Pappin, D. and Neefjes, J. (2000). Modulation of the major histocompatibility complex class II-associated peptide repertoire by human histocompatibility leukocyte antigen (HLA)-DO. *J. Exp. Med.* 191, 1127–1136.

131 van Ham, S. M., Tijn, E. P. M., Lillemeier, B. F., Grüneberg, U., van Meijgaarden, K. E., Pastoors, L., Verwoerd, D., Tulp, A., Canas, B., Rahman, D., Ottenhoff, T. H. M., Pappin, D. J. C., Trowsdale, J. and Neefjes, J. (1997). HLA-DO is a negative modulator of HLA-DM mediated MHC class II peptide loading. *Curr. Biol.* 7, 950–957.

132 Denzin, L. K., Sant'Angelo, D. B., Hammond, C., Surman, M. J. and Cresswell, P. (1997). Negative regulation by HLA-DO of MHC class II-restricted antigen processing. *Science* 278, 106–109.

133 Perraudeau, M., Taylor, P. R., Stauss, H. J., Lindstedt, R., Bygrave, A. E., Pappin, D. J., Ellmerich, S., Whitten, A., Rahman, D., Canas, B., Walport, M. J., Botto, M. and Altmann, D. M. (2000). Altered major histocompatibility complex class II peptide loading in H2-O-deficient mice. *Eur. J. Immunol.* 30, 2871–2880.

134 Brocke, P., Armandola, E., Garbi, N. and Hammerling, G. J. (2003). Down-modulation of antigen presentation by H2-O in B cell lines and primary B lymphocytes. *Eur. J. Immunol.* 33, 411–421.

135 Alfonso, C., Williams, G. S., Han, J. O., Westberg, J. A., Winqvist, O. and Karlsson, L. (2003). Analysis of H2-O influence on antigen presentation by B cells. *J. Immunol.* 171, 2331–2337.

136 Fallas, J. L., Tobin, H. M., Lou, O., Guo, D., Sant'Angelo, D. B. and Denzin, L. K. (2004). Ectopic expression of HLA-DO in mouse dendritic cells diminishes MHC class II antigen presentation. *J. Immunol.* 173, 1549–1560.

137 Glazier, K. S., Hake, S. B., Tobin, H. M., Chadburn, A., Schattner, E. J. and Denzin, L. K. (2002). Germinal center B cells regulate their capability to present antigen by modulation of HLA-DO. *J. Exp. Med.* 195, 1063–1069.

138 Boes, M., Cuvillier, A. and Ploegh, H. (2004). Membrane specializations and endosome maturation in dendritic cells and B cells. *Trends Cell Biol.* 14, 175–183.

139 Vogt, A. B., Spindeldreher, S. and Kropshofer, H. (2002). Clustering of MHC-peptide complexes prior to their engagement in the immunological synapse: lipid raft and tetraspan microdomains. *Immunol. Rev.* 189, 136–151.

140 Peters, P. J., Neefjes, J. J., Oorschot, V., Ploegh, H. L. and Geuze, H. J. (1991). Segregation of MHC class II molecules from MHC class I molecules in the Golgi complex for transport to lysosomal compartments. *Nature* 349, 669–676.

141 Escola, J.-M., Kleijmeer, M. J., Stoorvogel, W., Griffith, J. M., Yoshie, O. and Geuze, H. J. (1998). Selective enrichment of tetraspan proteins on the internal vesicles of multivesicular endosomes and on exosomes secreted by human B-lymphocytes. *J. Biol. Chem.* 273, 20 121–20 127.

142 Szollosi, J., Horejsi, V., Bene, L., Angelisova, P. and Damjanovich, S. (1996). Supramolecular complexes of MHC class I, MHC class II, CD20, and tetraspan molecules (CD53, CD81, and CD82) at the surface of a B cell line JY. *J. Immunol.* 157, 2939–2946.

143 Rubinstein, E., Le Naour, F., Lagaudriere-Gesbert, C., Billard, M., Conjeaud, H. and Boucheix, C. (1996). CD9, CD63, CD81, and CD82 are components of a surface tetraspan network connected to HLA-DR and VLA integrins. *Eur. J. Immunol.* 26, 2657–2665.

144 Tarrant, J. M., Robb, L., van Spriel, A. B. and Wright, M. D. (2003). Tetraspanins: molecular organisers of the leukocyte surface. *Trends Immunol.* 24, 610–617.

145 Kitadokoro, K., Bordo, D., Galli, G., Petracca, R., Falugi, F., Abrignani, S., Grandi, G. and Bolognesi, M. (2001). CD81 extracellular domain 3D structure: insight into the tetraspanin superfamily structural motifs. *EMBO J.* 20, 12–18.

146 Bromley, S. K., Iaboni, A., Davis, S. J., Whitty, A., Green, J. M., Shaw, A. S., Weiss, A. and Dustin, M. L. (2001). The immunological synapse and CD28-CD80 interactions. *Nat. Immunol.* 2, 1159–1166.

147 Mittelbrunn, M., Yanez-Mo, M., Sancho, D., Ursa, A. and Sanchez-Madrid, F. (2002). Dynamic redistribution of tetraspanin CD81 at the central zone of the immune synapse in both T lymphocytes and APC. *J. Immunol.* 169, 6691–6695.

148 Topilski, I., Harmelin, A., Flavell, R. A., Levo, Y. and Shachar, I. (2002). Preferential Th1 immune response in invariant chain-deficient mice. *J. Immunol.* 168, 1610–1617.

149 Chaturvedi, P., Hengeveld, R., Zechel, M. A., Lee-Chan, E. and Singh, B. (2000). The functional role of class II-associated invariant chain peptide (CLIP) in its ability to variably modulate immune responses. *Int. Immunol.* 12, 757–765.

150 Drbal, K., Angelisova, P., Rasmussen, A. M., Hilgert, I., Funderud, S. and Horejsi, V. (1999). The nature of the subset of MHC class II molecules carrying the CDw78 epitopes. *Int. Immunol.* 11, 491–498.

151 Irvine, D. J., Purbhoo, M. A., Krogsgaard, M. and Davis, M. M. (2002). Direct observation of ligand recognition by T cells. *Nature* 419, 845–849.

152 Li, Q. J., Dinner, A. R., Qi, S., Irvine, D. J., Huppa, J. B., Davis M. M. and Chakraborty, A. K. (2004). CD4 enhances T cell sensitivity to antigen by coordinating Lck accumulation at the immunological synapse. *Nat. Immunol.* 5, 791–799.

153 Wu, L. C., Tuot, D. S., Lyons, D. S., Garcia, K. C. and Davis, M.M. (2002). Two-step binding mechanism for T-cell receptor recognition of peptide MHC. *Nature* 418, 552–556.

154 Constant, S. L. and Bottomly, K. (1997). Induction of Th1 and Th2 CD4+ T cell responses: the alternative approaches. *Annu. Rev. Immunol.* 15, 297–322.

155 Anderson, H. A., Hiltbold, E. M. and Roche, P. A. (2000). Concentration of MHC class II molecules in lipid rafts facilitates antigen presentation. *Nat. Immunol.* 1, 156–162.

156 Hiltbold, E. M., Poloso, N. J. and Roche, P. A. (2003). MHC class II-peptide complexes and APC lipid rafts accumulate at the immunological synapse. *J. Immunol.* 170, 1329–1338.

157 Edidin, M. (2003). The state of lipid rafts: from model membranes to cells. *Annu. Rev. Biophys. Biomol. Struct.* 32, 257–283.

158 Karacsonyi, C., Knorr, R., Fulbier, A. and Lindner, R. (2004). Association of major histocompatibility complex II with cholesterol- and sphingolipid-rich membranes precedes peptide loading. *J. Biol. Chem.* 279, 34818–34826.

159 Hart, D. N. (1997). Dendritic cells: unique leukocyte populations which control the primary immune response. *Blood* 90, 3245–3287.

160 Kleijmeer, M. J., Morkowski, S., Griffith, J. M., Rudensky, A. Y. and Geuze, H. J. (1997). Major histocompatibility complex class II compartments in human and mouse B lymphoblasts represent conventional endocytic compartments. *J. Cell Biol.* 139, 639–649.

161 Kleijmeer, M., Ramm, G., Schuurhuis, D., Griffith, J., Rescigno, M., Ricciardi-Castagnoli, P., Rudensky, A. Y., Ossendorp, F., Melief, C. J., Stoorvogel, W. and Geuze, H. J. (2001). Reorganization of multivesicular bodies regulates MHC class II antigen presentation by dendritic cells. *J. Cell Biol.* 155, 53–63.

162 van Lith, M., van Ham, M., Griekspoor, A., Tjin, E., Verwoerd, D., Calafat, J., Janssen, H., Reits, E., Pastoors, L. and Neefjes, J. (2001). Regulation of MHC class II antigen presentation by sorting of recycling HLA-DM/DO and class II within the multivesicular body. *J. Immunol.* 167, 884–892.

163 Murk, J. L., Humbel, B. M., Ziese, U., Griffith, J. M., Posthuma, G., Slot, J. W., Koster, A. J., Verkleij, A. J., Geuze, H. J. and Kleijmeer, M. J. (2003). Endosomal compartmentalization in three dimensions: implications for membrane fusion. *Proc. Natl. Acad. Sci. U.S.A.* 100, 13332–13337.

164 Amigorena, S., Drake, J. R., Webster, P. and Mellman, I. (1994). Transient accumulation of new class II MHC molecules in a novel endocytic compartment in B lymphocytes. *Nature* 369, 113–120.

165 Turley, S. J., Inaba, K., Garrett, W. S., Ebersold, M., Unternaehrer, J., Steinman, R. M. and Mellman, I. (2000). Transport of peptide-MHC class II complexes in developing dendritic cells. *Science* 288, 522–527.

166 Chow, A., Toomre, D., Garrett, W. and Mellman, I. (2002). Dendritic cell maturation triggers retrograde MHC class II transport from lysosomes to the plasma membrane. *Nature* 418, 988–994.

167. Boes, M. and Ploegh, H. L. (2004). Translating cell biology in vitro to immunity in vivo. *Nature* 430, 264–271.
168. Boes, M., Bertho, N., Cerny, J., Op den Brouw, M., Kirchhausen, T. and Ploegh, H. (2003). T cells induce extended class II MHC compartments in dendritic cells in a Toll-like receptor-dependent manner. *J. Immunol.* 171, 4081–4088.
169. Montoya, M. C., Sancho, D., Bonello, G., Collette, Y., Langlet, C., He, H. T., Aparicio, P., Alcover, A., Olive, D. and Sanchez-Madrid, F. (2002). Role of ICAM-3 in the initial interaction of T lymphocytes and APCs. *Nat. Immunol.* 3, 159–168.
170. Bertho, N., Cerny, J., Kim, Y. M., Fiebiger, E., Ploegh, H. and Boes, M. (2003). Requirements for T cell-polarized tubulation of class II+ compartments in dendritic cells. *J. Immunol.* 171, 5689–5696.
171. Hmama, Z., Gabathuler, R., Jefferies, W. A., de Jong, G. and Reiner, N. E. (1998). Attenuation of HLA-DR expression by mononuclear phagocytes infected with Mycobacterium tuberculosis is related to intracellular sequestration of immature class II heterodimers. *J. Immunol.* 161, 4882–4893.
172. Flynn, J. L., Chan, J., Triebold, K. J., Dalton, D. K., Stewart, T. A. and Bloom, B. R. (1993). An essential role for interferon gamma in resistance to Mycobacterium tuberculosis infection. *J. Exp. Med.* 178, 2249–2254.
173. Pai, R. K., Convery, M., Hamilton, T. A., Boom, W. H. and Harding, C. V. (2003). Inhibition of IFN-gamma-induced class II transactivator expression by a 19-kDa lipoprotein from Mycobacterium tuberculosis: a potential mechanism for immune evasion. *J. Immunol.* 171, 175–184.
174. Chu, R. S., Askew, D., Noss, E. H., Tobian, A., Krieg, A. M. and Harding, C. V. (1999). CpG oligodeoxynucleotides down-regulate macrophage class II MHC antigen processing. *J. Immunol.* 163, 1188–1194.
175. Noss, E. H., Pai, R. K., Sellati, T. J., Radolf, J. D., Belisle, J., Golenbock, D. T., Boom, W. H. and Harding, C. V. (2001). Toll-like receptor 2-dependent inhibition of macrophage class II MHC expression and antigen processing by 19-kDa lipoprotein of Mycobacterium tuberculosis. *J. Immunol.* 167, 910–918.
176. Gehring, A. J., Dobos, K. M., Belisle, J. T., Harding, C. V. and Boom, W. H. (2004). Mycobacterium tuberculosis LprG (Rv1411c). A novel TLR-2 ligand that inhibits human macrophage class II MHC antigen processing. *J. Immunol.* 173, 2660–2668.
177. Fortune, S. M., Solache, A., Jaeger, A., Hill, P. J., Belisle, J. T., Bloom, B. R., Rubin, E. J. and Ernst, J. D. (2004). Mycobacterium tuberculosis inhibits macrophage responses to IFN-gamma through myeloid differentiation factor 88-dependent and -independent mechanisms. *J. Immunol.* 172, 6272–6280.
178. Geijtenbeek, T. B., Van Vliet, S. J., Koppel, E. A., Sanchez-Hernandez, M., Vandenbroucke-Grauls, C. M., Appelmelk, B. and Van Kooyk, Y. (2003). Mycobacteria target DC-SIGN to suppress dendritic cell function. *J. Exp. Med.* 197, 7–17.
179. Zhong, G., Fan, T. and Liu, L. (1999). Chlamydia inhibits interferon gamma-inducible major histocompatibility complex class II expression by degradation of upstream stimulatory factor 1. *J. Exp. Med.* 189, 1931–1938.
180. De Souza Leao, S., Lang, T., Prina, E., Hellio, R. and Antoine, J. C. (1995). Intracellular Leishmania amazonensis amastigotes internalize and degrade MHC class II molecules of their host cells. *J. Cell Sci.* 108, 3219–3231.
181. Moutaftsi, M., Mehl, A. M., Borysiewicz, L. K. and Tabi, Z. (2002). Human cytomegalovirus inhibits maturation and impairs function of monocyte-derived dendritic cells. *Blood* 99, 2913–2921.
182. Grigoleit, U., Riegler, S., Einsele, H., Laib Sampaio, K., Jahn, G., Hebart, H., Brossart, P., Frank, F. and Sinzger, C. (2002). Human cytomegalovirus induces a direct inhibitory effect on antigen presentation by monocyte-derived immature dendritic cells. *Br. J. Haematol.* 119, 189–198.
183. Johnson, D. C. and Hegde, N. R. (2002). Inhibition of the MHC class II antigen

presentation pathway by human cytomegalovirus. *Curr. Top. Microbiol. Immunol.* 269, 101–115.

184 Tomazin, R., Boname, J., Hegde, N. R., Lewinsohn, D. M., Altschuler, Y., Jones, T. R., Cresswell, P., Nelson, J. A., Riddell, S. R. and Johnson, D. C. (1999). Cytomegalovirus US2 destroys two components of the MHC class II pathway, preventing recognition by CD4+ T cells. *Nat. Med.* 5, 1039–1043.

185 Pollara, G., Speidel, K., Samady, L., Rajpopat, M., McGrath, Y., Ledermann, J., Coffin, R. S., Katz, D. R. and Chain, B. (2003). Herpes simplex virus infection of dendritic cells: balance among activation, inhibition, and immunity. *J. Infect. Dis.* 187, 165–178.

186 Neumann, J., Eis-Hubinger, A. M. and Koch, N. (2003). Herpes simplex virus type 1 targets the MHC class II processing pathway for immune evasion. *J. Immunol.* 171, 3075–3083.

187 Stumptner-Cuvelette, P., Morchoisne, S., Dugast, M., Le Gall, S., Raposo, G., Schwartz, O. and Benaroch, P. (2001). HIV-1 Nef impairs MHC class II antigen presentation and surface expression. *Proc. Natl. Acad. Sci. U.S.A.* 98, 12 144–12 149.

188 Ressing, M. E., van Leeuwen, D., Verreck, F. A., Gomez, R., Heemskerk, B., Toebes, M., Mullen, M. M., Jardetzky, T. S., Longnecker, R., Schilham, M. W., Ottenhoff, T. H., Neefjes, J., Schumacher, T. N., Hutt-Fletcher, L. M. and Wiertz, E. J. (2003). Interference with T cell receptor-HLA-DR interactions by Epstein-Barr virus gp42 results in reduced T helper cell recognition. *Proc. Natl. Acad. Sci. U.S.A.* 100, 11 583–11 588.

189 Molinari, M., Salio, M., Galli, C., Norais, N., Rappuoli, R., Lanzavecchia, A. and Montecucco, C. (1998). Selective inhibition of Ii-dependent antigen presentation by Helicobacter pylori toxin VacA. *J. Exp. Med.* 187, 135–140.

5
Antigen Processing and Presentation by CD1 Family Proteins

Steven A. Porcelli and D. Branch Moody

5.1
Introduction

T cells play in key role in cellular immune responses generated during infection, autoimmunity, tumor recognition and tissue transplantation. T cells are activated by antigens that are generated within cells and then bind to antigen presenting molecules for export to the surface of antigen presenting cells (APCs). For decades, it was thought that peptide antigens bound to MHC class I and II were the only structurally diverse antigens recognized by T cells. The discovery of CD1 antigen presenting molecules and their ability to display lipids to T cells greatly expands the range of compounds that can function as antigens for T cells to include lipids, glycolipids, lipopeptides and other small hydrophobic molecules. Despite the relatively recent discovery of the CD1 antigen presentation system, identification of T cells that recognize mammalian, bacterial and synthetic lipid antigens now suggest that the normal function of CD1-restricted T cells is to survey cells for changes in their lipid content that occur in response to infection and other stimuli. Here we summarize the molecular and cellular features of the CD1 antigen presentation system, and outline current views of how T cell recognition of CD1-presented lipid antigens may influence infectious, neoplastic and autoimmune diseases.

5.2
CD1 Genes and Classification of CD1 Proteins

Human CD1 proteins are encoded on chromosome 1, and thus map to a genetic locus that is unlinked to the human major histocompatibility complex (MHC) on chromosome 6. The CD1 locus consists of five homologous genes, CD1A, CD1B, CD1C, CD1D and CD1E [1]. Four of the genes encoded at this locus are expressed on the surface of antigen presenting cells as transmembrane glycoproteins, and these are designated CD1a, CD1b, CD1c and CD1d. The CD1E gene encodes the CD1e protein and is expressed as translational products of several differentially

Antigen Presenting Cells: From Mechanisms to Drug Development. Edited by H. Kropshofer and A. B. Vogt
Copyright © 2005 WILEY-VCH Verlag GmbH & Co. KGaA, Weinheim
ISBN: 3-527-31108-4

spliced mRNA transcripts, which include a soluble form and a transmembrane form that is retained intracellularly. However, CD1e is not known to be expressed at the cell surface, and at present it has no established role in antigen presentation [2]. Shortly after the molecular cloning of the human CD1 locus, the predicted products of the five human CD1 genes were divided into two groups based on their sequence homologies [3]. Group 1 proteins, which include CD1a, CD1b and CD1c, are most homologous to one another and are least homologous to CD1d, the sole group 2 isoform. The CD1e protein is not unambiguously classified according to homology as group 1 versus group 2, and its unique localization within cells suggest that it constitutes its own category (i.e., group 3 CD1).

Although there is some overlap in the functions of groups 1 and 2 CD1 proteins, emerging information about the separate functions of individual CD1 isoforms in immune responses has generally reinforced this classification scheme. For example, CD1a, CD1b and CD1c are coordinately expressed on dendritic cells and function in presentation of foreign lipids in host defense [4]. The group 2 protein, CD1d, mediates activation of a unique population of innate-like T lymphocytes known as NKT cells [5]. As detailed below, NKT cells play significant roles in a wide variety of immune reactions, including inflammation, autoimmunity and host defense.

CD1 orthologs have been found in all mammalian species examined to date, including mice, rats, guinea pigs, rabbits, sheep, dogs and monkeys [6–13]. CD1 proteins in non-human species are named according to the human isoform with which they show the highest level of homology. For example, guinea pigs express at least 7 CD1 proteins, including three orthologs of human CD1c, which are named CD1c1, CD1c2 and CD1c3. Muroid rodents (mice and rats) express orthologs only of CD1d, which stands in contrast to all other mammals, which have larger CD1 gene families. In humans, each of the human CD1 isoforms is differentially expressed on B cells, DCs and LCs. Detailed studies of each of the human isoforms have shown that these have different routes of intracellular trafficking, and contain antigen binding grooves that differ in their molecular specificity for antigen. This information, along with the diversification and retention of multiple CD1 genes in most mammalian species, supports the argument that each human CD1 isoform has a somewhat distinct function in the immune response and that each of these functions provides a survival advantage.

5.3
Structure and Biosynthesis of CD1 Proteins

It was appreciated from the very earliest studies of CD1 proteins that they are strikingly MHC class I-like in their overall protein structure [14]. In general, CD1 molecules are expressed as heterodimeric proteins composed of the CD1 heavy chain noncovalently associated with β2-microglobulin (β2m). CD1 heavy chains are glycoproteins with short cytoplasmic tails, a transmembrane region and three extracellular domains that, as for classical MHC class I molecules, are designated

α1, α2 and α3. In some cases, CD1d proteins have been detected as heavy chains that are not bound to β2m, though immunological functions of free CD1d heavy chains are not well understood [15, 16]. Because of the association with β2m and the obvious similarities of CD1 domain organization to classical MHC class I, CD1 proteins are often referred to as MHC class I-like molecules. However, based on their levels of amino acid homology, CD1, MHC class I and MHC class II are equally related to one another [17]. These three classes of antigen presenting molecules are thus believed to have diverged from a primordial ancestral protein family at a similar point in evolutionary time, and can be considered three distinct lineages of antigen presenting molecules.

Despite the strong overall structural similarity of CD1 to MHC class I proteins, the functions of CD1 proteins can in certain ways be more readily compared to MHC class II than class I. For example, in contrast to the ubiquitous expression of classical MHC class I molecules on nearly all cells of the body, both CD1 and MHC class II are typically expressed only on certain cell types, typically specialized antigen presenting cells. In addition, both are mainly focused on the presentation of antigens that gain access to endosomes, as opposed to MHC class I molecules, which mainly present antigens that gain access to the cytosol.

All human CD1 sequences contain a hydrophobic leader peptide at the N-terminus of the newly synthesized heavy chain, which signals co-translational insertion of CD1 heavy chains into membranes such that the α1, α2 and α3 domains are in the lumen of the endoplasmic reticulum and a short C-terminal tail protrudes into the cytoplasmic compartment. CD1b and CD1d proteins, and likely all other CD1 isoforms, associate with the protein folding chaperones calnexin and calreticulin in the ER [18–20]. There has been considerable speculation that CD1 proteins may initially fold around one or more self ligands available in the ER to stabilize the markedly hydrophobic ligand binding groove formed within the α1 and α2 domains of CD1 heavy chains. In support of this possibility, there is evidence that lipids can promote proper folding of CD1 proteins in vitro [21]. Also, endogenous phospholipids appear to be loaded onto CD1 within the secretory pathway, and such lipids might play a role in stabilizing and blocking the CD1 groove prior to encounter with exogenous lipids during trafficking through later stages of the secretory or endocytic compartments (Figure 5.2 below) [22–24].

A second event in ER assembly involves the association of CD1 heavy chains with β2m, which is generally necessary to promote cell surface expression of CD1 proteins at high levels [18, 25]. The functional importance of β2m association with CD1d has been shown in studies of β2m knockout mice, which have reduced development of CD1d-restricted NK T cells and decreased efficiency of CD1-restricted T cell activation [5, 26]. After exit from the ER, CD1 proteins rapidly traffic through the Golgi compartment, where they undergo processing and maturation of their N-linked glycans before appearing at the cell surface [27, 28].

5.3.1
Three-dimensional (3D) Structures of CD1 Proteins

X-ray crystal structures of CD1 proteins bound to lipid ligands now provide detailed insights into the 3D structure of CD1 proteins and helps to explain how CD1 heavy chains can simultaneously bind to β2m, lipid antigens and T cell antigen receptors (Figure 5.1) [29–32]. The most membrane-proximal domain of the CD1 heavy chain, α3, associates with β2m, and the 3D structure of these two domains is largely conserved among CD1 structures and with MHC class I. The α1 and α2 domains of the CD1 heavy function together as a single unit, the α1-α2 superdomain, which delimits a hollow cavity that functions as the CD1 antigen binding groove (Figure 5.1). The groove is situated above six β-strands that form a β-sheet platform and between two anti-parallel α-helices that form the lateral walls of the groove. The inner surface of the ligand binding grooves of all CD1 molecules studied thus far are lined almost exclusively by non-polar amino acids, so that the surface of the protein that contacts the ligands bound in the groove is hydrophobic and therefore well adapted for interactions with lipids.

CD1 antigen-binding grooves are larger and sequestered more deeply within the α1-α2 superdomain than those present in MHC class I and II. All known crystal structures of CD1 proteins show that the α1 and α2 domain helices are 4 to 6 Å higher above the β-sheet platform than are the two analogous antiparallel α-helices present in structures of MHC class I and II proteins. This basic difference in groove architecture results in part from bulky amino acids at positions 18, 40 and 49 of CD1 heavy chains [29]. The positioning of large amino acids just under the α1 helix of CD1 serves as a scaffold to elevate the α-helical regions of CD1 high above the floor of the groove. Thus, CD1 grooves can have almost twice the volume (~1300–2200 Å^3) than those seen in MHC class I proteins (~1000–1200 Å^3). Although 3D structures of all CD1 proteins are not yet known, bulky amino acids are present at positions 18, 40 and 49 in human and mouse CD1 proteins, suggesting that the deep grooves are a general structural feature that characterizes CD1 proteins (Figure 5.1).

Whereas peptides sit nearly atop MHC proteins, lipids are inserted deeply within CD1. The opening to the CD1 antigen binding groove can be thought of as a gap between the α1 and α2 helices (Figure 5.1). Compared to the approximately 11 Å wide gap in MHC class I and II grooves, the gap in CD1 proteins is narrower, ca. 7 Å. Furthermore, CD1 proteins studied to date have interdomain contacts between the α1 and α2 helices, which act like a zipper to cover part of the superior aspect of the groove. Thus, the narrow entrance to the groove can be thought of as a portal that connects the outer solvent-exposed face of the protein with the inner hydrophobic interior of the CD1 groove, which is sequestered from the solvent-exposed surface of the protein.

These structural features of CD1 explain the general molecular mechanism by which CD1 proteins bind amphipathic lipids and present them to T cells, as illustrated in the structure of CD1b bound to phosphatidylinositol (Figure 5.1). The alkyl chains of lipid and glycolipid antigens are inserted deeply within the CD1

5.3 Structure and Biosynthesis of CD1 Proteins

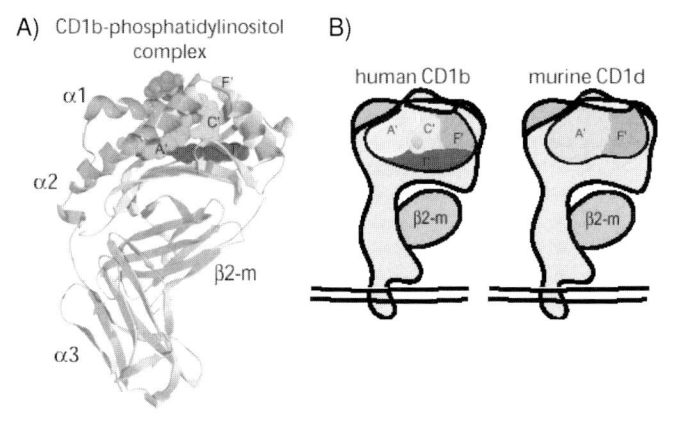

Figure 5.1 Structures of CD1 antigen binding grooves. Antigen binding grooves of CD1 proteins vary in size, architecture and ligand specificity. (A) The crystal structure of human CD1b bound to phosphatidylinositol shows that the two alkyl chains of the antigen are inserted into the A' and C' pockets, and the inositol moiety protrudes from the groove so that it can contact T cell antigen receptors [30]. (B) The grooves of CD1 proteins are composed of two or more discrete pockets. A schematic of the CD1b groove shows how it is divided into four pockets (named A', C', F' and T'), and illustrates the two portals located at the top of the F' pocket and the bottom of the C' pocket. The smaller mCD1d groove has two pockets and only one portal. (This figure also appears with the color plates.)

groove, where they make extensive Van der Waal's interactions with the non-polar side chains of the hydrophobic amino acids that line the interior of the groove. The more hydrophilic portions of the antigen, which are typically composed of carbohydrate, polyalcohol, peptide, phosphate or sulfate, protrude through the narrow portal present between the $\alpha 1$ and $\alpha 2$ helices. This mode of binding positions the hydrophilic caps of antigen to lie on the outer surface of the CD1 heavy chain, where they can directly contact T cell antigen receptors (TCRs) [30]. The interaction between TCRs and the hydrophilic caps of antigens has not been directly observed in crystal structures. However, the binding of TCRs to CD1-lipid complexes, and detailed studies of how antigen structure affects TCR-mediated recognition, provide strong evidence that the overall mechanism of CD1-restricted T cell activation involves the contact of TCR with a lipid ligand bound in the CD1 protein [33–38].

5.3.2
Molecular Features of CD1–Lipid Complexes

CD1 proteins show very low levels of allelic polymorphism, so that CD1 antigen binding groove structure does not vary significantly among individuals in a population. Yet, despite the nearly invariant structures of CD1 antigen binding grooves, many types of lipids are presented by CD1 proteins (Figure 5.2). Whereas the natural ligands for MHC class I and II grooves are essentially peptides or glycopeptides ranging from 8 to 22 amino acids in length, no such generic description

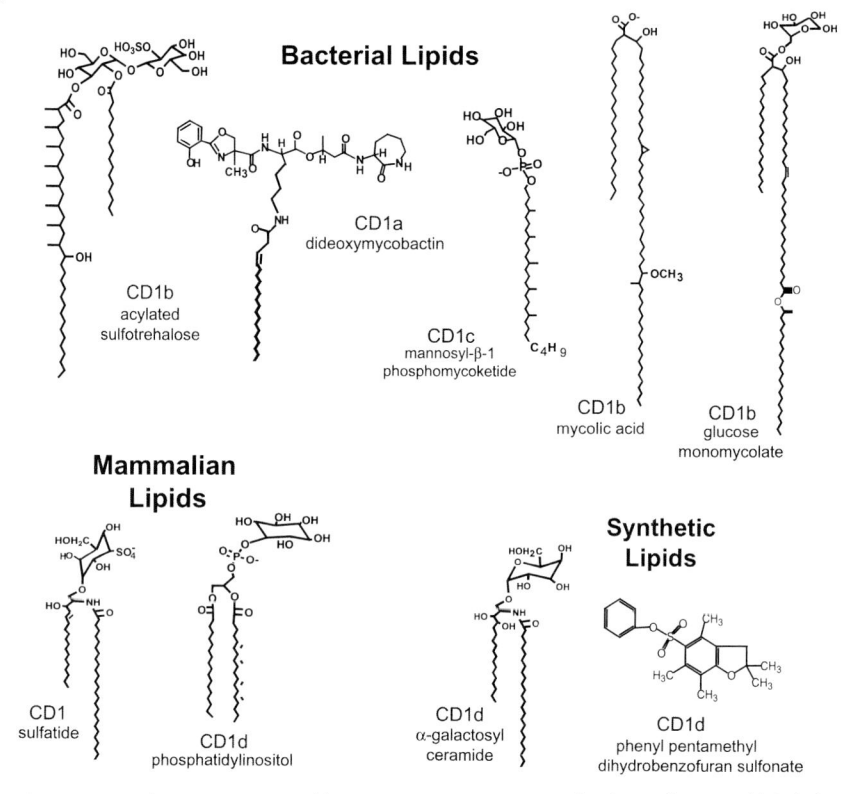

Figure 5.2 Lipid antigens presented by CD1. CD1a, CD1b, CD1c and CD1d proteins present various bacterial, mammalian and synthetic lipids to T cells. The structures of representative lipids are shown and labeled according to the CD1 isoform that mediates T cell activation, except for sulfatides, which can be presented by multiple CD1 isoforms.

could summarize the range of ligands that bind to CD1 molecules and can be presented by them for T cell recognition. Even a single CD1 isoform, such as CD1b, can present structurally diverse lipids that vary greatly in the size and shape, including mycolates, phosphatidylinositols, gangliosides, sulfatides and acyl sulfotrehaloses. This raises basic questions about how CD1 antigen binding grooves of nearly invariant structure could bind and present such chemically diverse lipid structures.

Much of the energy of binding of lipids with CD1 proteins comes from Van der Waal's interactions between the lipid anchors and the hydrophobic surface of the interior of the CD1 groove cavity. For example, CD1a uses four hydrogen bonds, no ionic interactions and 112 van der Waal's interactions to capture a bound sulfatide glycolipid ligand [32]. These hydrophobic interactions occur largely between the bland, repetitive structures of alkyl chains with the hydrophobic surface that lines the interior of the groove. Such interactions do not generally require precise positioning of any particular methylene unit at any point on the hydrophobic surface, in contrast to the hydrogen bond networks that anchor peptide side chains very precisely within the more shallow pockets of MHC class I and II. For CD1, binding is favored when the total number of hydrophobic interactions is increased, as would be the case when the lipid ligand fully occupies the groove. Further, openings in the base or side of the groove (the F' and C' portals; see below) may provide CD1 proteins greater flexibility to bind lipids that are variable in length and exceed the capacity of the groove by allowing antigens with long lipids tails to protrude from the interior of the groove, much as the open ends of the MHC class II groove allows the ragged ends of longer peptides to protrude [30, 39]. Thus, the CD1 system can use a relatively small number of proteins to capture many antigens by employing a series of less specific binding interactions, and thereby present a wide diversity of chemical classes of lipids and other hydrophobic small molecules.

5.3.3
CD1 Pockets and Portals

The molecular configuration of CD1 antigen binding grooves can be described by dividing them into either two or four contiguous antigen binding pockets. All CD1 structures studied to date have two pockets, named A' and F' following their initial description in the murine CD1d crystal structure [29]. The large CD1b groove contains these two pockets and two additional pockets, designated C' and T' (Figure 5.1) [30]. The A', C' and F' pockets derive their names from their side by side locations that correspond to the A, C and F pockets of MHC class I. The fourth pocket was named T' because it functions as a "tunnel" running along the bottom of the CD1b groove (Figure 5.1). In addition, CD1 proteins have either one or two portals, which are narrow gaps that connect the outer, solvent exposed surface of the protein to the inner, antigen binding surface of the groove. All CD1 proteins studied so far have a portal connecting the TCR contact surface to the groove, which is created by the ca. 7 Å wide gap between the $\alpha1$ and $\alpha2$ helices. This portal is known as the F' portal because it is located just above the F' pocket. CD1b has a second portal, which exits below the $\alpha2$ helix and connects the distal portion of the C' pocket to the outer surface $\alpha1$-$\alpha2$ superdomain. This C' portal is so named because it likely functions to guide long alkyl chains from the interior of the C' pocket to the outer surface of CD1 at a point that is distant from the TCR recognition surface (Figure 5.1).

The A' pockets of human and mouse CD1 proteins are buried within the $\alpha1$-$\alpha2$ superdomain, and they do not directly communicate with the outer surface of CD1. Thus, the central function of A' pockets is to anchor the alkyl moieties of lipid antigens within the groove. The shape of the A' pockets of human CD1 proteins is influenced by bulky amino acids that protrude from the roof and floor towards the center of the pocket. The resulting partial (mouse CD1d) or complete (human CD1a and CD1b) connection between the roof and floor forms a vertical pole, which transects the A' pocket. The key difference among A' pockets in various CD1 isoforms is the degree to which lipid ligands can wrap around the A' pole. For example, large lipids can wrap 360° around the pole in the A′ pocket of hCD1b or mCD1d and extend into other pockets. For this reason, CD1b has been referred to as a maze for alkyl chains because the largely interconnected pockets allow many different ways for lipids to fit within the groove [30, 31]. This contrasts with the A′ pocket of CD1a, which only partially encircles the pole and then abruptly terminates, so that its distal end is not connected to any other pocket. Thus, the A′ pocket of CD1a is an isolated, narrow tube that can accommodate antigens with lipid tails of a discrete length, leading to the description of the A′ pocket of CD1a as a molecular ruler for alkyl chains [32].

The antigen binding grooves in different CD1 isoforms differ in their overall volume, and therefore each is likely to show a preference for binding lipid antigens of differing sizes. For example, structural analysis of crystallized CD1 proteins estimate that the volumes of human CD1a, murine CD1d and human CD1b grooves are 1300, 1650, and 2200 Å^3, respectively [29, 30, 32]. The larger volume of the CD1b groove is accounted for by two additional pockets, C′ and T′, which are not yet known to exist in other CD1 proteins (Figure 5.1). The C′ pocket lies between the A′ and F′ pockets and guides an alkyl chain from the F′ portal to the C′ portal, located at the bottom of the groove, as depicted in the CD1b-phopshatidylinositol structure (Figure 5.1). The T′ tunnel of CD1b is at the bottom of the CD1b groove, where it connects the A′ and C′ pockets to form the A′T′F′ superchannel, which can bind particularly long alkyl chains. The larger volume of the CD1b groove, along with the C′ portal (which may allow particularly large lipids to escape from the interior of the groove), suggests that CD1b is specialized to bind antigens with particularly long alkyl chains. For example, a glucose monomycolate antigen with a C_{56} lipid moiety can fit within the CD1b groove, whereas a dideoxymycobactin lipopeptide with a C_{20} lipid can completely fill the CD1a groove (Figure 5.2) [31, 40].

5.4
Foreign Lipid Antigens Presented by Group 1 CD1

Human, mouse and guinea pig CD1 proteins have now been shown to present various different classes of lipid antigens to T cells (Figure 5.2). Antigens presented by the group 1 CD1 system have been found among mammalian cellular lipids such as sphingolipids [41–44], as well as specialized types of bacterial lipids, including mycolates [38, 45], lipoarabinomannan [46], mannosyl phosphomycoketides [47], dideoxymycobactins [40] and acylated sulfotrehaloses [48]. With increasing appreciation of the structural diversity of antigens presented by the CD1 system, it appears that the function of CD1-restricted T cells involves a survey of cells for alterations in the composition or structure of both endogenous self lipids and bacterial lipids introduced into cells as the result of infection. As outlined in Figure 5.2, three general types of lipids are presented by the CD1 system, which are self lipids derived from mammalian cells, foreign lipids from bacterial cell walls and synthetic lipids. These three classes mirror the three proposed functions of CD1-restricted T cells: host response to bacterial infection, immunoregulation of other immune effector cells through endogenously produced lipids, and modulation of immune responses using synthetic lipids as pharmacologic agents.

The antigen presenting function of CD1 proteins was first discovered through the use of human T cell clones that were responsive to lipid components of mycobacterial cell walls [4, 45, 46, 49] These early studies led to the identification of the first microbial lipid antigens presented by the group 1 CD1 system. These were *Mycobacterium tuberculosis* mycolic acids, which are α-branched, β-hydroxy fatty acids with unusually long alkyl chains (C_{70-86}) (Figure 5.2) [45]. Mycolic acids are so-called because they are only produced by mycobacteria and related species. These foreign lipids can be readily distinguished from all types of mammalian fatty acids based on their chemical structures, as mammalian fatty acids are much shorter (C_{12-22}) and lack α-branches and β-hydroxyl groups. Studies of the molecular specificity of CD1b-mediated T cell activation by a mycolyl lipid, known as glucose monomycolate, showed that these chemical features, which define these lipids as foreign, are also crucial for control of their recognition by T cells [38]. Other bacterial lipids presented by this system, such as lipoarabinomannan (LAM), mannosyl phosphomycoketide, didedeoxymycobactins and acylsulfotrehaloses, also differ structurally from all known mammalian lipids and can thus be considered foreign in the classical sense of the term [40, 46–48].

The display of foreign lipids bound to CD1 proteins on the surface of APCs that are either directly infected or near a site of infection is the molecular basis of the emerging hypothesis that CD1-restricted T cells play a significant role in host responses to infection [50]. Analysis of human T cells restricted by CD1a, CD1b and CD1c in vitro has found that these cells can inhibit growth of mycobacteria in cells, express anti-microbial proteins such as granulysin and secrete γ-interferon, which is an important Th1 effector cytokine during mycobacterial infection [51, 52]. Also, studies of clinical specimens have shown that CD1-restricted, lipid-reactive T cells are selectively expanded in the bloodstream of tuberculosis patients as compared to uninfected control subjects [47, 48, 53]. These studies suggest that human T cells recognize and respond to foreign lipids as part of the protective host response to mycobacterial infection.

5.5
Self Lipid Antigens Presented by CD1

Many examples have been reported of human and murine CD1-restricted T cells that respond to cells expressing CD1 proteins in the absence of infection or addition of other exogenous antigens. This apparent CD1-restricted autoreactivity is thought to reflect the presentation of self antigens, presumably lipid in nature, as a normal part of CD1 function [54, 55]. More recent studies have directly shown that CD1 proteins bind to endogenous self lipids and under certain circumstances can activate T cells. For example, crystal structures of human CD1a and CD1b proteins show how these human proteins bind to two different self lipids, sulfatide and phosphatidylinositol (Figure 5.1) [30, 32]. In addition, phosphatidylinositol-containing glycolipids have been eluted from cellular CD1d proteins, suggesting that such self lipids might play a role as non-antigenic chaperones that occupy the antigen-binding groove until more antigenic lipids can be loaded, analogous to the function of the MHC class II-associated invariant chain (Ii) in stabilizing and facilitating peptide loading of newly synthesized MHC class II molecules [22–24]. However, in other systems mammalian glycolipids, including phosphatidylinositol, gangliosides or sulfatides, lead to CD1-mediated T cell activation when they are applied to CD1-expressing antigen presenting cells at high concentrations [41, 43, 44, 56]. These studies show that mammalian lipids of normal structure can, under some circumstances, be sufficient to activate CD1-restricted T cells, raising the possibility that self-lipid antigens could mediate autoreactive T cell responses, including those that lead to autoimmune diseases.

5.6
Group 2 CD1-restricted T Cells

It is now well established that CD1d controls the development and functions of a subset of T lymphocytes known as Natural Killer T cells (NKT cells), which bear this name because they were initially defined by their coexpression of both T cell and NK cell markers (e.g., TCR$\alpha\beta$ and NK1.1) [57, 58]. A major population of cells in this phenotypic group undergo positive thymic selection by recognition of CD1d, which they recognize using antigen receptors consisting of an invariant T cell receptor (TCR) α chain (Vα14 rearranged to Jα18 in mice, and a strikingly homologous Vα24-Jα18 rearrangement in humans) [59]. These invariant TCRα chains pair with heterogeneous TCRβ chains, although these show a marked skewing of Vβ gene usage (especially Vβ8 in mice, and its closely homologous Vβ11 in humans) [59–61]. This population of cells, often referred to as "invariant" or "classic" CD1d-restricted NKT cells, is believed to be highly conserved in number, distribution and function in most mammals, including mice and humans [62, 63].

Many studies have pointed to a role for NKT cells as potent regulatory T cells that have the capacity to both initiate and shut down a wide variety of immune responses [58, 64]. For example, activation of CD1d-restricted NKT cells occurs early in various infectious disease models in mice, and such activation can lead to a so-called "adjuvant cascade" that reinforces innate immunity and promotes subsequent adaptive immunity [65, 66]. In addition, NKT cells have been implicated in immune surveillance for tumors in mice, as their deletion can promote tumor growth [67–71]. However, in the absence of a microbial or neoplastic trigger, many studies now suggest that the dominant effect of NKT cells may be to regulate the adaptive immune response in a way that prevents harmful autoimmunity. The evidence for this is particularly strong from mouse models of autoimmune disease, including experimental allergic encephalomyelitis (EAE) and type 1 diabetes mellitus [72–76]. For example, NOD mice, which spontaneously develop autoimmune diabetes, have fewer NKT cells than normal mice and also show a pronounced defect in the capacity of these cells to produce anti-inflammatory cytokines such as IL-4 [77, 78]. NOD mice that carry a homozygous CD1 knockout allele, which entirely lack NKT cells, show acceleration of diabetes development [79, 80]. Remarkably, similar NKT cell defects have been found in some studies of humans with type 1 diabetes and deficiencies of NKT cells have been documented in humans with other autoimmune diseases [81–85].

5.6.1
Antigens Recognized by Group 2 CD1-restricted T Cells

In an early attempt to discover CD1d-presented ligands, peptide phage display libraries were used to identify peptides that bound to recombinant murine CD1d. [86]. This study demonstrated that CD1d could, in principle, bind and present peptide ligands, although the frequency and physiologic significance of peptide presentation by CD1d is not yet known. In contrast, abundant evidence for the binding and presentation of lipid-based ligands by CD1d has emerged, thus recapitulating the central theme of the earlier studies on the lipid antigens presented by human group 1 CD1 proteins.

At present, the most extensively studied CD1d-presented antigens are synthetic glycosylceramides, and in particular a form of α-galactosylceramide (αGalCer) containing a phytosphingosine base and a long chain (C_{26}) fatty acid (Figure 5.2). This unusual glycolipid was initially identified as the active component of marine sponge extracts that possessed anti-tumor activity in mice [69]. Subsequent studies showed that αGalCer was a potent ligand for human and murine CD1d-restricted natural killer T cells (NKT cells) [87, 88]. While the identification of αGalCer as a CD1d-presented ligand and a powerful agonist for NKT cells has provided an extremely useful tool for the study of CD1d function, attempts to identify a natural homologue of αGalCer in either mammalian tissues or in relevant microbial pathogens have been unsuccessful.

While it is generally accepted that the frequent CD1d-restricted autoreactivity observed among murine NKT cells is likely to be due to recognition of self-antigens presented on CD1d, the precise identity of these putative natural self ligands remains unknown [89]. Some initial clues toward the identification of such self ligands may be provided by reports that CD1d-restricted NKT cells can be activated by natural and synthetic glycosyl phosphatidylinositol (GPI) [90] and other common phospholipids, including phosphatidylinositol (PI), phosphatidylethanolamine (PE) and phosphatidylglycerol (PG) [56, 91]. Together, these findings on the binding of endogenous lipids by CD1 have suggested a model in which some endogenous lipids might play a chaperone-like role during the biosynthesis of CD1 molecules. As recently described for PI in the case of CD1d molecules, the binding of these endogenous lipids is likely to occur in the ER, shortly after association of the CD1 heavy chain with the β2-microglobulin, [22, 24]. Such endogenously loaded chaperone lipids might serve to stabilize the molecule and protect the binding site until they are subsequently exchanged for other endogenous glycolipids, the identity of which may vary between different tissues [89, 92], or exogenous lipids internalized into endocytic compartments.

There has been scant evidence for the existence of natural microbial pathogen-associated antigens presented by CD1d, despite a considerable body of published work on the role of CD1d in various infectious diseases and disease models [93]. However, a recently published study found that a mycobacterial glycolipid known as phosphatidylinositol mannoside (PIM) was able to bind to CD1d, and that it activated a subset of human and murine CD1d-restricted NKT cells. [94]. Whether recognition of PIM by CD1d-restricted T cells contributes to the immune response against mycobacterial infection remains to be determined.

5.7
Tissue Distribution of CD1 Proteins

Analysis of the tissue distribution of CD1 molecules provides further support for the classification of CD1 molecules into at least two distinct groups. The distribution of group 1 CD1 molecules is relatively restricted, with strong expression limited mostly to immature cortical thymocytes and various specialized antigen-presenting cells, most notably myeloid lineage dendritic cells and a subset of B lymphocytes [95]. In contrast, CD1d is expressed on both hematopoietic cells and non-hematopoietic cell types in both human and mice [96, 97]. In humans, CD1d is expressed on B cells, macrophages, blood monocyte-derived dendritic cells and on dermal [98] and epidermal dendritic cells (Y. Dutronc & S.A. Porcelli, unpublished data). In addition, CD1d but not group 1 CD1 proteins are expressed by epithelial cells in the intestine [16, 99] and the skin [100], as originally suggested by the study of Canchis et al. [101]. A second important difference is that CD1d is constitutively expressed on monocyte precursors of myeloid DCs, whereas CD1a, CD1b and CD1c appear on monocytes only after they have received signals that induce differentiation to immature dendritic cells [4, 102, 103] The inducible and regulated nature of group 1 CD1 protein expression on myeloid DCs suggest that natural adjuvants could control the expression and functions of group 1 CD1 isoforms on this cell type [104].

5.8
Subcellular Distribution and Intracellular Trafficking of CD1

An additional aspect of the diversity of CD1 molecules resides in their divergent subcellular distribution. CD1 proteins rapidly traverse the secretory pathway to the cell surface, and then undergo a significant level of localization to the endocytic system, although the extent and precise positioning of each different CD1 isoform can vary substantially (Figure 5.3). The following sections summarize the differences in subcellular localization and trafficking that have been reported for the various CD1 isoforms.

5.8.1
Trafficking and Localization of CD1a

At steady state, human CD1a molecules are prominently expressed at the plasma membrane of blood monocyte-derived dendritic cells (DC) and epidermal Langerhans cells (LC). They are absent from the late endosomal and lysosomal compartments, but can be detected in the early/recycling endosomes [105]. Analysis of CD1a trafficking in human LC suggests that CD1a molecules present at the cell surface spontaneously internalize via clathrin-coated vesicles. Most likely, they are rapidly distributed to the early/sorting endosomes and delivered to early/recycling endosomes, and from this site they recycle back to the plasma membrane [106]. CD1a molecules are also present in Birbeck granules, rod-shaped structures specific to LC, which have been recently identified as part of the endosomal recycling compartment [107]. It is still unclear which mechanisms drive the intracellular trafficking of CD1a, since its short intracytoplasmic tail does not contain any recognizable sorting motif, in contrast to most other CD1 molecules. Possibly, CD1a simply follows the bulk flow of plasma membrane into endocytic vesicles in order to gain entry into the endosomal pathway [108], or CD1a may interact with other proteins at the cell surface that influence its trafficking. One candidate for such a role as an interacting protein is langerin, a recently identified C-type lectin specific to Langerhans cells, which functions as an endocytic receptor and colocalizes with CD1a in recycling endosomes and Birbeck granules [107, 109, 110].

5.8.2
Trafficking and Localization of CD1b

Human CD1b was the first CD1 protein for which the subcellular localization and mechanisms of intracellular trafficking were studied, thus establishing many of the general principles that apply in subtly different ways to other human CD1 isoforms and to CD1 proteins in nonhuman species. Human monocyte-derived DC express high levels of CD1b at their surface, but the molecule is also present in endosomes, lysosomes and the closely related MHC class II compartments (MIICs) of specialized antigen presenting cells [111, 112]. Pulse-chase experiments performed on B lymphoblastoid cells stably transfected with CD1b suggest that, after synthesis, CD1b molecules are transported rapidly to the cell surface via the secretory pathway, and from this site they are then internalized by clathrin-mediated endocytosis. This internalization is dependent on a tyrosine-based motif of the cytoplasmic tail (YXXΦ where Y is tyrosine, X is any amino acid, and Φ is a bulky hydrophobic residue), which is a canonical signal for interactions of transmembrane proteins with cytosolic adaptor protein (AP) complexes known as AP-1, AP-2, AP-3 and AP-4 [27]. These AP complexes are heterotetramers composed of two large subunits (α, γ, δ or ε paired with $\beta1$, $\beta2$, $\beta3$ or $\beta4$), one medium subunit ($\mu1$–$\mu4$) and one small subunit ($\sigma1$–$\sigma4$) [113, 114]. Internalization of CD1b from the plasma membrane to early recycling/sorting endosomes is most likely dependent on the AP-2 complex, whereas subsequent sorting to the late

Figure 5.3 Cellular distribution and trafficking of human and mouse CD1 molecules. Top left: *human CD1a*. After association with β2-microglobulin in the endoplasmic reticulum (ER), CD1a molecules are transported to the plasma membrane (PM) through the secretory pathway. CD1a spontaneously internalizes via clathrin-coated vesicles. After internalization, CD1a recycles to the plasma membrane through early/sorting and early/recycling endosomes (EE/RE). Top right: *human CD1b and mouse CD1d*. The cellular distributions of human CD1b and murine CD1d (mCD1d) molecules are similar. Human CD1d has been less well analyzed, and limited data suggest that it may more closely resemble human CD1c in its intracellular trafficking properties [116]. After expression at the cell surface, the interactions of the intracytoplasmic tails of CD1b and mCD1d with the cytosolic adaptor complex AP-2 mediate their internalization into clathrin-coated pits and delivery to early endosomes. From there, they access the late endosomes (LE), the lysosomes and the MHC Class II compartments (MIIC/Lys) through their interaction with another cytosolic adaptor, AP-3. CD1b and mCD1d molecules undergo several rounds of recycling to and from the cell surface, and over time accumulate in the LE and MIIC/Lys. Molecules that have been loaded with lipid antigens within the endocytic system escape to the cell surface and accumulate there for antigen presentation. Bottom left: *human CD1c*. CD1c molecules transit to the cell surface through the secretory pathway and are internalized via clathrin-coated vesicles through their interaction with AP-2. Since the cytoplasmic tail of CD1c fails to interact with AP-3, most of the molecules recycle to the cell surface through the EE/RE, and only a small fraction can be detected in the LE. Bottom right: *human CD1e*. Alternative splicing potentially leads to the generation of various forms of CD1e in the ER (see text). Only CD1e molecules with three lumenal α domains can associate with β2-microglobulin (β2-m) and leave the ER. In immature dendritic cells (DC), most CD1e molecules are present in the Golgi. After activation and maturation of DC, CD1e molecules accumulate in the late endosomes and lysosomes, where they are cleaved into a soluble form.

endosomal compartment may depend on interaction of CD1b tail with another adaptor protein, AP-3 [27, 111, 115]. Supporting this hypothesis is the observation that CD1b trafficking in AP-3 deficient cells derived from patients with Hermansky-Pudlak Syndrome type 2 is altered, showing increased surface expression and accumulation of CD1b molecules in the early endosomes with reduced localization to lysosomal compartments [116]. Compared to other group 1 CD1 proteins, only CD1b binds AP-3 efficiently [27, 116]. Notably, these features of trafficking and localization have been described for human CD1b, but may not apply in all cases to orthologues of this protein in other species. For example, the guinea pig expresses several proteins that appear to by closely related in sequence to human CD1b, but one of these, designated gpCD1b3, shows a pattern of cellular localization that is similar to that of human CD1a [117]. In this regard, the gpCD1b3 protein notably lacks the tyrosine residue in the cytoplasmic tail motif, and thus presumably is unable to interact with AP-2 and AP-3 complexes.

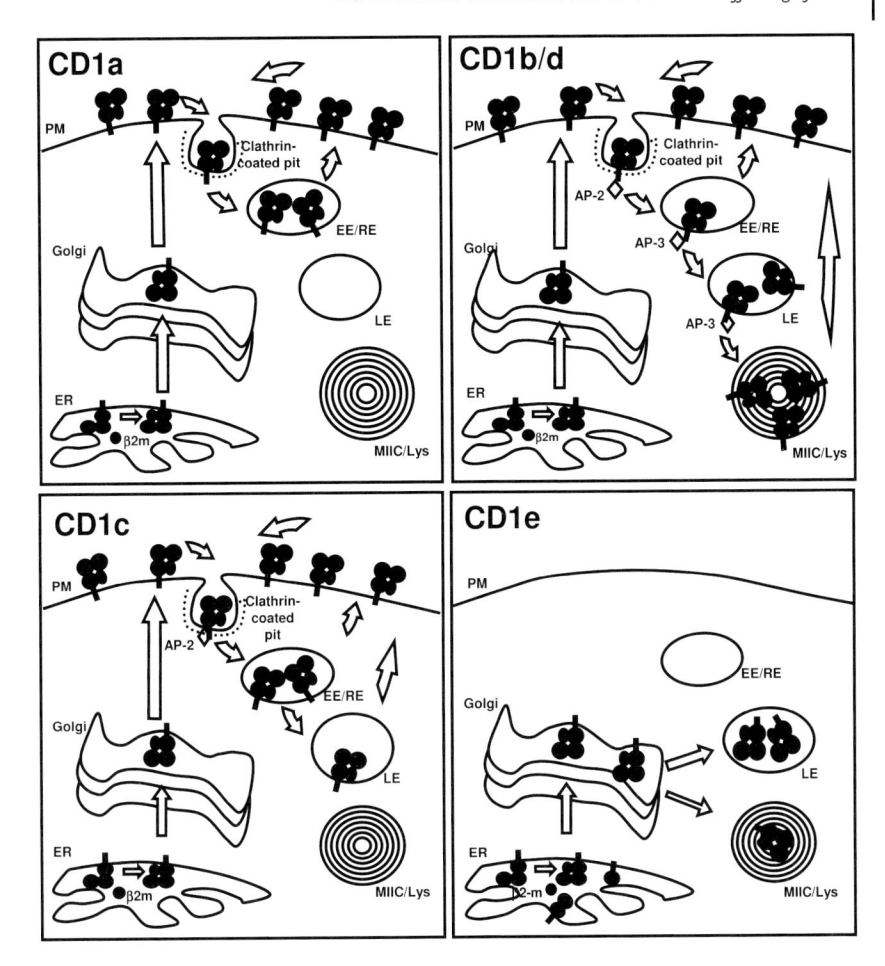

5.8.3
Trafficking and Localization of CD1c

The distribution of human CD1c partially overlaps those of CD1a and CD1b molecules. CD1c is expressed at the cell surface and most likely is internalized by clathrin-mediated endocytosis like CD1b. CD1c is abundantly present in the early/recycling compartment, but also colocalizes partially with CD1b in late endosomes [112]. Despite the presence of a tyrosine-based endosomal targeting motif very similar to that in CD1b, CD1c does not accumulate to any consistently appreciable extent in the MIIC compartment, probably because it cannot interact strongly with AP-3 [27, 112]. The different affinity of CD1b and CD1c for AP-3 may be linked to the identity of the amino acids adjacent to the YXXΦ motif, as the proline present at the C-terminus of the cytoplasmic tail of CD1b, but absent in CD1c, has been reported to strongly influence binding to AP-3 [116].

5.8.4
Trafficking and Localization of CD1d

The steady state localizations of human and murine CD1d molecules are similar to that of human CD1b, as all three proteins are expressed at the plasma membrane and in late endosomal and lysosomal compartments. However, one study has suggested that human CD1d may be less strongly directed to late endosomes than are human CD1b and murine CD1d [116]. Like CD1b and CD1c, CD1d molecules contain in their cytoplasmic tails a tyrosine-based internalization motif that mediates their internalization from the cell surface [28]. In similar fashion to CD1b, the cytoplasmic tail motif of murine CD1d interacts with both AP-2 and AP-3 complexes, whereas in vitro studies suggest that human CD1d may interact with AP-2 but not with AP-3 [116, 118]. Studies of AP-3 deficient mice have shown that the lysosomal trafficking attributable to this sorting complex is required for the normal thymic selection of CD1d-restricted NKT cells [118]. However, two studies have revealed the existence of additional trafficking mechanisms for delivery of human and mouse CD1d to MIIC compartments through their association with the MHC class II-associated invariant chain (Ii) or with MHC class II molecules themselves [28, 119]. Jayawardena-Wolf et al. [28] suggest that, after synthesis, CD1d molecules are quickly transported to the cell surface and progressively accumulate in the lysosomes and MIICs after several rounds of recycling between the plasma membrane and the endosomes. A fraction of murine CD1d co-precipitates with the Ii, implying that the two molecules associate in vivo. The role of Ii on CD1d trafficking is supported by the observation that, in cells expressing murine CD1d lacking the tyrosine-based targeting signal, transfection of Ii results in redistribution of the mutant CD1d from the cell surface to late endosomes and lysosomes. Conversely, Ii-deficient B cells and dendritic cells show an increased level of cell surface CD1d compared to normal cells [28].

Thus, while most CD1d appears to be rapidly transported to the cell surface after its release from the endoplasmic reticulum and then reinternalized to accumulate in the MIICs, a fraction might be delivered directly from the Golgi to the MIIC through its association with Ii. There, cleavage of Ii by resident proteases would allow loading of endosomal glycolipids and transfer to the cell surface, following a pathway similar to that followed by MHC class II molecules. This possibility has gained support by two studies implicating members of the cathepsin family of lysosomal proteases in the normal expression and function of mouse CD1d, although the mechanism for cathepsin involvement has not been fully resolved [120, 121]. Interestingly, Kang and Cresswell identified very similar mechanisms in their study of human CD1d, which demonstrated the formation of CD1d/Ii complexes that also contained MHC class II molecules [119]. These investigators further demonstrated that an association between human CD1d and MHC class II proteins could be detected even at the plasma membrane, suggesting that the preservation of this association all the way to the cell surface could potentially have effects on recycling of CD1d from the membrane and on its interactions with T cell receptors.

Taken together, these studies suggest the coexistence of two pathways of intracellular trafficking for CD1d. One pathway allows newly synthesized CD1d molecules to move directly to the cell surface where they subsequently internalize to reach endosomes and MIICs. A second pathway is followed by CD1d molecules that associate with Ii/MHC class II complexes in the ER. These would be predicted to enter the MIIC compartments directly as a result of intracellular sorting at the Golgi, and then subsequently be released to the cell surface after degradation of Ii and, apparently, in some cases still in association with MHC class II molecules. The former pathway appears to be quantitatively dominant, as the latter pathway is not sufficient to rescue the positive selection of NK T cells [122], and is most readily detected only when cytoplasmic tail-mediated sorting is abrogated [28]. While the currently published studies examined Ii association only for CD1d, preliminary studies indicate that this can also be demonstrated for human CD1b (M. Sugita and S.A. Porcelli, unpublished data). Thus, it is possible that a similar alternative trafficking pathway resulting in direct Golgi to MIIC transport exists for at least some group 1 CD1 proteins.

5.8.5
Trafficking and Localization of CD1e

Although the human CD1e gene and its transcripts were discovered more than two decades ago, only recently has CD1e protein been demonstrated to be expressed by myeloid lineage DCs [2]. Analysis of mRNA from monocyte-derived DCs identified more than 15 mRNA species encoding CD1e generated by alternative splicing. Two of these transcripts encode apparently full-length transmembrane proteins with 3 extracellular domains ($\alpha1$, $\alpha2$ and $\alpha3$) and a cytoplasmic tail of either 63 or 51 amino acids. Other transcripts encode forms of membrane-associated CD1e with only one ($\alpha3$ only) or two α domains ($\alpha2\alpha3$, $\alpha1\alpha3$), and yet others lack transmembrane domain sequences and appear to encode putative soluble forms of the protein.

Several antibodies were generated to study the cellular localization of the different CD1e isoforms expressed individually by transfection in HeLa and M10 cells. This study showed that only the membrane-associated isoforms with all three extracellular domains can leave the endoplasmic reticulum (ER), probably after they associate with the β2-microglobulin. These molecules are exported through the trans-Golgi network to an acidic compartment, most likely the late endosomes or lysosomes, where they are cleaved into a soluble form and accumulate. This intracellular traffic is controlled by a short sequence in the C-terminal end of the cytoplasmic tails containing two overlapping dilysine motifs (KKXK). Analyses of monocyte-derived DC and freshly isolated LCs confirm these data obtained with transfected cell lines. In immature DC, CD1e molecules are found mainly in the Golgi compartments and trans-Golgi network (TGN), whereas in LPS-treated DC, CD1e molecules accumulate in late endosomes and lysosomes. Notably, CD1e molecules seem never to reach the cell surface, which appears to exclude their direct participation in antigen presentation. Homologues of human CD1e have been identified in guinea pigs and in several other mammals, but their expression and intracellular distributions have not been studied [11]. However, the marked conservation of many predicted structural features of CD1e in humans and other mammals suggests that the unusual intracellular localization may be conserved, and thus may be associated with some important function of this molecule.

5.9
Antigen Uptake, Processing and Loading in the CD1 Pathway

As for the MHC class I and II pathways of peptide antigen presentation, the CD1 pathway is facilitated by a series of processes within antigen presenting cells that enhance the ability of these cells to generate complexes of specific lipid antigens with CD1 proteins. These processes include mechanisms for enhancing the uptake of lipid and glycolipid antigens and for delivering these to compartments in which CD1 molecules are abundantly concentrated. In addition, these compartments contain specific cofactors that assist in the formation of CD1/lipid complexes that form the targets of specific T cells.

5.9.1
Cellular Uptake of CD1-presented Antigens

Relatively little is known about how various lipids and glycolipids become associated with the surface of APCs such as dendritic cells, and through what routes these are subsequently internalized. However, two studies have identified cell surface receptors that appear to facilitate the cellular uptake and subsequent presentation of CD1-presented glycolipid antigens. These studies have identified two endocytic pattern recognition receptors in the C-type lectin family as potential antigen uptake receptors for the CD1 presentation pathway. In one case, the macrophage mannose receptor (MR) was shown to be involved in the binding and uptake of mycobacterial lipoarabinomannan (LAM), a complex glycolipid that was found to be presented to CD1b-restricted T cells [123]. Blockade of MR expressed on monocyte-derived dendritic cells could be shown to inhibit presentation of LAM to CD1b-restricted T cells in vitro, thus implicating this receptor as a major factor for LAM uptake. Consistent with the proposed role of MR, this receptor was demonstrated to colocalize with endocytosed LAM in late endosomal/lysosomal compartments of monocyte-derived dendritic cells. A similar role in antigen uptake has been attributed to the C-type lectin receptor known as langerin (CD207), which is highly expressed by human epidermal Langerhans cells that express several CD1 proteins, including very high levels of CD1a [109]. Langerin is expressed exclusively by Langerhans cells and plays a critical role in the formation of Birbeck granules, a unique endocytic compartment of these cells that has been proposed as a possible antigen loading compartment for CD1a molecules. The presentation of mycobacterial antigens to CD1a-restricted T cell clone could be effectively blocked by preincubation of APCs with anti-langerin antibodies, suggesting that this C-type lectin was involved in the uptake of antigen for appropriate processing and presentation by CD1a [110].

In addition to the likely effects of receptor-mediated uptake on the routing of CD1-presented lipid antigens to particular intracellular compartments, there is evidence that this may also be controlled to some extent by structural features of the lipids themselves. For example, a study on the uptake and intracellular trafficking of mycobacterial glucose monomycolates, a CD1b-presented antigen, revealed that the intracellular sorting of this antigen was strongly influenced by the length of the alkyl chains of the lipid [39]. Thus, a GMM molecule with a very long C80 mycolate chain was shown to accumulate significantly in dense late endosomal/lysosomal compartments in antigen presenting cells, whereas a GMM molecule with a C32 mycolate did not reach these compartments as efficiently. This correlated with a markedly higher efficiency of presentation to GMM specific T cells of the C80 form. Similarly, other CD1-presented lipid antigens show preferential localization to particular subcellular sites, such as sphingolipids, which accumulate in the Golgi compartment, and isoprenoid phospholipids, which are normally retained in the ER. Thus, regulated delivery of CD1 proteins to distinct intracellular compartments likely represents a means of controlling which types of lipids are ultimately presented to T cells.

5.9.2
Endosomal Processing of CD1-presented Antigens

In the MHC-restricted antigen presentation pathways, the term processing is generally used to refer to the proteolytic cleavage of protein antigens into the peptide fragments that eventually become associated with the MHC class I or class II molecules. Thus, processing can be defined generically as any process that modifies the covalent structure of an antigen to allow or enhance its presentation. Whether such modification or processing of lipid and glycolipid structures occurs at any biologically significant level for CD1-presented antigens is unclear. Initial studies of mycobacterial mycolyl glycolipids presented by human CD1b suggested that deglycosylation of these in the endosomes of human APCs was unlikely to occur in myeloid dendritic cells [124]. However, some of the earliest studies on CD1b-restricted presentation of mycobacterial LAM suggested that this very heavily glycosylated structure was likely to undergo enzymatic removal of a portion of its glycan structure prior to presentation to T cells [46]. Although there are no unequivocal examples of a requirement for enzymatic processing of natural CD1-presented antigens, the basic principle has been investigated using synthetic glycolipid ligands for CD1d-restricted NKT cells. Thus, a form of α-glycosyl ceramide bearing a Gal(α1→2)Gal disaccharide could not be presented to NKT cells unless it was first delivered to late endosomes or lysosomes where the terminal galactose was cleaved by a resident galactosidase [125]. Although not yet clearly shown for a natural glycolipid antigen, it seems likely that processing of glycolipid oligosaccharide or polysaccharide head groups by glycosidases will be involved in the generation of CD1-presented self antigens, such as gangliosides and other complex glycolipids.

5.9.3
Accessory Molecules for Endosomal Lipid Loading of CD1

As alluded to earlier, efficient presentation of lipid antigens by most CD1 molecules often requires endosomal localization of both the antigen and the CD1 protein. This implies that factors exist within the endosomes that facilitate the transfer of lipid antigens onto the CD1 protein to create complexes that are recognized by T cell receptors. In some cases, it appears likely that the acidic pH of endosomes and lysosomes is partly responsible for the enhanced lipid antigen loading of CD1 in these compartments. This is supported by studies showing that agents that neutralize the acidity of endosomes, such as the weak base chloroquine or the proton pump inhibitor concanamycin A, can reduce or prevent presentation of particular lipid antigens by CD1 proteins [4, 46, 87]. This effect is particularly evident with presentation of mycobacterial lipid antigens by human CD1b proteins, which are known to prominently localize to strongly acidified lysosomal and late endosomal compartments. The mechanism by which acidic pH facilitates lipid antigen loading onto CD1 is unknown, although existing data suggest that acidic conditions cause conformational changes in the CD1 structure that may facilitate access to the antigen binding groove [126, 127].

In addition to the physical properties of endosomes such as low pH, there clearly exist specific cofactors within these compartments that facilitate the loading of lipid antigens onto CD1 proteins. Three studies have demonstrated that several known endosomal lipid transport proteins (LTPs) function as such cofactors [128–130]. These proteins include the four members of the saposin family (saposins A, B, C and D), which are small dimeric proteins produced by endosomal proteolytic cleavage of a single prosaposin precursor protein. These proteins have been previously described as required cofactors for the activation of glycosphinglolipids for enzymatic processing and degradation in the lysosome [131–133]. Another protein involved in the processing of glycolipids in the lysosome, the GM2 activator protein [134–136], has also been implicated in CD1 function [130].

The role of the endosomal LTPs in antigen presentation by CD1 is most likely linked to their ability to promote the exchange of lipids between the membrane bilayer and the CD1 lipid binding groove. Studies of human T cells specific for endocytosed mycobacterial lipid antigens presented by CD1b showed a marked deficiency in presentation by cells lacking prosaposin expression. Experiments in which recombinant saposins were used to complement this defect revealed that only saposin C, and not other members of the saposin family, could reverse the defect in antigen presentation [128]. Studies of human and murine CD1d-restricted NKT cells likewise provided evidence for a significant role for saposins in lipid antigen loading. With human CD1d-restricted NKT cells, the absence of prosaposin, and thus of all four saposin molecules, eliminated the binding of an exogenous lipid antigen, αGalCer, to CD1d in the endosomal pathway. This defect could be reversed by reconstitution of prosaposin by transfection [129]. Murine NKT cells were shown to depend on saposins for their autoreactivity, which presumably reflects the presentation of endosomally loaded self lipids. Most strikingly, CD1d-restricted NKT cells failed to develop in mice that lacked saposins as a result of targeted deletion of the prosaposin gene [130].

Subcellular localization and biochemical studies provide further support for the importance of saposins in lipid antigen presentation by CD1. For example, saposins colocalize with CD1 molecules in late endosomes and lysosomes, which is consistent with their proposed role in lipid antigen loading in those compartments [128–130]. Furthermore, biochemical studies suggested that direct binding occurs between CD1b and saposin C in antigen presenting cells, and in vitro lipid solubilization studies showed that saposin C could extract mycobacterial LAM from an artificial membrane bilayer [128]. In vitro studies using mouse CD1d and purified saposins or GM2 activator protein also revealed the ability of these proteins to promote lipid exchange between liposomes and the CD1 protein. A surprising feature of the interactions between LTPs and CD1 proteins in the mouse CD1d studies is their apparent specificity in interacting with some but not all CD1d-glycolipid complexes, which appears to be dependent on the particular glycolipid bound to CD1d. Given that the lipid tails are sequestered in the CD1 groove, this suggests that the LTPs must perform some degree of recognition of the exposed head groups of the bound lipid or glycolipid ligands. Thus, it has been proposed that LTPs not only have a detergent-like function in promoting lipid exchange into the CD1 groove, but that they also may perform an editing function that involves binding to and removing certain lipids from CD1 while leaving other CD1-lipid complexes intact [130]. Such lipid-editing could have a substantial impact on the repertoire of antigens presented by CD1, similar to the influence of editing of MHC class II-bound peptides by HLA-DM on the selection and responses of conventional $CD4^+$ T cells [137].

5.9.4
Non-endosomal Loading of Lipids onto CD1 Molecules

In addition to specific protein cofactors within the endosomal system, it is also possible that lipid exchange proteins in other compartments of the cell or in the extracellular space could influence the expression and function of CD1 proteins. In fact, one example of a non-endosomal cofactor for lipid loading of CD1 has recently been proposed, which is the microsomal triglyceride transfer protein (MTP). This lipid transfer protein resides in the endoplasmic reticulum of hepatocytes and intestinal epithelial cells, and is essential for the lipidation of apolipoprotein B [138]. One study has found that deletion of the MTP gene alters the distribution of CD1d expression in murine hepatocytes, with increased levels in the endoplasmic reticulum, suggesting abnormal entry of CD1d into the secretory pathway. The absence of MTP was also associated with a reduction in the function of CD1d, in that cells lacking this protein were unable to activate CD1d-restricted

NKT cells [139]. These findings suggest that MTP may regulate lipid association with CD1d in the endoplasmic reticulum, thus altering its initial assembly and subsequent trafficking and function. Whether specific self or foreign antigens that can be recognized by T cells might be loaded onto CD1 in the ER remains to be determined.

Several studies also indicate that in some cases CD1 molecules can acquire lipid ligands at the cell surface. For example, some normal mammalian lipids can be loaded onto recombinant CD1 proteins at neutral pH [56], and mammalian glycosphingolipids that are presented to CD1-restricted T cells can associate with CD1 molecules at the cell surface [41, 42]. This may also be true for certain exogenous microbial lipid antigens, especially those that have relatively short lipid tails and are not efficiently sorted to endosomes [39]. It is currently unknown whether factors such as lipid exchange proteins in the extracellular environment might be involved in promoting the cell surface loading of CD1 molecules with exogenous lipids, and the functional importance, if any, of these putative non-endosomal pathways for lipid antigen loading onto CD1 remains unclear. It is probably significant that most examples of non-endosomal loading of CD1-presented antigens require relatively high concentrations of the exogenously added lipid antigen to induce specific T cell responses, suggesting that this pathway is likely to be less efficient than the endosomal loading pathway. These observations provide further support for the idea that an important mechanism by which APCs can control which lipids are presented by CD1 to T cells is by the selective sorting of lipids with intrinsically foreign structural features into endosomes, while lipids that are more characteristic of normal mammalian cell membranes may be largely excluded from such compartments [39]. Nevertheless, the non-endosomal pathway could promote the formation of CD1–self lipid complexes that are involved in the development and selection of CD1-restricted T cells, or required for the homeostasis and long-term survival of these T cells.

5.10
Conclusions

The CD1 system represents a surprising but extremely logical variation on the more familiar theme of MHC-restricted peptide antigen presentation. This system of MHC-related proteins provides a fundamentally different set of targets for immune recognition, and thus is likely to play many roles that complement or synergize with peptide antigen presentation. For group 1 CD1 proteins, there is clearly the potential to augment the protective immune response against pathogens that express lipids of intrinsically foreign structure. The advantage in having this additional level of recognition to augment the repertoire of peptide antigen specific T cells seems obvious, although the extent to which CD1-restricted T cells contribute to protective immunity in human infectious diseases remains to be determined. For group 2 CD1 proteins, currently available evidence indicates that these may generate signals for immunoregulation or innate immune responses mainly through the display of self lipid antigens. In either case, the mechanisms leading to the processing and loading of lipid antigens onto CD1 proteins are of fundamental importance to understanding and manipulating this unique antigen presentation pathway. The great challenge presented by the CD1 system is to understand the processes that allow the efficient handling of lipids in an environment that permits proteins and cells to function normally. As our insight and understanding into these processes advances, the potential of the CD1 system for augmenting immune responses or regulating harmful autoimmunity will likely become an area of increasing interest and importance.

Acknowledgments

The authors thank Dr. Carme Roura-Mir for creating images of CD1 proteins. D.B.M. is supported by grants from the Pew Foundation, the American College of Rheumatology Research and Education Foundation, the Cancer Research Institute, the Mizutani Foundation for Glycoscience and the NIH. S.A.P. is supported by the NIH and the Burroughs Wellcome Fund.

References

1 F. Calabi and C. Milstein, *Nature* **1986**, 323, 540–543.
2 C. Angenieux, J. Salamero, D. Fricker, J. P. Cazenave, B. Goud, D. Hanau, and S. H. de La, *J Biol. Chem.* **2000**, 275, 37 757–37 764.
3 F. Calabi, J. M. Jarvis, L. Martin, and C. Milstein, *Eur. J. Immunol.* **1989**, 19, 285–292.
4 S. Porcelli, C. T. Morita, and M. B. Brenner, *Nature* **1992**, 360, 593–597.
5 A. Bendelac, O. Lantz, M. E. Quimby, J. W. Yewdell, J. R. Bennink, and R. R. Brutkiewicz, *Science* **1995**, 268, 863–865.
6 F. Calabi, K. T. Belt, C. Y. Yu, A. Bradbury, W. J. Mandy, and C. Milstein, *Immunogenetics* **1989**, 30, 370–377.
7 A. Bradbury, F. Calabi, and C. Milstein, *Eur. J. Immunol.* **1990**, 20, 1831–1836.
8 P. F. Moore, M. D. Schrenzel, V. K. Affolter, T. Olivry, and D. Naydan, *Am. J. Pathol.* **1996**, 148, 1699–1708.
9 J. C. Woo and P. F. Moore, *Tissue Antigens* **1997**, 49, 244–251.
10 K. Kasai, A. Matsuura, K. Kikuchi, Y. Hashimoto, and S. Ichimiya, *Clin. Exp. Immunol.* **1997**, 109, 317–322.
11 C. C. Dascher, K. Hiromatsu, J. W. Naylor, P. P. Brauer, K. A. Brown, J. R. Storey, S. M. Behar, E. S. Kawasaki, S. A. Porcelli, M. B. Brenner, and K. P. LeClair, *J. Immunol.* **1999**, 163, 5478–5488.
12 K. Kashiwase, A. Kikuchi, Y. Ando, A. Nicol, S. A. Porcelli, K. Tokunaga, M. Omine, M. Satake, T. Juji, M. Nieda, and Y. Koezuka, *Immunogenetics* **2003**, 54, 776–781.
13 S. M. Hayes and K. L. Knight, *J. Immunol.* **2001**, 166, 403–410.
14 L. H. Martin, F. Calabi, F.-A. Lefebvre, C. A. G. Bilsland, and C. Milstein, *Proc. Natl. Acad. Sci. U.S.A.* **1987**, 84, 9189–9193.
15 H. S. Kim, J. Garcia, M. Exley, K. W. Johnson, S. P. Balk, and R. S. Blumberg, *J. Biol. Chem.* **1999**, 274, 9289–9295.

16 P. A. Bleicher, S. P. Balk, S. J. Hagen, R. S. Blumberg, T. J. Flotte, and C. Terhorst, *Science* **1990**, 250, 679–682.

17 S. A. Porcelli, *Adv. Immunol.* **1995**, 59, 1–98.

18 M. Sugita, S. A. Porcelli, and M. B. Brenner, *J. Immunol.* **1997**, 159, 2358–2365.

19 R. Huttinger, G. Staffler, O. Majdic, and H. Stockinger, *Int. Immunol.* **1999**, 11, 1615–1623.

20 S. J. Kang and P. Cresswell, *J. Biol. Chem.* **2002**, 277, 44 838–44 844.

21 A. Karadimitris, S. Gadola, M. Altamirano, D. Brown, A. Woolfson, P. Klenerman, J. L. Chen, Y. Koezuka, I. A. Roberts, D. A. Price, G. Dusheiko, C. Milstein, A. Fersht, L. Luzzatto, and V. Cerundolo, *Proc. Natl. Acad. Sci. U.S.A.* **2001**, 98, 3294–3298.

22 A. D. De Silva, J. J. Park, N. Matsuki, A. K. Stanic, R. R. Brutkiewicz, M. E. Medoł, and S. Joyce, *J. Immunol.* **2002**, 168, 723–733.

23 S. Joyce, A. S. Woods, J. W. Yewdell, J. R. Bennink, A. D. De Silva, A. Boesteanu, S. P. Balk, R. J. Cotter, and R. R. Brutkiewicz, *Science* **1998**, 279, 1541–1544.

24 J. J. Park, S. J. Kang, A. D. De Silva, A. K. Stanic, G. Casorati, D. L. Hachey, P. Cresswell, and S. Joyce, *Proc. Natl. Acad. Sci. U.S.A.* **2004**, 101, 1022–1026.

25 A. Bauer, R. Huttinger, G. Staffler, C. Hansmann, W. Schmidt, O. Majdic, W. Knapp, and H. Stockinger, *Eur. J. Immunol.* **1997**, 27, 1366–1373.

26 A. Bendelac, N. Killeen, D. R. Littman, and R. H. Schwartz, *Science* **1994**, 263, 1774–1778.

27 V. Briken, R. M. Jackman, S. Dasgupta, S. Hoening, and S. A. Porcelli, *EMBO J.* **2002**, 21, 825–834.

28 J. Jayawardena-Wolf, K. Benlagha, Y.-H. Chiu, R. Mehr, and A. Bendelac, *Immunity* **2001**, 15, 897–908.

29 Z. Zeng, A. R. Castano, B. W. Segelke, E. A. Stura, P. A. Peterson, and I. A. Wilson, *Science* **1997**, 277, 339–345.

30 S. D. Gadola, N. R. Zaccai, K. Harlos, D. Shepherd, J. C. Castro-Palomino, G. Ritter, R. R. Schmidt, E. Y. Jones, and V. Cerundolo, *Nat. Immunol.* **2002**, 3, 721–726.

31 T. Batuwangala, D. Shepherd, S. D. Gadola, K. J. Gibson, N. R. Zaccai, A. R. Fersht, G. S. Besra, V. Cerundolo, and E. Y. Jones, *J. Immunol.* **2004**, 172, 2382–2388.

32 D. M. Zajonc, M. A. Elsliger, L. Teyton, and I. A. Wilson, *Nat. Immunol.* **2003**, 4, 808–815.

33 S. Sidobre, O. V. Naidenko, B. C. Sim, N. R. Gascoigne, K. C. Garcia, and M. Kronenberg, *J. Immunol.* **2002**, 169, 1340–1348.

34 J. L. Matsuda, O. V. Naidenko, L. Gapin, T. Nakayama, M. Taniguchi, C. R. Wang, Y. Koezuka, and M. Kronenberg, *J Exp. Med.* **2000**, 192, 741–754.

35 K. Benlagha, A. Weiss, A. Beavis, L. Teyton, and A. Bendelac, *J Exp. Med.* **2000**, 191, 1895–1903.

36 E. P. Grant, M. Degano, J. P. Rosat, S. Stenger, R. L. Modlin, I. A. Wilson, S. A. Porcelli, and M. B. Brenner, *J. Exp. Med.* **1999**, 189, 195 205.

37 A. Melian, G. F. Watts, A. Shamshiev, G. de Libero, A. Clatworthy, M. Vincent, M. B. Brenner, S. Behar, K. Niazi, R. L. Modlin, S. Almo, D. Ostrov, S. G. Nathenson, and S. A. Porcelli, *J. Immunol.* **2000**, 165, 4494–4504.

38 D. B. Moody, B. B. Reinhold, M. R. Guy, E. M. Beckman, D. E. Frederique, S. T. Furlong, S. Ye, V. N. Reinhold, P. A. Sieling, R. L. Modlin, G. S. Besra, and S. A. Porcelli, *Science* **1997**, 278, 283–286.

39 D. B. Moody, V. Briken, T. Y. Cheng, C. Roura-Mir, M. R. Guy, D. H. Geho, M. L. Tykocinski, G. S. Besra, and S. A. Porcelli, *Nat. Immunol.* **2002**, 3, 435–442.

40 D. B. Moody, D. C. Young, T. Y. Cheng, J. P. Rosat, C. Roura-Mir, P. B. O'Connor, D. M. Zajonc, A. Walz, M. J. Miller, S. B. Levery, I. A. Wilson, C. E. Costello, and M. B. Brenner, *Science* **2004**, 303, 527–531.

41 A. Shamshiev, A. Donda, I. Carena, L. Mori, L. Kappos, and G. de Libero, *Eur. J Immunol.* **1999**, 29, 1667–1675.

42 A. Shamshiev, A. Donda, T. I. Prigozy, L. Mori, V. Chigorno, C. A. Benedict, L. Kappos, S. Sonnino, M. Kronenberg,

and G. de Libero, *Immunity*. **2000**, 13, 255–264.
43. A. Shamshiev, H. J. Gober, A. Donda, Z. Mazorra, L. Mori, and G. de Libero, *J. Exp. Med*. **2002**, 195, 1013–1021.
44. A. Jahng, I. Maricic, C. Aguilera, S. Cardell, R. C. Halder, and V. Kumar, *J. Exp. Med*. **2004**, 199, 947–957.
45. E. M. Beckman, S. A. Porcelli, C. T. Morita, S. M. Behar, S. T. Furlong, and M. B. Brenner, *Nature* **1994**, 372, 691–694.
46. P. A. Sieling, D. Chatterjee, S. A. Porcelli, T. I. Prigozy, R. J. Mazzaccaro, T. Soriano, B. R. Bloom, M. B. Brenner, M. Kronenberg, P. J. Brennan, and R. L. Modlin, *Science* **1995**, 269, 227–230.
47. D. B. Moody, T. Ulrichs, W. Muhlecker, D. C. Young, S. S. Gurcha, E. Grant, J. P. Rosat, M. B. Brenner, C. E. Costello, G. S. Besra, and S. A. Porcelli, *Nature* **2000**, 404, 884–888.
48. M. Gilleron, S. Stenger, Z. Mazorra, F. Wittke, S. Mariotti, G. Bohmer, J. Prandi, L. Mori, G. Puzo, and G. de Libero, *J. Exp. Med*. **2004**, 199, 649–659.
49. E. M. Beckman, A. Melian, S. M. Behar, P. A. Sieling, D. Chatterjee, S. T. Furlong, R. Matsumoto, J. P. Rosat, R. L. Modlin, and S. A. Porcelli, *J. Immunol*. **1996**, 157, 2795–2803.
50. S. H. Park and A. Bendelac, *Nature* **2000**, 406, 788–792.
51. S. Stenger, R. J. Mazzaccaro, K. Uyemura, S. Cho, P. F. Barnes, J. P. Rosat, A. Sette, M. B. Brenner, S. A. Porcelli, B. R. Bloom, and R. L. Modlin, *Science* **1997**, 276, 1684–1687.
52. S. Stenger, D. A. Hanson, R. Teitelbaum, P. Dewan, K. R. Niazi, C. J. Froelich, T. Ganz, S. Thoma-Uszynski, A. Melian, C. Bogdan, S. A. Porcelli, B. R. Bloom, A. M. Krensky, and R. L. Modlin, *Science* **1998**, 282, 121–125.
53. T. Ulrichs, D. B. Moody, E. Grant, S. H. Kaufmann, and S. A. Porcelli, *Infect. Immunol*. **2003**, 71, 3076–3087.
54. S. Porcelli, M. B. Brenner, J. L. Greenstein, S. P. Balk, C. Terhorst, and P. A. Bleicher, *Nature* **1989**, 341, 447–450.
55. S. H. Park, J. H. Roark, and A. Bendelac, *J. Immunol*. **1998**, 160, 3128–3134.
56. J. E. Gumperz, C. Roy, A. Makowska, D. Lum, M. Sugita, T. Podrebarac, Y. Koezuka, S. A. Porcelli, S. Cardell, M. B. Brenner, and S. M. Behar, *Immunity* **2000**, 12, 211–221.
57. A. Bendelac, M. N. Rivera, S. H. Park, and J. H. Roark, *Annu. Rev. Immunol*. **1997**, 15, 535–562.
58. D. I. Godfrey, K. J. Hammond, L. D. Poulton, M. J. Smyth, and A. G. Baxter, *Immunol. Today* **2000**, 21, 573–583.
59. O. Lantz and A. Bendelac, *J. Exp. Med*. **1994**, 180, 1097–1106.
60. P. Dellabona, E. Padovan, G. Casorati, M. Brockhaus, and A. Lanzavecchia, *J. Exp. Med*. **1994**, 180, 1171–1176.
61. S. Porcell, C. E. Yockey, M. B. Brenner, and S. P. Balk, *J. Exp. Med*. **1993**, 178, 1–16.
62. L. Brossay and M. Kronenberg, *Immunogenetics* **1999**, 50, 146–151.
63. M. Kronenberg, L. Brossay, Z. Kurepa, and J. Forman, *Immunol. Today* **1999**, 20, 515–521.
64. S. B. Wilson and T. L. Delovitch, *Nat. Rev. Immunol*. **2003**, 3, 211–222.
65. C. Carnaud, D. Lee, O. Donnars, S. H. Park, A. Beavis, Y. Koezuka, and A. Bendelac, *J. Immunol*. **1999**, 163, 4647–4650.
66. T. Kawamura, K. Takeda, S. K. Mendiratta, H. Kawamura, L. Van Kaer, H. Yagita, T. Abo, and K. Okumura, *J. Immunol*. **1998**, 160, 16–19.
67. J. Cui, T. Shin, T. Kawano, H. Sato, E. Kondo, I. Toura, Y. Kaneko, H. Koseki, M. Kanno, and M. Taniguchi, *Science* **1997**, 278, 1623–1626.
68. I. Toura, T. Kawano, Y. Akutsu, T. Nakayama, T. Ochiai, and M. Taniguchi, *J. Immunol*. **1999**, 163, 2387–2391.
69. E. Kobayashi, K. Motoki, T. Uchida, H. Fukushima, and Y. Koezuka, *Oncol. Res*. **1995**, 7, 529–534.
70. M. J. Smyth, N. Y. Crowe, and D. I. Godfrey, *Int. Immunol*. **2001**, 13, 459–463.
71. M. J. Smyth and D. I. Godfrey, *Nat. Immunol*. **2000**, 1, 459–460.
72. K. Miyamoto, S. Miyake, and T. Yamamura, *Nature* **2001**, 413, 531–534.
73. E. Pal, T. Tabira, T. Kawano, M. Taniguchi, S. Miyake, and T. Yamamura, *J. Immunol*. **2001**, 166, 662–668.

74 S. Sharif, G. A. Arreaza, P. Zucker, Q. S. Mi, J. Sondhi, O. V. Naidenko, M. Kronenberg, Y. Koezuka, T. L. Delovitch, J. M. Gombert, M. Leite-De-Moraes, C. Gouarin, R. Zhu, A. Hameg, T. Nakayama, M. Taniguchi, F. Lepault, A. Lehuen, J. F. Bach, and A. Herbelin, *Nat. Med.* **2001**, 7, 1057–1062.

75 S. Sharif, G. A. Arreaza, P. Zucker, Q. S. Mi, and T. L. Delovitch, *J. Mol. Med.* **2002**, 80, 290–300.

76 S. Hong, M. T. Wilson, I. Serizawa, L. Wu, N. Singh, O. V. Naidenko, T. Miura, T. Haba, D. C. Scherer, J. Wei, M. Kronenberg, Y. Koezuka, and L. Van Kaer, *Nat. Med.* **2001**, 7, 1052–1056.

77 L. D. Poulton, M. J. Smyth, C. G. Hawke, P. Silveira, D. Shepherd, O. V. Naidenko, D. I. Godfrey, and A. G. Baxter, *Int. Immunol.* **2001**, 13, 887–896.

78 J.-M. Gombert, T. Tancrede-Bohin, A. Hameg, M. do Carmo Leite-de-Moraes, A. P. Vicari, J.-F. Bach, and A. Herbelin, *Int. Immunol.* **1996**, 8, 1751–1758.

79 B. Wang, Y. B. Geng, and C. R. Wang, *J Exp. Med.* **2001**, 194, 313–320.

80 F. D. Shi, M. Flodstrom, B. Balasa, S. H. Kim, K. Van Gunst, J. L. Strominger, S. B. Wilson, and N. Sarvetnick, *Proc. Natl. Acad. Sci. U.S.A.* **2001**, 98, 6777–6782.

81 H. J. Van Der Vliet, B. M. Von Blomberg, N. Nishi, M. Reijm, A. E. Voskuyl, A. A. van Bodegraven, C. H. Polman, T. Rustemeyer, P. Lips, A. J. Van Den Eertwegh, G. Giaccone, R. J. Scheper, and H. M. Pinedo, *Clin. Immunol.* **2001**, 100, 144–148.

82 S. Kojo, Y. Adachi, H. Keino, M. Taniguchi, and T. Sumida, *Arthritis Rheum.* **2001**, 44, 1127–1138.

83 T. Maeda, H. Keino, H. Asahara, M. Taniguchi, K. Nishioka, and T. Sumida, *Rheumatology. (Oxford)* **1999**, 38, 186–188.

84 Y. Oishi, T. Sumida, A. Sakamoto, Y. Kita, K. Kurasawa, Y. Nawata, K. Takabayashi, H. Takahashi, S. Yoshida, M. Taniguchi, Y. Saito, and I. Iwamoto, *J. Rheumatol.* **2001**, 28, 275–283.

85 T. Sumida, A. Sakamoto, H. Murata, Y. Makino, H. Takahashi, S. Yoshida, K. Nishioka, I. Iwamoto, and M. Taniguchi, *J. Exp. Med.* **1995**, 182, 1163–1168.

86 A. R. Castano, S. Tangri, J. E. Miller, H. R. Holcombe, M. R. Jackson, W. D. Huse, M. Kronenberg, and P. A. Peterson, *Science* **1995**, 269, 223–226.

87 T. Kawano, J. Cui, Y. Koezuka, I. Toura, Y. Kaneko, K. Motoki, H. Ueno, R. Nakagawa, H. Sato, E. Kondo, H. Koseki, and M. Taniguchi, *Science* **1997**, 278, 1626–1629.

88 F. Spada, Y. Koezuka, and S. A. Porcelli, *J. Exp. Med.* **1998**, 188, 1–6.

89 Y. H. Chiu, J. Jayawardena, A. Weiss, D. Lee, S. H. Park, A. Dautry-Varsat, and A. Bendelac, *J. Exp. Med.* **1999**, 189, 103–110.

90 L. Schofield, M. J. McConville, D. Hansen, A. S. Campbell, B. Fraser-Reid, M. J. Grusby, and S. D. Tachado, *Science* **1999**, 283, 225–229.

91 J. Rauch, J. Gumperz, C. Robinson, M. Skold, C. Roy, D. C. Young, M. Lafleur, D. B. Moody, M. B. Brenner, C. E. Costello, and S. M. Behar, *J. Biol. Chem.* **2003**, 278, 47 508–47 515.

92 L. Brossay, S. Tangri, M. Bix, S. Cardell, R. Locksley, and M. Kronenberg, *J. Immunol.* **1998**, 160, 3681–3688.

93 M. Skold and S. M. Behar, *Infect. Immunol.* **2003**, 71, 5447–5455.

94 K. Fischer, E. Scotet, M. Niemeyer, H. Koebernick, J. Zerrahn, S. Maillet, R. Hurwitz, M. Kursar, M. Bonneville, S. H. Kaufmann, and U. E. Schaible, *Proc. Natl. Acad. Sci. U.S.A.* **2004**, 101, 10 685–10 690.

95 S. Porcelli, *Adv. Immunol.* **1995**, 59, 1–98.

96 L. Brossay, D. Jullien, S. Cardell, B. C. Sydora, N. Burdin, R. L. Modlin, and M. Kronenberg, *J. Immunol.* **1997**, 159, 1216–1224.

97 M. Exley, J. Garcia, S. B. Wilson, F. Spada, D. Gerdes, S. M. Tahir, K. T. Patton, R. S. Blumberg, S. Porcelli, A. Chott, and S. P. Balk, *Immunology* **2000**, 100, 37–47.

98 G. Gerlini, H. P. Hefti, M. Kleinhans, B. J. Nickoloff, G. Burg, and F. O. Nestle, *J. Invest Dermatol.* **2001**, 117, 576–582.

99 R. S. Blumberg, C. Terhorst, P. Bleicher, F. V. McDermott, C. H. Allan, S. B. Landau, J. S. Trier, and S. P. Balk, *J. Immunol.* **1991**, 147, 2518–2524.

100 B. Bonish, D. Jullien, Y. Dutronc, B. B. Huang, R. Modlin, F. M. Spada, S. A. Porcelli, and B. J. Nickoloff, *J. Immunol.* **2000**, 165, 4076–4085.

101 P. W. Canchis, A. K. Bhan, S. B. Landau, L. Yang, S. P. Balk, and R. S. Blumberg, *Immunology* **1993**, 80, 561–565.

102 W. Kasinrerk, T. Baumruker, O. Majdic, W. Knapp, and H. Stockinger, *J. Immunol.* **1993**, 150, 579–584.

103 H. Thomssen, M. Kahan, and M. Londei, *Cytokine* **1996**, 8, 476–481.

104 F. M. Spada, E. P. Grant, P. J. Peters, M. Sugita, A. Melian, D. S. Leslie, H. K. Lee, E. van Donselaar, D. A. Hanson, A. M. Krensky, O. Majdic, S. A. Porcelli, C. T. Morita, and M. B. Brenner, *J. Exp. Med.* **2000**, 191, 937–948.

105 M. Sugita, E. P. Grant, E. van Donselaar, V. W. Hsu, R. A. Rogers, P. J. Peters, and M. B. Brenner, *Immunity* **1999**, 11, 743–752.

106 J. Salamero, H. Bausinger, A. M. Mommaas, D. Lipsker, F. Proamer, J. P. Cazenave, B. Goud, S. H. de La, and D. Hanau, *J. Invest. Dermatol.* **2001**, 116, 401–408.

107 D. R. McDermott, U. Ziylan, D. Spehner, H. Bausinger, D. Lipsker, M. Mommaas, J. P. Cazenave, G. Raposo, B. Goud, S. H. de La Salle, J. Salamero, and D. Hanau, *Mol. Biol. Cell* **2002**, 13, 317–335.

108 S. Mayor, J. F. Presley, and F. R. Maxfield, *J. Cell Biol.* **1993**, 121, 1257–1269.

109 J. Valladeau, O. Ravel, C. Dezutter-Dambuyant, K. Moore, M. Kleijmeer, Y. Liu, V. Duvert-Frances, C. Vincent, D. Schmitt, J. Davoust, C. Caux, S. Lebecque, and S. Saeland, *Immunity* **2000**, 12, 71–81.

110 R. E. Hunger, P. A. Sieling, M. T. Ochoa, M. Sugaya, A. E. Burdick, T. H. Rea, P. J. Brennan, J. T. Belisle, A. Blauvelt, S. A. Porcelli, and R. L. Modlin, *J. Clin. Invest* **2004**, 113, 701–708.

111 M. Sugita, R. M. Jackman, E. van Donselaar, S. M. Behar, R. A. Rogers, P. J. Peters, M. B. Brenner, and S. A. Porcelli, *Science* **1996**, 273, 349–352.

112 V. Briken, R. M. Jackman, G. F. Watts, R. A. Rogers, and S. A. Porcelli, *J. Exp. Med.* **2000**, 192, 281–288.

113 M. Marks, H. Ohno, T. Kirchhausen, and J. Bonifacino, *Trends Cell Biol.* **1997**, 7, 124–128.

114 M. S. Robinson and J. S. Bonifacino, *Curr. Opin. Cell Biol.* **2001**, 13, 444–453.

115 R. M. Jackman, S. Stenger, A. Lee, D. B. Moody, R. A. Rogers, K. R. Niazi, M. Sugita, R. L. Modlin, P. J. Peters, and S. A. Porcelli, *Immunity* **1998**, 8, 341–351.

116 M. Sugita, X. Cao, G. F. Watts, R. A. Rogers, J. S. Bonifacino, and M. B. Brenner, *Immunity* **2002**, 16, 697–706.

117 K. Hiromatsu, C. C. Dascher, M. Sugita, C. Gingrich-Baker, S. M. Behar, K. P. LeClair, M. B. Brenner, and S. A. Porcelli, *Immunology* **2002**, 106, 159–172.

118 D. Elewaut, A. P. Lawton, N. A. Nagarajan, E. Maverakis, A. Khurana, S. Honing, C. A. Benedict, E. Sercarz, O. Bakke, M. Kronenberg, and T. I. Prigozy, *J. Exp. Med.* **2003**, 198, 1133–1146.

119 S. J. Kang and P. Cresswell, *EMBO J.* **2002**, 21, 1650–1660.

120 R. J. Riese, G. -P. Shi, J. Villadangos, D. Stetson, C. Driessen, A. M. Lennon-Dumenil, C. L. Chu, Y. Naumov, S. M. Behar, H. Ploegh, R. Locksley, and H. A. Chapman, *Immunity* **2001**, 15, 909–919.

121 K. Honey, K. Benlagha, C. Beers, K. Forbush, L. Teyton, M. J. Kleijmeer, A. Y. Rudensky, and A. Bendelac, *Nat. Immunol.* **2002**, 3, 1069–1074.

122 Y. H. Chiu, S. H. Park, K. Benlagha, C. Forestier, J. Jayawardena-Wolf, P. B. Savage, L. Teyton, and A. Bendelac, *Nat. Immunol.* **2002**, 3, 55–60.

123 T. I. Prigozy, P. A. Sieling, D. Clemens, P. L. Stewart, S. M. Behar, S. A. Porcelli, M. B. Brenner, R. L. Modlin, and M. Kronenberg, *Immunity* **1997**, 6, 187–197.

124 D. B. Moody, B. B. Reinhold, V. N. Reinhold, G. S. Besra, and S. A. Porcelli, *Immunol. Lett.* **1999**, 65, 85–91.

125 T. I. Prigozy, O. V. Naidenko, P. Qasba, D. Elewaut, L. Brossay, A. Khurana, T. Natori, Y. Koezuka, A. Kulkarni, and M. Kronenberg, *Science* **2001**, 291, 664–667.

126 W. A. Ernst, J. Maher, S. Cho, K. R. Niazi, D. Chatterjee, D. B. Moody, G. S. Besra, Y. Watanabe, P. E. Jensen, S. A. Porcelli, M. Kronenberg, and R. L. Modlin, *Immunity* **1998**, 8, 331–340.

127 J. S. Im, K. O. Yu, P. A. Illarionov, K. P. LeClair, J. R. Storey, M. W. Kennedy, G. S. Besra, and S. A. Porcelli, *J. Biol. Chem.* **2004**, 279, 299–310.

128 F. Winau, V. Schwierzeck, R. Hurwitz, N. Remmel, P. A. Sieling, R. L. Modlin, S. A. Porcelli, V. Brinkmann, M. Sugita, K. Sandhoff, S. H. Kaufmann, and U. E. Schaible, *Nat. Immunol.* **2004**.

129 S. J. Kang and P. Cresswell, *Nat. Immunol.* **2004**, 5, 175–181.

130 D. Zhou, C. Cantu, III, Y. Sagiv, N. Schrantz, A. B. Kulkarni, X. Qi, D. J. Mahuran, C. R. Morales, G. A. Grabowski, K. Benlagha, P. Savage, A. Bendelac, and L. Teyton, *Science* **2004**, 303, 523–527.

131 U. Bierfreund, T. Kolter, and K. Sandhoff, *Methods Enzymol.* **2000**, 311, 255–276.

132 K. Sandhoff, G. van Echten, M. Schroder, D. Schnabel, and K. Suzuki, *Biochem. Soc. Trans.* **1992**, 20, 695–699.

133 C. G. Schuette, B. Pierstorff, S. Huettler, and K. Sandhoff, *Glycobiology* **2001**, 11, 81R–90R.

134 C. S. Wright, Q. Zhao, and F. Rastinejad, *J. Mol. Biol.* **2003**, 331, 951–964.

135 C. S. Wright, S. C. Li, and F. Rastinejad, *J. Mol. Biol.* **2000**, 304, 411–422.

136 C. S. Wright, L. Z. Mi, and F. Rastinejad, *J. Mol. Biol.* **2004**, 342, 585–592.

137 H. Kropshofer, G. J. Hammerling, and A. B. Vogt, *Immunol. Today* **1997**, 18, 77–82.

138 D. A. Gordon, J. R. Wetterau, and R. E. Gregg, *Trends Cell Biol.* **1995**, 5, 317–321.

139 S. Brozovic, T. Nagaishi, M. Yoshida, S. Betz, A. Salas, D. Chen, A. Kaser, J. Glickman, T. Kuo, A. Little, J. Morrison, N. Corazza, J. Y. Kim, S. P. Colgan, S. G. Young, M. Exley, and R. S. Blumberg, *Nat. Med.* **2004**, 10, 535–539.

Part III
Antigen Presenting Cells' Ligands Recognized by T- and Toll-like Receptors

6
Naturally Processed Self-peptides of MHC Molecules

Harald Kropshofer and Sebastian Spindeldreher

6.1
Introduction

Almost 20 years have passed since pioneering work has revealed that most MHC class II molecules are constantly occupied by peptide antigens derived from self-proteins, termed self-peptides. The same phenomenon turned out to be true for MHC class I molecules.

These seminal findings gave rise to the idea that self-peptides may not solely be present on thymic antigen presenting cells (APCs) to allow for positive and negative selection of thymocytes, but may play a broader role in the adaptive branch of our immune system. Several years later it became clear that MHC class II-associated self-peptides on professional APCs, in particular dendritic cells (DCs), are critical for the survival of the $CD4^+$ helper T cell pool in the periphery and in the maintenance of peripheral tolerance against self-proteins.

Most recently, confocal microscopy technology disclosed that self-peptides apparently collaborate with foreign antigen derived peptides in triggering foreign antigen-specific helper T cells: there is evidence for focused co-localization of agonistic foreign peptides and non-agonistic self-peptides in the immunological synapse, encompassing the contact zone between an activated APC and a T cell. From these studies the concept emerges that self-peptide-MHC class II complexes favor rather than compete against recognition of foreign antigenic peptides.

Accordingly, to understand the pathogenesis of autoimmune diseases it may be essential to know the identity of the respective self-peptides that activate autoreactive helper or cytotoxic T cells, but is not sufficient in case foreign antigens do collaborate with self-peptides in triggering auto-reactivity. Finally, the knowledge of self-peptides derived from tumor-specific neo-self antigens may be critical for the design of more powerful vaccination strategies in the fight against cancer. Hence, MHC class II-bound self-peptides obviously play a considerably broader role in controlling helper T cell homeostasis than anticipated a decade ago. Yet, the naturally occurring self-peptides responsible for the aforementioned diverse tasks in health and disease remain to be identified in most instances.

Antigen Presenting Cells: From Mechanisms to Drug Development. Edited by H. Kropshofer and A. B. Vogt
Copyright © 2005 WILEY-VCH Verlag GmbH & Co. KGaA, Weinheim
ISBN: 3-527-31108-4

6.2
Milestone Events

In the early 1980s, immunologists realized that either peptide digests or synthetic peptides of model antigens, such as ovalbumin or hen egg lysozyme, in combination with the appropriate MHC class II molecules are sufficient to stimulate $CD4^+$ helper T cells [1–4]. At the same time it was established that protein antigens need to be physically altered or "processed" by the APC before being recognized by T cells [5–7]. From then on, laboratories attempted to characterize binding of a small variety of foreign antigens, either native proteins, protein fragments, or fully synthetic peptides, to intact MHC class II-bearing cells, membrane bilayer-embedded or solubilized MHC molecules [4, 8–10].

Most attempts using fresh APCs resulted in a high background of non-specific binding or poor T cell reactivity. This was because more than 90% of the sole MHC peptide sites were obviously pre-occupied with so-called "autologous" or "endogenous" peptides derived from the APC's own self-proteins [4, 11]. This view was supported by the first X-ray structural analysis of an MHC molecule, the human class I molecule HLA-A2: although no exogenous antigen was added, the peptide binding cleft of HLA-A2 was filled with "electron-dense material", which has been explained by the presence of a large variety of endogenous peptides [12]. This latter findings fuelled the awareness derived from protein studies, whereafter macrophages and other professional APCs cannot discriminate between self and non-self antigens [13, 14], because MHC molecules bind both types of antigens in the very same peptide binding cleft. The conclusion was that self-peptides apparently compete against foreign peptides at the level of antigen presentation to T cells. No wonder a race started in the mid-1980s to elucidate the identity of the postulated endogenous or autologous peptides that are constantly being generated by APCs and presented via MHC molecules.

Some of the burning questions at the time were:
1. Typical self-peptides are derived from which self-proteins?
2. How great is the diversity of self-peptides?
3. Different types of APCs harbor different sets of self-peptides?
4. What is the function of self-peptides in APCs of peripheral tissue?
5. Does the typical self-peptide repertoire contain epitopes triggering autoreactive T cells?

6.2.1
Nomenclature

6.2.1.1 Autologous Peptides

"Autologous peptides" is the original term used to define naturally processed MHC-associated peptides [15]. In the narrowest sense, autologous peptides are fragments from proteins derived from the same type of APC that expresses the MHC molecule the peptide has been found associated to. The term "self-peptides" is equivalent to the term "autologous peptides". It is used to underline the difference from "foreign peptides" that are derived from antigenic proteins of invading microorganisms.

6.2.1.2 Endogenous Peptides

Self-peptides purified from a particular type of APC can be derived either from the APC's own protein repertoire or from a protein that is expressed by another cell type of the same individual and has been taken up by the APC. In the former case the peptides are termed "endogenous", in the latter case they are termed "exogenous". Babbit et al. reported already in 1986 that the murine MHC class II molecule I-A^k was capable of binding the endogenous and autologous mouse lysozyme peptide 46–61 and the homologous exogenous xenogenic hen egg lysozyme peptide equally well [14]. This finding was also regarded as convincing evidence that MHC molecules do not distinguish between peptides representing self and non-self. Originally, it was believed that exogenous peptides are exclusively bound and presented by MHC class II molecules, whereas endogenous peptides can bind to both MHC class I and II molecules. Meanwhile, it became clear that APCs, such as DCs or macrophages, present exogenous peptides via MHC class I molecules, a phenomenon termed "cross-presentation" [16, 17].

6.2.1.3 Natural Peptides Ex Vivo and In Vitro

Table 6.1 lists the origin of naturally processed MHC class II-associated peptides derived from various cellular sources. Obviously, from the definitions given above, peptides derived from MHC molecules, the invariant chain, membrane, cytosolic or nuclear proteins are real "self-peptides" or "autologous peptides" with regard to human DCs [18], human peripheral blood mononuclear cells [26] or murine B cells [27]. However, peptides derived from serum proteins in preparations from cultured DCs or B cells are not of "endogenous" origin, as they are derived from the culture medium. Hence, exogenous peptides from serum proteins of cultured cells do not belong to the category of typical "self-peptides". However, the very same serum proteins, e.g., serum albumin or apolipoprotein, give rise to real "self-peptides" in the case of human peripheral blood mononuclear cells (Table 6.1). Here, "self" refers to the human body of the respective blood donor and not to a particular cell-type. In conclusion, it is essential to be aware of the context in order to use and interpret the aforementioned terms appropriately.

6 Naturally Processed Self-peptides of MHC Molecules

Table 6.1 Origin of class II MHC-associated natural peptides from different APC.

Origin	%				Ref.
	Human DC HLA-DR [18]	Human PBMC HLA-DR [26]		Murine B cells I-A^b [27]	
		Donor 1	Donor 2		
MHC molecules	3	2	4	5	18–23, 26, 27
Invariant chain	4	0	0	7	18–27
Membrane and vesicular proteins (e.g. receptors, cathepsins)	38	2	21	41	18–23, 26, 27
Cytosolic proteins (e.g. ribosomal proteins, actin)	24	64	8	36	18, 23, 26, 27
Nuclear proteins (e.g. histones, hnRNP)	7	14	4	0	18, 23, 26
Serum proteins (e.g. albumin, apolipoproteins)	24	18	63	11	18–23, 26, 27

6.2.2
Extra Electron Density Associated to MHC Molecules

Throughout the 1970s, possible X-ray crystallographic analysis of MHC molecules was the focus of numerous informal discussions. However, all attempts failed until Pamela Bjorkman in the laboratory of Don Wiley obtained suitable crystals of the MHC class I molecule HLA-A2 in 1979. Another 8 years passed before the three-dimensional structure of HLA-A2 was published [28], giving rise to a vastly growing immunological subspecialty, the study of antigen presentation. A deep groove formed by polymorphic parts of the MHC class I heavy chain was seen to be filled with slurry "extra electron density", apparently coming from a mixture of naturally processed endogenous peptides. Likewise, extra electron density was detected in the X-ray analysis of the first class II MHC molecule HLA-DR1 [29] (Figure 6.1). With refinement of the second MHC class I molecule, HLA-Aw68, it became clear that the peptide binding cleft contains substructures, so-called "specificity pockets" A–F, that were also filled with extra electron density [30]. The precise shape of these pockets turned out to differ between MHC alleles, thereby determining which set of peptides can fit into the peptide binding cleft.

Due to the high degree of heterogeneity of the set of bound self-peptides, their shapes remained decidedly uncertain. Realizing the complexity of peptide-sized structures in both the HLA-A2 and HLA-Aw68 crystals, and triggered by the

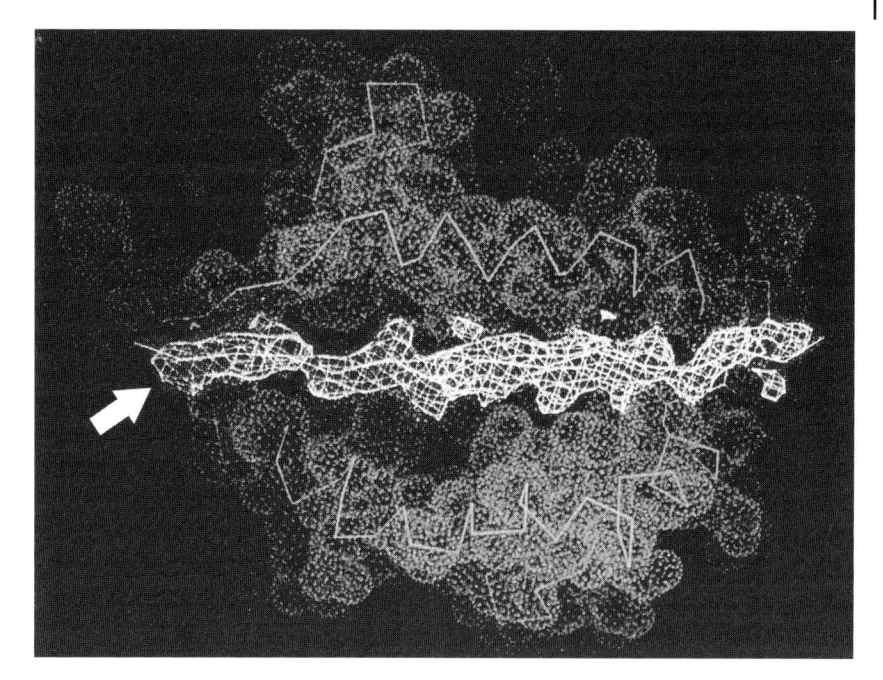

Figure 6.1 Extra electron density in the MHC class II HLA-DR1 crystal structure. Top view on the peptide binding domain reveals electron density (cf. arrow) that does not pertain to the class II MHC sequence. The rationale for this observation is the presence of a large variety of self-peptides. (Adapted from Ref. 29.)

increasing certainty that the groove-embedded material is central to immune recognition, numerous of immunologists, biologists and chemists set out to develop techniques to solve the mystery of the identity of the visualized self-peptides.

6.2.3
Acidic Peptide Elution Approach

In the mid-1980s, at least three independent lines of evidence indicated that self-peptides may constitutively occupy the binding site of MHC molecules:
1. Endogenously expressed murine lysozyme peptide 46–61 was found to bind with high affinity to murine MHC class II I-A^k molecules [14].
2. X-ray crystallographic structures of human MHC class I molecule HLA-A2 unraveled electron-dense material occupying the putative peptide binding groove [28].
3. Only about 5–10% of affinity-purified MHC class II molecules have the capacity to bind foreign antigenic peptides [11].

The pioneering work essential to studying natural self-peptides was performed in 1988 in the laboratory of Howard Grey. It was Soren Buus in collaboration with

Alessandro Sette who realized that I-A^d or I-E^d molecules do not lose putatively bound self-peptides when subjected to pH 10.5 treatment – a step included in their affinity purification protocol [15]. Therefore, they treated affinity-purified I-A^d and I-E^d molecules with acetic acid, pH 2.5, separated the low-molecular-weight (LMW) material from the MHC molecules by gel-filtration and fractionated the LMW pool by reversed-phase high-performance liquid chromatography (RP-HPLC) (Figure 6.2).

The broad peak that eluted at 23–43% acetonitrile in the HPLC fractionation contained material that displayed five critical features:

1. Material eluted from I-A^d inhibited binding of an I-A^d-restricted reporter peptide.
2. I-A^d-eluted material hardly bound to I-E^d.
3. The material lost its inhibitory capacity upon protease digestion.
4. The mean molecular weights were 3500 and 2000 for I-A^d and I-E^d, respectively, as judged by gel-filtration.
5. The broad RP-HPLC elution profile points to a high diversity, with regard to the hydrophobicity of the material.

```
Antigen Presenting Cells
          ↓
Detergent lysis:
Solubilization of MHC-peptide complexes
          ↓
Affinity chromatography:
Isolation of MHC-peptide complexes
          ↓
Acid elution:
Extraction of naturally processed peptides
          ↓
Gel filtration:
Separation of peptides from MHC
          ↓
RP-HPLC:
Fractionation of peptides
          ↓
T cell activation assay:
Detection of T cell-stimulating peptides
```

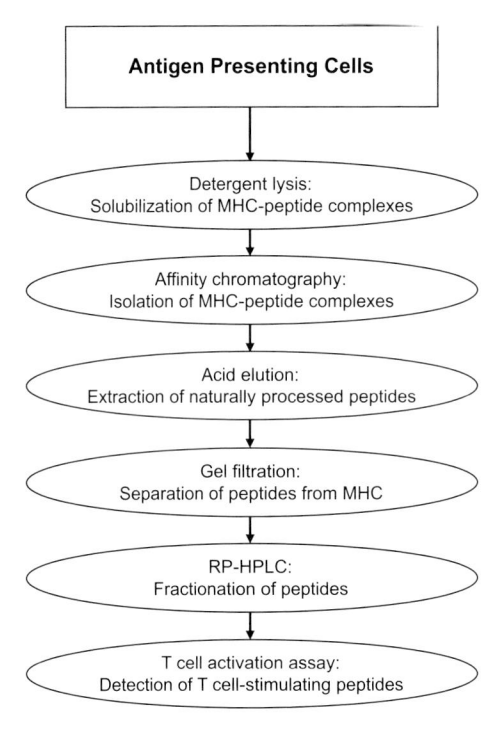

Figure 6.2 Strategy for isolating naturally processed MHC-associated peptides. Shown is the extraction method pioneered by Buus et al. [15] and optimized by Demotz et al. [31]. This technique was fundamental to numerous studies performed at later stages in both the MHC class I and class II field.

These results demonstrated for the first time that MHC molecules constitutively carry a heterogeneous set of naturally processed low-molecular-weight peptides that can be eluted by acid treatment. The two-step methodology introduced here, consisting of acid elution and RP-HPLC fractionation, was ground-breaking for both the MHC class II and MHC class I-field, and is still fundamental to the analysis of natural antigenic peptides almost two decades later.

6.2.4
First Natural Foreign Peptides on MHC Class II

Stephan Demotz and colleagues at Cytel were the first to apply successfully the acid elution/HPLC fractionation technique to characterize naturally processed length variants of a known foreign antigenic epitope, the I-E^d-restricted hen egg lysozyme-derived HEL(107-116) [31]. To detect the eluted peptides with maximal sensitivity, Demotz et al. introduced a T cell activation read-out: they prepared acid-eluted peptides from HEL-pulsed B lymphoma A20 cells and loaded them onto I-E^d purified from unpulsed A20 cells, incorporated the newly generated I-E^d-peptide complexes into planar membranes and measured IL-2 production of an I-E^d/HEL(107-116)-specific T cell hybridoma [31]. The significant extent of T cell stimulation proved that the mixture of eluted peptides contained the HEL(107-116) sequence and that functional I-E^d-HEL(107-116) complexes could be reconstituted by using acid-eluted peptides.

Beyond that, the T cell read-out of HPLC-fractionated natural peptides revealed that A20 lymphoma cells generated more than one length variant of HEL(107-116): from the HPLC retention time of synthetic HEL(107-116) and HEL(105-120) it was concluded that at least 2–3 length variants were eluted from A20 cells [31]. The microheterogeneity obtained upon HPLC analysis was later confirmed by sequence analysis of natural self-peptides eluting as nested sets of length variants of a single epitope [22].

Finally, by calibrating the T cell activation capacity per ng of synthetic HEL(107-118), the authors estimated that 10–40% of I-E^d molecules were originally occupied by HEL-derived naturally processed peptides. However, this astonishingly high number may rely on an over-estimation as, according current knowledge, it cannot be excluded that some natural HEL(107-116) length variants may be more potent than synthetic HEL(107-116) in activating T cells. Likewise, co-eluting self-peptides unrelated to HEL(107-116) may have contributed to T cell activation, according to the pseudodimer model (cf. Section 6.5.4).

6.2.5
First Natural Viral Epitopes on MHC Class I

A very similar strategy led to the isolation and sequence identification of the first endogenously processed MHC class I-restricted antigenic peptide by Grada van Bleek and Stanley Nathenson in 1990 [32]. They infected EL4 tumor cells with vesicular stomatitis virus (VSV) and fractionated the set of natural K^b-associated peptides. Compared with uninfected cells, material from [^{3}H]Tyr and [^{3}H]Leu metabolic radiolabeling experiments was sequenced and resulted in the octapeptide N(52-59) from the nucleocapsid protein of VSV that displayed allele-specific binding to K^b. This octapeptide turned out to be the major VSV antigenic determinant by both its sequence and biological activity, thereby supporting earlier reports showing that from a whole virus only one dominant peptide epitope is recognized per MHC allele [33, 34].

By an alternative strategy, where natural peptide extracts of whole tumor cells were used, the first sequences of influenza virus-specific T cell epitopes were elucidated [35]. From previous studies influenza nucleoprotein obviously contains two epitopes recognized by murine MHC class I molecules: the D^b-restricted 16-mer NP(365-380) and the K^d-restricted 12-mer NP(147-158).

Fractionation of the synthetic 12-mer and 16-mer by RP-HPLC revealed that several peptide truncation variants were far more potent in activating cytotoxic T cells recognizing these NP epitopes [35]. In both cases, one of these by-products appeared to be identical to the natural influenza NP epitope extracted from infected tumor cells. The respective sequences could be determined by a combination of mass spectrometry and comparison of their HPLC retention time with several synthetic length variants. Both natural influenza epitopes are nonamers: the D^b-restricted NP(366-374) and K^b-restricted NP(147-155).

Rough estimates indicated that 200–600 copies of both epitopes were present per infected cell [36]. Both nonamers were 10^2–10^3-fold more effective than their larger synthetic length variants [35, 36]. Likewise, a greater than 1000-fold lower off-rate for the natural NP(366-374) peptide as compared to the longer NP(Y365-380) variant has been described [37].

The same type of studies gave rise to the observation that most naturally processed peptides can only be extracted from cells that express the restricting MHC molecule [35, 38, 39]. A striking example is the K^b-restricted minor H antigen (H-4^b) that is at least 3000-fold enriched in K^b-transfected tumor cells compared with K^b-negative cells. The rationale for this finding is that peptides are rendered inaccessible to proteases and peptidases by binding to MHC molecules. Otherwise they are rapidly degraded.

6.2.6
Self-peptide Sequencing on MHC Class I: the First Anchor Motifs

The methods applied between 1987 and 1990 to characterize natural MHC ligands were limited to the identification of peptides for which specific T cells were available. Thus, novel epitopes for true self-peptides remained undiscovered. A major breakthrough in the characterization of class I MHC-associated peptide motifs was the approach of pooled sequencing of self-peptides eluted from purified MHC class I molecules by Edman degradation [40]. This methodology was developed in the laboratory of Rammensee and led to the elucidation of numerous allele-specific MHC class I ligand motifs based on naturally-processed self-peptides [40–42].

The idea to sequence the whole mixture of eluted self-peptides, rather than pursuing the conventional strategy to purify single peptides to homogeneity first, was driven by several considerations: (i) the main obstacle arguing against conventional techniques is the high heterogeneity of self-peptides bound to a single type of MHC molecule, rendering it difficult to obtain single peptides by RP-HPLC fractionation or alternative chromatographic steps; (ii) due to limitations in the sensitivity of the microsequencing technology, billions of cells are necessary to do sequence analysis; and (iii) even sharp peaks in HPLC fractionations may harbor more than a dozen different peptide sequences with similar degree of hydrophobicity. Thus, the concept arose that the radioactivity obtained by van Bleek and Nathenson in a previous report [32] in sequencing cycles 3 and 5 after metabolic labeling of EL4 cells with [^{3}H]Tyr following RP-HPLC fractionation of K^b-associated natural peptides and further analysis of the major VSV-induced peptide material may reflect the dominance of Tyr at positions 3 and 5 not only of the VSV nucleoprotein peptide N(52-59), as originally thought [32], but of several other K^b-bound self-peptides co-eluting with N(52-59).

Indeed, pool sequencing of the whole self-peptide mixture eluted from K^b molecules revealed a strong increase of Tyr and Phe in sequencing cycle 5 and a rise on Leu in cycle 8. In addition, a less pronounced increase was seen with Tyr in cycle 3 and Met in cycle 8 [40]. The result of these studies indicated that class I molecules follow stringent rules that are different for each individual class I allele. Allele-specific interactions rely on appropriately positioned anchor side-chains to fit into specificity pockets of the peptide binding groove [30]. The sum of allele-specific peptide-class I MHC interaction requirements defines the binding motif for a given MHC molecule, characterized by the number, spacing and specificity of anchors. Most of the motifs determined by pooled sequencing of naturally processed class I-associated self-peptides have anchor residues at positions P2, P3 or P5/P7 in addition to a C-terminal anchor [41–43].

The motifs deduced from pooled sequencing were subjected to refinement by in vitro peptide studies. Udaka et al. used octapeptide libraries and found that the K^b binding ligands do not solely rely on the major anchor residues, located at positions 5 and 9, but that all amino acids in octapeptides contribute to the overall binding affinity [44]. The combinatorial peptide library approach clearly showed that almost at each position particular amino acids are disfavored. These disfavored residues can prevent a peptide from binding to an MHC molecule even when the peptide bears allele-specific anchors at the appropriate positions [44].

6.2.7
First Murine MHC Class II-associated Self-peptides: Nested Sets

Only 5 months after the seminal report on pool sequencing of MHC class I self-peptides by Falk et al. [40], the group of Charles Janeway published the first de novo sequences of self-peptides eluted from MHC class II molecules [19]. The peptides were derived from a murine B cell lymphoma cell line and eluted from I-A^b and I-E^b molecules. Fortunately, as became clear later, Rudensky et al. did not chose the pooled sequencing technique: they used affinity-purified MHC class II molecules from 18 billion B cells, acid eluted the peptides, subjected them to RP-HPLC and sequenced them peak-by-peak via Edman microsequencing [19].

The conclusions drawn from the first set of 12 peptides differed in most instances from the characteristics of MHC class I-associated peptides described before:

1. The length varied from 13 to 17 residues, in a follow-up paper from the same group even a 22-mer was described [45].
2. The peptides displayed ragged ends.
3. It was difficult to define anchor motifs [19].

A first attempt to propose such motifs turned out to be unreliable, e.g. peptide binding to I-A^s does not solely depend on two anchors P1, P7 with the spacing i, i+6, as suggested originally [45], but on four anchors (P4, P6, P7 and P9), as shown almost a decade later [46].

Since the 1991 paper by Rudensky et al. was the first report involving classical sequence analysis, the immunological community was most interested in knowing to which parent proteins the self-peptides presented by MHC class II have to be assigned. In the context of I-A^b, only five parent proteins covering nine natural peptides were found [19, 20, 45]: the invariant chain, I-Eα chain, I-Aβ chain, IgV$_H$ chain and the murine leukemia virus envelope protein (MuLV) (Table 6.2). Likewise, the seven self-peptides eluted from I-E^b were derived from bovine serum albumin and MuLV envelope protein. Finally, again MuLV envelope protein, IgG2a and the transferrin receptor served as the sources for six self-peptides associated to I-A^s molecules (Table 6.2). Importantly, the same parent proteins gave rise to I-A^s-associated self-peptides when another B lymphoma line was investigated recently [46]. Even the "undefined" protein source, as described originally [20], was elucidated: it is the mosaic protein LR11 [46]. In conclusion, besides MHC molecules themselves and the accessory invariant chain, in particular membrane proteins from the endocytic pathway and serum albumin turned out to fuel the MHC class II processing pathway in generating autologous peptides. Most strikingly, only a few dominant epitopes and length variants thereof apparently constituted most of the natural peptide repertoire of cultured murine B cell lymphomas.

The first successful sequence analysis of MHC class II-associated peptides also allowed us to draw an interesting link between self-peptides presented by B cells and APCs in the thymus. The determinant recognized by the monoclonal antibody Y-Ae turned out to be the self-peptide Eα(52-68) bound to I-A^b [19, 47]. The remarkable feature of Y-Ae is that it readily stains medullary thymic epithelial cells (mTECs) but not cortical thymic epithelial cells (cTECs), albeit both types of thymic APCs express I-A^b and Eα molecules [48]. Thus, the Eα(52-68) self-peptide in the context of I-A^b is likely to contribute to negative selection of thymocytes through its presence on mTECs, whereas it is unlikely to trigger positive selection due to its absence on cTECs.

Table 6.2 The first class II MHC-associated self-peptides.

Allel	Parent protein	Length	Cell source	Ref.
Murine I-A^b	MuLV protein	13, 14	Lymphoma	19
	I-E α chain	15, 17	B cells	20
	Invariant Chain	15	Splenocytes	45
	I-A β chain	12	Spenocytes	45
	Ig V$_H$ chain	15	Splenocytes	45
Murine I-A^s	MuLV protein	15, 13	Lymphoma	19
	Transferrin receptor	16, 15	B cells	45
	IgG2a	16, 17	B cells	45
Human HLA-DR1	Unknown	16	EBV-transformed B cells	50

6.2.8
First Human MHC Class II-bound Self-peptides: Hydrophobic Motifs

To the astonishment of experts in this field, the first naturally processed peptides identified from class II MHC molecules did not display an apparent anchor motif, which was one of the hallmarks of both murine and human MHC class I-associated self-peptides. The situation changed when, about 6 months later, the first studies on human self-peptides from class II HLA-DR molecules were published. Our laboratory described the first human self-peptide sequence and named it "SP3", as the parent protein remained unknown [49]. SP3 was eluted from HLA-DR1 expressed by an EBV-transformed B lymphoblastoid cell line and was sequenced by Edman microsequencing (Table 6.2). According to the HPLC profile, it was one of the most abundant self-peptides. Systematic binding analysis with the synthetic counterpart of SP3 revealed a two-residue ligand motif for HLA-DR1 [49]: two bulky hydrophobic anchor residues with the relative spacing i, i+8 were sufficient for binding to HLA-DR1. Later studies revealed that SP3 belongs to the subset of high-affinity self-peptides in the context of HLA-DR1 [50]. In successive studies, it became clear that a bulky aliphatic or aromatic N-terminal anchor residue (P1 anchor) is of particular importance for binding to HLA-DR molecules [22, 51].

This phenomenon was also reflected when self-peptide mixtures eluted from different DR alleles were subjected to pooled sequencing [52]: in most instances, hydrophobic residues were enriched in cycles 2–4 of the microsequencing analysis. Due to the ragged ends of most self-peptide epitopes derived from class II MHC the pooled sequencing technology was unsuitable to define exact anchor positions. More refined anchor motifs were elucidated by in vitro binding analysis involving M13 phage peptide libraries [53]. Regarding HLA-DR1, the phage display approach confirmed and extended the findings deduced from the self-peptide

SP3: an aromatic residue was shown to be the optimal P1 anchor, an aliphatic residue was favorable at P9. In addition, Met or Leu can serve as a further anchor at P4. At P6, only very small amino acids, such as Gly, Ala or Pro, are allowed. Importantly, most HLA-DR1-associated self-peptides described in the meantime share only 2–3 of the characteristic features of the ligands motif described here, with only the P1 feature being obligatory. Accordingly, self-peptide SP3 can bind in at least two registers into the binding groove of HLA-DR1, with either Phe at position 5 or Ile at position 2 functioning as a P1 anchor (Figure 6.3).

Besides giving insight into anchor motifs of class II ligands, the initial Edman sequencing efforts gave us also clues as to what rules may govern antigen processing: in sequencing cycles 2 and 3, irrespective of the HLA-allele, there was a strong enrichment of Pro residues [52]. The rationale is that Pro acts as a stop signal for aminopeptidases. This means that the N-terminus of natural self-peptides is very frequently the result of successive trimming by exopeptidases, which is terminated by Pro residues [54].

Two additional reports, generated in the laboratory of Jack Strominger, were ground-breaking in defining further critical characteristics of class II HLA-DR associated self-peptides. Chicz et al. established a powerful combination of Edman microsequencing and matrix-assisted laser desorption mass spectrometry [22, 51]. They analyzed six HLA-DR alleles from a respective number of EBV-transformed human B cell lines and identified more than 200 self-peptide sequences. More than 85% of the peptides were of endogenous origin and most of them were derived from vesicular or membrane-spanning parent proteins. A similar distribution has been found with cultured human dendritic class and murine B cells but not with human PBMCs (Table 6.1). Strikingly, 20–25% of the self-peptides were derived from other MHC class I or II molecules, co-expressed by the respective B cell [54]. Moreover, numerous self-epitopes were represented by nested sets of N- and C-terminally truncated length variants, e.g. the set of Igκ chain peptides eluted from HLA-DR4 comprised 15-mers to 21-mers [51]. Overall, the shortest peptides were 13-mers, the longest were 30-mers; the majority attained a length of 15–17 residues.

Several epitopes were eluted from more than one HLA-DR allelic product, e.g. the HLA-A2-derived peptide A2(103-117) or the invariant chain peptide CLIP [22, 51]. The phenomenon of promiscuous binding to several HLA-DR allelic products is reminiscent of a few foreign epitopes, such as those derived from tetanus toxoid or influenza virus hemagglutinin [55].

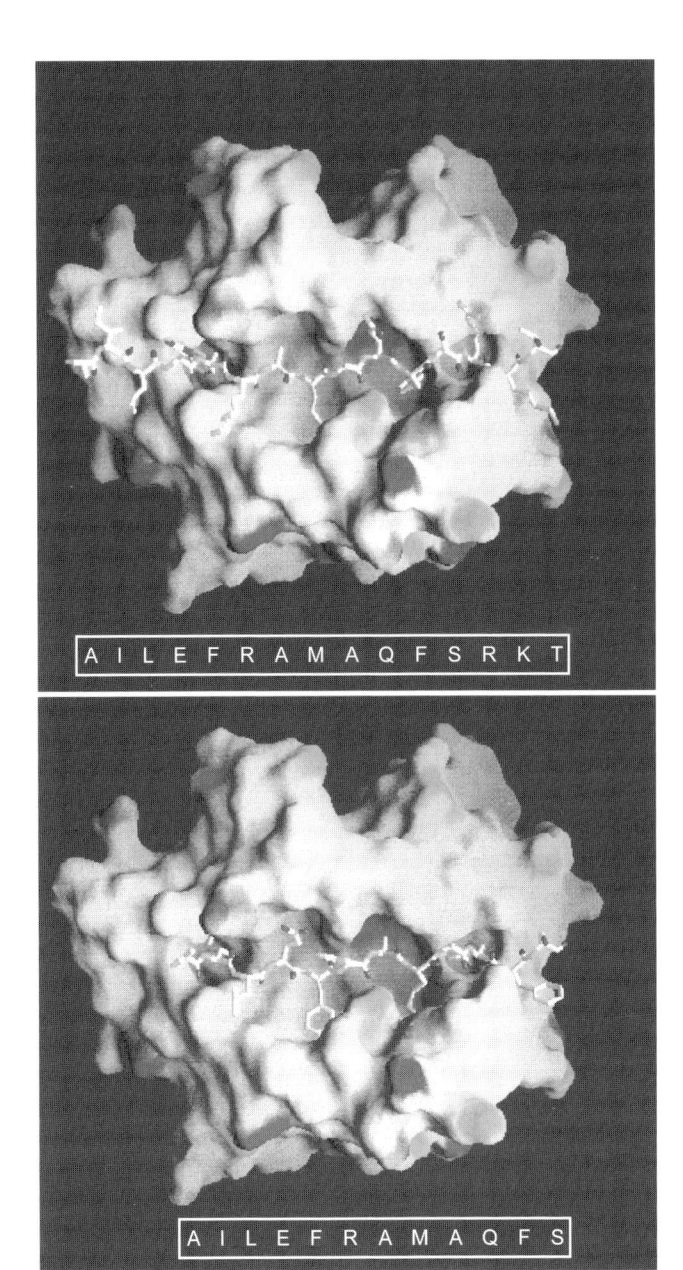

Figure 6.3 Computer modeling of the HLA-DR1:SP3 self-peptide complex. According to the X-ray crystal structure of the HLA-DR1 molecule, the self-peptide SP3 can bind into the peptide binding cleft in two different registers: either F-5 binds into the P1 pocket (upper panel) or I-2 (lower panel). The latter option foresees the hydrophobic residues L-3 and F-11 pointing towards the hydrophilic outer milieu, which is energetically disfavored. (Kindly provided by L. Mozyak, Harvard University, Boston, USA.) (This figure also appears with the color plates.)

6.3
Progress in Sequence Analysis of Natural Peptides

In the early 1990s when the first self-peptide sequencing attempts were envisaged, immunologists had to get used to at least two innovations: mass cell culture and HPLC technology. However, although several billions of cells had to be prepared, the heterogeneity of the self-peptide mixtures, the limitations in the resolution of 1D chromatography and the poor sensitivity of conventional Edman degradation rarely allowed identification of more than a dozen self-peptides per MHC allele.

One alternative to escape these limitations was the pooled sequencing technique, which was very successfully applied for the elucidation of anchor motifs in the MHC class I field. This technique failed in the MHC class II field due to the ragged ends of class II MHC-associated self-peptides.

The inclusion of mass spectrometry was the next step that led to an increase in the fidelity of the analysis and the number of peptides that could be identified. In particular, electrospray ionization tandem mass spectrometers directly coupled to RP-HPLC devices became powerful tools in the second half of the 1990s for sequencing natural peptides released from both MHC class I and class II molecules.

In the past 5 years, the objective was to increase the sensitivity of the whole methodology to such an extent that self-peptides from tissue samples rather than only peptides from cultured cells became accessible. To this end, sample preparation and peptide fractionation techniques were continuously optimized. In addition, HPLC separation and mass spectrometry were coupled in such a way that sample losses were minimized. The currently most advanced high end procedure is the so-called Multidimensional Protein Identification Technology (MudPIT). Compared with the original Edman technique, MudPIT reduced the cell requirements by about 1000- to 100-fold and increased the sequence output by about 30- to 50-fold. The following sections outline a few more details of the key sequencing technologies by describing a few key achievements attained with the respective technologies.

6.3.1
Edman Microsequencing

The heterogeneity of the self-peptide repertoire of MHC-associated peptides was so high that HLPC fractionation yielded only a handful of peptides that were pure enough for Edman sequencing. Beyond that, several milligrams of MHC molecules had to be eluted to gain sufficient self-peptide material. Therefore, only cells growing rapidly and easily in cell culture could by analyzed – mainly transformed B cell lines.

Shortly after the pooled sequencing attempts by Rammensee and colleagues, Jardetzky et al. published the first study on MHC class I-associated self-peptides, where a dozen individual self-peptides were identified by classical Edman sequencing and a binding motif was precisely elucidated [56]. As expected from a typical

MHC class I molecule that is responsible for the presentation of endogenous non-vesicular antigen sources, HLA-B27 displayed self-peptides derived from abundantly expressed proteins: ribosomal proteins, translation factors, cytosolic heat shock proteins and nuclear proteins. Apart from two apparently artificial HLA-B27 fragments, all peptides were nonamers and shared a common Arg anchor residue at position P2. Most of the peptides had a second positively charged anchor residues at P9.

These findings are relevant because the HLA-B27 genotype is strongly linked to the autoimmune disease ankylosing spondylitis [57]: knowing the ligand motif of HLA-B27 gives us a better chance of elucidating the autoantigenic peptides that contribute to this disease.

6.3.2
Electrospray Ionization Tandem Mass Spectrometry

In 1992, Hunt et al. pioneered the use of electrospray ionization tandem mass spectrometry (ESI-MS/MS) for sequence analysis of natural peptides eluted from MHC molecules [58]. This strategy relies on fractionation of peptide mixtures by microcapillary RP-HPLC and on-line coupling of the HPLC capillary with a mass spectrometer. The technique is not only able to determine the molecular mass, and therefore maximum length of each peptide component, it also allowed the analysis of HPLC fractions that contained more than one peptide. Sequence information can be obtained on sub-picomolar amounts of peptides by subjecting them to collision-activated dissociation on the triple-quadrupole mass spectrometer. Although, in principle, this technique is considerably more sensitive than the Edman technology, 200 µg to several mg of MHC protein were necessary – this corresponds to 10^9 to 10^{10} B cells [21, 58, 59].

Peptides eluted from the class I MHC molecule HLA-A2 may be representative of the value this technique has added to the field of naturally processed antigenic peptides [58, 59]. In human B lymphoblastoid cells, the existence of at least 200 distinct self-peptide species could be shown and eight of them could be sequenced [58]. They were all 9-mers, sharing a Leu at the P2 anchor position and Val, Leu or Ile at the P9 anchor position, which broadly agrees with the originally proposed HLA-A2 anchor motif deduced from pooled sequencing data [40] (Table 6.3). Analysis of a corresponding mutant B cell line that is defective in the transporter associated with antigen processing (TAP) revealed that HLA-A2 can bind self-peptides derived from the signal peptide domains of ER-resident proteins [59]. These peptides were usually larger than nine residues: 10- to 12-mers were described that were comparable to 9-mers with regard to their binding affinity.

Assuming that 10-mers or longer variants bind with the same P2 and P9 anchors into the peptide binding groove of HLA-A2 as 9-mers do, natural peptides can, obviously, protrude by 1 or 2 residues from the MHC class I groove. Systematic binding analysis involving synthetic peptide libraries have confirmed that 9-mers and 10-mers can bind with high affinity to HLA-A2 and that not only the two major anchor positions P2 and P9 but also residues at positions P3 to P7 contribute positively or negatively to the overall stability of HLA-A2 peptide complexes ([60]; Table 6.3).

The ESI-MS/MS technique also shed light on the natural peptide repertoire of class II MHC molecules. Again, Hunt et al. provided primary evidence that between 650 and 2000 distinct peptides are associated with the murine I-A^d molecule [21]. Sequences of nine self-peptides were described – all derived from secretory or integral membrane proteins. Interestingly, four of the five epitopes, where the parent proteins could be identified, were also described later in the context of human class II HLA-DR molecules [18, 23, 51, 61], albeit the sequence homology between the murine and human sequences of the respective epitopes was, in at least two cases, less than 70% (Table 6.4). These epitopes were derived from the murine I-E^d α chain that is highly homologous to the monomorphic HLA-DR α chain, the transferrin receptor, the exocytic protease inhibitor cystatin C and the invariant chain-derived peptide CLIP that has also been described as a self-peptide of I-A^b [19].

Table 6.3 Peptide binding motif of HLA-A*0201.

Report	Method	Anchor type[a]	P1	P2	P3	P4	P5	P6	P7	P8	P9	Length
Falk et al. 1991 [40]	Pooled sequencing	A1		L							V	9
		A2		M		E		V		K		
							K					
Hunt et al. 1992 [58, 59]	ESI-MS/MS	A1		L							V	9–12
											L	
											I	
Ruppert et al. 1993 [60]	Peptide binding assays	A1		L							V	9–10
		A2				Y	S	Y		A	P	
						F	T	F				
						W	G	W				
		R				D			R	D		
						E			K	E		
						R			H	R		

a The following anchor types have been defined: A1, primary anchor residue; A2, secondary anchor residue; R, residue with repulsory capacity.

Table 6.4 Reoccurring self-peptide epitopes on human and murine MHC class II.

Self-peptides from murine I-A^d [18]	Parent protein	Human analogues		
		Sequence	HLA class II allele	Ref.
ASFEAQGALANIAVDKA	E$^d\alpha$ [a]	GRFASFEAQGALAN	DR1/DR5	18
		SFEAQGALANIAVDKA	DQ7	61
VPQLNQMVRTAAEVAG	TfR [b]	XPELNKVARAAAEVAG	DR5/DRw52	23
DAYHSRAIQVVRARKQ	Cys-C [c]	DEYYRRLLRVLRAREQIV	DR8	51
KPVSQMRMATPLLMRPM	CLIP [d]	PPKPVSKMRMATPLLMQALP	DR1/DR2	51

(a) I-E$^d\alpha$.
(b) Transferrin receptor.
(c) Cystatin C.
(d) Class II MHC associated invariant chain peptide.

6.3.3
Automated Tandem Mass Spectrometry

ESI-MS/MS rendered the sequence analysis of MHC-associated peptide mixtures more reliable and increased the efficacy of sequencing. However, slow data acquisition and laborious manual interpretation of MS spectra limited the output of this technology. Technical improvements of the mass spectrometry devices, automated mass-spectral data acquisition and, in particular, data analysis by novel computer algorithms significantly increased the number of discernible peptide sequences. The most important example is the algorithm SEQUEST, which uses protein and nucleotide sequence databases to perform cross-correlation analyses of experimental and theoretically generated tandem mass spectra [62]. An alternative and very frequently employed algorithm is MASCOT.

The group of Alexander Rudensky used this strategy and discerned 128 self-peptides displayed by the murine class II MHC molecule I-A^b in activated B cells and macrophages [27]. This is about ten-fold more sequences than in the equivalent attempt by conventional Edman sequencing a decade earlier [19]. The novel finding, which became possible through the large number of identified distinct epitopes, was that 21% of peptides were derived from cytosolic proteins [27]. Moreover, these cytosolic self-peptides are constitutively displayed by splenic and thymic DC, thereby tolerizing self-reactive T cells. The same self-epitopes were hardly presented by resting B cells or thymic cortical epithelial cells [27].

6.3.4
MAPPs: MHC-associated Peptide Proteomics

The latest progress in mass spectrometry-based analysis of naturally processed peptides relies on the optimization of two critical steps: sample preparation and high-resolution fractionation of enormously complex peptide mixtures. Solubilization of natural MHC-peptide complexes from tissues or cells requires detergents. Most suitable detergents, however, interfere with the performance of HPLC and ESI-MS/MS. Therefore, a detergent depletion step based on ultrafiltration had to be introduced prior to HPLC fractionation. The inclusion of the so-called "Multidimensional Protein Identification Technology (MudPIT)" led to further improvements [63]. MudPIT separates complex peptide mixtures by 2D HPLC (Figure 6.4). The separation needle is packed with a strong cation exchange material directly followed by C18 RP material that extends up to the tip of the needle. The peptides leaving the needle tip are sprayed directly into the orifice on an ion trap mass spectrometry device. In the total set-up, dead volumes are minimized, thereby increasing the sensitivity and resolution of the separation and reducing sample loss. Finally, a rapidly scanning ion-trap MS replaces the former ESI-MS/MS device; sequence determination is carried out with the computer algorithm SEQUEST.

MAPPs technology is a high-throughput approach yielding several hundred peptide sequences from 1–5 µg purified HLA-DR molecules [18]. Compared with previous attempts, at least 100-fold less cell material is sufficient. Consequently, MAPPS allows sequence analysis of self-peptide repertoires from low-abundance cell types, blood samples, tissue biopsies or micro dissections. Applying MAPPs, the first comprehensive self-peptide analysis of human monocyte-derived DCs

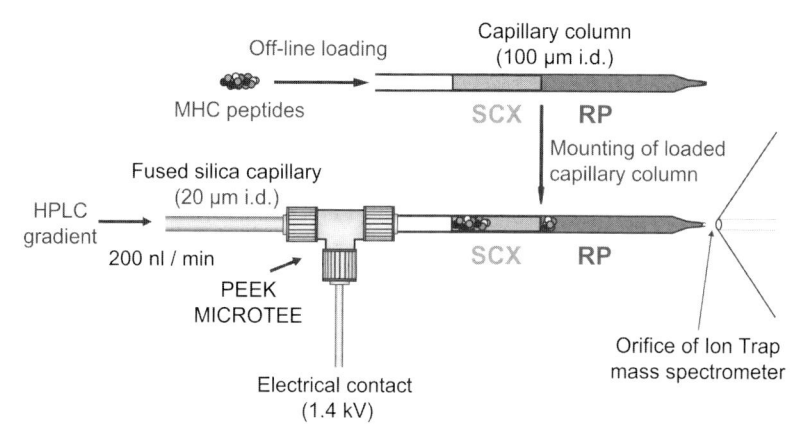

Figure 6.4 Multidimensional protein identification technology (MudPIT). This method is based on two-dimensional liquid chromatography followed by electrospray mass spectrometry (2D-LC-MS/MS) and can also be used to sequence complex peptide mixtures, e.g., MHC-associated peptides. Peptide mixtures are separated on a cation-exchange matrix (SCX) combined with a reversed-phase (RP) matrix. The capillary column is connected directly to the orifice of an ion trap mass spectrometer.

was feasible [18]. Only 5×10^6 DCs were necessary to identify 215 self-peptide sequences derived from 55 parent proteins. Some 38% of the peptides originated from transmembrane and vesicular proteins of the endocytic pathway and, surprisingly, 31% from nuclear and cytosolic proteins (Table 6.1). More than 80% of the self-peptide epitopes were represented in both immature DCs and after LPS-induced maturation. Interestingly, epitopes from leukocyte elastase, inter-α-trypsin inhibitor and nicotin nucleotide pyrophosphorylase were only present in immature DCs, whereas epitopes from mannose receptor, TNFα, aminopeptidase N and invariant chain-derived peptide CLIP were strongly enriched in mature DCs [18]. The latter observation gave rise to the intriguing finding that mature DCs utilize surface CLIP-HLA complexes to antagonize T_H1 cell polarization (cf. Section 6.5.4).

6.4
Natural Class II MHC-associated Peptides from Different Tissues and Cell-types

Initially, only naturally processed peptides extracted from lymphoblastoid B cells grown in culture could be studied. With increasing sensitivity of the sequencing techniques the self-peptide repertoire of other professional and non-professional APCs became accessible. Very recently, MAPPs technology has allowed us to identify self-peptides from ex-vivo sources, e.g. APCs from blood or the thymus. The following sections present an overview of these studies.

6.4.1
Peripheral Blood Mononuclear Cells

When human peripheral blood leukocytes are subjected to Ficoll density gradient centrifugation, a fraction strongly enriched in monocytes, NK cells, B and T lymphocytes can be obtained, denoted as "peripheral blood mononuclear cells (PBMCs)". In PBMCs, mainly B cells and monocytes carry class II MHC-associated self-peptides. Regarding the expression level, resting B cells and monocytes carry 10- to 50-fold less class II MHC molecules than B lymphoblastoid cell lines (i.e. about $1–5 \times 10^4$ molecules per cell). From the conventional amount of blood to be obtained from a single donor, which is up to 500 ml, only about 1–5 µg of HLA-DR molecules can be purified.

Analysis of the self-peptide repertoire of PBMCs from several healthy blood donors [26] revealed considerable variability in the frequency of peptides derived from endogenous cytosolic or exogenous serum proteins (Table 6.1). With regard to serum proteins, at least four parent proteins have been repeatedly found (Table 6.5): serum albumin, apolipoprotein A-II, α_1-antitrypsin and β_2-microglobulin. Interestingly, the epitope HSA(444-457) has been described several times before: in murine splenocytes and thymic APCs [64], in human splenocytes [65] and in human EBV-transformed B cells [66]. Likewise, β_2-microglobulin(24-35) has been eluted from HLA-DR6 molecules of EBV B cells [67]. A rationale for the presentation of the HSA(444-457) epitope by both peripheral and central APCs could be that thymic tolerance induction by deletion of HSA(444-457)-reactive thymocytes is essential for the maintenance of self-tolerance against most abundant serum proteins, such as serum albumin.

Table 6.5 Major serum protein epitopes eluted from PBMC [26].

Serum protein	Sequence (shortest length variant)	Residue numbering	Truncation variants	Previous source Tissue	Ref.
Serum albumin	TPTLVEVSRNLGK	444–457	8	Spleen, thymus	66–68
Apolipoprotein A-II	SKEQLTPLIKKAGTEL	68–83	5	–	–
α_1-Antitrypsin	KAVLTIDEKGTEAA	359–372	4	–	–
β_2-Microglobulin	TPKIQVYSRHPA	24–35	5	–	–

6.4.2
Myeloid Dendritic Cells

Myeloid dendritic cells (mDCs) can be purified from peripheral blood cells by usage of antibodies recognizing the mDC-specific surface marker BDCA-1 [68]. The MAPPs technology enabled us to sequence eight prominent epitopes from seven parent proteins (Table 6.6). It was six serum proteins and one endocytic membrane protein, HLA-DP. Strikingly, two of the eight self-epitopes derived from serum albumin and α_1-antitrypsin were identical to the self-epitopes found in PBMCs (cf. Tables 6.5 and 6.6). As it is most likely B cells that give rise to both self-peptides in the PBMC fraction, B cells and myeloid DCs share a common processing program with regard to two of the most abundant serum proteins, serum albumin and α_1-antitrypsin. This may be a very critical aspect for the maintenance of peripheral self-tolerance against major serum self-proteins. Resting B cells alone may not suffice and, therefore, may require help from myeloid DCs in tolerizing CD4$^+$ helper T cells that specifically recognize the outlined self-epitopes from serum albumin and α_1-antitrypsin.

Table 6.6 Prominent HLA-DR-associated self-epitopes eluted from peripheral blood myeloid DCs.

Donor protein	Sequence	Residue numbering	Previous source	
			Tissue, cell-type	Ref.
Complement C1R	GDRWILTAAHTLYPK	493–507	–	–
HLA-DP α chain	MFYDLDKKETVWH	62–75	–	–
α1-Antitrypsin	KAVLTIDEKGTEA	359–371	PBMC	Table 6.5
α1-Acid glycoprotein 1	AHLLILRDTKTYMLA	117–131	–	–
Ficolin 1	GNHQFAKYKSFKVADE	213–228	–	–
Apolipoprotein J	KNPKFMETVAEKALQ	426–440	–	–
Albumin	LGEYKFQNALLVRYT	442–436	–	–
	TPTLVEVSRNLGKVG	444–458	Spleen, Thymus, PBMC	65–67

6.4.3
Medullary Thymic Epithelial Cells

The experimental demonstration of acquired self-tolerance and clonal deletion of self-reactive lymphocytes were important milestones in attaining our current understanding of how the adaptive immune system fights foreign invaders but prevents deleterious self-reactivity [69]. Tolerance to tissue-specifically expressed self-proteins has been ascribed to extrathymic (peripheral) tolerance mechanisms, whereas the thymus was viewed as the organ of central tolerance induction focusing on thymic and blood-borne self-antigens [70]. However, the recent past has provided increasing evidence that there is a pool of genes, covering 5–10% of all currently known genes, that are ectopically expressed in thymic APCs, in particular in medullary thymic epithelial cells (mTECs) [71]. These findings suggest that mTEC may be able to present self-peptides not only derived from thymus-specific endogenous self-proteins or blood-borne self-antigens but also from tissue-specific antigens that are aberrantly expressed in mTECs [72].

As yet, only a handful of reports have addressed the question of which naturally expressed self-peptides are actually presented by mTECs or other thymic APCs. As mentioned above (cf. Table 6.4), Murphy et al. described a natural epitope in mice that was only expressed in the thymic medulla, and not in the cortex [48]. This self-peptide turned out to be the I-A^b- and I-A^d-restricted epitope Eα(52-68) [16–18]. In another report, length variants of the naturally processed β_2-microglobulin epitope β_2-m(46-60) were described to be presented by various murine tissues, including the thymus of BALB/c mice [73, 74]. The length variant

β_2-m(50-63) was found in the thymus and in the spleen of C3H/HeJ mice, associated to I-E^k and, likewise, on I-E^s of ΔY mice [64]. The major conclusion from the latter study was that a few self-peptides of unknown origin are tissue-specifically distributed, as they were only found on splenic APC, but there were no fundamental differences discernible with regard to the presence of the most abundant self-peptides in spleen or thymus [64].

When medullary thymic epithelial cells (mTECs) were isolated from murine thymus, about 2×10^6 highly purified mTECs were analyzed by MAPPs technology. In accordance with previous findings described above, serum proteins, such as serum albumin, β_2-microglobulin or apolipoprotein E, were parent proteins of very abundant self-epitopes (Table 6.7). In total, serum proteins accounted for roughly 25% of the 134 self-peptide species. Some 55% of the self-peptides were derived from typical endogenous self-proteins that are known from B cells, macro-

Table 6.7 Self-peptides from medullary thymic epithelial cells (mTEC).

Origin	Parent protein	Previous reports	
		Cell type	Ref.
mTEC	CLIP	B cells, macrophages	27
	A(α) β chain	B cells, macrophages	27
	GAPDH	B cells, macrophages	27
	Aspartat-aminotransferase	B cells, macrophages	27
	Calnexin	–	–
Serum	Serum albumin	Splenocytes, thymus	65
	Apolipoprotein E	–	–
	Complement C3	–	–
	β2-Microglobulin	B cells	27
	Serum amyloid A-1	–	–
B lymphocytes	L-Plastin	–	–
Heart, liver	Cytochrome c oxidase	–	–
Liver, monocytes	Carboxylesterase 1	–	–
Liver, intestine	Cadherin-17	–	–
Liver, intestine	Carboxylesterase 2	–	–
B lymphocytes	IgM μ chain	–	–
Liver	Monooxygenase R	–	–

phages or other APCs (Table 6.7). Among this subset, the invariant chain-derived CLIP was the most prominent self-peptide. In striking contrast to other APCs, about 20% of the peptides eluted from I-A^d from mTECs originated from cells normally found in extra-thymic tissues (Table 6.7). Most of these parent proteins are expressed in the liver or intestine. Beyond that, the membrane-bound form of the IgM μ chain or L-Plastin are expressed in B lymphocytes. In summary, we conclude from the emerging evidence that promiscuously expressed genes of mTECs give rise to ectopically expressed tissue-specific self-peptides: Thymocytes recognizing these self-peptides on the surface of mTECs with high avidity are prone to deletion for the benefit of self-tolerance.

6.4.4
Splenic APCs

A reoccurring question in research on natural MHC-associated peptides was whether the self-peptides found on APCs in cell culture would really reflect the self-peptide repertoire of the same type of APC in vivo. To address this question, I-A^b-associated self-peptides were prepared from splenocytes of a B6 mouse and subjected to high-throughput MAPPs analysis.

The data obtained from this ex vivo approach can be compared with that obtained from lymphoblastoid B cells in vitro [19, 20, 27] (Table 6.2). It turned out that most self-proteins constituting the self-peptide repertoire of B lymphoblastoid cells in culture, such as the invariant chain, I-A^b, apolipoprotein B, immunoglobulins, interleukin receptors and the transferrin receptor [27], are also serving as parent proteins for the very abundant I-A^b-bounded self-peptides of splenic APCs in vivo [26].

However, with the exception of the invariant chain CLIP or the transferrin receptor, the epitopes presented on splenocytes were different from the ones eluted from lymphoblastoid B cells. This is, most likely, because the activity and type of proteases and H2-M, which are the major contributors to endocytic antigen processing [75], are different in vivo than in vitro. Another notable aspect relates to the protein half-lives in naive B cell versus virus-transformed, and hence activated, B cells. Rapidly dividing transformed B cells may have a higher turnover in certain proteins, which may alter the sequence of proteolytic processing steps and, thereby, generate another set of immunodominant self-peptides in relation to slowly dividing naive B cells that are dominant in the spleen. Thus, it is obviously advisable for future attempts to study naturally processed peptides ex vivo, in particular in those cases where we want to improve our knowledge on self-antigens that drive life-threatening diseases.

6.4.5
Tumor Cells

The relevance of T cell mediated anti-tumor immunity has been demonstrated in both animal models and human cancer therapy [76, 77]. The identification of

MHC class I-restricted tumor epitopes has generated increasing interest in immunotherapy for cancer. However, therapeutic strategies that have focused on the use of only MHC class I-associated tumor peptides were merely transiently effective in eliminating cancer cells in patients [78]. Novel strategies aim at prolonging tumor antigen presentation by the use of DC and the inclusion of MHC class II-restricted tumor antigens. Given the importance of $CD4^+$ T cells in anti-tumor immunity, it is critical to identify class II MHC-restricted self-peptides originating from tumor cells [79]. Several approaches have been developed for this purpose, including a genetic targeting expression system [79], biochemical purification using tumor cell lysates [80] or self-peptide elution from tumor cells [81].

The value of identifying natural peptides from class II MHC molecules on tumor cells is exemplified by the melanoma cell line FM3 [81]. FM3 was established from metastatic melanoma of a 70-year old woman. It was shown to express several common melanoma antigens, including Melan-A, MAGE-1, MAGE-3, tyrosinase and gp100 [82]. When HLA-DR molecules were purified from FM3 cells and self-peptides eluted, two 16-mers were identified that were derived from parent proteins known to be over-expressed in different types of tumors [81]: gp100(44-59) and annexin II(208-223). The gp100 protein is a lineage restricted melanocyte marker and several melanoma antigens are known in the context of recognition by HLA-A2-restricted T cells. Annexin II, however, is a ubiquitously expressed protein that is over-expressed in tumors. Both gp100(44-59) and annexin II (208-223) sensitised T cells for effective HLA-DR4-restricted recognition of melanoma cells [83]. In a one study, the self-peptide annexin II(208-223) was presented by autologous DC and induced $CD4^+$ T helper cells, which responded to melanoma cells over-expressing annexin II by secreting large amounts of type 1 cytokines [84]. Although this annexin II-derived self-peptide is a candidate epitope to be incorporated into tumor vaccines against malignant melanoma, it remains to be established whether it will function as a good tumor rejection antigen and how it can be used effectively in active immunotherapy. In particular, intracellular delivery of MHC class I and II tumor self-peptides into DCs may be a promising, novel approach to be tested in forthcoming clinical settings.

6.4.6
Autoimmunity-related Epithelial Cells

In inflammatory conditions, such as autoimmunity and organ transplantation, peripheral epithelial cells are induced to express class II MHC molecules [85, 86]. In both cases, self-antigens can either be presented by epithelial cells themselves or by professional APCs. Whatever the case, epithelial cells become the unique targets of the pathogenic mechanisms in organ-specific autoimmune diseases and transplant rejection [87]. A role for class II MHC molecules in epithelium is postulated either in the triggering of the initial autoimmune reaction or in the perpetuation of the response. This is indicated by the class II MHC-restricted reactivity to antigen presented by autologous epithelial cells and not by conventional APCs, as shown by $CD4^+$ T cells isolated from autoimmune glands [88]. Thus, the self-peptide repertoire displayed at the surface of class II-expressing epithelial cells must be relevant to the understanding of the autoreactive T cell response. The rat insular neuroendocrine epithelial cell line RINm5F has been widely used for human studies of autoimmune diabetes and, therefore, has been chosen to explore the self-peptide repertoire [89]. RINm5F transfected with the human genes encoding HLA-DR4, HLA-DM and the invariant chain revealed a remarkably heterogeneous pool of self-peptides from the cell surface, secretory vesicles and the cytosol [90]. In marked contrast, a lymphoblastoid B cell line displayed a very restricted and homogeneous peptide repertoire, where a few epitopes from surface molecules were predominant [90]. Although many of the parent proteins found with RINm5F epithelial cells were ubiquitous, some self-proteins, e.g., GD3 ganglioside synthase, neprilysin or carboxypeptidase H, were specific of neuroendocrine cells. Importantly, carboxypeptidase H peptides have been proposed as putative autoantigens in type I diabetes mellitus [91]. Thus, class II molecules expressed on epithelial cells could lead to the display of altered sets of self-peptides capable of stimulating otherwise silent T cells.

The two most critical chaperones of the class II antigen processing pathway, HLA-DM and the invariant chain (Ii), turned out to govern the composition of the self-peptide repertoire of RINm5F cells [92]. In the absence of HLA-DM and Ii, HLA-DR4 was mostly associated to peptides from cytosolic proteins. This is consistent with preferential loading in the ER which is fuelled with cytosolic self-peptides via the TAP transporter [93]. Expression of Ii alone forced the association of CLIP to HLA-DR4, but also of several other self-peptides of endogenous and exogenous sources [92]. This agrees with a switch of the loading location from the ER to the endocytic pathway. HLA-DM expression resulted in a similar composition of the self-peptide repertoire, although the sequences from the respective transfectants were very different from the transfectants discussed before. These data support the notion that the reported variability in the relative expression of class II MHC, Ii and HLA-DM in autoimmune tissues relates to the maintenance of autoimmunity [94]. Low or absent HLA-DM expression can prevent the presentation of dominant epitopes, thereby allowing new or cryptic endogenous epitopes to be presented at the surface of endocrine cells and to potentially stimulate non-tolerant T cell pools. In summary, more accurate knowledge of the expression of Ii, HLA-DM and HLA-DR molecules in target organs of autoimmune diseases is crucial to identify epithelial cell-specific self-peptides and to understand the dynamics in the presentation of autoepitopes.

6.5
The CLIP Story

In 1992, Peter Cresswell and colleagues analyzed human TxB hybrid cell lines defective in antigen processing and identified a naturally processed peptide derived from the class II MHC-associated chaperone Ii [24]. This class II MHC-associated Ii peptide was named "CLIP"– and it became the most frequently cited natural peptide. The striking observation was that in these types of mutant APCs, CLIP covered 70–80% of the HLA-DR-associated self-peptide repertoire [24, 25]. However, the murine equivalent of CLIP has already been described as a very abundant self-peptide in wild-type APCs: by Rudensky et al. in murine B cells [19] and by Chicz et al. in human EBV-transformed B cells [22]. As soon as it became clear that it is the CLIP region that is mainly responsible for the interaction of intact Ii with the peptide binding groove of any class II MHC molecule [95, 96] it was easy to understand why CLIP is binding promiscuously to class II MHC proteins [97]. Furthermore, as Ii and class II MHC molecules are widely co-expressed, CLIP is a key representative in the vast majority of naturally processed self-peptide repertoires of class II MHC.

6.5.1
CLIP in APCs Lacking HLA-DM

Class II MHC null APCs, such as 721.174 or T2, that displayed the $CLIP^{high}$ phenotype upon transfection with the genes for class II heterodimers [24, 25] were widely defective in antigen presentation; however, they could be fully rescued upon transfection with the genes encoding the non-classical MHC molecule HLA-DM [98, 99]. Indeed, HLA-DM could be shown to release CLIP from class II-CLIP complexes [100] and exert this task in a true catalytic fashion [101]. HLA-DM-mediated CLIP release is accomplished by binding of HLA-DM to class II dimers, induction of conformational changes, preferentially affecting the P1 specificity pocket [102] and by stabilizing the empty binding cleft in a chaperone-like fashion [103].

As HLA-DM is only capable of releasing peptides of moderate or low kinetic stability [104], the sequence of CLIP had to be evolutionarily adapted to the polymorphism of the class II alleles, so that critical anchor residues, such as P1 and P4 or P6, do not fit too tightly into the specificity pockets. Otherwise, a long-lasting CLIP release step would become limiting in antigen presentation and prevent the host from mounting a cellular immune response. Met-91 (P1), Ala-94 (P4) and Pro-96 (P6) are highly conserved in all known species and appear to contribute exquisitely to promiscuous but low-stability binding in the context of the multitude of allelic variants of class II binding grooves [105]. Conversely, the evolution of further allelic class II variants is restricted by the conserved CLIP sequences, since the incompatibility with CLIP removal will be disfavored by natural selection mechanisms.

6.5.2
Flanking Residues and Self-release of CLIP

A closer look at the naturally occurring CLIP length variants in the human system reveals that there are two principal variants: CLIP(long), consisting of 21- to 26-mers, and CLIP(short), consisting of 14- to 19-mers [22, 51, 66]. X-ray structural analysis of the HLA-DR3-CLIP crystal shows that CLIP(short) contains the residues critical for interacting with the peptide binding groove [105]. According to this crystal structure, the N-terminal flanking residues of CLIP, positions 81–89, have no defined folding, as they remained invisible. However, CLIP(long) dissociated rapidly from HLA-DR molecules at endocytic pH, whereas CLIP(short) displayed a lower off-rate [101, 106]. These findings suggested that the N-terminal 9 residues of CLIP, which are positioned outside the 9-mer region occupying the binding cleft, are functionally relevant.

The critical impact of flanking residues is reminiscent of naturally occurring hen-egg-lysozyme (HEL) derived epitopes: C-terminal Trp-residues enhance the immunogenicity of HEL(52-63), resulting in altered T cell receptor variable region usage [107]. Likewise, the N-terminal flanking region of CLIP augments the immunogenic potential of the cryptic "self" epitope from the C-erb oncogene (Her2/neu), leading to protective antitumor immunity of the chimeric tumor antigen [108].

Although most class II allelic products depend on HLA-DM with respect to CLIP removal, no necessity for HLA-DM or the murine counterpart H2-DM is apparent for DRB1*0401, I-A^k or I-A^d [109, 110]. Likewise, loading with conventional self-peptides other than CLIP has been demonstrated with several HLA-DM-defective mutants [25, 111]. The rationale is that, at endosomal pH, the N-terminal residues 81–89 facilitate rapid release of CLIP(long) [111]. Furthermore, CLIP(81-89) catalyzes the release of CLIP(short) and a subset of other self-peptides. Flanking residues of CLIP, in particular Lys-83, Lys-86 and Pro-87, are thought to interact with an effector site outside the binding-cleft. This would be consistent with an allosteric mode of action [111]. Interestingly, a lateral HLA-DM-interacting surface on HLA-DR that includes acidic and hydrophobic HLA-DR residues near the N-terminus of the peptide has been mapped [112]. Hence, from these parallels one could envisage that class II molecules bear a binding-site outside the groove close to the P1 pocket that allows induction of conformational changes in and around the P1 pocket, thereby initiating the release of bound peptides.

According to this model, HLA-DM would be an optimized effector molecule that has release capacities superior to the original effector, the N-terminal flanking residues of CLIP.

6.5.3
CLIP in Tetraspan Microdomains

The density of MHC class II on the surface of APCs is a critical parameter in activating helper T cells. Being aware that cognate antigenic peptides can only be a minority in relation to the multitude of endogenous self-peptides, such as CLIP, the question emerges as to how T cells manage to track down the few copies of cognate peptide. At least 200–300 class II-peptide complexes are estimated to be required to fully activate a naive $CD4^+$ T cell so that it can acquire effector functions in immune defense [113, 114]. However, it remained open whether abundant self-peptides, such as CLIP, can be excluded from the contact zone where cognate T cell receptors recognize cognate antigenic peptide.

Until recently, it was believed that supra-molecular assemblies of monospecific T cell receptors on the T cell surface drive cluster formation of cognate class II-MHC peptide complexes on APCs [115]. Novel evidence suggests that professional APCs actively organize the lateral distribution of class II MHC-peptide complexes on their own – before they emerge in immunological synapses with T cells – by segregation of class II molecules into membrane microdomains [116]. B cells and some other APCs concentrate class II-peptide complexes in detergent-resistant cholesterol- and glycosphingolipid-enriched microdomains at the cell surface, denoted as "lipid rafts" [117, 118]. This type of microdomains can accumulate in synapses and thereby favor antigen presentation at low doses of antigen; however, lipid rafts do not seem to be equally important in all types of APCs [119, 120].

Microdomains formed by so-called tetraspanins seem to be more broadly distributed across leukocytes [121]. Tetraspanins, such as CD9, CD53, CD81 or CD82 are proteins that contain four membrane-spanning domains and have a strong tendency to build up so-called tetraspan webs by homo- and hetero-dimerization with each other through their luminal stalk subdomains. Class II MHC molecules segregate in tetraspan microdomains already in endosomal compartments, preferentially bound to CD82 and CD63 and to the co-stimulator CD86 [120].

In developing DC, class II molecules loaded with cognate antigen co-localize with CD86 in membrane domains of transport vesicles and remain co-associated upon arrival on the surface [122]. Hence, tetraspan microdomains appear to facilitate clustering of proteins relevant for class II-restricted antigen presentation assemblies during transport to the cell surface and until synapses with T cells are generated. Strikingly, tetraspan microdomains revealed to carry a rather narrow and selected set of self-peptides [120]. This is probably due to the presence of the peptide editor HLA-DM. In mature DCs tetraspan microdomains display a strong enrichment of CLIP, although it may attain an abundance of less than 10% in the total self-peptide repertoire [116]. In agreement with this observation, the tetraspanin CD82 co-precipitates readily with HLA-DR-CLIP complexes in DCs [18] and in human B lymphoblastoid cell lines [120].

When we explored the total self-peptide repertoire of tetraspan microdomains, affinity-purified from human peripheral blood mononuclear cells [26], we found a strong enrichment of epitopes from nuclear and cytosolic proteins, in particular several ribonuclear proteins (hnRNP) (Figure 6.5). Strikingly, none of the self-peptides were derived from serum, plasma membrane or endocytic proteins. However, most self-peptides found in tetraspan microdomains displayed another interesting feature: several pairs of self-peptides originated from regions adjacent to each other in the respective parent protein. Table 6.8 shows two examples, derived from LckBP1 and hnRNP A2/B1. As one peptide of each pair has an apparent class II binding motif, whereas the other peptide lacks such a motif, it is tempting to speculate that the "null" peptide is a processing by-product protected from degradation by neighboring class II molecules or tetraspanins in the same microdomain. Future studies will have to expand on the peptide repertoire of tetraspanin domains of other APCs.

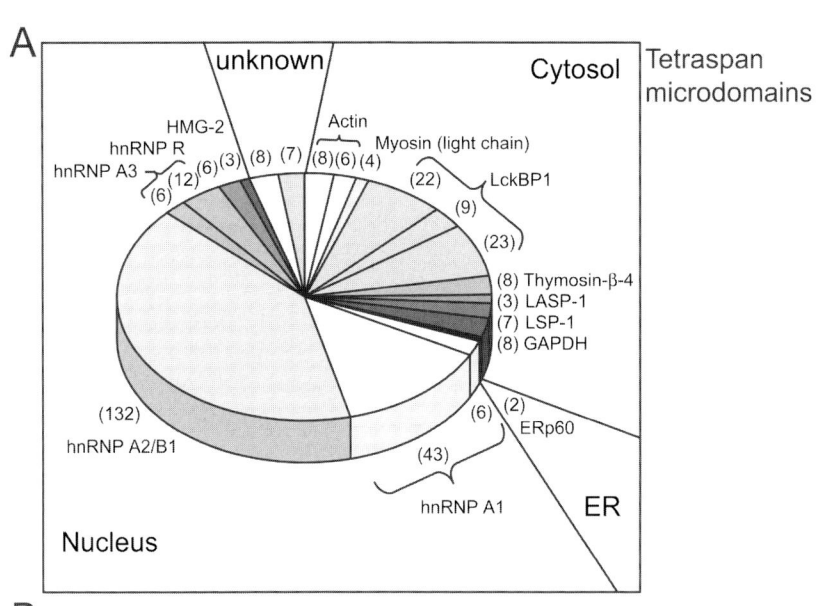

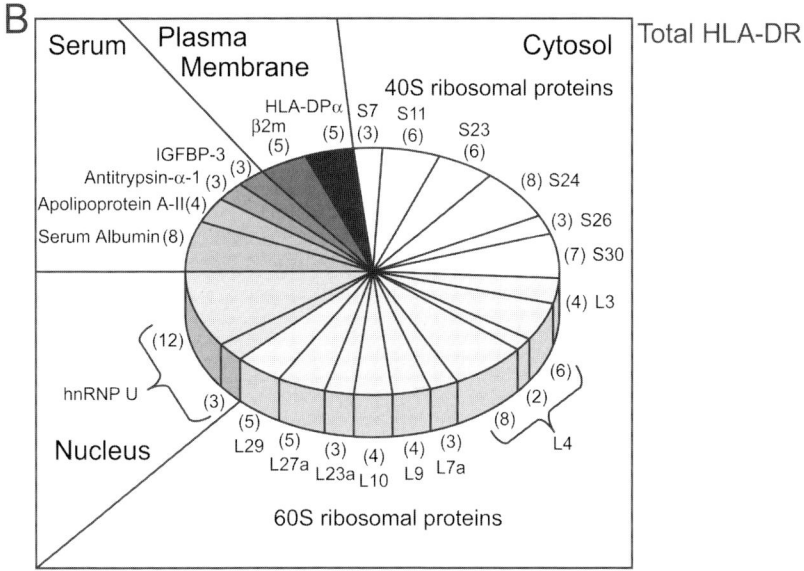

Figure 6.5 Tetraspan microdomains carry a select self-peptide set. Indicated are the parent proteins of the natural peptides found exclusively in tetraspan microdomains (A) or bound to HLA-DR molecules but not to microdomains (B). Self-peptides were purified from peripheral blood mononuclear cells. Also indicated is the number of identified length variants of a given epitope in the respective self-protein.

Table 6.8 Self-peptides in tetraspan microdomains of PBMC [26].

Protein name	Sequence	Epitope
Hematopoietic cell-specific LYN substrate 1 (LckBP1)	TIEGSGRTEHINIHQLRNK	42–60
	EGSGRTEHINIHQL	44–57
	GSGRTEHINIHQL	45–57
	TEHINIHQL	49–57
	TEHINIHQLR	49–58
	TEHINIHQLRNK	49–60
	RNKVSEEHDVLR	58–69
	RNKVSEEHDVLRK	58–70
	NKVSEEHDVLR	59–69
	VSEEHDVLR	61–69
	VSEEHDVLRK	61–70
	VSEEHDVLRKK	61–71
Heterogeneous nuclear ribonucleo-protein A2/B1 (hnRNP A2/B1)	GIKEDTEEHHL	118–128
	GIKEDTEEHHLR	118–129
	GIKEDTEEHHLRD	118–130
	GIKEDTEEHHLRDYF	118–132
	GIKEDTEEHHLRDYFEEY	118–135
	GIKEDTEEHHLRDYFEEYGK	118–137
	FEEYGKIDTIEIITDRQSGKKRGFGFVT	132–159
	EEYGKIDTIEIITDRQSGKKRGFGFVT	133–159
	GKIDTIEIITDRQSGKK	136–152
	GKIDTIEIITDRQSGKKRGFG	136–156
	GKIDTIEIITDRQSGKKRGFGFVT	136–159
	GKIDTIEIITDRQSGKKRGFGFVTF	136–160

6.5.4
CLIP as an Antagonist of T_H1 Cells

Many investigators have proposed that endogenous self-peptides may be critical for the extraordinary sensitivity of T cells for foreign antigen [123–125]. This view is supported by the fact that T cells rely on the weak interaction with MHC–self-peptide complexes in the thymus to progress to full maturity [126] and in the periphery to guarantee survival [125]. In addition, along with agonist ligands, large quantities of class II MHC-self-peptide complexes, attaining up to 20% relative abundance, accumulate in the immunological synapse [124, 127]. One of the models attempting to explain how self-peptides may influence T cell responses is the heterodimer model [123, 127]: MHC-agonist peptide and MHC-self-peptide complexes form heterodimers that induce dimerization of TCRs, thereby facilitating T cell activation. Until most recently, it remained open whether each self-peptide can influence T cell responsiveness in a similar manner or whether distinct self-peptides modulate T cell activity in distinct way. CLIP is the first defined self-peptide that could be demonstrated to have an impact on the quality of T cell activation [18]. During maturation, DCs strongly up-regulate class II-CLIP complexes on the surface, apparently driven by down-regulation of the abundance and catalytic activity of HLA-DM. Class II-CLIP complexes attain an abundance of up to 10% of the total class II-self-peptide pool on the surface of mature DC [18]. CLIP and cognate antigen co-segregate in immunological synapses of DCs and naive $CD4^+$ T cells, which is in accordance with class II-CLIP complexes being constituents of tetraspan microdomains. The increase in class II-CLIP complexes on mature DC led to a significant shift in the polarization of naive T cells [18]: CLIP reduced the number of T_H1 cells secreting IFN-γ and promoted the secretion of the T_H2 cytokine IL-4. Likewise, APCs from H2-DM mice favored polarization of T_H2 cells, whereas APCs from wild-type mice gave rise to more T_H1 cells [18]. Mechanistically, CLIP may interfere with signaling relevant for T_H1 polarization. This would be equivalent to the situation when low doses of agonist or low-affinity agonists give rise to low-avidity APC-T cell interactions, thereby favoring T_H2 polarization.

Alternatively, CLIP may promote the formation of a particular type of TCR dimers, as described in the pseudodimer model [128]. Compared to other self-peptides, CLIP may play a special role in this scenario as it contributes to positive selection of T cells in the thymus and, thus, may preferentially attract those peripheral T cells that have been positively selected by thymic MHC-CLIP complexes. In the presence of exogenous antigenic peptide, these T cells may undergo T_H2 polarization. Class II MHC alleles linked to autoimmune diseases, such as rheumatoid arthritis, juvenile dermatomyositis, autoimmune hepatitis and Graves disease, form unstable class II-CLIP complexes [18]. As most autoimmune diseases rely on T_H1 cells, a lack of CLIP may be accomplished by a low capacity to counterbalance T_H1 polarization. Thus, the $CLIP^{low}$ phenotype may increase the risk for autoimmunity, whereas the $CLIP^{high}$ phenotype may confer protection. This mechanism may normally prevent the organism from excessively strong T_H1 responses, but maintain balance in the task to combat foreign invaders without losing tolerance against the set of self-antigens.

6.6
Outlook: Natural Peptides as Diagnostic or Therapeutic Tools

Recent advances in genome sequencing have provided us with the complete genome sequence of both pathogens and the host homo sapiens (cf. http://www.ncbi.nlm.nih.gov/Genomes). In parallel, the ongoing refinement of MHC allele-specific ligand binding motifs in combination with the sophistication of computational tools allowed us to set-up algorithms for the identification of T cell epitopes. As a consequence, expectations are high that we may be able to predict those natural self-peptide sequences that confer susceptibility to autoimmune diseases, would be suitable for immunotherapy of tumors or may serve as biomarkers for certain diseases and the severity of a malignancy.

However, it is still the lack of fundamental data on the cooperative or interference interactions of various adjacent and non-adjacent amino acids and our ignorance on proteolytic processing of antigenic peptides that limit our current prediction algorithms. One way out of this situation is to expand the field of high-throughput sequencing of naturally processed peptides. In particular, the MAPPs approach is a powerful tool that will pave the way to generate comprehensive self-peptide sequence databases that can be exploited for diagnostic and therapeutic purposes. One obstacle in the usage of self-protein sequences from the genome sequencing approach is that, as yet, in most instances only unmodified sequences are available in database searches. However, in both the oncology and autoimmune fields, mutations and post-translational modifications, such as phosphorylation, glycosylation or citrullination, play a critical role. Future mass-spectrometry-based search algorithms will have to take this into account.

As the MAPPs technology is sensitive enough to generate self-peptide profiles from peripheral blood samples, it becomes feasible to search for novel self-peptide markers that are linked to clinical events. An obvious field that could benefit from such a diagnostic innovation is the autoimmune disease area. Rheumatoid arthritis (RA) would be particular amenable to self-peptide profiling, as not only blood samples but also synovial fluid from RA-patients' joints could be used in an individualized manner. Apart from diagnostic purposes, the same strategy could give us more direct insight into disease-related alterations of putative autoantigenic self-peptides and the dynamics of epitope spreading at later stages of RA progression. The great advantage of this type of biomarker search is that it is strictly focused from the beginning: it is well established that certain MHC class II alleles function as susceptibility or severity factors in RA and other autoimmune diseases, and we are persuaded that it is some critical self-peptides that trigger autoreactive T cells. Hence, it is only matter of time before we will identify them via self-peptide profiling.

Another aspect that could be addressed by high-throughput self-peptide analysis is the phenomenon of naturally antagonized T cells that can be found in late phases of the immune response to HIV-1, Epstein-Barr Virus or melanoma antigens. This is most likely due to antagonist self-peptides [129]. In support of this view, many self-peptides that induce positive selection in the thymus exhibit antagonist activity for mature T cells. Thus, antagonist self-peptides might influence immune responses as a rule, rather than as an exception. Antagonist self-peptides may vary in their number and potency towards different epitopes and MHC alleles, thereby explaining why certain MHC alleles have an inhibitory effect on immune responses. The identification of this type of antagonist self-peptides could be of particular value in exploring infectious diseases and in oncology.

A key challenge facing researchers in oncology is the need to supplement current cancer vaccines with tumor-specific class II MHC-restricted tumor antigens. As yet, only a dozen helper T cell epitopes have been defined, most of which are connected to malignant melanoma. Pilot studies are underway using DCs and necrotic tumor cells in vitro to mimic the situation in vivo. MAPPs technology will be suitable for identifying more class II-restricted tumor antigens by sequencing those naturally processed peptides from DCs that are newly presented upon engaging tumor cells. Mixtures of naturally processed helper T cell epitopes may prove to be critical in conditioning DCs so that they are potent enough to trigger a long-lasting cytotoxic anti-tumor response by CD8$^+$ T cells. Future clinical trials in cancer patients may benefit from the availability of an increasing number of class I- and class II-MHC-restricted, tumor-specific self-peptides; in particular, at the beginning of a century when individualized health care is no longer a foreign word but the focus.

References

1 Watts, T. H., Brian, A. A., Kappler, J. W., Marrack, P. and McConnell, H. M. (1984): Antigen presentation by supported planar membranes containing affinity-purified I-Ad. *Proc. Natl. Acad. Sci. U.S.A.* 81: 7564–7568.

2 Allen, P. M., Strydom, D. J. and Unanue, E. R. (1984): Processing of lysozyme by macrophages: identification of the determinant recognized by two T-cell hybridomas. *Proc. Natl. Acad. Sci. U.S.A.* 81: 2489–2493.

3 Guillet, J. G., Lai, M. Z., Briner, T. J., Smith, J. A. and Gefter, M. L. (1986): Interaction of peptide antigens and class II major histocompatibility complex antigens. *Nature* 324: 260–262.

4 Babbitt, B. P., Allen, P. M., Matsueda, G., Haber, E. and Unanue, E. R. (1985): Binding of immunogenic peptides to Ia histocompatibility molecules. *Nature* 317: 359–361.

5 Ziegler, K. and Unanue, E. R. (1981): Identification of a macrophage antigen-processing event required for I-region-restricted antigen presentation to T lymphocytes. *J. Immunol.* 127: 1869–1875.

6 Thomas, J. W., Danho, W., Bullesbach, E., Fohles, J. and Rosenthal, A. S. (1981): Immune response gene control of determinant selection. III. Polypeptide fragments of insulin are differentially recognized by T but not by B cells in insulin immune guinea pigs. *J. Immunol.* 126: 1095–1100.

7 Chesnut, R. W., Colon, S. M. and Grey, H. M. (1982): Requirements for the processing of antigens by antigen-presenting B cells. I. Functional comparison of B cell tumors and macrophages. *J. Immunol.* 129: 2382–2388.

8 Clark, R. B., Chiba, J., Zweig, S. E. and Shevach, E. M. (1982): T-cell colonies recognize antigen in association with specific epitopes on Ia molecules. *Nature* 295: 412–414.

9 Shimonkevitz, R., Colon, S., Kappler, J. W., Marrack, P. and Grey, H. M. (1984): Antigen recognition by H-2-restricted T cells. II. A tryptic ovalbumin peptide that substitutes for processed antigen. *J. Immunol.* 133: 2067–2074.

10 Buus, S., Sette, A., Colon, S. M., Jenis, D. M. and Grey, H. M. (1986): Isolation and characterization of antigen-Ia complexes involved in T cell recognition. *Cell* 47: 1071–1077.

11 Buus, S., Sette, A. and Grey, H. M. (1987): The interaction between protein-derived immunogenic peptides and Ia. *Immunol. Rev.* 98: 115–141.

12 Bjorkman, P. J., Saper, M. A., Samraoui, B., Bennett, W. S., Strominger, J. L. and Wiley, D. C. (1987): Structure of the human class I histocompatibility antigen, HLA-A2. *Nature* 329: 506–512.

13 Unanue, E. R. (1981): The regulatory role of macrophages in antigenic stimulation. Part Two: symbiotic relationship between lymphocytes and macrophages. *Adv. Immunol.* 31: 1–136.

14 Babbitt, B. P., Matsueda, G., Haber, E., Unanue, E. R. and Allen, P. M. (1986): Antigenic competition at the level of peptide-Ia binding. *Proc. Natl. Acad. Sci. U.S.A.* 83: 4509–4513.

15 Buus, S., Sette, A., Colon, S. M. and Grey, H. M. (1988): Autologous peptides constitutively occupy the antigen binding site on Ia. *Science* 242: 1045–1047.

16 Bevan, M. J. (1976): Cross-priming for a secondary cytotoxic response to minor H antigens with H-2 congenic cells which do not cross-react in the cytotoxic assay. *J. Exp. Med.* 143: 1283–1288.

17 Rock, K. L., Gamble, S. and Rothstein, L. (1990): Presentation of exogenous antigen with class I major histocompatibility complex molecules. *Science* 249: 918–921.

18 Rohn, T. A., Boes, M., Wolters, D., Spindeldreher, S., Muller, B., Langen, H., Ploegh, H., Vogt, A. B. and Kropshofer, H. (2004): Upregulation of the CLIP self peptide on mature dendritic cells antagonizes T helper type 1 polarization. *Nat. Immunol.* 5: 909–918.

19 Rudensky, A., Preston-Hurlburt, P., Hong, S. C., Barlow, A. and Janeway, C. A., Jr. (1991): Sequence analysis of peptides bound to MHC class II molecules. *Nature* 353: 622–627.

20 Rudensky, A., Preston-Hurlburt, P., al-Ramadi, B. K., Rothbard, J. and Janeway, C. A., Jr. (1992): Truncation variants of peptides isolated from MHC class II molecules suggest sequence motifs. *Nature* 359: 429–431.

21 Hunt, D. F., Michel, H., Dickinson, T. A., Shabanowitz, J., Cox, A. L., Sakaguchi, K., Appella, E., Grey, H. M. and Sette, A. (1992): Peptides presented to the immune system by the murine class II major histocompatibility complex molecule I-Ad. *Science* 256: 1817–1820.

22 Chicz, R. M., Urban, R. G., Lane, W. S., Gorga, J. C., Stern, L. J., Vignali, D. A. and Strominger, J. L. (1992): Predominant naturally processed peptides bound to HLA-DR1 are derived from MHC-related molecules and are heterogeneous in size. *Nature* 358: 764–768.

23 Newcomb, J. R. and Cresswell, P. (1993): Characterization of endogenous peptides bound to purified HLA-DR molecules and their absence from invariant chain-associated alpha beta dimers. *J. Immunol.* 150: 499–507.

24 Riberdy, J. M., Newcomb, J. R., Surman, M. J., Barbosa, J. A. and Cresswell, P. (1992): HLA-DR molecules from an antigen-processing mutant cell line are associated with invariant chain peptides. *Nature* 360: 474–477.

25 Sette, A., Ceman, S., Kubo, R. T., Sakaguchi, K., Appella, E., Hunt, D. F., Davis, T. A., Michel, H., Shabanowitz, J. and Rudersdorf, R. (1992): Invariant chain peptides in most HLA-DR molecules of an antigen-processing mutant. *Science* 258: 1801–1804.

26 Spindeldreher, S. (2005): Multimolekulare Antigenpräsentationskomplexe in Membran-Mikrodomänen von B-Lymphozyten und Dendritischen Zellen. (Ruprecht-Karls Universität, Heidelberg, PhD thesis).

27 Dongre, A. R., Kovats, S., deRoos, P., McCormack, A. L., Nakagawa, T., Paharkova-Vatchkova, V., Eng, J., Caldwell, H., Yates, J. R., 3rd and Rudensky, A. Y. (2001): In vivo MHC class II presentation of cytosolic proteins revealed by rapid automated tandem mass spectrometry and functional analyses. *Eur. J. Immunol.* 31: 1485–1494.

28 Bjorkman, P. J., Saper, M. A., Samraoui, B., Bennett, W. S., Strominger, J. L. and Wiley, D. C. (1987): The foreign antigen binding site and T cell recognition regions of class I histocompatibility antigens. *Nature* 329: 512–518.

29 Brown, J. H., Jardetzky, T. S., Gorga, J. C., Stern, L. J., Urban, R. G., Strominger, J. L. and Wiley, D. C. (1993): Three-dimensional structure of the human class II histocompatibility antigen HLA-DR1. *Nature* 364: 33–39.

30 Garrett, T. P., Saper, M. A., Bjorkman, P. J., Strominger, J. L. and Wiley, D. C. (1989): Specificity pockets for the side chains of peptide antigens in HLA-Aw68. *Nature* 342: 692–696.

31 Demotz, S., Grey, H. M., Appella, E. and Sette, A. (1989): Characterization of a naturally processed MHC class II-restricted T-cell determinant of hen egg lysozyme. *Nature* 342: 682–684.

32 Van Bleek, G. M. and Nathenson, S. G. (1990): Isolation of an endogenously processed immunodominant viral pep-

tide from the class I H-2Kb molecule. *Nature* 348: 213–216.
33 Bodmer, H. C., Pemberton, R. M., Rothbard, J. B. and Askonas, B. A. (1988): Enhanced recognition of a modified peptide antigen by cytotoxic T cells specific for influenza nucleoprotein. *Cell* 52: 253–258.
34 Oldstone, M. B., Whitton, J. L., Lewicki, H. and Tishon, A. (1988): Fine dissection of a nine amino acid glycoprotein epitope, a major determinant recognized by lymphocytic choriomeningitis virus-specific class I-restricted H-2Db cytotoxic T lymphocytes. *J. Exp. Med.* 168: 559–570.
35 Rotzschke, O., Falk, K., Deres, K., Schild, H., Norda, M., Metzger, J., Jung, G. and Rammensee, H. G. (1990): Isolation and analysis of naturally processed viral peptides as recognized by cytotoxic T cells. *Nature* 348: 252–254.
36 Falk, K., Rotzschke, O., Deres, K., Metzger, J., Jung, G. and Rammensee, H. G. (1991): Identification of naturally processed viral nonapeptides allows their quantification in infected cells and suggests an allele-specific T cell epitope forecast. *J. Exp. Med.* 174: 425–434.
37 Cerundolo, V., Elliott, T., Elvin, J., Bastin, J., Rammensee, H. G. and Townsend, A. (1991): The binding affinity and dissociation rates of peptides for class I major histocompatibility complex molecules. *Eur. J. Immunol.* 21: 2069–2075.
38 Falk, K., Rotzschke, O. and Rammensee, H. G. (1990): Cellular peptide composition governed by major histocompatibility complex class I molecules. *Nature* 348: 248–251.
39 Griem, P., Wallny, H. J., Falk, K., Rotzschke, O., Arnold, B., Schonrich, G., Hammerling, G. and Rammensee, H. G. (1991): Uneven tissue distribution of minor histocompatibility proteins versus peptides is caused by MHC expression. *Cell* 65: 633–640.
40 Falk, K., Rotzschke, O., Stevanovic, S., Jung, G. and Rammensee, H. G. (1991): Allele-specific motifs revealed by sequencing of self-peptides eluted from MHC molecules. *Nature* 351: 290–296.
41 Rammensee, H. G., Falk, K. and Rotzschke, O. (1993): Peptides naturally presented by MHC class I molecules. *Annu. Rev. Immunol.* 11: 213–244.
42 Rammensee, H. G., Friede, T. and Stevanoviic, S. (1995): MHC ligands and peptide motifs: first listing. *Immunogenetics* 41: 178–228.
43 Madden, D. R. and Wiley, D. C. (1992): Peptide binding to the major histocompatibility complex molecules. *Curr. Opin. Struct. Biol.* 2: 300–304.
44 Udaka, K. (1996): Decrypting class I MHC-bound peptides with peptide libraries. *Trends Biochem. Sci.* 21: 7–11.
45 Rudensky, A. and Janeway, C. A., Jr. (1993): Studies on naturally processed peptides associated with MHC class II molecules. *Chem. Immunol.* 57: 134–151.
46 Kalbus, M., Fleckenstein, B. T., Offenhausser, M., Bluggel, M., Melms, A., Meyer, H. E., Rammensee, H. G., Martin, R., Jung, G. and Sommer, N. (2001): Ligand motif of the autoimmune disease-associated mouse MHC class II molecule H2-A(s). *Eur. J. Immunol.* 31: 551–562.
47 Rudensky, A., Rath, S., Preston-Hurlburt, P., Murphy, D. B. and Janeway, C. A., Jr. (1991): On the complexity of self. *Nature* 353: 660–662.
48 Murphy, D. B., Lo, D., Rath, S., Brinster, R. L., Flavell, R. A., Slanetz, A. and Janeway, C. A., Jr. (1989): A novel MHC class II epitope expressed in thymic medulla but not cortex. *Nature* 338: 765–768.
49 Kropshofer, H., Max, H., Muller, C. A., Hesse, F., Stevanovic, S., Jung, G. and Kalbacher, H. (1992): Self-peptide released from class II HLA-DR1 exhibits a hydrophobic two-residue contact motif. *J. Exp. Med.* 175: 1799–1803.
50 Kropshofer, H., Vogt, A. B., Moldenhauer, G., Hammer, J., Blum, J. S. and Hammerling, G. J. (1996): Editing of the HLA-DR-peptide repertoire by HLA-DM. *EMBO J.* 15: 6144–6154.
51 Chicz, R. M., Urban, R. G., Gorga, J. C., Vignali, D. A., Lane, W. S. and Strominger, J. L. (1993): Specificity and promiscuity among naturally processed pep-

tides bound to HLA-DR alleles. *J. Exp. Med.* 178: 27–47.

52 Kropshofer, H., Max, H., Halder, T., Kalbus, M., Muller, C. A. and Kalbacher, H. (1993): Self-peptides from four HLA-DR alleles share hydrophobic anchor residues near the NH2-terminal including proline as a stop signal for trimming. *J. Immunol.* 151: 4732–4742.

53 Hammer, J., Takacs, B. and Sinigaglia, F. (1992): Identification of a motif for HLA-DR1 binding peptides using M13 display libraries. *J. Exp. Med.* 176: 1007–1013.

54 Nelson, C. A., Vidavsky, I., Viner, N. J., Gross, M. L. and Unanue, E. R. (1997): Amino-terminal trimming of peptides for presentation on major histocompatibility complex class II molecules. *Proc. Natl. Acad. Sci. U.S.A.* 94: 628–633.

55 Marshall, K. W., Liu, A. F., Canales, J., Perahia, B., Jorgensen, B., Gantzos, R. D., Aguilar, B., Devaux, B. and Rothbard, J. B. (1994): Role of the polymorphic residues in HLA-DR molecules in allele-specific binding of peptide ligands. *J. Immunol.* 152: 4946–4957.

56 Jardetzky, T. S., Lane, W. S., Robinson, R. A., Madden, D. R. and Wiley, D. C. (1991): Identification of self peptides bound to purified HLA-B27. *Nature* 353: 326–329.

57 Colbert, R. A. (2004): The immunobiology of HLA-B27: variations on a theme. *Curr. Mol. Med.* 4: 21–30.

58 Hunt, D. F., Henderson, R. A., Shabanowitz, J., Sakaguchi, K., Michel, H., Sevilir, N., Cox, A. L., Appella, E. and Engelhard, V. H. (1992): Characterization of peptides bound to the class I MHC molecule HLA-A2.1 by mass spectrometry. *Science* 255: 1261–1263.

59 Henderson, R. A., Michel, H., Sakaguchi, K., Shabanowitz, J., Appella, E., Hunt, D. F. and Engelhard, V. H. (1992): HLA-A2.1-associated peptides from a mutant cell line: a second pathway of antigen presentation. *Science* 255: 1264–1266.

60 Ruppert, J., Sidney, J., Celis, E., Kubo, R. T., Grey, H. M. and Sette, A. (1993): Prominent role of secondary anchor residues in peptide binding to HLA-A2.1 molecules. *Cell* 74: 929–937.

61 Khalil-Daher, I., Boisgerault, F., Feugeas, J. P., Tieng, V., Toubert, A. and Charron, D. (1998): Naturally processed peptides from HLA-DQ7 (alpha1*0501-beta1*0301): influence of both alpha and beta chain polymorphism in the HLA-DQ peptide binding specificity. *Eur. J. Immunol.* 28: 3840–3849.

62 Yates, J. R., 3rd, Eng, J. K., McCormack, A. L. and Schieltz, D. (1995): Method to correlate tandem mass spectra of modified peptides to amino acid sequences in the protein database. *Anal. Chem.* 67: 1426–1436.

63 Washburn, M. P., Wolters, D. and Yates, J. R., 3rd. (2001): Large-scale analysis of the yeast proteome by multidimensional protein identification technology. *Nat. Biotechnol.* 19: 242–247.

64 Marrack, P., Ignatowicz, L., Kappler, J. W., Boymel, J. and Freed, J. H. (1993): Comparison of peptides bound to spleen and thymus class II. *J. Exp. Med.* 178: 2173–2183.

65 Gordon, R. D., Young, J. A., Rayner, S., Luke, R. W., Crowther, M. L., Wordsworth, P., Bell, J., Hassall, G., Evans, J. and Hinchliffe, S. A. (1995): Purification and characterization of endogenous peptides extracted from HLA-DR isolated from the spleen of a patient with rheumatoid arthritis. *Eur. J. Immunol.* 25: 1473–1476.

66 Verreck, F. A., van de Poel, A., Drijfhout, J. W., Amons, R., Coligan, J. E. and Konig, F. (1996): Natural peptides isolated from Gly86/Val86-containing variants of HLA-DR1, -DR11, -DR13, and -DR52. *Immunogenetics* 43: 392–397.

67 Davenport, M. P., Quinn, C. L., Chicz, R. M., Green, B. N., Willis, A. C., Lane, W. S., Bell, J. I. and Hill, A. V. (1995): Naturally processed peptides from two disease-resistance-associated HLA-DR13 alleles show related sequence motifs and the effects of the dimorphism at position 86 of the HLA-DR beta chain. *Proc. Natl. Acad. Sci. U.S.A.* 92: 6567–6571.

68 Penna, G., Sozzani, S. and Adorini, L. (2001): Cutting edge: selective usage of

chemokine receptors by plasmacytoid dendritic cells. *J. Immunol.* 167: 1862–1866.
69. Schwartz, R. H. and Müller, D. in *Fundamental Immunology*. (ed. Paul, W. R.) 901–934 (Lippincott Williams & Wilkins, Philadelphia, 2003).
70. Mathis, D. and Benoist, C. (2004): Back to central tolerance. *Immunity* 20: 509–516.
71. Kyewski, B. and Derbinski, J. (2004): Self-representation in the thymus: an extended view. *Nat. Rev. Immunol.* 4: 688–698.
72. Derbinski, J., Schulte, A., Kyewski, B. and Klein, L. (2001): Promiscuous gene expression in medullary thymic epithelial cells mirrors the peripheral self. *Nat. Immunol.* 2: 1032–1039.
73. Guery, J. C. and Adorini, L. (1995): Dendritic cells are the most efficient in presenting endogenous naturally processed self-epitopes to class II-restricted T cells. *J. Immunol.* 154: 536–544.
74. Guery, J. C., Sette, A., Appella, E. and Adorini, L. (1995): Constitutive presentation of dominant epitopes from endogenous naturally processed self-beta 2-microglobulin to class II-restricted T cells leads to self-tolerance. *J. Immunol.* 154: 545–554.
75. Watts, C. (2004): The exogenous pathway for antigen presentation on major histocompatibility complex class II and CD1 molecules. *Nat. Immunol.* 5: 685–692.
76. Greenberg, P. D. (1991): Adoptive T cell therapy of tumors: mechanisms operative in the recognition and elimination of tumor cells. *Adv. Immunol.* 49: 281–355.
77. Rosenberg, S. A. (2001): Progress in human tumour immunology and immunotherapy. *Nature* 411: 380–384.
78. Lau, R., Wang, F., Jeffery, G., Marty, V., Kuniyoshi, J., Bade, E., Ryback, M. E. and Weber, J. (2001): Phase I trial of intravenous peptide-pulsed dendritic cells in patients with metastatic melanoma. *J. Immunother.* 24: 66–78.
79. Wang, R. F. (2001): The role of MHC class II-restricted tumor antigens and CD4+ T cells in antitumor immunity. *Trends Immunol.* 22: 269–276.
80. Pieper, R., Christian, R. E., Gonzales, M. I., Nishimura, M. I., Gupta, G., Settlage, R. E., Shabanowitz, J., Rosenberg, S. A., Hunt, D. F. and Topalian, S. L. (1999): Biochemical identification of a mutated human melanoma antigen recognized by CD4(+) T cells. *J. Exp. Med.* 189: 757–766.
81. Halder, T., Pawelec, G., Kirkin, A. F., Zeuthen, J., Meyer, H. E., Kun, L. and Kalbacher, H. (1997): Isolation of novel HLA-DR restricted potential tumor-associated antigens from the melanoma cell line FM3. *Cancer Res.* 57: 3238–3244.
82. Kirkin, A. F., thor Straten, P. and Zeuthen, J. (1996): Differential modulation by interferon gamma of the sensitivity of human melanoma cells to cytolytic T cell clones that recognize differentiation or progression antigens. *Cancer Immunol. Immunother.* 42: 203–212.
83. Li, K., Adibzadeh, M., Halder, T., Kalbacher, H., Heinzel, S., Muller, C., Zeuthen, J. and Pawelec, G. (1998): Tumour-specific MHC-class-II-restricted responses after in vitro sensitization to synthetic peptides corresponding to gp100 and Annexin II eluted from melanoma cells. *Cancer Immunol. Immunother.* 47: 32–38.
84. Heinzel, S., Rea, D., Offringa, R. and Pawelec, G. (2001): The self peptide annexin II (208-223) presented by dendritic cells sensitizes autologous CD4+ T lymphocytes to recognize melanoma cells. *Cancer Immunol. Immunother.* 49: 671–678.
85. Bottazzo, G. F., Pujol-Borrell, R., Hanafusa, T. and Feldmann, M. (1983): Role of aberrant HLA-DR expression and antigen presentation in induction of endocrine autoimmunity. *Lancet* 2: 1115–1119.
86. Feldmann, M., Londei, M. and Essery, G. (1986): Autoimmune disease: synergy of HLA class II overexpression and abrogation of self tolerance. *Pathol. Biol. (Paris)* 34: 779–782.
87. Markmann, J., Lo, D., Naji, A., Palmiter, R. D., Brinster, R. L. and Heber-Katz, E.

(1988): Antigen presenting function of class II MHC expressing pancreatic beta cells. *Nature* 336: 476–479.

88 Bach, J. F., Koutouzov, S. and van Endert, P. M. (1998): Are there unique autoantigens triggering autoimmune diseases? *Immunol. Rev.* 164: 139–155.

89 Gazdar, A. F., Chick, W. L., Oie, H. K., Sims, H. L., King, D. L., Weir, G. C. and Lauris, V. (1980): Continuous, clonal, insulin- and somatostatin-secreting cell lines established from a transplantable rat islet cell tumor. *Proc. Natl. Acad. Sci. U.S.A.* 77: 3519–3523.

90 Muntasell, A., Carrascal, M., Serradell, L., Veelen Pv, P., Verreck, F., Koning, F., Raposo, G., Abian, J. and Jaraquemada, D. (2002): HLA-DR4 molecules in neuroendocrine epithelial cells associate to a heterogeneous repertoire of cytoplasmic and surface self peptides. *J. Immunol.* 169: 5052–5060.

91 Castano, L., Russo, E., Zhou, L., Lipes, M. A. and Eisenbarth, G. S. (1991): Identification and cloning of a granule autoantigen (carboxypeptidase-H) associated with type I diabetes. *J. Clin. Endocrinol. Metab.* 73: 1197–1201.

92 Muntasell, A., Carrascal, M., Alvarez, I., Serradell, L., van Veelen, P., Verreck, F. A., Koning, F., Abian, J. and Jaraquemada, D. (2004): Dissection of the HLA-DR4 peptide repertoire in endocrine epithelial cells: strong influence of invariant chain and HLA-DM expression on the nature of ligands. *J. Immunol.* 173: 1085–1093.

93 Busch, R., Cloutier, I., Sekaly, R. P. and Hammerling, G. J. (1996): Invariant chain protects class II histocompatibility antigens from binding intact polypeptides in the endoplasmic reticulum. *EMBO J.* 15: 418–428.

94 Louis-Plence, P., Kerlan-Candon, S., Morel, J., Combe, B., Clot, J., Pinet, V. and Eliaou, J. F. (2000): The down-regulation of HLA-DM gene expression in rheumatoid arthritis is not related to their promoter polymorphism. *J. Immunol.* 165: 4861–4869.

95 Vogt, A. B., Stern, L. J., Amshoff, C., Dobberstein, B., Hammerling, G. J. and Kropshofer, H. (1995): Interference of distinct invariant chain regions with superantigen contact area and antigenic peptide binding groove of HLA-DR. *J. Immunol.* 155: 4757–4765.

96 Stumptner, P. and Benaroch, P. (1997): Interaction of MHC class II molecules with the invariant chain: role of the invariant chain (81-90) region. *EMBO J.* 16: 5807–5818.

97 Sette, A., Southwood, S., Miller, J. and Appella, E. (1995): Binding of major histocompatibility complex class II to the invariant chain-derived peptide, CLIP, is regulated by allelic polymorphism in class II. *J. Exp. Med.* 181: 677–683.

98 Denzin, L. K., Robbins, N. F., Carboy-Newcomb, C. and Cresswell, P. (1994): Assembly and intracellular transport of HLA-DM and correction of the class II antigen-processing defect in T2 cells. *Immunity* 1: 595–606.

99 Fling, S. P., Arp, B. and Pious, D. (1994). HLA-DMA and DMB genes are both required for MHC class II/peptide complex formation in antigen-presenting cells. *Nature* 368: 554–558.

100 Sloan, V. S., Cameron, P., Porter, G., Gammon, M., Amaya, M., Mellins, E. and Zaller, D. M. (1995): Mediation by HLA-DM of dissociation of peptides from HLA-DR. *Nature* 375: 802–806.

101 Kropshofer, H., Vogt, A. B. and Hammerling, G. J. (1995): Structural features of the invariant chain fragment CLIP controlling rapid release from HLA-DR molecules and inhibition of peptide binding. *Proc. Natl. Acad. Sci. U.S.A.* 92: 8313–8317.

102 Pashine, A., Busch, R., Belmares, M. P., Munning, J. N., Doebele, R. C., Buckingham, M., Nolan, G. P. and Mellins, E. D. (2003): Interaction of HLA-DR with an acidic face of HLA-DM disrupts sequence-dependent interactions with peptides. *Immunity* 19: 183–192.

103 Kropshofer, H., Arndt, S. O., Moldenhauer, G., Hammerling, G. J. and Vogt, A. B. (1997): HLA-DM acts as a molecular chaperone and rescues empty HLA-DR molecules at lysosomal pH. *Immunity* 6: 293–302.

104 Kropshofer, H., Hammerling, G. J. and Vogt, A. B. (1997): How HLA-DM edits

the MHC class II peptide repertoire: survival of the fittest? *Immunol. Today* 18: 77–82.

105 Ghosh, P., Amaya, M., Mellins, E. and Wiley, D. C. (1995): The structure of an intermediate in class II MHC maturation: CLIP bound to HLA-DR3. *Nature* 378: 457–462.

106 Urban, R. G., Chicz, R. M. and Strominger, J. L. (1994): Selective release of some invariant chain-derived peptides from HLA-DR1 molecules at endosomal pH. *J. Exp. Med.* 180: 751–755.

107 Carson, R.T., Vignali, K. M., Woodland, D. L. and Vignali, D. A. A. (1997): Recognition of MHC class II-bound peptide flanking residues enhances immunogenicity and results in altered TCR V region usage. *Immunity* 7: 387–399.

108 Hess, A. D., Thoburn, C., Chen, W., Miura, Y. and Van der Wall, E. (2001): The N-terminal flanking region of the invariant chain peptide augments the immunogenicity of a cryptic "self" epitope from a tumor-associated antigen. *Clin. Immunol.* 101: 67–76.

109 Avva, R. R. and Cresswell, P. (1994): In vivo and in vitro formation and dissociation of HLA-DR complexes with invariant chain-derived peptides. *Immunity* 1: 763–774.

110 Brooks, A. G., Campbell, P. L., Reynolds, P., Gautam, A. M. and McCluskey, J. (1994): Antigen presentation and assembly by mouse I-Ak class II molecules in human APC containing deleted or mutated HLA DM genes. *J. Immunol.* 153: 5382–5392.

111 Kropshofer, H., Vogt, A. B., Stern, L. J. and Hammerling, G. J. (1995): Self-release of CLIP in peptide loading of HLA-DR molecules. *Science* 270: 1357–1359.

112 Doebele, R. C., Busch, R., Scott, H. M., Pashine, A. and Mellins, E. D. (2000): Determination of the HLA-DM interaction site on HLA-DR molecules. *Immunity* 13: 517–527.

113 Demotz, S., Grey, H. M. and Sette, A. (1990): The minimal number of class II MHC-antigen complexes needed for T cell activation. *Science* 249: 1028–1030.

114 Harding, C. V. and Unanue, E. R. (1990): Quantitation of antigen-presenting cell MHC class II/peptide complexes necessary for T-cell stimulation. *Nature* 346: 574–576.

115 Bromley, S. K., Burack, W. R., Johnson, K. G., Somersalo, K., Sims, T. N., Sumen, C., Davis, M. M., Shaw, A. S., Allen, P. M. and Dustin, M. L. (2001): The immunological synapse. *Annu. Rev. Immunol.* 19: 375–396.

116 Vogt, A. B., Spindeldreher, S. and Kropshofer, H. (2002): Clustering of MHC-peptide complexes prior to their engagement in the immunological synapse: lipid raft and tetraspan microdomains. *Immunol. Rev.* 189: 136–151.

117 Anderson, H. A., Hiltbold, E. M. and Roche, P. A. (2000): Concentration of MHC class II molecules in lipid rafts facilitates antigen presentation. *Nat. Immunol.* 1: 156–162.

118 Hiltbold, E. M., Poloso, N. J. and Roche, P. A. (2003): MHC class II-peptide complexes and APC lipid rafts accumulate at the immunological synapse. *J. Immunol.* 170: 1329–1338.

119 Huby, R., Chowdhury, F. and Lombardi, G. (2001): Rafts for antigen presentation? *Nat. Immunol.* 2: 3.

120 Kropshofer, H., Spindeldreher, S., Rohn, T. A., Platania, N., Grygar, C., Daniel, N., Wolpl, A., Langen, H., Horejsi, V. and Vogt, A. B. (2002): Tetraspan microdomains distinct from lipid rafts enrich select peptide-MHC class II complexes. *Nat. Immunol.* 3: 61–68.

121 Hemler, M. E. (2003): Tetraspanin proteins mediate cellular penetration, invasion, and fusion events and define a novel type of membrane microdomain. *Annu. Rev. Cell. Dev. Biol.* 19: 397–422.

122 Turley, S. J., Inaba, K., Garrett, W. S., Ebersold, M., Unternaehrer, J., Steinman, R. M. and Mellman, I. (2000): Transport of peptide-MHC class II complexes in developing dendritic cells. *Science* 288: 522–527.

123 Janeway, C. A., Jr. (2001): How the immune system works to protect the host from infection: a personal view. *Proc. Natl. Acad. Sci. U.S.A.* 98: 7461–7468.

124 Wulfing, C., Sumen, C., Sjaastad, M. D., Wu, L. C., Dustin, M. L. and Davis, M. M. (2002): Costimulation and endogenous MHC ligands contribute to T cell recognition. *Nat. Immunol.* 3: 42–47.

125 Stefanova, I., Dorfman, J. R. and Germain, R. N. (2002): Self-recognition promotes the foreign antigen sensitivity of naive T lymphocytes. *Nature* 420: 429–434.

126 Goldrath, A. W. and Bevan, M. J. (1999): Selecting and maintaining a diverse T-cell repertoire. *Nature* 402: 255–262.

127 Irvine, D. J., Purbhoo, M. A., Krogsgaard, M. and Davis, M. M. (2002): Direct observation of ligand recognition by T cells. *Nature* 419: 845–849.

128 Li, Q. J., Dinner, A. R., Qi, S., Irvine, D. J., Huppa, J. B., Davis, M. M. and Chakraborty, A. K. (2004): CD4 enhances T cell sensitivity to antigen by coordinating Lck accumulation at the immunological synapse. *Nat. Immunol.* 5: 791–799.

129 Vukmanovic, S., Neubert, T.A., Santori, F.R. (2003): Could TCR antagonism explain the association between MHC genes and disease? *Trends Mol. Med.* 9: 139–146.

7
Target Cell Contributions to Cytotoxic T Cell Sensitivity

Tatiana Lebedeva, Michael L. Dustin and Yuri Sykulev

7.1
Introduction

Functioning of the immune system is mediated by multiple cell-cell contacts. In particular, antigen recognition by T cells requires interaction of immune receptors on T cells with their ligands on target cells and antigen-presenting cells (APC) and leads to large-scale redistribution of immune receptors and formation of specialized intercellular junction, termed immunological synapse (IS). While redistribution of immune receptors on T cells is well documented, initial or preexisting distribution of these molecules is less well understood. It is also not clear whether the preexisting distribution may change upon activation and differentiation of T cells. Even less is known about initial distribution of immune receptors on target cells and APC. In particular, it is unclear whether the redistribution of the immune receptors on target cells and APC may be an active process or whether these receptors just follow a pattern of redistribution of their counter receptors on T cells.

In this chapter we focus on the distribution of major histocompatibility complex (MHC) and adhesion molecules on the surface of target cells and APC. Based on our own and other published data we propose that, in addition to well-documented role of intercellular adhesion molecule 1 (ICAM-1, CD54) to function as an adhesion receptor, productive engagement of ICAM-1 leads to the induction of cell surface and intracellular molecular events that facilitate antigen presentation to T cells.

7.2
Intercellular Adhesion Molecule 1 (ICAM-1)

7.2.1
Adhesion Molecules on the Surface of APC and Target Cells

Antigen recognition by T cells requires conjugate formation between the T cell and target cell or APC. Upon recognition of specific target cells, CTL adhesion to the target cell precedes antigen recognition (Spits et al., 1986). In fact, CTL-target cell conjugates could be formed in the absence of antigen, but their contacts with CTL do not lead to target cell lysis. In addition, CTL may utilize regulated changes in adhesion to control immune recognition events (Dustin and Springer, 1989).

Among others, adhesion molecule ICAM-1 plays a key role in CTL interactions with target cells. Since ICAM-1 and MHC-I are the ligands for LFA-1 (lymphocyte function associated antigen-1, CD11a/CD18) and TCR (T cell receptor), respectively, which segregate upon productive CTL-target encounter, these systems are generally thought to cooperate at a distance due to the molecular segregation at the interface (Somersalo et al., 2004; Springer, 1990 ; Stinchcombe et al., 2001).

7.2.2
ICAM-1 Structure and Topology on the Cell Surface

ICAM-1 is composed of five Ig-like domains, a transmembrane segment, and 28 amino acids cytoplasmic tail (Staunton et al., 1990). Electron microscopy studies revealed a fixed 90° bend between domains 3 and 4 such that monomers are L-shaped (Kirchhausen et al., 1995; Staunton et al., 1990). It is expressed at low basal level on various cells and could be significantly up-regulated in response to activation or immune challenge, augmenting immune response in inflamed tissues (Dustin et al., 1986).

Oligomerization and clustering of adhesion receptors promote adhesion by favoring rebinding of receptors and ligands after dissociation. This can be a very significant effect since adhesion molecule interactions tend to have fast monomeric dissociation rates. Cell surface fluorescent resonance energy transfer (FRET) with bivalent monoclonal antibodies (mAb) revealed that ICAM-1 molecules may self-associate as well as form clusters with other cell surface molecules (Bacso et al., 2002; Bene et al., 1994). Biochemical approaches supported the organization of ICAM-1 into non-covalent homodimers on the cell surface (Miller et al., 1995; Reilly et al., 1995). Consistent with these data, analysis of the crystal structure of the three C-terminal domains (D3-D5) of ICAM-1 shows that this molecule can form dimers. Domain 4 (D4) of ICAM-1 has a 16 amino acid residue disordered region enabling formation of the energetically favorable D4-D4 dimer (Yang et al., 2004). The ICAM-1 D1-D2 structure also suggested a D1-D1 dimer interface on the side of D1 opposite the LFA-1 binding surface (Casasnovas et al., 1998). All these data suggest that at high ICAM-1 cell surface density D1-D5 ICAM-1 molecules form D4-D4 dimers that could be brought together by D1-D1 interactions, resulting in W-shaped tetramers capable of further propagation into linear arrays. In this structure the LFA-1 binding site in D1 and Mac-1 binding site in D3 are favorably oriented and readily accessible for ligation of LFA-1. The ICAM-1 clustering offers a plausible model for IS formation. The linear array, created by ICAM-1 and zigzagging between the two cells, can likely bend and close to form a circle, i.e., adhesion ring. Could this molecular ring be related to formation of the adhesion ring of the immunological synapse? The adhesion ring in the immunological synapse involves thousands of LFA-1·ICAM-1 interactions and has a thickness of several microns and so it would have to incorporate many of these molecular scale rings. Nonetheless, the ability of LFA-1·ICAM-1 interactions to form linear arrays may account for the sharp dividing line between the cSMAC and pSMAC (Grakoui et al., 1999). In fact, CTL also form antigen-independent micron scale adhesion rings in response to high surface densities of ICAM-1 in the planar glass-supported bilayer (Somersalo et al., 2004).

Early electron microscopy studies of integrins showed globular structures supported on two straight "legs" connected to the membrane, with a total height of 20 nm (Carrell et al., 1985). Subsequent studies supported the notion that the globular domain contains the ligand binding sites, which are then 20 nm from the membrane. The crystal structure of integrins revealed unexpected conformational flexibility. One crystal structure of an integrin showed a "genuflected" conformation in which "knees" in the leg domains are folded back to position the ligand binding domain within 7.5 nm of the membrane (Xiong et al., 2001), similar to the size of the TCR. The ICAM-1 D1 may be up to 18 nm from the membrane when this angle between domain 5 and the membrane is 45° or as close as 7.5 nm when D5 is perpendicular to the membrane surface (Kirchhausen et al., 1995). Thus the LFA-1–ICAM-1 interaction may be able to reach 40 nm to initiate interactions, but may then be able to pull the membranes to within less than 15 nm to allow nearby TCR–pMHC interaction. This possibility would not have seemed likely (Shaw and Dustin, 1997) until the surprising crystal structure was published, making it possible to explain how LFA-1–ICAM-1 interactions directly facilitate TCR-pMHC binding. A caveat to this idea is evidence that the genuflected form of integrins have a very low affinity for ligand. However, this evidence was obtained for a different integrin, not LFA-1. Thus, it is possible that LFA-1 may have a structural difference to allow high-affinity ICAM-1 binding in the genuflected conformation.

7.2.3
ICAM-1 as Co-stimulatory Ligand and Receptor

Usually, the signal received by a T cell via TCR binding to MHC is designated as signal 1, while the signal received via accessory or co-stimulatory molecules is referred to as signal 2. This model is typically focused on events in the T cells where signal 2 is mainly mediated by co-stimulatory molecules such as CD28 (Johnson and Jenkins, 1994; Mueller et al., 1990; Schwartz et al., 2002; Slavik et al., 1999), but also by adhesion molecules such as CD2 (Bierer and Hahn, 1993;

Davis et al., 2003) and others. The APC can also receive signals that have important immunoregulatory functions, including activating signals through CD40 and the CD28 ligands CD80 and CD86. ICAM-1 is expressed on both the T cell and APC and may also function as a co-stimulatory ligand on APC signaling through LFA-1 to the T cells, and also as a co-stimulatory receptor on the signaling directly into the APC. Since the APC often express LFA-1, it is also possible for ICAM-1 to act as a co-stimulatory receptor on certain T cells. Thus, this versatile adhesion system may provide information to the T cell and APC at several levels to facilitate MHC-II or MHC-I restricted immune responses. The role of ICAM-1 as a co-stimulatory receptor may vary in regard to different subsets of T cells involved in intercellular contacts. For example, memory T cells express ICAM-1, which is absent from naïve T cells (Buckle and Hogg, 1990).

MHC-II restricted antigen presentation is limited to professional APC that also have the capacity to express intermediate or high levels of B7, depending upon their state of activation. $CD4^+$ T cells, thus, seldom encounter antigens in the total absence of B7 ligand, but different levels of co-stimulation may make the difference between tolerance and a productive response. In contrast, virtually all nucleated cells express MHC-I. Most of these cells lack expression of B7 family proteins such that MHC-I restricted presentation often takes place in the absence of CD28 ligands. Under these conditions overall expression of ICAM-1 and its up-regulation by inflammatory stimuli would provide an important source of co-stimulation to $CD8^+$ T cells through engagement of LFA-1. High level of ICAM-1 expression on APC provides effective co-stimulation and enhances the activation of $CD8^+$ T cells isolated from spleen; the extent of this enhancement is more profound at lower peptide concentration (Oh et al., 2003). In addition, LFA-1-ICAM-1 co-engagement, even at a very high level of signal 1, stimulates greater $CD8^+$ T cell proliferation than the TCR-mediated signal alone. LFA-1-ICAM-1 interactions are thought to act specifically by increasing cell contact and, thus, TCR avidity and, consequently, the TCR occupancy (Bachmann et al., 1997). Cytotoxic T lymphocytes (CTL) induced with high levels of ICAM-1 kill tumor cells expressing low levels of tumor-associated antigen very effectively. ICAM-1 is more effective in co-stimulation of IL-2 production by $CD8^+$ T cells, than by $CD4^+$ T cells (Deeths and Mescher, 1999). Thus, under conditions of B7-independent co-stimulation of $CD8^+$ T cells, the level of ICAM-1 signal becomes a crucial factor for initiation of CTL response to MHC-I antigens presented by non-professional APC. ICAM-1 co-stimulation may also facilitate activation of $CD4^+$ T cells (Gaglia et al., 2000). All these data support the notion that ICAM-1 plays an important role as a co-stimulatory ligand for LFA-1 in T-cell signaling in addition to its role as an adhesion ligand.

7.2.4
ICAM-1-mediated Signaling

Various membrane receptors signal via dimerization. Similarly, ICAM-1 dimerizes and/or oligomerizes upon ligation with LFA-1 or with other multivalent ligands such as fibrinogen to initiate intracellular signaling. Experimentally, anti-ICAM-1 antibodies can be used to trigger signaling. ICAM-1-mediated signaling events include phosphorylation of several proteins, including Src family kinases p53/p56 lyn and pp60 src, actin-binding protein p85 cortactin, focal adhesion kinase (FAK), paxillin, as well as transient phosphorylation of 34kD cdc2 cyclin-dependent protein kinase, a regulator of cell cycle (Greenwood et al., 2002; Hubbard and Rothlein, 2000). The 28 amino acid cytoplasmic tail of ICAM-1 lacks intrinsic enzymatic activity. The IKKY(485)RLQ sequence in the ICAM-1 cytoplasmic domain bears weak similarity to the immunoreceptor tyrosine inhibition motifs (ITIM) that generate docking sites for protein phosphatases with Src homology 2 (SH2) domains. Upon engagement with multivalent ligand ICAM-1 becomes phosphorylated at Y485 and can bind SH2-containing phosphatase-2 (SHP-2) (Pluskota et al., 2000). SHP-2 dephosphorylates certain substrates, but can also serve as scaffolding protein mediating recruitment of Grb-2, which binds to SOS and promotes activation of Ras, initiating the Raf-1/Mek1/ERK pathway. Once activated, these mitogen activated protein kinases (MAPK) are involved in the phosphorylation of transcription factors such as c-fos [via ERK (Sano et al., 1998)], c-jun [via JNK (Etienne et al., 1998)], or ATF-2 [via p38 family MAPK proteins (Wang and Doerschuk, 2001)]. Phosphorylated c-fos and c-jun form AP-1 (activating protein–1), a transcription factor involved in cell growth and differentiation (Koyama et al., 1996) (Figure 7.1).

The physiological consequences of increased kinase activity and transcription factor activation include increased production of proteins such as cytokines and cell surface proteins. Cross-linking of ICAM-1 on HUVEC with anti-ICAM antibody followed by secondary antibody results in increased mRNA and protein production of IL-8 and RANTES, chemokines that enhance leukocyte trafficking. Expression of IL-1 is also increased following ICAM-1 engagement. Elevated expression of the IL-1 gene appeared dependent upon the activation of AP-1 within the LPS responsive enhancer region. Engagement of ICAM-1 on lymphocytes also increases expression of MHC-II, IL-1 receptor, VCAM-1 and ICAM-1, which all may facilitate antigen presentation by target cells. Remarkably, under the same conditions the expression levels of the accessory molecule B7 (Poudrier and Owens, 1994) or endothelial adhesion molecule E-selectin (Lawson et al., 1999) were not upregulated by ICAM-1 signaling directly, although IL-1 produced in response to ICAM-1 cross-linking will up-regulate E-selectin to enhance leukocyte–endothelial interactions.

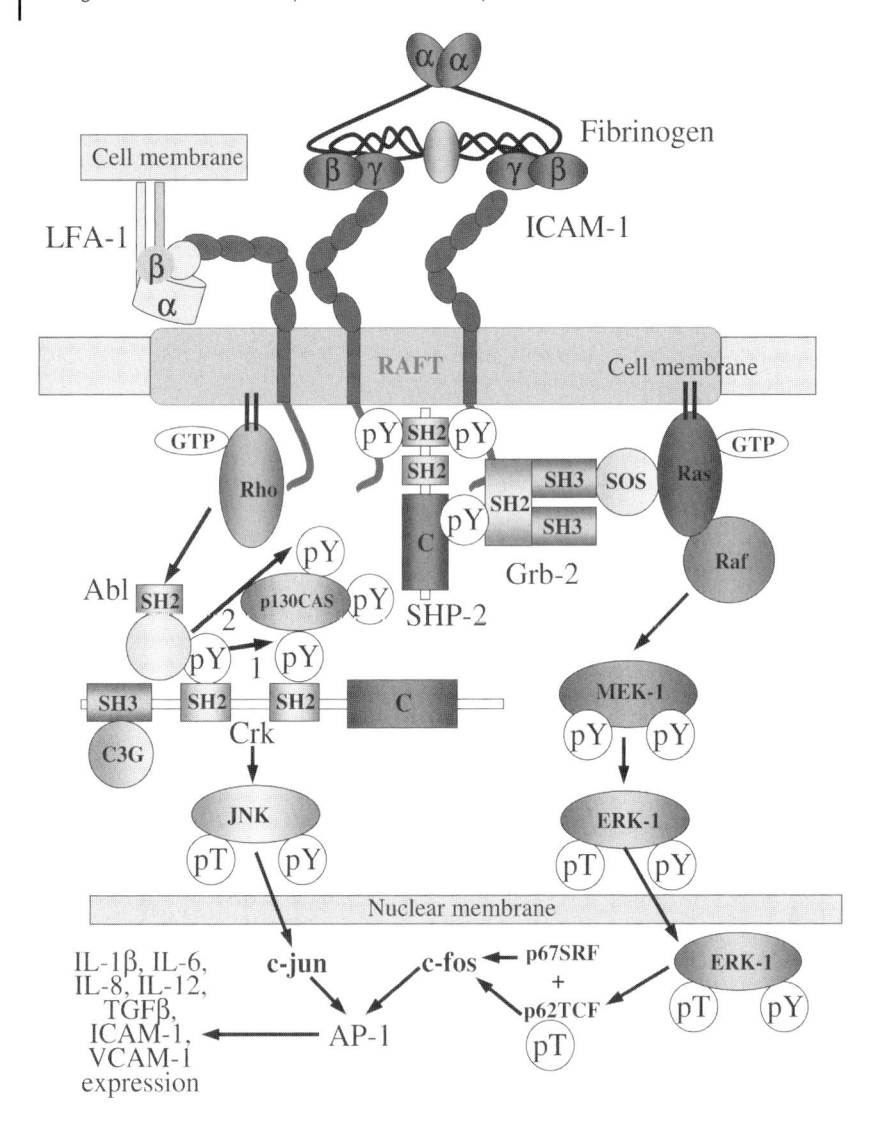

Figure 7.1 Productive engagement of ICAM-1 activates several signaling pathways. ICAM-1 cross-linking with multivalent ligands (such as fibrinogen) results in ICAM-1 recruitment to lipid rafts and phosphorylation of both ICAM-1 (at Y485) and SHP-2, Src homology 2 domain (SH2)-containing tyrosine phosphatase (Pluskota et al., 2000). SHP-2 recruitment to rafts is usually mediated by high levels of membrane cholesterol and interaction with annexin II, Ca^{2+}-dependent phospholipid-binding protein (Burkart et al., 2003). SHP-2 contains two N-terminal SH2 domains, catalytic domain (C) and a short C-terminal tail. Even though pY485 containing amino acid sequence does not resemble ITIM (immune receptor tyrosine-based inhibitory motif) consensus motif, association of SHP-2 and phosphorylated ICAM-1 occurs via SH2 domain of SHP-2. ICAM-1 is associated with SHP-2 on endothelial cells, Raji B cell line and human kidney fibroblast line 293 (Pluskota et al., 2000). Along with its

enzymatic activity, SHP-2 serves as adaptor molecule; being phosphorylated, it interacts with Grb2 (growth factor receptor binding protein 2) and via its connection with SOS (guanidine exchange factor for Ras) can activate Ras, small G-protein. Both signaling through the phosphatase catalytic domain of SHP-2 and signaling through Ras are necessary for activation of mitogen-activated protein kinase (MAPK). The activated Ras associates with the serine/threonine kinase Raf. Ras is localized to the plasma membrane due to its prenylation, while Raf is recruited to membrane by association with Ras. Its localization at the membrane results in activation and subsequent phosphorylation of the dual specific MAPK–ERK kinase (MEK) that, in turn, phosphorylates ERK (extracellular signal regulated kinase) on both tyrosine (pY) and threonine (pT) residues. Phosphorylated ERK dimerizes and exposes a signal peptide that mediates its transfer to the nucleus. Inside the nucleus ERK phosphorylates p62TCF (ternary complex factor) that then associates with p67SRF (serum response factor) to form active transcription factor complex to promote transcription of c-fos gene. Initially it was found that ICAM-1–SHP-2 interactions are necessary for cellular survival, but they are implicated in regulation of cytokines and growth factors expression during cell proliferation and differentiation as well as in upregulation of some membrane proteins, including immune receptors. Another signal transduction route activated upon ICAM-1 engagement is associated with activation of RhoA-family G-proteins (Etienne et al., 1998; Thompson et al., 2002) that activates downstream Abl tyrosine kinase capable of autophosphorylation. Abl initially phosphorylates Crk-accosiated substrate, CAS, at a single tyrosine residue (1) and at the same time binds to SH2 domain of adaptor protein Crk (CT10 regulator of kinase), contributing to additional phosphorylation of tyrosine residues of p130CAS (2). The Crk-based scaffold also provides a binding site for C3G, guanidine exchange factor protein. Assembly of Crk, p130CAS and C3G (Etienne et al., 1998) leads to activation of JNK (c-jun N-terminal kinase) that, similar to ERK, can translocate to nucleus and induce transcription of c-jun gene. C-Fos and c-Jun proteins together form activator protein complex-1 (AP-1), which controls expression of cytokines and genes encoding other proteins, ICAM-1 and VCAM-1, in particular (Koyama et al., 1996; Lawson et al., 1999; Poudrier and Owens, 1994; Sano et al., 1998). ICAM-1 molecules lacking the intracellular domain can not activate Rho proteins upon cross-linking (Greenwood et al., 2003). ICAM-1-associated signaling is not limited to the above pathways. Engagement of ICAM-1 also induces activation of Src-family kinases, in particular, p53Lyn upstream PLCγ. This cascade is responsible for Ca^{2+} influx and phosphorylation of cortactin p85 and other actin-associated proteins involved in cytoskeletal rearrangement (Etienne-Manneville et al., 2000). Src-family kinase pathway is also implicated in activation of p38 MAPK via ezrin phosphorylation and is responsible for generation of reactive oxygen species and nitric oxide (Wang and Doerschuk, 2001; Wang et al., 2003), suggesting an ICAM-1 role in mediating inflammation. The ability of ICAM-1 to induce various cell responses could depend on the nature of its ligand, cell type, ICAM-1 expression level and molecular distribution on the cell surface, suggesting that ICAM-1 mediated signaling is involved in modulation of immune response by a wide spectrum of various effects. (This figure also appears with the color plates.)

7.2.5
Role of ICAM-1 in Endothelial Response to Leukocytes

Leukocyte–endothelial interactions are vital for appropriate immune responses because they control the access of leukocytes to the tissue. ICAM-1 is an important adhesion molecule on endothelial cells that has significant basal expression, but is strongly up-regulated by inflammatory cytokines. ICAM-1 also participates in a second level of regulation of the leukocyte endothelial interaction by regulating the cytoskeletal response of the endothelial cells to leukocyte adhesion (Barreiro et al., 2002; Wojciak-Stothard et al., 1999). In culture models, LFA-1 expressing activated leukocytes elicits a dramatic response from endothelial cells, involving the formation of a cup-like structure by endothelial cell membrane projections that reach around the leukocyte. The signaling pathway responsible for this response involves RhoA on the endothelial cells and the mechanism of protrusion generation to embrace the leukocyte includes ezrin interactions with the cytoplasmic domains of ICAM-1 and VCAM on the endothelial cells. This system has led to insights in ICAM-1 signaling that will be discussed below and also may be a model for the active role of the APC in engaging T cells.

7.2.6
ICAM-1 Association with Lipid Rafts

Plasma membrane lipids can form highly ordered specialized microdomains, named lipid rafts. Rafts represent a fraction of cell membrane enriched in cholesterol, sphingolipids (mainly GM1 ganglioside) and phospholipids with long saturated fatty acyl chains, which are insoluble at 4 °C in non-ionic detergents, such as Triton X-100 or Brij, and, thus, can be separated from the bulk of the plasma membrane by floatation on sucrose gradients (Brown and Rose, 1992; Simons and Ikonen, 1997). Analysis of lipid raft functions confirms that they are important constituents of plasma membrane, serving as active sites of membrane protrusion (Naslavsky et al., 2004), protein sorting and a platform for signaling (Brown and London, 1998; Edidin, 2003). Disruption of rafts blocks some signaling processes in T lymphocytes, particularly Ca^{2+} mobilization, but also stimulates some tyrosine phosphorylation, leading to the concept that membrane domains are involved in initiating signaling and maintaining off states (Dustin, 2002).

At rest ICAM-1 molecules are primarily raft-excluded, whereas after cross-linking they are mostly found in the raft fraction on sucrose gradients (Amos et al., 2001; Tilghman and Hoover, 2002a). This raises the possibility that ICAM-1 recruitment into rafts initiates intracellular signaling and may enhance antigen presentation.

Consistent with this notion, we have shown that productive ICAM-1 engagement results in the recruitment of Src kinases and concomitant clustering of the raft marker GM1 and MHC-I-ICAM-1 assemblies to the area of initial target cell-CTL contact (Lebedeva et al., 2004). The significance of ICAM-1-mediated signaling and subsequent large-scale molecular redistribution on the surface of target cells is demonstrated by experiments in which disruption of raft integrity or blocking of Src kinase activity decreased the ability of the target cells to present viral peptides to CTL, especially at lower peptide abundance (Lebedeva et al., 2004). Productive engagement of ICAM-1 molecules also leads to their recruitment to rafts in other cell types that could serve as target cells for CTL. For instance, antibody-mediated cross-linking of ICAM-1 on brain endothelial cells induces its partitioning into raft fractions and ICAM-1 co-immunoprecipitation with Src kinases (Tilghman and Hoover, 2002b).

Since neither transmembrane nor cytoplasmic tail of ICAM-1 contains characteristics that could target it to membrane rafts, it is thought that cross-linking of ICAM-1 brings together transmembrane segments, increasing their affinity to the liquid-ordered environment in the lipid rafts. Treatment with cytochalasin B (Tilghman and Hoover, 2002a) does not block ICAM-1 redistribution to the raft fraction after its cross-linking. This suggests that actin polymerization is not required for ICAM-1 recruitment to lipid rafts. However, earlier studies have shown ICAM-1 association with α-actinin, a linker between actin cytoskeleton and cell membrane (Carpen et al., 1992). Ezrin, another linker between integral membrane proteins and actin cytoskeleton, binds directly to ICAM-1 in the presence of phosphatidylinositol 4,5-biphosphate, strongly suggesting a regulatory role of phosphoinositide signaling pathways in anchoring ICAM-1 (Heiska et al., 1998). Thus, ICAM-1 is linked to the actin cytoskeleton independent of association with membrane rafts. The cytoplasmic domain of ICAM-1 binds the β-tubulin and glyceroaldehyde-3-phosphate dehydrogenase, a microtubule bundling enzyme (Federici et al., 1996). ICAM-1 engineered to express a glycosylphosphatidylinositol (GPI) anchor in place of its transmembrane and cytoplasmic domain still binds to LFA-1 and can function in antigen presentation (Greenwood et al., 2003); however, GPI anchored proteins also activate src family kinases upon rafts clustering such that the GPI anchor may simply replace the normal signaling mediated by the ICAM-1 transmembrane and cytoplasmic domains with a related set of signals characteristic of GPI anchored proteins. Thus, ICAM-1 facilitates antigen presentation by initiating Src kinase-dependent signaling that leads to increase of cytokine expression and the level of immune receptors on target and APC; this signaling also facilitates recruitment of the immune receptors into rafts and rafts coalescence, resulting in the accumulation of these receptors at the T cell-target cell contact area. Although these events seem to be independent on actin cytoskeleton, the cytoskeletal changes induced by ICAM-1 ligation may result in optimized membrane contact between target cells and CTL.

7.3
Major Histocompatability Complex (MHC)

7.3.1
MHC Molecules

MHC molecules that display fragments of pathogens or tumor-specific proteins serve as natural ligands for antigen-specific receptor (TCR) on T cells. There are two classes of MHC, MHC-I and MHC-II, which present antigenic peptides to two different subpopulations of T cells, $CD8^+$ and $CD4^+$, respectively. These proteins share similar molecular structure, but present peptides of different origin. Peptides associated with MHC-I are derived from proteins produced inside the cell, while MHC-II-bound peptides are derived from proteins taken up into the cell. This difference may reflect the profile of MHC-I and MHC-II distribution between tissues: MHC-II is expressed only on professional APC, most of which exercise phagocytic activity, while MHC-I is present on the surface of all nucleated cells.

Measuring diffusion characteristics of MHC-I molecules, Edidin and colleagues first concluded that these molecules are organized in clusters on the cell surface (Edidin, 1988; Wier and Edidin, 1988). The value of the rotational diffusion coefficient of HLA-A2 MHC-I on B lymphoblastoid JY cells suggests that there are, on average, about 25 molecules of HLA-A2 per cluster, which could reach 240 molecules per cluster on HeLa cells (Tang and Edidin, 2001). In accord with these data, analysis of the hierarchy of MHC-I clustering with electron microscopy showed the existence of small-scale (2–10 nm) MHC-I clusters as well as larger-scale clusters (up to hundreds of nanometers) (Damjanovich et al., 1995).

7.3.2
Molecular Associations of MHC-I Molecules

Molecular homo-association between different MHC-I molecules on T and B lymphoblastoid cell lines was confirmed by fluorescence energy transfer between FITC- or TRITC-labeled Fab fragments of W6/32 mAb bound to non-polymorphic domain of human MHC-I (Bene et al., 1994). MHC-I clusters appeared to be stable over dozens of minutes and were not caused by random collision of the monomers. For instance, homo-association of MHC-I was detected on the different B cell lines (regardless of MHC-I expression level), as well as on transformed fibroblasts, lymphoblasts and activated lymphocytes, epithelial (melanoma) and colon carcinoma cells, but not on the resting lymphocytes and normal fibroblasts.

Receptor clustering at the cell surface may facilitate intermolecular cross talk and influence the pattern of organization of signaling proteins associated with the receptors. While actual organizing forces of these non-random molecular assemblies are not completely understood, several mechanisms may contribute to MHC clustering. Similar to ICAM-1, MHC molecules are concentrated in lipid rafts of various cells of lymphoid and non-lymphoid origin. On average 2–50% of the cell surface MHC could be found in lipid rafts. Lipid rafts of an individual cell are not uniform and may form highly ordered and semi-ordered domains concentrating different types of proteins. Thus, the variability in the amount of MHC molecules in lipid rafts may arise from variations in cholesterol and/or ganglioside level due to cultivating conditions, cell type and stage of activation, or depending on the concentration and type of detergents used for the isolation of rafts. The level of cholesterol, an essential raft component, influences membrane fluidity and may change the balance of raft-included and raft-excluded proteins. Thus, elevation of cholesterol concentration in the membrane of JY cells is accompanied by a higher degree of MHC-I clustering, as evident from the enhanced intermolecular energy transfer efficiency (Bodnar et al., 1996). MHC-I, MHC-II and CD48 molecules also form clusters with IL2-R (CD25) in cholesterol-dependent manner on the surface of human T lymphoma. Disruption of lipid rafts with filipin or depletion of membrane cholesterol with MCD results in the blurring of cluster boundaries and an apparent dispersion of clusters for all four proteins (Vereb et al., 2000). We have also found that almost 50% of cell-surface HLA-A2 MHC-I resides in lipid rafts on JY cells (Lebedeva et al., 2004). Moreover, cholesterol extraction with MCD led to partial release of HLA-A2 from lipid rafts to a fraction of soluble membranes, affecting about 60% of raft-associated HLA-A2. Clustering of MHC-II is also thought to depend on their recruitment to lipid rafts. Up to 50% of cell surface MHC-II molecules are included into the fraction of membrane rafts on B cells (Anderson et al., 2000). In contrast, MHC II do not localize within rafts on human myelomonocytes, but could be recruited to the rafts after cross-linking; this process is accompanied by intracellular proteins phosphorylation that depends on the activity of Src family kinases (Huby et al., 1999). Thus, evidently, lipid rafts can be utilized as a platform for protein clustering on the cell surface. Localization of transmembrane proteins that are not classical receptor molecules in lipid rafts may provide an opportunity to initiate signaling followed by their productive engagement. Indeed, cross-linking of either MHC-I or MHC-II by multivalent ligands induces phosphorylation of several intracellular proteins within-in the very first minute after ligation. The ability of MHC molecules to "report" the binding by activation of intracellular signaling could be utilized to enhance antigen presentation.

Another platform for protein clustering is based on multiple intermolecular associations with tetraspan superfamily (TM4SF) molecules. TM4SF proteins may serve as adaptors that regulate the assembly of protein complexes in the cell membrane due to their well-recognized ability to associate with each other and other transmembrane proteins. Both MHC-I and MHC-II are found co-aggregated with TM4SF. In particular, a fraction of MHC-II proteins is associated with three proteins of the TM4SF family, CD53, CD81, and CD82, on JY cells. In addition, two other B cell surface molecules, CD20 and MHC-I, are in close vicinity to each other and to MHC-II and the TM4SF proteins (based on FRET experiments). The efficiency of the FRET from CD20, CD53, CD81, and CD82 to HLA-DR suggests that all these molecules are in a single type of complex with the HLA-DR molecules (Szollosi et al., 1996). Association of MHC-II molecules with TM4SF microdomains has been recently confirmed by biochemical analysis of plasma membranes derived from dendritic cells (DC) and B cells (Kropshofer et al., 2002). In these clusters, HLA-DR and HLA-DP are associated with CD9, CD81, CD82 and peptide "editor" molecule HLA-DM. Up to 10% of cell surface MHC-II is confined to TM4SF microdomains. TM4SF associations with MHC-I, MHC-II and other Ig superfamily proteins are sensitive to 1% Triton X-100 treatment, suggesting that these are not liquid ordered domains like rafts and may be more fluid than the raft-associated MHC-II fractions.

While raft-associated MHC-II molecules show a highly diverse self-peptide repertoire that is very similar to the peptide repertoire of MHC-II molecules derived from the non-raft fraction, TM4SF-based MHC-II clusters are marked by the presence of three dominant peptides, whereas most other self-peptides, including abundant peptide such as CLIP, are absent (Kropshofer et al., 2002). This suggests that the spatial organization of peptide–MHC-II complexes in the plasma membrane could be important for the T cell activation potential. If this subset of self-peptides was particularly important for positive selection in the thymus then replacement of a few of these peptides with foreign peptides during antigen presentation could result in very sensitive foreign antigen recognition based on recent models for synergy of foreign and self-MHC complexes. Because both types of microdomains differ from each other with regard to the peptide repertoire, it is tempting to propose that they may evoke different types of T cell response; for example, they may determine whether naïve T cells differentiate into type 2 or type 1 cells.

The physiological significance of pMHC-II clustering is consistent with the findings, showing that soluble, agonist pMHC-II ligands are nonstimulatory as monomers and minimally stimulatory as dimers. In contrast, trimeric or tetrameric agonist ligands that engage multiple TCRs for a sustained duration are potent stimuli. MHC-II-driven formation of TCR clusters seems required for effective activation and helps to explain the specificity and sensitivity of T cells (Boniface et al., 1998; Cochran et al., 2000).

Molecular clustering could also arise from the concentration of molecules in the transport vesicles delivering cargo to the cell surface. This mechanism seems to be a plausible explanation of the origin of MHC-I patches (Tang and Edidin, 2001) and is supported by analysis of changes of apparent size of HLA-A2 patches on the cell membrane under conditions of inhibition of vesicles traffic. Some MHC-I molecules could cluster already in the endoplasmic reticulum (ER) (Pentcheva and Edidin, 2001). This MHC-I fraction is not bound to the TAP; instead, it appears clustered after peptide loading and is associated with the transport receptor BAP31, which shuttles between the ER and the *cis*-Golgi (Spiliotis et al., 2000). Similarly, MHC-II–TM4SF clusters are found in the intracellular compartments as well as on the plasma membrane (Kropshofer et al., 2002). The TM4SF microdomains are reminiscent of clusters containing MHC-II and co-stimulatory molecule CD86 (B7-2); the latter are located in transport vesicles of murine dendritic cells. Through these vesicles, MHC-II–CD86 clusters are transported from internal loading compartments to the cell surface, where they remain clustered (Turley et al., 2000). Protein association with lipid rafts also could occur during intracellular transport. For example, the GPI-linked protein, PLAP, fully soluble in Triton X-100 after biosynthesis and for the time of residence in the ER compartment, becomes insoluble traveling between the ER and medial Golgi (Garcia et al., 1993). This suggests that insolubility is not an inherent property of GPI-linked and other proteins, but is acquired after exposure to the lipid environment that proteins encounter entering the Golgi complex, where liquid ordered rafts begin to form. Similarly, in the endoplasmic reticulum, MHC-II complexes are cold 1% Triton X-100 soluble and acquire their insolubility during transport through Golgi apparatus (Poloso et al., 2004).

In addition to TM4SF proteins, MHC-I and/or MHC-II clusters contain other transmembrane proteins, such as ICAM-1, insulin receptor, IL-2 receptor. Detection of ICAM-1 in MHC-I and MHC-II enriched clusters has a functional significance. In fact, ICAM-1–MHC-I co-clustering in lipid rafts facilitates efficiency of the presentation of viral peptides displayed on HLA-A2 of B lymphoblastoid cells to CTL (Lebedeva et al., 2004). Co-clustering of MHC-I and ICAM-1 within lipid rafts was also observed on the surface of LS-174-T colon carcinoma cell line (Bacso et al., 2002). Treatment of the cells with γ-interferon enhances heteroassociation between MHC-I, ICAM-1 and GM1 due to elevated cell surface level of these molecules and an increase of their partitioning in rafts (Bacso et al., 2002). We propose that changing of the pattern of the cell surface molecular assemblies may enhance recognition of carcinoma cells by CTL.

7.3.3
Association of MHC-I and ICAM-1

ICAM-1–MHC-I association has been found in both raft and non-raft membrane fractions of B lymphoblastoid cell line (Lebedeva et al., 2004). Moreover, disruption of the integrity of isolated rafts does not reverse ICAM-1–MHC-I association. Consistent with these findings, association between TM4SF molecules CD9 and CD81 and $\alpha 3\beta 1$ integrins on the surface of fibrosarcoma cell line HT1080 and epidermoid carcinoma cells A431 is independent of raft environment, even though these molecular assemblies could be located in rafts (Claas et al., 2001). These data suggest that molecular aggregation and recruitment into rafts might have different functional implications. For instance, co-clustering of ICAM-1 and MHC-I facilitates initial detection of a small number of cognate pMHC on target cells by CTL, regardless of association with rafts, while signaling initiated by productive engagement of raft-included ICAM-1 and/or MHC-I leads to coalescence of rafts and raft-associated ICAM-1–MHC-I assemblies at the CTL-target cell interface, enhancing antigen presentation to CTL. We propose that ICAM-1–MHC-I interactions provide a link between the initial step of antigen recognition and immunological synapse formation.

7.3.4
Could APC and Target Cells Play an Active Role in Ag Presentation?

The total number of any allelic variant of MHC protein on the cell surface of potential target cells is measured by hundreds of thousands; the level of MHC molecules on professional APC, e.g., dendritic cells, is even higher. At the same time, the number of cognate pMHC complexes recognizable by T cells, say, the number of MHC molecules displaying virus- or tumor-associated peptides, typically varies from a few to a few hundred per cell (Christinck et al., 1991; Demotz et al., 1990; Harding and Unanue, 1990; Irvine et al., 2002; Kageyama et al., 1995; Sykulev et al., 1996), amounting to a fraction of a percent of the total MHC. Thus, identification of such a small number of pMHC complexes by T cells on the surface of target cells is equivalent to finding a needle in a haystack.

When antigen concentration is limited, signaling through the TCR bound to a rare pMHC may be quite weak, requiring an amplification mechanism triggered by the initial recognition event. While it is well established that T cells possess mechanisms to amplify initial signals, whether target cells and APC can actively contribute to the amplification mechanism is only beginning to be understood. In particular, APC and target cells can modulate the efficiency of antigen presentation and influence the outcome of immune response by changing the composition of the molecular assemblies on the cell surface in response to inflammatory stimuli or ligation of MHC or accessory molecules. There are several molecular features that can foster efficiency of antigen presentation.

7.3.5
Identical pMHCs are Clustered in the Same Microdomain

As described above, there are several mechanisms to concentrate pMHC on the plasma membrane and to facilitate recognition of cognate pMHC by TCR. But it is difficult to imagine how identical pMHC can reside in the same microdomain. This drawback could be bypassed in several ways. At any given time the spectrum of synthesized and degraded proteins in the cell is limited. Thus, it is possible that the same peptides could be loaded to MHC-I molecules that are already clustered in the ER (Pentcheva and Edidin, 2001) and that these clusters could be preserved until they appear on the cell surface. In a similar scenario, MHC-II molecules are thought to be clustered in a specialized vesicles (Turley et al., 2000), where they are loaded with a limited set of peptides derived from an endocytosed protein and are subsequently released at the cell surface remaining to be clustered. Yet another example is given by clustering of MHC-II associated with TM4SF microdomains that present only limited peptide repertoire on the surface of human transformed B cells and DCs (Kropshofer et al., 2002). These domains are likely to originate from MHC-II compartments where TM4SF members (CD63 and CD82) are associated with HLA-DR, HLA-DO and HLA-DM, suggesting a role for TM4SF proteins in loading peptides onto MHC-II molecules (Hammond et al., 1998). Thus, MHC-II complexes in this compartment are likely to be loaded with a distinct set of peptides compared to MHC-II occurring in vesicles devoid of TM4SF proteins.

7.3.6
Identical pMHC can be Recruited to the Same Microdomain During Target Cell–T Cell Interaction

Another scenario suggests that agonist pMHC are recruited to the IS, where they accumulate in antigen-dependent manner. At limited peptide concentration both relevant and irrelevant pMHC are accumulated in the center of IS; it is believed that pMHC are recruited into rafts and the rafts' aggregation results in a rapid delivery of the pMHCs to the interface, facilitating their accumulation at the site of the T cell contact with APC (Hiltbold et al., 2003). At the stage of mature IS, agonist pMHC segregate toward the center of the IS while irrelevant pMHC are excluded to the periphery of the center of the IS. The presence of both types of pMHC in the cSMAC could stabilize weak TCR–pMHC interactions by direct binding of TCR with either pMHC (Wulfing et al., 2002). Thus, higher avidity of TCR–pMHC interaction is gained by multivalent engagement of co-clustered self- and agonist pMHC complexes.

7.3.7
Co-clustering of MHC and Accessory Molecules

Clustered pMHC complexes may be co-aggregated with accessory or co-stimulatory molecules that could be beneficial for effective antigen presentation. The presence of such assemblies on the cell surface could form scaffolds that facilitate intercellular interactions. Recruitment of these assemblies into rafts may further promote receptor oligomerization and, therefore, increase the receptors avidity. In fact, interactions between MHC-I and ICAM-1 are enhanced as a result of raft aggregation (Lebedeva et al., 2004). In addition, other co-stimulatory molecules can facilitate these interactions. For instance, NKG2D, a co-stimulatory receptor for T cells, contributes to the formation of an antigen-independent ring junction formed between CTL and glass-supported lipid bilayer containing ICAM-1 at lower surface densities (Somersalo et al., 2004). MHC-I association with tetraspan family proteins also increases the ability of target cells to induce cytotoxic response. In particular, epithelial membrane protein 2 (EMP2) regulates the surface expression of some membrane proteins, notably those destined for lipid raft microdomains. Overexpression of EMP2 in target cells increased their susceptibility to CTL cytotoxicity by increasing the surface levels of MHC-I, ICAM-1, and GM1 glycolipids that reside with EMP2 in a lipid raft membrane compartment (Wadehra et al., 2003). All the above data suggest that pre-existing MHC-containing assemblies on the target cell membrane can dramatically influence the outcome of T cell–target cell encounters.

7.3.8
Role of Cytoskeleton

Analysis of T cells interaction with supported lipid bilayers containing MHC and adhesion molecules (Grakoui et al., 1999) as well as earlier experiments on pre-treatment of B cells with inhibitors of actin cytoskeleton (Wulfing et al., 1998; Wulfing et al., 2002) suggest that B cells do not actively contribute to the dynamics of molecular segregation in the IS.

Some recent data force a re-evaluation of a possible role of the APC cytoskeleton. Pre-treatment of B cells with PMA, an activator of protein kinase C isoforms, leads to tighter T-B cells contacts (Wulfing et al., 1998). This may have been because of changes on ICAM-1 on the B cell or increased LFA-1 activity on the B cell binding to ICAM-1 on the T cell. Either way it suggests that active regulation of adhesion in the B cell can contribute to the avidity of the T cell–B cell contact, but PMA is an artificial way to induce this activation. In another example, GPI-anchored D^b molecules were defective in mediating CTL lysis of transfected lymphoma cells, compared to the same cells expressing intact MHC proteins. This disparity could result from differences between GPI-D^b and wild-type-D^b assemblies and transport or from differences in their cell membrane topology that affect $CD8^+$ CTL recognition of pMHC-I complex (Cariappa et al., 1996). In contrast to the results of Wulfing et al , Gordy et al found that pre-treatment of CH27 cells with latrunculin B caused a significant reduction in the formation of B cell–T cell conjugates and abolished enrichment of PKC-Θ at the c-SMAC in T cells (Gordy et al., 2004). Treatment of dendritic cells with cytochalasin D did significantly impair antigen presentation by mature dendritic cells (Al-Alwan et al., 2001a). The expression of the actin bundling protein fascin, which plays a role in the formation of long denditic processes by mature dendritic cells, was important for the active role of the dendritic cell cytoskeleton in antigen presentation based on experiments with anti-sense oligonucleotides (Al-Alwan et al., 2001b). More recently, the actin cytoskeleton regulatory small G proteins Rac1 and Rac2 were shown to play an important role in antigen presentation by mature dendritic cells (Benvenuti et al., 2004). In this study dendritic cells were demonstrated to very actively embrace T cells. The receptors responsible for this response of the dendritic cells were not demonstrated, but the response is reminiscent of the ICAM-1-mediated response of endothelial cells to leukocyte adhesion described above. A distinct role of cytoskeleton in antigen presentation by dendritic cells was suggested by experiments in which maturation of bone-marrow-derived DCs induces the formation of tubular MHC-II-positive compartments. The recognition of cognate pMHC on DC by T cells provides a signal that leads to intracellular transport of MHC-II to the cell surface. Polarized MHC-II-positive tubules form within minutes, and point toward T cells. Loading DC with antigen is not sufficient to induce tubulation; an additional signal through Toll-like receptor is required to induce T-cell-polarized endosomal tubulation (Boes et al., 2002; Boes et al., 2004). A specific enhancing role of these provocative tubules for antigen presentation has yet to be demonstrated.

7.4 Conclusion

There is compelling evidence that the APC play an active role facilitating antigen presentation to T cells. Molecular events involved in this process may be cell type specific and may depend on the stage of cellular differentiation. This explains why various target cells may present the same antigen with different efficacy. This difference is manifested the most in recognition by $CD8^+$ cytotoxic T cells that can target virtually any cell, but are initially activated by professional APC. Similar to T cells, adhesion molecules on target cells and APC primarily mediate cell–cell contact, an essential step in antigen recognition by T cells. It is becoming increasingly evident, however, that they also trigger intracellular signaling that ultimately leads to accumulation of immune receptor, MHC molecules in particular, at the contact area with T cell, thereby increasing the effective concentration of the immune receptors recognizable by respective counter receptors on T cells. In this event, adhesion molecules on target cells and APC appear to function as co-stimulatory molecules. This mechanism seems reminiscent of co-stimulation in T cells and is especially important at the condition of limited antigen concentration. On both T cells and target cells, immune receptors undergo large-scale rearrangement, resulting in accumulation of the immune receptors at the cell–cell interface. In addition, the receptors on T cells are segregated to form very well organized molecular assemblies. Whether active molecular segregation occurs on target cell membranes has yet to be determined.

References

Al-Alwan, M. M., Rowden, G., Lee, T. D., West, K. A. (2001a): The dendritic cell cytoskeleton is critical for the formation of the immunological synapse. *J. Immunol.* 166(3): 1452–1456.

Al-Alwan, M. M., Rowden, G., Lee, T. D., West, K. A. (2001b): Fascin is involved in the antigen presentation activity of mature dendritic cells. *J. Immunol.* 166(1): 338–345.

Amos, C., Romero, I. A., Schultze, C., Rousell, J., Pearson, J. D., Greenwood, J., Adamson, P. (2001): Cross-linking of brain endothelial intercellular adhesion molecule (ICAM)-1 induces association of ICAM-1 with detergent-insoluble cytoskeletal fraction. *Arterioscler. Thromb. Vasc. Biol.* 21(5): 810–816.

Anderson, H. A., Hiltbold, E. M., Roche, P. A. (2000): Concentration of MHC class II molecules in lipid rafts facilitates antigen presentation. *Nat. Immunol.* 1(2): 156–162.

Bachmann, M. F., McKall-Faienza, K., Schmits, R., Bouchard, D., Beach, J., Speiser, D. E., Mak, T. W., Ohashi, P. S. (1997): Distinct roles for LFA-1 and CD28 during activation of naive T cells: adhesion versus costimulation. *Immunity* 7(4): 549–557.

Bacso, Z., Bene, L., Damjanovich, L., Damjanovich, S. (2002): INF-gamma rearranges membrane topography of MHC-I and ICAM-1 in colon carcinoma cells. *Biochem. Biophys. Res. Commun.* 290(2): 635–640.

Barreiro, O., Yanez-Mo, M., Serrador, J. M., Montoya, M. C., Vicente-Manzanares, M., Tejedor, R., Furthmayr, H., Sanchez-Madrid, F. (2002): Dynamic interaction of VCAM-1 and ICAM-1 with moesin and ezrin in a novel endothelial docking structure for adherent leukocytes. *J. Cell Biol.* 157(7): 1233–1245.

Bene, L., Balazs, M., Matko, J., Most, J., Dierich, M. P., Szollosi, J., Damjanovich, S. (1994): Lateral organization of the ICAM-1 molecule at the surface of human lymphoblasts: a possible model for its co-distribution with the IL-2 receptor, class I and class II HLA molecules. *Eur. J. Immunol.* 24(9): 2115–2123.

Benvenuti, F., Hugues, S., Walmsley, M., Ruf, S., Fetler, L., Popoff, M., Tybulewicz, V. L., Amigorena, S. (2004): Requirement of Rac1 and Rac2 expression by mature dendritic cells for T cell priming. *Science* 305(5687): 1150–1153.

Bierer, B. E., Hahn, W. C. (1993): T cell adhesion, avidity regulation and signaling: a molecular analysis of CD2. *Semin. Immunol.* 5(4): 249–261.

Bodnar, A., Jenei, A., Bene, L., Damjanovich, S., Matko, J. (1996): Modification of membrane cholesterol level affects expression and clustering of class I HLA molecules at the surface of JY human lymphoblasts. *Immunol. Lett.* 54(2–3): 221–226.

Boes, M., Cerny, J., Massol, R., Op den Brouw, M., Kirchhausen, T., Chen, J., Ploegh, H. L. (2002): T-cell engagement of dendritic cells rapidly rearranges MHC class II transport. *Nature* 418(6901): 983–988.

Boes, M., Cuvillier, A., Ploegh, H. (2004): Membrane specializations and endosome maturation in dendritic cells and B cells. *Trends Cell Biol.* 14(4): 175–183.

Boniface, J. J., Rabinowitz, J. D., Wulfing, C., Hampl, J., Reich, Z., Altman, J. D., Kantor, R. M., Beeson, C., McConnell, H. M., Davis, M. M. (1998): Initiation of signal transduction through the T cell receptor requires the multivalent engagement of peptide/MHC ligands [corrected]. *Immunity* 9(4): 459–466.

Brown, D. A., London, E. (1998): Functions of lipid rafts in biological membranes. *Annu. Rev. Cell Dev. Biol.* 14: 111–136.

Brown, D. A., Rose, J. K. (1992): Sorting of GPI-anchored proteins to glycolipid-enriched membrane subdomains during transport to the apical cell surface. *Cell* 68(3): 533–544.

Buckle, A.-M., Hogg, N. (1990): Human memory T cells express intercellular adhesion molecule-1 which can be increased by interleukin 2 and interferon-gamma. *Eur. J. Immunol.* 20: 337–341.

Burkart, A., Samii, B., Corvera, S., Shpetner, H. S. (2003): Regulation of the SHP-2 tyrosine phosphatase by a novel cholesterol- and cell confluence-dependent mechanism. *J. Biol. Chem.* 278(20): 18 360–18 367.

Cariappa, A., Flyer, D. C., Rollins, C. T., Roopenian, D. C., Flavell, R. A., Brown, D., Waneck, G. L. (1996): Glycosylphosphatidylinositol-anchored H-2Db molecules are defective in antigen processing and presentation to cytotoxic T lymphocytes. *Eur. J. Immunol.* 26(9): 2215–2224.

Carpen, O., Pallai, P., Staunton, D. E., Springer, T. A. (1992): Association of intercellular adhesion molecule-1 (ICAM-1) with actin-containing cytoskeleton and alpha-actinin. *J. Cell Biol.* 118(5): 1223–1234.

Carrell, N. A., Fitzgerald, L. A., Steiner, B., Erickson, H. P., Phillips, D. R. (1985): Structure of human platelet membrane glycoproteins IIb and IIIa as determined by electron microscopy. *J. Biol. Chem.* 260: 1743–1749.

Casasnovas, J. M., Stehle, T., Liu, J. H., Wang, J. H., Springer, T. A. (1998): A dimeric crystal structure for the N-terminal two domains of intercellular adhesion molecule-1. *Proc. Natl. Acad. Sci. U.S.A.* 95(8): 4134–4139.

Christinck, E. R., Luscher, M. A., Barber, B. H., Williams, D. B. (1991): Peptide binding to class I MHC on living cells and quantitation of complexes required for CTL lysis. *Nature* 352: 67–70.

Claas, C., Stipp, C. S., Hemler, M. E. (2001): Evaluation of prototype transmembrane 4 superfamily protein complexes and their relation to lipid rafts. *J. Biol. Chem.* 276(11): 7974–7984.

Cochran, J. R., Cameron, T. O., Stern, L. J. (2000): The relationship of MHC-peptide binding and T cell activation probed using chemically defined MHC class II oligomers. *Immunity* 12(3): 241–250.

Damjanovich, S., Vereb, G., Schaper, A., Jenei, A., Matko, J., Starink, J. P., Fox,

G. Q., Arndt-Jovin, D. J., Jovin, T. M. (1995): Structural hierarchy in the clustering of HLA class I molecules in the plasma membrane of human lymphoblastoid cells. *Proc. Natl. Acad. Sci. U.S.A.* 92(4): 1122–1126.

Davis, S. J., Ikemizu, S., Evans, E. J., Fugger, L., Bakker, T. R., van der Merwe, P. A. (2003): The nature of molecular recognition by T cells. *Nat. Immunol.* 4(3): 217–224.

Deeths, M. J., Mescher, M. F. (1999): ICAM-1 and B7-1 provide similar but distinct costimulation for CD8+ T cells, while CD4+ T cells are poorly costimulated by ICAM-1. *Eur. J. Immunol.* 29(1): 45–53.

Demotz, S., Grey, H. M., Sette, A. (1990): The minimal number of class II MHC-antigen complexes needed for T cell activation. *Science* 249: 1028–1030.

Dustin, M. L. (2002): Membrane domains and the immunological synapse: keeping T cells resting and ready. *J. Clin. Invest.* 109(2): 155–160.

Dustin, M. L., Rothlein, R., Bhan, A. K., Dinarello, C. A., Springer, T. A. (1986): Induction by IL-1 and interferon, tissue distribution, biochemistry, and function of a natural adherence molecule (ICAM-1). *J. Immunol.* 137: 245–254.

Dustin, M. L., Springer, T. A. (1989): T cell receptor cross-linking transiently stimulates adhesiveness through LFA-1. *Nature* 341: 619–624.

Edidin, M. (1988): Function by association? MHC antigens and membrane receptor complexes. *Immunol. Today* 9(7–8): 218–219.

Edidin, M. (2003): The state of lipid rafts: from model membranes to cells. *Annu. Rev. Biophys. Biomol. Struct.* 32: 257–283.

Etienne, S., Adamson, P., Greenwood, J., Strosberg, A. D., Cazaubon, S., Couraud, P. O. (1998): ICAM-1 signaling pathways associated with Rho activation in microvascular brain endothelial cells. *J. Immunol.* 161(10): 5755–5761.

Etienne-Manneville, S., Manneville, J. B., Adamson, P., Wilbourn, B., Greenwood, J., Couraud, P. O. (2000): ICAM-1-coupled cytoskeletal rearrangements and transendothelial lymphocyte migration involve intracellular calcium signaling in brain endothelial cell lines. *J. Immunol.* 165(6): 3375–3383.

Federici, C., Camoin, L., Hattab, M., Strosberg, A. D., Couraud, P. O. (1996): Association of the cytoplasmic domain of intercellular-adhesion molecule-1 with glyceraldehyde-3-phosphate dehydrogenase and beta-tubulin. *Eur. J. Biochem.* 238(1): 173–180.

Gaglia, J. L., Greenfield, E. A., Mattoo, A., Sharpe, A. H., Freeman, G. J., Kuchroo, V. K. (2000): Intercellular adhesion molecule 1 is critical for activation of CD28-deficient T cells. *J. Immunol.* 165(11): 6091–6098.

Garcia, M., Mirre, C., Quaroni, A., Reggio, H., Le Bivic, A. (1993): GPI-anchored proteins associate to form microdomains during their intracellular transport in Caco-2 cells. *J. Cell Sci.* 104 (Pt 4): 1281–1290.

Gordy, C., Mishra, S., Rodgers, W. (2004): Visualization of antigen presentation by actin-mediated targeting of glycolipid-enriched membrane domains to the immune synapse of B cell APCs. *J. Immunol.* 172(4): 2030–2038.

Grakoui, A., Bromley, S. K., Sumen, C., Davis, M. M., Shaw, A. S., Allen, P. M., Dustin, M. L. (1999): The immunological synapse: a molecular machine controlling T cell activation. *Science* 285: 221–227.

Greenwood, J., Amos, C. L., Walters, C. E., Couraud, P. O., Lyck, R., Engelhardt, B., Adamson, P. (2003): Intracellular domain of brain endothelial intercellular adhesion molecule-1 is essential for T lymphocyte-mediated signaling and migration. *J. Immunol.* 171(4): 2099–2108.

Greenwood, J., Etienne-Manneville, S., Adamson, P., Couraud, P. O. (2002): Lymphocyte migration into the central nervous system: implication of ICAM-1 signalling at the blood-brain barrier. *Vascul Pharmacol.* 38(6): 315–322.

Hammond, C., Denzin, L. K., Pan, M., Griffith, J. M., Geuze, H. J., Cresswell, P. (1998): The tetraspan protein CD82 is a resident of MHC class II compartments where it associates with HLA-DR, -DM, and -DO molecules. *J. Immunol.* 161(7): 3282–3291.

Harding, C. V., Unanue, E. R. (1990): Quantitation of antigen-presenting cell MHC class II/peptide complexes necessary for T-cell stimulation. *Nature* 346: 574–576.

Heiska, L., Alfthan, K., Gronholm, M., Vilja, P., Vaheri, A., Carpen, O. (1998): Association of ezrin with intercellular adhesion molecule-1 and -2 (ICAM-1 and ICAM-2). Regulation by phosphatidylinositol 4,5-bisphosphate. *J. Biol. Chem.* 273(34): 21 893–21 900.

Hiltbold, E. M., Poloso, N. J., Roche, P. A. (2003): MHC class II-peptide complexes and APC lipid rafts accumulate at the immunological synapse. *J. Immunol.* 170(3): 1329–1338.

Hubbard, A. K., Rothlein, R. (2000): Intercellular adhesion molecule-1 (ICAM-1) expression and cell signaling cascades. *Free. Radic. Biol. Med.* 28(9): 1379–1386.

Huby, R. D., Dearman, R. J., Kimber, I. (1999): Intracellular phosphotyrosine induction by major histocompatibility complex class II requires co-aggregation with membrane rafts. *J. Biol. Chem.* 274(32): 22 591–22 596.

Irvine, D. J., Purbhoo, M. A., Krogsgaard, M., Davis, M. M. (2002): Direct observation of ligand recognition by T cells. *Nature* 419(6909): 845–849.

Johnson, J. G., Jenkins, M. K. (1994): Monocytes provide a novel co-stimulatory signal to T cells that is not mediated by the CD28/B7 interaction. *J. Immunol.* 152(2): 429–437.

Kageyama, S., Tsomides, T. J., Sykulev, Y., Eisen, H. N. (1995): Variations in the number of peptide-MHC class I complexes required to activate cytotoxic T cell responses. *J. Immunol.* 154: 567–576.

Kirchhausen, T., Staunton, D. E., Springer, T. A. (1995): Location of the domains of ICAM-1 by immunolabeling and single-molecule electron microscopy. *J. Leukocyte Biol.* 53: 342–346.

Koyama, Y., Tanaka, Y., Saito, K., Abe, M., Nakatsuka, K., Morimoto, I., Auron, P. E., Eto, S. (1996): Cross-linking of intercellular adhesion molecule 1 (CD54) induces AP-1 activation and IL-1beta transcription. *J. Immunol.* 157(11): 5097–5103.

Kropshofer, H., Spindeldreher, S., Rohn, T. A., Platania, N., Grygar, C., Daniel, N., Wolpl, A., Langen, H., Horejsi, V., Vogt, A. B. (2002): Tetraspan microdomains distinct from lipid rafts enrich select peptide-MHC class II complexes. *Nat. Immunol.* 3(1): 61–68.

Lawson, C., Ainsworth, M., Yacoub, M., Rose, M. (1999): Ligation of ICAM-1 on endothelial cells leads to expression of VCAM-1 via a nuclear factor-kappaB-independent mechanism. *J. Immunol.* 162(5): 2990–2996.

Lebedeva, T., Anikeeva, N., Kalams, S. A., Walker, B. D., Gaidarov, I., Keen, J. H., Sykulev, Y. (2004): MHC-I-ICAM-1 association on the surface of target cells: implications for antigen presentation to cytotoxic T lymphocytes. *Immunology* 113(4): 460–471.

Miller, J., Knorr, R., Ferrone, M., Houdei, R., Carron, C. P., Dustin, M. L. (1995): Intercellular adhesion molecule-1 dimerization and its consequences for adhesion mediated by lymphocyte function associated-1. *J. Exp. Med.* 182(5): 1231–1241.

Mueller, D. L., Jenkins, M. K., Chiodetti, L., Schwartz, R. H. (1990): An intracellular calcium increase and protein kinase C activation fail to initiate T cell proliferation in the absence of a co-stimulatory signal. *J. Immunol.* 144(10): 3701–3709.

Naslavsky, N., Weigert, R., Donaldson, J. G. (2004): Characterization of a nonclathrin endocytic pathway: membrane cargo and lipid requirements. *Mol. Biol. Cell* 15(8): 3542–3552.

Oh, S., Hodge, J. W., Ahlers, J. D., Burke, D. S., Schlom, J., Berzofsky, J. A. (2003): Selective induction of high avidity CTL by altering the balance of signals from APC. *J. Immunol.* 170(5): 2523–2530.

Pentcheva, T., Edidin, M. (2001): Clustering of peptide-loaded MHC class I molecules for endoplasmic reticulum export imaged by fluorescence resonance energy transfer. *J. Immunol.* 166(11): 6625–6632.

Pluskota, E., Chen, Y., D'Souza, S. E. (2000): Src homology domain 2-containing tyrosine phosphatase 2 associates with intercellular adhesion molecule 1 to regulate cell survival. *J. Biol. Chem.* 275(39): 30 029–30 036.

Poloso, N. J., Muntasell, A., Roche, P. A. (2004): MHC Class II Molecules Traffic into Lipid Rafts during Intracellular Transport. *J. Immunol.* 173(7): 4539–4546.

Poudrier, J., Owens, T. (1994): CD54/intercellular adhesion molecule 1 and major histocompatibility complex II signaling induces B cells to express interleukin 2 receptors and complements help provided through CD40 ligation. *J. Exp. Med.* 179(5): 1417–1427.

Reilly, P. L., Woska, J. R., Jr., Jeanfavre, D. D., McNally, E., Rothlein, R., Bormann, B. J. (1995): The native structure of intercellular adhesion molecule-1 (ICAM-1) is a dimer. Correlation with binding to LFA-1. *J. Immunol.* 155(2): 529–532.

Sano, H., Nakagawa, N., Chiba, R., Kurasawa, K., Saito, Y., Iwamoto, I. (1998): Cross-linking of intercellular adhesion molecule-1 induces interleukin-8 and RANTES production through the activation of MAP kinases in human vascular endothelial cells. *Biochem. Biophys. Res. Commun.* 250(3): 694–698.

Schwartz, J. C., Zhang, X., Nathenson, S. G., Almo, S. C. (2002): Structural mechanisms of costimulation. *Nat. Immunol.* 3(5): 427–434.

Shaw, A., Dustin, M. (1997): Making the T cell receptor go the distance: a topological view of T cell activation. *Immunity* 6: 361–369.

Simons, K., Ikonen, E. (1997): Functional rafts in cell membranes. *Nature* 387(6633): 569–572.

Slavik, J. M., Hutchcroft, J. E., Bierer, B. E. (1999): CD28/CTLA-4 and CD80/CD86 families: signaling and function. *Immunol. Res.* 19(1): 1–24.

Somersalo, K., Anikeeva, N., Sims, T. N., Thomas, V. K., Strong, R. K., Spies, T., Lebedeva, T., Sykulev, Y., Dustin, M. L. (2004): Cytotoxic T lymphocytes form an antigen-independent ring junction. *J. Clin. Invest.* 113(1): 49–57.

Spiliotis, E. T., Osorio, M., Zuniga, M. C., Edidin, M. (2000): Selective export of MHC class I molecules from the ER after their dissociation from TAP. *Immunity* 13: 841–851.

Spits, H., van Schooten, W., Keizer, H., van Seventer, G., Van de Rijn, M., Terhorst, C., de Vries, J. E. (1986): Alloantigen recognition is preceded by nonspecific adhesion of cytotoxic T cells and target cells. *Science* 232: 403–405.

Springer, T. A. (1990): Adhesion receptors of the immune system. *Nature* 346: 425–433.

Staunton, D. E., Dustin, M. L., Erickson, H. P., Springer, T. A. (1990): The arrangement of the immunoglobulin-like domains of ICAM-1 and the binding sites for LFA-1 and rhinovirus. *Cell* 61: 243–254.

Stinchcombe, J. C., Bossi, G., Booth, S., Griffiths, G. M. (2001): The immunological synapse of CTL contains a secretory domain and membrane bridges. *Immunity* 15(5): 751–761.

Sykulev, Y., Joo, M., Vturina, I., Tsomides, T. J., Eisen, H. N. (1996): Evidence that a single peptide-MHC complex on a target cell can elicit a cytolytic T cell response. *Immunity* 4: 565–571.

Szollosi, J., Horejsi, V., Bene, L., Angelisova, P., Damjanovich, S. (1996): Supramolecular complexes of MHC class I, MHC class II, CD20, and tetraspan molecules (CD53, CD81, and CD82) at the surface of a B cell line JY. *J. Immunol.* 157(7): 2939–2946.

Tang, Q., Edidin, M. (2001): Vesicle trafficking and cell surface membrane patchiness. *Biophys. J.* 81(1): 196–203.

Thompson, P. W., Randi, A. M., Ridley, A. J. (2002): Intercellular adhesion molecule (ICAM)-1, but not ICAM-2, activates RhoA and stimulates c-fos and rhoA transcription in endothelial cells. *J. Immunol.* 169(2): 1007–1013.

Tilghman, R. W., Hoover, R. L. (2002a): E-selectin and ICAM-1 are incorporated into detergent-insoluble membrane domains following clustering in endothelial cells. *FEBS Lett.* 525(1–3): 83–87.

Tilghman, R. W., Hoover, R. L. (2002b): The Src-cortactin pathway is required for clustering of E-selectin and ICAM-1 in endothelial cells. *FASEB J.* 16(10): 1257–1259.

Turley, S. J., Inaba, K., Garrett, W. S., Ebersold, M., Unternaehrer, J., Steinman, R. M., Mellman, I. (2000): Transport of peptide-MHC class II complexes in developing dendritic cells. *Science* 288(5465): 522–527.

Vereb, G., Matko, J., Vamosi, G., Ibrahim, S. M., Magyar, E., Varga, S., Szollosi, J., Jenei, A., Gaspar, R., Jr., Waldmann, T. A., Damjanovich, S. (2000): Cholesterol-dependent clustering of IL-2Ralpha and its colocalization with HLA and CD48 on T lymphoma cells suggest their functional association with lipid rafts. *Proc. Natl. Acad. Sci. U.S.A.* 97(11): 6013–6018.

Wadchra, M., Su, H., Gordon, L. K., Goodglick, L., Braun, J. (2003): The tetraspan protein EMP2 increases surface expression of class I major histocompatibility complex proteins and susceptibility to CTL-mediated cell death. *Clin. Immunol.* 107(2): 129–136.

Wang, Q., Doerschuk, C. M. (2001): The p38 mitogen-activated protein kinase mediates cytoskeletal remodeling in pulmonary microvascular endothelial cells upon intracellular adhesion molecule-1 ligation. *J. Immunol.* 166(11): 6877–6884.

Wang, Q., Pfeiffer, G. R., 2nd, Gaarde, W. A. (2003): Activation of SRC tyrosine kinases in response to ICAM-1 ligation in pulmonary microvascular endothelial cells. *J. Biol. Chem.* 278(48): 47 731–47 743.

Wier, M., Edidin, M. (1988): Constraint of the translational diffusion of a membrane glycoprotein by its external domains. *Science* 242(4877): 412–414.

Wojciak-Stothard, B., Williams, L., Ridley, A. J. (1999): Monocyte adhesion and spreading on human endothelial cells is dependent on Rho-regulated receptor clustering. *J. Cell Biol.* 145(6): 1293–1307.

Wulfing, C., Sjaastad, M. D., Davis, M. M. (1998): Visualizing the dynamics of T cell activation: intracellular adhesion molecule 1 migrates rapidly to the T cell/B cell interface and acts to sustain calcium levels. *Proc. Natl. Acad. Sci. U.S.A.* 95(11): 6302–6307.

Wulfing, C., Sumen, C., Sjaastad, M. D., Wu, L. C., Dustin, M. L., Davis, M. M. (2002): Costimulation and endogenous MHC ligands contribute to T cell recognition. *Nat. Immunol.* 3(1): 42–47.

Xiong, J. P., Stehle, T., Diefenbach, B., Zhang, R., Dunker, R., Scott, D. L., Joachimiak, A., Goodman, S. L., Arnaout, M. A. (2001): Crystal structure of the extracellular segment of integrin alpha Vbeta3. *Science* 294(5541): 339–345.

Yang, Y., Jun, C. D., Liu, J. H., Zhang, R., Joachimiak, A., Springer, T. A., Wang, J. H. (2004): Structural basis for dimerization of ICAM-1 on the cell surface. *Mol. Cell* 14(2): 269–276.

8
Stimulation of Antigen Presenting Cells: from Classical Adjuvants to Toll-like Receptor (TLR) Ligands
Martin F. Bachmann and Annette Oxenius

8.1
Synopsis

The immune system has long been viewed as based on two distinct pillars, a non-specific, rapidly responding innate system and a more slowly but antigen-specifically responding adaptive system. Only recently it has emerged that the innate and adaptive immune systems are intimately linked, cross-regulating each other. In particular, no relevant adaptive immune response will be mounted unless the innate immune system has been alarmed through pathogen-associated molecular patterns (PAMPs). Important sensors for such PAMPs are toll-like receptors (TLRs). This chapter discusses the role of TLRs in regulating adaptive immune responses and outlines ways how current vaccine strategies attempt to integrate these novel insights into immunology.

8.2
Pathogen-associated Features that Drive Efficient Immune Responses

Live viruses, bacteria and parasites usually induce rapid and strong immune responses. Indeed, most pathogens are cleared by the host's immune system in due time and the harm caused by the infection remains limited. What are the reasons for the potent immune responses raised by pathogens? Several features account for the rapid induction of protective immunity (Figure 8.1): (i) Live pathogens replicate in the host. This leads to prolonged exposure of the immune system to pathogen-derived antigens, a requirement for the induction of optimal T cell responses and for the generation of effector T cells [1–3]. (ii) Pathogens are large compared to isolated proteins. This facilitates phagocytosis by macrophages and DCs, enhancing antigen presentation and T cell responses. Thus, the simple fact that pathogens are relatively large serves as a clue for the immune system that they are potentially dangerous. In addition, the surfaces of most pathogens, in particular of bacteria, have the potential to directly activate components of the complement cascade, leading to opsonization and enhanced antigen presentation

Antigen Presenting Cells: From Mechanisms to Drug Development. Edited by H. Kropshofer and A. B. Vogt
Copyright © 2005 WILEY-VCH Verlag GmbH & Co. KGaA, Weinheim
ISBN: 3-527-31108-4

[4, 5]. (iii) Most pathogens, in particular viruses and bacteria, exhibit highly repetitive surfaces, exhibiting many copies of a single protein. Such highly organized antigens efficiently cross-link B cell receptors of specific B cells, facilitating their activation [6]. Moreover, repetitive surfaces are preferentially decorated by components of the complement cascade [5], facilitating phagocytosis. (iv) Last, but not least, live pathogens are strong stimulators of the innate immune system, leading to extensive activation of innate cells, such as macrophages, dendritic cells (DCs) and natural killer cells [7]. In addition, pathogen replication often results in the death of host cells, leading to the release of substances not usually present in the extracellular environment. Such substances include heat-shock proteins, β-defensins and oxidized lipids, which are also recognized by receptors expressed on innate cells [8].

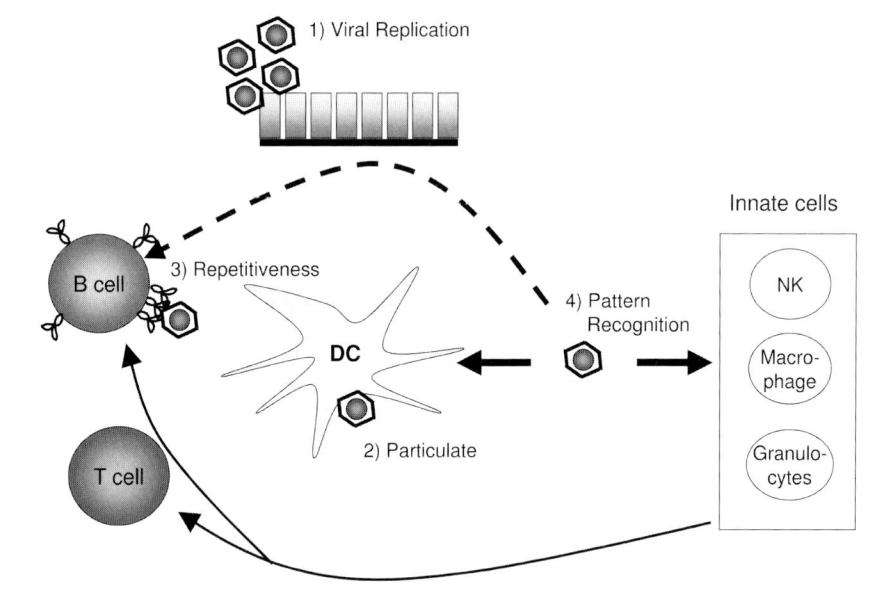

Figure 8.1 Key features of pathogens that render them highly immunogenic.

8.3
Composition and Function of Adjuvants

Although the concept of pathogen-derived pattern recognition is relatively recent [9–12], the power of the innate immune system to enhance adaptive immune responses has been used by immunologists for centuries for the development of vaccines. Indeed, the historically most successful vaccines are live, attenuated ones, carrying all the features of pathogens, including their ability to activate the innate immune system. In contrast to live pathogens, isolated pathogen components or recombinant subunit vaccines are usually unable to efficiently activate innate cells and fail to induce B and, in particular, T cell responses in the absence of helper substances, so-called adjuvants ("adiuvare" means "to help" in Latin). In essence, adjuvants attempt to give the antigens a "flavor of pathogen" by mimicking some of their characteristics. This is achieved in several ways. On the one hand, most adjuvants create an antigen depot. This is an important feature, since soluble antigens exhibit a short half-life and are rapidly eliminated from the host. In the absence of a depot, an antigen therefore triggers the immune system only for a short period of time, leading to abortive immune responses. In addition, antigen depots usually result in the formation of antigen aggregates, facilitating phagocytosis (features 1 and 2 in Figure 8.1). A typical member of depot-forming adjuvants is Alum (aluminum hydroxide), the most widely used adjuvant in humans. In fact, in most countries, Alum remains the only adjuvant licensed for use in humans. A more aggressive form of a depot-forming adjuvants is incompletes Freund's adjuvants (IFA), a water in oil emulsion, which causes prolonged antigen presentation but also exhibits major local side reactions. IFA is therefore only exceptionally used in humans. Whether the induction of cell death by adjuvants, and subsequent release of heat-shock proteins and other endogenous ligands for innate receptors is important for the generation of immune responses remains to be elucidated.

An option to specifically enhance B cell responses is to make the antigen of choice highly organized, leading to efficient cross-linking of B cell receptors [13] (Feature 3 in Figure 8.1). In fact, most viral vaccines are highly repetitive and this is an important reason for their strong immunogenicity for B cells. Recently, it has been shown that antigens of choice can be rendered highly organized by displaying them on the repetitive surface of virus-like particles. Such vaccines are even able to break B cell unresponsiveness and induce the production of self-specific antibodies [14, 15].

The situation is different for the generation of strong effector T cell responses, as is important for therapeutic vaccination against cancer or chronic viral infections. Here, formation of antigen depots or rendering antigens highly organized alone is not sufficient, and innate stimuli have to be added to the adjuvants formulations (feature 4 of Figure 8.1). That T cell responses remain ineffective in the absence of "non-specific" stimuli has been known for some time and was once termed the "immunologist's dirty little secret" [9]. Indeed, "dirt" it is that stimulates the innate immune system for enhancement of adaptive responses. Extracts from mycobacteria, for example, have been mixed with IFA to generate complete Freund's adjuvants (CFA), a much more potent adjuvant than the "clean" IFA. Indeed, immunization with antigen in CFA results in formation of local granulomas in the tissue. Such granulomas keep the immune system chronically stimulated and maintain effector T cell responses for extended periods, in a way similar to what is seen in chronic tuberculosis. LPS and bacterial cell wall components also enhance adaptive immune responses [16] and even debris from lysed insect cells dramatically enhanced cytotoxic T cell responses against recombinant proteins [17]. The molecular basis of the perception of this "dirt" and the target cells involved in recognition have only been elucidated recently. In first approximation, the following two statements summarize current knowledge:

1. Key receptors for the "dirt", which also has been termed "pathogen-associated molecular patterns" (PAMPs), belong to the family of toll-like receptors (TLRs). These receptors are evolutionary old and are already found in insects, where they are also involved in host defense.
2. The key cell responsible for sensing the PAMPs and enhancing T cell responses is the dendritic cell (DC) (Figure 8.1).

Not unexpectedly, there are exceptions to these rules. PAMPs, such as LPS or bacterial DNA, stimulate not only DCs but also B cells and, in some species, macrophages and other leukocytes. Thus, PAMPs may directly facilitate B cell responses by stimulating B cell activation or, indirectly, by activating DCs, resulting in enhanced Th cell responses. The relative importance of the two mechanisms remains to be further investigated. Similarly, activation of macrophages through PAMPs results in enhanced clearance of pathogens. Whether activation of macrophages also affects adaptive immunity through increased T cell responses remains to be seen.

For these reasons, attempts to manipulate the innate immune system to enhance vaccine induced immune responses have focused around the DC and TLRs. In the next sections, the structure and function of TLRs and their ligands will be described as well as current strategies to harness their potential for optimizing vaccination strategies.

8.4
TLR Protein Family in Mammals

Molecular identification of the mechanisms involved in the activation of innate immune responses after stimulation with adjuvant substances or triggered by natural PAMPs has advanced the field of innate immune activation tremendously. Within the past few years it has been appreciated that multiple receptors, expressed on various cells participating in innate and also adaptive immune responses, can recognize defined chemical structures that are unique to viruses, bacteria or fungi. As already discussed, many of these receptors belong to the family of Toll-like receptors (TLRs). TLRs are evolutionary conserved receptors and the first TLR "Toll" was identified in the fruit fly *Drosophila melanogaster*. Toll is important for embryogenesis as well as the control of fungal infections in the adult fly [18]. Subsequent database searches revealed that mammalian genomes also contained genes that showed homology to the insect Toll and, to date, 13 such mammalian receptors have been identified (10 in humans and 12 in mice) [19–21]. Table 8.1 lists the various TLRs and their ligands.

Table 8.1 TLRs: Expression and main natural PAMP ligands.

Receptor	Species	Localization	Expression in DC subsets*		Natural ligands	Pathogen	Ref.
			mDC	pDC			
TLR1[a]	H	S	++	+	Triacyl lipopeptides	Bacteria and	84
	M		++	++	Soluble factors	mycobacteria	34
						Neisseria meningitidis	
TLR2	H	S	++	–	Lipoprotein/	Multiple pathogens	85
	M		++	++	lipopeptides		
					PGN	Gram-positive bacteria	25, 86
					Lipoteichoid acid	Gram-positive bacteria	86
					Lipoarabinomannan	Mycobacteria	36
					Glycoinositolphospholipids	Trypanosoma cruzi	38
					Glycolipids	Treponema maltophilum	87
					Gr.B streptococcal sol. factors	Group B streptococci	37
					Porins	Neisseria	32
					Atypical LPS	Leptospira interrogans	30
					Atypical LPS	Porphyr. gingivalis	31
					Atypical LPS	Legionella pneumophila	29
					Atypical LPS		29
					Zymosan	Bortedella fungi	88
TLR3	H	E	++	–	dsRNA	Viruses	50
	M		++/–	–			
TLR4	H	S	–	–	LPS	Gram-negative bacteria	24
	M		++/–	++/–	Fusion protein	RSV	89
					Env protein	MMTV	90
TLR5	H	S	+	–	Flagellin	Bacteria	41
	M		++/–	+			
TLR6[a]	H	S	++	++	Diacyl lipopeptides	Mycoplasma	28
	M		++	+	Lipoteichoic acid	Gram-positive bacteria	86
					Zymosan	Fungi	88
TLR7	H	E	+/–	++	ssRNA	Viruses	51, 52
	M		++/–	++			
TLR8	H	E	++	–	ssRNA	Viruses	51

Table 8.1 Continued.

Receptor	Species	Localization	Expression in DC subsets*		Natural ligands	Pathogen	Ref.
			mDC	pDC			
TLR9	H	E	–	++	Unmethylated CpG DNA	Multiple pathogens	54
	M		++	++			
TLR10	H		+	+	n.d.	n.d.	
TLR11	M				n.d	Uropathogenic bacteria	44
TLR12	M				n.d.	n.d.	
TLR13	M				n.d.	n.d.	

* Freshly isolated cells; "+" and "–" denote relative mRNA expression levels [57].
a Dimerizes with TLR2.
S: cell surface. E: endosomal compartments (including ER for TLR9). ER: endoplasmic reticulum. H: human. L: cellular localization. M: mouse. MMTV: mouse mammary tumor virus. mDC: mycloid DC (in mice these contain $CD4^+$, $CD8^+$ or DN subsets). pDC: plasmacytoid DC. PGN: peptidoglycan. Porphyr.: Porphyromonas gingivalis. RSV: respiratory syncytial virus.

The general structure of the TLRs consists of an extracellular domain that contains 10–25 tandem copies of a leucine-rich motif (LRR). This domain is responsible for the ligand binding. The cytoplasmic domain contains a conserved region of about 200 amino acids, which is termed the Toll/IL-1R (TIR) domain. TLR signaling is mediated by the TIR domains (reviewed in Refs. 8, 10, 22 and 23). The various TLRs and their ligands are described individually in the following sections.

8.4.1
TLR4

The first mammalian TLR that was identified was TLR4: Two mutant mouse strains (C3H/HeJ and C57BL/10ScCr) that were unresponsive to lipopolysaccharide (LPS) were shown to have mutations within the TLR4 gene [24]. LPS is one of the oldest known and best characterized PAMPs; it is composed of polysaccharides and the lipid A portion that is embedded in the cell membrane of Gram-negative bacteria. It is the lipid A component that constitutes the PAMP. LPS binding to TLR4 is facilitated by CD14, a glycosylphosphatidylinositol (GPI)-anchored protein; CD14 binds and retains LPS on the cell surface of mammalian cells. The initial observation made in the mutant mouse strains (i.e., that LPS signaling required the presence of functional TLR4) was further corroborated by the generation of TLR4-deficient mice [25]. Furthermore, humans that carry mutations in their TLR4 genes have impaired responsiveness to LPS [26]. Endogenous host-derived ligands have also been described for TLR4. These putative ligands include heat-shock proteins and extracellular matrix degradation products. However, this remains a controversial topic due to potential LPS contaminations.

8.4.2
TLR2

Gram-positive bacteria do not have LPS and hence different bacterial constituents are recognized by other TLRs. The cell wall of Gram-positive bacteria contains a thick, porous peptidoglyan (PGN) layer. This is a polymer of alternating N-acetylmuramic acid (NAM) and N-acetylglucosamine (NAG). Long strands of this alternating polymer may be linked by L-alanine, D-glutamic acid, L-lysine, D-alanine tetrapeptides to NAM. Embedded in the PGN layer are lipoproteins and lipoteichoic acids (LTA); these polysaccharides extend through the entire PGN layer and appear on the cell surface. Constituents of the PGN signal via TLR2 [25, 27]. TLR2 has further been shown to bind several other PAMPs such as mycoplasma lipopeptides (MALP-2) [28], lipoproteins derived from various pathogens, atypical LPS [29–31], porins [32] and zymosan [33].

The remarkably broad ligand specificity of TLR2 can be explained by the fact that TLR2 forms functional heterodimers with either TLR1 or TLR6 [28, 33, 34]. TLR1/TLR2 heterodimers preferably recognize triacylated lipoproteins (most bacterial lipoproteins) whereas TLR6/TLR2 heterodimers are activated by diacylated lipopeptides (as in PGN, and MALP) [28, 35]. Furthermore, several other ligands are dependent on TLR2 for signaling but do not depend on the presence of either TLR1 or TLR6. These ligands include mycobacterial lipoarabinomannan [36], atypical LPS from *Legionella* [29], *Leptospira* [30], *Bortadella* [29] and *Porphyromonas* [31] and group B streptococcal soluble factors [37]. A protozoan molecule that is recognized by TLR2 is glucosylphosphatidylinositol (GPI) from *Trypanosoma cruzi* [38]. This suggests that TLR2 might heterodimerize with other TLR molecules than TLR1 and 6 or even non-TLR molecules. Additionally, TLR2 might also homodimerize in vivo. TLR-2 mediated signaling, comparable to TLR4-mediated signaling, is also facilitated by the presence of CD14, which augments the efficiency of ligand recognition.

8.4.3
TLR5

Most bacteria possess flagella, oranelles that are important for bacterial motility. Flagella are composed of flagellin subunits that are polymerized to make up a long filament (up to 20 μm long) which is anchored in the bacterial cell wall. Bacteria may possess a single polar flagellum, several polar, bipolar or nonpolar flagella. Flagellin shows potent proinflammatory activity [39, 40] and TLR5 has been identified as the receptor for flagellin [41]. The precise TLR5 recognition site within flagellin was mapped to a conserved and functionally important cluster of 13 amino acids. Only monomeric but not the filamentous molecule exposes the recognition site and hence stimulates TLR5 [42]. In humans polymorphisms in TLR5 are reportedly associated with differential susceptibility to infection with *Legionella pneumophila* [43].

8.4.4
TLR11

Very recently another TLR was identified, TLR11, which mediates specific recognition of uropathogenic bacteria [44]. Most uropathogenic bacteria are strains of *Escherichia coli* (uropathogenic *E. coli*, or UPEC). TLR11 is expressed on macrophages and liver, kidney and bladder epithelial cells and might play an important role for control of infections within the urogenital system.

8.4.5
TLR12 and TLR13

In mice, two novel TLR have been identified, i.e. TLR12 and TLR 13. The ligands for these receptors have, so far, not been identified [23].

8.4.6
Nucleic Acids as PAMPs

As opposed to the ligands discussed above, which represent biochemically unique structures of bacteria and fungi, nucleic acids of pathogens have a structure largely identical to host nucleic acid. The initial observation that bacterial DNA is a potent adjuvant and stimulates innate immunity [45] has spawned an intensive search for bacterial DNA receptors. It was demonstrated that short oligonucleotides that contain unmethylated CG motifs (CpG) were the active component in bacterial DNA [46]. Interestingly, unmethylated CG motifs are under-represented in the genome of eukaryotic cells, offering an explanation for the ability of TLR9 to distinguish between prokaryotic and eukaryotic DNA.

Not only bacterial DNA is sensed by cells of the innate and adaptive immunity but also dsRNA and ssRNA. Since all of these nucleic acid species might also exist in large or small quantities in mammalian cells, an additional feature of PAMP recognition has to be operational: a different subcellular localization of nucleic acid between host and infecting or ingested pathogen.

8.4.6.1 TLR3
dsRNA (both natural and synthetic analogues such as poly I: C) induces potent type I IFN production. Type I interferon (IFN-α/β) production by virus-infected cells is a crucial early event during viral infection and represents a first line of defense. IFN α/β leads to induction of an "antiviral state" (by down-regulation of host cell protein synthesis) not only in the virus infected cells but also in sur-

rounding non-infected cells (reviewed in Refs. 47–49). dsRNA is an intermediate of virus replication and is not produced by mammalian host cells (at least not in large quantities); hence it represents a virus-associated molecular pattern. TLR3, which is expressed intracellulary in endosomal compartments, was found to recognize dsRNA and to transmit signals to activate NF-kappaB and the IFN-beta promoter [50]. Thus, the receptor for extracellular dsRNA is TLR3, while there is an additional intracellular system activated by dsRNA: PKR (IFN-inducible dsRNA-dependent protein kinase, see below). The relative importance of the two systems probably differs between pathogens and a detailed understanding of their respective functions requires further investigation.

8.4.6.2 TLR7 and TLR8

Mouse TLR7 and human TLR7 and 8 were initially shown to recognize imidazoquinolines, which are guanosine-based compounds that are structurally similar to nucleic acid. The natural ligand for TLR7 and 8 was only recently identified: ssRNA [51, 52]. In TLR7-deficient mice, ssRNA viruses such as Vesicular Stomatitis Virus (VSV) or Influenza Virus (Flu) were unable to stimulate IFNα/β production in the absence of replication. In contrast, viruses with a dsDNA genome (such as Herpes Simplex Viruses) were not dependent on TLR7 signaling for type I IFN production. TLR7 signaling appears to occur in endosomal compartments, since inhibition of endosomal acidification inhibited TLR7 signaling.

8.4.6.3 TLR9

Bacterial unmethylated DNA as well as oligonucleotides containing specific non-methylated CpG motifs (CpG-ODN) are mitogenic for B cells and other leukocytes and can directly activate macrophages and dendritic cells (DCs), leading to upregulation of co-stimulatory molecules and cytokine production (reviewed in Ref. 53). Comparable to TLR7 and 8, activation of APCs was preceded by non-specific endocytosis of CpG-ODN and required endosomal acidification. Subsequently, the intracellular receptor for CpG-ODN turned out to be TLR9. Accordingly, TLR9-deficient mice are refractory to immunostimulatory CpG-DNA [54]. TLR9 exhibits a rather unique subcellular localization; in resting DCs, TLR9 is localized in the endoplasmatic reticulum. Upon activation, TLR9 is rapidly relocalizes to the endosomal compartment, where the interaction with CpGs occurs [55].

8.4.7
Compartmentalization of Sensing Renders the Nucleic Acid PAMPs

Nucleic acids are common chemical structures to host cells and pathogens. For instance, ssRNA is very abundant in host cells and does not usually lead to TLR7/8 activation. However, host ssRNA is predominantly located within the nucleus and cytoplasm whereas ssRNA viruses may at least partly end up in endosomal compartments after viral uptake. Hence, the intracellular compartmentalization of ssRNA, rather than ssRNA structure itself, is the key determinant of TLR7/8-mediated cell activation.

Similarly, some dsRNA species may inevitably exist within the cytoplasm of host cells, either due to secondary structures of ssRNA molecules or due to partial complementarity of different ssRNA species. TLR3 recognizes dsRNA, however, due to its localization within endosomal compartments; TLR3 is not activated by cellular dsRNA unless it is presented to a cell such that endosomal uptake is involved [56]. Recognition of double stranded RNA in the cytoplasm, indicative of viral infection, is mediated by a TLR-independent system: dsRNA binding proteins (DRBPs), which induce production of IFNα/β in virtually all cell types (see below).

8.5
TLR Signaling

Binding of the respective ligands to TLRs induces an intracellular signaling cascade that eventually leads to the activation of innate immune defenses. This includes the production of proinflammatory cytokines and the upregulation of co-stimulatory molecule expression. Thus, TLR signaling not only activates inflammatory pathways of the innate immune response, which lead to recruitment of neutrophils and activation of macrophages (enhancing phagocytes and inducing microbicidal mechanisms), but is also critically involved in the activation of adaptive immune responses. Upregulation of co-stimulatory molecules, in particular on DCs, is required for the activation of naïve T cells to become differentiated effector cells. To activate pathogen-specific T cells in secondary lymphoid organs, tissue-resident immature DCs need to undergo a maturation process, during which they upregulate co-stimulatory molecules and cytokine production, increase the half-life of MHC expression, downregulate phagocytic activity and migrate from the peripheral site of pathogen contact to the secondary lymphoid organs. This DC migration is facilitated by cell surface chemokine receptors (such as CCR7) that are expressed after TLR-mediated DC activation. Furthermore, TLR activation induces chemokine secretion, including IL-8, CXCL1, CCL2, CCL4, CCL5, CCL7, CCL8 and CCL13. These chemokines enable the activation and recruitment of leukocytes by modulation of cell surface integrin-affinity, which allow leukocyte adhesion to endothelial surfaces followed by extravasation into the tissue interstitium (reviewed in Ref. 57).

The following section summarizes the key signaling events that lead from TLR ligand binding to induction of a gene transcription program.

8.5.1
Signal Transduction Across the Membrane

Upon ligand binding and dimerization, TLRs undergo a conformational change that leads to the recruitment of downstream signaling molecules. These downstream adaptor molecules have been largely identified in recent years. Four adaptor proteins are implicated in TLR signaling: MyD88 (myeloid differentiation factor 88), MAL/TIRAP (MyD88-adaptor-like/TIR-associated protein), TRIF (Toll-receptor-associated activator of interferon) and TRAM (Toll-receptor-associated molecule). Binding of the adaptor molecules to the activated TLRs allows signal transduction from the TIR (Toll/interleukin-1 receptor homologous region) domains, leading to activation of a series of protein kinases and, eventually, to activation of transcription factors that control the transcription of genes involved in inflammation and APC activation (Figure 8.2)

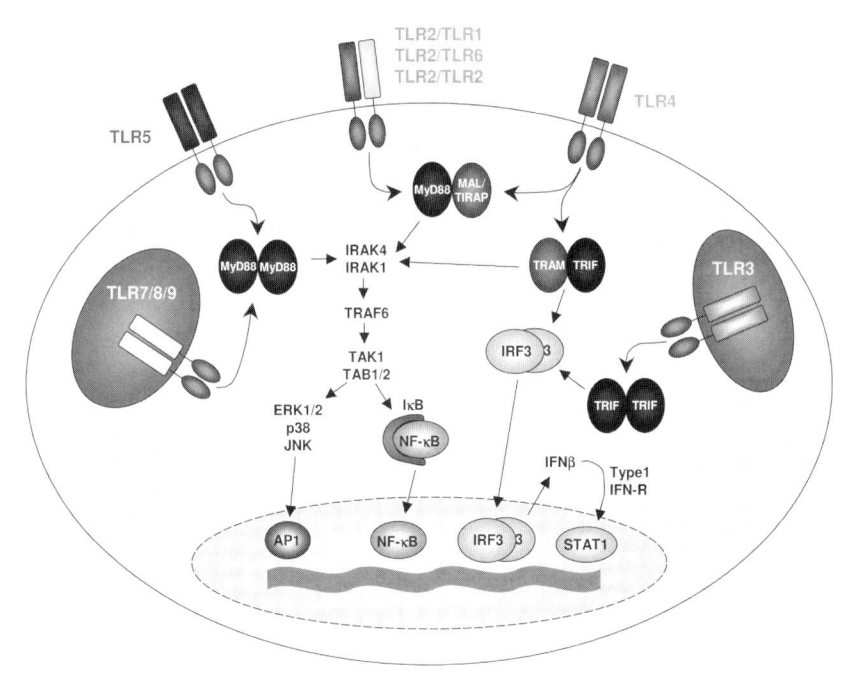

Figure 8.2 TLR signaling cascades. (This figure also appears with the color plates.)

8.5.2
MyD88-dependent Pathways

MyD88 is involved in signaling of most TLRs (TLR 1, 2, 4–9) (reviewed in Ref. 22). It is composed of an N-terminal death domain and a C-terminal TIR domain; homodimerization of MyD88 is required for signal transduction. TIR domains mediate interaction with other TIR-domain carrying proteins, such as TLRs, while death domains interact with death domains on other proteins. In this way, MyD88 relays TLR-signaling to down-stream molecules expressing death domains. An important death domain carrying target of MyD88 is IRAK4 (IL-1R-associated kinase 4), which is recruited by MyD88 binding to the signaling complex [58, 59]. IRAKs contain an N-terminal death domain and a central Ser/Thr-kinase domain. In TLR signaling, IRAK4 and its kinase activity are upstream of IRAK1 activity. After IRAK activation the TRAF6 (TNF-receptor-associated factor 6) is engaged: TRAF6 contains an N-terminal coiled coil domain (which is required for downstream signaling) and a conserved C-terminal domain (required for self-association and for interaction with IRAKs) [22]. TRAF6-mediated activation of the transcription factors NF-κB and AP1 (activator protein 1) requires TAK1 (transforming growth factor-β activated kinase) and TAB1/2 (TAK1-binding proteins). TAK1 belongs to the family of mitogen-activated protein kinase kinase kinase (MAPKKK) [60] and it is activated by binding of TAB1 and linked to TRAF6 via TAB2 [61, 62]. Finally, activation of TAK1 leads to the activation of IKKs (IκB kinases), which degrade the inhibitors (IκB) associated with NF-κB in the cytoplasm. Subsequent release of NF-κB from its bound inhibitors allows its translocation to the nucleus. Activation of TAK1 also results in the activation of MAPKs (mitogen-activated protein kinases), including JNK (JUN N-terminal kinase), ERK1/2 (extracellular signal-regulated kinase) and p38 kinase [22, 23].

An alternative player in the MyD88-dependent pathway is the adaptor molecule TIRAP (TIR-domain-containing adaptor protein or MyD88-adaptor-like protein, MAL). TIRAP can dimerize with MyD88 for initiation of downstream signals from TLR4 and TLR2 [63, 64] but not for other TLRs.

8.5.3
MyD88-independent Pathways

TLR4-dependent LPS signaling was not completely abolished in MyD88-deficient macrophages. Instead, NFκB-activation was induced, albeit with delayed kinetics [65] and activation of IFN-inducible genes, such as glucocorticoid-attenuated response gene 16 (GARG16), CXC-chemokine ligand 10 (CXCL10) and immunoresponsive gene 1 (IRG1), was even normal [66]. This observation resulted in the discovery of MyD88-independent TLR signaling pathways. Subsequent experiments demonstrated that induction of IFNβ production was completely MyD88-independent, explaining the remaining upregulation of IFN-inducible genes [67].

MyD88-independent TLR-4 signaling is initiated by a different TIR-domain containing adaptor: TRIF (TIR-containing adaptor molecule inducing IFNβ or TICAM, TIR-domain-containing molecule1) [68]. In addition, a fourth TIR-containing adaptor protein is required: TRAM (TRIF-related adaptor molecule or TIR-domain-containing molecule 2, TICAM) [69–71]. TRAM associates with TRIF in TLR4 signaling and mediates production of IFNβ and activation of IFN-inducible genes [70, 72].

In TLR3 signaling, TRIF is used as the only adaptor molecule. TRIF activation by TLR3 or activation of TRIF/TRAM by TLR4 signals leads to the activation of

IRF3 (interferon-regulatory factor 3) [73]. IRF3 is constitutively expressed in various cells and upon activation of its C-terminal domain the formation of IRF3 dimers is induced, allowing translocation to the nucleus, where it activates type I IFN gene transcription. Binding of IFNβ to the type I IFN receptor results in activation of the transcription factor STAT-1 (signal transducer and activator of transcription-1) [48].

In summary, MyD88-dependent signaling is important for all TLR-mediated production of inflammatory cytokines. TLR3 and TLR4 signaling leads to a MyD88-independent but TRIF-dependent induction of type I IFN production that is linked to upregulation of co-stimulatory molecule expression (Figure 8.3). However, TLR7 and TLR9-mediated MyD88-dependent signaling also leads to production of type I IFNs; the players involved in this pathway are still to be determined.

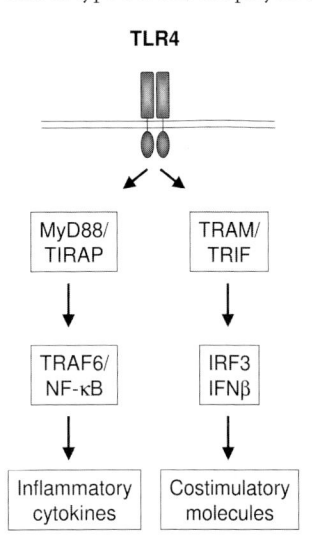

Figure 8.3 Different signaling pathways lead to the production of inflammatory cytokines or upregulation of co-stimulatory markers.

8.6
TLR-independent Recognition of PAMPs: Nods, PKR and Dectin-1

To date several other known TLR-independent mechanisms are induced by PAMPs and lead to induction of innate defense mechanisms, including cellular activation and cytokine production. The following section briefly describes a selection of some of these mammalian TLR-independent pathogen sensing mechanisms.

8.6.1
Nods

TLR-independent recognition of certain bacterial PAMPs is mediated via Nod (nucleotide-binding oligomerization domain) family members (reviewed in Ref.

74). The nucleotide-binding site/leucine-rich repeat (NBS/LRR) proteins are cytoplasmic proteins that can sense microbial motifs. For two members of this family (Nod1 and Nod2) PAMPs have been identified. Both Nod1 and Nod2 recognize distinct molecular motifs of PGN: Nod1 recognizes a naturally occurring muropeptide of PGN containing the unique terminal amino acid diaminopimelic acid (DAP), which is mainly found in the PGN of Gram-negative bacteria [75]. Nod2 senses muramyl dipeptide, a common small fragment of PGN, which renders Nod2 a more general detector of bacterial PGN than Nod1 [76, 77]. Nod proteins have three distinct domains: a central nucleotide-binding oligomerization domain (Nod), a C-terminal ligand-recognition domain (LRD) and an N-terminal effector-binding domain (EBD). In Nod1 and 2, the EBD domain is reminiscent of a caspase-recruitment domain (CARD), mediating homophilic protein interactions [78]. Caspase activation results in active interleukin-1 converting enzyme (ICE), triggering the production of interleukin-1. The LRD of Nods contains leucine-rich repeats (LRRs), a motif that is also present in the extracellular domain of TLRs. Hence, a similar mechanism seems responsible for ligand recognition by Nods and TLRs. Nods initiate cellular activation but also cell death by multiprotein complexes known as inflammasomes [79] that eventually lead to NF-κB and inflammatory caspase activation.

Intracytoplasmic detection of bacteria seems to be relevant for general immune control and homeostasis since mutations in Nod2 have been associated with Crohn's disease, a chronic inflammatory disease [80]. Thus, the cytoplasmic Nod-mediated sensing system complements well the TLR-mediated surface/endosomal sensing system of bacterial PAMPs.

8.6.2
PKR (IFN-inducible dsRNA-dependent Protein Kinase)

DsRNA binding proteins (DRBPs) contain dsRNA binding domains (DRBDs) and have been shown to interact with as little as 11 bp of dsRNA, independent of the RNA sequence. The cytoplasmic dsRNA-dependent protein kinase PKR is involved in dsRNA sensing, signaling and host defense against virus infections. PKR may mediate multiple effects; a predominant one in host defense is increased IFN production and activity (reviewed in Refs. 81 and 82). In contrast to TLR3, which exhibits a restricted expression pattern, DRBPs and PKR are expressed in virtually all cell types, leading to rapid production of IFNα/β by infected host cells.

8.6.3
Dectin-1

Recently, dectin-1, a C-type lectin-like transmembrane receptor that is expressed on macrophages and other innate immune cells, such as neutrophils and dendritic cells, was found to be a pattern-recognition receptor for β-glucans expressed by fungi [83]. Dectin-1 contains a cytoplasmic ITAM-like (immunoreceptor tyrosine-based activation motif) motif that is necessary for proinflammatory cytokine production and macrophage activation. Dectin-1 might therefore represent an innate receptor on immune cells sensing fungal pathogens.

8.7
Therapeutic Potential of TLRs and their Ligands

Given the manyfold biological activities of TLR-ligands it is not surprising that they are broadly used to manipulate the immune system. Essentially, TLR-ligands are employed in two distinct ways (Figure 8.4). They are used in monotherapies, i.e. their potential to non-specifically stimulate the immune system is utilized to induce local or systemic secretion of cytokines etc. Alternatively, TLR ligands are co-administered with antigens to enhance the adaptive immune response against these antigens. The history of the discovery of CpGs and their subsequent development for use in humans may be used to exemplify the two strategies. The mycobacterium Bovis strain BCG and BCG extracts were known to induce regression of tumors in experimental animal models if injected locally. In an attempt to purify the component with anti-tumor activity, bacterial extracts were fractionated and individual components assessed for activity. Unexpectedly, the DNA faction was most active [45]. Detailed analysis demonstrated that short oligonucletotides exhibited the highest activity, in particular those with a palindromic CG motif such as GACGTC [91]. The first CpG-motif was discovered. Subsequent experiments focused on use of CpGs as a monotherapy, in particular by injecting the DNA oligonucleotides into tumors and searching for anti-tumor activity. Later, it was found that CpGs may be used for non-specific activation of white blood cells, putting the immune system into an alarm state and creating a Th1 environment. Consequently, CpGs are now developed for prophylactic use against biological warfare to induce non-specific protection against viral, bacterial and parasitic infections. At the same time, it became evident that CpGs may also enhance immune responses against co-administered antigens [92, 93], making them a potent adjuvant, in particular for the induction of T cell responses [94].

Such dual use of TLR ligands in monotherapies or adjuvant formulations occurs in many instances; derivatives of LPS engaging TLR4 are, for example, used both as monotherapies against pathogens and tumors and as adjuvants for enhancement of immune responses [16, 95]. Imidazole quinolines, ligands for TLR7/8, were developed as drugs against genital warts and it was only recently that a link to TLRs could be made [96]. Experimental use of imidazole quinolines as adjuvant is only now beginning.

The use of TLR ligands in adjuvant formulations faces several problems. (i) Since TLR ligands activate the immune system non-specifically, they are inherently toxic. (ii) Many TLR-ligands are unstable in vivo. (iii) TLR expression may be limited to specialized cell types in vivo and targeted delivery to these cells is desirable. In most instances, these cells will be antigen presenting cells, in particular DCs. The problem of TLR-ligand toxicity has been solved elegantly for LPS. By

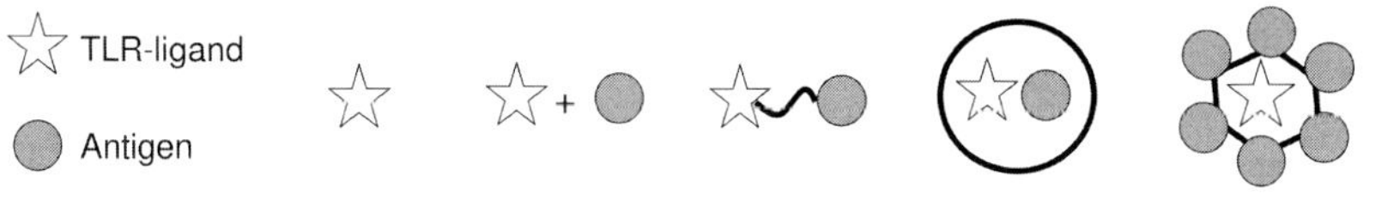

	Mono-therapy	Mixed	Linked	Micro-Particles	Virus-like particels
☆ TLR-ligand ● Antigen	☆	☆ + ●	☆∼●	(☆●)	(☆ surrounded by ●)
Non-specific activation	++	++	+	+/-	+/-
Amounts required	high	high	intermediate	low	low
Side-effects	high	high	intermediate	low	low
DC targeting	-	-	-	+	++
TLR-ligand stabilisation	-	-	-	+	+
T cell response	-	+	++	+++	+++
B cell response	-	+	++	++	++

Figure 8.4 TLR ligands and their modes of application.

separating the polysaccharides from the lipid A moiety and subsequent modification of lipid A, it was possible to reduce or eliminate the pyrogenic potential of LPS from its adjuvant function, creating safe and efficacious vaccine formulations [e.g. monophosphoryl lipid A (MLA or MPL), OM-174] [16, 97]. However, in most instances, such a separation will not be possible, since the non-specific ability of the TLR-ligands to activate innate immune cells is in fact responsible for their adjuvant function. Hence, eliminating the non-specific stimulation equals eliminating the adjuvant property.

An alternative strategy to eliminate or reduce the toxicity of TLR ligands is to co-deliver them in close proximity to the antigens (Figure 8.4). This may mean chemically coupling or recombinantly fusing the ligand to the antigen, or co-packaging them into micro- or nano-particles, liposomes, iscoms or virosomes [98, 99]. An additional recent strategy is to package TLR-ligands, in particular CpGs, into virus-like particles [100]. This not only enhanced the adjuvants properties of the CpGs but also eliminated the side-effects of the CpGs, such as splenomegaly. An important reason for the reduced side-effects was that much lower amounts of CpGs were required for adjuvants activity [100]. Packaging TLR-ligands into micro- or nanoparticles, virosomes or virus-like particles further overcomes the limitations of TLR-ligand stability and targeting to dendritic cells. Specifically, packaging of TLR-ligands protects them from degradation by enzymes present in the tissue and serum of the vaccinated host or even present in the cell culture medium. Single stranded RNA, for example, is only able to trigger TLR-7/8 if complexed to cationic lipids [51, 52], protecting it from degradation by RNAse. Without this protective coat, RNA fails to induce a response in DCs. As already discussed, this appears to be the clue that is used by the immune system to distinguish between infectious and non-infectious RNA: only RNA protected by a viral coat or a bacterial membrane and cell wall will reach the endosome and be able to trigger TLR7/8 [51, 52].

An alternative strategy to overcome stability issues with natural TLR-ligands is to develop synthetic, small molecular versions of them. Imidazole quinolines, which trigger TLR7/8, or fully synthetic lipid A derivatives are good examples of this trend. Such molecular analogues of natural TLR-ligands will also be easier to manufacture, and clinical development will also be easier for well-defined, fully synthetic molecules.

The potential for the various TLR ligands to co-stimulate T and B cell responses may differ and probably depends on the antigen used for immunization and the adjuvants they are combined with. Moreover, different TLR-ligands may co-stimulate different effector responses – some may favor $CD8^+$ T cell responses, while others preferably induce $CD4^+$ T cell responses or activate B cells. We have recently determined in mice the efficiency of various TLR-ligands to enhance $CD8^+$ T cell responses induced by VLPs [101]. CpGs, engaging TLR9, were by far the most effective enhancers of $CD8^+$ T cell responses, followed by poly (I: C) (TLR3) and single stranded RNA (TLR7/8). In contrast, ligands for other TLRs (TLR 2,4,5) exhibited little potential to enhance $CD8^+$ T cell responses. Whether similar rules govern activation of $CD4^+$ T cells and B cells remains to be determined.

8.8
Conclusion

Taken together, these considerations suggest that TLR-ligands are optimally used for vaccination if they are linked or better packaged together with the antigen. Nanoparticles or virus-like particles are preferred, since these small particles are efficiently taken up by professional antigen presenting cells, in particular DCs, making sure that the TLR-ligand is co-delivered with the antigen into the same DC, leading to an optimal synergy between the innate recognition system, the DC and the T cells. If the antigen is presented in a highly organized fashion on virus-like particles, a potent B cell response is additionally induced, coming close to an immune response as induced by live attenuated vaccines.

References

1 Kundig, T.M., A. Shahinian, K. Kawai, H.W. Mittrucker, E. Sebzda, M.F. Bachmann, T.W. Mak, and P.S. Ohashi. 1996. Duration of TCR stimulation determines costimulatory requirement of T cells. *Immunity* 5: 41–52.

2 Iezzi, G., K. Karjalainen, and A. Lanzavecchia. 1998. The duration of antigenic stimulation determines the fate of naive and effector T cells. *Immunity* 8: 89–95.

3 Storni, T., C. Ruedl, W.A. Renner, and M.F. Bachmann. 2003. Innate immunity together with duration of antigen persistence regulate effector T cell induction. *J. Immunol.* 171: 795–801.

4 Carroll, M.C., and M.B. Fischer. 1997. Complement and the immune response. *Curr. Opin. Immunol.* 9: 64–69.

5 Matsushita, M., and T. Fujita. 2001. Ficolins and the lectin complement pathway. *Immunol. Rev.* 180: 78–85.

6 Bachmann, M.F., and R.M. Zinkernagel. 1997. Neutralizing antiviral B cell responses. *Annu. Rev. Immunol.* 15: 235–270.

7 Janeway, C.A., Jr., and R. Medzhitov. 2002. Innate immune recognition. *Annu. Rev. Immunol.* 20: 197–216.

8 Ulevitch, R.J. 2004. Therapeutics targeting the innate immune system. *Nat. Rev. Immunol.* 4: 512–520.

9 Janeway, C.A., Jr. 1989. Approaching the asymptote? Evolution and revolution in immunology. *Cold Spring Harb. Symp. Quant. Biol.* 54 Pt 1: 1–13.

10 Beutler, B., K. Hoebe, X. Du, and R.J. Ulevitch. 2003. How we detect microbes and respond to them: the Toll-like receptors and their transducers. *J. Leukoc. Biol.* 74: 479–485.

11 Akira, S., K. Takeda, and T. Kaisho. 2001. Toll-like receptors: critical proteins linking innate and acquired immunity. *Nat. Immunol.* 2: 675–680.

12 Medzhitov, R. 2001. Toll-like receptors and innate immunity. *Nat. Rev. Immunol.* 1: 135–145.

13 Bachmann, M.F., and M.R. Dyer. 2004. Therapeutic vaccination for chronic diseases: a new class of drugs in sight. *Nat. Rev. Drug Discov.* 3: 81–88.

14 Chackerian, B., D.R. Lowy, and J.T. Schiller. 2001. Conjugation of a self-antigen to papillomavirus-like particles allows for efficient induction of protective autoantibodies. *J. Clin. Invest.* 108: 415–423.

15 Jegerlehner, A., A. Tissot, F. Lechner, P. Sebbel, I. Erdmann, T. Kundig, T. Bachi, T. Storni, G. Jennings, P. Pumpens, W. Renner, and M.F. Bachmann. 2002. Molecular assembly system that renders antigens of choice highly repetitive for induction of protective B cell responses. *Vaccine* 20: 25–30.

16 Persing, D.H., R.N. Coler, M.J. Lacy, D.A. Johnson, J.R. Baldridge, R.M. Hershberg, and S.G. Reed. 2002. Taking toll: lipid A mimetics as adjuvants and immunomodulators. *Trends Microbiol.* 10: S32–37.

17 Bachmann, M.F., T.M. Kundig, G. Freer, Y. Li, C.Y. Kang, D.H. Bishop, H. Hengartner, and R.M. Zinkernagel. 1994. Induction of protective cytotoxic T cells with viral proteins. *Eur. J. Immunol.* 24: 2228–2236.

18 Lemaitre, B., E. Nicolas, L. Michaut, J.M. Reichhart, and J.A. Hoffmann. 1996. The dorsoventral regulatory gene cassette spatzle/Toll/cactus controls the potent antifungal response in Drosophila adults. *Cell* 86: 973–983.

19 Tabeta, K., P. Georgel, E. Janssen, X. Du, K. Hoebe, K. Crozat, S. Mudd, L. Shamel, S. Sovath, J. Goode, L. Alexopoulou, R.A. Flavell, and B. Beutler. 2004. Toll-like receptors 9 and 3 as essential components of innate immune defense against mouse cytomegalovirus infection. *Proc. Natl. Acad. Sci. U.S.A.* 101: 3516–3521.

20 Beutler, B., K. Hoebe, and L. Shamel. 2004. Forward genetic dissection of afferent immunity: the role of TIR adapter proteins in innate and adaptive immune responses. *C.R. Biol.* 327: 571–580.

21 Beutler, B., and M. Rehli. 2002. Evolution of the TIR, tolls and TLRs: func-

tional inferences from computational biology. *Curr. Top. Microbiol. Immunol.* 270: 1–21.

22. Akira, S., and K. Takeda. 2004. Toll-like receptor signalling. *Nat. Rev. Immunol.* 4: 499–511.

23. Beutler, B. 2004. Inferences, questions and possibilities in Toll-like receptor signalling. *Nature* 430: 257–263.

24. Poltorak, A., X. He, I. Smirnova, M.Y. Liu, C. Van Huffel, X. Du, D. Birdwell, E. Alejos, M. Silva, C. Galanos, M. Freudenberg, P. Ricciardi-Castagnoli, B. Layton, and B. Beutler. 1998. Defective LPS signaling in C3H/HeJ and C57BL/10ScCr mice: mutations in Tlr4 gene. *Science* 282: 2085–2088.

25. Takeuchi, O., K. Hoshino, T. Kawai, H. Sanjo, H. Takada, T. Ogawa, K. Takeda, and S. Akira. 1999. Differential roles of TLR2 and TLR4 in recognition of Gram-negative and Gram-positive bacterial cell wall components. *Immunity* 11: 443–451.

26. Arbour, N.C., E. Lorenz, B.C. Schutte, J. Zabner, J.N. Kline, M. Jones, K. Frees, J.L. Watt, and D.A. Schwartz. 2000. TLR4 mutations are associated with endotoxin hyporesponsiveness in humans. *Nat. Genet.* 25: 187–191.

27. Wetzler, L.M. 2003. The role of Toll-like receptor 2 in microbial disease and immunity. *Vaccine* 21(Suppl 2): S55–60.

28. Takeuchi, O., T. Kawai, P.F. Muhlradt, M. Morr, J.D. Radolf, A. Zychlinsky, K. Takeda, and S. Akira. 2001. Discrimination of bacterial lipoproteins by Toll-like receptor 6. *Int. Immunol.* 13: 933–940.

29. Girard, R., T. Pedron, S. Uematsu, V. Balloy, M. Chignard, S. Akira, and R. Chaby. 2003. Lipopolysaccharides from Legionella and Rhizobium stimulate mouse bone marrow granulocytes via Toll-like receptor 2. *J. Cell Sci.* 116: 293–302.

30. Werts, C., R.I. Tapping, J.C. Mathison, T.H. Chuang, V. Kravchenko, I. Saint Girons, D.A. Haake, P.J. Godowski, F. Hayashi, A. Ozinsky, D.M. Underhill, C.J. Kirschning, H. Wagner, A. Aderem, P.S. Tobias, and R.J. Ulevitch. 2001. Leptospiral lipopolysaccharide activates cells through a TLR2-dependent mechanism. *Nat. Immunol.* 2: 346–352.

31. Hirschfeld, M., J.J. Weis, V. Toshchakov, C.A. Salkowski, M.J. Cody, D.C. Ward, N. Qureshi, S.M. Michalek, and S.N. Vogel. 2001. Signaling by toll-like receptor 2 and 4 agonists results in differential gene expression in murine macrophages. *Infect. Immun.* 69: 1477–1482.

32. Massari, P., P. Henneke, Y. Ho, E. Latz, D.T. Golenbock, and L.M. Wetzler. 2002. Cutting edge: Immune stimulation by neisserial porins is toll-like receptor 2 and MyD88 dependent. *J. Immunol.* 168: 1533–1537.

33. Ozinsky, A., D.M. Underhill, J.D. Fontenot, A.M. Hajjar, K.D. Smith, C.B. Wilson, L. Schroeder, and A. Aderem. 2000. The repertoire for pattern recognition of pathogens by the innate immune system is defined by cooperation between toll-like receptors. *Proc. Natl. Acad. Sci. U.S.A.* 97: 13 766–13 771.

34. Wyllie, D.H., E. Kiss-Toth, A. Visintin, S.C. Smith, S. Boussouf, D.M. Segal, G.W. Duff, and S.K. Dower. 2000. Evidence for an accessory protein function for Toll-like receptor 1 in anti-bacterial responses. *J. Immunol.* 165: 7125–7132.

35. Akira, S. 2003. Mammalian Toll-like receptors. *Curr. Opin. Immunol.* 15: 5–11.

36. Means, T.K., E. Lien, A. Yoshimura, S. Wang, D.T. Golenbock, and M.J. Fenton. 1999. The CD14 ligands lipoarabinomannan and lipopolysaccharide differ in their requirement for Toll-like receptors. *J. Immunol.* 163: 6748–6755.

37. Henneke, P., O. Takeuchi, J.A. van Strijp, H.K. Guttormsen, J.A. Smith, A.B. Schromm, T.A. Espevik, S. Akira, V. Nizet, D.L. Kasper, and D.T. Golenbock. 2001. Novel engagement of CD14 and multiple toll-like receptors by group B streptococci. *J. Immunol.* 167: 7069–7076.

38. Campos, M.A., I.C. Almeida, O. Takeuchi, S. Akira, E.P. Valente, D.O. Procopio, L.R. Travassos, J.A. Smith, D.T. Golenbock, and R.T. Gazzinelli. 2001. Activation of Toll-like receptor-2 by glycosylphosphatidylinositol anchors from

a protozoan parasite. *J. Immunol.* 167: 416–423.

39 Steiner, T.S., J.P. Nataro, C.E. Poteet-Smith, J.A. Smith, and R.L. Guerrant. 2000. Enteroaggregative *Escherichia coli* expresses a novel flagellin that causes IL-8 release from intestinal epithelial cells. *J. Clin. Invest.* 105: 1769–1777.

40 Eaves-Pyles, T., K. Murthy, L. Liaudet, L. Virag, G. Ross, F.G. Soriano, C. Szabo, and A.L. Salzman. 2001. Flagellin, a novel mediator of Salmonella-induced epithelial activation and systemic inflammation: I kappa B alpha degradation, induction of nitric oxide synthase, induction of proinflammatory mediators, and cardiovascular dysfunction. *J. Immunol.* 166: 1248–1260.

41 Hayashi, F., K.D. Smith, A. Ozinsky, T.R. Hawn, E.C. Yi, D.R. Goodlett, J.K. Eng, S. Akira, D.M. Underhill, and A. Aderem. 2001. The innate immune response to bacterial flagellin is mediated by Toll-like receptor 5. *Nature* 410: 1099–1103.

42 Smith, K.D., E. Andersen-Nissen, F. Hayashi, K. Strobe, M.A. Bergman, S.L. Barrett, B.T. Cookson, and A. Aderem. 2003. Toll-like receptor 5 recognizes a conserved site on flagellin required for protofilament formation and bacterial motility. *Nat. Immunol.* 4: 1247–1253.

43 Hawn, T.R., A. Verbon, K.D. Lettinga, L.P. Zhao, S.S. Li, R.J. Laws, S.J. Skerrett, B. Beutler, L. Schroeder, A. Nachman, A. Ozinsky, K.D. Smith, and A. Aderem. 2003. A common dominant TLR5 stop codon polymorphism abolishes flagellin signaling and is associated with susceptibility to legionnaires' disease. *J. Exp. Med.* 198: 1563–1572.

44 Zhang, D., G. Zhang, M.S. Hayden, M.B. Greenblatt, C. Bussey, R.A. Flavell, and S. Ghosh. 2004. A toll-like receptor that prevents infection by uropathogenic bacteria. *Science* 303: 1522–1526.

45 Tokunaga, T., H. Yamamoto, S. Shimada, H. Abe, T. Fukuda, Y. Fujisawa, Y. Furutani, O. Yano, T. Kataoka, and T. Sudo. 1984. Antitumor activity of deoxyribonucleic acid fraction from Mycobacterium bovis BCG. I. Isolation, physicochemical characterization, and antitumor activity. *J. Natl. Cancer Inst.* 72: 955–962.

46 Krieg, A.M., A.K. Yi, S. Matson, T.J. Waldschmidt, G.A. Bishop, R. Teasdale, G.A. Koretzky, and D.M. Klinman. 1995. CpG motifs in bacterial DNA trigger direct B-cell activation. *Nature* 374: 546–549.

47 Bogdan, C. 2000. The function of type I interferons in antimicrobial immunity. *Curr. Opin. Immunol.* 12: 419–424.

48 Taniguchi, T., and A. Takaoka. 2002. The interferon-alpha/beta system in antiviral responses: a multimodal machinery of gene regulation by the IRF family of transcription factors. *Curr. Opin. Immunol.* 14: 111–116.

49 Le Page, C., P. Genin, M.G. Baines, and J. Hiscott. 2000. Interferon activation and innate immunity. *Rev. Immunogenet.* 2: 374–386.

50 Alexopoulou, L., A.C. Holt, R. Medzhitov, and R.A. Flavell. 2001. Recognition of double-stranded RNA and activation of NF-kappaB by Toll-like receptor 3. *Nature* 413: 732–738.

51 Heil, F., H. Hemmi, H. Hochrein, F. Ampenberger, C. Kirschning, S. Akira, G. Lipford, H. Wagner, and S. Bauer. 2004. Species-specific recognition of single-stranded RNA via toll-like receptor 7 and 8. *Science* 303: 1526–1529.

52 Diebold, S.S., T. Kaisho, H. Hemmi, S. Akira, and C. Reis e Sousa. 2004. Innate antiviral responses by means of TLR7-mediated recognition of single-stranded RNA. *Science* 303: 1529–1531.

53 Krieg, A.M. 2002. CpG motifs in bacterial DNA and their immune effects. *Annu. Rev. Immunol.* 20: 709–760.

54 Hemmi, H., O. Takeuchi, T. Kawai, T. Kaisho, S. Sato, H. Sanjo, M. Matsumoto, K. Hoshino, H. Wagner, K. Takeda, and S. Akira. 2000. A Toll-like receptor recognizes bacterial DNA. *Nature* 408: 740–745.

55 Leifer, C.A., M.N. Kennedy, A. Mazzoni, C. Lee, M.J. Kruhlak, and D.M. Segal. 2004. TLR9 is localized in the endoplas-

mic reticulum prior to stimulation. *J. Immunol.* 173: 1179–1183.

56. Kariko, K., H. Ni, J. Capodici, M. Lamphier, and D. Weissman. 2004. mRNA is an endogenous ligand for Toll-like receptor 3. *J. Biol. Chem.* 279: 12 542–12 550.

57. Iwasaki, A., and R. Medzhitov. 2004. Toll-like receptor control of the adaptive immune responses. *Nat. Immunol.* 5: 987–995.

58. Wesche, H., W.J. Henzel, W. Shillinglaw, S. Li, and Z. Cao. 1997. MyD88: an adapter that recruits IRAK to the IL-1 receptor complex. *Immunity* 7: 837–847.

59. Muzio, M., J. Ni, P. Feng, and V.M. Dixit. 1997. IRAK (Pelle) family member IRAK-2 and MyD88 as proximal mediators of IL-1 signaling. *Science* 278: 1612–1615.

60. Yamaguchi, K., K. Shirakabe, H. Shibuya, K. Irie, I. Oishi, N. Ueno, T. Taniguchi, E. Nishida, and K. Matsumoto. 1995. Identification of a member of the MAPKKK family as a potential mediator of TGF-beta signal transduction. *Science* 270: 2008–2011.

61. Shibuya, H., K. Yamaguchi, K. Shirakabe, A. Tonegawa, Y. Gotoh, N. Ueno, K. Irie, E. Nishida, and K. Matsumoto. 1996. TAB1: an activator of the TAK1 MAPKKK in TGF-beta signal transduction. *Science* 272: 1179–1182.

62. Takaesu, G., S. Kishida, A. Hiyama, K. Yamaguchi, H. Shibuya, K. Irie, J. Ninomiya-Tsuji, and K. Matsumoto. 2000. TAB2, a novel adaptor protein, mediates activation of TAK1 MAPKKK by linking TAK1 to TRAF6 in the IL-1 signal transduction pathway. *Mol Cell* 5: 649–658.

63. Horng, T., G.M. Barton, R.A. Flavell, and R. Medzhitov. 2002. The adaptor molecule TIRAP provides signalling specificity for Toll-like receptors. *Nature* 420: 329–333.

64. Fitzgerald, K.A., E.M. Palsson-McDermott, A.G. Bowie, C.A. Jefferies, A.S. Mansell, G. Brady, E. Brint, A. Dunne, P. Gray, M.T. Harte, D. McMurray, D.E. Smith, J.E. Sims, T.A. Bird, and L.A. O'Neill. 2001. Mal (MyD88-adapter-like) is required for Toll-like receptor-4 signal transduction. *Nature* 413: 78–83.

65. Kawai, T., O. Adachi, T. Ogawa, K. Takeda, and S. Akira. 1999. Unresponsiveness of MyD88-deficient mice to endotoxin. *Immunity* 11: 115–122.

66. Kawai, T., O. Takeuchi, T. Fujita, J. Inoue, P.F. Muhlradt, S. Sato, K. Hoshino, and S. Akira. 2001. Lipopolysaccharide stimulates the MyD88-independent pathway and results in activation of IFN-regulatory factor 3 and the expression of a subset of lipopolysaccharide-inducible genes. *J. Immunol.* 167: 5887–5894.

67. Yoneyama, M., W. Suhara, Y. Fukuhara, M. Fukuda, E. Nishida, and T. Fujita. 1998. Direct triggering of the type I interferon system by virus infection: activation of a transcription factor complex containing IRF-3 and CBP/p300. *EMBO J.* 17: 1087–1095.

68. Yamamoto, M., S. Sato, K. Mori, K. Hoshino, O. Takeuchi, K. Takeda, and S. Akira. 2002. Cutting edge: a novel Toll/IL-1 receptor domain-containing adapter that preferentially activates the IFN-beta promoter in the Toll-like receptor signaling. *J. Immunol.* 169: 6668–6672.

69. Yamamoto, M., S. Sato, H. Hemmi, S. Uematsu, K. Hoshino, T. Kaisho, O. Takeuchi, K. Takeda, and S. Akira. 2003. TRAM is specifically involved in the Toll-like receptor 4-mediated MyD88-independent signaling pathway. *Nat. Immunol.* 4: 1144–1150.

70. Fitzgerald, K.A., D.C. Rowe, B.J. Barnes, D.R. Caffrey, A. Visintin, E. Latz, B. Monks, P.M. Pitha, and D.T. Golenbock. 2003. LPS-TLR4 signaling to IRF-3/7 and NF-kappaB involves the toll adapters TRAM and TRIF. *J. Exp. Med.* 198: 1043–1055.

71. Bin, L.H., L.G. Xu, and H.B. Shu. 2003. TIRP, a novel Toll/interleukin-1 receptor (TIR) domain-containing adapter protein involved in TIR signaling. *J. Biol. Chem.* 278: 24 526–24 532.

72. Oshiumi, H., M. Sasai, K. Shida, T. Fujita, M. Matsumoto, and T. Seya. 2003. TIR-containing adapter molecule (TICAM)-2, a bridging adapter recruiting to toll-like receptor 4 TICAM-1 that

induces interferon-beta. *J. Biol. Chem.* 278: 49 751–49 762.

73 Yamamoto, M., S. Sato, H. Hemmi, K. Hoshino, T. Kaisho, H. Sanjo, O. Takeuchi, M. Sugiyama, M. Okabe, K. Takeda, and S. Akira. 2003. Role of adaptor TRIF in the MyD88-independent toll-like receptor signaling pathway. *Science* 301: 640–643.

74 Royet, J., and J.M. Reichhart. 2003. Detection of peptidoglycans by NOD proteins. *Trends Cell Biol.* 13: 610–614.

75 Chamaillard, M., M. Hashimoto, Y. Horie, J. Masumoto, S. Qiu, L. Saab, Y. Ogura, A. Kawasaki, K. Fukase, S. Kusumoto, M.A. Valvano, S.J. Foster, T.W. Mak, G. Nunez, and N. Inohara. 2003. An essential role for NOD1 in host recognition of bacterial peptidoglycan containing diaminopimelic acid. *Nat. Immunol.* 4: 702–707.

76 Girardin, S.E., I.G. Boneca, J. Viala, M. Chamaillard, A. Labigne, G. Thomas, D.J. Philpott, and P.J. Sansonetti. 2003. Nod2 is a general sensor of peptidoglycan through muramyl dipeptide (MDP) detection. *J. Biol. Chem.* 278: 8869–8872.

77 Inohara, N., Y. Ogura, A. Fontalba, O. Gutierrez, F. Pons, J. Crespo, K. Fukase, S. Inamura, S. Kusumoto, M. Hashimoto, S.J. Foster, A.P. Moran, J.L. Fernandez-Luna, and G. Nunez. 2003. Host recognition of bacterial muramyl dipeptide mediated through NOD2. Implications for Crohn's disease. *J. Biol. Chem.* 278: 5509–5512.

78 Inohara, N., and G. Nunez. 2003. NODs: intracellular proteins involved in inflammation and apoptosis. *Nat. Rev. Immunol.* 3: 371–382.

79 Martinon, F., and J. Tschopp. 2004. Inflammatory caspases: linking an intracellular innate immune system to autoinflammatory diseases. *Cell* 117: 561–574.

80 Girardin, S.E., J.P. Hugot, and P.J. Sansonetti. 2003. Lessons from Nod2 studies: towards a link between Crohn's disease and bacterial sensing. *Trends Immunol.* 24: 652–658.

81 Samuel, C.E. 2001. Antiviral actions of interferons. *Clin. Microbiol. Rev.* 14: 778–809, table of contents.

82 Saunders, L.R., and G.N. Barber. 2003. The dsRNA binding protein family: critical roles, diverse cellular functions. *FASEB J.* 17: 961–983.

83 Herre, J., A.S. Marshall, E. Caron, A.D. Edwards, D.L. Williams, E. Schweighoffer, V. Tybulewicz, E.S.C. Reis, S. Gordon, and G.D. Brown. 2004. Dectin-1 utilizes novel mechanisms for yeast phagocytosis in macrophages. *Blood* 104: 4038–4045.

84 Takeuchi, O., S. Sato, T. Horiuchi, K. Hoshino, K. Takeda, Z. Dong, R.L. Modlin, and S. Akira. 2002. Cutting edge: role of Toll-like receptor 1 in mediating immune response to microbial lipoproteins. *J. Immunol.* 169: 10–14.

85 Aliprantis, A.O., R.B. Yang, M.R. Mark, S. Suggett, B. Devaux, J.D. Radolf, G.R. Klimpel, P. Godowski, and A. Zychlinsky. 1999. Cell activation and apoptosis by bacterial lipoproteins through toll-like receptor-2. *Science* 285: 736–739.

86 Schwandner, R., R. Dziarski, H. Wesche, M. Rothe, and C.J. Kirschning. 1999. Peptidoglycan- and lipoteichoic acid-induced cell activation is mediated by toll-like receptor 2. *J. Biol. Chem.* 274: 17 406-17 409.

87 Opitz, B., N.W. Schroder, I. Spreitzer, K.S. Michelsen, C.J. Kirschning, W. Hallatschek, U. Zahringer, T. Hartung, U.B. Gobel, and R.R. Schumann. 2001. Toll-like receptor-2 mediates Treponema glycolipid and lipoteichoic acid-induced NF-kappaB translocation. *J. Biol. Chem.* 276: 22041–22047.

88 Underhill, D.M., A. Ozinsky, A.M. Hajjar, A. Stevens, C.B. Wilson, M. Bassetti, and A. Aderem. 1999. The Toll-like receptor 2 is recruited to macrophage phagosomes and discriminates between pathogens. *Nature* 401: 811–815.

89 Kurt-Jones, E.A., L. Popova, L. Kwinn, L.M. Haynes, L.P. Jones, R.A. Tripp, E.E. Walsh, M.W. Freeman, D.T. Golenbock, L.J. Anderson, and R.W. Finberg. 2000. Pattern recognition receptors TLR4 and CD14 mediate response to

respiratory syncytial virus. *Nat. Immunol.* 1: 398–401.

90. Rassa, J.C., J.L. Meyers, Y. Zhang, R. Kudaravalli, and S.R. Ross. 2002. Murine retroviruses activate B cells via interaction with toll-like receptor 4. *Proc. Natl. Acad. Sci. U.S.A.* 99: 2281–2286.

91. Yamamoto, S., T. Yamamoto, S. Shimada, E. Kuramoto, O. Yano, T. Kataoka, and T. Tokunaga. 1992. DNA from bacteria, but not from vertebrates, induces interferons, activates natural killer cells and inhibits tumor growth. *Microbiol. Immunol.* 36: 983–997.

92. 12. Chu, R. S., O. S. Targoni, A. M. Krieg, P. V. Lehmann, and C. V. Harding. 1997. CpG oligodeoxynucleotides act as adjuvants that switch on T helper 1 (Th1) immunity. *J. Exp. Med.* 186: 1623–1629.

93. Roman, M., E. Martin-Orozco, J.S. Goodman, M.D. Nguyen, Y. Sato, A. Ronaghy, R.S. Kornbluth, D.D. Richman, D.A. Carson, and E. Raz. 1997. Immunostimulatory DNA sequences function as T helper-1-promoting adjuvants. *Nat. Med.* 3: 849–854.

94. Krieg, A.M. 2003. CpG motifs: the active ingredient in bacterial extracts? *Nat. Med.* 9: 831–835.

95. Meraldi, V., R. Audran, J.F. Romero, V. Brossard, J. Bauer, J.A. Lopez, and G. Corradin. 2003. OM-174, a new adjuvant with a potential for human use, induces a protective response when administered with the synthetic C-terminal fragment 242-310 from the circumsporozoite protein of Plasmodium berghei. *Vaccine* 21: 2485–2491.

96. Hemmi, H., T. Kaisho, O. Takeuchi, S. Sato, H. Sanjo, K. Hoshino, T. Horiuchi, H. Tomizawa, K. Takeda, and S. Akira. 2002. Small anti-viral compounds activate immune cells via the TLR7 MyD88-dependent signaling pathway. *Nat. Immunol.* 3: 196–200.

97. Brandenburg, K., B. Lindner, A. Schromm, M.H. Koch, J. Bauer, A. Merkli, C. Zbaeren, J.G. Davies, and U. Seydel. 2000. Physicochemical characteristics of triacyl lipid A partial structure OM-174 in relation to biological activity. *Eur. J. Biochem.* 267: 3370–3377.

98. O'Hagan, D.T., and N.M. Valiante. 2003. Recent advances in the discovery and delivery of vaccine adjuvants. *Nat. Rev. Drug Discov.* 2: 727–735.

99. Gursel, I., M. Gursel, K.J. Ishii, and D.M. Klinman. 2001. Sterically stabilized cationic liposomes improve the uptake and immunostimulatory activity of CpG oligonucleotides. *J. Immunol.* 167: 3324–3328.

100. Storni, T., C. Ruedl, K. Schwarz, R.A. Schwendener, W.A. Renner, and M.F. Bachmann. 2004. Nonmethylated CG motifs packaged into virus-like particles induce protective cytotoxic T cell responses in the absence of systemic side effects. *J. Immunol.* 172: 1777–1785.

101. Schwarz, K., T. Storni, V. Manolova, A. Didierlaurent, J.C. Sirard, P. Rothlisberger, and M.F. Bachmann. 2003. Role of Toll-like receptors in costimulating cytotoxic T cell responses. *Eur. J. Immunol.* 33: 1465–1470.

Part IV
The Repertoire of Antigen Presenting Cells

9
Evolution and Diversity of Macrophages
Nicholas S. Stoy

9.1
Evolution of Macrophages: Immunity without Antigen Presentation

9.1.1
Introduction

With the rapid advances in DC research over recent years it might appear that many traditional roles ascribed to macrophages are being whittled away. It is now accepted that DCs are the most potent 'professional' APCs and unique in their ability to present antigen to naive T cells. Like macrophages, DCs are capable of phagocytosis, cytokine and chemokine production. Nevertheless, increasing knowledge of DC function should clarify, not detract from, an understanding of macrophage involvement in immune responses. A major challenge (and a preoccupation of this and the next chapter) is to integrate the contributions of these two key cells. Macrophages and DCs are usually complementary, not competitive in their functions. Far from being redundant, macrophages are essential for the coordination and implementation of both innate and adaptive immunity, being consistently located in the very tissue microenvironments where immune responses are played out. Innate immunity does not require antigen presentation at all, yet macrophages are crucially involved in it. The innate immune system, which antedates the adaptive immune system in evolution, can be studied 'in isolation' in invertebrates. Phagocytosis and the ability to display immunological responses are early and conserved components of macrophage function and are carried forward into the mammalian immune system. Antigen presentation is just one facet of the macrophage repertoire, and a more recent addition.

9.1.2
Drosophila: a Window into Innate Immunity

Most information about the invertebrate immune system, to date, has come from *Drosophila melanogaster*. *Drosophila* plasmatocytes (a subset of hemocytes) [1] are macrophage-like cells with two important functions, firstly phagocytosis of apoptotic cells and secondly protection from infection. The clearance of apoptotic cells during ontogeny is well demonstrated by the microchaete, a simple mechanosensory organ of *Drosophila*, which loses a glial cell by apoptosis shortly after birth. Thereafter, axonal growth coincides with the removal of apoptotic fragments by 'mobile macrophages', identified by their ability to take up injected indian ink [2]. For normal *Drosophila* nervous system morphogenesis, apoptotic cell removal is as necessary as apoptosis itself. If the plasmatocyte scavenger receptor Croquemort is inhibited, or if hemocytes are deleted, nervous system defects occur. However, plasmatocytes themselves are not thought to be responsible for promoting apoptosis in the developing nervous system [3]. Croquemort shares homology with the mammalian class B scavenger receptor, CD36, which, together with the vitronectin and phosphatidylserine receptors, engulfs apoptotic cells [1]. Cell proliferation, cell differentiation and programmed cell-death are indispensable requirements for orderly development in both invertebrates and vertebrates. It is essential that no tissue-damaging molecules are produced during phagocytosis of apoptotic remnants. Phosphatidylserine, thought to be an important molecule identifying apoptotic cells for phagocytosis, is exposed on apoptotic cell membranes in both mammals and *Drosophila*. Possible homologues of genes encoding a specific phosphatidylserine receptor in mammals have been identified in *Drosophila*. On a note of caution, it has recently been found that, in mammals, a combination of other receptors may be just as effective in removing apoptotic cells [4], as may another non-phosphatidylserine receptor (Draper) in *Drosophila* [5]. Some key molecules required for the development and/or preservation of neuronal function in *Drosophila* can also be produced by mammalian macrophages, where they usually exhibit antiinflammatory phenotypes. For example, the cytokine TGF-β is inducible in mammalian macrophages during phagocytosis of apoptotic bodies and is predominantly antiinflammatory (Figure 9.2), whilst in *Drosophila* other members of the TGF-β superfamily, bone morphogenetic proteins (BMPs) and activins, regulate protein expression during growth, differentiation and morphogenesis [6].

In mammals, too, TGF-β is a pleotropic cytokine involved in growth and development. When produced by macrophages, it enhances phagocytosis, as well as downregulating iNOS, toxic oxygen and nitrogen intermediates and antagonizing proinflammatory cytokines, such as IL-12 [7]. By contrast, in *Drosophila*, the TGF-β pathway may be upregulated as part of a defensive immune response to sepsis/injury. When comparing mammals and *Drosophila*, the functional equivalence of downstream transcriptional activators and repressors cannot always be assumed [8]. Despite these differences, TGF-β family ligands, on binding to their surface receptor complexes, universally activate intracellular signaling molecules (receptor-activated Smads), which, in turn, complex with co-Smad (Smad4) and then transfer to the nucleus to interact with their target genes (Figure 9.2). In mammals, Smad1, Smad5 and Smad8 mediate BMP signaling and Smad2 and Smad3 mediate TGF-β/activin signaling. Smads modulate gene transcription in conjunction with coactivators or corepressors. One well-described coactivator is CBP [9]. An example of the antiinflammatory role of mammalian TGF-β is its dis-

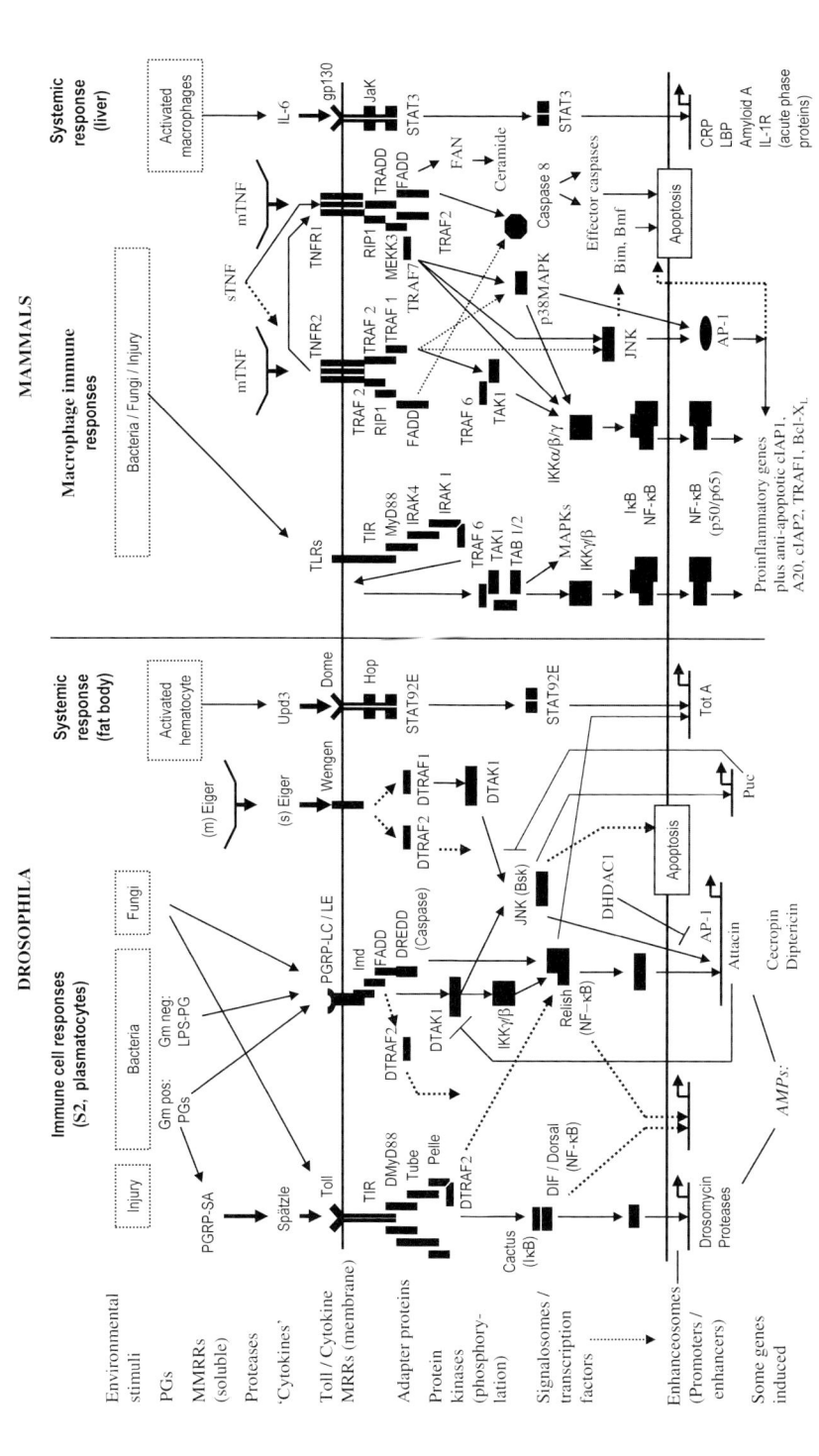

Figure 9.1 Comparison of some immune response signal-transduction pathways in Drosophila and mammals. Differences as well as similarities are apparent, as indicated in the text. JNK – AP-1 pathways mediate apoptosis and are countered by NF-κB-activating pathways in both Drosophila and mammals. In Drosophila, AP-1 attracts HDAC1 to the NF-κB-activated promoter of Attacin-A and inhibits it, therefore acting as a repressor of transcription. Sustained-response, Relish (NF-κB)-dependent, genes limit the activity of the JNK pathway by directing TAK1 to proteosomal degradation. Such 'cross-talk' mechanisms, acting together or sequentially, control the extent/duration of (innate) proinflammatory responses.

ruption of LPS/NF-κB signaling in microglia ('brain macrophages') through mechanisms involving smad3/CBP and MAPK, thereby contributing to neuroprotection [10]. In murine (RAW 264.7) macrophages, TGF-β suppresses LPS-induced NF-κB activation by enhancing ERK activation, which induces a MAPK phosphatase and thus inhibits the p38 MAPK drive to NF-κB [11, 12]. Similarly, in colonic epithelium, TGF-β inhibits infection-triggered TLR-2/MyD88/p38 MAPK-dependent signaling, by inducing MAPK phosphatase-1 (MKP-1), which deactivates p38 MAPK (Figure 9.2) [13]. Although Smad pathways are functionally disparate between *Drosophila* and mammals, the Smads themselves and Smad receptors (both type I and II) tend to be highly conserved [9]. The molecule cAMP is an intracellular second messenger essential for many antiinflammatory responses in mammalian macrophages but also has a pivotal role in preserving neuronal function and memory in *Drosophila* [14]. The cAMP/PKA/CREB pathway is well described in both mammalian macrophages and in *Drosophila* Steiner 2 (S2) cells, a primary haemocyte-derived macrophage-like cell culture [15], but the regulation of cAMP response element-dependent genes is different between *Drosophila* and vertebrates [16, 17]. Another protein conserved between *Drosophila* and mammals is TORC1, which may act to amplify the magnitude of CREB responses in the absence of extracellular stimuli and so modulate the expression of CREB-responsive genes during development [18]. It is anticipated that many more molecules concerned with homeostasis during *Drosophila* ontogeny will be discovered in the phenotypes of *Drosophila* plasmatocytes and have homologues in mammalian macrophages, providing insights into molecular mechanisms counteracting inflammation and maintaining self-tolerance.

On environmental challenge (injury or sepsis), innate immunity in *Drosophila* provides a robust defense [19–22]. An exoskeleton, epithelial barriers and phagocytosis by plasmatocytes are primary mechanisms averting bacterial sepsis. Just as the Croquemort phagocytic receptor recognizes apoptotic cells, so does the scavenger receptor, CI, (resembling the mammalian class A scavenger receptor) recognize both Gram-negative and Gram-positive bacteria [15]. Phagocytosis of microbes by S2 cells leads to lysosomal proteolysis by cysteine proteases of the cathepsin family [23], thus rendering potentially immunogenic microbial molecules incapable of further stimulating cell surface receptors. Additional to any conserved role in innate tolerance, a wider spectrum of activity of these cysteine proteases in mammals suggests they have evolved further to undertake processing of antigen for MHC II presentation in adaptive immunity. *Drosophila* cathepsins are apparently not inducible by immune stimuli. *Drosophila* S2 cells, as well as cells in the fat body (corresponding to the mammalian liver), can be stimulated through the *Drosophila* Toll receptor to adopt an antimicrobial phenotype. Most members of the now well-categorized *Drosophila* Toll family are exclusively concerned with developmental processes, but Toll itself also functions as an immunomodulatory receptor, linked to an NF-κB analogue (Dorsal/Dif). Another *Drosophila* immunomodulatory pathway (Imd) is linked to a different NF-κB analogue (Relish). These two pathways are activated by a range of bacteria and fungi to induce numerous anti-microbial peptides (AMPs), comprising a humoral innate immune defense upon which *Drosophila* is heavily reliant (Figure 9.1) [24].

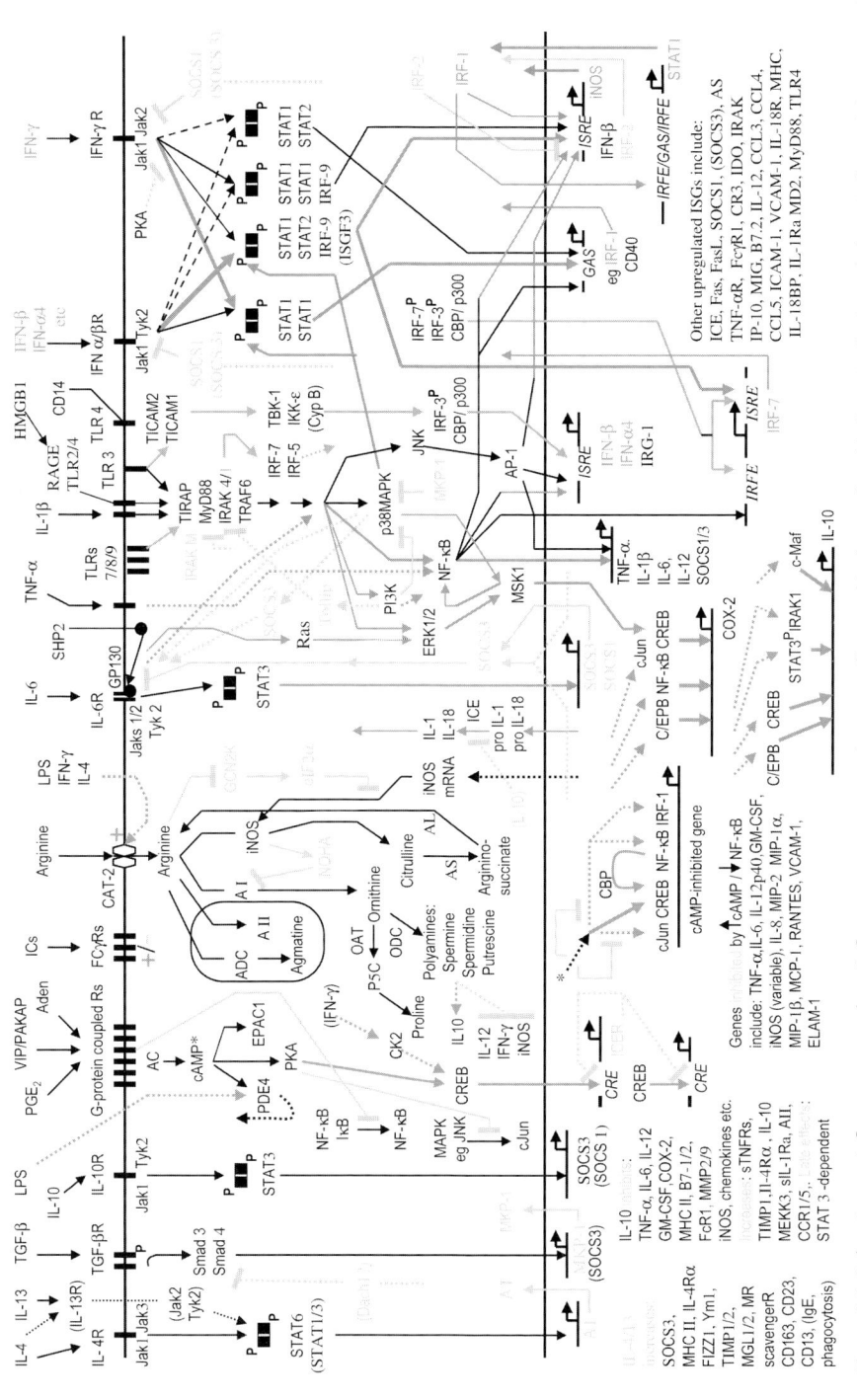

Figure 9.2 Balancing proinflammatory and antiinflammatory responses in macrophages. Some of the better-known intracellular transduction/signalling pathways and their interactions are shown. These control upregulation, downregulation, and the limits and sequencing of innate and adaptive immune responses, initiated by a variety of external stimuli. Most are discussed in the text, but this figure is also intended as a 'reference' diagram, of possible help in the interpretation of future research findings. (Green lines: facilitatory/proinflammatory; red lines: inhibitory/antiinflammatory; blue lines: pathways involving IRFs; dotted lines: detailed pathways omitted for clarity or because unknown.) (This figure also appears with the color plates.)

The Imd pathway triggers also MAPK signaling, which rapidly activates a short burst of JNK-responsive genes, which are in fact downregulated as a result of Relish-responsive genes becoming activated. Although JNK and Relish are activated from a common branch point (TAK1), the known JNK-activated genes do not produce AMPS. Rather, the JNK-responsive gene, Punch, could contribute to the stress response and wound healing, through enhanced dopamine and melanin production, respectively [25]. The TLRs in mammals act as microbial molecular recognition receptors (MMRRs) [20], often referred to as pattern recognition receptors (PRRs). Their ligands are highly conserved microbial-associated molecules, MAMs, (more restrictively referred to as pathogen associated molecular patterns, PAMPs). TLR stimulation often activates NF-κB (along with other signaling pathways) to promote rapid induction of proinflammatory genes. The situation is somewhat different in Drosophila. Interposed between the microbial macromolecules and the two Drosophila NF-κB pathways are peptidoglycan (PG) recognition molecules, which either activate Toll indirectly, through an intervening cytokine-like molecule, Spätzle (PPR-S) [26], induced exclusivly in hemocytes, or activate the Imd pathway via a transmembrane portion [27]. There are over a dozen PG recognition molecules in Drosophila able to detect conserved microbial components, thus putatively distinguishing 'self' from 'nonself', although only PGRP-SA, PGRP-LC and PGRP-LE are currently known 'receptors' for bacterial PGs in immune responses. Discrimination of 'self'/'non-self' in the mammalian innate immune system is largely a function of membrane-located MMRRs. [28]. Whereas TLRs transduce microbial recognition into immune responses through intracellular signaling pathways, at least one peptidoglycan-recognition molecule provides an interesting functional parallel with downregulatory mammalian decoy receptors. Like some soluble mammalian decoy receptors, which are secreted extracellularly and neutralize immunoactive molecules (discussed in 9.2.6.4), Drosophila extracellular PGRP-SC1B may be involved in inhibiting the immunostimulatory properties of some PGs [29]. Drosophila can generate a homologue of the mammalian proinflammatory (and apoptosis-inducing) cytokine TNF, [30, 31] and can also express NOS, both of which, in mammalian macrophages, are inducible as part of the proinflammatory response. The TNF-α-like system in Drosophila consists of a ligand (Eiger) a receptor (Wengen) and intracellular signaling through JNK (Figure 9.1). Although not directly required for AMP production, the JNK-induced gene Punch codes for a rate-limiting enzyme in the synthesis of tetrahydrobiopterin, an essential cofactor for NOS function. The Drosophila (D)NOS, itself, is not inducible but is expressed additionally as short spliced variants, which inhibit DNOS1 by forming heterodimers with the full length enzyme. Apart from being involved in organ development and synaptogenesis, DNOS has a role in Drosophila's response to hypoxia and NO may be microbicidal in Drosophila phagocytes, just as in mammalian macrophages [25, 32, 33]. The systemic response in mammals through which acute phase proteins, such as CRP and serum amyloid A, are secreted by hepatocytes is dependent on (macrophage) proinflammatory cytokine release at the site of initial infection/injury, on circulating IL-6 as a signaling link to the liver and on the operation of the Jak-STAT intracellular signaling pathway to activate acute phase protein genes in the liver. As shown in Figure 9.1, all the basic machinery is present in *Drosophila* to produce an acute phase response in the fat body, the equivalent of the mammalian liver. In addition to Jak-STAT signaling, generation of the acute phase *Drosophila* 'marker' protein totA may also require the participation of Relish (NF-κB) [34–36]. The only Jak known in *Drosophila* is Hopscotch and this may upregulate a protein resembling the C3 component of complement, another conserved element in humoral innate immunity [37].

Drosophila is useful in studying microbial responses, whether to its own natural pathogens or to typical mammalian pathogens [38]. Amongst the former are the protozoan parasite *Octosporea* [39] and the Gram-negative bacterial genus *Erwinia* [40], each of which trigger very different specific immune (hemocyte) responses. Amongst the latter is *Listeria monocytogenes*, which is phagocytosed by S2 cell and requires the virulence factor listeriolysin O (LLO) and vacuolar acidification for vacuolar escape into the cytoplasm, just as in mammalian macrophages [41]. Analogous to its virulence in murine macrophages, *Mycobacterium marinum* in *Drosophila* blocks vacuolar acidification and fails to elicit an immune response (AMPs) [42]. Knowledge of the whole *Drosophila* genome allows microarray analysis of the full immune response. One such study tracks an immune response to septic penetrating injury of the thorax by combined *Escherichia coli* and *Micrococcus luceus*. Of 13197 genes investigated, 230 were induced and 170 repressed, representing the whole co-ordinated innate response through pathogen recognition, phagocytosis, NF-κB activation and production of reactive oxygen species. The extreme stimulus applied in this case triggers both antifungal and antibacterial responses and expression of all AMPS, but in less extreme conditions *Drosophila* distinguishes between natural pathogens and mounts more specific immune responses, with identifiable 'signatures' of gene expression [43]. A genome-wide RNA interference screen of the response of *Drosophila* S2 cells to LPS has shed light on the mechanism of NF-κB activation, revealing that LPS exerts a suppressive effect on a caspase inhibitor gene named defense repressor 1, thereby activating the caspase 8 homolog, Dredd, which cleaves Relish and releases its activated NF-κB domain for nuclear translocation (Figure 9.1) [44].

Possession of an immune system gives *Drosophila* an evident survival advantage. As a general thesis, immune competence may have been achieved through the evolutionary selection of molecules required for normal development but with coincidental properties useful for immunity. The complexity of molecular signaling during *Drosophila* ontogeny contrasts with the relative simplicity of the immune response, suggesting that certain strategic molecules, some perhaps in mutated form, may have been co-opted into microbial defense. Coordination of signaling pathways during *Drosophila* development is exemplified by the complex regulation of embryonic dorsal-ventral patterning in which a series of maternally-transcribed genes influence early embryogenesis. Several maternal proteins constitute a proteolytic cascade, which activates the Toll receptor ligand Spätzle, and yet others activate BMP and Jak/STAT signaling pathways, all ultimately regulating the relative and absolute levels of Cactus and Dorsal, which control NF-κB translocation to the nucleus [45]. In plasmatocytes the same Spätzle, Toll and Cactus/Dorsal (NF-κB) pathway is diverted to the production of AMPs and in this situation Spätzle is cleaved to its activated form by proteases, activated when PGRP-SA binds fungal or bacterial molecules (PGs). PGRPs may themselves have derived from an active enzyme (lysozyme) by substitution of one amino acid, negating the enzyme activity whilst retaining the binding capability [29]. Mammalian TLRs may have evolved as far back as plants which contain the TIR motif of mammalian TLRs and the IL-1 receptor – so it is a moot point as to whether developmental genes arose before or after immunological genes [20]. Comparing *Drosophila* and mammalian macrophage phenotypes, the former demonstrate much less requirement for multiple signaling pathways, or transcription factor cooperation, in order to switch on genes fully. In mammals, presumably from further evolutionary pressures, proinflammatory molecules are more tightly regulated and often require coordinated activation of several separate pathways and transcription factors, complexing together in the 'enhancesome' region of the gene promoter, to achieve maximum gene activation. For example, full IFN-β transduction requires three signaling cascades to activate the transcription factors NF-κB, AP-1 (cJun) and IRF (Figure 9.2) [25]. Even to trigger TLR activity in mammals can require several essential accessory molecules (TLR4 associates with CD14, LBP and MD2) [46]. Despite the simpler, less-regulated signaling-transduction in *Drosophila*, sequential and coordinated activation of signaling pathways are readily apparent, as in the response to sepsis, during which a rapid short burst of JNK-triggered genes gives way to more sustained Dorsal/Dif (Gram-positive) or imd (Gram-negative)-triggered genes (the AMP response), overlapping a transient expression of Jak-STAT induced genes [47]. The control of early and late response genes in mammals is considered in more detail in 9.2.6. The TNF-like molecule, Eiger, is mainly expressed in the nervous system, where it initiates JNK-induced (caspase-independent) apoptosis. Apoptosis may be blocked by inhibiting the JNK pathway in S2 cells but is also partially blocked by caspase inhibitors. Programmed cell-death pathways are either intrinsically regulated by caspases, under genetic control (entirely so in *Caenorhabditis elegans*) or by extrinsic mechanisms, through cell surface receptors, of which TNF/TNFR/JNK signaling is a prime example [31]. Whilst the single TNF/TNFR system of *Drosophila* is directed towards programmed cell death in the control of normal morphogenesis, many of the more than 20 members of the mammalian TNF/TNFR superfamily proteins are expressed in the immune system [30]. With its ability to trigger apoptosis, but also induce NF-κB-responsive proinflammatory genes, TNF-α appears to have a contradictory role in the mammalian innate immune response. Speculatively, a pathway that can be switched between microbial killing and apoptosis may provide flexibility in regulating the duration of immune responses in the more sophisticated mammalian immune system. The divergent roles of TNF in development and in antimicrobial immunity are regulated separately in *Drosophila*. The JNK signaling pathway and apoptosis is controlled by DTRAF1. NF-κB signaling and AMP production is controlled by DTRAF2 through the Toll pathway [48]. In mammals TRAF4 most closely corresponds to DTRAF1 and both are most prevalent in neuronal and epithelial precursor cells and neither play much, if any, part in NF-κB signaling. Mammalian TRAF 6 is the closest functional counterpart of DTRAF2 in activating NF-κB [49], but signal regulation is far more intricate in mammals. For example, TNF interaction with TNF-R1, which contains a death domain, favors apoptosis, in contrast to signaling initiated by TNF-R2 (Figure 9.1). However, TNF receptor stimulation also activates NF-κB, which induces anti-apoptotic proteins alongside proinflammatory ones. Inhibition of activated NF-κB, for whatever reason, sensitizes cells to TNF-α-induced cell death. Negative regulation of JNK during NF-κB-mediated responses is conserved in mammals. Conversely, prolonged activation of the JNK pathway signals cell death [50]. Overall, and regardless of any speculations, *Drosophila* provides insights into an immense range of autonomous macrophage functions, ranging from antiinflammatory uptake of apoptotic cells to full-blown pathogen-lethal immune responses.

9.1.3
Evolution of Adaptive Immunity: Macrophages in a New Context

Although antigen presentation is not a feature of the invertebrate innate immune system, a conserved MHC-like region is present in the *Drosophila* genome and the *Drosophila* proteosome can generate peptides suitable for loading onto MHC molecules [51]. It is an estimated 540 million years ago that adaptive immunity first emerged, with the jawed vertebrates – a dramatic change that must have had a major impact on the subsequent course of immune system evolution. However, it is necessary to go back much further, in excess of 800 million years, before the divergence of deuterostomes and protostomes, to find a common ancestor of mammals on the one hand and *Drosophila* and nematodes on the other. Unsurprisingly, therefore, comparisons of invertebrate and mammalian innate immune systems have revealed marked difference as well as similarities. For a contemporary view of the isolated mammalian innate immune system, the SCID and nude (athymic) mouse models, which lack functional T cells, provide an alternative to extrapolations from *Drosophila*.

The advantages of the adaptive immune system are discussed in detail elsewhere in this book, but in considering the contributions of macrophages to the combined innate and adaptive systems several conceptual points should be emphasized:

1. The adaptive immune system is grafted onto the innate immune system and cannot function without it. The adaptive immune system simply modulates the responses of the innate immune system in various ways.
2. The 'master-switch', determining immunity versus tolerance, lies in the innate immune system. Once determined, polarity is faithfully transmitted right through the adaptive immune system, both afferently and efferently. Therefore, the adaptive immune system only amplifies a pre-existing response. Any changes in polarity must be initiated in the innate immune system and by implication require an environmental input. Macrophage and DC phenotypes are programmed by the local microenvironment, so

the fundamental basis of immune responses lies in the modulation of the microenvironment of non-lymphoid tissues.
3. The innate immune system recognizes generic microbial molecular patterns and the adaptive immune system increases the specificity of the response by focussing it on unique microbial identifiers (antigens) and then generating immunological memory.
4. The adaptive immune system is usually programmed at a distance from immunologically active peripheral tissues, introducing the requirement for new types of cells. These are DCs for information transfer and T cells for fine-tuning of the specificity and diversity of the response and for amplification of the signal strength. Lymphoid organs (nodes and spleen), with controlled access and egress, are specialized for these functions.
5. The stand-alone innate immune system of the invertebrate, as exemplified by *Drosophila*, involves a phagocytic macrophage-like cell in two ways: first, disposal of unwanted self in the form of apoptotic cells and second disposal of pathogens, achieved either by ingestion and destruction (a cellular response) or by secretion of antimicrobial molecules (as a form of 'humoral' response). Macrophages employ 'hard-wired' germline-encoded receptors that directly distinguish apoptotic cells (self) from pathogen-specific three-dimensional molecular conformations or 'patterns' (non-self). By contrast, the adaptive immune system must interpret coded information (antigenic peptides) presented to it to decipher (noninfectious) self from (infectious) non-self.
6. To contribute to a tissue adaptive response macrophages must be able to process and present antigen on MHC II molecules with the required co-stimulation. Macrophages are adept at internalizing, degrading and presenting exogenous protein-derived antigens under the right conditions, but not endogenously processed protein. IFN-γ is required for MHC II expression and antigen presentation. Treatment of macrophages with GM-CSF, however, encourages endogenous peptide occupation of MHC II molecules [52]. There is evidence that non-TLR receptors may also assist macrophages to function as APCs, as elaborated in more detail in 9.2.4.
7. If macrophages are considered to be sentinels of the innate immune system and DCs of the adaptive immune system, one measure of the balance between innate and adaptive responses would the ratio of macrophages to DCs generated in a tissue. However, this definitive statement depends on a strict functional definition of the two cell types. Currently, distinguishing the two either functionally or by surface markers is not always easy (as discussed in 9.2.3).
8. There is a growing realization that canonical cell types, such as Th1 and Th2, may be a gross over-simplification. Rather, immune cells integrate information presented to multiple receptors at the cell surface and transduce the information through multiple signaling pathways to induce multiple genes (often interacting), resulting in a diverse range of (effector) outcomes that often change over time.

9. Microbial killing mechanisms induced in macrophage are essentially the same whether triggered by TLRs or T cells. Humoral immunity and differential apoptosis of immune-activated cells contribute to both innate and adaptive immunity.
10. The above statements outline the normal operation of the immune system Autoimmunity, cancer and other diseases affecting the immune system should be interpretable as specific abnormalities operating in the same framework. The adaptive immune system is complex and more likely to fail operationally, resulting in autoimmunity or allergy. Conversely, the innate immune system initiates and sustains most chronic inflammatory conditions, including autoimmunity. Autoimmunity and cancer expression is modulated by both innate and adaptive immunity

9.2
Diversity of Macrophages in Mammalian Tissues

9.2.1
Classifying Heterogeneity

Macrophage phenotypic diversity is identifiable through variations in morphology, expression of cell-surface markers and, most importantly, through function. In all these respects there is considerable overlap with DCs. Indeed, a recent textbook dedicated to the macrophage states that "it is probably obsolete to define DCs and macrophages as separate subtypes of APCs: rather it may be appropriate to consider them as an interactive network, sensitive to intercellular crosstalk and extremely responsive to environmental factors" [53]. Other researchers have claimed that DCs and macrophages could lie at the extremes of a continuous phenotypic spectrum [54]. In considering macrophage diversity, therefore, fundamental questions still demand an answer. What are the defining characteristics that distinguish macrophages from DCs and at what stage, in what manner and under what conditions do macrophages and DCs part phenotypic company in their development from a common marrow progenitor cell?

Morphology cannot be used to separate the two cell types reliably because macrophages and DCs can look deceptively similar. Macrophage morphology is extremely variable and is substantially influenced by the physical constraints of tissue architecture [55]. Kupffer cells, located in hepatic sinusoids, are elongated and can appear highly dendritic in the neonatal period. Muscle macrophages are spindle-shaped, lying parallel to muscle fibers. Microglial cells probably show the greatest morphological diversity, ranging from highly ramified, radially or longitudinally branched, to compact [56]. Veiled macrophages in afferent lymphatics may simulate veiled DCs in appearance but fail to display accepted DC markers [57]. Aluminum hydroxide, used as an adjuvant, is taken up by macrophages as persistent crystalline inclusions but, although the cells are unchanged morphologically, they take on the function and phenotypic markers of DCs [58]. Thus morphology, at best, gives a static and sometimes misleading impression of cell types and, in particular, gives no idea of sequential changes in phenotype or functional interactions with the tissue microenvironment. It might be anticipated that cell surface markers would constitute a more reliable method for distinguishing macrophages from DCs. The problem here is that expression of combinations of surface molecules change over time in both cell types as well as exhibiting marked variations according to the developmental stimuli to which the cells are subjected, as now discussed.

9.2.2
Phenotypic Manipulations and Transdifferentiations: Routes to and from Macrophages

It is now broadly accepted that most new tissue macrophages, of whatever final phenotype, are recruited from circulating uncommitted monocytes [59]. After being released from bone marrow, monocytes probably remain in the circulation for less than 24 h, before migrating into tissues, apparently at random, to differentiate in response to local conditions. Resident tissue macrophages are widely distributed. Injury or inflammation, however, substantially enhances monocyte recruitment through the local release of chemoattractants and upregulation of endothelial adhesion molecules. At least two distinct subsets of circulating blood monocytes are identifiable from expressed chemokine receptors. A CCR2-expressing subset is actively recruited to inflamed tissue and a CX_3CR1 subset to non-inflamed tissues, and both can differentiate into DCs [59].

Microenvironmental factors putatively involved in the 'maturation' and 'differentiation' of macrophages and DCs in vivo have been mainly investigated in vitro, driven, not least, by the search for 'therapeutic' DCs. Subjecting blood-derived monocytes to various cocktails of cytokines and other biologically active molecules results in the generation of cells with the behavioral characteristics (and accepted surface markers) of either macrophages or DCs. When cultured in serum with GM-CSF for a week, human monocytes (bearing the marker CD14), differentiate, without proliferation, into typical macrophages ($CD14^+$), but if the culture medium also contains IL-4 the resulting cells have the morphological and functional characteristics of immature DCs ($CD14^-$). Replacing GM-CSF/IL-4 with macrophage colony-stimulating factor (M-CSF) at this stage converts the immature DCs into macrophages [60, 61]. A combination of GM-CSF and IL-4 (or IL-13) allows monocytes to differentiate to DCs, but inclusion of IL-10 in the culture medium blocks DC differentiation and produces macrophage phenotypes. IL-10 is only effective in promoting macrophages if present in the first few days of culture and cannot reverse the differentiation to DCs if added at later times [62, 63]. Culture of monocytes in serum-free conditions clearly indicates that the primary action of GM-CSF or IL-3 on monocytes is to downregulate their M-CSF receptors, so pushing differentiation towards immature DCs rather than macrophages. The M-CSF receptor, encoded by the gene c-fms, is obligatory for macrophage differentiation from monocytes and acts at an early stage [64]. Engagement of the monocyte 'marker' CD11b (Mac-1) downregulates the expression of Foxp1, a transcription factor that inhibits c-fms transcription. The monocyte-specific expression of c-fms is regulated by several other transcription factors (PU.1, AML1, C/EBP), which bind to adjacent regions of the promoter. Deficiency of c-fms causes global depletion of macrophages, whilst CD11b has an important role in monocyte migration, differentiation, phagocytosis and oxidative burst [65]. Activated monocytes themselves secrete M-CSF, which can therefore act in an autocrine manner via M-CSFR. Moreover, IL-10 has a synergistic effect with M-CSF to promote macrophage differentiation from monocytes by enhancing the upregulation of M-CSFR. The resultant macrophages show enhanced priming for effector function, in that they express higher levels of FcγRI, FcγRII and FcγRIII, augmented phagocytosis and increased generation of reactive oxygen species when stimulated with zymosan, and of IL-6 when stimulated with LPS [66]. Microenvironmental IL-6 itself (or the soluble IL-6R which, unusually, potentiates IL-6) upregulates M-CSFR on monocytes, skewing their development towards macrophages. Newly arriving monocytes can apparently trigger IL-6 release from stromal cells such as fibroblasts [67]. M-CSF is intimately involved in establishing and regulating resting tissue macrophages during normal organogenesis, again suggesting an important physiological role for macrophages as scavengers of apoptotic cells [55]. Monocyte-derived DCs may provide another source of M-CSF, and IL-10 can induce DC to macrophage transdifferentiation by upregulating DC c-fms (M-CSFR), enabling, at least in principal, a reversal of the balance between the two cell types [68]. An antiinflammatory ('resting') phenotype is supported in M-CSF-treated macrophages by the induction of a phosphatase (MKP-1) that inhibits the Raf/MEK/ERK (MAPK) pathway without inducing apoptosis [69].

It is instructive to contrast the requirements for macrophage culture in vitro with those for DCs [70, 71]. DCs, with phenotypes resembling Langerhans cells (LCs) (as found in epidermis and mucosa), can be obtained from adult peripheral blood monocytes, without cell division, in the presence of GM-CSF/IL-4 (and serum), by addition of TGF-β1 [72]. Alternatively, 'mature' DCs, bearing the marker CD83$^+$, may be derived from monocytes by including TNF-α in the initial GM-CSF/IL-4 culture medium, or, even more effectively, by adding TNF-α at day five. One of the most effective ways of producing mature DCs (CD83$^+$) from monocytes is by initial culture with GM-CSF and IL-4 (to produce immature DCs) followed by exposure to a 'monocyte conditioned medium', which contains variable quantities of TNF-α IL-1β IL-6, IFN-α and PGE$_2$ [73, 74]. Monocytes express IL-3R, and when cultured with IL-3 and IL-4 yield a subset of DCs producing less IL-12 and supporting Th2 responses [75]. Conversely, IL-3 and IFN-β-treated monocytes acquire a mature DC phenotype and exhibit potent Th1 polarizing signals despite low secretion of IL-12 [76]. In addition to the monocyte pool a small number of circulating mononuclear cells coexpressing CD14 and the progenitor marker CD34 have been identified, and these (like monocytes) are able to traverse endothelial layers and differentiate into tissue DCs. CD14$^+$CD34$^+$ cells could be direct precursors of DCs and circumvent the alternative monocyte (CD14$^+$, CD34$^-$) pathway. Cord blood, or bone marrow, are other sources of CD34$^+$ progenitor cells from which DCs will differentiate in vitro if they are cultured with GM-CSF or IL-3, with the later addition of TNF-α [77]. Blood monocytes exposed to type I interferons after incubation with GM-CSF develop rapidly into plasmacytoid DCs, secreting IL-15 and skewing T cells towards Th1 [78]. Even activated human neutrophils can escape apoptosis and adopt the phenotype of DCs if they are cultured with IFN-γ or GM-CSF [79].

Production of macrophages from M-CSF-stimulated monocytes does not occur universally. The M-CSF route to macrophages can be diverted to osteoclasts (bone resorptive cells) if the surface receptor RANK, which is induced by M-CSF, interacts with its ligand RANKL, found primarily on osteoblasts and on some T cell subsets. Osteoclasts exhibit various macrophage markers but are distinguished by multinucleation, resulting from cell-fusion, and become TRAP positive. Induction with GM-CSF, by contrast, yields myeloid DCs on addition of soluble RANKL [80]. Marrow-derived osteoclast precursor cells (in murine cultures containing M-CSF and serum) respond directly to IL-3, to inhibit, irreversibly, RANKL-induced osteoclast differentiation in favor of macrophages [81]. The phenotype of osteoclasts is extensively modulated by cytokines in an analogous way to macrophages. The process requires the expression of RANKL and may be induced through functional NMDA receptors, which activate NF-κB, and is enhanced by TNF-α and inhibited by IL-4 [82]. Apart from its role as an excitatory neurotransmitter, glutamate may have a signaling role in bone homeostasis [83]. Functional NMDA is expressed by osteoclasts [84] and NMDA receptors are expressed by osteoclast precursor cells, including the murine macrophage cell line RAW 264.7 [84], but to what extent NMDA receptors are present on other macrophages/microglia appears to be largely unexplored. Recent work has revealed that the c-Fos element of the AP-1 transcription factor may provide an essential intracellular switch between osteoclast and macrophage differentiation and, more significantly, may have a general suppressive effect on macrophage and DC proinflammatory activation [85]. Two elements, c-Fos and c-Jun, make up the AP-1 transcription factor with binding sites in the promoter regions of several proinflammatory genes. Early expression of AP-1 in murine peritoneal macrophages may help induce iNOS but overexpression of c-Fos is suppressive towards iNOS and TNF-α gene transcription, by mechanisms that do not involve AP-1 directly [86]. In DCs, elevated c-Fos suppresses IL-12p70 and, consequently, moves T cells towards the antiinflammatory end of the T cell repertoire [87, 88]. The plasticity of murine bone marrow monocytoid progenitors is well-displayed by expanding them with the growth factor Flt3-L and then subjecting them to various cytokines at various times during incubation. In this way, immature monocytoid cells can give rise to macrophages, osteoclasts, DCs or microglia. Macrophages can be derived at all stages simply by adding M-CSF. Cells in late cultures display macrophage markers and, from these, microglial cells will differentiate if glial-cell conditioned medium is added [89].

Other aspects of macrophage heterogeneity extensively explored in vitro are phenotypic reversals and 'transdifferentiations'. In the light of such studies, the concepts of 'maturation' and 'differentiation', which imply set sequencing and fixed terminal phenotypes, fall into some disarray. For example, upregulation of proinflammatory cytokines (IL-12, IL-15 and TNF-α) in DCs, which is promoted by TNF-α, is transient, reversing to a predominantly antiinflammatory phenotype (IL-10, TGF-β) on withdrawal of TNF-α, but is reinducible again to a proinflammatory phenotype on re-exposure [90]. IL-4/GM-CSF-treated blood monocytes tend to revert from DCs to adherent macrophage-like cells unless terminally differentiated by LPS [78]. The antiinflammatory phenotype which may be adopted by macrophages after phagocytosing apoptotic cells can be reversed to a proinflammatory one, either by exposure to anti-TGF-β1 antibody or if PGE_2 is neutralized with indomethacin [91]. Stable GM-CSF-responsive progenitor cell lines, derived from fetal liver or adult bone marrow of some mouse strains, and then transformed with c-Myb into confer immortality, can differentiate either into macrophage-like cells in response to TNF-α and IL-4 or into DCs when IFN-γ is substituted for IL-4. Myb-transformed cells differentiate into macrophages under the influence of M-CSF, as would be expected [54]. Microglia differentiate readily into DCs in vitro and have been described, on the basis of immunochemistry, as uncommitted myeloid progenitors of both immature DCs and macrophages [92]. Resting microglia transdifferentiate to immature DCs on exposure to GM-CSF. In brain inflammation, as during EAE or *Toxoplasma encephalitis*, brain DCs (CD11c$^+$) secrete IL12p70 while microglia (CD11c$^-$) secrete TNF-α GM-CSF and NO. Both cell types can act as APCs, producing IFN-γ and IL-2 on contact with primed naive T cells [93, 94]. Microglial/DC transdifferentiation is of more than academic interest because there is definite evidence for routes by which antigen can be transferred from brain parenchyma to cervical lymph nodes [95]. Macrophage into DC conversion may be a much more general phenomenon than currently appreciated, with far-reaching consequences in vivo. Murine Peyer's patch macrophages (F4/80$^+$ CD34$^+$ MHC II$^+$) can produce DCs when cultured in GM-CSF and IL-4 [96]. The murine cell line RAW264.7 only requires LPS stimulation to differentiate into DCs, a transdifferentiation that may depend on the second-messenger phosphatidic acid [97]. Thus, not only can LPS convert immature DCs into mature DCs but also macrophages into DCs. DCs can be generated from alveolar macrophages (AMs) and blood monocytes [98]. Peritoneal macrophages separated by adherence methods can take on DC characteristics on culture with GM-CSF and IL-4. Remarkably, monocytes or immature DCs can assume the phenotypes of either macrophages or endothelial cells in vitro when cultured in a suitable endothelial (angiogenic) growth medium [99]. Loading of murine aortic smooth muscle cells with cholesterol transform a proportion of them into 'foamy macrophage'-like cells, although these do not appear to be activated to a proinflammatory phenotype [100].

Preadiocytes, which populate adipose tissue throughout life, can transdifferentiate to macrophages. This is not surprising since preadipocytes share more features in common with macrophages than with mature adipocytes. Transplanting murine preadipocytes into the peritoneal cavity efficiently produces phagocytic cells with the gene expression profile of macrophages (F4/80$^+$ CD11b$^+$ CD45$^+$), resembling peritoneal macrophages already present. Cell-to-cell contact between preadipocytes and peritoneal macrophages is partially responsible for the phenotypic changes and the peritoneal microenvironment, by implication, contributes the rest [101]. In obesity, bone marrow-derived macrophages also accumulate in adipose tissue and are largely responsible for a state of chronic low-grade inflammation (including elevated TNF-α and iNOS), which may mediate many of the associated metabolic consequences. Adipokines, such as leptin, activate endothelial cells to recruit monocytes, while fat tissue growth is associated with monocyte recruitment and an accumulation of resident macrophages, suggesting that weight gain may be another consequence of inflammation [102, 103].

9.2.3
Function-related 'Markers' in Macrophages and DCs

Even in the face of mounting evidence of microenvironmentally-controlled, phenotypic plasticity and interchangeability, the different attributes of macrophages and DCs might still be expected to correlate with the expression of unique markers by which the two cell types might be distinguished. The essential attributes of DCs, as originally defined, are inducible mobility, ability to transport antigen from peripheral tissues to lymphoid tissues and to stimulate an immune response in naive T cells. The defining functional characteristics of macrophages are rather the opposite in that they are relatively immobile, adherent cells (a property used experimentally to segregate them) with a tendency to stay in peripheral tissues, where they display heterogeneity (both morphological and functional) until their final apoptotic demise and removal. During inflammatory responses macrophages are able to present antigen effectively to memory T cells. The terminology that has developed to describe DCs – 'immature', 'semi-mature' and 'mature,' – may simply reflect the dramatic, easily recognizable, phenotypic changes that accompany the transition from a resting state to a fully mobile, functional APC, as embodied in the so-called LC paradigm [104]. For DCs, 'maturity' equates most closely to the ability to stimulate naive T cells. The terminology of macrophages has developed differently, emphasizing the various induced activation states of, presumably, 'mature' but relatively non-mobile, tissue-located cells. Specific markers are perhaps most likely to correlate with the different behavior of macrophages and DCs. As DCs mature they should be directed towards lymphoid tissue. One approach has been to compare chemokine receptor expression. Monocyte chemokine receptors are responsible for recruitment to areas of inflammation but thereafter their expression is readily modified by the local microenvironment, including cytokines and LPS [105]. Whilst both macrophages and DCs express similar (pro-)inflammatory chemokine receptors in the 'immature' state, only DCs require an 'exit strategy' from peripheral tissues. This is thought to be provided in part by CCR7, expressed on mature activated DCs, which migrate towards CCL19 and CCL21, constitutively expressed in lymphatics/lymphoid tissue [106]. Proinflammatory chemokine receptors (upregulated by IL-10 and in some cases by steroids) [107] prime monocytes for recruitment to proinflammatory microenvironments. On in vitro stimulation by the outer membrane protein

A (a TLR2 ligand) of Klebsiella pneumonia, or with LPS, or IFN-γ, both immature DCs and macrophages acquire CCR7 and downregulate CCR1, CCR2 and CCR5 (inflammatory chemokine receptors). At the same time they upregulate CXCL8 (IL-8), CCL2 (MCP-1), CCL3 (MIP-1α) and CCL5 (RANTES), which can recruit further cells. On expression of CCR7, macrophages do not lose their phenotypic markers, but display proinflammatory cytokines, and neither do CCR7-expressing macrophages migrate via lymphatics directly to CCL21 (at least not in vivo). Therefore, CCR7 expression alone is insufficient to induce macrophage lymphatic migration [90]. Macrophages undoubtedly do reach lymphoid tissue, in vivo, but it is suspected that this could be via blood in response to CCL 21 expression near or on high endothelial venules [108]. Even mature LCs which express CCR7 can fail to migrate to lymphoid tissue if the 'chemokine switch' [105] operates incompletely and they continue to express aberrant CCR6 (which is attracted to the proinflammatory chemokine CCL20) as in the disease histiocytosis [109]. A further requirement for proinflammatory DC migration appears to be the presence of microenvironmental PGE_2 during DC maturation, because PGE_2 is essential for coupling CCR7 to intracellular signal transduction mechanisms, which employ cAMP and PKA [110]. Since PGE_2 is a product of proinflammatory macrophages, the resultant DC migration is a good example of the two cell types working synergistically [111, 112]. The downregulation of proinflammatory cytokine secretion by PGE_2, via elevation of intracellular cAMP and PKA, is described for both macrophages and DCs, constituting a mechanism for limiting the time course of innate proinflammatory immune cell activation both in peripheral tissues (macrophages and DCs) and lymphoid tissue (primarily DCs). In response to LPS, DCs generate COX-2 at 4–6 h and PGE_2 (but not the proinflammatory PGD_2) at 6–18 h, allowing DCs enough time to reach lymphoid tissue before proinflammatory, Th1-directing, cytokines (notably IL-12) are curtailed [113, 114]. Consequently, it can be surmised that in vitro generation of DCs using proinflammatory cytokine mixtures incorporating PGE_2 over several days inevitably runs the risk of producing Th2-directing DCs of little relevance to tumor control [115]. In sharp contrast to their response to proinflammatory microenvironments, 'resting' macrophages, in the steady state, are biased towards secreting antiinflammatory cytokines, one of which, TGF-β, is essential for DC migration under such conditions. TGF-β promotes survival of, and expands, a subset of human $CD16^+$ ($Fc\gamma RIII^+$) blood monocytes having the greatest propensity to become migratory DCs, as tested in vitro by their ability to pass from the ablumenal to luminal side of an endothelial monolayer ('reverse transmigration'). In the absence of exposure to TGF-β, or alternatively simply to particulates that can be phagocytosed, monocytes that have traversed the monolayer do not retraverse, but develop instead into tissue macrophages [116]. Notably, however, the same $CD16^+$ monocyte subset and $CD16^+$ DCs (belonging to a subset identified by a monoclonal antibody, M-DC8) can also be activated by LPS to become potent TNF-α producers and such DCs have been found in inflamed mucosa-associated lymphoid tissue in Crohn's disease [117]. TNF-α may paradoxically be a survival factor for DCs, rescuing them from the deleterious effects of reactive oxygen intermediates [118]. Melanin granules, taken up by both macrophages and DCs, excite no proinflammatory response. Melanin in DCs can easily be tracked to regional nodes when injected into the skin. Using this approach in transgenic mice, it is found that TGF-β signaling is essential for DC migration under 'resting' steady state conditions [119]. Since loaded macrophages can also be found in lymph nodes, a transfer of melanin by phagocytosis of short-lived DCs by macrophages located in lymphoid areas is another possibility. Uptake of apoptotic cells exerts an antiinflammatory effect on macrophages and DCs alike, but on exposure to apoptotic cells immature DCs upregulate CCR7 without maturation, which is of possible relevance to steady-state transfer of apoptotic material to lymph nodes [120]. Whilst informing as to the circumstances of DC migration, these observations do not explain why LCs can traffic antigen from skin to regional nodes independently of CCR7/CCL21 signaling in the steady state [121] and, conversely, why macrophages, which can also express CCR7, do not migrate in the proinflammatory state. Additional mechanisms to guide DCs to lymphoid tissue must exist. Amongst candidate molecules, attention has recently focused on CD38, which is present on precursor monocytes, then becomes downregulated on immature DCs only to be upregulated again on mature DCs. Upregulation of CD38 can be achieved by LPS or IFN-γ and is dependent on NF-κB activation. Functionally, CD38 increases cyclic ADP ribose and intracellular calcium and stimulates IL-12 secretion and T cell proliferation, in keeping with mature DC function [122]. More remotely, CD38 regulates humoral immunity, but this is now thought to be an indirect effect of CD38 promoting both recruitment of monocytes to inflamed tissues and the migration of mature DCs to draining lymph nodes. It may be that that necrotic cells at sites of inflammation release the CD38 substrate NAD^+, which is then available to drive CD38-dependent pathways that activate chemokine responsiveness, such as the attraction between CCR7 and CCL19/21 or between CCR2 and CCL2 [123]. A signaling pathway has been identified in DCs for upregulation of CCR7. This depends on surface expression of TREM-2/DAP12. The pathway triggered is independent of NF-κB and p38 MAPK and it is thought that it may amplify DC responses to pathogens [124]. DAP12, which has several associated molecules (MLD-1, TREM-1 and TREM-2), is expressed on monocytes/macrophages and DCs and has a high constitutive expression in lung. Following BCG infection of C57BL/6 mice, MDL-1 and TREM-1 are both induced, but only in the absence of IFN-γ (which downregulates them) and TNF-α specifically upregulates MDL-1 expression. Because the kinetics of expression of the DAP12-associated molecules correlate with that of TNF-α and IFN-γ, they may be of importance in generating proinflammatory responses [125].

The balance between innate and adaptive immunity, reflected in the balance between activated macrophages and activated ('mature') DCs, will necessarily have a far-reaching effect on the initiation and progression of immune responses. Apart from the inevitable delay in instructing the adaptive response, several other constraints can tend to anchor, and modulate, the immune response within the confines of the innate immune system. Using *Salmonella typhimurium* it has been shown that whole bacteria (viable or non-viable) and, to a lesser extent LPS alone, can block the development of DCs from influxing proinflammatory monocytes.

The LPS-mediated part of this effect is lost in TLR4 knock-out mice. The failure of DCs to migrate to local nodes results in poor antigen presentation to T cells. Monocytes initially exposed to the bacterium cannot be induced to differentiate into DCs even when exposed to GM-CSF [126]. On intracellular infection with *Mycobacterium tuberculosis* (MTB), monocyte differentiation to DCs under the influence of IFN-α is diverted in favor of an 'immunoprivileged' macrophage-like cell, thus constituting a potential mechanism for reduction of an adaptive response against the infection. The macrophage-like phenotype retains CD14, fails to acquire CD1, partially expresses B7.2 but does not upregulate B7.1 or MHC I or II. The cytokines TNF-α and IL-10 are induced, but not IL-12 [127]. In another recent murine study, MTB antigen did not arrive in mediastinal nodes, in $CD11c^+$, $DEC205^+$ DCs, until two weeks after intratracheal inoculation [128]. Further aspects of the relationship between macrophages and DCs in tuberculosis will be discussed in 9.2.4. Microbial infection is often accompanied by increased apoptosis. The integration of signals from combined exposure to apoptotic cells and TLR ligands (such as LPS) generates a different macrophage phenotype from that of either stimulus individually. The combination enhances early TNF-α and proinflammatory chemokines, but reduces late TNF-α, IL-10 and IL-12, while increasing late TGF-β. In this way, the output of proinflammatory cytokines from TLR stimulation could be profoundly modified by the build up of apoptotic cells associated with infection, limiting the extent and duration of an immune response, with a knock-on effect on DCs. IFN-γ, however, can override the downregulatory effects of apoptotic cells [129]. Uptake of apoptotic cells by macrophages impedes their ability to present exogenous antigen to autologous T cells and is associated with occupation of MHC II molecules by apoptotic DNA, downregulation of co-stimulatory molecules and increased secretion of TGF-β. Such mechanisms would tend to prevent a proinflammatory adaptive immune response [130]. Infection is commonly accompanied also by necrosis, and TLR ligands could even promote this. Macrophage, and neutrophil, necrosis, which may occur early, could combine with bacteria to activate DCs and load them with potentially antigenic material [131]. Increases in bioactive mediators such as adenosine and PGE_2 during proinflammatory innate immune responses might influence whether incoming monocytes adopt the phenotype of macrophages or DCs. Both adenosine and PGE_2 elevate intracellular cAMP and give rise to cells intermediate between macrophages and DCs, with good antigen presenting abilities, but reduced expression of CCR7, thereby blocking migration of these cells to lymph nodes, and weighting the immune response towards innate rather than adaptive immunity [132]. Doubt has recently been cast on the exact role of IL-12 production by DCs, assumed in the past to direct the adaptive response towards Th1. In fact, in a proinflammatory setting, IL-12 from DCs may have most impact on the innate response (through stimulating IFN-γ release from NK and other cells) because IL-12 production is superior in immature DCs and is rapidly reduced on DC maturation, regardless of the specific maturation stimulus. Even mature DCs with upregulated CD40 fail to liberate proinflammatory cytokines in vitro when stimulated with CD40L [133]. In a mildly antiinflammatory microenvironment, when late apoptotic cells are co-cultured with both macrophages and immature DCs (rather than with macrophages alone), the effect of the DCs is to reduce secretion of IL-6 and MIP-2 (murine equivalent of the chemokine IL-8), while at the same time inhibiting phagocytosis by DCs and enhancing phagocytosis by macrophages. These results introduce the notion that extrinsic factors, notably contact with immature DCs, may significantly hold in check the production of some proinflammatory molecules by macrophages engulfing apoptotic cells [134]. IL-6 is amongst cytokines produced early in the proinflammatory response of macrophages to TLR stimulation, but increasingly the evidence points to the need to reappraise the function of this enigmatic cytokine. It has only recently been reported that when monocyte-derived DCs are 'matured' by LPS, or other proinflammatory stimulus, in the presence of GM-CSF and IL-4, the inclusion of IL-6 skews differentiation towards phenotypically mature but functionally impaired DCs. Specifically, IL-6 inhibits NF-κB activation, reduces CCR7 transcription (partly via autocrine IL-10) and decreases production of TNF-α and IP-10 [135]. At the same time, during an innate response, IL-6 can oppose the suppressor activity of regulatory T cells. The downregulatory effects of IL-6 on DCs must tend to keep an innate response from propagating to the adaptive immune system. Chronic inflammation in skin alters the phenotype of LCs, which upregulate CCR7 and migrate to lymphoid tissue, but fail to express co-stimulatory molecules or DC-LAMP and behave as immature LCs with poor T cell stimulating abilities. Finally, there is considerable debate about DC function and distribution in the lung, examined in a later section (9.2.5). In summary, anchoring immune responses in the innate immune system can result either from lack of transmission of information to the adaptive immune system and/or from blunting of the response of DCs to microbial infection, but currently little appears to be known about what controls the balance between these two cell types and between innate and adaptive immunity.

9.2.4
Macrophage Phenotypic Diversity in Response to Microbial Challenge

An outstanding recent achievement in immunology has been the discovery and characterization of the mammalian TLRs (currently totaling 11), along with their ligands and signaling pathways [136, 137]. Amongst membrane receptors of macrophages and DCs, the TLRs are major transducing receptors but, in addition, there are other MMRRs, which often contribute to phagocytosis, as well as transduction. Recently, putative cytosolic MMRRs have also been described. Despite lacking the remarkable antigen specificity and versatility of T and B cell adaptive responses (achieved by receptor gene rearrangements and clonal expansion), the innate immune system still has an impressive range of germline encoded, fixed-specificity, MMRRs that accurately perform the primary discrimination between self and non-self molecules, across the whole spectrum of potential pathogens. The innate system for recognizing highly conserved microbial molecules has been honed for reliability by evolution, but occasional confusion of self/non-self ligand recognition could predispose to autoimmunity, and in this respect dsDNA signaling through TLR9 has attracted particular attention [138]. All membrane-bound receptors, including TLRs, probably follow the same principles of forming dynamic complexes in lipid rafts (microdomains), as described for the immunological synapse. For stimulation with LPS, the complex may include some or all of the following: CD14 (which first binds LPS), CD11b/CD18, HSP70, HSP90 and CXCR4, as well as soluble adaptor proteins, such as MD-2, and components of the intracellular signaling pathway, such as the adaptor protein MyD88 and the signaling kinase IRAK, which is recruited to the complex on ligand engagement (Figures 9.1 and 9.2) [138–140]. Of the non-TLR MMRRs, the C-type lectin family is growing in size and significance, although knowledge of the corresponding in vivo ligands is lagging behind. In general, C-type lectins bind conserved glycan ligands through carbohydrate recognition domains, but 'non-classical' C-type lectins can also bind non-carbohydrate ligands [141]. The group includes the mannose receptor (MR), Dectin-1 and DC-SIGN. MR is expressed on macrophages and DCs and is involved in the uptake of pathogens, as diverse as HIV, *Candida*, *Leishmania* and *Trypanosoma*. As a result of recognizing surface polysaccharides, MRs may have an additional role in enhancing antigen presentation since, in DCs, internalization of peptide ligands, attached to MR, massively increases the presentation of MHC II-restricted peptides to antigen-specific T cells [142]. Similarly, the mycobacterial non-peptide glycolipid LAM can be delivered by MRs efficiently into the endosomal compartment for presentation on CD1b to T cells. Therefore, MRs may be an important mechanism for presenting antigen to T cells in the adaptive amplification or suppression of macrophage responses [143]. In murine macrophages, MR is upregulated by the antiinflammatory cytokines, IL-13 and IL-4, and by PGE_2. Upregulation of MR by IL-13 is through a pathway involving PPARγ and phospholipase A_2. Type I interferons (IFNα/β) upregulate MR, but they are downregulated by IFN-γ [144]. Cross linking by bacterial Man-LAM of MR on immature monocyte-derived DCs triggers an antiinflammatory response with increased IL-10, IL-1Ra, IL-1RII, CCL17 and CCL22 and reduced IL-12 [145]. MBL is a soluble C-type lectin belonging to the collectins and is a component of the 'humoral' innate immune system, in that it activates complement and opsonizes target cells and microbes. MBL may opsonize apoptotic cells for phagocytosis and recognizes various bacteria, viruses, fungi and protozoa, and some forms of LPS [146] with relatively high specificity. The outcome is not always favorable since MBL, or other collectins, may introduce viable pathogens into macrophages. Genetically, low MBL appears protective against leishmaniasis and tuberculosis [147], but, in this context, other collectins, surfactant A and D, rather than MBL, enhance MR-mediated uptake of *Mycobacterium avium* by macrophages. Dectin 1 is the main macrophage receptor for recognition of fungal β-glucans, a role previously ascribed to the complement receptor CR3 (Cd11/CD18b). Dectin 1 interacts with both non-opsonized and opsonized yeast particles, and is expressed at high levels in AMs [148]. Most recently, the C-type lectin receptor, DC-SIGN, has gained prominence as a major MMRR on DCs, and, despite its name, is expressed on at least some macrophage subsets (AMs), and is of relevance to HIV and mycobacterial pathogenesis, as discussed below. Compared to the TLRs much less is known about the specificity, inter-relationships, and internal signaling aspects of C-type lectins. Dectin-1, for example, is associated with a proinflammatory (TNF-α-producing) response to zymosan, is rapidly expressed on IL-4/IL-13 stimulated macrophages and more protractedly on GM-CSF or TGF-β-stimulated macrophages. Dectin-1 is downregulated by IL-10, LPS and steroids, but not IFN-γ [149, 150]. Another important group of non-TLR MMRRs is the scavenger receptors. Scavenger receptor A binds microbes and apoptotic cells and can even bind the lipid A moiety of LPS without eliciting proinflammatory responses. In general, many 'non-TLRs' fail to excite proinflammatory responses on ligand binding, are more prominently displayed on resting macrophages and are downregulated on proinflammatory activation. The outcome of innate recognition of diverse microbial-associated molecules is governed by variations in the conformation of the stimulating molecule(s) ('molecular patterns'), by effects of combinatorial MMRR stimulation, by the prevailing status of receptor expression and receptor-linked intracellular signaling pathways, by effects over time of repeated/continuous MMRR stimulation (characteristically associated with tolerance), and by 'host' genetic variability, which can alter the final macrophage or DC phenotype through its effects on any component of the system. Although relevant to all MMRRs, these general properties are discussed here with particular reference to TLR4 and TLR2 and their ligands. The classical transducer for LPS is TLR4. The resultant proinflammatory response, endotoxicity, is a reliable pointer to TLR4 activation. Enterobacteria, such as *E. coli* and *Yersinia*, stimulate TLR4 to a much greater extent than non-enterobacteria such as *Borrelia*, as has been assessed by the output of proinflammatory chemokines. Endotoxic potential correlates with variations in the lipid A moieties of microbial LPS. The specificity of TLR4 for LPS has been investigated by transfecting TLR4 alone into a human monocytic cell line (THP-1) and measuring NF-κB-driven transcriptional activity. TLR4 is functional in the transfected cells but the effects of LPS are further augmented by cotransfection of MD2 and CD14, as might be expected. Significantly, neither TLR1, TLR2 nor TL6, nor combinations of these, in transfected cells are able to respond to LPS [151]. Stimulation of the human CD14/TLR4/MD-2 receptor complex by LPS is reported to differ from that of the mouse in that the lipid A portion of LPS must be bound to polysaccharide to be effective [152]. Expression of TLR2 is higher on macrophages than all other cells [153]. Amongst TLR2 agonists are PG, lipotechoeic acid and lipoproteins, initiating responses to Gram-positive bacteria, spirochaetes and mycobacteria, but the purest TLR2 stimulant to emerge is the synthetic Pam3Cys, which induces rapid expression of proinflammatory genes, such as TNF-α, in a human monocytic cell line [154]. The LPS of *Porphyromonas gingivalis* responsible for periodontitis appears to act primarily through TLR2. In vivo combinatorial stimulation of MMRRs clustered together in the membrane is the rule, extending the scope and versatility of TLRs. A well-documented example is the combination of TLR2, TLR6, CD14 and Dectin-1, all of which collaborate in the recognition of zymosan (a β-glucan-containing yeast particle) and elicit proinflammatory cytokines (TNF-α, IL-12) and phagocytosis [139].

Synergism amongst TLRs is perhaps almost inevitable because of their proximity on the cell surface and because responses to most are funneled through MyD88, an adapter shared by most TLRs, including TLR2, TLR4 and TLR6 (but not TLR3), which then signals via IRAK1/4 to TRAF6 (Figures 9.1 and 9.2). Another Myd88 adapter-like protein (Mal/TIRAP), associated with TLR2, TLR4 and TLR6, probably functions proximally to MyD88 and both are required for TLR2 and TLR4 signaling. In addition, Mal/TIRAP interacts with downstream TRAF6 to activate NF-κB through MAPKs and the p65 NF-κB subunit [155]. Mal/TIRAP is analogous to TRIF, which is a component of TLR3 signal transduction via TRAF 6. In MyD88 or TIRAP deficient mice, bacterial stimulation fails to produce proinflammatory cytokines (TNF-α, IL-1β, IL12p40 and IL-6) but slower MyD88-independent pathways remain intact, through which NF-κB can be activated and through which LPS/TLR4 signaling and/or TLR3 signaling (possibly via TRIF) trigger IFN-β and thereby type I IFN-inducible genes [136, 137]. Type 1 interferons can also be induced through MyD88-dependent pathways linked to TLR7 and TLR9.

An influential variable modulating TLR signaling is ligand concentration. Low ligand concentration and/or short exposure tend to upregulate TLR pathways, whereas high ligand concentration and/or prolonged exposure tend to downregulate them, priming for initiation and termination of proinflammatory responses, respectively. LPB binding of LPS is important in that it controls the concentration of ligand available at receptors. At low concentrations of whole bacteria, LPS, lipoproteins, double-stranded RNA, CpG DNA and Pam3Cys, all readily upregulate macrophage surface TLR2 expression and, for LPS and dsRNA, this is independent of MyD88. Thus, TLR2 expression constitutes a remarkably sensitive marker for a whole variety of microbial products and may boost the initial reaction to infection [153]. Additionally, there may be a minimum threshold for direct MyD88-linked TLR4 signaling because low concentrations of LPS appear to signal independently of MyD88 and bias towards Th2, rather than Th1 obtained at higher concentrations [156]. Both Myd88-dependent and -independent pathways can increase cell surface expression of the co-stimulatory molecules required for adaptive responses. Factors underlying tolerance induction on prolonged/repeated ligand exposure are considered in 9.2.6.

To summarize, microbial receptors are 'in the front line' when it comes to generating macrophage heterogeneity. Evidently, microbial recognition activates both TLR and non-TLR signaling. Mycobacterial LAM, for example, signals through LBP/CD14/TLR2 to produce TNF-α and IL-8, whereas mycobacterial ManLAM signals through MR or DC-SIGN to produces IL-10, and inhibits LPS-induced IL-12 and TNF-α [157]. With some exceptions, signaling initiated through TLRs tends to be proinflammatory while signaling initiated through MR or DC-SIGN is antiinflammatory. Tissue macrophages are uniquely placed to act as innate transducers and integrators of these competing signals, as well as being potent effectors of the resultant intracellular processing – delivering either a net pro- or anti-inflammatory response.

Recognition of microbial molecular patterns is but one part of the macrophage response to infection. Much information has now accumulated about how macrophages respond to microbial infections that target them specifically. Infection of macrophages (and DCs) by the Gram-positive facultative intracellular (cytosolic) bacterium *Listeria monocytogenes* is optimally cleared through gene expression linked to TLR2/MyD88-signaling. The innate response includes increases in TNF-α, IL-12, NO and co-stimulatory molecules and promotes a Th1 adaptive response. Additional immunity may be provided by heterodimerization of TLR2 with TLR1 or TLR6, recognition of *Listeria* flagellin by TLR5 and/or bacterial non-methylated CpG-containing DNA by TLR9. Thus, TLR2-deficient macrophages show reduced TNF, IL-12, NO and co-stimulatory CD40 and CD86 expression on *Listeria* infection, but these responses are virtually abolished if the deficit is widened to MyD88 [158]. A recent mouse DNA microarray study of *Listeria*-infected bone marrow-derived macrophages indicates that, when *Listeria* reaches the cytosol, a cluster of 'late response' genes is induced that is MyD88-independent. Significantly, these include IFN-responsive genes (IRGs), leading to IFN-β gene transcription, amongst many others. The late response appears to be initiated by an intracellular innate MMRR belonging to a family of receptors termed NOD-LRRs (nucleotide-binding oligomerization domain-leucine-rich repeats). IRGs are distinct from TLR-dependent late response genes that are induced, for example, by the lipoteichoic acid of *Staphylococcus aureus*, although these also induce type I IFN. *Listeria* mutants lacking LLO fail to escape the vacuolar compartment into the cytosol and induce an early gene cluster, consistent with TLR/NF-κB signaling, but not the late IRG cluster [159]. The discovery of NODs points to another facet of macrophage diversity, surveillance of the cytosol for microbial infection that might otherwise spread from cell to cell without further engagement of TLRs. Currently, the best described NODs are NOD1 and NOD2. The former appears to be an upstream regulator of NF-κB and can be triggered in epithelial cells by enterobacteria, such as *Shigella flexneri*. The latter is found in myeloid cells and can detect any PG. These putative intracellular innate MRRs have homology with plant disease-resistant proteins, some of which are cytosolic and induce cell death. There are further functional similarities between intracellular NODs and the extracellular free PG-recognition proteins of *Drosophila* [160]. A human mutation in which NOD2 is defective is found in some cases of Crohn's disease and is associated with a marked reduction in the antiinflammatory cytokine IL-10, as well as significant reductions in TGF-β, TNF-α, IL 6 and IL-8 on stimulation of peripheral blood mononuclear cells with PG [161]. Stimulation of NOD2-defective cells with *Bacteroides* or heat-killed *Salmonella* reduces IL-10 and TGF-β but not TNF-α output, through NOD2 recognition of a motif present in all PGs (MDP) and found in the cell walls of both bacteria. These results have been interpreted as showing that intact NOD2 is necessary for effective production of IL-10 and TGF-β and may be part of an antiinflammatory network counterbalancing the proinflammatory pathways induced by TLR activation. More fundamentally, NOD2 may be important for intracellular pathogen surveillance and initiation of immune responses [160]. The latest twist to this unfolding story is that the role of PG as a proteotypic TLR2 ligand has been challenged, in which case intracellular NODs could become the only known PG-recognition molecules [162].

There are many unanswered questions concerning NODs but, doubtless, clarification of the role of these fascinating macrophage proteins will follow soon. It should perhaps be pointed out that, even in the total absence of TLR signaling via TLR2 or 4, or absence of MyD88, antigenic information about *Listeria* can still be transmitted to the adaptive immune system and *Listeria*-specific protective CD4$^+$ and CD8$^+$ T cells generated [163].

Leishmania is a protozoan obligate intracellular parasite of macrophages. Its survival depends on the failure of infected macrophages to mount an effective proinflammatory response against internalized leishmanial amastigotes, coupled with inefficient antigen presentation of leishmanial antigens by infected macrophages and an impaired ability of infected macrophages to respond to external proinflammatory stimuli [164–166]. Indeed, in an extreme scenario, the parasite might evade all immune responses and replicate unchecked, spreading by macrophage lysis to other macrophages and eventually to all macrophage-containing organs, such as spleen, liver and bone marrow. Visceral leishmaniasis is the closest clinical equivalent to this absolute failure of immunity. If *Leishmania* is tracked through the stages of an immune response (innate and adaptive) it becomes clear that any deviation from a null response depends on changes in macrophage phenotype generated following the initiation of infection. A minimum condition for elimination of *Leishmania* from macrophages is the induction of iNOS. Therefore, *Leishmania* survival will correlate with programming by the parasite of a macrophage phenotype suppressive of iNOS whereas elimination will correlate with all other immune responses that singly or together upregulate iNOS to the required threshold. In a very resistant host this could be achieved by innate immunity alone, whereas in a very susceptible host it may not be achievable even with combined innate and adaptive immunity. Although, hypothetically, an innate proinflammatory response can result in elimination, in practice a Th1 (IFN-γ-producing) adaptive response is usually required because *Leishmania* potently subverts macrophage proinflammatory activity. The influence of genetic variability of *Leishmania* itself is most apparent at the level of the macrophage phenotype, whereas host genetics impacts on many facets of the immune response. The experimental evidence will now be reviewed with reference to this simple framework. Following uptake of the parasite into macrophages, promastigote-containing phagosomes fuse with lysosomes to form the parasitiophorus vacuole, and promastigote lipophosphoglycan (LPG) is shed, inhibiting both the respiratory burst and lysosomal enzyme hydrolysis. LPG, which coats the promastigote, and derived PG can resist lysis by complement, resist toxic radicals, impair macrophage signaling pathways and modulate cytokines and NO. Upregulation of iNOS is part of the early intrinsic macrophage response to *L. major* infection, occurring within 24 h in resistant C57BL/6 mice and driven by *Leishmania*-triggered IFN-α/β, rather than IFN-γ [167]. In human cutaneous leishmaniasis iNOS expression is highest in localized lesions containing few parasites and lowest in more diffuse lesions with a higher parasite burden [168]. Phagocyte NADPH oxidase may be an additional requirement to clear *L. major* from spleen, but in cutaneous leishmaniasis iNOS, alone, suffices [169]. In general, high levels of iNOS are associated with proinflammatory macrophages and NF-κB activation, whereas low levels of iNOS are associated with resting or antiinflammatory macrophages One of the first responses of macrophages in resistant C57BL/6 mice is LPG signaling through TLR2/MyD88, which induces NF-κB and TNF-α, indicating a proinflammatory response [170]. TLR4 stimulation also helps control Leishmania through the early induction of iNOS, whereas TLR4 deficiency increases intracellular arginase in Leishmania infection [171]. Not only does NO kill Leishmania and promote macrophage activation-induced apoptosis, but iNOS also diverts L-arginine away from the arginase pathway, which produces polyamines required for parasite growth and survival (Figure 9.2). Arginase inhibition, either by antiinflammatory cytokines or specific inhibitors, is a limiting factor on Leishmania growth [172, 173]. L. donovani induces elevated ceramide in host macrophages, which downregulates PKC and ERK intermediates, thereby reducing activation of key proinflammatory transcription factors such as AP-1 and NF-κB, involved in NO generation [174]. LPG prevents phosphorylation of proinflammatory signaling kinases, p38 MAPK, JNK, and ERK 1/2, and degradation of IκBα in naive macrophages [175], and both LPG and IL-10 downregulate PKC in activated macrophages. Another crucial and sustained macrophage response following phagocytosis of amastigotes is suppression of IL-12 production [176]. Parasite-related disruption of proinflammatory macrophage signaling is again involved and, in infection with L. mexicana, LPS fails to stimulate IL-12 and a protective Th1 response because of expression of leishmanial cysteine peptidase B, which cleaves JNK and ERK [177]. IL-12 suppression is generalized across both susceptible (BALB/c) and resistant (C57BL/6) mouse strains and occurs in macrophages activated by LPS, by CD40 cross-linkage or by cognate interaction with T cells and, interestingly, can be mimicked by the phagocytosis of non-macrophage-activating inert particles. In vitro, uptake of opsonized or unopsonized L. mexicana amastigotes into bone-marrow-derived macrophages, or into resident or inflammatory peritoneal macrophages, fails in all cases to trigger NO, or induce IL-12, regardless of whether the mouse strain is resistant or susceptible. Another mechanism of IL-12 suppression is engagement by Leishmania of CR3 or FcγR, which antagonizes IL-12 upregulation on cognate interaction with T cells [178]. Leishmanial PGs selectively inhibit synthesis of biologically active IL-12 in activated murine (BALB/c) macrophages (without reducing TNF-α or IL-6) [179]. Despite expression on infected macrophages, CD40 is progressively prevented from interacting with CD40L on T cells (to generate IFN-γ) because of interference by Leishmania with p38 MAPK phosphorylation, in the signaling pathway which links CD40/CD40L ligation to IL-12 gene induction [180]. Suppressed IL-12 favors parasite survival, as does upregulation of macrophage IL-10 by Leishmania. Higher resistance to Leishmania is displayed by IL-10 'knock-out' than wild type BALB/c. A reciprocal relationship between IL-10 and IL-12 is well known [181] and, furthermore, IL-10-producing macrophages are refractory to activation by IFN-γ. Infected BALB/c IL-10$^{-/-}$ mice produced more IFN-γ and NO and less IL-12 p40 and IL-12R [182]. Although IL-10 can be part of a Th2 response, its synthesis is directly and dramatically increased in macrophages by amastigote infection alone. Host IgG-opsonized amastigotes are ligands of macrophage FcγR and directly induce IL-10 expression [183]. Similarly adherent spleen cell TGF-β is increased to high levels early in *Leishmania* infection and may contribute to the immunosuppression of visceral leishmaniasis [184]. Previous studies have suggested that MHC I and II expression is impaired in *Leishmania*-infected macrophages but, at the single cell level, surface expression of MHC II-peptide complexes is found to be intact, but cannot support the sustained TCR signaling required for IFN-γ production [166]. Either way, reduced functional MHC and co-stimulatory molecules in *Leishmania*-infected macrophages protects the parasite. DNA microarrays have recently been used to analyze gene expression in BALB/c bone marrow macrophages infected with *L. chagasi* and reveal downregulation of many proinflammatory genes along with upregulation of some but not all of the genes associated with an antiinflammatory phenotype. Of note, under these particular conditions, the macrophage 'survival' phenotype for *L. chagasi* lies somewhere between LPS- and alternatively-activated macrophages and in particular the balance of IL-10, TNF and arginase are unaltered [185]. An important consideration in the immune response to *Leishmania* is the impact of secondary effects from the immediate macrophage microenvironment. If intrinsic innate macrophage defenses have failed, successful elimination of *Leishmania* can only be achieved through microenvironmental changes taking place in the immediate vicinity of infected macrophages (including cell–cell contact), sufficient to activate macrophages to the required proinflammatory phenotype. Neutrophils may provide a vehicle for amastigotes to enter macrophages. In addition, co-culture of *Leishmania*-infected BALB/c macrophages with dead exudate neutrophils increases parasite replication, concomitantly with enhanced production of PGE_2 and TGF-β. The latter induces ornithine decarboxylase (ODC in Figure 9.2) in macrophages, leading to increased polyamine synthesis and possibly enhanced growth and replication of *L. major*. Under the same conditions, the more resistant mouse strain C57BL/6 shows more effective clearance, mediated by a serine protease (neutrophil elastase) and macrophage TNF-α. Strikingly, early depletion of neutrophils renders BALB/c resistant to *L. major* and shifts the adaptive immune response towards Th1 [186]. NK cells are an integral component of the innate immune system, in close contact with infected macrophages. NO may stimulate IFN-γ release in the early IFN-α/β response, but the main stimulus for INF-γ release, IL-12, is missing, at least from infected macrophages themselves. Furthermore, a transient *Leishmania*-triggered upregulation of SOC3 strongly inhibits macrophage activation (Figure 9.2) [187]. DCs constitute another local influence and, unlike macrophages, can secrete IL-12 on uptake of *Leishmania* [176], possibly utilizing DC-SIGN as a receptor. DCs are the vital link to the adaptive response to leishmaniasis. However, the same (genetic) factors that cause failure of the isolated innate response militate against a successful outcome of an adaptive response. In the complex inter-relationship between tissue macrophages and DCs it is now known that there are two waves of antigen transfer by DCs to lymph nodes. The first does not carry any viable parasites (either in BALB/c or C57BL/6) and, although producing IL-12, these DCs trigger a Th2 adaptive response in adoptive transfer experiments. The second wave does carry viable *Leishmania* [188]. Nevertheless, the impaired innate responsiveness of BALB/c mice carries through as a Th2 bias in the adaptive response compared to Th1-orientated C57BL/6 mice. In fact, Th2 bias in BALB/c may depend partly on impaired expression of IL-12R [189] and partly on a greater suppression of APC IL-12 by PGE_2 than is found in Th1-biased mice. Increased PGE_2 production by BALB/c macrophages on LPS challenge depends on disproportionate upregulation of downstream enzymes in PGE_2 synthesis [190]. BALB/c spleen cells reportedly have a greater number of receptors for PGE_2, making them more sensitive to PGE_2-suppression of a proinflammatory response when stimulated by *Staphylococcus* [191]. Following infection of BALB/c mice with *L. donovani*, splenic macrophages can produce TNF-α but not IL-12 whereas DCs (with surrounding clusters of mononuclear cells) can produce an early transient burst of IL-12 (resolved by day 3) and stimulate an innate response from neighboring NK cells and the beginnings of a Th1 response from neighboring T cells, which produce IL-2 (a co-stimulator with IL-12 of NK cells) [192]. DC migration to lymphoid tissue is compromised by the downregulation of CCR7. Displaying only attenuated adaptive responses, BALB/c remain highly susceptible to *Leishmania* infection. In the more resistant C57BL/6 strain, innate uptake into macrophages produces TNF-α, but not IL-12. Release of amastigotes from macrophages or direct uptake of promastigotes by DCs provides sequential triggering of DCs (LCs in the skin) to produce IL-12 and MHC II molecules. IL-12 stimulates local release of IFN-γ from NK cells. DCs are able to promote a specific adaptive Th1 immune response in local nodes by an MHC II-restricted mechanism, which depends on antigen presentation and IL-12 co-stimulation [193]. Comparing different mouse strains, none exhibit production of IL-12 from macrophages on infection, but DCs from all strains can process *Leishmania* promastigotes (the mobile precursors of amastigotes), and amastigotes, to present antigen, together with co-stimulatory molecules (e.g. CD40), and IL-12 as required for a primary T cell response [176, 194]. An intrinsic, genetically-determined, partial defect in APC IL-12 signaling in BALB/c mice, rather than a deficit of (DC) IL-12, per se, seems to account for an IL-4–dominated (Th2) adaptive response to *Leishmania* [190]. Increased NO production, derivable from macrophage iNOS, is found in resistant strains on infection with *Leishmania* while the response in the BALB/c strain shows predominance of PGE_2 and TGF-β [195]. In keeping with this, ablation of functional IL-12R renders the C57BL/6 strain susceptible to large ulcerative cutaneous lesions akin to BALB/c, even in the presence of effective IL-23 (IL-12p40/IL-12p19) mediated-signaling [196]. Following subcutaneous infection of BALB/c mice with *L. major*, IL-4 increases rapidly in draining lymph nodes and downregulates IL-12R β2-chain on $CD4^+$ T cells, a reaction that fails to occur in more resistant C57BL/6 mice [197]. Th2 cells recognizing a *Leishmania*-derived epitope are found at high frequency in BALB/c. Spleen DCs from naive BALB/c express less CD40 and STAT4 than C57BL/6, interpreted as showing that BALB/c DCs have a less 'mature' phenotype than in Th1-orientated C57BL/6, and so are less resistant to intracellular infections such as *Listeria* and *Leishmania* [198]. Cutaneous leishmaniasis in BALB/c mice is lessened by daily intraperitoneal injections of IL-12 and is attenuated in PI3K knock-out mice, in which endogenous IL-12 and Th1-responsiveness is increased; conversely, C57BL/6 mice are rendered more susceptible by treatment with IL-12 antibodies [199–201]. All these observations underline the contribution of a vigorous Th1 adaptive response acting on infected macrophages for parasite killing. Successful elimination of *Leishmania*, in fact, seems to require a concerted immune response since even elevated levels of iNOS and IL-12 are not enough to clear the parasite in mice with a defective Fas/FasL apoptotic pathway, presumably because of reduced susceptibility of infected macrophages to apoptosis [202, 203]. Aging assists immunity to *Leishmania* and senescent BALB/c mice become increasingly more able to counter the infection by exhibiting a more effective release of IL-12 which, however, has to be primed by previous infection(s) as resistance fails to develop in senescent 'specific pathogen-free' BALB/c [204].

A consequence of inaccessibility of *Leishmania* in macrophages may be an overexuberant, but ineffectual, response to antigen, as found in the non-healing inflammatory destructive lesions of mucosal leishmaniasis. As in murine strains, host genetic factors must be implicated in the various clinical syndromes. Mucosal leishmaniasis is dominated by much increased IFN-γ and TNF-α in an exaggerated Th1 response driven by very few *Leishmania* in the lesions. There is reduced secretion of IL-10 from peripheral blood mononuclear cells in response to leishmanial antigens [205]. Finally, the enigmatic role of TNF-α in *Leishmania* must be considered. Although this pleotropic cytokine is part of the innate proinflammatory response of macrophages, including a role in upregulation of iNOS, it may have much more significant impact on the adaptive arm of the immune system in *Leishmania* infection. In fact, TNF-α is not well suppressed by leishmania in macrophages and so, presumably, confers an evolutionary advantage to the parasite. Even though TNF-α is responsible for activation-induced apoptosis of macrophages by an autocrine mechanism operating through TNFR (early at 3–6 h) and by induction of iNOS (late at 12–24 h) [206], these manifestations may be effectively countered by the prevailing antiinflammatory phenotype of *Leishmania*-infected macrophages [165]. A more fundamental protection from apoptosis in infected macrophages may be afforded by repression of cytochrome c release from mitochondria, an effect independent of cytokine balance or NF-κB and present in both susceptible and resistant strains alike [207]. Whilst macrophages are relatively protected from apoptosis, activated IFN-γ-secreting Th1 cells would be subject to TNF-α-induced apoptosis. Experiments with anti-TNF-α antibodies tend to support this view in that IFN-γ and IL-10 from lymph node cells are increased in infected animals. Mice lacking TNF-α receptor function (TNFRp55$^{-/-}$) eventually control the parasite through proinflammatory macrophage responses, even though parasite burden is initially increased, but at the expense of an expansion and persistence of active inflammatory lesions because of failure to eliminate lesional T cells by apoptosis [208]. Thus, in TNF-α deficiency it is evident that reduced iNOS, initially, might be offset by increased availability of IFN-γ from T cell longevity, later. Interestingly, intact Fas/FasL pathways and iNOS (and the availability of IFN-γ) may be more critical to *Leishmania* elimination from macrophages than TNF-α, whereas TNFα is more important than

Fas/FasL in curtailing T cell activity in inflammatory foci [202, 208]. Nevertheless, even the resistant strain C57BL/6 can succumb to leishmania in the absence of TNF-α under some experimental conditions [209], and TNF-α antibodies reduce iNOS in macrophages, though not in neutrophils, and contribute to increased lesion size in murine cutaneous leishmaniasis [210]. Notably, (in C57BL/6 mice) iNOS function must be maintained to prevent recrudescence of 'cured' murine *Leishmania* that has been found to persist in other cell types, such as fibroblasts or atypical DCs (CD11c+) in lymph nodes, with low expression of MHC II, co-stimulatory molecules, and CD86, combined with expression of markers normally associated with macrophages (CD11b and F4/80). BALB/c is much less effective in controlling amastigotes, which readily accumulated in CD11c$^+$ giant cells [211].

Mycobacteria are intracellular pathogens and, like *Leishmania*, are able to survive for long periods in (alveolar) macrophages, their major habitat. Elimination of the infection from macrophages depends on the integration of pro- and antiinflammatory signaling mechanisms, which mediate immunity and tolerance (immunosuppression), respectively. Innate modulation of this balance results from bacterial coat–immune cell receptor interactions and from intracellular bacterial–phagosomal interactions. Communication between innate and adaptive systems depends on interactions between mycobacteria and DCs, eventually feeding back as antigen-specific T cell responses directed onto infected macrophages. A redundancy of receptors recognizing mycobacterial capsular components ensures efficient uptake by macrophages and explains why selective receptor blockade does not alter survival and growth of mycobacteria. The multiple macrophage receptors available include MRs, CR3, scavenger receptors [212] and DC-SIGN. The latter is expressed on, and could affect entry into, AMs expressing an antiinflammatory phenotype and more specifically alternatively-activated (IL-4 or IL-13 stimulated) macrophages. DC-SIGN is up-regulated by IL-4 and IL-13 and down-regulated by TNF-α and LPS and has even been proposed as a good marker for this macrophage phenotype [213]. Since DC-SIGN is now thought to be the major receptor for uptake of mycobacteria into DCs it could well assist entry into both antiinflammatory macrophages and DCs. The main tolerogenic bacterial wall component interacting with DCs is ManLAM, a mycobacteria-specific mannosylated lipoglycan present on mycobacteria, ranging from MTB to *M. bovis* (BCG) and *M. avium*, and a key ligand of DC-SIGN, through which mycobacteria can gain entry into DCs [214]. ManLAM can cross-link MRs expressed by macrophages and immature DCs, and has been shown, in DCs, to be capable of inducing an immunosuppressive phenotype (IL-10, low IL-12, IL-1R antagonist and IL-1R type II) [145]. Thus, mycobacteria would appear capable of triggering an antiinflammatory phenotype on first encounter with macrophages. ManLAM does not induce proinflammatory cytokines (TNF-α, IL-8) whereas lipomannan (the core of LAM) does [157]. In addition to non-TLR receptors, TLR2 can elicit IL-6 and IL-10 in DCs in response to whole MTB, but has highlighted the complexity of bacterial coat–receptor interactions because ManLAM from virulent MTB fails to activate TLR2- or TLR4-transfected cells, in contrast to fast-growing mycobacteria, which can bring about activation through TLR2. Conversely, MTB can induce IL-12 in the absence of both TLR2 and TLR4 [215, 216]. Like *Leishmania*, mycobacteria suppress IL-12 much more in macrophages than DCs, but infected macrophages and DCs both produce TNF-α and IL-10. In transgenic mice expressing human IL-10 linked to the MHC II promoter, susceptibility to *M. avium* is much increased in association with suppression of mycobacterially-induced TNF, NO and IL-12p40. Apoptosis is also reduced, but Th1 effector responses and IFN-γ production remain normal. Clinically, these transgenic mice harbor increased mycobacteria, have more viable macrophages and exhibit increased morbidity (hepatic fibrosis) and mortality [217]. In a more physiological setting, uptake of another intracellular bacterium, *Chlamydia pneumoniae*, into human peripheral blood monocytes actively induces IL-10, and again inhibits apoptosis and IL-12. Virulent mycobacteria induce less macrophage apoptosis than attenuated strains such as *M. avium*. Production of IL-10 by mycobacterial-infected macrophages inhibits apoptosis by suppressing TNF-α. Although MTB infection itself causes minimal apoptosis, the cell wall and secreted 'virulence factor', 19 kDa glycolipoprotein, operating through TLR2, does increase apoptosis and reduce MTB survival. In the short-term this may facilitate the spread of residual viable TB to other cells but can also trigger host-protective T cell responses emanating from APC uptake of apoptotic cells [218]. Whatever the mode and consequences of mycobacterial access to macrophages, in vivo, during murine pulmonary MTB infection, murine AMs contain numerous bacteria and few DCs [219]. Once mycobacteria have gained entry, survival depends on quelling any initial innate response and consequentially delaying and subduing any specific adaptive response. Macrophage phagosomes infected with MTB remain non-acidified and unfused with lysosomes, conditions conducive to bacterial growth. Unlike macrophages, DCs are not permissive to the growth and replication of virulent MTB and the mycobacterial phagosome remains isolated from the biosynthetic pathways of DCs. Nevertheless, infected DCs can prepare and present mycobacterial antigens [220]. IFN-γ promotes phagosomal maturation to phagolysosomes. Further information about the phagosomal microenvironment in macrophages can be inferred from the secondary effects of immunological activation on the transcriptosome of MTB, used as an ingenious bioprobe. Of particular interest, IFN-γ-stimulated (NO-producing) macrophages evoke an MTB/phagosomal microenvironment that is nitrosative, oxidative, inhibitory to aerobic respiration and hostile to the pathogen, though the transcriptosome demonstrates the remarkable ability of MTB to adapt [221]. Binding of ManLam to DC-SIGN prevents mycobacteria and even LPS from inducing DC maturation, of obvious survival value to a pathogen ultimately eliminated by an adaptive, IFN-γ-mediated, response. The resultant phenotype in DCs includes IL-10, which limits upregulation of IL-12 and of co-stimulatory molecules and prevents Th1 responses [222]. The impairment of DC potency may explain why SCID mice, in which only an innate response is possible, show the same limited resistance to infection as wild-type mice up to 8 weeks post infection [223]. In relation to the adaptive response, TLR2 knock-out mice, paradoxically, exhibit an enhanced tissue-damaging proinflammatory response in chronic infection that is associated with increased T cells and elevation of IFN-γ and TNF-α, and must be driven by non-TLR2-related mechanisms. In TLR2-deficient C57BL/6 mice, bacterial burden increases and a chronic pneumonia develops alongside an ineffective granulomatous response [224]. In MTB infection of MyD88-deficient mice there is no significant effect on TNF-α, IL-12 and NF-κB, suggesting possible alternative TLR2-mediated pathways. TNF-α signaling via TNF-αR still operates in these circumstances to induce NF-κB [225]. Infection of MyD88-deficient mice with M. bovis produces low levels of proinflammatory cytokines and NO, but co-stimulatory capacity, MHC II expression and Th1 responses remain intact. The antiinflammatory bias of the innate response in macrophages prevents bacterial clearance from lungs, which show increased mycobacterial burden and confluent pneumonia, but T-cell priming and IFN-γ release is sufficient to clear infection from liver and spleen [226]. With intact TLR-signaling, antigen presentation is reduced in MTB infection, favoring bacterial survival. Reduced antigen presentation is similar to the situation in *Leishmania*, but in MTB infection is ascribed to decreased expression of MHC II molecules, dependent on the mycobacterial 19 kDa lipoprotein, signaling through TLR2. In the earlier stages of infection the same mycobacterial component can activate microbicidal innate immunity and it has been suggested that, as MHC II inhibition begins, existing mycobacterial peptide–MHC II complexes are stabilized, prolonging presentation as a 'freeze frame' [227]. Another MTB TLR2 ligand (LprG) can inhibit MHC II antigen-processing in primary human macrophages but requires long-term exposure [228]. MTB (or its 19 kDa protein) inhibits, in an MyD88 (TLR)-dependent manner, various IFN-γ-mediated macrophage responses, with suppression of genes involved in MHC II antigen-processing, antigen presentation and recruitment of T cells [229]. M. avium infection of BALB/c macrophages also inhibits the induction of IFN-γ-inducible genes by downregulating IFN-γR with resultant impairment of Jak-STAT signaling [230].

In vivo, the phenotype of MTB mycobacterial-infected macrophages probably tends to lie somewhere between a pure pro- and pure antiinflammatory macrophage phenotype. Although mycobacterial-infected adherent macrophages prepared in vitro by GM-CSF do not secrete IL-12 they can still produce another cytokine of the same family, IL-23 (IL12.p40/p19 heterodimer), rendering them capable of stimulating Th1/NK cells [231]. As in *Leishmania*, IL-10 secretion, inhibition of apoptosis and reduced MHC II, CD86 and CD40 expression in tuberculosis are antiinflammatory features of the macrophage phenotype, weighting the immune balance in favor of mycobacterium survival, but, on final apoptosis, infected macrophages, replete with intracellular bacteria, can be assimilated into DCs which express IL-12, MHC I/II and co-stimulatory molecules and provide a means of cross-presentation to CD8 and CD4 T cells for the generation of antigen-specific immunity. Direct processing of MTB by DCs adds to the Th1 response [232]. However, the delay in arrival of DCs into mediastinal nodes following MTB lung infection has been commented on in Section 9.2.3 [127]. Genetic factors influence the immunological balance. Unsurprisingly, Th2-orientated BALB/c mice are readily susceptible to infection with *M. avium* from combined failure of proinflammatory macrophage activation (T cell-independent) and adaptive (CD4$^+$ T cell-dependent) responses, analogous to mechanisms described in *Leishmania*. IL-12 protects against mycobacterium infection in pathogen-free female BALB/c mice [233]. Granulomata are a feature of chronic MTB infection and contain infected activated macrophages, driving a persistent but only partially effective Th1 adaptive response, represented by infiltrating T cells [213]. The role of TNF-α in relation to intracellular pathogens is only now being clarified. MTB-induced TNF-α in RAW 264.7 cells is largely dependent on TLR-signaling [234]. Paradoxically, although widely regarded as a proinflammatory cytokine, TNF-α may additionally be a negative regulator of an over-exuberant Th1/IFN-γ response, as previously mentioned. TNF-α restricts the number of proliferating T cells and their activation during mycobacterial infection. In TNF-α deficient C57BL/6 mice, the granulomatous lung lesions are subjected to an enhanced, uncontrolled Th1 adaptive response with overproduction of IFN-γ and IL-12 and lung tissue destruction through target cell (macrophage)/bystander apoptosis and necrosis. The pathology is reduced by depletion of $CD4^+$ and $CD8^+$ T cells, which reduce IFN-γ, or by reconstitution of a wild-type phenotype using TNF-α gene transfer [235]. The importance of TNF-α in causing T cell apoptosis and contracting the population of activated T cells at the end of an immune response is illustrated by the increased survival of $CD8^+$ cells in anti-TNF-α antibody treated mice. $CD8^+$ T cells, normally deleted in the liver, survive in the antibody treated mice and recirculate into the periphery [236]. In the clinical arena, anti-TNF-α used in the treatment of rheumatoid arthritis has resulted in the disturbing side-effect of systemic autoimmunity. Signaling through the p55 TNFR in the EAE model is proinflammatory but is countered by signaling through p75 TNFR which brings about regression of disease by promoting apoptosis of T cells [237].

Considering that macrophages are a major reservoir for HIV persistence and replication, there will, predictably, be interactions with other pathogens sharing the same cell. In terms of the global threat from HIV it is unfortunate that interactions with *Leishmania* and tuberculosis appear more synergistic than antagonistic [238, 239]. Synergism between HIV and 'opportunistic' infection undoubtedly occurs primarily in macrophages but the secondary ramifications are discernible throughout the immune system [240]. Chemokine receptor expression and proinflammatory/antiinflammatory balance are two aspects of the macrophage phenotype crucial to both HIV progression and its interaction with other pathogens. Progression of HIV involves complex interactions, between macrophages, DCs and T cells, that are not yet fully understood. Coverage here is limited to a few general observations. Viral tropism for macrophages is present at disease initiation and is maintained throughout HIV infection, although the optimum chemokine co-receptor for HIV uptake changes as HIV envelope proteins alter with disease progression. It can be surmised that, during the early latent phase of HIV infection, the virus is taken up by, and replicates in, macrophages, which display a predominantly antiinflammatory phenotype. Later progression to AIDS seems to coincide with a more proinflammatory macrophage phenotype, notably including TNF-α, which supports increased viral replication and correlates, paradoxically, with a decrease in the ratio of Th1/Th2. In early HIV, macrophage tropic (M tropic) HIV uses CCR5 as a co-receptor, which, together with the primary receptor CD4, binds the viral envelope protein, gp120. The antiinflammatory cytokine, M-CSF, is strongly induced in HIV-infected macrophages and upregulates CD4 and CCR5 by an autocrine/paracrine circuit. Not only does M-CSF enhance HIV replication but it also promotes macrophage growth and differentiation. Expression of CCL3, which is chemoattractant to T cells and macrophages, is also increased [241]. Apoptotic cell uptake by macrophages vigorously enhances HIV replication, at the same time inducing the antiinflammatory cytokine TGF-β and suppressing proinflammatory cytokines such a GM-CSF, IL-1β and TNF-α. Notably, apoptotic T cells may be taken up using the macrophage vitronectin receptor, which can, additionally, interact with the Tat protein of HIV to increase TGF-β and HIV replication [242]. Thus, antiinflammatory macrophages are, in various ways, conducive to HIV survival and replication. This may be a relatively silent phase in HIV disease clinically, although chronically elevated TGF-β may be an important factor in the excessive fibrosis of HIV renal nephropathy [243]. The role of IL-10 in HIV is more difficult to ascertain. It is an upregulator of CCR5 in human monocyte-derived macrophages and an even stronger upregulator of CCR5 in human microglia, so potentially increases HIV load by enhancing viral entry [244]. With regard to HIV replication, the effect of IL-10 is more contentious, with possible suppression at lower concentrations and enhancement at higher concentrations. As HIV progresses to AIDS, virus tropism switches to CCXR4, originally identified on T cells but in fact also readily expressed on macrophages. Thus, CCR5-tropic HIV (R5) gives way to CXCR4-tropic (X4) HIV, with an intermediate type (R5X4), which can infect macrophages through either receptor. These changes can all occur in the context of an existing antiinflammatory microenvironment. Indeed, the cytokines IL-10 and TGF-β may encourage receptor progression by increasing surface expression of CXCR4 on human monocyte-derived macrophages, whereas proinflammatory GM-CSF and the Th2 cytokines IL-4 and IL-13 reduce it [245]. Increased absolute surface expression of CXCR4 over CCR5 seems to be a major determinant of faster disease progression, since, at the intracellular level, a common gp120/gp41 mechanism in both X4 and R5 infected cells triggers identical caspase-dependent and -independent cytopathic effects on T cells [246].

There are several reasons for suggesting that macrophages assume a proinflammatory phenotype in symptomatic HIV (AIDS):

1. Proinflammatory macrophages are well documented in the AIDs dementia complex and there is evidence of increased NO production by brain macrophages [247]. Brain IFN-γ is elevated in seropositive HIV-1, independent of the presence of opportunistic infection or neuropsychiatric symptoms [248]. Microglia can behave very much as macrophages elsewhere, with TLR expression and the ability to present antigen when appropriately stimulated, including CNS antigen. HIV-1 Tat is a protein required for replication but is not functional in resting macrophages. However, it becomes elevated in inflammation and inhibits cAMP-mediated pathways by blocking adenyl cyclase, so potentiating the proinflammatory drive [249]. IFN-γ and

TLR stimulation can induce IDO, which is elevated in some strains and correlates with neuropathology [250].

2. The proinflammatory macrophage phenotype is associated with an accelerated replication rate of HIV, a feature of AIDS. Upregulation of HIV-1 by TNF-α involves NF-κB-responsive sites in the viral long terminal by which transcription is increased. Concurrently, elevation of TNF-α secretion in HIV-1 infected monocyte-derived macrophages increases anti-apoptotic [Bcl-2 and Bcl-x(L)] proteins and decreases pro apoptotic proteins (Bax and Bad) – partly dependent on an intact NF-κB pathway, and favoring viral persistence [251]. Increased COX-2 driven by NF-κB is part of the proinflammatory response but provides feed-forward autocrine inhibition of HIV-1 replication, through cAMP elevation and activation of PKA [246, 252].

3. There is evidence of a Th1 adaptive response, which implies a preceding proinflammatory innate immune response [253]. The elevated monocyte/macrophage activation marker, HLADR, correlates with CD4$^+$ T cell depletion [254]. IFN-γ downregulates both coreceptors and chemotaxis but cannot prevent entry of HIV into monocytes [255]. A study of IL-10 promoter alleles suggests that decreased IL-10 in the later stages of infection could contribute to progression and is consistent with increased Th1 responsiveness [256]. DCs provide the essential conduit between innate immunity (macrophage phenotypes) and adaptive immunity (T cell phenotypes). HIV in the periphery can 'hitch a ride' on DCs, attaching externally to the C-type lectin DC SIGN, for direct delivery (in trans) to CD4 and coreceptors on T cells [257]. As with macrophages, HIV can be internalized by DCs which may act as a reservoir for the virus in lymphoid tissue. Infected DCs and macrophages may respond differently to cytokines. Notably, IL10 inhibits X4 replication in macrophages but increases it in DCs, as well as enhancing DC CXCR4 expression, thus possibly contributing to disease progression [258]. New research suggests that, early in the disease, a genetically determined balance may be set, essentially between innate and adaptive immunity, which controls the rate of loss of T cells, many of which may not be directly infected, since there is a generalized T cell activation [259]. In this context, a reduced TH1/Th2 ratio with disease progression results not from a Th1 to Th2 polarity switch but from a decline in the availability of IFN-γ-producing activated Th1 cells, resulting from preferential expression of CXCR4 on Th1 cells and activation-induced cell death. Productively HIV-infected cells show a much increased output of IFN-γ, which is consistent with preferential Th1 viral replication and associated upregulation of CXCR4 and CCR5 on Th1 cells compared to Th2. There is a commitment of T cells to the Th1 lineage, despite the differential effects of T cell apoptosis on the size and composition of the T cell pool. The apoptosis rate of Th1 is slightly increased despite upregulation of the anti-apoptotic protein Bcl-2 [260]. Susceptibility to (HIV) activation-induced apoptosis is most apparent in TNF-α- and IFN-γ-producing T cells of both CD4$^+$

and CD8$^+$ subsets [261]. Although Th1 and Th2 T cells express similar levels of CD4 and bind recombinant gp120 equally, Th2 cells are much more resistant to apoptosis than Th1 [262].

4. Superadded infections clearly demonstrate that there is a correlation between increased HIV expression and microbially-induced proinflammatory macrophage phenotypes. Indeed, the proinflammatory effects of TLR stimulation during opportunistic infection may be a more critical determinant of progression to AIDS than changes in the virus itself [238, 263]. However, in this situation, macrophage phenotypes (whether pro- or antiinflammatory) will be influenced by a daunting number of variables, including the prevailing microenvironment [264], variability in molecular cell-wall structures within and between pathogens, relative concentrations of pathogen, order of presentation of pathogens, phenotype of macrophages(or DCs) at presentation, receptor types engaged (TLRs, lectins, others) and mode of presentation (acute or continuing, sequential, overlapping) [265]. A few examples illustrate the complexities. Stimulation of TLRs of macrophages infected with HIV may potentiate viral expression either directly (in cis), through induction of transcription factors (NF-κB), or indirectly (in trans) through induction of proinflammatory cytokines such as TNF-α or IL-1β acting in an autocrine or paracrine manner [266]. TLR4 stimulation by LPS is a prototypical proinflammatory stimulator of macrophages and upregulates cyclin T1 required for HIV Tat protein to mediate viral transcription. HIV itself also upregulates cyclin T1 in macrophages [267]. Against this must be offset the fact that LPS (TLR4 stimulation) causes sustained downregulation of CCR5 in monocyte-derived macrophages, with the overall effect of inhibiting HIV [268]. To complicate the situation further, HIV expression is, paradoxically, enhanced in LPS tolerance, in which proinflammatory cytokines and NF-κB are downregulated. One candidate mediator for this effect is c-Myb, which is overexpressed in LPS tolerance and also upregulates HIV expression [269]. AMs are the major reservoir of HIV in lung infection. TB induces high levels of CXCR4 in AMs as part of the innate immune response to MTB. Expression of CXCR4 correlates with increased HIV X4 uptake and replication. At the same time, TB-infected macrophages produce increased chemokines, including CCL4 (MIP-1β) and CCL5 (RANTES), which block CCR5 and so reduce R5 uptake, if only transiently [270, 271]. Any reduction in uptake appears to be more than countered because RANTES may have the adverse effect of inducing a proinflammatory signaling cascade, which enhances virus replication. In AMs, the balance between inhibition of excessive responses to harmless particles and essential responses to significant TLR agonists (TLR2, 4 or 9) is largely maintained by IL-10. It is, therefore, of relevance that IL-10 sustained expression of CCR1 and CCR5 is reversible by various TLR ligands that inhibit IL-10-induced CCR1/CCR5 expression [272]. MTB tends to induce an overall antiinflammatory macrophage phenotype, as discussed earlier, but phagocytosis of MTB into HIV-infected macrophages increases

output of TNF-α as well as IL-10 and enhances MTB growth, probably as a result of elevated TNF-α [273]. Monocyte-derived macrophages, preinfected with TB before HIV exposure, show acute suppression of HIV replication by unknown factors, since CCR5 binding chemokines and IL-10 are only marginally affected [265]. The much less virulent *M. avium* readily induces a proinflammatory response in macrophages, which includes TNF-α, IL-1β, IL-6, GM-CSF and upregulation of IDO [274]. In compromised HIV-infected macrophages, *M. avium* is an opportunistic pathogen that increases both HIV and CCR5 expression, along with the expected activation of NF-κB and TNF-α [263]. Paradoxically, the less virulent 'opportunistic' infections, such as *M. avium* and *Pneumocystis carinii*, may be more likely to elicit proinflammatory enhancement of HIV replication and viraemia than perhaps MTB or *Leishmania*, both of which tend to reinforce an antiinflammatory macrophage phenotype. In *Leishmania*, the observed LPG-induced activation of HIV-1 LTR transcription is through a cAMP/PKA dependent pathway, as well as being NF-κB-dependent [275]. Of considerable current interest among candidate receptors for HIV-1 is DC-SIGN, which binds gp120 and, in combination with CD4/CCR5, facilitates a more efficient entry of HIV (in cis) into CD4-expressing AMs, particularly when the CD4/CCR5 receptor combination is expressed at low levels [276]. Furthermore, as DC-SIGN is now recognized as the leading receptor for mycobacterial cell wall glycolipid, ManLAM (discussed earlier), interactive effects on intracellular signaling pathways must be anticipated, notably from increased IL-10 and reduced LPS-mediated DC maturation and Th1 responsiveness [238]. Measurement of plasma cytokines in HIV, TB and HIV/TB-coinfected patients indicates that immune activation is highest (IFN-γ, IL-18 and IL-12) in patients with advanced HIV infection, low CD4$^+$ T cell counts and TB co-infection [277].

In conclusion, HIV can survive and replicate in macrophages of either antiinflammatory or proinflammatory phenotype, though replication is more rapid in proinflammatory macrophages. HIV-infected monocytes/macrophages are a major vehicle of viral persistence and dissemination. Progression to AIDS is associated with changes in viral tropism and a switch to proinflammatory macrophage phenotypes. Concurrent infections by opportunistic pathogens, particularly those inducing proinflammatory changes in macrophages, may increase HIV expression.

9.2.5
Interactions between Tissue Microenvironments and Macrophages Generate Diversity

It has been traditional do classify macrophage heterogencity according to perceived differences in macrophage morphology and function in different mammalian organs – lung, gut, brain, skin, liver, spleen, kidney, bone. Alternatively, it might reasonably be assumed that steady-state phenotypes of resident macrophages in any tissue will correlate with the normal resting microenvironment and that changes in phenotype of both resting and monocyte-derived macrophages will occur predictably in response to specific perturbations of the system. The complexities and diversity of macrophage–tissue interactions are illustrated by looking at aspects of three very different microenvironments: liver, lung and tumor.

Kupffer cells (KCs), the resident macrophages of the liver, constitute the major portion of fixed macrophages in the body. Removal of senescent cells by KCs is a non-inflammatory process, mediated in part by a C type lectin receptor. KCs are strategically placed to clear LPS, most of which enters the body from the gut, but KCs, like macrophages elsewhere, can generate the whole gamut of proinflammatory cytokines, as well as reactive oxygen/nitrogen intermediates, and are consequently able to cause proinflammatory liver damage [278, 279]. KCs are the main source of proinflammatory cytokines in the liver. The normal liver microenvironment would, of necessity, be one of LPS tolerance. Scavenger receptors help clearance of LPS arriving from the gut, and speculatively the effect on KCs could be antiinflammatory. LBP could also help neutralize LPS by transferring it to lipoproteins, such as high-density lipoprotein. In normal liver, expression of CD14 is low and largely confined to KCs. Nevertheless, TLR4 signaling by the KC network is critically required for the containment of acute *Salmonella* infection, and *Salmonella* LPS triggers a full proinflammatory innate and adaptive immune response [280]. When LPS tolerance is vigorously induced prior to virulent *Salmonella* challenge, proinflammatory cytokine output is markedly decreased but this may be more than compensated for by markedly increased numbers of KCs, increased phagocytosis and clearance of *Salmonella* from the blood, together with increased neutrophil recruitment [281]. Systemic infection with *Listeria monocytogenes* has given insight into the interaction between KCs and hepatocytes. Following intravenous injection, the vast majority of *Listeria* are cleared by hepatocytes (60% in 10 min), but KCs are the principal source of proinflammatory cytokines (TNF-α, IL-1β, IL-12) on *Listeria* infection. In addition, early secretion from KCs of IL-6 (within the first hour) activates STAT3 intracellular signaling in hepatocytes and induces the synthesis of acute phase proteins such as CRP, fibrinogen, α_1 proteinase inhibitor and α_1 acid glycoprotein (Figures 9.1 and 9.2). Depletion of KCs, or IL-6 deficiency, largely abrogates the hepatocyte response, though hepatocytes may have some potential for autocrine/paracrine stimulation from self-generated IL-6 [282]. IL-6R is mainly expressed on hepatocytes, monocytes/macrophages and some lymphocytes. Acute phase proteins are part of the humoral innate immune system. CRP enhances opsonization and clearance of apoptotic cells by macrophages, which are induced to express TGF-β [283] but, conversely, CRP may be proinflammatory when binding to macrophage Fc receptors and may contribute to atherosclerosis [284]. Fibrinogen appears to be capable of stimulating TLR4 in the extravascular compartment, with chemokine release attracting other immune cells to sites of inflammation [285].

Infection with *Schistosoma mansoni* (SM), a helminth that colonizes portal veins, elicits an antiinflammatory phenotype in murine KCs, characterized by IL-10, IL-6, IL-4 and IL-13, but not IL-12 or IL-18. The associated T cell response is dominated by IL-4, IL-5 and IL-13 but little IFN-γ. The KC phenotype in this instance instructs an antigen-specific Th2 response and the pathological changes are those of hepatic fibrosis [286]. In chronic SM infection in humans, increased IL-4 and IL-13 are associated with egg granulomata. The ability of stimulated peripheral blood mononuclear cells to produce IL-5 and IL-13 correlates with the degree of hepatic fibrosis. TGF-β is also elevated during the evolution of hepatic fibrosis in SM infection. In the severest fibrosis TNF-α and IL-10 tend to be elevated and IFN-γ low. Although the T cell response is clearly Th2 [287], the KC phenotype driving the adaptive response on sustained stimulation by SM could be compensated proinflammatory activation in which IL-10 has suppressed IL-12 whilst PGE$_2$ has suppressed iNOS, through elevation of cAMP. In support of this view, LPS-induced iNOS in rat KCs can be markedly decreased by prior, or early, cAMP elevation, preventing the degradation of IκBα, so keeping NF-κB sequestered in the cytosol [288]. A recent study in a reversible model of liver injury, in which macrophages can be selectively depleted, either during distinct injury or recovery phases, indicates the existence of functionally distinct subpopulations of macrophages in the same tissue. In this model, depletion during advanced fibrosis results in less scarring and fewer myofibroblasts, but depletion during recovery is associated with a failure of matrix degradation [289].

The effects of alcohol on KCs are complex and, in some respects, contradictory. CD14 expression on KCs is elevated by alcohol, increasing sensitivity to LPS [290]. There is much evidence that LPS is a cofactor in toxin/alcohol-induced liver injury, which is decreased experimentally by oral nonabsorbable antibiotics, colectomy or a germ-free bowel, or more directly by inhibition of the proinflammatory response of KCs [291]. In the early stages of alcohol-induced liver disease, fatty accumulation appears to be related to PGE$_2$, which is a product of COX-2 induction in proinflammatory macrophages. PGE$_2$ exerts its effect through EP2/EP4 receptors expressed on hepatocytes, thereby increasing the intracellular second messenger cAMP, leading to triglyceride accumulation. The trigger to increased proinflammatory activation and raised PGE$_2$ could be alcohol-related increases in LPS absorption from the gut and/or increased LPS sensitivity [292]. The toxic effects of alcohol appear to be associated with an enhanced response of KCs to LPS rather than an impairment of the known inhibitory feedback loop of cAMP/PKA on TNF-α production. Alcohol substantially decreases the rise in cAMP and PKA activation normally found with PGE$_2$ stimulation, but this is compensated for in some way 'downstream', leaving the degree of TNF-α suppression unaltered [293]. Despite the increased likelihood of LPS-induced damage with alcohol, and the impression that the main fibrogenic mediators are TNF-α, IL-1β and ET-1 [294], the effect of acute alcohol intake in 'binge' quantities is found to be immunosuppressive in peritoneal macrophages, with TLR4 and TLR3 signaling impaired at several points and proinflammatory cytokine expression reduced [295]. PGE$_2$, acting in an autocrine/paracrine manner through EP2 and EP4 receptors, inhibits fibrogenic mediators but enhances the anti-fibrinogenic cytokines IL-10 and IL-6 [294]. The following sequence is suggested for the development of alcoholic liver disease. The process is initiated by KCs, which subsequently activate hepatic stellate cells (HSCs). Because of the chronicity of the input to KCs, feedback downregulatory pathways become activated but are ineffective and so KCs reach a new equilibrium with only partial compensation of the proinflammatory phenotype. Thus, KCs and the microenvironment exhibit a mixture of competing pro- and antiinflammatory mediators (TNF-α, PGE$_2$, IL-12, IL10; see Figure 9.3).

The proinflammatory microenvironment driving the fibrogenic process is increased by hypoxia and cell necrosis. The cytotoxic actions of TNF-α are enhanced by exogenous stressors and there is increased activation of proinflammatory MAPK cascades. Fibrogenesis is seen as a protracted process consequent upon low grade chronic innate immune activation [296]. However, the interplay of KCs and HSCs is the key to understanding liver fibrogenesis. HSCs make up 15% of cells in normal liver but proliferate when activated and are the main effector cells of hepatic fibrosis (of which irreversible cirrhosis is the end stage). Both KCs and HSCs are activated through all stages of alcoholic liver disease [297, 298]. Increased TGF-β following alcoholic liver injury is one of the factors promoting fibrogenesis [295], presumably a chronically-activated but thwarted attempt at healing. IL-10 reverses fibrosis by downregulating proinflammatory overactivity, allowing the liver microenvironment to revert to a normal resting ('steady') state. Failure of HSCs to continue to secrete sufficient IL-10 may be the prelude to irreversible liver cirrhosis [299].

The lung mucosa, like gut, faces constant microbial exposure. Despite this it has strong 'immunosuppressive' properties, to which alveolar epithelium and AMs contribute. In the 'resting' state, AMs are antiinflammatory, secreting IL-10 and expressing IL-10 receptors (IL-10Rs). In humans, AMs may even be the major source of local IL-10 and, compared to the epithelium, the major respondent, since alveolar epithelium appears to lack both IL-10 output and IL-10Rs [300]. In C57BL/6 mice, normal alveolar epithelium can produce IL-10 constitutively and influences the AM phenotype through AM-expressed IL-10Rs. AMs are not universally immunosuppressive because, on appropriate microbial challenge, they retain a vigorous proinflammatory response. Ligands for TLR2, TLR4 and TLR9 all inhibit IL-10-mediated STAT3 signaling, with loss of IL-10-induced chemokines and loss of suppression of TNF-α [272]. Equally, phagocytosis of allergen-containing pollen starch granules by rat AMs through C-type lectins and β_2-integrins evokes rapid iNOS upregulation and release of NO [301]. In the normal lung of Lewis rats, low concentrations of NO are produced by a minority of interstitial macrophages, whilst AMs are inactive, but on in vitro incubation with heat-killed *Listeria*, rat AMs readily upregulate iNOS [302]. In IL-10 knockout mice, iNOS rises to higher levels on airway OVA challenge, following sensitization, than in wild-type mice [303]. LPS activates NF-κB nuclear translocation in human AMs, resulting in the release of proinflammatory NF-κB-dependent cytokines such as IL-6, IL-8 and TNF-α [304]. AMs are essential for maximal NF-κB expression in the lung on LPS challenge and AM depletion results in failure to mount an innate

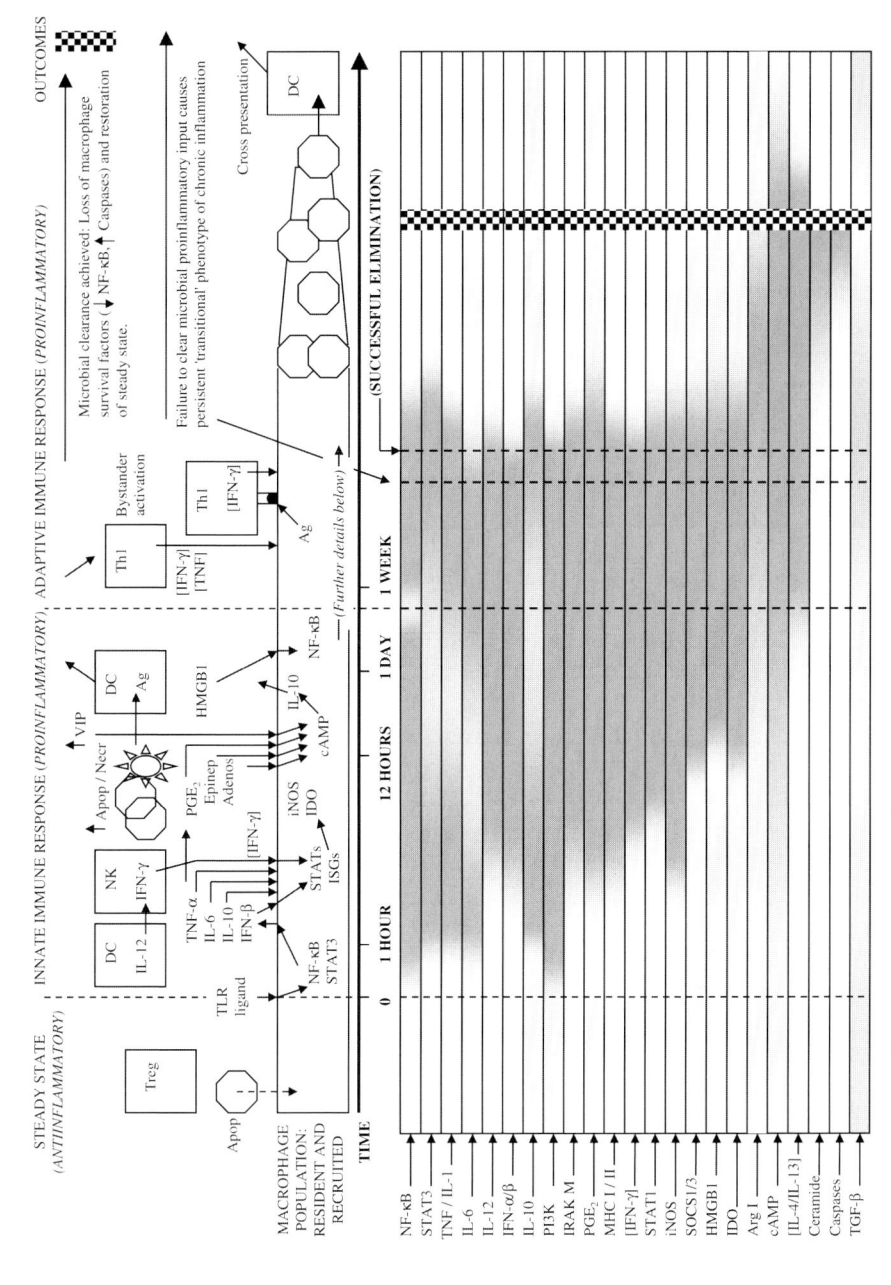

9.2 Diversity of Macrophages in Mammalian Tissues | 287

Figure 9.3 A 'macrophage-centric' impression of an immune response. Schematic representation of kinetics and sequencing of some key macrophage phenotypic changes during a prototypical immune response to de novo TLR stimulation. The macrophage phenotype changes in response to the initial microbial stimulus, alterations in the microenvironment and the activation status of adjacent cells. For intracellular pathways involved see also Figure 9.2. Early IL-10 may be triggered by LPS, mediated by activated intranuclear IRAK1, phosphorylating STAT3. The constitutive transcription factor c-Maf also acts on the IL-10 promoter and is required for LPS-induced IL-10 expression and for enhanced IL-10 expression following LPS and IL-4.

response, with reduction of proinflammatory cytokines, chemokines and impaired neutrophil attraction [305]. Resting AMs have a reduced activity of the Fos/Jun transcription factor AP-1 [306], but this is upregulated by exposure to GM-CSF [307]. GM-CSF is expressed constitutively in normal lung and amplifies the LPS/NF-κB and AP-1 pathways responsible for release of proinflammatory cytokines (including TNF-α and GM-CSF itself), chemokines (responsible for neutrophil recruitment) and matrix metalloproteases [308]. A role for GM-CSF in macrophage differentiation in vitro has been alluded to earlier. Much of the differentiation of AMs from monocytes has been attributed to the effects of GM-CSF operating through the PU.1 transcription factor, which induces a wide selection of genes controlling cell surface receptors (including TLRs), phagocytosis, surfactant clearance, cell adhesion, collectin-binding and proinflammatory cytokines [309, 310]. Bleomycin produces an AM phenotype secreting TGF-β and inducing pulmonary fibrosis [311]. Lack of GM-CSF is implicated in the pathogenesis of pulmonary fibrosis because GM-CSF knockout mice are susceptible to the condition [312]. Conversely, GM-CSF enhances production of PGE$_2$ by upregulating COX-2 in macrophages, fibroblasts and alveolar epithelial cells. PGE$_2$ is potently antifibrogenic and inhibits bleomycin-induced fibrosis [313]. The potency of GM-CSF in switching macrophage phenotypes is dramatically illustrated in the entirely different setting of macrophage infection by *Leishmania*, genetically manipulated to express GM-CSF. As discussed earlier, Th2-biased BALB/c mice normally have difficulty clearing *Leishmania* from macrophages and an antiinflammatory response is induced. When intracellular *Leishmania* are engineered to produce and secrete GM-CSF, BALB/c-strain macrophages show a dramatic reversal of phenotype to one dominated by proinflammatory cytokines (IL-1β, IL-6, IL-18) and chemokines (RANTES/CCL5, MIP-1α/CCL3, MIP-1β/CCL4, MIP-2/CXCL2 and MCP-1/CCL2) and deliver much-enhanced parasite killing [314]. Airway instillation of liposome-encapsulated dichloromethylene diphosphonate, presumed to be taken up selectively by AMs, suppresses the activation of intrapulmonary NF-κB in an acute inflammatory lung model, triggered by immune complexes. There are corresponding reductions in TNF-α and MIP-2 and the model suggests that AM NF-κB may be essential for the initiation of lung inflammation [315].

Lung DCs have similarities in their behavior to AMs in both the resting state and on exposure to provocations. Resting DCs are functionally immature, secreting IL-10 [316]. On exposure to respiratory antigen (OVA), murine pulmonary DCs reach bronchial nodes at 24 h and produce transient increases in IL-10. These DCs are described as 'mature' in that they express MHC II and co-stimulatory molecules and can induce regulatory (antiinflammatory) T cells, secreting IL-10 and IL-4 [317]. Alternatively, if OVA exposure is combined with stimulation by proinflammatory cytokines, particularly TNF-α and GM-CSF together, DCs express IL-12, MHC II and co-stimulatory molecules and then usually instruct mixed Th1/Th2 adaptive responses [316]. Lung eosinophilia is said to be dependent on antigen presentation by DCs to Th2 cells, as demonstrated by the injection of OVA-pulsed mature DCs into the airways of mice and rechallenge with inhaled OVA at 2 weeks. The injected cells can be tracked to draining lymph nodes but there is no firm evidence, on the basis of markers used, that effector lung APCs, with which memory CD4$^+$ or CD8$^+$ Th1 and Th2 react, must necessarily be DCs, particularly as the generation of (primary) T cell responses is strictly compartmentalized to lymph nodes [318]. Lung immunology highlights incongruities between DC and AM identification markers and function. The first difficulty is whether the antigen-bearing DCs, which migrate to regional nodes, and the APCs, which present antigen to memory T cells in the lung, are in fact both DCs [319, 320], or whether the latter are AMs [321]. Looking at markers and function of human lung DCs, in vitro, they appear to express some features of immature DCs (expression of CCR1 and CCR5 ability to endocytose dextran through MRs, low co-stimulatory molecules, CD40 CD80 and CD86, poor expression of differentiation marker CD83 and no CD1a) but also resemble mature DCs in their high expression of MHC-II and a powerful ability to stimulate T cell proliferation in mixed lymphocyte reactions, despite their low (but effective) expression of co-stimulatory molecules [322]. Unfortunately, in vitro studies give no indication of DC kinetics or site of T cell interactions but it has also been reported that T cell proliferation is confined to nodes and specifically does not occur when memory T cells are activated to produce effector cytokines in non-lymphoid tissues, such as lung [321]. A second difficulty is actually distinguishing AMs from DCs, particularly if both are proposed to be immunocompetent APCs in non-lymphoid lung tissue, and this again depends on the interpretation of cell markers. A subset of APCs with high MHC II expression and the ability to stimulate naive T cells can be isolated from BAL fluid and further subdivided into CD1a$^+$ and CD1a$^-$ groups. CD1a$^+$ behave as mature LCs (potent T cell stimulators as measured by T cell proliferation in mixed lymphocyte reactions, but poor cytokine producers and S100$^+$). The CD1a$^-$ cells have a lower capacity to drive T cell proliferation but express IL-1, IL-6 and TNF. Unlike monocytes these cells have dendritic morphology and low CD14, suggesting they may be monocytes in the process of differentiating to DCs but, on the basis of other cell markers, could equally well be differentiating towards mature macrophages [323]. DC precursors (like monocytes) are said to be attracted to lungs using expressed chemokine receptors, such as CCR1, CCR2, CCR5 and CCR6, some of which may be stimulus specific, such as CCR1 and CCR5 in bacterial infection [324]. What is not clear is to what extent DCs are pre-committed circulating precursors or differentiate in situ from monocytes [325]. Some studies have relied on minimal or ambivalent markers to distinguish AMs from DCs and there are very few studies that attempt to look at the kinetics of both AMs and DCs concurrently. A final difficulty is that both AMs and DCs are in very close proximity in the lung [326]. Indeed, it has been postulated that AMs provide a suppressive counterbalance to the activation of DCs and that it could be NO from macrophages that keep DCs in an immature state [327]. Resident AMs have been considered to be immunosuppressive to local T cell activation and to dampen effector responses to antigens, as measured in vitro by T cell proliferation and effects of macrophage-depletion [328]. These hypotheses must be accepted with caution. Macrophage depletion may increase the response to antigen by other means, such as increased availability of antigen to DCs. Paradoxically, AM depletion increases the Th2 response and IgE production in the lungs of presensitized OVA-challenged mice [329], whereas, arguably, removal of a suppressive cell population should release a proinflammatory response. On the basis of AM depletion and adoptive transfer studies in OVA sensitized and challenged BALB/c mice it has, on the contrary, been suggested that AMs are inherently Th1-promoting and important in antagonizing ongoing Th2-dependent allergic responses [330]. Perhaps the effects of AM depletion in a bacterial pneumonia may give a more representative picture of the biological function of AMs. In *Pseudomonas* lung infection, depletion initially limits the severity of the proinflammatory response, but at the expense of increased bacterial load and ultimately increases morbidity and mortality through impaired recruitment of neutrophils and failure to eliminate the bacterium [331].

It may be concluded that AMs, DCs and associated cells, most importantly lung epithelium [332, 333], act in a concerted way in response to environmental changes, reinforcing either antiinflammatory, proinflammatory or transitional responses. This view is consistent with a simpler interpretation of AM–DC interactions, that AMs may buffer rather than polarize DCs. For example, if AMs are in an antiinflammatory state, as in LPS tolerance or during the downregulation of a proinflammatory response, they would tend to stabilize the DC population in, or move it towards, an immature (antiinflammatory) state. Relevant antiinflammatory signaling molecules involved in this paracrine 'crosstalk' between DC and AMs might be TGF-β, IL-10 and PGE$_2$. Therefore, instead of suppressing DCs, it seems more likely that AMs (and probably all macrophages) normally operate in parallel with them, being antiinflammatory in the resting state but proinflammatory on TLR stimulation, or when exposed to proinflammatory molecules, an important principle in understanding the contribution of AMs to asthma pathogenesis, considered in the following chapter (10.2).

Tumor cells often produce molecules that recruit [334] and sustain 'tumor associated macrophages' (TAMs). The inflammatory profile of the tumor environment is shaped largely by unregulated production by tumor cells of cytokines such as IL-10, TGF-β and IL-4, and, additionally, PGE$_2$, all molecules which suppress proinflammatory responses [335]. Moreover, it is just such secreted molecules that ensure tumor growth and survival. Tumor-produced VEGF and M-CSF and several chemokines assist monocyte/macrophage recruitment and survival [336]. Natural tumor resolution is associated with a vigorous innate proinflammatory response, supporting a Th1 adaptive response. TAMs at the invasive margins of tumors can express co-stimulatory molecules and have been observed in cell-to-cell contact with CD4$^+$ and CD8$^+$ cells, suggesting they are fully functional APCs [337]. Exposure of macrophages to PGE$_2$ increases intracellular cAMP [338, 339] and inhibits IL-1 and TNF-α output. Exposure of T cells to PGE$_2$ is permissive of IL-4 and slightly enhances IL-5 secretion, but inhibits IL-2 and IFN-γ. Tumor cell-derived PGE$_2$ inhibits TNF-α release by augmenting macrophage IL-10 synthesis [340]. In turn, a TAM phenotype biased towards IL-10 exhibits autocrine reduction in IL-12 and reduced activation of NF-κB [341], perhaps in part from increased nuclear translocation of NF-κB (inhibitory) p50 homodimer. TAMs express antiinflammatory cytokines, IL-1 receptor anatgonist, high levels of MR and scavenger receptors, arginase in preference to iNOS and are poor APCs. In fact, TAMs from suppressive tumor microenvironments have many features associated with 'type 2' antiinflammatory macrophages (M2), and may provide insights into the mechanisms by which proinflammatory (M1) responses are downregulated [342]. Malignant ascites from ovarian carcinoma has been found to contain an IL-10, TGF-β_2-producing HLA-DR-negative subset of monocytes, which inhibits T cell proliferation and IFN-γ production and is unable to secrete IL-12 or TNF-α [343]. In mice, cytotoxic $CD8^+$ T cell adaptive responses against tumors may be specifically inhibited by macrophages of an M2 phenotype, expressing high arginase activity [344]. In an antiinflammatory microenvironment there is little impetus for the differentiation of mature DCs from monocytes or macrophages. As would be predicted, DCs are usually scarce in tumor microenvironments and locked into an 'immature' phenotype [345]. M-CSF and IL-6, released from renal carcinoma cells, inhibit normal DC development from $CD34^+$ progenitors and direct them instead towards a macrophage phenotype [346]. Injected human antigen-loaded DCs fail to localize effectively to tumors or nodes affected by cancer [347]. Therapeutic efforts may be aimed at reversing the antiinflammatory tumor microenvironment [348]. This is the rationale for intra-tumoral delivery of GM-CSF, which has been shown, in mice, to convert (F4/80) TAMs into $CD11c^+$ $CD80^+$ MHC II expressing DCs that activate NK cells and T-cells. The proinflammatory innate/adaptive response is further improved, as is the prognosis, by additional delivery of IL-12 and co-stimulators [349]. Another way to redress the macrophage functional balance is by inhibition of COX-2 (overexpressed in human lung cancer). Specific COX-2 inhibitors prevent PGE2-mediated induction of IL-10 and restore IL-12 in APCs [350]. A paradox in tumor immunology is that a chronic proinflammatory drive, whether from infection or other causes, finally yields a tumor to which there is little innate or adaptive immune response. Escape of a tumor from effective immunosuveillance is most readily explained through a process of 'immunoediting' [348] taking place in what has been dubbed the 'Darwinian' microenvironment of the tumor [336]. Thus, abnormal cancer cells expressing antigens most capable of eliciting a Th1 response are successfully removed, leaving a less proinflammatory cancer cell phenotype, which is mirrored in the phenotypes of influxing immune cells. The phenotypes of TAMs change the situation further by promoting fibrosis and/or angiogenesis (aiding tumor spread). In the hypoxic conditions of many tumors, macrophages upregulate a hypoxia-inducible transcription factor, which increases expression of hypoxia-inducible genes, including VEGF and CXCL8 implicated in angiogenesis [336, 337]. In support of the principle of immunoediting, tumors arising in SCID or nude mice are more immunogenic and less tumorogenic [348]. In a setting of chronic inflammation macrophages may be pushed into a transitional phenotype through the induction of regulatory programs, and so may secrete IL-10 as well as IL-12 and TNF-α and have increased cAMP and reduced NF-κB activation. In summary, TAMs illustrate well the varied phenotypes and functions of macrophages from M1 at one end of the continuum through to M2 at the other. It is increasingly recognized that TAMs play a decisive role in determining tumor resolution or progression, as has been formulated most eloquently in the 'macrophage balance' hypothesis [351].

9.2.6
Sequential and Regulatory Changes in Macrophage Phenotypes: Limiting Pro- and Antiinflammatory Responses

TLR-triggered proinflammatory responses should run a finite course and should automatically be downregulated to the steady state when pathogen destruction eliminates TLR ligands. The proinflammatory response to pathogens tends to be incremental, with staged expression of early and late innate response genes. The innate response may be further amplified by the more specific adaptive response. To prevent inappropriately large or persistent responses, upper limits to proinflammatory activation have evolved. Conversely, lower limits to antiinflammatory responses (involving active expression of other subsets of genes) are equally necessary and these may take the form of priming the system to react rapidly to further proinflammatory stimuli and/or reversing excessive or protracted antiinflammatory phenotypes. Examples of the prevention of profound antiinflammatory phenotypes might be the priming effect of IL-4 on Dectin 1 expression [147] and the IL-4-induced enhancement of bioactive IL-12p70 in DCs and macrophages during microbial infection, or in the presence of ongoing inflammation [352]. Effective Th-1 immunity, required to control the fungus *Candida albicans*, depends on the presence of endogenous IL-4. The early IFN-γ response (possibly from NK cells of the innate immune system) is actually greater in IL-4 knock-out mice, but in later infection IL-4 is necessary to prime the system for a continuing effective IL-12-dependent Th1 response [353]. Surprisingly, even the direct IL-12 antagonist, IL-10, has been shown to be required for optimal co-stimulation of an antifungal Th1 response to *Candida* [354]. In septic peritonitis, mice depleted of KC-derived IL-10 carry an increased bacterial load which they fail to clear because of reduced IFN-γ production, rather than because of changes in other relevant cytokines. Depletion of IL-10 before infection results in increased mortality, but neutralization of IL-10 several hours after the onset of infection appears beneficial – in line with the view that IL-10 primes for the proinflammatory response at the onset of infection but, as the initial surge in IL-12 declines, a reciprocal rise in IL-10 will be predominantly antiinflammatory (Figure 9.3) [355]. The antiinflammatory cytokines L-4 and IL-13 contribute to the downregulation of a proinflammatory microenvironment in rheumatoid joints, but the same two cytokines prime for increased IL-12p70 release from synovial fluid macrophages [356]. IL-4 priming of DCs may be a factor promoting Th1 and cytotoxic T cell responses to tumors [357]. Such results do not support an invariant inverse relationship between IL-4 and IFN-γ. On the contrary, IL-4 can substitute for IFN-γ in priming macrophages for IL-12p70 production and, particularly in the presence of GM-CSF, IL-4 can increase IFN-γ by upregulating IL-12. The inference must be that a proinflammatory response might just as effectively be initiated in the innate immune system in the presence of antiinflammatory (Th2) feedback as in the steady ('resting') state. An explanation for the actions of IL-4 and IL-13 may be found in the IL-4R type I (IL-4Rα/γ_c) and IL-4R type II (IL-4Rα/IL-13Rα1) receptors expressed by macrophages and DCs. Although IL-13 binds preferentially to IL-4R type II, IL-4 and IL-13 can both activate STAT-6 through IL-R4α present in both type I and II receptor heterodimers. At least for murine DCs in vitro, it appears that IL-4 and IL-13 can synergize with GM-CSF to activate STAT-6 and bring about DC maturation (upregulation of MCH II and co-stimulatory molecules) via IL-4R type II, while IL12p70 in response to microbial products is mediated by IL-4 (but not IL-13), acting through IL-4R type I and Jak3 [358]. In general, mechanisms preventing overshoot of the antiinflammatory response are poorly understood.

The macrophage is a major effector of both pro- and antiinflammatory responses. During a typical immune response, macrophage phenotypes change sequentially. Initial direct proinflammatory gene expression (in cis) under the influence of the innate immune system is later modulated by integrated signals from both innate and adaptive immune systems. Expression of these additional genes often occurs in trans, following autocrine/paracrine receptor stimulation by released cytokines, or in response to other proinflammatory molecules in the microenvironment. Induction of proinflammatory genes is invariably offset by induction of downregulatory genes, since failure to regulate the upper limit or duration of immune responses could provoke septic shock, chronic inflammation or even autoimmunity. Conversely, failure to control the lower limit could manifest as post-sepsis immunosuppression. Prolonged expression of antiinflammatory macrophage phenotypes mediates tolerance, as in the case of the much-researched LPS tolerance. LPS tolerance has been defined as "a reduced capacity of the host (in vivo) or of cultured macrophage/monocyte (in vitro) to respond to LPS activation following first exposure to this stimulus" [359]. The multiple mechanisms that produce LPS tolerance are, probably, similar to those limiting, or reducing, proinflammatory macrophage activity during any immune response [359, 360]. Some sequential aspects of macrophage pro- and antiinflammatory activation during an immune response are shown schematically in Figure 9.3 and some of the mechanisms responsible are now discussed. Many operate entirely within the confines of the innate immune system.

9.2.6.1 Pre-TLR and TLR Regulation of Immune Responses

In LPS tolerance CD14 (membrane and soluble), LPB and MD2 can be upregulated, but not consistently in many cases, and the net outcome of these interrelated factors in 'tuning' immune responses is not known [359]. Early research into mechanisms of tolerance investigated whether downregulation of TLR receptors was responsible. Though this may occur it is not now thought to be the most important factor. Indeed, TLR2 expression is increased in LPS tolerance. Furthermore, Pam3Cys TLR2 stimulation promotes an antiinflammatory-biased innate cytokine profile (abundant IL-10 and reduced elevation of IL-12) and a Th2 adaptive response, suggesting that signaling through TLR2 can, under some circumstances, counteract TLR4 signaling in vivo [361]. Another TLR with antiinflammatory properties is TLR9, which responds to CpG-DNA of bacterial and viral origin, and which could account for the antiinflammatory effects of probiotics in a murine model of experimental colitis. Because microbial DNA is normally immunostimulatory, the route of absorption, through the small intestine, may be critical to an antiinflammatory probiotic action [362].

9.2.6.2 Signal Transduction in the Regulation of Immune Responses

The LPS, CD14/TLR4, MyD88, IRAK, TRAF 6, MAPK/NF-κB signaling sequences are rapidly curtailed after LPS exposure by a pathway involving PI3K and a PKC isoform, which inactivates IRAK 1, and is operative within 30 min of LPS stimulation (Figures 9.2 and 9.3). PI3K is constitutively expressed in innate immune cells (macrophages/DCs) so is activated alongside the primary proinflammatory signaling pathways and is immediately available to limit the expression of proinflammatory mediators such as TNF-α and IL-12 [363, 364]. Additionally, there is a slower inactivation of an alternative LPS, IRAK, NF-κB signaling pathway, commencing at the CD18/CD11b (CR3) receptor (but not involving PI3K). This alternative mechanism of downregulation may be of importance in CD14 negative DCs, as well as in macrophages, which express CR3 [365]. As well as inhibiting innate immune responses to LPS, the negative regulatory effects of PI3K induction by TLR4 stimulation are carried forward into the adaptive immune system, because PI3K regulates IL-12 output from DCs [199]. Although *Porphyromonas gingivalis* LPS signals primarily through TLR2, it again activates the PI3K pathway and, if this is blocked experimentally, there is an excessive elevation of IL-12, and a concomitant severe reduction of IL-10 [366]. Stimulation of rat KCs either by Gram-positive associated LTA (via TLR2/6) or Gram-negative associated LPS (via TLR4) triggers expression of IL-10 and IL-6, which is specifically dependent on PI3K and Jak2 signaling and counterbalances proinflammatory cytokines produced by other pathways, setting the balance of pro- and antiinflammation at the very earliest stages of an immune response [367, 297].

A further means of suppressing TLR4 and TLR2 signaling in monocytes and macrophages (but probably not DCs) involves IRAK-M, which inhibits activation of the NF-κB pathway by preventing the dissociation of the TLR/MyD88/IRAK complex necessary for downstream engagement of TRAF6 [368]. Induction of IRAK-M occurs within 2 h of stimulation of monocytes with TNF-α, itself released as a result of TLR2 or TLR4 stimulation, or following exposure to NO [369]. The tolerance that rapidly develops following Pam3Cys stimulation of TLR2 (in human MonoMac6 cells) is thought to be predominantly the result of increased IRAK-1 degradation and/or disruption of the NF-κB signaling pathway, reducing NF-κB nuclear translocation [154]. Tolerance and cross-tolerance affecting TLR signaling are by no means absolute and are countered, for example, by IFN-γ and GM-CSF, which rescue IRAK 1, allowing other proinflammatory inputs (such as CpG DNA-induced IFN-γ) to alter the balance away from tolerance [370].

The early response to LPS through TLR4 and NF-κB, amongst other transcription factors, is transient, as indicated by the decline of TNF-α and IL-1β after a few hours. During the initial surge, TNF-α can act in a positive feedback loop to stimulate further NF-κB activation. However, another target gene strongly induced by NF-κB is that of I-κB, which binds NF-κB and restricts its cytoplasmic availability for nuclear translocation and so constitutes a potent negative feedback regulator of all NF-κB-dependent genes [371]. Upregulation of the p50 subunit of NF-κB also plays a part in LPS tolerance by forming homodimers that cannot bind to I-κB but can translocate to the nucleus, where they engage and block the NF-κB sites of gene promoters [372]. In resting macrophages, PKA phosphorylation may account for a suppressive constitutive binding of p50/p50 homodimer to DNA to reduce oversensitivity to proinflammatory activation in the absence of an external stimulus [373]. An inhibitory spliced variant of MyD88 may be induced by LPS in murine macrophages and have a role in LPS tolerance [374]. Intracellular signal transduction responsible for NF-κB activation may be reduced in LPS tolerance but in vitro studies show that continuous stimulation of different TLRs provokes sustained and additive activation of (RAW264) macrophage proinflammatory gene promoters [375], emphasizing that innate tolerance, at least as expressed extracellularly in the tissue microenvironment, must be defined in terms of cell output and not intracellular pathways. EKLF is a transcription factor operating on the IL12p40 promoter, which activates it in resting macrophages but represses it in proinflammatory activated macrophages by inhibiting NF-κB nuclear localization. Other promoters of proinflammatory cytokines and chemokines could be regulated similarly [376]. The early response to LPS is transient but a second surge of proinflammatory activity is triggered, with an onset 8 to 12 h after LPS exposure, and includes a novel cytokine, HMGB1, which reaches a prolonged plateau from 18 to 24 h after LPS and could, therefore, be a putative mediator of septic shock [377] (Figures 9.2 and 9.3). CD14, TLR4 and selected cytokines (TNF, IL-1 and IFN-γ, presumably acting in trans) are implicated in signaling pathways leading to HMGB1 release from LPS-stimulated macrophages, but HMGB1 can also enter the microenvironment passively from necrotic cells [378, 379]. HMGB1 may signal in part through RAGE, a receptor for advanced glycated end products, amyloid β-peptide and S100 proteins (all ligands characterized by β-sheets and fibrils). Recent research on RAGE indicates that its engagement may be responsible for prolonged activation of NF-κB in models of innate sepsis but appears to have little impact on adaptive immunity, other than its effects on proinflammation at the level of effector cells (macrophages). Thus, in EAE, activation of NF-κB and the initiation of the inflammatory response can occur independently of RAGE, but RAGE has a role in its perpetuation [380].

LPS stimulation through TLR 4, or double-stranded (viral) RNA through TLR3 (or other routes), can upregulate IFN-β in macrophages and DCs. The transduction pathways have been considered to involve NF-κB and MAPKs and are enhanced by NO in RAW 264.7 macrophages [381]. However, the induction of IFN-β following TLR stimulation may, in fact, be mainly independent of the adapters MyD88 and Mal/TIRAP which lead to NF-κB. Instead, two additional adapters, TICAM-1 and TICAM-2, may be required to activate IRF- and IFN-inducible genes (Figure 9.2). Not only does this allow variable control of the different pathways, and of IFN-β signaling, but also mediates MyD88-independent DC maturation [382]. As the immune response evolves, IFN-β can signal, in trans, through IFNR1 to induce a second wave of genes that are Jak/STAT dependent by virtue of possessing ISRE-binding sites. These genes include IP-10, iNOS, RANTES and MCP-5, as well as co-stimulatory molecules, such as CD80, CD86 and CD40, which are essential for participation of APCs in adaptive responses (Figure 9.2) [383, 384]. Induction of iNOS may be much more effectively achieved by this pathway than by NF-κB acting in the first phase of the response to TLR stimulation. Furthermore, it seems likely that Jak/STAT signaling might be a necessity for iNOS induction, because this is impaired in STAT1 deficiency and abolished in Type 1 IFN receptor knock-out mice [383]. Feed-forward activation of NF-κB by TNF-α, operating in trans through TNFR1, could act synergistically with STAT1 on the iNOS promoter. Yet other interferon-responsive genes could be activated directly through IRF3. In murine resident AMs, autocrine/paracrine feed-forward amplification through IFN-β and STAT1 is relatively inoperative compared to peritoneal macrophages, regardless of whether stimulation is applied through TLR2, TLR3 or TLR4. The early gene response (TNF-α, CCL3, CCL4 and CCL5) is preserved, as is STAT1 activation, by exogenous IFN-β. One explanation is that continuous exposure to low concentrations of LPS at mucosal surfaces induces partial tolerance, manifest as a failure to produce bioactive IFN-β. A second burst of a proinflammatory response may then be held in check unless IFNα/β or IFN-γ is provided by other proinflammatory cells (epithelial cells or recruited monocytes), activated above a certain threshold of microbial infection [385]. IFN-α is typically produced in viral infections in conjunction with IFN-β. IFN-α is also produced in large amounts by plasmacytoid DCs [386]. Recent research indicates that IFN-α-producing murine DCs are mainly of myeloid origin, casting further doubts on the reliability of lineage markers [387].

The type II interferon, IFN-γ, is foremost amongst cytokines boosting innate immunity and is copiously released by NK cells in proximity to proinflammatory macrophages, in the presence of IL-12. In adaptive immunity, IFN-γ is the defining cytokine of a Th1 response, further boosting proinflammatory macrophage activation [388]. It is often stated that IFN-γ primes for TLR responses and, indeed, it is often present early in immune reactions, macrophages themselves controversially being one possible source [389]. There is no doubt that IFN-γ rapidly primes macrophages for full proinflammatory activation on subsequent exposure to LPS or TNF-α [390]. However, a purist interpretation of an immune response generated, de novo, requires TLR stimulation to precede IFN-γ upregulation. Short co-stimulation with a TLR agonist and IFN-γ in the murine RAW 264.7 macrophages causes early amplification of IFN-γ signaling attributed to p38 MAPK phosphorylation of STAT1 and to a transient increase in TLR expression that resolves in 24 h [391]. In intracellular murine macrophage infection by *Chlamydia pneumoniae*, TLR stimulation is thought to upregulate IFN-γ both directly, via an NF-κB pathway, and (perhaps more certainly) indirectly through IFNα/β

upregulation and STAT1 activation [392]. Both IFN α/β and IFN-γ can bring about sequential enhancement of a TLR-initiated response by STAT1 phosphorylation. For IFN-γ, phosphorylated STAT1 dimers bind to the γ activation site (GAS) of IFN-γ-inducible genes, including class II transactivator and IFR genes. STAT1-dependent pathways are supplemented by other non-STAT-dependent pathways in IFN-γ signaling, using various signaling intermediates, including MAPKs (Figure 9.2). These non-STAT pathways and their functions are not yet well defined. Up to 500 genes are estimated to be regulated by IFN-γ [393].

Jak/STAT pathways are triggered by interferons, and by many other cytokines, but their activation is often not sufficient for full responses and freely diffusible STATS show overlap in their target genes [394]. Important regulators of Jak/STAT pathways are members of the SOCS family of proteins. The mechanism of SOCS regulation of TLR signaling has recently been defined [395]. Macrophage stimulations through TLRs, IFN-γ, IL-6 or GM-CSF are all targets of inhibition by SOCS. For example, prolonged co-stimulation of RAW 267.4 macrophages with a TLR agonist and IFN-γ or preincubation with TLR agonist before IFN-γ elicits 'tolerance' through TLR-induced expression of at least three SOCS proteins, SOCS1 SOCS2 and CIS, which mediate feedback inhibition directed to IFN-γ signaling [391]. SOCS inhibition shows no absolute pathway specificity but, nevertheless, SOCS1 is the most efficient inhibitor of IFN-γ signaling. Macrophages subjected to prolonged LPS stimulation or to intracellular infection by *Listeria* exhibit both p38 MAPK-dependent STAT1 phosphorylation and p38 MAPK-dependent induction of SOCS3. While STAT1 enhances the immediate response to IFN-γ, with increased output of IFN-γ responsive genes, SOCS3 accounts for a later reduction in STAT1 phosphorylation and downregulation of IFN-γ-responsive genes. SOCC3 protein is induced to maximal levels about 6 h after LPS stimulation, probably with the involvement of NF-κB. Reduction of STAT1 phosphorylation is evident at 4 h, reaching a maximum effect after 8-12 h. Similar kinetics apply to *Listeria* infection. SOCS 3 inhibition affects not only IFN-γ. For example SOCS3 also inhibits Jak2 phosphorylation that is required for GM-CSF signal transduction through STAT5 [396, 397]. The balance of activation and suppression of IFN-γ responsive genes (including MHC II) likely changes over time, depending on the relative balance of TLR ligands and interferon concentrations in the immediate microenvironment. Because of its potency in amplifying proinflammatory immune responses, tight regulation of IFN-γ is mandatory. Recent evidence suggests IFN-γ potently inhibits gene transcription in macrophages by inducing ICER through a novel signaling pathway, utilizing casein kinase 2 and CREB. ICER behaves as an early response gene with maximal expression within 3–6 h (Figure 9.2) [398]. Other mechanisms by which IFN-γ is inhibited include IFN-γ-induced antiinflammatory genes, such as the decoy receptors IL-1Ra and IL-18BP, and enhanced activation-induced apoptosis in macrophages by mechanisms that include upregulation of caspase 1 and increased sensitivity to Fas-mediated apoptosis [399].

SOCS3 is rapidly and strongly induced in macrophages by LPS, TNF-α, IL-1, IL-6 and IL-10. An intact MAPK pathway is required and SOCS3 is produced by new protein synthesis, unlike STATS which are constitutively and immediately available [400]. In addition, p38 MAPK rapidly inhibits the IL-6/STAT 3 pathway by a more direct route, independently of SOCS3 and not requiring protein synthesis (Figure 9.2) [401]. IL-6 is an early response cytokine, with binding sites for NF-κB and AP-1 in its promoter, and is readily induced by low concentrations of LPS [402]. In the past it has usually been considered proinflammatory in its actions because it is induced early alongside other proinflammatory cytokines (IL-1, TNF α), but this bland description of such a pleotropic cytokine is no longer tenable. The IL-6R incorporates gp130, through which it is linked to Jak/STAT3 signaling. IL-6 responsiveness in macrophages is negatively regulated by SOCS3, which specifically prevents STAT3 activation in response to gp-130 mediated signals. In the absence of SOCS3, IL-6 action is unregulated and prolonged, with the induction of different gene sets, which may be antiinflammatory, similar to IL-10-STAT3-induced genes, or may be genes normally induced by interferons and STAT1. Therefore, SOC3 in macrophages can be seen as blocking a fundamental action of IL-6, which is to inhibit (LPS) proinflammatory signaling, thereby inhibiting proinflammatory cytokine production (TNF-α, IL-12). Hypothetically, IL-6 antiinflammatory activity in macrophages should be most evident at the onset of an immune response before the build-up of SOC3, but alternative pathways, possibly including 38p MAPK, inhibit IL-6 STAT3 signaling within minutes [403, 404]. With such effective silencing it is currently difficult to discern any role for IL-6 in macrophages! Perhaps the major impact of IL-6 is on adjacent, or more distant, cells, particularly hepatocytes in the generation of acute phase proteins. Intriguingly a soluble form of IL-6R (sIL-6R) may acquire IL-6 and then act, in trans, by attaching to isolated gp130 molecules, universally expressed on all cell membranes (unlike conventional IL-6R). One process that can be explained by this mechanism is thermal activation of L-selectin adhesion of T cells, via a Jak triggered T cell response [405]. Very recently IL-6 has been shown to suppress the maturation of DCs [135].

IL-10 is one of the most powerful antiinflammatory cytokines, regulating the upper limits and duration of macrophage proinflammatory activation during immune responses, but, unlike IL-6, its receptor lacks gp130 and its STAT3 signaling pathway is therefore insensitive to inhibition by SOCS3 [406]. The IL10R1 subunit of the IL10 receptor is upregulated on activated macrophages, making them the major target for the cytokine. IL-10 inhibits the expression of MHC II, co-stimulatory molecules, NO, cytokines, chemokines and adhesion molecules, actions previously attributed to its interference with the function of transcription factors, such as NF-κB or AP-1. It now seems more likely, from microarray experiments, that IL-10 exerts virtually all its effects by activating STAT3, but the exact mechanism by which STAT3 represses activated genes still requires clarification [407]. Although most IL10-responsive genes are repressed by STAT3, IL-10 is a strong inducer of both SOCS3 and SOCS1 and the latter is involved in negative regulation of IL-10 signaling [408]. It is of great interest that glucocorticoids, which generate a similar range of immunosuppression as IL-10, may in fact share a common mode of action, through the glucocorticoid-induced leucine zipper

(GILZ), which is upregulated in macrophages by both IL-10 and glucocorticoids, binds directly to the p65 subunit of NF-κB and inhibits its action [409]. The reciprocal autocrine relationship between IL-12 and IL-10 in macrophages still requires an explanation at the molecular level. The IL-12p40 promotor may, in fact, be more sensitive to the transcription factor c-rel than to p50/p65 NF-κB and, significantly, IL-10 has been shown to reduce the level of c-rel in RAW 264.7 nuclei, by inhibiting IκBα. A further observation from microarray experiments is that c Maf is induced by IL-10, both beeing reinforced by a positive feedback loop. While potently activating IL-10 and IL-4 expression, c Maf, at the same time, selectively inhibits IL-12p40 and IL12p35 genes [410]. Another proinflammatory cytokine negatively regulated by IL-10 is IL-1β. LPS can stimulates IL-10 production in macrophages through a Jak3 pathway and IL10 then acts through an autocrine feedback circuit to downregulate proinflammatory cytokines [411]. LPS possibly induces IL-10 through nuclear transfer of modified IRAK1 and phosphorylation of STAT3 (serine 727) on the IL-10 promoter.

9.2.6.3 Regulation of Immune Responses by Cytokines and other Bioactive Molecules

Bioactive molecules generated in the proinflammatory microenvironment have a major effect in defining the upper limit of the proinflammatory macrophage response. Two examples are given. Firstly, COX-2 induction occurs in the early response to TLR stimulation under the control of various transcription factors, including NF-κB. Apart from the proinflammatory effects of PGD_2, another released product of COX-2 is PGE_2, which exerts a strong anti-inflammatory feedback, operating through specific membrane receptors which increase adenyl cyclase activity and elevate intracellular cAMP. The overall outcome of elevated cAMP is an increase in many antiinflammatory mediators, notably IL-10 and TGF-β and reduction in many proinflammatory molecules, notably TNF-α and IL-12. Elevation of cAMP blocks p38 MAPK phosphorylation and thereby downregulates IL-12p40mRNA and IL-12p70 in murine peritoneal macrophages [412]. Elevated cAMP reduces GM-CSF by a mechanism not involving IL-10 [409]. Autocrine reciprocal regulation of IL-12 and IL-10 in macrophages, independently of IFN-γ, is well known [413]. The second example is the downregulation of macrophage proinflammatory responses by the inflammatory product adenosine, operating through adenosine receptors, which can occur entirely within the confines of the innate immune system. Stimulation of the A_{2B} receptor of bone-marrow-derived BALB/c macrophages increases cAMP and inhibits IFN-γ-induced expression of MHC II, proinflammatory cytokines and NO. The A_{2B} receptor is itself up-regulated by IFN-γ [414]. In human macrophages the A_{2A} receptor mediates similar antiinflammatory functions, most significantly reducing IL-12 and increasing IL-10. The A_{2A} receptor is upregulated by IL-1β and TNF-a but downregulated by IFN-γ [415]. The macrophage A_{2A} receptor has been described as an 'angiogenic switch' because of a broad synergism between TLR2, TLR4, TLR7, TLR9 and $A_{2A}R$ to upregulate vascular endothelial growth factor expression and downregulate

TNF-α in murine macrophages, effects hypothesized to promote wound healing in a proinflammatory setting [416].

9.2.6.4 Regulation of Immune Responses by Decoys

Decoy receptors prevent access of ligands to their normal signaling pathways, and may act extracellularly, at the cell membrane or intracellularly. First described for the IL-1/IL-18 receptor family, and later for TNF-α receptors, decoys moderate proinflammatory signals. However, other decoys can reduce the activity of antiinflammatory cytokines, such as IL-10 or IL-13 [417]. Thus, such receptors should produce graduated negative feedback, tending to keep immune responses within upper and lower limits. This is particularly well described for inducible soluble receptors. With IL-1RII, its availability is increased by antiinflammatory agents, such as IL-4 and glucocorticoids, but inhibited by IFN-γ. A more general regulator of proinflammatory responses is D6, a promiscuous decoy and scavenger of many inflammatory chemokines (ligands of CCR1 through to CCR5) and located in lymphatic and placental endothelium [418]. Another recently described decoy receptor is sRAGE [380].

9.2.6.5 Regulation of Immune Responses by the Adaptive Immune System

Responses generated in the adaptive immune system amplify the innate responses that instruct them, whether these be pro- or antiinflammatory. The main innate signal for Th1 amplification of (proinflammatory) immunity is IL-12, and this is tightly regulated to avoid excessive magnitude or duration of action. Factors opposing IL-12 'overshoot' may include IL-10, IFN-α, IFN β, ligation of FcγR, complement or scavenger receptors, PGE_2 and corticosteroids, as well as several 'proinflammatory' mediators such as TNF-α and some chemokines. It is important that activated T cells should not trigger IL-12 overproduction indiscriminately through CD40–CD40L interaction with bystander APCs in the tissues, but this is prevented by the requirement for initial APC activation by microbial stimulus during an innate response before co-stimulatory amplification can become operative in the adaptive response [419]. IFN-γ is well established as the prototypical T cell-derived (Th1) proinflammatory cytokine acting on macrophages. Recent research suggests that an APC-triggered T-cell response is a 'one-off' process and restimulation is ineffective unless new T cells are recruited [420]. Another T-cell-derived cytokine secreted by activated memory $CD4^+$ T cells is IL-17, and it, too, has the capability of interacting with macrophages to augment the output of proinflammatory cytokines (TNF-α and IL-1β) and eicosanoids. The IL-12-related cytokine, IL-23, from proinflammatory activated macrophages may be a major instigator of the IL-17-producing T cell phenotype [421]. At the opposite (antiinflammatory) end of the spectrum, antigen-specific regulatory T cells can also receive instruction from the innate immune system and, in turn, augment an antiinflammatory microenvironment by suppressing proinflammatory T cells, restraining DC maturation and APC function and promoting an antiinflammatory (IL-10 secreting) phenotype [422]. The novel cytokine IL27, produced mainly by macrophages and DCs, is an inhibitor of both Th1 and Th2 responses but, paradoxically, helps to initiate Th1 responses. The IL-27R (WSX-1 combined with gp130) is mainly expressed on naive T cells and NK cells and its ligand, IL-27, can activate at least four Jak/STAT pathways (Jak1 and STAT1, STAT3, STAT4 and STAT5) and exerts a modulating effect, which depends on the prevailing T cell phenotype [423]. In naive murine $CD4^+$ T cells the Th1-specific transcription factor T bet and its gene product IL-12Rβ2 are strongly induced by IL-27 [424]. In activated Th1 cells, as following IL-12 stimulation, STAT4 will be operative already and in this case IL-27 might inhibit the response by additional activation of STAT1 and STAT3. The STAT3 inhibitory pathway, in particular, also mediates the downregulatory effects of IL-6 and IL-10, as already discussed. Conversely, GATA 3 is a crucial transcription factor for Th2 responses and is inhibited by IL-27. GATA 3 is itself induced by IL-4 through a STAT6 pathway and reciprocally inhibits IL-12Rβ2 and STAT4 expression, which are involved in the development of Th1 responsiveness. Although much of the detail is not yet known, these preliminary findings suggest that IL-27 is not a primary determinant of the polarity of an immune response, but is a strong regulator of both excessive pro- and excessive antiinflammatory activity. Accordingly, WSX knock-out mice demonstrate unregulated combined Th1 and Th2 hyperactivity in response to *Toxoplasma gondii* [425] or *Trypanosoma cruzi* infection [426] and in the case of *Leishmania* a predominant initial unregulated Th2 hyperactivity [427]. Excessive Th2 responsiveness encourages parasitemia while excessive Th1 responsiveness causes progressive and eventually fatal proinflammatory liver damage. Further interactions between macrophages and T cells, particularly in the generation of tolerance, are considered later, but it notable that TLR3 and TLR9 are now known to be expressed on activated $CD4^+$ T cells and, on binding their respective ligands (dsRNA and CpG DNA), prolong T cell survival [428].

9.2.6.6 Regulation of Immune Responses by Apoptosis

The effects of apoptosis are apparent at every stage of an immune response, affecting its amplitude and/or duration [429]. Apoptotic mechanisms modulating immune responses include apoptosis of effector macrophages, secondary effects on macrophage phenotypes from uptake of apoptotic cells and feed-forward effects of innate cytokines on T cells, such as the requirement for IL-12 to prevent Th1 apoptosis [430]. TNF-α and FasL directly trigger intracellular death pathways, as do glucocorticoids. The balance of survival factors versus apoptotic factors in immune cells is complex, depending on the expression and interplay of multiple pro- and anti-apoptotic genes.

9.2.6.7 Interaction of Regulatory Mechanisms during Immune Responses

In conclusion, all these mechanisms fit a model based on simple principles. Firstly, innate immunity instructs adaptive immunity, implying that the proinflammatory/antiinflammatory balance (or polarity) of an immune response is programmed in the innate immune system. Secondly, specific primary signaling pathways link biologically significant inputs to appropriate biologically significant outputs. The decision as to what constitutes a significant input/output must be intuitive, but based on the principles of evolution might include TLR stimulation leading to effector mechanisms for microbial killing, such as NO; apoptosis leading to antiinflammatory removal of cells or immune surveillance; and tissue damage (proinflammation) leading to tissue repair (antiinflammation). The most direct or prominent signaling route(s) between these chosen inputs and outputs must then be defined pragmatically as the primary signaling (transduction) pathway(s). Proinflammatory and antiinflammatory responses are either amplified by positive feedback (direct or indirect) or reduced by negative feedback (direct or indirect), making up subsidiary or secondary signaling pathways. Negative feedback mechanisms, for example, are initiated by the primary signaling pathways and serve to reduce (dampen) or reverse them. When acting within the same cell (intracrine, in cis), such opposing mechanisms may be contemporaneous with the build-up of innate responses. Delayed or sequential responses (acting in trans) may be directed back onto the same cell (autocrine) or onto adjacent cells (paracrine) or even distant cells, via blood (endocrine). Adaptive tolerance and immunity are viewed as special cases of delayed feedback. Some of the regulating molecules involved in secondary signaling, such as those working through gp130-containing receptors, are particularly versatile and can dampen both pro- and antiinflammatory immune responses in both the innate and adaptive immune systems. Macrophages are central to the successful operation of immune signaling networks because they can receive primary inputs and deliver primary outputs (through their diverse phenotypes) and contain all the necessary primary and secondary signaling apparatus. They are, for the same reasons, often involved in immune system failures, such as sustaining an inappropriate proinflammatory output in autoimmunity or an inappropriate antiinflammatory output in cancer.

9.2.7
Macrophage Diversity: an Overview

The determinants of all macrophage phenotypic diversity, including differentiation and maturation, activation and 'deactivation', proinflammatory, 'transitional' and antiinflammatory status, survival and apoptosis, are the induced and sequential expression of gene products. Heterogeneity is the result of gene transcription programs induced by multivarious environmental stimuli, providing a huge range of phenotypes. Only necrotic cell-death might be excluded because this is simple chemical disruption of the cell. The dynamics of the tissue microenvironment accounts for inducible gene expression in precursor monocytes and determines the multiple phenotypes of macrophages and DCs and their subtypes. Functional macrophage diversity is caused by changes in tissue microenvironments that include the transduction of inputs from cytokines, hormones and other bioactive molecules, as well as exogenous inputs from the wider environment, notably infection. Macrophages are seen primarily as sentinels of the innate immune system, intercepting and reacting directly to microbial stimuli [431], and DCs primarily as sentinels of the adaptive immune system. Therefore, the ratio of the two cell types produced at any one time in any non-lymphoid tissue (though difficult to measure, as activated DCs usually emigrate) will determine the initial balance of innate and adaptive immunity. Macrophage input necessarily precedes output but the output can in turn profoundly and directly alter the shaping of the microenvironment – the activated cell is both a transducer and effector, with perhaps more potential to change a local microenvironment than any other immune cell. Much of traditional 'polarized' immunology is currently unraveling because inputs and outputs of immune cells are far more complex and varied than previously envisaged. Intracellular transduction pathways integrate a multiplicity of signals and generate a multiplicity of outputs [136, 139]. Even the boundaries between macrophages and DCs are blurred with redundancy of function where their phenotypes overlap. DCs and macrophages may sometimes develop in the same conditions from common precursors or one may differentiate from the other. With the advent of total genome arrays it will be particularly interesting to discover to what extent the diversity of macrophage outputs generated by innate and adaptive inputs overlap and to what extent they differ. In the current state of uncertainty, macrophage phenotypes are probably best specified in terms of as many of the relevant parameters as possible. The convention is developing of prefixing the responding cell type with some of the important inputs, such as TNF-DCs or IFN-DCs [432] and of extrapolating back from patterns of multiple gene expression ('signatures') to putative inputs [433]. From macrophage diversity emerge recurring and unifying themes. Macrophages in leishmaniasis and cancer share many common phenotypic features, as do macrophages in listeriosis and organ-specific autoimmunity. Increasingly, insights into one condition are serving to illuminate another.

Acknowledgments

I thank the library staff of Epsom General Hospital, part of the Epsom and St Helier NHS Trust, for their unfailing support over the past seven years. This chapter is dedicated to my wife, Liz.

References

1 Meister, M. (2004): Blood cells of *Drosophila*: cell lineages and role in host defence. *Curr. Opin. Immunol.* 16: 10–15.

2 Fichelson, P., Gho, M. (2003): The glial cell undergoes apoptosis in the microchaete lineage of *Drosophila*. *Development* 130: 123–133.

3 Sears H.C., Kennedy C.J., Garrity P.A. (2003): Macrophage-mediated corpse engulfment is required for normal *Drosophila* CNS morphogenesis. *Development* 130: 3557–3565.

4 Böse, J., Gruber, A. D., Helming, L., Schiebe, S., Wegener, I., Hafner, M., Beales, M., Köntgen, F., Lengeling, A. (2004): The phosphatidylserine receptor has essential functions during embryogenesis but not in apoptotic cell removal. *J. Biol.* 3: 15.

5 Manaka, J., Kuraishi, T., Shiratsuchi, A., Nakai, Y., Higashida, H., Henson, P., Nakanishi, Y. (2004): Draper-mediated and phosphatidylserine-independent phagocytosis of apoptotic cells by *Drosophila* hemocytes/macrophages. *J. Biol. Chem.* 279: 48 466–48 476.

6 Parker L., Stathakis D.G., Arora K. (2004): Regulation of BMP and activin signaling in *Drosophila*. *Prog. Mol. Subcell. Biol.* 34: 73–101.

7 Letterio, J.J., Roberts, A.B. (1998): Regulation of immune responses by TGF-β. *Annu. Rev. Immunol.* 16: 137–161.

8 Hyman, C.A., Bartholin, L., Newfeld, S.J., Wotton, D (2003): *Drosophila* TGIF proteins are transcriptional activators. *Mol. Cell. Biol.* 23: 9262–9274.

9 Newfeld, S. J., Wisotzkey, R. G., Kumar, S. (1999): Molecular evolution of a developmental pathway: phylogenetic analyses of transforming growth factor-β family ligands, receptors and Smad signal transducers. *Genetics* 152: 783–795.

10 Le, Y., Iribarren, P., Gong, W., Cui, Y., Zhang, X., Wang, J.M. (2004): TGF-β1 disrupts endotoxin signaling in microglial cells through Smad3 and MAPK pathways. *J. Immunol.* 173: 962–968.

11 Watanabe, T., Jono, H., Han, J., Lim, D. J., Li, J-D. (2004): Synergistic activation of NF-κB by nontypeable Haemophilus influenzae and tumor necrosis factor α. *Proc. Natl. Acad. Sci. U.S.A.* 101: 3563–3568.

12 Xiao Y. Q., Malcolm, K., Worthen, G. S., Gardai, S., Schiemann, W. P., Fadok, V. A., Bratton, D. L., Henson, P. M. (2002): Cross-talk between ERK and p38 MAPK mediates selective suppression of pro-inflammatory cytokines by transforming growth factor-β. *J. Biol. Chem.* 277: 14 884–14 893.

13 Jono, H., Xu, H., Kai, H., Lim, D.J., Kim, Y.S., Feng, X.-H., Li, J-D. (2003): Transforming growth factor-β-Smad pathway negatively regulates nontypeable Haemophilus influenzae-induced MUC5AC mucin transcription via mitogen-activated protein kinase (MAPK) phosphatase-1 dependent inhibition of p38 MAPK. *J. Biol. Chem.* 278: 27 811–27 819.

14 Korzus, E. (2003): The relation of transcription to memory formation. *Acta Biochim. Pol.* 50: 775–782.

15 Rämet, M., Pearson, A., Manfruelli, P., Li, X., Koziel, H., Göbel, V., Chung, E., Krieger, M., Ezekowitz, R. A. B. (2001): *Drosophila* scavenger receptor CI is a pattern recognition receptor for bacteria. *Immunity* 15: 1027–1038.

16 Poels, J., Franssens, V., Van Loy, T., Martinez, A., Suner, M.-M., Dunbar, S.J., De Loof, A., Vanden Broeck, J. (2004): Isoforms of cyclic AMP response element binding proteins in *Drosophila* S2 cells. *Biochem. Biophys. Res. Commun.* 320: 318–324.

17 Horiuchi, J., Jiang, W., Zhou, H., Wu, P., Yin, J. C. P. (2004): Phosphorylation of conserved casein kinase sites regulates CREB DNA binding in *Drosophila*. *J. Biol. Chem.* 279: 12 117–12 125.

18 Iourgenko, V., Zhang, W., Mickanin, C., Daly, I., Jiang, C., Hexham, J. M., Orth, A. P., Miraglia, L., Meltzer, J., Garza, D., Chirn, G-W., McWhinnie, E., Cohen, D., Skaelton, J., Terry, R., Yu, Y., Bodian, D., Buxton, F. P., Zhu, J., Song, C.,

Labow, M. A. (2003): Identification of a family of cAMP response element-binding protein coactivators by genome-scale functional analysis in mammalian cells. *Proc. Natl. Acad. Sci. U.S.A.* 100: 12 147–12 152.

19 Hoffmann J. A. (2003): The immune response of *Drosophila*. *Nature* 426:33–38.

20 Beutler, B. (2004): Innate immunity: an overview. *Mol. Immunol.* 40: 845–859.

21 Naitza, S., Ligoxygakis, P. (2004): Antimicrobial defenses in *Drosophila*: the story so far. *Mol. Immunol.* 40: 887–896.

22 Leclerc, V., Reichart, J.M. (2004): The immune response of *Drosophila* melanogaster. *Immunol. Rev.* 198: 59–71.

23 Kocks, C., Maehr, R., Overkleeft, H. S., Wang, E. W., Lackshmanan, K., Lennon-Duménil, A.-M., Ploegh, H. L., Kessler, B. M. (2003): Functional proteomics of the active cysteine protease content in *Drosophila* S2 cells. *Mol. Cell. Proteom.* 2: 1188–1197.

24 Hedengren-Olcott, M., Olcott, M.C., Mooney, D.T., Ekengren, S., Geller, B. L., Taylor, B. J. (2004): Differential activation of the NF-κB-like factors Relish and Dif in *Drosophila* melanogaster by fungi and gram-positive bacteria. *J. Biol. Chem.* 279: 21 121–21 127.

25 Silverman, N., Zhou, R., Erlich, R. L., Hunter, M., Bernstein, E., Schneider D., Maniatis, T. (2003): Immune activation of NF-κB and JNK requires *Drosophila* TAK1. *J. Biol. Chem.* 278: 48 928–48 934.

26 Hu, X., Yagi, Y., Tanji, T., Zhou, S., Ip, Y. T. (2004): Multimerization and interaction of Toll and Spätzle in *Drosophila*. *Proc. Natl. Acad. Sci. U.S.A.* 101: 9369–9374.

27 Dziarski, R. (2004): Peptidoglycan recognition proteins (PGRPs). *Mol. Immunol.* 877–886.

28 Steiner, H. (2004): Peptidoglycan recognition proteins: on and off switches for innate immunity. *Immunol. Rev.* 198: 83–96.

29 Mellroth, P, Karlsson, J., Steiner, H. (2003): A scavenger function for *Drosophila* peptidoglycan recognition protein. *J. Biol. Chem.* 278: 7059–7064.

30 Igaki, T., Kanda, H., Yamamoto-Goto, Y., Kanuka, H., Aigaki, T., Miura, M. (2002): Eiger, a TNF superfamily ligand that triggers the *Drosophila* JNK pathway. *EMBO J.* 21: 3009–3018.

31 Kaupilla, S., Maaty, W. S. A.,m Chen, P., Tomar, R.S., Eby, M.T., Chapo, J., Chew, S., Rathore, N., Zachariah, S., Sinha, S. K., Abrams, J. M., Chaudhary, P. M. (2003): Eiger and its receptor, Wengen, comprise a TNF-like system in *Drosophila*. *Oncogene* 22: 4860–4867.

32 Staziv, Y., Regulski, M., Kuzin, B., Tully, T., Enikolopov, G. (2001): The *Drosophila* nitric-oxide synthase gene (dNOS) encodes a family of proteins that can modulate NOS activity by acting as dominant negative regulators. *J. Biol. Chem.* 276: 42 241–42 251.

33 Foley, E., O'Farrell, P.H.O. (2002): Nitric oxide contributes to induction of innate immune responses to gram-negative bacteria in *Drosophila*. *Genes Develop.* 17: 115–125.

34 Agaisse, H., Petersen, U.-M., Boutros, M., Mathey-Prevot, B., Perrimon, N. (2003): Signaling role of hemocytes in *Drosophila* Jak/STAT-dependent response to septic injury. *Devlop. Cell* 5: 441–450.

35 Bach, E.A., Vincent, S., Zeidler, M. P., Perrimon, N. (2003): A sensitized genetic screen to identify novel regulators and components of the *Drosophila* Janus kinase/signal transducer and activator of transcription pathway. *Genetics* 165: 1149–1166.

36 Bodian, D. L., Leung, S., Chiu, H., Govind, S. (2004): Cytokines in *Drosophila* hematopoiesis and cellular immunity. *Prog. Mol. Subcell. Biol.* 34: 27–36.

37 Lagueux, M., Perrodou, E., Levashina, E. A., Capovilla, M., Hoffmann, J. A. (2000): Constitutive expression of a complement-like protein in Toll and JAK gain-of-function mutants of *Drosophila*. *Proc. Natl. Acad. Sci. U.S.A.* 97: 114 27–11 432.

38 Steinert, M., Leippe, M., Roeder, T. (2003): Surrogate hosts: protozoa and invertebrates as models for studying pathogen-host interactions. *Int. J. Med. Microbiol.* 293: 321–332.

39 Roxström-Lindquist, K., Terenius, O., Faye, I. (2004): Parasite-specific immune response in adult *Drosophila* melanogaster: a genomic study. *EMBO Rep.* 5: 207–212.

40 Basset, A., Khush, R. S., Braun, A., Gardan, L., Boccard, F., Hoffmann, J. A., Lemaitre, B. (2000): The phytopathogenic bacteria Erwinia caratovora infects *Drosophila* and activates an immune response. *Proc. Natl. Acad. Sci. U.S.A.* 97: 3376–3381.

41 Cheng, L. W., Portnoy, D. A. (2003): *Drosophila* S2 cells: an alternative infection model for *Listeria monocytogenes*. *Cell. Microbiol.* 5: 875–885.

42 Dionne MS, Ghori N, Schneider DS (2003): *Drosophila* melanogaster is a genetically tractable model host for *Mycobacterium marinum*. *Infect. Immunity* 71: 3540–3550.

43 De Gregorio, E., Spellman, P.T., Rubin, G. M., Lemaitre, B. (2001): Genome-wide analysis of the *Drosophila* immune response by using oligonucleotide microarrays. *Proc. Natl. Acad. Sci. U.S.A.* 98: 12590–12595.

44 Foley, E., O'Farrell, P.H. (2004): Functional dissection of an innate immune response by a genome-wide RNAi screen. *PloS Biol.* 2: 1091–1106.

45 Lopes, E. S., Araujo, H. M. (2004): The maternal JAK/STAT pathway of *Drosophila* regulates embryonic dorsal-ventral patterning. *Brazil. J. Med. Biol. Res.* 37: 1811–1818.

46 Mushegian, A., Medzhitov, R. (2001): Evolutionary perspective on innate immune recognition. *J. Cell. Biol.* 155: 705–710.

47 Boutros, M., Agaisse, H., Perrimon, N. (2002): Sequential activation of signaling pathways during innate immune responses in *Drosophila*. *Develop. Cell* 3: 711–722.

48 Cha, G-H., Cho, K. S., Lee, J. H., Kim, M., Kim, E., Park, J., Lee, S. B., Chung, J. (2003): Discrete functions of TRAF1 and TRAF2 in *Drosophila* melanogaster mediated by c-Jun N-terminal kinase and NF-κB-dependent signaling pathways. *Mol. Cell. Biol.* 23: 7982–7991.

49 Wajant, H., Scheurich, P. (2004): Analogies between *Drosophila* and mammalian TRAF pathways. *Prog. Mol. Subcell. Biol.* 34: 47–72.

50 Park, J. M., Brady, H., Ruocco, M. G., Sun, H., Williams, DA., Lee, S. J., Kato Jr., T., Richards, N., Chan, K., Mercurio, F., Karin, M., Wasserman, S.A. (2004): Targeting of TAK1 by the NF-κB protein Relish regulates the JNK-mediated immune response in *Drosophila*. *Genes Develop.* 18: 584–594.

51 Danchin, E., Vitiello, V., Richard, O., Gouret, P., McDermott, M. F., Pantarotti, P. (2004): The major histocompatability complex origin. *Immunol. Rev.* 198: 216–232.

52 Brazil, M. I., Weiß, S., Stockinger, B. (1997): Excessive degradation of intracellular protein in macrophages prevents presentation in the context of major histocompatability complex class II molecules. *Eur. J. Immunol.* 27: 1506–1514.

53 Paglia, P., Colombo, M. P. (2002): Macrophages as antigen-presenting cells: relationship to dendritic cells and use in vaccination studies. In: Burke, B., Lewis, C. E. (Eds.), *The Macrophage*: Second edition, Oxford University Press, Oxford, UK. 103–137.

54 Banyer, J. L., Hapel, A. J. (1999): Myb-transformed haematopoietic cells as a model for monocyte differentiationinto dendritic cells and macrophages. *J. Leukoc. Biol.* 66: 217–223.

55 Cecchini, M. G., Dominguez, M. G., Mocci, S., Wetterwald, A., Felix, R., Fleisch, H., Chisholm, O., Hofstetter, W., Pollard, J. W., Stanley, E.R. (1994): Role of colony stimulating factor-1 in the establishment and regulation of tissue macrophages during postnatal development of the mouse. *Development* 120: 1357–1372.

56 Matyszak, M. K., Perry, V. H. (2002): Macrophages in the central nervous system. In: Burke, B., Lewis, C. E. (Eds.), *The Macrophage*: Second edition, Oxford University Press, Oxford, UK. 523–547.

57 Ruwhof, C., Canning, M. A., Grotenhuis, K., De Wit, H. J., Florencia, Z. Z., De Haan-Meulman, M., Drexhage, H.A.

(2002): Accessory cells with a veiled morphology and movement pattern generated from monocytes after avoidance of plastic adherence and of NADPH oxidase activation. A comparison with GM-CSF/IL-4-induced monocyte-derived dendritic cells. *Immunobiology* 205: 247–266.

58 Rimaniol, A. C., Gras, G., Verdier, F., Capel, F., Grigoriev, V. B., Porcheray, F., Sauzeat, E., Fournier, J. G., Clayette, P., Siegrist, C. A., Dormont, D. (2004): Aluminium hydroxide adjuvant induces macrophage differentiation towards a specialized antigen-presenting cell type. *Vaccine* 22: 3127–3135.

59 Geissmann, F., Jung, S., Littman, D.R. (2003): Blood monocytes consist of two principal subsets with distinct migratory properties. *Immunity* 19: 71–82.

60 Zhou, L.-J., Tedder, T. F. (1996): $CD14^+$ blood monocytes can differentiate into functionally mature $CD38^+$ dendritic cells. *Proc. Natl. Acad. Sci. U.S.A.* 93: 2588–2592.

61 Palucka, K. A., Taquet, N., Sanchez-Chapuis, F., Gluckman, J. C. (1998): Dendritic cells as the terminal stage of monocyte differentiation. *J. Immunol.* 160: 4587–4595.

62 Buelens, C., Verhasselt, V., De Groote, D., Thielemans, K., Goldman, M., Willems, F. (1997): Interleukin-10 prevents the generation of dendritic cells from human peripheral blood mononuclear cells cultured with interleukin-4 and granulocyte/macrophage-colony-stimulating factor. *Eur. J. Immunol.* 27: 756–762.

63 Allavena, P., Piemonti, L., Longoni, D., Bernasconi, S., Stoppacciaro, A., Ruco, L., Mantovani, A. (1998): IL-10 prevents the differentiation of monocytes to dendritic cells but promotes their maturation to macrophages. *Eur. J. Immunol.* 28: 359–369.

64 Suzuki, H., Katayama, N., Ikuta, Y., Mukai, K., Fujieda, A., Mitani, H., Araki, H., Miyashita, H., Hoshino, N., Nishikawa, H., Nishii, K., Minami, N., Shiku, H. (2004): Activities of granulocyte-macrophage colony-stimulating factor and interleukin-3 on monocytes. *Am. J. Haematol.* 75: 179–189.

65 Shi, C., Zhang, X., Chen, Z., Sulaiman, K., Feinberg, M. W., Ballantyne, C. M., Jain, M. K., Simon, D. I. (2004): Integrin engagement regulates monocyte differentiation through the forkhead transcription factor Foxp1. *J. Clin. Invest.* 114: 408–418.

66 Hashimoto, S.-i., Yamada, M., Motoyoshi, K., Akagawa. (1997): Enhancement of macrophage colony-stimulating factor-induced growth and differentiation of human monocytes by interleukin-10. *Blood* 89: 315–321.

67 Chomarat, P., Banchereau, J., Davoust, J., Palucka, A. K. (2000): IL-6 switches the differentiation of monocytes from dendritic cells to macrophages. *Nat. Immunol.* 1: 510–514.

68 Rieser, C., Ramoner, R., Böck, G., Deo, Y. M., Hölti, L., Bartsch, G., Thurnher, M. (1998): Human monocyte-derived dendritic cells produce macrophage colony-stimulating factor: enhancement of c-fms expression by interleukin-10. *Eur. J. Immunol.* 28: 2283–2288.

69 Valledor, A. F., Xaus, J., Marquès, L., Celada, A. (1999): Macrophage colony-stimulating factor induces the expression of mitogen-activated protein kinase phosphatase-1 through a protein kinase C-dependent pathway. *J. Immunol.* 163: 2452–2462.

70 Palucka, K., Banchereau, J., Caux, C., Dezutter-Dambuyant, C., Liu, Y.-J. (2002): Isolation and propagation of human dendritic cells. *Methods Microbiol.* 32: 591–620.

71 Krause, S. W., Rehli, M., Andreesen, R. (2002): Isolation, characterization and cultivation of human monocytes and macrophages. *Methods Microbiol.* 32: 767–784.

72 Geissmann F, Prost C, Monnet J-P, Dy M, Brousse N, Hermine O (1998): Transforming growth factor $\beta1$, in the presence of granulocyte/macrophage colony-stimulating factor and interleukin 4, induces differentiation of human peripheral blood monocytes into dendritic Langerhans cells. *J. Exp. Med.* 187: 961–966.

73 Reddy, A., Sapp, M., Feldman, M., Subklewe, M., Bhardwaj, N. (1997): A monocyte conditioned medium is more effective than defined cytokines in mediating the terminal maturation of human dendritic cells. *Blood* 90: 3640–3646.

74 Sato, M., Takayama, T., Tanaka, H., Konishi, J., Suzuki, T., Kaiga, T., Tahara, H. (2003): Generation of mature dendritic cells fully cabale of T helper type1 polarization using OK-432 combine with prostaglandin E$_2$. *Cancer Sci.* 94: 1091–1098.

75 Ebner, S., Hofer, S., Nguyen, V. A., Furhapter, C., Herold, M., Fritsch, P., Heufler, C., Romani, N. (2002): A novel role for IL-3: human monocytes cultured in the presence of IL-3 and IL-4 differentiate into dendritic cells that produce less IL-12 and shift Th cell responses toward a Th2 cytokine pattern. *J. Immunol.* 168: 6199–6207.

76 Buelens, C., Bartholomé, E. J., Amraoui, Z., Boutriaux, M., Salmon, I., Thielemans, K., Willems, F., Goldman, M. (2002): Interleukin-3 and interferon β cooperate to induce differentiation of monocytes into dendritic cells with potent helper T-cell stimulatory properties. *Blood* 99: 993–998.

77 Caux, C. (1998): Pathways of development of human dendritic cells. *Eur. J. Dermatol.* 8: 375–384.

78 Santini, S. M., Lapenta, C., Logozzi, M., Parlato, S., Spada, M., Di Pucchio, Belardelli, F. (2000): Type 1 interferon as apowerful adjuvant for monocyte-derived dendritic cell development and activity in vitro and in Hu-PBL-SCID mice. *J. Exp. Med.* 191: 1777–1788.

79 Iking-Konert, C., Csekö, C., Wagner, C., Stegmaier, S., Andrassy, K., Hänsch, G. M. (2001): Transdifferentiation of polymorphonuclear neutrophils: acquisition of CD83 and other functional characteristics of dendritic cells. *J. Mol. Med.* 79: 464–474.

80 Miyamoto, T., Suda, T. (2003): Differentiation and function of osteoclasts. *Keio J. Med.* 52: 1–7.

81 Khapli, S. M., Mangashetti, L. S., Yogesha, S. D., Wani, M. R. (2003): IL-3 acts directly on osteoclast precursors and irreversibly inhibits receptor activator of NF-κB ligand-induced osteoclast differentiation by diverting the cells to macrophage lineage. *J. Immunol.* 171: 142–151.

82 Hinoi, Takarada, T., Yoneda, Y. (2004): Glutamate signaling system in bone. *J. Pharmacol. Sci.* 94: 215–220.

83 Espinosa, L., Itzstein, C., Cheynel, H., Delmas, P.D., Chenu, C. (1999): Active NMDA glutamate receptors are expressed by mammalian osteoclasts. *J. Physiol.* 518: 47–53.

84 Merle, B., Itzstein, C., Delmas, P. D., Chenu, C. (2003): NMDA glutamate receptors are expressed by osteoclast precursors and involved in the regulation of osteoclastogenesis. *J. Cell. Biochem.* 90: 424–436.

85 Matsuo, K., Ray, N. (2004): Ostoclasts, mononuclear phagocytes, and c-fos: new insights into osteoimmunology. *Keio J. Med.* 53: 78–84.

86 Okada, S., Obata, S., Hatano, M., Tokuhisa, T. (2003): Dominant-negative effect of thr c-fos family gene products on inducible NO synthase expression in macrophages. *Int. Immunol.* 15: 1275–1282.

87 Dillon, S., Agrawal, A., Van Dyke, T., Landreth, G., McCauley, L., Koh, A., Maliszewski, C., Akira, S., Pulendran, B. (2004): A Toll-like receptor 2 ligand stimulates Th2 responses in vivo, via induction of extracellular signal-regulated kinase mitigen-activated protein kinase and c-Fos in dendritic cells. *J. Immunol.* 172: 4733–4743.

88 Fox, S. W., Haque, S. J., Lovibond, A. C., Chambers, T. J. (2003): The possible role of TGF-β-induced suppressors of cytokine signaling expression in osteoclast/macrophage lineage commitment in vitro. *J. Immunol.* 170: 3679–3687.

89 Servet-Delprat, C., Arnaud, S., Jurdic, P., Nataf, S., Grasset, M.-F., Soulas, C., Domenget, C., Destaing, O., Rivollier, A., Perret, M., Dumontel, C., Hanau, D., Gilmore, G. L., Belin, M.-f., Rabourdin-Combe, C., Mouchiroud, G. (2002): Flt3$^+$ macrophage precursors commit sequentially to osteoclasts, dendritic

cells and microglia. *BMC Immunol.* 3: 15.

90 Nelson, E. L., Strobl, S., Subleski, J., Prieto, D., Kopp, W. C., Nelson, P. J. (1999): Cycling of human dendritic cell effector phenotypes in response to TNF-α: modification of the current 'maturation' paradigm and implications for in vivo immunoregulation. *FASEB J.* 13: 2021–2030.

91 Fadok, V. A., Bratton, D. L., Konowai, A., Freed, P. W., Westcott, J.Y., Henson, P. M. (1998): Macrophages that have ingested apoptotic cells in vitro inhibit proinflammatory cytokine production through autocrine/paracrine mechanisms involving TGF-β, PGE$_2$, and PAF. *J. Clin. Invest.* 101: 890–898.

92 Santambrogio, L., Belyanskaya, S.L., Fischer, F.R., Cipriani, B., Brosnan, C.F., Ricciardi-Castagnoli, P., Stern, L.J., Strominger, J.L., Riese, R. (2001): Developmental plasticity of CNS microglia. *Proc. Natl. Acad. Sci. U.S.A.* 98: 6295–6300.

93 Fischer, H.-G., Reichmann, G. (2001): Brain dendritic cells and macrophages/microglia in central nervous system inflammation. *J. Immunol.* 2001: 2717–2726.

94 Fischer, H.-G., Bonifas, U., Reichmann, G. (2000): Phenotype and functions of brain dendritic cells emerging during chronic infection of mice with *Toxoplasma gondii*. *J. Immunol.* 164: 4826–4834.

95 Karman, J., Ling, C., Sandor, M., Fabry, Z. (2004): Dendritic cells in the initiation of immune responses against central nervous system-derived antigens. *Immunol. Lett.* 92: 107–115.

96 Makala. L.H.C., Reyes, J.C.S., Nishikawa, Y., Tsushima, Y., Xaun, X., Huang, X., Ngasawa, H. (2003): A comparison of the phenotype of dendritic cells derived from discrete Peyer's patch macrophages of non-infected and *Toxoplasma gondii* infected mice. *J. Vet. Med. Sci.* 65: 591–597.

97 Lee, Y.-N., Lee, H.-Y., Lee, Kang, H.-K., Kwak, J.-Y., Bae, Y.-S. (2004): Phosphatidic acid positively regulates LPS-induced differentiation of RAW264.7 murine macrophage cell line into dendritic-like cells. *Biochm. Biophys. Res. Commun.* 318: 839–845.

98 Suzuki, Y., Yanagawa, H., Nishioka, Y., Nishimura, N., Takeuchi, E., Sone, S. (2001): Efficient generation of dendritic cells from alveolar and pleural macrophages as well as blood monocytes in patients with lung cancer. *Lung Cancer* 34: 195–205.

99 Havermann, K., Pujol, B.F., Adamkiewicz. (2003): In vitro transformation of monocytes and dendritic cells into endothelial like cells, in Moldovan, N.I. (Ed.) *Novel Angiogenic Mechanisms: Role of Circulating Progenitor Endothelial Cells*, Kluwer Academic/Plenum Publishers, 47–57.

100 Rong, J.X., Shapiro, M., Trogan, E., Fisher, E.A. (2003): Transdifferentiation of mouse aortic smooth muscle cells to a macrophage-like state after cholesterol loading. *Proc. Natl. Acad. Sci. U.S.A.* 100: 13 531–13 536.

101 Charrière, G., Cousin, B., Arnaud, E., André, M., Bacou, F., Pénicaud, L., Casteilla, L. (2003): Preadipocyte conversion to macrophage. Evidence of plasticity. *J. Biol. Chem.* 278: 9850–9855.

102 Weisberg, S.P., McCann, D., Desai, M., Rosenbaum, M., Leibel, R.L., Ferrante Jr., A.W. (2003): Obesity is associated with macrophage accumulation in adipose tissue. *J. Clin. Invest.* 112: 1796–1808.

103 Curat, C. A., Miranville, A., Sengenes, C., Diehl, M., Tonus, C., Busse, R., Bouloumie, A. (2004): From blood monocytes to adipose tissue-resident macrophages: Induction of diapedesis by human mature adipocytes. *Diabetes* 53: 1285–1292.

104 Wilson, S.W., Villadangos, J.A. (2004): Lymphoid organ dendritic cells: beyond the Langerhans cells paradigm. *Immunol. Cell. Biol.* 82: 91–98.

105 Allavena, P., Sica, A., Vecchi, A., Locati, M., Sozzani, S., Mantovani, A. (2000): The chemokine receptor switch paradigm and dendritic cell migration: its significance in tumor tissues. *Immunol. Rev.* 177: 141–149.

106 Forster, R., Schubel, A., Bretfeld, D., Kremmer, E., Renner-Muller, I., Wolf, E., Lipp, M. (1999): CCR7 coordinates the primary immune response by establishing functional microenvironments in secondary lymphoid organs. *Cell* 99: 23–33.

107 Sozzani, S., Ghezzi, S., Iannolo, G., Luini, W., Borsatti, A., Polentarutti, N., Sica, A., Locati, M., Mackay, C., Wells, T. N. C., Biswas, P., Vincenzi, E., Poli, G., Mantovani, A. (1998): Interleukin 10 increases CCR5 expression and HIV infection in human monocytes. *J. Exp. Med*. 187: 439–444.

108 Jeannin, P., Magistrelli, G., Herbault, N., Goetsch, L., Godfroy, S., Charbonnier, P., Gonzalez, A., Delneste, Y. (2003): Outer membrane protein A renders dendritic cells and macrophages responsive to CCL21 and triggers dendritic cells to migrate to secondary lymphoid organs. *Eur. J. Immunol*. 33: 326–333.

109 Fleming, M.D., Pinkus, J.L., Alexander, S.W., Tam, C., Loda, M., Sallan, S.E., Nichols, K.E., Carpentieri, D.F., Pinkus, G.S., Rollins, B.J. (2003): Coincident expression of the chemokine receptors CCR6 and CCR7 by pathologic Langerhans cells in Langerhans cell histiocytosis. *Blood* 101: 2473–2475.

110 Luft, T., Jefford, M., Luetjens, P., Toy, T., Hochrein, H., Masterman, K.-A., Maliszewski, C., Shortman, K., Cebon, J., Maraskovsky. (2002): Functionally distinct dendritic cell (DC) populations induced by physiologic stimuli: prostaglandin E2 regulates the migratory capacity of specific DC subsets. *Blood*: 100: 1362–1372.

111 Scandella, E., Men, Y., Gillessen, S., Förster, R., Groettrup, M. (2002): Prostaglandin E_2 is a key factor for CCR7 surface expression and migration of monocyte-derived dendritic cells. *Blood* 100: 1345–1361.

112 Scandella, E., Men, Y., Legler, D.F., Gillessen, S., Prickler, L., Ludewig, B., Groettrup, M. (2004): CCL9/CCL21-triggered signal transduction and migration of dendritic cells requires prostaglandin E_2. *Blood* 103: 1596–1601.

113 J-J. Lee, M. Takei, S. Hori, Y. Inoue, Y. Harada, R. Tanosaki, Y Kanda, M. Kami. A. Makimoto, S.Mineishi, H. Kawai, A. Shimosaka, Y. Heike, Y. Ikarashi, H. Wakasugi, Y. Takaue, T-J. Hwang, H-J. Kim, T. Kakizoe (2002): The role of PGE_2 in the differentiation of dendritic cells: how do dendritic cells influence T-cell polarization and chemokine receptor expression? *Stem Cells* 20: 448–459.

114 Fogel-Petrovic, M., Long, J. A., Knight, D. A., Thompson, P. J., Upham, J. W. (2004): Activated human dendritic cells express inducible cyclo-oxygenase and synthesize prostaglandin E_2 but not prostaglandin D2. *Immunol. Cell. Biol*. 82: 47–54.

115 Kubo, S., Takahashi, H. K., Takei, M., Iwagaki, H., Yoshino, T., Tanaka, N., Mori, S., Nishibori, M. (2004): Eprostanoid (EP)2/EP4 receptor-dependent maturation of human monocyte-derived dendritic cells and induction of helper/T2 polarization. *J. Pharmacol. Exp. Ther*. 309: 1213–1220.

116 Randolph, G.J., Sanchez-Schmitz, G., Liebman, R.M., Shökel, K. (2002): The CD16+ (FcγRIII+) subset of human monocytes preferentially becomes migratory dendritic clls in a model tissue setting. *J. Exp. Med*. 196: 517–527.

117 de Baey, A., Mnede, I., Baretton, G., Greiner, A., Hartl, W.H., Baeurle, P.A., Diepolder, H.M. (2003): A subset of human dendritic cells in the T cell area of mucosa-associated lymphoid tissue with a high potential to produce TNF-α. *J. Immunol*. 170: 5089–5094.

118 Um, H. D., Cho, Y-H., Kim, D. K., Shin, J-R, Lee.Y-J., Choi, K-S., Kang, J-M., Lee, M-G. (2004): TNF-α suppresses dendritic cell death and the production of reactive oxygen intermediates induced by plasma withdrawal. *Exp. Dermatol*. 13: 282–288.

119 Hemmi, H., Yoshino, M., Yamazaki, H., Naito, M., Iyoda, T., Omatsu, Y., Shimoyama, S., Letterio, J.J., Nakabayashi, T., Tagaya, H., Yamane, T., Ogawa, M., Nishikawa, S.-I., Ryoke, K., Inaba, K., Hayashi, S.-I., Kunisada, T. (2001): Skin antigens in the steady state are traf-

ficked to the regional lymph nodes by transforming growth factor-β1-dependent cells. *Int. Immunol.* 13: 695–704.

120 Verbovetski, I., Bychkov, H., Trahtemberg, U., Shapira, I., Hareuveni, M., Ben-Tal, O., Kutikov, I., Gill, O., Mevorach, D. (2002): Opsonoization of apoptotic cells by autologous iC3b facilitates clearance by immature dendritic cells, downregulates DR and CD86 and upregulates CC chemokine receptor 7. *J. Exp. Med.* 196: 1553–1561.

121 Yoshino, M., Yamazaki. H., Nakano, H., Kakiuchi, T., Ryoke, K., Kunisada, T., Hayashi, S.-I. (2003): Distinct antigen trafficking from skin in the steady and active state. *Int. Immunol.* 15: 773–779.

122 Fedele, G., Frasca, L., Palazzo, R., Ferrero, E., Malavasa, F., Ausiello, C.M. (2004): CD38 is expressed on human mature monocyte-derived dendritic cells and is functionally involved in CD83 expression and IL-12 induction. *Eur. J. Immunol.* 34: 1342–1350.

123 Höpken, U.E., Lipp, M. (2004): All roads lead to Rome: triggering dendritic cell migration. *Immunity* 20: 247–253.

124 Bouchon, A., Hernandez-Munain, C., Cella, M., Colonna, M. (2001): A DAP12-mediated pathway regulates expression of CC chemokine receptor 7 and maturation of human dendritic cells. *J. Exp. Med.* 194: 1111–1122.

125 Aoki, N., Zganiacz, A., Margetts, P., Xing, Z. (2004): Differential regulation of DAP12 and molecules associated with DAP12 during host responses to mycobacterial infection. *Infect. Immun.* 72: 2477–2483.

126 Rotta, G., Edwards, E.W., Sangaletti, S., Bennett, C., Ronzoni, S., Colombo, M.P., Steinman, R.M., Randolph, G.J., Rescigno, M. (2003): Lipopolysaccharide or whole bacteria block the conversion of inflammatory monocytes into dendritic cells in vivo. *J. Exp. Med.* 198: 1253–1263.

127 Mariotti, S., Teloni, R., Iona, E., Fattorini, L., Romagnoli, G., Gagliardi, M. C., Orefici, G., Nisini, R., (2004): *Mycobacterium tuberculosis* diverts alpha interferon-induced monocyte differentiation from dendritic cells into immunoprivileged macrophage-like cells. *Infect. Immun.* 72: 4385–4392.

128 García-Romo, G. S., Pedroza-Gonzáles, A., Aguilar-León, Orozco-Estevez, H., D., Lambrecht, B. N., Estrada-García, I., Flores-Romo, L. (2004): Airways infection with virulent *Mycobacterium tuberculosis* delays the influx of dendritic cells and the expression of costimulatory molecules in mediastinal lymph nodes. *Immunology* 112: 661–668.

129 Lucas, M., Stuart, L.M., Savill, J., Lacy-Hulbert, A. (2003): Apoptotic cells and innate immune stimuli combine to regulate macrophage cytokine secretion. *J. Immunol.* 2003: 2610–2615.

130 Filaci, G., Contini, P., Fravega, M., Fenoglio, D., Azzarone, B., Julien-Giro Fiocca R., Boggio, M., Nechi, V., De Lerma Barbaro, A., Merlo, A., Rizzi Ghio, M., Setti, M., Puppo, F., Zanetti, M., Indiveri, F. (2003): Apoptoic DNA binds to HLA classII molecules inhibiting antigen presentation and participating in the development of anti-inflammatory functional behaviour of phagocytic macrophages. *Hum. Immunol.* 64: 9–20.

131 Kirschnek, S., Scheffel, J., Heinzmann, U., Häcker, G. (2004): Necrosis-like cell death induced by bacteria in mouse macrophages. *Eur. J. Immunol.* 34: 1461–1471.

132 Giordano, D., Magaletti, D.M., Clark, E.A., Beavo, J.A. (2003): Cyclic nucleotides promote monocyte differentiation toward a DC-SIGN(+) (CD209) intermediate cell and impair differentiation into dendritic cells. *J. Immunol.* 2003: 6421–6430.

133 Tschoep, K., Manning, T.C., Harlin, H., George, C., Johnson, M., Gajewski, T.F. (2003): Disparate functions of immature and mature human myeloid dendritic cells: implications for dendritic cell-based vaccines. *J. Leukoc. Biol.* 74: 69–80.

134 Takahashi, M., Kurosaka, K., Kobayashi, Y. (2004): Immature dendritic cells reduce proinflammatory cytokine production by a coculture of macrophages and apoptotic cells in a cell-to-cell con-

135 Hegde, S., Pahne, J., Smola-Hess, S. (2004): Novel immunosuppressive properties of interleukin-6 in dendritic cells: inhibition of NF-κB binding activity and CCR7 expression. *FASEB J.* 18: 1439–1441.

136 Akira, S., Takeda, K. (2004): Toll-like receptor signalling. *Nat. Rev. Immunol.* 4: 499–511.

137 Yamamoto, M., Takeda, K., Akira, S. (2004): TIR domain-containing adaptors define the specificity of TLR signaling. *Mol. Immunol.* 40: 861–868.

138 O'Neill, L. A. J. (2004): After the Toll rush. *Science* 303: 1481–1482.

139 Underhill, D.M. (2003): Toll-like receptors: networking for success. *Eur. J. Immunol.* 33: 1767–1775.

140 Triantafilou, M., Miyake, K., Golenbock, D.T., Triantafilou, K. (2002): Mediators of innate immune recognition of bacteria concentrate in lipid rafts and facilitate lipopolysaccharide-induced cell activation. *J. Cell Sci.* 115: 2603–2611.

141 Marshall, A.S.J., Gordon, S. (2004): C-type lectins on the macrophage cell surface – recent findings. *Eur. J. Immunol.* 34: 18–24.

142 Agnes, M. C., Tan, A., Jordansa, R., Geluk, A., Roep, B. O., Ottenhoff, T., Drijfhout, J. W., Koning, F. (1998): Strongly increased efficiency of altered peptide ligands by mannosylation. *Int. Immunol.* 10: 1299–1304.

143 Prigozy, T.I., Sieling, P.A., Clemens, D., Stewart, P.L., Behar, S.M., Porcelli, S.A., Brenner, M.B., Modlin, R.L., Kronenberg, M. (1997): The mannose receptor delivers lipoglycan antigens to endosomes for presentation to T cells by CCD1b molecules. *Immunity* 6: 187–197.

144 Coste, A., Dubourdeau, M., Linas, M.D., Cassaing, S., Lepert, J.-C., Balard, P., Chalmeton, S., Bernad, J., Orfila, C., Séguéla, J.-P., Pipy, B. (2003): PPARγ promotes mannose receptor gene expression in murine macrophages and contributes to the induction of this receptor by IL-13. *Immunity* 19: 329–339.

145 Chieppa, M., Bianchi, G., Doni, A., Del Prete, A., Sironi, M., Laskarin, G., Monti, P., Piemonte, L., Biondi, A., Mantovani, A., Introna, M., Allavena, P. (2003): Cross-linking of the mannose receptor on monocyte-derived dendritic cells activates an anti-inflammatory immunosuppressive program. *J. Immunol.* 171: 4552–4560.

146 Devyatyarova-Johnson, M., Rees, I. H., Robertson, B. D., Turner, M. W., Klein, N. J., Jack, D. L. (2000): The lipopolysaccharide structures of *Salmonella* enterica serovar typhimurium and Neisseria gonorrhoea determine the attachment of human mannose-binding lectin to intact organisms, *Infect. Immun.* 68: 3894–3899.

147 Fraser, I.P., Stuart, L., Ezekowitz, R.A.B. (2004): TLR-independent pattern recognition receptors and anti-inflammatory mechanisms. *J. Endotox. Res.* 10: 120–124.

148 Herre, J., Gordon, S., Brown, G.D. (2004): Dectin-1 and its role in the recognition of β-glucans by macrophages. *Mol. Immunol.* 40: 869–876.

149 Willment, J. A., Lin, H.-H., Reid, D.M., Taylor, P.R., Williams, D.L., Wong, S.Y.C., Gordon, S., Brown, G.D. (2003): Dectin-1 expression and function are enhanced on alternatively activated and GM-CSF-treated macrophages and are negatively regulated by IL-10, dexamethasone, and lipopolysaccharide. *J. Immunol.* 171: 4569–4573.

150 Adachi, Y., Ishii, T., Ikeda, Y., Hoshino, A., Tamura, H., Aketagawa, J., Tanaka, S., Ohno, N. (2004): Characterization of b-glucan recognition site on C-type lectin, dectin 1. *Infect. Immun.* 72: 4159–4171.

151 Duenas, A. I., Orduna, A., Crespo, M. S., Garcia-Rodriguez, C. (2004): Interaction of endotoxins with Toll-like receptor 4 correlates with their endotoxic potential and may explain the proinflammatory effect of Brucella spp. LPS. *Int. Immunol.* 16: 1467–1475.

152 Muroi, M., Tanamoto, K.-i. (2002): The polysaccharide portion plays an indespensible role in *Salmonella* lipopolysaccharide-induced activation of NF-κB

through human Toll-like receptor 4. *Infect. Immun.* 70: 6043–6047.

153 Nilsen, N., Nonstad, U., Khan, N., Knetter, C.F., Akira. S., Sundan, A., Espevik, T., Lien, E. (2004): Lipopolysaccharide and double-stranded RNA upregulate Toll-like receptor 2 independently of myeloid differentiation factor 88. *J. Biol. Chem.* 279: 39 727–39 735.

154 Siedlar, M., Frankenberger, M., Benkhart, E., Espevik, T., Quirling, M., Brand, K., Zembala, M., Ziegler-Heitbrock, L. (2004): Tolerance induced by the lipopeptide Pam$_3$Cys is due to ablation of IL-1R-associated kinase. *J. Immunol.* 173: 2736–2745.

155 Mansell, A., Brint, E., Gould, J.A., O'Niell, L.A., Hertzog, P.J. (2004): Mal interacts with TNF receptor associated factor (TRAF)-6 mediated NF-κB activation by Toll-like receptor (TLR)-2 and TLR4. *J. Biol. Chem.* 279: 37 227–37 230.

156 Eisenbarth, S.C., Piggotti, D.A., Huleatt, J.W., Visintin, I., Herrick, C.A., Bottomly, K. (2002): Lipopolysaccharide-enhanced, Toll-like receptor 4-dependent T helper type 2 responses to inhaled antigen. *J. Exp. Med.* 196: 1645–1651.

157 Vignal, C., Guérardel, Y., Kremer, L., Masson, M., Legrand, D., Mazurier, J., Elass, E. (2003): Lipomannans, but not lipoarabinomannans, purified from Mycobacterium chelonae and Mycobacterium kanasii induce TNF-α and IL-8 secretion by a CD14-Toll-like receptor 2-dependent mechanism, *J. Immunol.* 171: 2014–2033.

158 Torres, D., Barrier, M., Bihl, F., Quesniaux, V. J. F., Maillet, I., Akira, S., Ryffel, B., Erard, F. (2004): Toll-like receptor 2 is required for optimal control of Listeria monocytogenes infection. *Infect. Immun.* 72: 2131–2139.

159 McCaffrey, R. L., Fawcett, P., O'Riordan, M., Lee, K-D., Havell, E. A., Brown, P. O., Portnoy, D. A. (2004): A specific gene expression program triggered by Gram-positive bacteria in the cytosol. *Proc. Natl. Acad. Sci. U.S.A.* 101: 11 386–11 391.

160 Girardin, S. E., Philpott, D. J. (2004): The role of peptidoglycan recognition in innate immunity. *Eur. J. Immunol.* 34: 1777–1782.

161 Netea, M. G., Kullberg, B. J., de Jong, D. J., Franke, B., Sprong, T., Naber, T. H. K., Drenth, J. P. H., Van der Meer, J. W. M. (2004): NOD2 mediates anti-inflammatory signals induced by TLR ligands: implications for Crohn's disease. *Eur. J. Immunol.* 34: 2052–2059.

162 Travassos, L. H., Girardin, S. E., Philpott, D. J., Blanot, D., Nahori, M. A., Werts, C., Bonecca, I. G. (2004): Toll-like receptor 2-dependent bacterial sensing does not occur via peptidoglycan recognition. *EMBO Rep.* 5: 1000–1006.

163 Kursar, M., Mittrucker, H. W., Koch, M., Kohler, A., Herma, M., Kaufmann, S. H. (2004): Protective T cell response against intracellular pathogens in the absence of Toll-like receptor signaling via myeloid differentiation factor 88. *Int. Immunol.* 16: 415–421.

164 Awasthi, A., Mathur, R. K., Saha, B. (2004): Immune response to *Leishmania* infection. *Indian J. Med. Res.* 119: 238–258.

165 Goto, H., Lindoso, J. A. L. (2004): Immunity and immunosuppression in experimental visceral Leishmaniasis. *Brazil. J. Med. Biol. Res.* 37: 615–623.

166 Meier, C. L., Svensson, M., Kaye, P. M. (2003): *Leishmania*-induced inhibition of macrophage antigen presentation analyzed at the single-cell level. *J. Immunol.* 171: 6706–6713.

167 Diefenbach A, Schindler H, Donhauser N, Lorenz E, Laskay T, MacMicking J, Röllinghoff M, Gresser I, Bogdan C. (1998): Type 1 interferon (IFNα/β) and type 2 nitric oxide synthase regulate the innate immune response to a protozoan parasite. *Immunity* 8: 77–87.

168 Qadoumi, M., Becker, I., Donhauser, N., Röllinghoff, M., Bogdan, C. (2002): Expression of inducible nitric oxide synthase in skin lesions of patients with American cutaneous Leishmaniasis. *Infect. Immun.* 70: 4638–4642.

169 Blos, M., Schleicher, U., Soares Rocha, F. J., Meißner, U., Röllinghoff, M., Bogdan, C. (2003): Organ-specific and stage-dependent control of *Leishmania major* infection by inducible nitric oxide

synthase and phagocyte NADPH oxidase. *Eur. J. Immunol.* 33: 1224–1234.

170 De Veer, M. J., Curtis, J. M., Baldwin, T. M., DiDonato, J. A., Sexton, A., McConville, M. J., Handman, E., Schofield, L. (2003): MyD88 is essential for clearance of *Leishmania*: possible role for lipophosphoglycan and Toll-like receptor 2 signaling. *Eur. J. Immunol.* 33: 2822–2831.

171 Kropf, P., Freudenberg, M. A., Modolell, M., Price, H. P., Herath, S., Antoniazi, S., Galanos, C., Smith, D. F., Muller, I. (2004): Toll-like receptor 4 contributes to efficient control of infection with the protozoan parasite *Leishmania major*. *Infect. Immun.* 72: 1920–1928.

172 Iniesta, V., Gómez-Nieto, L. C., Corraliza, I. (2001): The inhibition of arginase bN$^{\omega}$-hydroxy-L-arginine controls the growth of *Leishmania* inside macrophages. *J. Exp. Med.* 193: 777–783.

173 Melby, P. C., (2002): Recent developments in Leishmaniasis. *Curr. Opin. Infect. Dis.* 15: 485–490.

174 Ghosh, S., Bhattacharyya, S., Sirkar, M., Shankar Sa, G., Das, T., Majumdar, D., Roy, S., Majumdar, S. (2002): *Leishmania donovani* suppresses activation in host macrophages via ceramide generation: involvement of extracellular signal-regulated kinase. *Infect. Immun.* 70: 6828–6838.

175 Privé, C., Descoteaux, A. (2000): *Leishmania donovani* promastigotes evade the activation of mitogen-activated protein kinases p38, c-Jun N-terminal kinase, and extracellular signal-regulated kinase-1/2 during infection of naive macrophages. *Eur. J. Immunol.* 30: 2235–2244.

176 Marovich, M. A., McDowell, M. A., Thomas, E. K., Nutman, T. B. (2000): IL-12p70 production by *Leishmania major*-harboring human dendritic cells is a CD40/CD40 ligand-dependent process. *J. Immunol.* 164: 5858–5865.

177 Cameron, P., McGachy, A., Anderson, M., Paul, A., Coombs, G.H., Mottram, J. C., Alexander, J., Plevin, R. (2004): Inhibition of lipopolysaccharie-induced macrophage IL-12 production by *Leishmania mexicana* amastigotes: the role of cysteine peptidases and the NF-κB signaling pathway. *J. Immunol.* 173: 3297–3304.

178 Weinheber, N., Wolfram, M., Harbecke, D., Aebischer, T. (1998): Phagocytosis of *Leishmania mexicana* amastigotes by macrophages leads to a sustained suppression of IL-12 production. *Eur. J. Immunol.* 28: 2467–2477.

179 Piedrafita, D., Proudfoot, L., Nikolaev, A. V., Xu, D., Sands, W., Feng, G-J., Thomas, E., Brewer, J., Ferguson, M. A. J., Alexander, J., Liew, F. Y. (1999): Regulation of macrophage IL-12 synthesis by *Leishmania* phosphoglycans. *Eur. J. Immunol.* 29: 235–244.

180 Awasthi, A., Mathur, R., Khan, A., Joshi, B. N., Jain, N., Sawant, S., Boppana, R., Mitra, D., Saha, B. (2003): CD40 signaling is impaired in *L. major*-infected macrophages and is rescued by p38MAPK activator establishing a host-protective memory T cell response. *J. Exp. Med.* 197: 1037–1043.

181 Zhou, L., Nazarian, A.A., Smale, S.t. (2004): Interleukin-10 inhibits interleukin-12 p40 gene transcription by targeting a late event in the activation pathway. *Mol. Cell. Biol.* 24: 2385–2396.

182 Padigel, U. M., Alexander, J., Farrell, J.P. (2003): The role of interleukin-10 in susceptibility of BALB/c mice to infection with *Leishmania mexicana* and Leishmania amazonensis. *J. Immunol.* 171: 3705–3710.

183 Kane, M. M., Mosser, D. M. (2001): The role of IL-10 in promoting disease progression in leishmaniasis. *J. Immunol.* 166: 1141–1147.

184 Rodrigues, V. Jr., Santana da Silva, J., Campos-Neto, A. (1998): Transforming growth factor β and immunosuppression in experimental visceral leishmaniasis. *Infect. Immun.* 66: 1233–1236.

185 Rodriguez, N. E., Chang, H. K., Wilson, M. E. (2004): Novel program of macrophage gene expression induced by phagocytosis of *Leishmania chagasi*. *Infect. Immun.* 72: 2111–2122.

186 Ribiero-Gomes, F. L., Otero, A. C., Gomes, N. A., Moniz-de-Souza, M. C. A., Cysne-Finkelstein, Arnholdt, A. C., Calich, V. L., Coutinho, S. G., Lopes,

M. F., DosReis, G. A. (2004): Macrophage interactions with neutrophils regulate *Leishmania* major infection. *J. Immunol.* 172: 4454–4462.

187 Bertholet, S., Dickensheets, H. L., Sheikh, F., Gam, A. A., Donnelly, R. P., Kenney, R. T. (2003): *Leishmania donovani*-induced expression of suppressor of cytokine signaling 3 in human macrophages: a novel mechanism for intracellular parasite suppression of activation. *Infect. Immun.* 2095–2101.

188 Misslitz, A. C., Bonhagen, K., Harbecke, D., Lippuner, C., Kamradt, T., Aebischer, T. (2004): Two waves of antigen-containing dendritic cells in vivo in experimental *Leishmania major* infection. *Eur. J. Immunol.* 34: 715–725.

189 Galbiati, F., Rogge, L., Adorini, L. (2000): IL-12 receptor regulation in IL-12-deficient BALB/c and C57BL/6 mice. *Eur. J. Immunol.* 30: 29–37.

190 Kuroda, E., Yamashita, U. (2003): Mechanisms of enhanced macrophage-mediated prostaglandin E_2 production and its suppressive role in Th1 activation in Th2-dominant BALB/c mice. *J. Immunol.* 170: 757–764.

191 Kuroda, E., Sugiura, T., Zeki, K., Yoshida, Y., Yamashita, U. (2000): Sensitivity difference to the suppressive effect of prostaglandin E_2 among mouse strains: a possible mechanism to polarize Th2 type response in BALB/c mice. *J. Immunol.* 164: 2386–2395.

192 Gorak, P. M. A., Engwerda, C. R., Kaye, P. M. (1998): Dendritic cells, but not macrophages, produce IL-12 immediately following *Leishmania donovani* infection. *Eur. J. Immunol.* 28: 687–695.

193 von Stebut, E., Belkaid, Y., Jakob, T., Sacks, D. L., Udey, M. C. (1998): Uptake of *Leishmania major* amastigotes results in activation and interleukin 12 release from murine skin-derived dendritic cells: implications for the initiation of anti-*Leishmania* immunity. *J. Exp. Med.* 188: 1547–1552.

194 Konecny, P., Stagg, A. J., Jebbari, H., English, N., Davidson, R. N., Knight, S. C. (1999): Murine dendritic cells internalize *Leishmania major* promastigotes, produce il-12p40 and stimulate primary T cell proliferation in vitro. *Eur. J. Immunol.* 29: 1803–1811.

195 Soares, M. B. P., David, J. R., Titus, R. G. (1997): An in vitro model for infection with *Leishmania major* that mimics the immune response in mice. *Infect. Immun.* 65: 2837–2845.

196 Chakir, H., Campos-Neto, A., Mojibian, M., Webb, J. R. (2003): IL12Rβ2-deficient mice of a genetically resistant background are susceptible to *Leishmania major* infection and develop a parasite-specific Th2 immune response. *Microbes Infect.* 5: 241–249.

197 Himmelrich, H., Parra-Lopez, C., Tacchini-Cottier, F., Louis, J. A., Launois, P. (1998): The IL-4 rapidly produced in BALB/c mice after infection with *Leishmania major* down-regulates IL-12 receptor β2-chain expression on $CD4^+$ T cells resulting in a state of unresponsiveness to IL-12. *J. Immunol.* 161: 6156–6163.

198 Liu, T., Matsuguchi, T., Tsuboi, N., Yarima, T., Yoshikai, Y. (2002): Differences in expression of Toll-like receptors and their reactivities in dendritic cells in BALB/c and C57BL/c mice. *Infect. Immun.* 70: 6638–6645.

199 Fukao, T., Tanabe, M., Terauchi, Y., Ota, T., Matsuda, S., Asano, T., Kadowaki, T., Takeuchi, T., Koyasu, S. (2002): PI3K-mediated negative feedback regulation of IL-12 production in DCs. *Nat. Immunol.* 3: 875–881.

200 Heinzl, F. P., Schoenhaut, D. S., Rerko, R. M., Rosser, L. E., Gately, M. K. (1993): Recombinant interleukin 12 cures mice infected with *Leishmania major*. *J. Exp. Med.* 177: 1505–1509.

201 Sypek, J. P., Chung, C. L., Mayor, S. H., E., Subramanyam, J. M., Goldman, S. J., Sieburth, D. S., Wolf, S. F., Schaub, R. G. (1993): Resolution of cutaneous Leishmamniasis: interleukin 12 initiates a protective T helper 1 immune response. *J. Exp. Med.* 177: 1797–1802.

202 Chakour, R., Guler, R., Bugnon, M., Allenbach, C., Garcia, I., Mauël, J., Louis, J., Tacchini-Cottier, F. (2003): Both the Fas ligand and inducible nitric oxide synthase are needed for control of parasite replication within the lesions in

mice infected with *Leishmania major* whereas the contribution of tumor necrosis factor is minimal. *Infect. Immun.* 71: 5287–5295.
203 Huang, F.-P., Xu, D., Esfandiari, E.-O., Sands, W., Wei, X-q., Liew, F. Y. (1998): Mice defective in Fas are highly susceptible to *Leishmania major* infection despite elevated IL-12 synthesis, strong Th1 responses, and enhanced nitric oxide production. *J. Immunol.* 160: 4143–4147.
204 Ehrchen, J., Sindrilaru, A., Grabbe, S., Schonlau, F., Schlesiger, C., Sorg, C., Scharffetter-Kochanek, K., Sunderkotter, C. (2004): Senescent BALB/c mice are able to develop resistance to *Leishmania* infection. *Infect. Immun.* 72: 5106–5114.
205 Bacellar, O., Lessa, H., Schriefer, A., Machado, P., Ribeiro de Jesus, A., Dutra, W. O., Gollob, K.J., Carvalho, M. (2002): Up regulation of TI I-1 type responses in mucosal leishmaniasis patients. *Infect. Immun.* 70: 6734–6740.
206 Xaus, J., Comalada, M., Valledor, A.F., Lloberas, J., López-Soriano, Argilés, J. M., Bogdan, C., Celada, A. (2000): LPS induces apoptosis in macrophages mostly through autocrine production of TNF-α. *Blood* 95: 3823–3831.
207 Akarid, K., Arnoult, D., Micic-Polianski, J., Sif, J., Estaquier, J., Ameisen, J. C. (2004): *Leishmania major*-mediated prevention of programmed cell death induction in infected macrophages is associated with repression of mitochondrial release of cytochrome c. *J. Leukoc. Biol.* 76: 95–103.
208 Kanaly, S. T., Nashleanas, M., Hondowicz, B., Scott, P. (1999): TNF receptor p55 is required for elimination of inflammatory cells following control of intracellular pathogens. *J. Immunol.* 163: 3883–3889.
209 Wilhelm, P., Ritter, U., Labbow, S., Donhauser, N., Röllinghoff, M., Bogdan, C., Körner, H. Rapidly fatal leishmaniasis in resistant C57BL/c mice lacking TNF. *J. Immunol.* 166: 4012–4019.
210 Fonseca, S. G., Ramao, P. R., Figueiredo, F., Morais, R. H., Lima, H. G., Ferreira, S. H., Cunha, F. Q. TNF-α mediates the induction of nitric oxide synthase in macrophages but not neutrophils in experimental cutaneous leishmaniasis. *Eur. J. Immunol.* 33: 2297–2306.
211 Muraille, E., De Trez, C., Pajak, B., Torrentera, F. A., De Baetselier, P., Leo, O., Carlier, Y. (2003): Amastigote load and cell surface phenotype of infected cells from lesions and lymph nodes of susceptible and resistant mice infected with *Leishmania major*. *Infect. Immun.* 71: 2704–2715.
212 Zimmerli, S., Edwards, S., Ernst, J.D. (1996): Selective receptor blockade during phagocytosis does not alter the survival and growth of *Mycobacterium tuberculosis* in human macrophages. *Am. J. Respir. Cell Mol. Biol.* 15: 760–770.
213 Puig-Kroger, A., Serrano-Gomez, D., Caparros, E., Dominguez-Soto, A., Relloso, M., Colmenares, M., Martinez-Munoz, L., Longo, N., Sanchez-Sanchez, N., Rincon, M., Rivas, L., Sanchez-Mateos, P., Fernandez-Ruiz, E., Corbi, A. L. (2004): Regulated expression of the pathogen receptor dendritic cell-specific intercellular adhesion molecule 3 (ICAM-3)-grabbing nonintegrin in THP-1 leukemic cells, monocytes, and macrophages. *J. Biol. Chem.* 279: 25 680–25 688.
214 Tailleux, L., Schwartz, O., Herrman, J. L., Pivert, E., Jackson, M., Amara, A., Legres, Gluckman, J. C., L., Lagrange, P. H., Gicquel, B., Neyrolles, O. (2003): DC-SIGN is the major *Mycobacterium tuberculosis* receptor on human dendritic cells. *J. Exp. Med.* 197: 121–127.
215 Jang, S., Uematsu, S., Akira, S., Salgame, P. (2004): IL-6 and IL-10 induction from dendritic cells in response to *Mycobacterium tuberculosis* is predominantly dependent on TLR2-mediated recognition. *J. Immunol.* 173: 3392–3397.
216 Quesniaux, V., Fremond, C., Jacobs, M., Parida, S., Nicolle, D., Yeremeev, V., Bihl, F., Erard, F., Botha, T., Drennan, M., Soler, M-N., Le Bert, M., Schnyder, B., Ryffel, B. (2004): Toll-like receptor pathways in the immune response to mycobacteria. *Microbes Infect.* 6: 946–959.

217 Feng, C. G., Kullberg, M. C., Jankovic, D., Cheever, A. W., Caspar, P., Coffman, R. L., Sher, A. (2002): Transgenic mice expressing human interleukin-10 in the antigen-presenting cell compartment show increased susceptibility to infection with *Mycobacterium avium* associated with decreased macrophage effector function and apoptosis. *Infect. Immun.* 70: 6672–6679.

218 López, M., Sly, L. M., Luu, Y., Young, D., Cooper, H., Reiner, N. E. (2003): The 19-kDA *Mycobacterium tuberculosis* protein induces macrophage apoptosis through Toll-like receptor-2. *J. Immunol.* 170: 2409–2416.

219 Pedroza-Gonzáles, A., García-Romo, G. S., Aguilar-León, D., Calderon-Amador, J., Hurtado-Ortiz, R., Orozco-Estevez, H., Lambrecht, B. N., Estrada-García, I., Hernández-Pando, R., Flores-Romo, L. (2004): In situ analysis of lung antigen-presenting cells during murine pulmonary infection with virulent *Mycobacterium tuberculosis*. *Int. J. Exp. Pathol.* 85: 135–145.

220 Tailleux, L., Maeda, N., Nigou, J., Gicquel, B., Neyrolles, O. (2003): How is the phagocyte lectin keyboard played? Master class lesson by *Mycobacterium tuberculosis*. *Trends Microbiol.* 11: 259–263.

221 Schnappinger, D., Ehrt, S., Voskuil, M. I., Liu, Y., Mangan, J. A., Monahan, I. M., Dolganov, G., Efron, B., Butcher, P. D., Nathan, C., Schoolnik, G. K. (2003): Transcriptional adaptation of *Mycobacterium tuberculosis* within macrophages: insights into the phagosomal environment. *J. Exp. Med.* 198: 693–704.

222 Geijtenbeek, T. B. H., van Vliet, S. J., Koppel, E. A., Sanchez-Hernandez, M., Vandenbroucke-Grauls, C. M. J. E., Appelmelk, B., van Kooyk. (2003): Mycobacteria target DC-SIGN to suppress dendritic cell function. *J. Exp. Med.* 197: 7–17.

223 Doherty, T. M., Sher, A. (1997): Defects in cell-mediated immunity affect chronic, but not innate, resistance of mice to *Mycobacterium avium* infection. *J. Immunol.* 158: 4822–4831.

224 Drennan, M. B., Nicolle, D., Quesniaux, V. J. F., Jacobs, M., Allie, N., Mpagi, J., Fremond, C., Wagner, H., Kirschning, C., Ryffel, B. (2004): Toll-like recptor 2-deficient mice succumb to *Mycobacterium tuberculosis* infection. *Am. J. Pathol.* 164: 49–57.

225 Sugawara, I., Yamada, H., Mizuno, S., Takeda, K., Akira, S. (2003): Mycobacterial infection in MyD88-deficient mice. *Microbiol. Immunol.* 47: 841–847.

226 Nicolle, D. M., Pichon, X., Bouchot, A., Maillet, I., Erard, F., Akira, S., Ryffel, B., Quesniaux, V. F. J. (2004): Chronic pneumonia despite adaptive immune response to Mycobacterium bovis BCG in MyD88-deficient mice. *Lab. Invest.* 84: 1305–1321.

227 Noss, E. H., Pai, R. K., Sellati, T. J., Radolf, J. D., Belisle, J., Golenbock, D. T., Boom, W. H., Harding, C. V. (2001): Toll-like receptor 2-dependent inhibition of macrophage class II MHC expression and antigen processing by 19-kDa lipoprotein of Mycobaccterium tuberculosis. *J. Immunol.* 167: 910–918.

228 Gehring, A. J., Dobos, K. M., Belisle, J. T., Harding, C. V., Boom, W.H. (2004): *Mycobacterium tuberculosis* LprG (Rv1411c): a novel TLR-2 ligand that inhibits human macrophage class II MHC antigen processing. *J. Immunol.* 173: 2660–2668.

229 Pai, R. K., Pennini, M. E., Tobian, A. A., Canaday, D.H., Boom, W.H., Harding, C. V. (2004): Prolonged toll-like receptor signaling by *Mycobacterium tuberculosis* and its 19-kilodalton lipoprotein inhibits gamma interferon-induced regulation of selected genes in macrophages. *Infect. Immun.* 72: 6603–6614.

230 Hussain, S., Zwilling, B. S., Lafuse, W. P. (1999): *Mycobacterium avium* infection of mouse macrophages inhibits IFN-γ Janus kinase-STAT signaling and gene induction by down-regulation of the IFN-γ receptor. *J. Immunol.* 163: 2041–2048.

231 Verreck, F. A. W., de Boer, T., Langenberg, D. M. L., Hoeve, M. A., Kramer, M., Vaisberg, E., Kastalein, R., Kolk, A., de Waal-Malefyt, R., Ottenhoff, T. (2004): Human IL-23-producing type 1

macrophages promote but IL-10-producing type 2 macrophages subvert immunity to (myco)bacteria. *Proc. Natl. Acad. Sci. U.S.A.* 101: 4560–4565.

232 Winau, F., Kaufmann, S. H. E., Schaible, U. E. (2004): Apoptosis paves the detour path for CD8 T cell activation against intracellular bacteria, *Cell. Microbiol.* 6. 599–607.

233 Silva, R. A., Pais, T. F., Appelberg, R. (1998): Evaluation of IL-12 in immunotherapy and vaccine design in experimental *Mycobacterium avium* infections. *J. Immunol.* 161: 5578–85.

234 T. K. Means, B.W. Jones, A.B. Schromm, B A. Shurtleff, J.A. Smith, J. Keane, D.T. Golenbock, S.N.Vogel, M.J. Fenton (2001): Differential effects of Toll-like receptor antagonists on *Mycobacterium tuberculosis*-induced macrophage responses, *J. Immunol.* 166: 4074–4082.

235 Zganiacz, A., Santosuosso, Wang, J., Yang, T., Chen, L., Anzulovic, M., Alexander, S., Gicquel, B., Wan, Y., Bramson, J., Inman, M., Xing, Z. (2004): TNF-α is a critical negative regulator of type 1 immune activation during intracellular bacterial infection. *J. Clin. Invest.* 113: 401–413.

236 Murray, D. A., Crispe, I. N. (2004): TNF-α controls intrahepatic T cell apoptosis and peripheral t cell numbers. *J. Immunol.* 173: 2402–2409.

237 Kassiotis, G., Kollias, G. (2001): Uncoupling the proinflammatory from the immunopsupresive properties of tumor necrosis factor (TNF) at the p55 TNF receptor level: implications for pathogenesis and therapy of autoimmune demyelination. *J. Exp. Med.* 193: 427–434.

238 Kaufmann, S. H. E., Schaible, U. E. (2003): A dangerous liason between two major killers: Mycobacterium tuberulosis and HIV target dendritic cells through DC-SIGN. *J. Exp. Med.* 197: 1–5.

239 Olivier, M., Badaro, R., Medrano, F. J., Moreno. J. (2002): The pathogenesis of *Leishmania*/HIV co-infection: cellular and immunological mechanisms. *Ann. Trop. Med. Parasitol.* 97: S79–S98.

240 Levy, J. A. (2001): Thje importance of the innate immune system in controlling HIV infection and disease. *Trends Immunol.* 22: 312–316.

241 Kutza, J., Crim, L., Feldman, S., Hayes, M. P., Gruber, M., Beeler, J., Clouse, K.A. (2000): Macrophage colony-stimulating factor antagonists inhibit replication of HIV-1 in human macrophages. *J. Immunol.* 164: 4955–4960.

242 Lima, R. G., Van Weyenbergh, J., Saraiva, E. M. B., Barral-Netto, M., Galvao-Castro, B., Bou-Habib, D. C. (2002): The replication of human immunodeficiency virus type 1 in macrophages is enhanced after phagocytosis of apoptotic cells. *J. Infect. Dis.* 185: 1561–1566.

243 Yamamoto, T., Noble, N. A., Miller, D. E., Gold, L. I., Hishida, A., Nagase, M., Cohen, A. H., Border, W. A. (1999): Increased levels of transforming growth factor-β in HIV-asociated nephropathy. *Kidney Int.* 55: 579–592.

244 Wang, J., Roderiquez, G., Oravecz, T., Norcross, M. A. (1998): Cytokine regulation of human immunodeficiency virus type 1 entry and replication in human monocytes/macrophages through modulation of CCR5 expression. *J. Virol.* 72: 7642–7647.

245 Wang, J., Crawford, K., Yuan, M., Wang, H., Gorry, P. R., Gabuzda, D. (2002): Regulation of CC chemokine receptor 5 and CD4 expression and human immunodeficiency virus type 1 replication in human macrophages and microglia by T helper type 2 cytokines. *J. Infect. Dis.* 185: 885–897.

246 Blanco, J., Barretina, J., Clotet, B., Esté, J. A. (2004): R5 HIV gp120-mediated cellular contacts induce the death of single CCR5-expressing CD4 T cells by a gp41-dependent mechanism. *J. Leukoc. Biol.* 76: 804–811.

247 Boven, L. A., Gomes, L., Hery, C., Gray, F., Verhoef, J., Portegies, P., Tardieu, M., Nottet, H. S. L. M. (1999): Increased peroxynitrite activity in AIDS dementia complex: implications for the neuropathogenesis of HIV-1 infection. *J. Immunol.* 162: 4319–4327.

248 Shapshak, P., Duncan, R., Minagar, A., Rodriguez de la Vega., Stewart, R.V.,

Goodkin, K. (2004): Elevated expression of IFN-γ in the HIV-1 infected brain. *Front. Biosci.* 9: 1073–1081.

249 Grant, R. S., Naif, H., Thuruthyil, S. J., Nasr, N., Littlejohn, T., Takikawa, O., Kapoor, V. (2000): Induction of indoleamine 2,3-dioxygenase in primary human macrophages by HIV-1. *Redox. Rep.* 5: 105–107.

250 Minghetti, L., Visentin, S., Patrizio, M., Franchini, L., Ajmone-Cat, M. A., Levi, G. (2004): Multiple actions of the human immunodeficiency virus type-1 Tat protein on microglial cell functions. *Neurochem. Res.* 29: 965–978.

251 Guillemard, E., Jacquemot, C., Aillet, F., Schmitt, N., Barre-Sinoussi, F., Israel, N. (2004): Human immunodeficiency virus 1 favors the persistence of infection by activating macrophages through TNF. *Virology* 329: 371–380.

252 Hayes, M. M., Lane, B. R., King, S. R., Markovitz, D. M., Coffey, M. J. (2002): Prostaglandin E2 inhibits replication of HIV-1 in macrophages through activation of protein kinase A. *Cell. Immunol.* 215: 61–71.

253 Dalgleish, A. G., O'Byrne, K. J. (2002): Chronic immune activation and inflammation in the pathogenesis of AIDS and cancer. *Adv. Cancer Res.* 84: 231–276.

254 Gascon, R. L., Narvaez, A. B., Zhang, R., Khan, J.O., Hecht, F. M., Herndier, B.G., McGrath, M. S. (2002): Increased HLA-DR expression on peripheral blood monocytes in subsets of subjects with primary HIV infection is associated with elevated CD4 T-cell apoptosis. *J. Acquir. Immune Defic. Syndr.* 30: 146–153.

255 Creery, D., Weiss, W., Lim, W. T., Aziz, Z., Angel, J.B., Kumar, A. (2004): Downregulation of CXCR-4 and CCR-5 expression by interferon-γ is associated with inhibition of chemotaxis and human immunodeficiency virus (HIV) replication but not HIV entry into human monocytes. *Clin. Exp. Immunol.* 137: 156–165.

256 Shin, H. D., Winkler, C., Stephens, J. C., Bream, J., Young, H., Goedert, J. J., O'Brien, T. R., Vlahov, D., Buchbinder, S., Giorgi, J., Rinaldo, C., Donfield, S., Willoughby, A., O'Brien, S.J., Smith, M.W. (2000): Genetic restrictions of HIV-1 pathogenesis to AIDS by promoter alleles of IL-10. *Proc. Natl. Acad. Sci. U.S.A.* 97: 14 467–14 472.

257 Soilleux, E. J. (2003): DC-SIGN (dendritic cell-specific ICAM-grabbing non-integrin) and DC-SIGN-related (DC-SIGNR): friend or foe? *Clin. Sci.* 104: 437–446.

258 Ancuta, P., Bakri, Y., Chomont, N., Hocini, H., Gabuzda, D., Haeffner-Cavaillon, N. (2001): Opposite effects of IL-10 on the ability of dendritic cells and macrophages to replicate primary CXCR4-dependent HIV-1 strains. *J. Immunol.* 166: 4244–4253.

259 Deeks, S. G., Kitchen, C. M. R., Liu, L., Guo, H., Gascon, R., Varváez, A. B., Hunt, P., Martin, J. N., Khan, J. O., Levy, J., McGrath, M. S., Hecht, F. M. (2004): Immune activation set point during early HIV infection predicts subsequent CD4[+] T cell changes independent of viral load. *Blood* 104: 942–947.

260 Bahbouhi, B., Landay, A., Al-Harthi, L. (2004): Dynamics of cytokine expression in HIV productively infected primary CD4[+] T cells. *Blood* 103: 4581–4587.

261 Ledru, E., Herve, L., Garcia, S., Debord, T., Gougeon, M.-L. (1998): Differential susceptibility to activation-induced apoptosis among peripheral Th1 subsets: correlation with Bcl-2 expression and consequences for AIDS pathogenesis. *J. Immunol.* 160: 3194–3206.

262 Accornero, P., Radrizzani, M., Delia, D., Gerosa, F., Kurrie, R., Colombo, M. P. (1997): Differential susceptibility to HIV-gp120-sensitized apoptosis in CD4[+] T-cell clones with different T-helper phenotypes: role of CD95/CD95L interactions. *Blood* 89: 558–569.

263 Wahl, S. M., Greenwell-Wild, T., Peng. G., Hale-Donze, H., Doherty, T. M., Mizel, D., Orenstein, J. M. (1998): *Mycobacterium avium* complex augments macrophage HIV-1 production and increases CCR5 expression. *Proc. Natl. Acad. Sci. U.S.A.* 95: 12 574–12 579.

264 Kedzierska, K., Crowe, S. M., Turville, S., Cunningham, A. L. (2003): The

influence of cytokines, chemokines and their receptors on HIV-1 replication in monocytes and macrophages. *Rev. Med. Virol.* 13: 39–56.

265 Goletti, D., Carrara, S., Vincenti, D., Giacomini, E., Fattorini, L., Garbuglia, A. R., Capobianchi, M. R., Alonzi, T., Fimia, G. M., Federico, M., Poli, G., Coccia, E. (2004): Inhibition of HIV-1 replication in monocyte-derived macrophages by *Mycobacterium tuberculosis*. *J. Infect. Dis.* 189: 624–633.

266 Báfica, A., Scanga, C. A., Schito, M., Chaussabel, D., Sher, A. (2004): Influence of coinfecting pathogens on HIV expression: evidence for a role of Toll-like receptors. *J. Immunol.* 172: 7229–7234.

267 Liou, L. Y., Herrmann, C. H., Rice, A.P. (2004): HIV-1 infection and regulation of Tat function in macrophages. *Int. J. Biochem. Cell Biol.* 36: 1767–1775.

268 Franchin, G., Zybarth, G., Dai, W. W., Dubrovsky, L., Reiling, N., Schmidtmayerova, H., Bukrinsky, M., Sherr, B. (2000): Lipopolysaccharide inhibits HIV-1 infection of monocyte-derived macrophages through direct and sustained down-regulation of CC chemokine receptor 5. *J. Immunol.* 164: 2592–2601.

269 Báfica, A., Scanga, C. A., Equils, O., Sher, A. (2004): The induction of Toll-like receptor tolerance enhances rather than suppresses HIV-1 gene expression in transgenic mice. *J. Leukoc. Biol.* 75: 460–466.

270 Gross, E., Amelia, C. A., Pompucci, L., Franchin, G., Sherry, B., Schmidtmayerova, H. (2003): Macrophages and lymphocytes differentially modulate the ability of RANTES to inhibit HIV-1 infection. *J. Leukoc. Biol.* 74: 781–790.

271 Hoshino, Y., Tse, D. B., Rochford, G., Prabhakar, S., Hosino, S., Chitkara, N., Kuwabara, K., Ching, E., Raju, B., Gold, J. A., Borkowsky, W., Rom, W. N., Pine, R., Weiden, M. (2004): *Mycobacterium tuberculosis*-induced CXCR4 and chemokine expression leads to preferential X4 HIV-1 replication in human macrophages. *J. Immunol.* 172: 6251–6258.

272 Fernandez, S., Jose, P., Avdiushko, M. G., Kaplan, A. M., Cohen, D. A.
(2004): Inhibition of IL-10 receptor function in alveolar macrophages by Toll-like receptor agonists. *J. Immunol.* 172: 2613–2620.

273 Imperiali, F. G., Zaninoni, A., La Maestra, L., Tarsia, P., Blasi, F., Barcellini, W. (2001): Increased *Mycobacterium tuberculosis* growth in HIV-1-infected human macrophages: role of tumour necrosis factor-α. *Clin. Exp. Immunol.* 123: 435–442.

274 Hayashi, T., Rao, S. P., Takabayashi, K., Van Uden, J. H., Kornbluth, R. S., Baird, S, M., Taylor, M. W., Carson, D. A., Catanzaro, A., Raz, E. (2001): Enhancement of innate immunity against *Mycobacterium avium* infection by immunostimulatory DNA is mediated by indoleamine 2,3-deoxygenase. *Infect. Immun.* 69: 6156–6164.

275 Bernier, R., Turco, S. J., Olivier, M., Tremblay, M. (1995): Activation of human immunodeficiency virus type 1 in monocytoid cells by the protozoan parasite *Leishmania donovani*. *J. Virol.* 69: 7282–7285.

276 Lee, D., Leslie, G., Soilleux, E., O'Doherty U., Baik, S., Levroney, E., Flummerfelt, K., Swiggard, W., Coleman, N., Malim, M., Doms, R.W. (2001): cis expression of DC-SIGN allows for more efficient entry of human and simian immunodeficiency virus. *J. Virol.* 75: 12 028–12 038.

277 Subramanyam, S., Hanna, L. E., Venkatesan, P., Sankaran, K., Narayanan, P. R., Swaminathan, S. (2004): HIV alters plasma and M. tuberculosis-induced cytokine production in patients with tuberculosis. *J. Interferon Cytokine Res.* 24: 101–106.

278 Seki, E., Tsutsui, H., Nakano, H., Tsuji, N. M., Hoshino, K., Adachi, O., Adachi, K., Futatsugi, S., Kuida, K., Takeuchi, O., Okamura, H., Fujimoto, J., Akira, S., Nakanishi, K. (2001): Lipopolysaccharide-induced IL-18 secretion from murine Kupffer cells independently of myeloid differentiation factor 88 that is critically involved in induction of production of IL-12 and IL-1β *J. Immunol.* 166: 2651–2657.

279 Aldini, R., Marangoni, A., Guardigli, M., Sambri, V., Giacani, L., Montagnani, M., Roda, A., Cevenini, R. (2004): Chemiluminescence detaction of reactive oxygen species in isolated Kupffer cells during phagocytosis of Treponema Pallidum. *Comparative Hepatol.* 3(suppl 1): 541.

280 Vazquez-Torres, A., Vallance, B. A., Bergman, M. A., Finlay, B. B., Cookson, B. T., Jones-Carson, J., Fang, F. C. (2004): Toll-like receptor4 dependence of innate and adaptive immunity to *Salmonella*: importance of Kupffer cell network. *J. Immunol.* 172: 6202–6208.

281 Lehner, M. D., Ittner, J., Buncschuh, D. S., van Rooijen, N., Wendel, A., Hartung, T. (2001): Improved innate immunity to endotoxin tolerant mice increases resistance to *Salmonella enterica* serovar *typhimurium* infection despite attenuated cytokine response. *Infect. Immun.* 69: 463–471.

282 Gregory, S. H., Wing, E. J., Danowski, K. L., van Rooijen, N., Dyer, K. F., Tweardy, D. J. (1998): IL-6 produced by Kupffer cells induces STAT protein activation in hepatocytes early during the course of systemic Listerial infections. *J. Immunol.* 160: 6056–6061.

283 D. Gershov, S. Kim, N. Brot, K.B. Elkon (2000): C-reactive protein binds to apoptotic cells, protects the cells from assembly of the terminal complement components, and sustains an antiinflammatory innate immune response: implications for systemic autoimmunity, *J. Exp. Med.* 192: 1353–1363.

284 Venugopal, S. K., Devaraj, S., Jialal, L. (2005): Effect of C-reactive protein on vascular cells: evidence for a proinflammatory, proatherogenic role. *Curr. Opin. Nephrol. Hypertens.* 14: 33–37.

285 Smiley, S. T., King, J. A., Hancock, W.W. (2001): Fibrinogen stimulates macrophage chemokine secretion through Toll-like receptor 4. *J. Immunol.* 167: 2887–2894.

286 Hayashi, N., Matsui, K., Tsutsui, H., Osada, Y., Mohamed, R. T., Nakano, H., Kashiwamura, S-i, Hyodo, Y., Takeda, K., Akira, S., Hada, T., Higashino, K., Kojima, S., Nakanishi, K. (1999): Kupffer cells from *Schistosoma mansoni*-infected mice participate in the prompt type 2 differentiation of hepatic T cells in response to worm antigens. *J. Immunol.* 163: 6702–6711.

287 Magalhães, A., Miranda, D. G., Miranda, R. G., Araújo, M. I., Almeida de Jesus, A., Silva, A., Santana, L. B., Pearce, E., Carvalho, E. M., Ribiero de Jesus. (2004): Cytokine profile associated with human chronic schistosomiasis mansoni. *Mem. Inst. Oswaldo Cruz* 99 (suppl 1): 21–26.

288 Mustafa, S. B., Olson, M. S. S. (1998): Expression of nitric oxide synthase in rat Kupffer cells is regulated by cAMP. *J. Biol. Chem.* 273: 5073–5080.

289 Duffield, J. S., Forbes, S. J., Constandinou, C. M., Clay, S., Partolina, M., Vuthoori, S., Wu, S., Lang, R., Iredale, J. P. (2005): Selective depletion of macrophages reveals distinct, opposing roles during liver injury and repair. *J. Clin. Invest.* 115, 56–65.

290 Wheeler, M. D., Thurman, R. G. (2003): Up-regulation of CD14 in liver caused by acute ethanol involves oxidant-dependent AP-1 pathway. *J. Biol. Chem.* 278: 8435–8441.

291 Su, G. L. (2002): Lipopolysaccharide in liver injury: molecular mechanisms of Kupffer cell activation. *Am. J. Physiol. Liver Physiol.* 283: G256–G265.

292 Enomoto, N., Ikejima, K., Yamashina, S., Enomot, A., Nishiura, T., Nishimura, T., Brenner, D. A., Schemmer, P., Bradford, B. U., Rivera, C. A., Zhong, Z., Thurman, R. G. (2000): Kupffer cell-derived prostaglandin E_2 is involved in alcohol-induced fat accumulation in rat liver. *Am. J. Physiol. Gastrointest. Liver Physiol.* 279: G100–G106.

293 Aldred, A., Nagy, L. E. (1999): Ethanol dissociates hormone-stimulated cAMP production from inhibition of TNF-a production in rat Kupffer cells. *Am. J. Physiol. Liver Physiol.* 276: G98–G106.

294 Treffkorn, L., Schreibe, R., Maruyama, T., Dieter, P. (2004): PGE_2 exerts its effects on the LPS-induced release of TNF-α, ET-1, IL-1α, IL-6 and IL-10 via the EP2 and EP4 receptor in rat liver

macrophages. *Prostaglandins Lipid Mediat.* 74: 113–123.

295 Pruett, S. B., Schwab, C., Zheng, Q., Fan, R. (2004): Suppression of innate immunity by acute ethanol administration: a global perspective and a new mechanism beginning with inhibition of signaling through TLR3. *J. Immunol.* 173: 2715–2724.

296 Friedman, S. L. (2000): Molecular regulation of hepatic fibrosis, an integrated cellular response to tissue injury. *J. Biol. Chem.* 275: 2247–2250.

297 Mann, D. A., Smart, D. E. (2002): Transcriptional regulation of hepatic stellate cell activation. *Gut* 50: 891–896.

298 Chedid, A., Arain, S., Synder, A., Mathurin, P., Capron, F., Naveau, S. (2004): The immunology of fibrogenesis in alcoholic liver disease. *Arch. Pathol. Lab. Med.* 128: 1230–1238.

299 Wang, S. C., Ohata, M., Schrum, L., Rippe, R. A., Tsukamoto, H. (1998): Expression of interleukin-10 by in vitro and in vivo activated hepatic stellate cells. *J. Biol. Chem.* 273: 302–308.

300 Lim, S., Caramori, G., Tomita, K., Jazrawi, E., Oates, T., Chung, K. F., Barnes, P.J. Adcock, I.M. (2004): Differential expression of IL-10 receptor by epithelial cells and alveolar macrophages. *Allergy* 59: 505–514.

301 Currie, A. J., Stewart, G. A., McWilliam, A. S. (2000): Alveolar macrophages bind and phagocytose allergen-containing pollen starch granules via C-type lectin and integrin receptors: implications for airway inflammatory disease. *J. Immunol.* 164: 3878–3886.

302 Liu, H.-W., Anand, A., Bloch, K., Christiani, D., Kradin, R. (1997): Expression of inducible nitric oxide synthase by macrophages in rat lung. *Am. J. Respir. Crit. Care Med.* 156: 223–228.

303 B.T. Ameredes, R. Zamora, K.F. Gibson, T.R. Billiar, B.Dixon-McCarthy, S. Watkins, W.J. Calhoun (2001): Increased nitric oxide production by airway cells of sensitized and challenged IL-10 knockout mice, *J. Leukoc. Biol.* 70: 730–736.

304 Carter, A. B., Monick, M. M., Hunninghake, G.W. (1998): Lipopolysaccharide-induced NF-κB activation and cytokine release in human alveolar marophages is PKC-independent and TK- and PC-PLC-dependent. *Am. J. Respir. Cell Mol. Biol.* 16: 384–391.

305 Koay A, Gao X, Washington MK, Parman KS, Sadikot RT, Blackwell TS, Christman JW. (2002): Macrophages are necessary for maximal nuclear factor-κB activation in response to endotoxin. *Am. J. Respir. Cell. Mol. Biol.* 26: 572–578.

306 Monick, M. M., Carter, A. B., Gudmundsson, G., Geist, L.J., Hunninghake, G. W. (1998): Changes in PKC isoforms in human alveolar macrophages compared with blood monocytes. *Am. J. Physiol.* 275: L389–L397.

307 Flaherty, D. M., Monick, M. M., Carter, A. B., Peterson, M.W., Hunninghake, G. W. (2001): GM-CSF increases AP-1 DNA binding and Ref-1 amounts in human alveolar macrophages. *Am. J. Respir. Cell Mol. Biol.* 25: 254–259.

308 Bozinovski, S., Jones, J.E., Viahos, R., Hamilton, J. A., Anderson, G. P. (2002): Granulocyte/macrophage-colony-stimulating factor (GM-CSF) regulates lung innate immunity to lipopolysaccharide through Akt/Erk activation of NFκB and AP-1 in vivo. *J. Biol. Chem.* 277: 42808–42814.

309 B.C. Trapnell, J.A. Whitsett, GM-CSF regulates pulmonary surfactant homeostasis and alveolar macrophage-mediated innate host defense, *Annu. Rev. Physiol.* 64 (2002) 775–802.

310 Y. Shibata, L.A. Foster, J.F. Bradfield, Q.N. Myrvik (2000): Oral administration of chitin down-regulates serum IgE levels and lung eosinophilia in the allergic mouse, *J. Immunol.* 164: 1314–1321.

311 Daniels, C. E., Wilkes, M. C., Edens, M., Kottorn, T. J., Murphy, S. J., Limper, A. H., Leof, E. B. (2004): Imatinib mesylate inhibits the profibrogenic activity of TGF-β and prevents bleomycin-mediated lung fibrosis. *J. Clin. Invest.* 114: 1308–1316.

312 Moore, B. B., Coffey, M. J., Christensen, P., Sitterding, S., Ngan, R., Wilke, C.A., McDonald, R., Phare, S. M., Peters-Golden, M., Paine 3rd, R., Toews, G. B. (2000): GM-CSF regulates bleomycin-induced pulmonary fibrosis via a prosta-

glandin-dependent mechanism. *J. Immunol.* 165: 4032–4039.
313 Charbeneau, R. P., Christensen, P. J., Chrisman, C. J., Paine 3rd, R., Toews, G. B., Peters-Golden, M., Moore, B. B. (2003): Impaired synthesis of prostaglandin E_2 by lung fibroblasts and alveolar epithelial cells from GM-CSF –/– mice: implications for fibroproliferation. *Am. J. Physiol. Lung Cell Mol. Physiol.* 284: L1103–L1111.
314 Dumas, C., Muyombwe, A., Roy, G., Matte, C., Ouellette, M., Olivier, M., Papadopoulou, B. (2003): Recombinant *Leishmania major* secreting biologically active Granulocyte-macrophage colony-stimulating factor survives poorly in macrophages in vitro and delays disease development in mice. *Infect. Immun.* 71: 6499–6509.
315 Lentsch, A. B., Czermak, B. J., Bless, N. M., Van Rooijen, N., Ward, P.A. (1999): Essential role of alveolar macrophages in intrapulmonary activation of NF-κB. *Am. J. Respir. Cell Mol. Biol.* 20: 692–698.
316 Stumbles, P. A., Thomas, J. A., Pimm, C. L., Lee, P. T., Venaille, T. J., Proksch, S., Holt, P. G. (1998): Resting respiratory tract dendritic cells preferentially stimulate T helper cell type 2 (Th2) responses and require obligatory cytokine signals for induction of Th1 immunity. *J. Exp. Med.* 188: 2019–2031.
317 Akbari, O., R. H. DeKruyff, R. H., Umetsu, D. T. (2001): Pulmonary dendritic cells producing IL-10 mediate tolerance induced by respiratory exposure to antigen, *Nat. Immunol.* 2: 725–731.
318 Lambrecht, B. N., De Veerman, M., Coyle. A. J., Gutierrez-Ramos, J.-C., Thielemans, K., Pauwels, R. A. (2000): Myeloid dendritic cells induce Th2 responses to inhaled antigen, leading to eosinophiloic airway inflammation. *J. Clin. Invest.* 106: 551–559.
319 Lambrecht, B. N. (2001): Allergen uptake and presentation by dendritic cells. *Curr. Opin. Allergy Clin. Immunol.* 1: 51–59.
320 Lambrecht, B. N., Hammad, H. (2002): Myeloid dendritic cells make it to the top. *Clin. Exp. Allergy* 32: 805–810.
321 Harris, N. L., Watt, V., Ronchese, F., Le Gros, G. (2002): Differential T cell function and fate in lymph node and non-lymphoid tissues. *J. Exp. Med.* 195: 317–326.
322 Cochand, L., Isler, P., Songeon, F., Nicod, P. (1999): Human lung dendritic cells have an immature phenotype with efficient mannose receptors. *Am. J. Respir. Cell. Mol. Biol.* 21: 547–554.
323 van Haarst, J. M. W., Verhoeven, G. T., de Wit, H. J. D., Hoogsteden, H. C., Debets, R., Drexhage, H. A. (1996): $CD1a^+$ and $CD1a^-$ accessory cells from human bronchoalveolar lavage differ in allostimulatory potential and cytokine production. *Am. J. Respir. Cell. Mol. Biol.* 15: 752–759.
324 Stumbles, P. A., Strickland, D. H., Pimm, C. L., Proksch, S. F., Marsh, A. M., McWilliam, A. S., Bosco, A., Tobagus, I., Thomas, J. A., Napoli, S., Proudfoot, A. E. I., Wells, T. N. C., Holt, P. (2001): Regulation of dendritic cell recruitment into resting and inflamed airway epithelium: use of alternative chemokine receptors as a function of inducing stimulus. *J. Immunol.* 167: 228–234.
325 Lambrecht, B. N., Carro-Muino, I., Vermaelen, K., Pauwels, R. A. (1999): Allergen-induced changes in bone marrow progenitor and airway dendritic cells in sensitized rats. *Am. J. Respir. Cell. Mol. Biol.* 20: 1165–1174.
326 Holt, P. G., Oliver, J., Bilyk, N., McMenamin, C., McMenamin, P. G., Kraal, G., Thepen, T. (1993): Downregulation of the antigen presenting cell function(s) of pulmonary dendritic cells in vivo by resident alveolar macrophages. *J. Exp. Med.* 177: 397–407.
327 Bingisser, R. M., Holt, P. G. (2001): Immunomodulating mechanisms in the lower respiratory tract: nitric oxide mediated interactions between alveolar macrophages, epithelial cells, and T cells. *Swiss Med. Wkly* 131: 171–179.
328 Strickland, D. H., Thepen, T., Kees, U. R., Kraal, G., Holt, P. G. (1993): Regulation of T-cell function in lung tissue by pulmonary alveolar macrophages. *Immunology* 80: 266–272.

329 Thepen, T., McMenamin, C., Girn, B., Kraal, G., Holt, P. G. (1992): Regulation of IgE production in pre-sensitized animals: in vivo elimination of alveolar macrophages preferentially increases IgE response to inhaled allergen. *Clin. Exp. Allergy* 22: 1107–1114.

330 Tang, C., Inman, M. D., van Rooijen, N., Yang, P., Matsumoto, K. and O'Byrne, P. M. (2001): Th type 1-stimulating activity of lung macrophages inhibitsTh2-mediated allergic airway inflammation by an IFN-γ-dependent mechanism. *J. Immunol*. 166: 1471–1481.

331 Kooguchi, K., Hashimoto, S., Kobayashi, A., Kitamura, Y., Kudoh, I., Wiener-Kronish, J., Sawa, T. (1998): Role of alveolar macrophages in initiation and regulation of inflammation in *Pseudomonas aeruginosa* pneumonia. *Infect. Immun*. 66: 3164–3169.

332 Holtzman MJ, Morton JD, Shornick LP, Tyner JW, O'Sullivan MP, Antao A, Lo M, Castro M, Walter MJ (2001): Immunity, inflammation, and remodeling in the airway epithelial barrier: epithelial-viral-allergic paradigm. *Physiol. Rev*. 82: 19–46.

333 M.J. Walter, N. Kajiwara, P. Karanja, M. Castro, M.J. Holtzman (2001): Interleukin-12p40 production by barrier epithelial cells during airway inflammation, *J. Exp. Med*. 193: 339–351.

334 Mantovani, A., Allavena, P., Sozzani, S., Vecchi, A., Locati, M., Sica, A. (2004): Chemokines in the recruitment and shaping of the leukocyte infiltrate of tumors. *Semin. Cancer Biol*. 14: 155–160.

335 Kim, J., Modlin, R. L., Moy, R. L., Dubinett, S. M., McHugh, T., Nickoloff, B. J., Uyemura, K. (1995): IL-10 production in cutaneous basal and squamous cell carcinomas. A mechanism for evading the local T cell immune response. *J. Immunol*. 155: 2240–2247.

336 Mantovani, A., Allavena, P., Sica, A. (2004): Tumour-associated macrophages as a prototypic type II polarized phagocytic population: role in tumour progression. *Eur. J. Cancer* 40: 1660–1667.

337 Ohno, S., Suzuki, N., Ohno, Y., Inagawa, H., Soma, G-I., Inoue, M. (2003): Tumor-associated macrophages: foe or accomplice of tumors? *Anticancer Res*. 23: 4395–4410.

338 Betz, M, Fox, B. S. (1991): Prostaglandin E_2 inhibits production of Th1 lymphokines but not of the Th2 lymphokines. *J. Immunol*. 146: 108–113.

339 Uotila, P. (1996): The role of cyclic AMP and oxygen intermediates in the inhibition of cellular immunity in cancer. *Cancer Immunol. Immunother*. 43: 1–9.

340 Kambayashi, T., Alexander, H.R., Fong, M., Strassmann, G. (1995): Potential involvement of IL-10 in suppressing tumor-associated macrophages. Colon-26-derived prostaglandin E_2 inhibits TNF-α release via a mechanism involving IL-10. *J. Immunol*. 154: 3383–3390.

341 Sica A, Saccani A, Bottazzi B, Polentarutti N, Vecchi A, van Damme J, Mantovani A (2000): Autocrine production of IL-10 mediates defective IL-12 production and NF-κB activation in tumor-associated macrophages. *J. Immunol*. 164: 762–767.

342 Mantovani, A., Sozzani, S., Locati, M., Allavena, P., Sica, A. (2002): Macrophage polarization: tumor-associated macrophages as a paradigm for polarized M2 mononuclear phagocytes. *Trends Immunol*. 23: 549–555.

343 Loercher, A. E., Nash, M. A., Kavanagh, J. J., Platsoucas, C. D., Freedman, R. S. (1999): Identification of an IL-10-producing HLA-DR-negative monocyte subset in the malignant ascites of patients with ovarian carcinoma that inhibits cytokine protein expression and proliferation of autologous T cells. *J. Immunol*. 163: 6251–6260.

344 Liu, Y., Van Ginderachter, J. A., Brys, L., De Baetselier, P., Raes, G., Geldhof, A. B. (2003): Nitric-oxide independent CTL suppression during tumor progression: association with arginase-producing (M2) myeloid cells. *J. Immunol*. 170: 5064–5074.

345 Mantovani, A., Schioppa, T., Biswas, S. K., Marchesi, F., Allavena, P., Sica, A. (2003): Tumor-associated macrophages and dendritic cells as prototypic type II

polarized myeloid populations. *Tumori* 89: 459–468.

346 Menetrier-Caux, C., Montmain, M.C., Dieu, M. C., Bain, C., Favrot, M. C., Caux, C., Blay, Y.C. (1998): Inhibition of the differentiation of dendritic cells from CD34$^+$ progenitors by tumor cells: role of interleukin-6 and macrophage colony-stimulating factor. *Blood* 92: 4778–4791.

347 Morse, M. A., Coleman, R. E., Akabani, G., Niehaus, N., Coleman, D., Lyerly, H. K. (1999): Migration of human dendritic cells after injection in patients with metastatic malignancies. *Cancer Res.* 59: 56–58.

348 Ikeda, H., Chamoto, K., Tsuji, T., Suzuki, Y., Wakita, D., Takeshima, T., Nishimura, T. (2004): The critical role of type-1 innate and acquired immunity in tumor immunotherapy. *Cancer Sci.* 95: 697–703.

349 Li, Q., Pan, P-Y., Gu, P., Xu, D., Chen, S-H. (2004): Role of immature myeloid Gr-1+ cells in the development of antitumor immunity. *Cancer Res.* 64: 1130–1139.

350 Stolina M, Sharma S, Lin Y, Dohadwala M, Gardner B, Luo J, Zhu L, Kronenberg M, Miller PW, Portanova J, Lee JC, Dubinett SM (2000): Specific inhibition of cyclooxygenase 2 restores antitumor reactivity by altering the balance of IL-10 and IL-12 synthesis. *J. Immunol.* 164: 361–370.

351 Mantovani, A., Bottazzi, B., Colotta, F., Sozzani, S., Ruco, L. (1992): The origin and function of tumor-associated macrophages. *Immunol. Today* 13: 265–270.

352 Hochrein, H., O'Keefe, M., Luft, T., Vandenbeele, S., Grumont, R. J., Maraskovsky, E., Shortman, K. (2000): Interleukin (IL)-4 is a major regulatory cytokine governing bioactive IL-12 production by mouse and human dendritic cells. *J. Exp. Med.* 192: 823–833.

353 A. Mencacci, G. Del Sero, E. Cenci, C.Fe d'Ostiani, A. Bacci, M. Montagnoli, M. Kopf, L. Romani, (1998): Endogenous interleukin 4 is required for development of protective CD4$^+$ helper type 1 cell response to *Candida albicans*. *J. Exp. Med.* 187: 307–317.

354 Mencacci A, Cenci E, Del Sero G, Fé d'Ostiani C, Mosci P, Trinchieri G, Adorini L, Romani L (1998): IL-10 is required for development of protective Th1 responses in IL-12-deficient mice upon *Candida albicans* infection. *J. Immunol.* 161: 6228–6237.

355 Emmanuilidis, K., Weighardt, H., Maier, S., Gerauer, K., Fleischmann, T., Zheng, X. X., Holzmann, B., Heidecke, C. D. (2001): Critical role of Kupffer cell-derived IL-10 for host defense in septic peritonitis. *J. Immunol.* 167: 3919–3927.

356 Mottonen, M., Isomaki, P., Luukkainen, R., Lassila, O. (2002): Regulation of CD154-induced interleukin-12 production in synovial fluid macrophages. *Arthr. Res.* 4: R9.

357 Schuler, T., Kammertoens, T., Preiss, S., Debs, P., Noben-Trauth, N., Blankenstein, T. (2001): Generation of tumour-associated cytotoxic T lymphocytes requires interleukin 4 from CD8+ T cells. *J. Exp. Med.* 194: 1767–1775.

358 Lutz, M. B., Schnare, M., Menges, M., Rossner, S., Rollinghoff, M., Schuler, G., Gessner, A. (2002): Differential functions of IL-4 receptor types I and II for dendritic cell maturation and IL-12 production and their dependency on GM-CSF, *J. Immunol.* 169: 3574–3580.

359 Fan, H., Cook, J.A. (2004): Molecular mechanisms of endotoxin tolerance. *J. Endotox. Res.* 10: 71–83.

360 Dobrovolskaia MA, Vogel SN (2002): Toll receptors, CD14, and macrophage activation and deactivation by LPS. *Microbes Infect.* 4: 903–914.

361 Dillon, S., Agrawal, A., Van Dyke, T., Landreth, G., McCauley, L., Koh, A., Maliszewski, C., Akira, S., Pulendran, B. (2004): A Toll-like receptor 2 ligand stimulates Th2 responses in vivo, via induction of extracellular signal-regulated kinase mitigen-activated protein kinase and c-Fos in dendritic cells. *J. Immunol.* 172: 4733–4743.

362 Rachmilewitz, D., Katakura, K., Karmeli, F., Hayashi, T., Reinus, C., Rudensky, B., Akira, S., Takeda, K., Lee, J., Takabayashi, K., Raz, E. (2004): Toll-

363 Fukao, T., Koyasu, S. (2003): PI3K and negative regulation of TLR signaling. *Trends Immunol.* 24: 358–363.

364 Guha, M., Mackman, N. (2002): The phosphatidylinositol 3-kinase-Akt pathway limits lipopolysaccharide activation of signaling pathways and expression of inflammatory mediators in human monocytic cells. *J. Biol. Chem.* 35: 32 124–32 132.

365 Noubir, S., Hmama, Z., Reiner, N. E. (2004): Dual receptors and distinct pathways mediate interleukin-1 receptor-associated kinase degradation in response to lipopolysaccharide. *J. Biol. Chem.* 279: 25 189–25 195.

366 Martin, M., Schifferle, R.E., Cuesta, N., Vogel, S. N., Katz, J., Michalek, S. M. (2003): Role of the phosphatidylinositol 3 kinase-Akt pathway in the regulation of IL-10 and IL-12 by *Porphyromonas gingivalis* lipopolysaccharide. *J. Immunol.* 171: 717–725.

367 Dahle, M. K., Overland, G., Myhre, A. E., Stuestol, J. F., Hartung, T., Krohn, C. D., Mathiesen, O., Wang, J. E., Aasen, A. O. (2004): The phosphatidyl 3-kinase/protein kinase B signaling pathway is activated by lipoteichoic acid and plays a role in Kupffer cell production of interleukin-6 (IL-6) and IL-10. *Infect. Immun.* 72: 5704–5711.

368 Kobayashi K, Hernandez LD, Galán JE, Janeway CA Jr, Medzhitov R, Flavell RA (2002). IRAK-M is a negative regulator of Toll-like receptor signaling. *Cell* 110: 191–202.

369 Del Fresno, C., Gomez-Garcia, L., Caveda, L., Escoll, P., Arnalich, F., Zamora, R., Lopez-Collazo (2004): Nitric oxide activates the expression of IRAK-M via the release of TNF-α in human monocytes. *Nitric Oxide* 10: 213–220.

370 Dalpke, A., Heeg, K. (2002): Signal integration following Toll-like receptor triggering. *Crit. Rev. Immunol.* 22: 217–250.

371 Hanada, T., Yoshimura, A. (2002): Regulation of cytokine signaling and inflammation, *Cytokine Growth Factor Rev.* 13: 413–421.

372 Bohuslav, J., Kravchenko, V. V., Parry, G. C. N., Erlich, J. H., Gerondakis, S., Mackman, N., Ulevitch, R. J. (1998): Regulation of an essential innate response by the p50 subunit of NF-κB. *J. Clin. Invest.* 102: 1645–1652.

373 Guan, H., Hou, S., Ricciardi, R. P. (2005): DNA binding of repressor NF-kB p50/p50 depends on phosphorylation of Ser^{337} by PKAc. *J. Biol. Chem.* 280: 9957–9962.

374 Janssens, S., Burns, K., Tschopp, J., Beyaert, R. (2002): Regulation of interleukin-1- and lipopolysaccharide-induced NF-κB activation of alternative splicing of MyD88. *Curr. Biol.* 12: 467–471.

375 Hume, D. A., Underhill, D. M., Sweet, M. J., Ozinsky, A. O., Liew, F. Y., Aderem, A. (2001): Macrophages exposed continuously to lipopolysaccharide and other agonists that act via toll-like recptors exhibit a sustained and additive activaion state. *BMC Immunol.* 2: 11.

376 Luo, Q., Ma, X., Wahl, S. M., Bieker, J. J., Crossley, M., Montaner, L. J. (2004): Activation and repression of interleukin-12 p40 transcription by erythroid Kruppel-like factor in macrophages. *J. Biol. Chem.* 279: 18 451–18 456.

377 Wang, H., Bloom, O., Zhang, M., Vishnubhakat, J. M., Ombrellino, M., Che, J., Frazier, A., Yang, H., Ivanova, S., Borovikova, L., Manogue, K. R., Faist, E., Abraham, E., Andersson, J., Andersson, U., Molina, P. E., Abumrad, N. N., Sama, A., Tracey, K. J. (1999): HMG-1 as a late mediator of endotoxin lethality in mice. *Science* 285: 248–251.

378 Chen, G., Li, J., Ochani, M., Rendon-Mitchell, B., Qiang, X., Susaria, S., Ulloa, L., Yang, H., Fan, S., Goyert, S.M., Wang, P., Tracey, K.J., Sama, A.E., Wang, H. (2004): Bacterial endotoxin stimulates macrophages to release HMB1 partly through CD14- and TNF-dependent mechanisms. *J. Leukoc. Biol.* 76: 76: 994–1001.

379 Andersson, U., Tracey, K. J. (2003): HMGB1 in sepsis. *Scand. J. Infect. Dis.* 35: 577–584.

380 Liliensiek, B., Weigand, M.A., Bierhaus, A., Niklas, W., Kasper,M., Hofer, S., Plachky, J., Gröne, H.-J., Kurschus, F.G., Schmidt, A.M., Du Yan, S., Martin, E., Schleicher, E., Stern, D.M., Hämmerling, G.J., Nawroth, P.P., Arnold, B. (2004): Receptor for advanced glycation end products (RAGE) regulates sepsis but not the adaptive immune response. *J. Clin. Invest.* 113: 1641–1650.

381 Jacobs, A. T., Ignarro, L. J. (2003): Nuclear factor-κB and mitogen-activated protein kinases mediate nitric oxide-enhanced transcriptional expression of interferon-β. *J. Biol. Chem.* 278: 8018–8027.

382 Seya, T., Oshiumi, H., Sasai, M., Akazawa, T., Matsumoto, M. (2005): TICAM-1 and TICAM-2: toll-like receptor adapters that participate in induction of type 1 interferons. *Int. J. Biochem. Cell Biol.* 37: 524–529.

383 Herzog, P.J., O'Neill, L.A., Hamilton, J.A. (2003): The interferon in TLR signaling: more than just antiviral. *Trends Immunol.* 24: 534–538.

384 Beutler, B., Hoebe, K., Shamel, L. (2004): Forward genetic dissection of afferent immunity: the role of TIR adapter proteins in innate and adaptive immune responses. *C.R. Biol.* 327: 571–580.

385 Punturieri, A., Alviani, R. S., Polak, T., Copper, P., Sonstein, J., Curtis, J. L. (2004): Specific engagement of TLR4 and TLR3 does not lead to IFN-b-mediated innate signal amplification and STAT1 phosphorylation in resident murine alveolar macrophages. *J. Immunol.* 173: 1033–1042.

386 A.K. Paluka, J. Banchereau, P. Blanco, V. Pascual (2002): The interplay of dendritic cell subsets in systemic lupus erythematosus, *Immunol. Cell. Biol.* 80: 484–488.

387 Karsunky, H., Merad, M., Mende, I., Manz, M. G., Engleman, E. G., Weissman, I. L. (2005): Developmental origin of interferon-α-producing dendritic cells from haemopoeitic precursors. *Exp. Haematol.* 33: 173–181.

388 Schroder, K., Hertzog, P. J., Ravasi, T., Hume, D. A. (2004): Interferon-γ: an overview of signals, mechanisms and functions. *J. Leukoc. Biol.* 75: 163–189.

389 Schleicher, U., Hesse, A., Bogdan, C. (2004): Minute numbers of contaminant $CD8^+$ T cells or $CD11b^+CD11c^+$ NK cells are the source of IFN-γ in IL-12/I-18-stimulated mouse macrophage populations. *Blood* 105: 1319–1328.

390 Ma, J., Chen, T., Mandelin, J., Ceponis, A., Miller, N. E., Hukkanen, M., Ma, G. F., Konttinen, Y. T. (2003): Regulation of macrophage activation. *CMLS Cell Mol. Life Sci.* 60: 2334–2346.

391 Dalpke, A. H., Eckerle, S., Frey, M., Heeg, K. (2003): Triggering of Toll-like receptors modulate IFN-g signaling: involvement of serine 727 STAT1 phosphorylation and suppressors of cytokine signaling. *Eur. J. Immunol.* 33: 1776–1787.

392 Rothfuchs, A. G., Trumstedt, C., Wigzell, H., Rottenberg, M. E. (2004): Intracellular bacterial infection-induced IFN-γ is critically but not solely dependent on Toll-like receptor 4-myeloid differentiation factor 88-IFN-$\alpha\beta$-STAT1 signaling. *J. Immunol.* 172: 6345–6353.

393 Ramana, C. V., Gil, M. P., Schreiber, R. D., Stark, G. R. (2002): Stat1-dependent and -independent pathways in IFN-γ-dependent signaling. *Trends Immunol.* 23: 96–101.

394 Kerr, I. M., Costa-Pereira, A. P., Lillemeier, B. F., Strobl, B. (2003): Of JAKs, STATs, blind watchmakers, jeeps and trains. *FEBS Lett.* 546: 1–5.

395 Baetz, A., Frey, M., Heeg, K., Dalpke, A. H. (2004): Suppressor of cytokine signaling (SOCS) proteins indirectly regulate Toll-like receptor signaling in innate immune cells. *J. Biol. Chem.* 279: 54 708–54 715.

396 Stoiber, D., Kovarik, P., Cohney, S., Johnston, J. A., Steinlein, P., Decker, T. (1999): Lipopolysaccharide induces in macrophages the synthesis of the supprssor of cytokine signaling 3 and suppresses signal transduction in

396 ...response to the activating factor IFN-γ. *J. Immunol.* 163: 2640–2647.
397 Stoiber, D., Stockinger, S., Steinlein, P., Kovarik, J., Decker, T. (2001): *Listeria monocytogenes* modulates macrophage cytokine responses through STAT serine phosphorylation and the induction of suppressor of cytokine signaling 3. *J. Immunol.* 166: 466–472.
398 Mead, J. R., Hughes, T. R., Irvine, S. A., Singh, N. N., Ramji, D. P. (2003): Interferon-g stimulates the expression of the inducible cAMP early repressor in macrophages through the activation of casein kinase 2. *J. Biol. Chem.* 278: 17741–17751.
399 Mühl, H., Pfeilschifter, J. (2003): Anti-inflammatory properties of interferon-γ. *Int. Immunopharmacol.* 3: 1247–1255.
400 Terstegen, L., Gatsios, P., Bode, J. G., Schaper, F., Heinrich, P. C., Graeve, L. (2000): The inhibition of interleukin-6-dependent STAT activation by mitogen-activated protein kinases depends on tyrosine 759 in the cytoplasmic tail of glycoprotein 130. *J. Biol. Chem.* 275: 18810–18817.
401 Ahmed, S. T., Ivashkiv, L. B. (2000): Inhibition of IL-6 and IL-10 signaling and Stat activation by inflammatory and stress pathways. *J. Immunol.* 165: 5227–5237.
402 Schilling, D., Thomas, K., Nixdorff, K., Vogel, S. N., Fenton, M. J. (2002): Toll-like receptor 4 and Toll-IL-1 receptor domain-containing adapter protein (TIRAP)/myeloid differentiation protein 88 adapter-like (Mal) contribute to maximal IL-6 expression in macrophages. *J. Immunol.* 169: 5874–5880.
403 Johnston, J. A., O'Shea, J. J. (2003): Matching SOCS with function. *Nat. Immunol.* 4: 507–509.
404 Yasukawa, H., Ohishi, M., Mori, H., Marakami, M., Chinen, T., Aki, D., Hanada, T., Takeda, K., Akira, S., Hoshijima, M., Hirano, T., Chien, K. R., Yoshimura, A. (2003): IL-6 induces an anti-inflammatory response in the absence of SOCS3 in macrophages. *Nat. Immunol.* 4: 551–556.

405 Rose-John, S., Neurath, M. F. (2004): IL-6 trans signaling: the heat is on. *Immunity* 20: 2–4.
406 Niemand, C., Nimmesgern, A., Haan, S., Fischer, P., Schaper, F., Rossaint, R., Heinrich, P. C., Müller-Newen, G. (2003): Activation of STAT3 by IL-6 and IL-10 in primary human macrophages is differentially modulated by suppressor of cytokine signaling 3. *J. Immunol.* 170: 3263–3272.
407 Williams, L., Bradley, L., Smith, A., Foxwell, B. (2004): Signal transducer and activator of transcription 3 is the dominant mediator of the anti-inflammatory effects of IL-10 in human macrophages. *J. Immunol.* 172: 567–576.
408 Ding, Y., Chen, D., Tarcsafalvi, A., Su, R., Qin, L., Bromberg, J. S. (2003): Suppressor of cytokine signaling 1 inhibits IL-10-mediated immune responses. *J. Immunol.* 170: 1383–1391.
409 Berrebi, D., Bruscoli, S., Cohen, N., Foussat, A., Migliorati, G., Bouchet-Delbos, L., Maillot, M-C., Portier, A., Couderc, J., Galanaud, P., Peuchmaur, M., Riccardi, C., Emilie, D. (2003): Synthesis of glucocorticoid-induced leucine zipper (GILZ) by macrophages: an anti-inflammatory and immunosuppressive mechanism shared by glucocorticoids and IL-10. *Blood* 101: 729–738.
410 Cao, S., Liu, J., Chesi, M., Bergsagel, P. L., Ho, I-C., Donnelly, R. P., Ma, X. (2002): Differential regulation of IL-12 and IL-10 gene expression in macrophages by the basic leucine zipper transcription factor c-Maf fibrosarcoma. *J. Immunol.* 169: 5715–5725.
411 Kim, H.-J., Hart, J., Knatz, N., Hall, M. W., Wewers, M. D. (2004): Janus kinase 3 downregulates lipopolysaccharide-induced IL-1b-converting enzyme activation by autocrine IL-10. *J. Immunol.* 172: 4948–4955.
412 Feng, W. G., Wang, Y. B., Zhang, J. S., Wang, X. Y., Li, C. L., Chang, Z. L. (2002): cAMP elevators inhibit LPS-induced IL-12 p40 expression by interfering with phosphorylation of p38 MAPK in murine peritoneal macrophages. *Cell Res.* 12: 331–337.

413 Brewington, R., Chatterji, M., Zoubine, M., Miranda, R.N., Norimatsu, M., Shnyra, A. (2001): IFN-γ-independent autocrine cytokine regulatory mechanisms in reprogramming of macrophage responses to bacterial lipopolysaccharide. *J. Immunol.* 2001: 392–398.

414 Xaus, J., Mirabet, M., Lloberas, J., Soler, C., Lluis, C., Franco, R., Celada, A. (1999): IFN-g up-regulates the A2B adenosine receptor expression in macrophages: a mechanism of macrophage deactivation. *J. Immunol.* 162: 3607–3614.

415 Khoa, N. D., Montesinos, M. C., Reiss, A. B., Delano, D., Awadallah, N., Cronstein, B. N. (2001): Inflammatory cytokines regulate function of adenosine A_{2A} receptors in human monocytic THP-1 cells. *J. Immunol.* 167: 4026–4032.

416 Pinhal-Enfield, G., Ramanathan, M., Hasko, G., Vogel, S.N., Salzman, A.L., Boons, G.-J, Leibovich, S.J. (2003): An angiogenic switch in macrophages involving synergy between Toll-like receptors 2, 4, 7, and 9 and adenosine A_{2a} receptors. *Am. J. Pathol.* 163: 711–721.

417 Mantovani, A., Bonecchi, R., Martinez, F. O., Galliera, E., Perrier, P., Allavena, P., Locati, M. (2003): Tuning of innate immunity and polarized responses by decoy receptors. *Int. Arch. Allergy Immunol.* 132: 109–115.

418 Mantovani, A., Locati, M., Polentarutti, N., Vecchi, A., Garlanda, C. (2004): Extracellular and intracellular decoys in the tuning of inflammatory cytokines and Toll-like receptors: the new entry TIR8/SIGIRR. *J. Leukoc. Biol.* 75: 738–742.

419 Schulz, O., Edwards, A. D., Schito, M., Aliberti, J., Manickasingham, S., Sher, A., Reis e Sousa, C. (2000): CD40 triggering of heterodimeric IL-12p70 production by dendritic cells in vivo requires a microbial priming signal. *Immunity* 13: 453–462.

420 Corbin, G. A., Harty, J. T. (2005): T cells undergo rapid ON/OFF but not ON/OFF/ON cycling of cytokine production in response to antigen. *J. Immunol.* 174: 718–726.

421 Stamp, L.K., James, M.J., Cleland, L.G. (2004): Interleukin-17: the missing link between T-cell accumulation and effector cell actions in rheumatoid arthritis. *Immunol. Cell. Biol.* 82: 1–9.

422 Misra, N., Bayry, J., Lacroix-Desmazes, S., Kazatchkine, M.D. Kaveri, S.V. (2004): Human $CD4^+CD25^+$ T cells restrain the maturation and antigen-presenting function of dendritic cells. *J. Immunol.* 172: 4676–4680.

423 Villarino, A. V., Huang, E., Hunter, C. A. (2004): Understanding the pro- and anti-inflammatory properties of IL-27. *J. Immunol.* 173: 715–720.

424 Lucas, S., Ghilardi, N., Li, J., de Sauvage, F. J. (2003): IL-27 regulates IL-12 responsiveness of naïve $CD4^+$ T cells through Stat1-dependent and -independent mechanisms. *Proc. Natl. Acad. Sci. U.S.A.* 100: 15 047–15 052.

425 Villarino, A., Hibbert, L., Lieberman, L., Wilson, E., Mak, T., Yoshida, H., Kastelein, R. A., Saris, C., Hunter, C. A. (2003): The IL-27R (WSX-1) is required to suppress T cell hyperactivity during infection. *Immunity* 19: 645–655.

426 Hamano, S., Himeno, K., Miyazaki, Y., Ishii, K., Yamanaka, A., Takeda, A., Zhang, M., Hisaeda, H., Mak, T. W., Yoshida, H. (2003): WSX-1 is required for resistance to *Trypanosoma cruzi* infection by regulation of proinflammatory cytokine production. *Immunity* 19: 657–667.

427 Yoshida, H., Hamano, S., Senaldi, G., Covey, T., Faggioni, R., Mu, S., Xia, M., Wakeham, A. C., Nishina, H., Potter, J., Saris, C. J. M., Mak, T. W. (2001): WSX-1 is required for the initiation of Th1 responses and resistance to *L. major* infection. *Immunity* 15: 569–578.

428 Gelman, A.E., Zhang, J., Choi, Y., Turka, L.A. (2004): Toll-like receptor ligands directly promote activated $CD4^+$ T cell survival. *J. Immunol.* 15: 6065–6073.

429 Oberholzer, C., Oberholzer, A., Clare-Salzler, M., Moldawer, L.L. (2001): Apoptosis in sepsis: a new target for therapeutic exploration. *FASEB J.* 15: 879–892.

430 Marth T, Zeitz M, Lúdvíksson BR, Strober W, Kelsall BL (1999): Extinction of IL-12 signaling promotes Fas-mediated apoptosis of antigen-specific T cells. *J. Immunol.* 162: 7233–7240.

431 Hume, D. A., Ross, I. L., Himes, S. R., Sasmono, R. T., Wells, C. A., Ravasi, T. (2002): The mononuclear phagocyte system revisited. *J. Leukoc. Biol.* 72: 621–627.

432 Banchereau, J., Pascual, V., Palucka, A. K. (2004): Autoimmunity through cytokine-induced dendritic cell activation. *Immunity* 20: 539–550.

433 Bennett, L., Palucka, A. K., Arce, E., Cantrell, V., Borvak, J., Banchereau, J., Pascual, V. (2003): Interferon and granulopoiesis signatures in systemic lupus erythematosus blood. *J. Exp. Med.* 197: 681–685.

10
Macrophages – Balancing Tolerance and Immunity
Nicholas S. Stoy

10.1
Balancing Tolerance and Immunity

10.1.1
Introduction

The existence of innate and adaptive immunity is well established, but less is known about the generation and maintenance of tolerance. Immune tolerance has usually been ascribed to T cell-processing mechanisms located either centrally (in the thymus) or peripherally (in secondary lymphoid tissue). However, more recently, an instructive role for DCs in shaping tolerance has emerged. In the paradigm presented here, innate (peripheral) tolerance is proposed as a more formal counterpart to innate immunity. By definition, innate tolerance is a primary response expressed in peripheral non-lymphoid tissue microenvironments and must be independent of any T cell adaptive response. It depends essentially on the ability of macrophages to display variable antiinflammatory, as well as proinflammatory, phenotypes in response to local tissue microenvironments. Macrophages are deemed to play a crucial role in both inflammation and immunomodulation at the local level, and are assumed to be reliable quantifiers of ongoing tissue inflammation. DCs, as well as several 'non-professional' immunologically-activated cells, also contribute to the polarization of the local microenvironment, but DCs characteristically emigrate following activation and antigen capture. The relevance of opposing mechanisms of innate tolerance and innate immunity to overall immunity, and to the pathogenesis of inflammatory conditions, is examined. Towards the end of the chapter, asthma and systemic lupus erythematosus are analyzed in some detail to illustrate how the paradigm might prove useful in unraveling complex interactions between macrophages, DCs, NKs, natural T cells, T cells, B cells, and eosinophils. Hopefully, the paradigm will go some way towards bridging the gap between observational studies and mathematical modeling of the mammalian immune system, as well as indicating how macrophage-based treatment strategies of immune disorders might offer some unique advantages.

Antigen Presenting Cells: From Mechanisms to Drug Development. Edited by H. Kropshofer and A. B. Vogt
Copyright © 2005 WILEY-VCH Verlag GmbH & Co. KGaA, Weinheim
ISBN: 3-527-31108-4

In vivo, it is now almost axiomatic that innate immunity instructs and activates adaptive immunity [1]. When inflammation is initiated by the innate immune system, proinflammatory activated macrophages characteristically promote [2] and reflect a microenvironment in which DCs process antigen derived from pathogens [3], or necrotic cells [4, 5]. In addition to transporting antigenic material, DCs carry specific polarizing and activating signals to secondary lymphoid tissue [6, 7]. Thus, through what may be termed 'innate functional response linkage' (IFRL) [8], a proinflammatory macrophage phenotype is usually linked to a Th1 response [9, 10]. More recently, macrophages with antiinflammatory phenotypes have been described [11–14], often associated with antiinflammatory or 'steady-state' microenvironments in which DCs are still able to take up antigen but receive different polarizing and activating signals [15–20]. In this case, IFRL describes a sequence linked to Th2 or regulatory T cell responses [10, 21, 22]. An apparent contradiction to IFRL, which will be considered in detail, is 'inflammation' in asthma, since in that condition proinflammatory responses in the lung may correlate more closely with Th2 than Th1 adaptive responses [23].

The default setting of macrophages (and microenvironment) at mucosal boundaries is generally antiinflammatory [24–26], antigen being processed in an antiinflammatory manner from the outset [27]. Extrapolating from extensive in vitro studies it has been suggested that, through 'sampling' the local microenvironment, macrophages assume phenotypes similar to other immunologically active cells. Uncommitted monocytes recruited to immunologically active sites, transform into activated macrophages of similar phenotype to those already present [28]. Likewise, DCs can differentiate from uncommitted monocytes, responding to the prevailing microenvironment by developing pro- or antiinflammatory phenotypes [29–32].

It is hypothesized that tissue macrophages are ideally placed to integrate patterns of input from local microenvironments. The innate immune system can function independently, generating innate immunity or innate tolerance, but its operation is usually further modified by T cell responses [33]. Innate tolerance, it is argued, instructs adaptive tolerance.

10.1.2
Macrophage Phenotypes: Effects on Immunity and Tolerance

Proinflammatory macrophages are presumed to be principal effector cells of both innate immunity (as through TLR stimulation) and adaptive immunity (as through MHC II/antigen/TCR engagement and co-stimulation). The macrophage proinflammatory phenotype consists of a range of outputs, including pathogen-toxic and/or tissue-damaging molecules, adhesion molecules, MHC and co-stimulatory molecules and various macrophage signaling molecules, which, directly or indirectly, modulate the existing phenotype. Despite NF-κB activation, implying eventual apoptosis, anti-apoptotic mechanisms prolong proinflammatory macrophage survival [34, 35]. IL-12 from proinflammatory macrophages, or other cells, stimulates local NK cells and T cells to secrete IFN-γ, amplifying the proinflammatory response in the tissues as well as providing anti-apopototic signaling to activated Th1 cells [36]. IL-12 produced by migrating DCs promotes polarization towards Th1 in lymphoid tissue [37]. However, recent information suggests that mature DCs, which are derived in vitro from monocytes (regardless of how the maturation process is achieved), rapidly secrete IL-12p70 initially, but then lose this capability. The time course of these changes is critical to polarizing T cells in lymphoid tissue and, in vitro, the process may take 10 days. In any event the data have led to the suggestion that one of the main functions of DCs could simply be to amplify the initial innate response in the tissues [38].

By contrast, antiinflammatory macrophages are both producers of, and may be induced by, antiinflammatory molecules [8, 39–41]. So-called 'alternatively activated' macrophages, specifically induced in vitro by IL-4 or IL-13, exhibit upregulated antigen-presenting capacity (MHC II) [42], novel chemokine expression [43], and anti-proliferative effects on T cells [44]. Some molecules from antiinflammatory macrophages, particularly TGF-β and proline [45], promote angiogenesis and fibrosis [43], of importance in tissue healing [46–48]. The most complete and authoritative functional classification of macrophage phenotypes to date recognizes innate activation, classical (IFN-γ-induced) activation, innate/acquired deactivation, alternative (IL-4 or IL-13-induced) activation and humoral activation [28]. Innately activated, innately deactivated and humorally activated macrophages are all capable of displaying antiinflammatory (or non-inflammatory) properties and so could be mediators of innate tolerance. In addition, 'transitional' macrophages, with mixed proinflammatory and antiinflammatory phenotypes, may arise during induction and resolution of inflammatory responses. It is proposed that an immune response, in vivo, always begins with a change in the balance of innate tolerance and innate immunity. The change may then be communicated to the adaptive immune system, which responds predictably. The initial innate response also programs into the system later modulatory processes, through which the magnitude and duration of both innate and adaptive responses are altered. Contrary to common assumptions, the magnitude and kinetics of the antigen-specific T cell response to bacterial infection may be programmed from peripheral tissues within the first day, well before the height of the proinflammatory response and independently of both duration of infection or quantity of antigen, although persistence of infection is required to drive continued T cell expansion [49]. If stimulation of the innate immune system persists, various competing mechanisms may eventually be found operating at the same time in the same tissue microenvironment, presenting a dynamic balance of cells, cytokines and other bioactive molecules, of great apparent complexity [50]. Any paradigm attempting to explain immunological responses in peripheral tissues impacted by the environment (gut, lung, skin) must accommodate not only the full range of observed macrophage phenotypes but also temporal sequencing, positive and negative feedback modulation of inflammation and mechanisms switching the polarity of immune responses.

Innate tolerance actively opposes innate immunity. Both innate tolerance and innate immunity are graded responses, lying on the same continuum. Closely related to the 'innate tolerance–innate immunity continuum', and identifiable in the same tissue microenvironment, is an 'antiinflammatory–proinflammatory continuum'. As discussed in 9.2.4, innate immunity is a phylogenetically ancient (and originally 'stand-alone') mechanism based on the recognition of highly conserved microbial-associated molecules by pattern recognition receptors, which in the case of some well-categorized TLRs activate immediate proinflammatory effector mechanisms to bring about an immune response [51]. Inflammation, although a powerful effector of immune responses (both innate and adaptive), is relatively non-specific and results in tissue damage [52], itself a potential enhancer of inflammation. Tolerizing signals are necessary, at every stage of an immune response, to prevent over-exuberant responses leading to damaging sequelae such as endotoxic (LPS-induced) shock, chronic inflammation and autoimmunity.

10.1.3
Concept of Innate (Peripheral) Tolerance

As a general principal, it appears that immunological tolerance, regardless of its origin, exerts an antiinflammatory (or non-inflammatory) influence on tissue microenvironments. For example, mucosally-induced (oral) tolerance and tolerance to self-T cells are both associated with reduced local proinflammatory activity. Similarly, peripheral T cell tolerance ('adaptive tolerance'), whether from T cell elimination, anergy or the production of phenotypically antiinflammatory T cells, manifests in the tissues either as a total lack of T cell response or as a suppressive (or neutral) influence on tissue proinflammatory responses. Thus, T cell tolerance reinforces, or has no effect on, antiinflammatory or non-activated macrophage phenotypes, but may exert a downregulatory influence on proinflammatory macrophage phenotypes. As discussed, tolerance to tumor cells, which themselves often secrete antiinflammatory molecules, can often be traced to a failure to induce proinflammatory macrophage responses, or to their active suppression [53, 54]. Transplant tolerance and tolerance to the developing fetus both depend on a lack of activated proinflammatory macrophages in the local microenvironment, since these are recognized as major effectors of rejection [55]. In keeping with these observations, it is postulated that innate tolerance is specifically associated with non-inflammatory or antiinflammatory microenvironments. Like innate immunity, innate tolerance is induced as a direct reaction to external (environmental) and/or internal immune system perturbations. As well as opposing immunity, innate tolerance enhances or maintains local resistance to tissue-damaging inflammation. In the context of infection, innate tolerance can be protective of both 'host' and 'pathogen' [56], but the balance can become unstable in either direction. Innate tolerance may initially increase pathogen virulence by enabling evasion or neutralization of an effective immune response, but, eventually, innate or adaptive immunity may be triggered and the balance redressed. Stimuli, other than microbial, may trigger the same mechanisms as mediate innate tolerance. The definition of innate tolerance is proposed to encompass not only stabilization and enhancement of antiinflammatory macrophage phenotypes but also downregulation of classically-activated macrophages [57], so long as the mechanism does not involve the adaptive immune system.

10.1.4
Concept of Adaptive Tolerance

In the 'resting' or 'steady state' [20], macrophages in peripheral tissues display non-inflammatory or antiinflammatory phenotypes, characteristically secreting IL-10 and/or TGF-β. Under such circumstances, DC priming and antigen acquisition results in the induction of subsets of antiinflammatory T cells, secreting IL-10/IL4 (Th2), IL-10 (T_R1) or TGF-β (Th3 or CD4$^+$CD25$^+$) [18, 28, 58–61] and can express the same T cell receptors as their Th1 counterparts [21, 62, 63]. In vitro, DCs prepared in an IL-10-dominated microenvironment secrete IL-10 and produce IL-4-secreting Th2 cells [64]. IL-10 causes a persistent anergy in human CD4$^+$T cells [65] whilst antigen-specific T cell anergy is imparted by IL-10-primed immature DCs [66]. The presence of TGF-β in the innate microenvironment may reinforce tolerance through the adaptive immune system, judging from studies in which TGF-β-treated APCs induce TGF-β-secreting CD4$^+$ T cells and delete antigen-specific APCs through a Fas/FasL pathway [67–69]. CD4$^+$CD25$^+$ regulatory T are able to inhibit other T cells (notably CD8$^+$T cell proliferation), perforin, granzyme B and IFN-γ production [70]. Antiinflammatory T cells are probably relatively short-lived and deleted after a few cycles of cell proliferation [71], implying that a continuous input of tolerizing antigen (foreign and/or self) from the innate system into the adaptive system is likely to be necessary for the maintenance of antigen-specific T cell tolerance and for deletion of antigen-specific T cells [72]. Adoptive transfer studies have confirmed that repetitive presentation of very low doses of antigen, including self-antigen, exclusively by DCs, can cause clonal deletion (or anergy) of antigen-specific T cells [16, 73]. In the 'maturation' hypothesis [74–77], DCs, when continuously exposed in the 'immature' state to 'physiological' mixtures of low dose self and/or non-self antigens, behave as tolerogenic DCs, but on exposure to microbial wall stimulants and interferons in a pro-inflammatory microenvironment they rapidly 'mature' to immunogenic DCs. Once instructed in lymphoid tissues, Th1- and Th2-polarized T cells tend to 'run true-to-form', in vivo, in their tissue effector roles, regardless of the polarity of the microenvironment [78], in keeping with the important proposition that the polarity switch between tolerance and immunity lies not in the adaptive but in the innate immune system. This does not exclude natural T cells, including probably regulatory T cells, having a primary polarizing role within the innate immune system.

Macrophages and DCs express receptors with overlapping functions. These receptors are usually induced and activated by the same environmental or endogenous stimuli. Some receptors are identical, such as TLRs, and some are different but with overlapping functions, such as MRs on macrophages and DEC205 on DCs. The paradigm presented here proposes that macrophages and DCs function so closely together in tissue microenvironments [79] that they can be amalgamated into a single 'macrophage–DC functional unit' (MDU; see Figure 10.1).

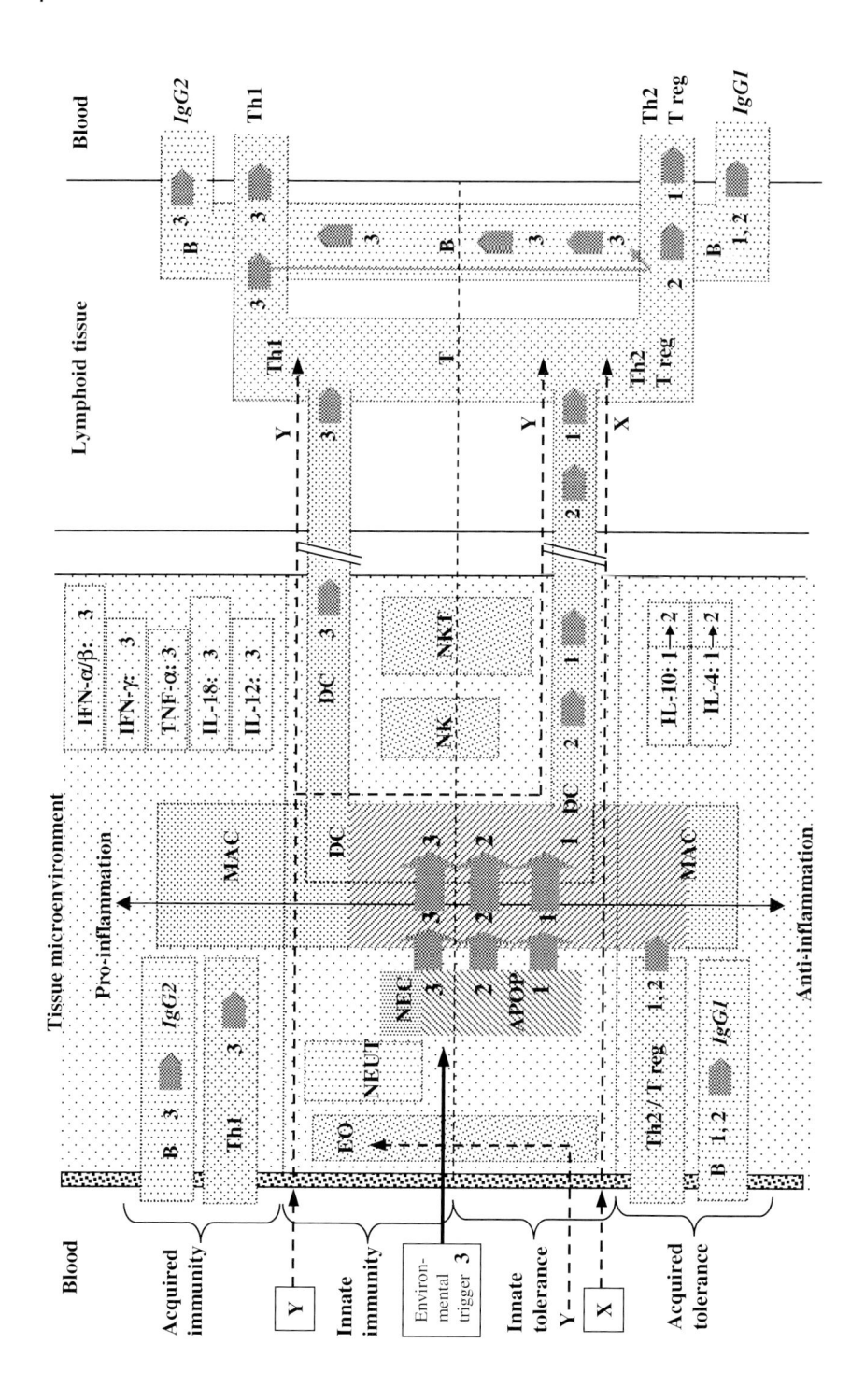

336 | *10 Macrophages – Balancing Tolerance and Immunity*

Figure 10.1 A qualitative/semi-quantitative representation to analyse immune responses, with specific reference to macrophages and the sequential events affecting macrophage phenotypes. Here, lupus is analysed: the pre-disease heightened state of innate tolerance (sequence 1+2) is overwhelmed following an additional environmental input (+ sequence 3), precipitating clinical lupus. Note that the final microenvironmental cytokine balance is a mixture of 'type 1' and 'type 2' with (in this particular case) an excess of 'type 1'. Semi-quantitative features of the model are the representation of concentrations (of cells, cytokines, etc.) by areas, including hypothetical maximum concentrations (dashed area outlines) and concentrations at a particular stage in a specific immune response. The ratio of macrophage to DC concentration may vary; indeed, to explore different functional outcomes, all cells which fail to migrate may, arbitrarily, be defined as macrophages or, alternatively, any cells which react with naive T cells, even if they fail to migrate, may, arbitrarily, be defined as DCs. Macrophages and DCs constitute the MDU, which includes adaptive as well as innate responses; the ratio of the MDU committed to innate versus adaptive immunity is another potential variable. The scheme can be manipulated to model the steady state (equivalent to sequence 1 in the lupus example), any other proinflammatory environmentally triggered response (infection, organ-specific autoimmunity), LPS tolerance, cancer and 'allergic' asthma (X) and/or 'proinflammatory asthma' (Y) (see dotted lines and eosinophil area). NK and NKT may be incorporated in sequences and other cells, such as mast cells and effector cells (eg goblet cells, fibroblasts, and smooth muscle cells in asthma) may be added.

The MDU has a dual role: the first is information collection, integration and transduction into a local effector output (mainly implemented by macrophages) and the second is information transfer to lymphoid tissue (mainly implemented by DCs). The identification of each cell type or their roles need not be absolute and partly on the vagaries of cell surface markers used to identify them. Indeed, as hinted earlier, a purely functional definition of macrophages and DCs may be worth considering. In the steady state the MDU constitutively expresses receptors capable of initiating responses to either anti- or proinflammatory stimuli. The properties of the MDU provide the basis for an elegant, flexible and powerful system of innate and adaptive immune surveillance, which generates tolerance in the steady state. Macrophages and DCs are continuously sampling the microenvironment to assemble a pro- or antiinflammatory phenotype in the MDU. Stimulation of macrophages and DCs depends on cell-surface receptor concentrations and ligand affinity and on the concentration of ligands in the microenvironment, resulting in a hierarchical pattern of response (as is already recognized for T cell epitopes). At the same time as the innate pro- or antiinflammatory polarity of the MDU is established, receptor-mediated uptake of selected material from the microenvironment provides a mechanism for multiple antigen processing and antigen-presentation in lymphoid tissue. Whereas anti- and proinflammatory macrophages mainly stay in the tissues, antiinflammatory (immature) and proinflammatory (mature) activated DCs mainly migrate to lymphoid organs. All the different grouped antigens taken up by one DC are transferred, tagged with either an anti- or proinflammatory polarizing signal, to lymphoid tissue to instruct the appropriate T cell response. In the steady state, a continuous trickle of self antigens derived from apoptosed cells will be presented by tolerogenic DCs to the adaptive immune system [80]. This will include uptake by DCs of apoptosed macrophages, together with their phagocytosed contents and antigenic material, providing the basis for tolerogenic cross-presentation of antigen to the adaptive immune system. Adjuvants work by biasing the polarization of the MDU either towards proinflammation or antiinflammation, usually corresponding to changes in the balance between innate immunity and innate tolerance [81].

The processes by which tolerogenic T cells are generated in lymphoid tissue are not entirely clear. Some investigators have found that a low concentration of antigen is an important determinant of tolerogenicity [82] but, in addition, antiinflammatory cytokines are considered to have a polarizing role. The two points of view are reconciled if low dose antigen fails to trigger maturity in DCs, keeping them in an antiinflammatory mode. Under steady-state conditions, T cell activation fails to produce IFN-γ and is followed by early T cell deletion [71]. By comparison, proinflammatory microenvironments containing TNF-α, IL-12 and/or other proinflammatory cytokines (interferons) usually bring about maturation of DCs to a Th1-inducing phenotype. In particular IL-12, augmented by (innate) IFN-γ primes peripheral blood myeloid DCs (CD1a$^+$ CD11c$^+$) for a strong Th1 adaptive response, as would occur in vivo when they encounter viruses, bacteria or other Th1-polarizing infections. However, plasmacytoid DCs (CD1a$^-$ CD11c$^-$) do not behave in this 'classic' manner, as judged from in vitro studies. They secrete primarily IFNα/β (on viral infection), but this cytokine does not apparently promote DC maturation, supports an IL-10 secreting phenotype and may influence myeloid DC polarization towards T reg [83]. With the exception, perhaps, of total elimination of proinflammatory T cell clones through 'exhaustion' (on exposure to continuing high antigen concentrations), a usual prerequisite for induction of adaptive T cell tolerance would appear to be an antiinflammatory microenvironment in lymphoid tissue, as would be found in the immediate surroundings of phenotypically antiinflammatory DCs during cognate antigen interaction with T cells. As indicated above, related groups of antigens presented by a single tolerogenic DC would then induce related groups of antigen-specific tolerogenic T cells. Influencing T cells towards the same outcome in response to multiple epitopes from the same pathogen has been used as a strategy in attempting to formulate effective anti-*Leishmania* vaccines.

A further important consideration in adaptive tolerance is how DC-primed T cells affect, and are affected by, the microenvironment of the peripheral tissues, the actual 'battleground' of immunological responses. As yet there appears to be continuing uncertainty about cytokine profiles and types of APC (macrophages, DCs or indeed other 'non-professional' APCs) that present antigen to antigen-specific memory T cells in tissues in vivo, the so-called effector limb of DTH [69, 84, 85]. Furthermore, the presence or absence of proinflammatory cytokines such as IL-15 and IL-12 may regulate nonspecific (antigen-independent) bystander activation of memory T cells in tissues [86], thus providing amplification of immune responses. The paradigm assumes that 'committed' memory T cells (programmed as Th1, Th2, Th3 or T reg) can all be recruited to immunologically active tissues by chemokines, but will usually behave as terminally differentiated cells [87], subsets of which are selected for specific and nonspecific (bystander) activation by the prevailing microenvironmental conditions, so reinforcing the response that has generated them, whether this be pro- or antiinflammatory. In this view, the Th1 phenotypes enhances proinflammation and T reg phenotypes enhance antiinflammation but the predetermined polarity of T cells is not changed by the local microenvironment [73]. Antigen-specific TCR engagement of Th1 cells with macrophages will promote an increased availability of IFN-γ and IL-12 and other proinflammatory molecules. Similarly, antigen-specific TCR engagement of regulatory T cell subsets with macrophages could increase IL-4 and promote an alternatively activated macrophage phenotype. Antigen non-specific T cells, activated through bystander mechanisms, may easily outnumber activated antigen specific T cells. Regulatory T cells introduced into a tissue microenvironment could also suppress innate immune responses directly through the release of antiinflammatory cytokines (IL-10, TGF-β) [88]. In these respects memory T cells do not display (or display to a much lesser extent) the plasticity displayed by recruited monocytes/macrophage [89], DCs and probably some other cells such as eosinophils, all of which may receive direct instruction from the microenvironment to develop either anti- or proinflammatory phenotypes, and presumably then exert a decisive influence on the programming and modulation of subsequent T cell adaptive responses.

On microbial challenge tissue DCs may be the earliest cells to synthesize bioactive Th1-inducing IL-12 as they do not require IFN-γ co-stimulation. However, the paradigm predicts that macrophages too are readily activated to a proinflammatory phenotypes, including IL-12 secretion, because the necessary combination of proinflammatory cytokines rapidly becomes available in acute inflammatory microenvironments. Sources of initial IL-12 could include macrophage and/or DC membrane-stored IL-12, or IL-12 released from neutrophils, leading to the rapid release of IFN-γ from local NK or T cells [6]. The cytokine/chemokine osteopontin (OPN) is a product of proinflammatory activated macrophages. It increases IL-12 and reduces IL-10 secretion and aids macrophage recruitment and differentiation [90, 91]. At the same time OPN promotes migration of DCs to lymph nodes [92], constituting one putative mechanism of IFRL, tying macrophages and DCS into the early shaping of Th1 polarization. Environmental stimulation of OPN release from macrophages and NK cells of the innate immune system is provoked by *Mycobacterium tuberculosis* infection of macrophages. Conversely, transgenic OPN 'knock-out' mice react to some viral or bacterial infections with reduced IL-12 and increased IL-10-production and fail to mount an effective Th1 (IFN-γ) response [90, 93, 94]. IL-15 is another proinflammatory macrophage cytokine of the innate immune system that may help prime DCs for antigen-presentation to T cells [95], as part of IFRL. In these ways, the build-up of proinflammatory tissue macrophages and DCs would occur in parallel, as required by the paradigm. Conversely, the same sort of microenvironmental linkages are presumed to occur in the generation of innate and acquired tolerance, except that the cell phenotypes in this case are antiinflammatory. Absence of IL-12 during DC priming in tissues is probably as critical to polarization of T cells away from Th1 and towards Th2 and T reg as is the presence of innately-generated suppressive cytokines, such as IL-10 and TGF-β. Ratios of IL-12/IL-10 or IL-12/PGE$_2$ occurring in the tissues may be useful measures of the balance between innate tolerance and innate immunity, as well as predicting proinflammatory versus antiinflammatory T cell polarization in draining lymph nodes and predicting (in mice) the ratio of B cell-generated IgG2/IgG1 isotypes. An important component of adaptive tolerance is the production of relatively ineffective antibodies of the IgG1 isotype (in mice). The innate cytokine IL-12 plays an important part in directing isotype switching to complement fixing IgG2 antibodies and protective antibody immunity in infection [96, 97].

Finally, the paradigm takes full account of temporal changes in macrophage function with changing microenvironmental inputs. In contrast to macrophages, DCs are usually considered to undergo maturation or terminal differentiation after tissue microenvironmental stimulation but this may simply reflect the fact that they promptly migrate to lymphoid tissue, where they are considered to have reached a definitive state of maturation, whether mature or semi-mature [75], and are not subject to further modifying stimuli. Based on IFRL, the adaptive immune system in vivo serves only to amplify, or sustain, (innate) tissue effector responses, without changing their polarity (whether this be pro- or antiinflammatory). T cells modulate inflammation by providing feedback onto macrophage receptors. T cell signals, which may operate through cytokine release or direct cell–cell contact, are integrated, along with other microenvironmental inputs, into the final macrophage phenotype (Figures 9.3 and 10.1). Indeed, the division between innate and adaptive inputs in determining macrophage phenotypes may be hard to decipher because the former initiate, and usually overlap, the latter and because macrophages and T cells not uncommonly employ the same biologically active molecules (IL-10, TGF-β, TNF-α, OPN [98] and others). In the present paradigm, 'switching' of a tissue microenvironment from anti- to proinflammatory (or vice versa) originates in macrophages rather than T cells (Figure 10.1). Thus, macrophages are both potent modulators of environmental stimuli and potent local effectors of inflammatory responses. Macrophage self-regulatory molecules may act immediately (usually reinforcing a proinflammatory phenotype as part of an innate immune response) or in a 'delayed direct' manner (either delayed reinforcement or delayed inhibition [99, 100] of a proinflammatory phenotype) or in a 'delayed indirect' manner (involving other intermediate cells, such as T cells [101] or B cells [102]). A growing number of molecules produced initially by macrophages in innate proinflammatory responses are now thought to be directly instrumental in the later downregulation of these same responses; PGE$_2$ and OPN may be in this category, comprising autocrine or paracrine mechanisms of innate tolerance. Thus, PGE$_2$ is one of the NF-κb-dependent products of proinflammatory macrophage activation that can later stimulate increased IL-10 by raising cAMP via PGE$_2$ receptors [99, 103, 104]. Although downregulatory for proinflammatory cytokines, PGE$_2$ is, however, potently anti-fibrotic, and so, presumably, does not contribute to the healing phenotype of antiinflammatory macrophages. As already indicated, OPN clearly promotes the proinflammatory pathway by increasing the IL-12/IL-10 ratio and its synthesis is increased by concomitant induction of iNOS (in RAW 264.7 macrophages). Conversely, OPN exerts negative feedback onto iNOS [105] and, furthermore, it may be chemotactic and stimulant to fibroblasts in the antiinflammatory phase of an immune response [90, 106]. Rheumatoid synovial fibroblasts themselves produce large amounts of OPN [107]. There are numerous further examples of tolerogenic feedback instigated by the innate immune system. For example, proinflammatory activated macrophages and other APCs express a molecule, PDL-1, which could be induced (in a STAT1-dependent manner) as part of an innate response to microbial products, including LPS (via TLR4), or in response to IFN-γ (from NK cells), but which is inhibitory to peripheral T cell activation when it interacts with PD-1, a co-stimulatory molecule expressed by activated effector T cells in peripheral tissues. This negative feedback loop is reinforced by IFN-γ-secreting activated T cells and, presumably, suppresses over-exuberant adaptive responses. Another macrophage molecule, PD-L2, is only inducible (in a STAT6-dependent manner) through interaction of proinflammatory activated macrophages with IL-4 or Th2 cells, and so, presumably, could be a downregulator of excessive Th2 activation during the effector phase of an immune response or could perhaps be a reciprocal modulator of macrophage function [108]. Although IDO induction occurs in proinflammatory activated macrophages and DCs, it may nevertheless be downregulatory. Currently there are more hypotheses than answers in this area, but tryptophan depletion, T cell anergy and imparting of an apoptotic signal to activated Th1 cells, all associated with IDO upregulation, may have regulatory significance [109–111]. In general, therefore, 'switching-off' proinflammatory responses in macrophages takes place either if the initial stimulus resolves, as in pathogen elimination by a successful proinflammatory response [112], or if tolerance to the stimulus develops. Whatever the mechanism, changes in T cell phenotypes in vivo are usually linked via IFRL with preceding changes in macrophage phenotypes (Figure 10.1). This applies even to the determination of differential apoptosis between T cell subsets, to activation/deactivation of cytotoxic $CD8^+$ T cells through 'cross presentation' of antigen (originating from one cell and presented by another, [113], to 'infectious' spread of tolerance to other T cells [114], or to secondary effects of T cells on the function of other cells, particularly B cells. Although mechanisms still need much clarification, generation of regulatory T cells [115] (and lack of support for Th1 [36]) is a sequential adaptive response to innate tolerance, and the outcome is stabilization, restoration or enhancement of the antiinflammatory ('suppressor') macrophage phenotype. The absence of proinflammatory cytokines in the 'priming' tissue microenvironment is evidently crucial to the induction of adaptive as well as innate tolerance.

10.1.5
Innate Tolerance: Receptors, Responses and Mechanisms

Like innate immunity, innate tolerance employs germline encoded macrophage receptors but, unlike innate immunity, these when stimulated induce antiinflammatory phenotypes. Primary stimulation of macrophages must be distinguished from stimulation via the adaptive immune system. By definition, innate tolerance cannot be mediated by cells of the adaptive immune system, namely T and B cells, specifically activated by the response. However, recently described natural T cells and a subset of 'innate' B cells are exceptional in having verifiable claims to the innate immune system. The definition excludes antiinflammatory T cell cytokines acting on macrophages, but not the same cytokines (IL-4, IL-13 and TGF-β) if produced by macrophages through primary stimulation [116–119]. Although B7/CD28 blockade effectively and reciprocally induces anergic T cells and antiinflammatory macrophages [55] it too lies outside the definition of innate tolerance. Also excluded are immunoglobulins from antigen-specific adaptive B cell responses, which, in the presence of LPS or other co-stimulatory molecules, can elicit rapid and potent antiinflammatory signals by engaging a subset of monocyte/macrophage FCγ receptors [102, 120], thereby preventing, or reversing, IL-12 production (either directly or through IL-10 induction) [121–124]. Macrophage receptors involved in innate tolerance respond to environmentally derived stimuli, exactly as demonstrated for receptors involved in innate immunity [125]. MMRRs for innate tolerance and innate immunity may even be reciprocally regulated, e.g. increased TLR expression with proinflammatory stimuli, decreased with antiinflammatory; increased expression of MRs with antiinflammatory stimuli (IL-10 and IL-4) [126–128], decreased with proinflammatory [129]. Another possibility is that a single receptor may direct pro- or antiinflammatory intracellular signaling depending on the stimulus, as exemplified by CD14, which can participate in TLR signaling, as described, but can also, and perhaps dominantly, be a receptor for apoptoic cells. In the latter role it has been described as an 'apoptotic cell pattern recognition receptor' and could help terminate proinflammatory responses through activating antiinflammatory pathways [130]. Antiinflammatory macrophage phenotypes are already known to be induced by some microbial-related ligands, including the receptor-binding B unit of cholera toxin [131–133], the similar non-toxic AB complex of *E. coli* [134] excretory/secretory material from live adult filarial parasites [135–137] and Gram-negative flagellin [138]. Both extracellular and intracellular infections may elicit innate tolerance. These include the intracellular macrophage pathogens *Leishmania*, *Trypanosoma cruzi*, MTB, *Plasmodium* and *Chlamydia*. In both *Leishmania* [139] and *Chlamydia*, secretion of TGF-β and/or IL-10 from infected macrophages is a manifestation of innate tolerance. In splenic infection by *Leishmania donovani*, the parasites localize to the macrophage-rich red pulp and secrete TGF-β, in the demonstrable absence of co-localizing T cells, as part of a primary macrophage immunosuppressive response [140]. *Toxoplasma* is an obligate intracellular protozoon parasite, which in SCID mice increases macrophage secretion of IL-10 and the active form of TGF-β, indicating clearly that T cells are not required for this response [141]. In IL-10 'knock-out' mice there is a vigorous immune response with elevated IL-12 and IFN-γ causing considerable 'bystander' damage to the liver and increased mortality [142]. When mediating innate tolerance, TGF-β may act in an autocrine way to increase its own output and may stimulate IL-10 production and downregulation of IL-12R [143]. Intracellular infection of macrophages by *Toxoplasma* suppresses NF-κB nuclear translocation by enhancing IκBα phosphorylation and inhibits the output of the proinflammatory cytokines, IL-12 and TNF-α, in response to LPS. [144]. Induction of innate tolerance is instrumental to the survival of MTB. Inoculation of macrophages with MTB provokes a short burst of TNF-α, maximal at about 8 h, but this is followed by a sustained secretion of IL-10, downregulation of TNF-α and increased output of sTNFR2, thereby silencing an apoptotic signal to the infected macrophage [145]. Macrophages taken from acute helminth infections have proinflammatory phenotypes, supporting a Th1 response, whereas macrophages from chronic infaction have suppressed IL-12 and NO, but secrete higher levels of IL-6 and PGE$_2$. These APCs express MHC II and co-stimulatory CD40 and B7-2 and bias T cells toward IL-4 secretion (Th2). Intact STAT6 signaling is essential for expression of this antiinflammatory phenotype [146]. The filarial nematode (helminth) *Brugia malaya* produces a tolerant phenotype in responding peritoneal macrophages, which biases the adaptive response towards Th2, an effect dependent on macrophage output of TGF-β rather than IL-10 [147, 148]. *Brugia malaya* produces a homolog of MIF-1 that recruits macrophages to the site of infection and induces in them Ym1, a reliable marker of a tolerogenic antiinflammatory macrophage phenotype. Among other cells recruited by the MIF-1 homolog are eosinophils [149].

The maintenance of innate tolerance and its restoration after an immune response may crucially depend on phagocytosis of apoptotic cells [150], a process requiring the collaboration of a series of receptors to assist tethering [151, 152]. Of these, the phosphatidylserine receptor has been given prominence [153, 154], perhaps to an unwarranted degree. Recent evidence from phosphatidyl receptor knock-out mice suggests that apoptotic cells can be engulfed quite satisfactorily without it [155]. Apoptotic cell-uptake satisfies the criteria for innate tolerance because an antiinflammatory phenotype (increased TGF-β and/or IL-10 and reduced TNF-α) is characteristically induced [156–159].

LPS tolerance is included in innate tolerance because the macrophage phenotype can be directly skewed in an antiinflammatory direction without the intervention of the adaptive immune system [160, 161]. Downregulation of IL-12 [162, 163] and reciprocal upregulation of IL-10 occurs in LPS-tolerant macrophages and is accompanied by induction of cytosolic arginase I [164, 165], also a marker of alternatively activated macrophages [39]. Arginase I competes with iNOS for available arginine (Figure 9.2) [166]. Survival of antiinflammatory macrophages is supported by the anti-apoptotic effects of arginase and IL-10 [147]. TGF-β, a product of antiinflammatory macrophages, suppresses iNOS directly [167] and increases arginase activation [168]. Macrophages (particularly AMs) may produce endogenous IL-4 and IL-13, both of which could act in an autocrine manner, conceivably within the boundaries of innate tolerance [41, 116, 117, 119, 169]. IL-13 induces arginase in macrophages, by increasing cAMP [166], and (like IL-10) can reduce NF-κB and proinflammatory molecules [170–172]. Very high arginase/iNOS ratios occur with exposure of macrophages to IL-4, IL-13 and IL-10 acting synergistically [173], as might be encountered in adaptive (alternative) activation. On its own, LPS coinduces arginase and iNOS but addition of IFN-γ pushes the balance more towards iNOS [174]. Upregulation of arginase can be triggered by pathogens, such as *Leishmania*, and correlates with innate tolerance. Stimulation of macrophages in vitro by the trypanasomal antigen cruzipain increases IL-10, TGF-β and arginase and reduces iNOS [175]. Schistosoma induces arginase 1, and the iNOS/arginase ratio is inversely correlated to the degree of granulomatous and fibrotic change in the liver. Trauma is associated with release of TGF-β, Th2 cytokines and catecholamines, which synergistically induce arginase, and probably contribute to post-traumatic immunosuppression [176]. Since arginine is a unique substrate for both arginase and iNOS, factors regulating the two enzymes greatly influence the dynamic balance of macrophage effector functions. Apart from direct enzyme competition for substrate there is additional post-translational inhibition of iNOS mRNA by arginine metabolites (Figure 9.2) [177, 178]. When macrophages are in antiinflammatory mode the balance is in favor of arginase, with production of molecules that are implicated in tissue healing and repair, such as polyamines and (pro-fibrotic) proline. Measurements of the relative expression of arginase 1 and iNOS after LPS stimulation of rat peritoneal macrophages in vitro demonstrates that iNOS first appears at around 2 h, and is maximal at 8–12 h, but arginase appears later, at 4 h, and is maximal at 12 h. Similarly, in vivo injection of LPS rapidly induces iNOS in lungs and spleen, maximal as early as 2 to 6 h whereas the induction of arginase, though strong, is delayed, reaching a maximum at 10 h, and so is ideally sequenced to downregulate NO production in immune responses. Genetic factors regulating the iNOS/arginase ratio impact markedly on the ability of macrophages to control infection or to reject tumors (both requiring a high NO output). BALB/c macrophages are poor producers of NO but are high in arginase 1, which is triggered by Jak/STAT6 signaling following IL-4 and/or IL-13 receptor stimulation [179–180]. The iNOS/arginase ratio in macrophages is already receiving therapeutic attention. Greater than 95% wild-type BALB/c mice succumb to metatstaic mammary carcinoma (4T1) after removal of the primary tumor, but greater than 60% of STAT6-deficient mice survive under the same circumstances. Ablation of the Jak/STAT6 pathway is highly effective in promoting (M1) macrophage phenotypes and tumor rejection [181]. In infection, too, STAT6 knock-out mice achieve a much greater bacterial clearance in peritoneal sepsis [182]. In the paradigm, in the absence of infection, antiinflammatory macrophages, with a low iNOS/arginase ratio, are induced innately and reinforced adaptively.

The phenomenon of LPS tolerance supports the general view that low dose stimulation by constant bombardment with putative antigens may maintain the gut and airways in the 'default' antiinflammatory state, through continuous induction of non-specific innate tolerance. LPS/antigen-tolerant APCs are markedly impaired in their ability to induce IFN-γ from NK or T cells in the adaptive phase of the immune responses, [161, 183] a clear demonstration of IFRL. Of interest, neonatal tolerance can occur in the absence of functional Th2 or Th1 responsiveness [184], and maintenance of oral tolerance specifically requires neither Th1 nor Th2 cells [185, 186]. However, if neonatal adaptive responses are provoked they are of adaptive tolerance with expression of IL4 [187]. Assuming most mucosal surfaces at any one time are set to 'antiinflammation' (innate tolerance), it is possible to postulate a general 'bystander' downregulatory effect on localized proinflammatory responses exerted both afferently in lymph nodes and efferently through regulatory T cells entering the tissues [188]. Tolerance of the lung or gut to the normal (external) environment is viewed as a dynamic process in which innate immunity or tolerance is continually instructing the adaptive immune system so that the setting of overall tolerance versus immunity is itself 'acquired' from the environment.

Like arginase, the recently-described molecules, Ym1 and Fizz1, are highly expressed in alternatively-activated macrophages and stimulated by the Th2 cytokines IL-4 and/or IL-13, through a STAT6 pathway [14, 189]. Furthermore, Ym1 and Fizz1 may be directly inducible in (parasitic) infection, as by *Trichinella spiralis* [190], *Trypanosoma congalese* [191] and *Fasciola hepatica* [192], giving a clue that Ym1 might be a reliable indicator of innate (as well as adaptive) tolerance. Perhaps this explains why nematode-elicited macrophages in vivo exhibit a greater elevation of Ym1 than achieved by IL-4 and IL-10 stimulation of macrophages in vitro [44] and why selective upregulation of Ym1 and Fizz1 in APCs is now recognized as a generalized feature of nematode infection [193].

A highly effective way of ending an immune response, and a means to restore a tolerogenic microenvironment, is proinflammatory macrophage apoptosis. It may, additionally, serve a last immune function by killing intracellular microbes for cross-presentation to the adaptive immune system. Activation-induced cell death of macrophages both removes a potent proinflammatory source and potentially converts it into an antiinflammatory one through uptake and processing of apoptotic cells. However, macrophages tend to be resistant to apoptosis, despite producing copious cytotoxic molecules. Constitutive [194], and upregulated, expression of NF-κB is a survival factor for macrophages and renders them resistant to TNF-α- and FasL-mediated apoptosis occurring through pathways involving caspases 2 and/or 8. Migration inhibitory factor (MIF) aids macrophage longevity and sustains global proinflammatory function through suppressing activation-induced p53-dependent apoptosis [195]. Although MIF was earlier described as a pituitary-secreted hormone it is produced in high local concentrations by LPS and exotoxin-stimulated macrophages as well as by macrophages exposed to TNFα and IFN-γ [196]. Conversely, proinflammatory stimuli such as LPS increase intracellular ceramide, which in turn inhibits IKK and several of the protein kinase signaling pathways responsible for activating NF-κB and/or AP-1 and required for sustaining proinflammatory gene transcription (including iNOS and COX-2) [197]. The mode of action of ceramide is not entirely known, and may, to some extent, be cell-specific, but in T cells it activates caspase-2 and -8 upstream of mitochondria [198]. Differentiation of human monocytes into macrophages induces rapid downregulation of caspase-8 as well as upregulation of Bfl-1, an anti-apoptotic member of the Bcl-2 family. Monocytes are relatively short-lived unless they differentiate into macrophages [199]. Sphingomyelinase catalyzes the conversion of sphingomyelin into ceramide, which is pro-apoptotic but is counteracted by sphingosine-1-phosphate, which promotes macrophage survival. Apoptosis of bone–marrow-derived macrophages after withdrawal of M-CSF can be reduced by ceramide-1-phosphate, which inhibits sphingomyelinase and prevents the build-up of ceramide. Protracted survival of AMs in vivo may depend on constitutive activity of ERK MAPK and the PI3K-regulated effector Akt maintaining a high level of ceramidase activity, which converts ceramide into the harmless product sphingosine, which is found at high concentration in AMs [200–202]. Caspases are not indispensable for macrophage activation-induced death because apoptosis still occurs by a caspase-independent route when macrophages are stimulated in the presence of a pan-caspase inhibitor [203].

The paradigm lays great emphasis on cytokines and other biologically active molecules having a determining role in setting the polarity of the microenvironment at every stage of an immune response. In microenvironments already exhibiting inflammation, PGE_2 (from proinflammatory activated macrophages) could, for example, downregulate IL-12 production in DCs, even in the presence of IFN-γ, thereby polarizing towards Th2 and promoting immunosuppression in macrophages [204]. This would occur through both direct autocrine and T-cell-mediated mechanisms. Exposure to LPS [205] or GM-CSF [206] induces COX-2 along with other NFκ-B-dependent proinflammatory molecules and chemokines in stimulated monocytes [207]. Subsequently, PGE_2 increases IL-10 in macrophages, constituting a delayed tolerogenic feedback loop. PGE2 enhances Th2 at induction and activation. Nevertheless, in the basal state, as discussed, PGE_2 may prime the Th1 pathway, as evidenced from in vitro experiments in which DCs are incubated with GM-CSF and IL-4, with or without addition of PGE_2 [208]. The molecule PGE_2 is turning out to have diverse innate tolerogenic functions, including a role in the age-related reduction of Th1 immunity. Enhanced induction of COX-2 may be a consequence of increased NF-κB activation, which occurs with aging as a result of accentuated cytoplasmic I-κB degradation. LPS produces greater elevations of ceramide in macrophages of old mice, suggesting an alternative explanation for increased PGE_2 not involving NF-κB [209]. Ceramide stimulates JNK/AP-1-mediated activation of the C/EBPb gene, which is a promoter of COX-2 transcription. Ceramide also enhance p38 MAPK-mediated COX-2 induction via CREB [210]. PGE_2 downregulates phagocytosis. Overall, antiinflammatory macrophage phenotypes displayed in innate tolerance correlate well with increased intracellular cAMP [211, 212], whereas proinflammatory, but not antiinflammatory [213, 214], macrophage phenotypes often correlate with increased NF-κB activation/nuclear translocation [215].

In the context of innate tolerance, mention should be made of the endocrine influences, exogenous and endogenous, of glucocorticoids, which directly induce an antiinflammatory phenotype in macrophages and/or downregulate a proinflammatory phenotype. An important contribution of glucocorticoids to innate tolerance may be their ability to enhance the nonphlogistic uptake of apoptotic leukocytes, including spent neutrophils [216, 217]. By enhancing apoptosis, glucocorticoids counterbalance the proinflammatory innate immune system mediator, macrophage migratory inhibitory factor (MIF) [218, 219]. Cell-specific stimulation of the glucocorticoid receptor increases caspases 9/3 by an intracellular signaling pathway involving PKC and the generation of acid sphingomyelinase and ceramide [220, 221]. Not all GC actions are overtly tolerogenic and a recent report suggests that they can synergize with TNF-α to upregulate TLR2 expression [222]. UV light can promote innate tolerance by directly suppressing the proinflammatory phenotype of skin macrophages (reduced IL-12, increased IL-10) through engagement of the macrophage complement receptor, CD11b, by iC3b deposited in UV-irradiated skin. Although macrophages secrete TNF-α under these conditions, UV enhances the tolerogenic phenotype [223]. In experimental silicosis, AMs in the acute phase display a transitional phenotype with upregulated iNOS offset by upregulation of arginase-1, Fizz1 and Ym1/2. These studies revealed, unexpectedly, that in the later fibrotic stages there is no association with M2 macrophages. Nevertheless, there is a reduction of silica-induced lung fibrosis in IL-10 knock-out mice. The pro-fibrotic effect of IL-10, in vivo, probably results from increased expression of macrophage pro-fibrotic TGF-β together with a reduction in antifibrotic PGE_2, from downregulation of COX-2. Net increased fibrosis in the absence of IL-10 apparently depends on influxing Th1 $CD4^+$ T cells, bringing into question some established views [224, 225]. In a wider context, these studies indicate that tolerance may be the net outcome of a delicately balanced microenvironment, with elements of both innate and adaptive immune systems. In chronic inflammatory conditions, macrophages appear to develop partially suppressed proinflammatory phenotypes.

Induction or maintenance of innate tolerance by viruses is an effective 'strategy' for evasion of innate immunity [226]. An estimated one-third of the adenovirus genome is used to counteract innate and adaptive antiviral immune responses [227]. The ways in which viruses neutralize cell immunity are numerous, and include virus-induced release of soluble receptors for proinflammatory cytokines (viroceptors), homologues of IL-18-binding protein and of IL-10 [228], a GM-CSF inhibitory factor, interference with TNF-α-related cytolysis/pro-inflammation [229], induction of TGF-β1 by infected cells [228], chemokine homologs that block endogenous chemokine binding [229], prevention of MHC I antigen presentation [227], downregulation of MHC I and MHC II expression and blockade of complement activation [228]. The induction of IFNα/β by viruses stimulates an immediate innate cytotoxic and, more importantly, IFN-γ response from NK cells, within hours of primary infection, and a delayed response from adaptive CTLs, as well as negatively regulating IL-12 expression [230]. The presence of dsRNA in virally infected cells strongly induces Jak1/Tyk2 STAT pathways (which mediate actions of IFNα/β) and upregulates NF-κB and MAPK, contributing to the host immune responses [230, 231]. DCs may become locked into a tolerogenic (immature) mode. For example, the highly pathogenic Lassa virus produces a high level of tolerance in both MCs and DCs, neither of which are activated. The related Mopeia virus does not cause the same severe immunosuppression because it strongly activates infected macrophages, but not DCs, to produce IFNα/β along with HLA and co-stimulatory molecules [232]. Viruses demonstrate well how immune tolerance can be maintained by mechanisms interfering with the apoptosis, which ultimately governs the presence and balance of all immunologically active cells in tissue microenvironments. Both apoptosis and prevention of apoptosis can be advantageous to viruses [233]. Adenoviruses and Herpes viruses oppose apoptosis by interfering with intrinsic apoptotic mechanisms in the host cell and by blocking extrinsic mechanisms such as the death receptors Fas and TRAIL, which are targeted by cytotoxic T cells and NK cells. Viable viral target cells are required for viral replication and persistence [226, 233, 234]. Conversely, apoptosis may assist the spread of viruses in apoptotic blebs to other phagocytic cells without exciting a proinflammatory response. Apoptosis is construed in the paradigm as an antiinflammatory, tolerogenic, process and prevention of apoptosis may likewise be associated with antiinflammatory phenotypes. Against this, apoptosis, which is particularly promoted by interferon α/β synergizing with TNF-α, may be an important host defense mechanism favoring immunity by restricting viral replication and survival [235].

The correlation between antiinflammatory cell phenotypes and innate tolerance is not absolute. While proinflammatory changes are usually associated with innate immunity (HIV perhaps being an exception), antiinflammatory changes or lack of inflammation are not always reliable indicators of innate tolerance since they may also accompany innate immunity. Perhaps the simplest example of non-inflammatory innate immunity is protection against infection by epithelial or endothelial barriers [236], as in original concepts of immune privilege. Barrier immunity may be reinforced by AMPs of the defensin/cathelicidin families, which provide a constitutive antimicrobial presence, as when secreted onto the ocular surface epithelium or skin, but are produced more abundantly by (and are chemotactic for) neutrophils during proinflammatory responses [237]. Microbial interactions with MBL, other collectins, such as surfactant-associated proteins and with the complement system can all bring about innate immunomodulatory effects on macrophage phenotypes and DC programming, both pro- and antiinflammatory. By binding to microbial carbohydrate moieties serum MBL can aid in the physical elimination of potential pathogens without necessarily eliciting a proinflammatory response [238]. One possibility is a direct interaction of MBL with macrophages to promote opsonisation of bacteria and viruses. More significantly, MBL-bound to carbohydrate moieties expressed on common bacteria, such as *Staphlococcus aureus*, can activate the complement system, through MBL-associated serine proteases, independently of anti-bodies and C1q, to produce opsonic C3 fragments, which assist phagocytosis [239]. Both MBL and C1q opsonize apoptotic cells and facilitate their uptake into both macrophages and immature DCs. C1q may be especially important in inducing an antiinflammatory DC phenotype, to the extent that IL-10, TNF-α and IL-6 are induced, but not IL-12. It has been speculated that C1q opsonization may even assist in the non-inflammatory removal necrotic debris [240, 241]. Several other opsonizers (CRP, amyloid A, surfactants) may remove cells and microbes nonphlogistically. Phagocytosis of apoptotic neutrophils by AMs (but not peritoneal macrophages) can be dramatically enhanced by surfactant A opsonization, but not by MBL and C1q [242]. However, surfactant A significantly enhances the uptake of C1q-coated particles by AMs [243]. There remains an unresolved debate as to what determines whether the phagocytic uptake of apoptotic cells will be anti- or proinflammatory. Macrophages and immature DCs are important sources of C1q and other early components of complement but the significance of this is unknown. Upregulation of MR, both macrophage-associated and in soluble form, is induced by IL-10 and/or IL-4 [126], as would be found in antiinflammatory (or 'transitional') microenvironments of either innate or adaptive origin. Gram-positive bacteria display surface carbohydrate patterns that can bind directly to several membrane-anchored lectin and lectin-like molecules (including MR) on macrophages and DCs, some of which have been identified with antiinflammatory phenotypic modulation [244]. Carbohydrates on *Shistosoma* eggs may be taken up by MR of macrophages and DCs to produce, sequentially, innate tolerance and the anticipated adaptive responses, Th2, IgE and IgG1 (IFRL) [245]. Finally, there is growing interest in natural antibodies, produced by a subset of B cells (B-1) which secrete germline encoded (innate) antigen-specific IgM that recognizes microbial and viral coat antigens and leads to complement activation via the lectin pathway [246, 247].

Clearly, many complex balances exist between host and pathogen during the innate phase of an immune response, even prior to macrophage involvement. In attempting to determine the overall outcome, two simple questions may be posed: firstly, is the response antiinflammatory/non-inflammatory (best assessed by the activation phenotype of local tissue macrophages) and, secondly, does it favor the persistence/replication of the pathogen? If the answer to both questions is 'yes' it is likely that the response involves innate tolerance rather than innate immunity.

10.1.6
Incorporating NK and NT Cells into the Innate Tolerance/Innate Immunity Paradigm

The possible inclusion of NK and NT cells in innate tolerance deserves special consideration because of their firmly established role in the innate immune system. The cytotoxic and IFN-γ-producing roles of NK cells are well known but the activating signals for NK cells have to compete against ligands of a series of inhibitory immunoglobulin-like NK receptors that bind to and protect cells expressing MHC I [248]. Cells lacking MHC I ('missing self') are preferentially targeted for lysis and/or cytokine production by NK cells [249]. Proinflammatory signals for recruitment, proliferation, activation and survival of NK cells include several chemokines and combinations of the cytokines IL-2, IL-12, IL-18, IL-15 [250, 251], IFN-γ and IFNα/β, mostly derivable from activated macrophages and/or DCs. Combinations of IL-15 and IL-12 induce a powerful proinflammatory drive through induction of STAT4 responsive genes [252]. It has recently been discovered that bacterially-activated murine DCs rapidly secrete a combination of IL-2 and IFNα/β that strongly activates NK cell IFN-γ output without the requirement for an adaptive response (these results assume non-contaminated cell purification). Human DCs too secrete IL-2 in an IL-15-dependent manner [253]. In the absence of IL-12/IL-18 or IL-2/IFNα/β, inhibitory ligands dominate over excitatory ligands, but when these proinflammatory cytokines are present the outcome is in favor of NK activation and IFN-γ production [254]. IL-12 can directly stimulate the perforin gene promoter (and cytolysis) via STAT4 [255] and IL-15 can also activate perforin-dependent cytolysis [256]. IL-15 and IL-18 synergistically upregulate IFN-γ transcription in NK cells by STAT and NF-κB-mediated pathways respectively and IL-15, but not IFN-γ or IFNα/β upregulate IL-12Rβ2 on NK cells [257]. In CMV viral infection of C57BL/6 mice, the IL-12/STAT4 pathway is responsible for IFN-γ expression while IFNα/β/STAT1 accounts for induction of IL-15-mediated cytotoxicity [258]. The clear implication is that NK cells are strongly activated to proinflammatory cytokine release and cytotoxicity, when macrophages or DCs are already activated to proinflammatory phenotypes in local tissue proinflammatory microenvironments [251], but, in a prevailing antiinflammatory microenvironment, NK cells, if present, may mediate a much more selective surveillance of putative target cells for cytolysis. Cells would be selected for elimination because of downregulated MHC I molecules, one cause of which could be intracellular infection [259, 260]. Another exciting recent discovery is that human NK cells possess TLRs that respond to CpG and dsRNA with IFN-γ and TNF-α release and cytotoxicity that is effective against tumor cells or immature DCs [260]. The cytokine output of NK cells (ranging from IFN-γ, TNF-α, GM-CSF and IL-8 to IL-10, IL-5, IL-4 and IL-13) appears to correlate well with microenvironmental conditions, as mirrored by local macrophage phenotypes. For example, IL-12 promotes IFN-γ and IL-10 in NK cells whereas IL-4 stimulates IL-5 and IL-13 [261, 262]. However, intraepithelial NK cells in the rat gut spontaneously secrete an unusual combination of IL 4 and/or IFN-γ [263]. An important means of curtailing immune responses is the suppression of NK cell IFN-γ output by PGE$_2$, which downregulates NK IL-15 receptor function, so blocking continued stimulation by IL-15 produced by macrophages [264]. Contributing to regulation of proinflammatory responses might be the cytolysis of 'immature' dendritic cells that have failed to take up bacteria or antigen and which are preferentially targeted by activated NK cells. Activated (maturing) DCs, which express IL-12, TNF-α and CCR-7 are relatively protected by upregulated MHC I [265]. The most tangible expression of innate tolerance by NK-cells is their subversion by some microbes and tumor cells. Mechanisms of interaction with viruses include blocking of NK activating receptors by viral proteins, viral components acting as ligands for inhibitory receptors and virally-induced modulation of cytokines to an antiinflammatory mode [266]. Tumor cells usually have an antiinflammatory cytokine profile that fails to activate NK cells but, with addition of IL-12 and with a coating of IgG, tumor cells are successfully lysed by NK cells in vivo in BALB/c mice [267]. *Leishmania* provides a very good illustration of the dynamic balance of innate tolerance and innate immunity for it is now known that live *Leishmania* promastigotes can directly activate NK cells to produce IFN-γ without any requirement for antigen presentation [268]. On the contrary, *Leishmania* induces a tolerogenic response from its main host cell, the macrophage, as already discussed. The overall picture of NK-mediated tolerance is of highly regulated phenotypes, providing surveillance of abnormal cells in an antiinflammatory microenvironment but responding to a proinflammatory microenvironment with a surge of reinforcing cytokines and cytotoxicity [249, 269].

NT cells are a heterogeneous group that includes both $\alpha\beta$ and γd T cells and are thought to possess their own innate recognition pathways for lipid/glycolipid components of microbial cell walls or infected cell membranes. Responses range from proinflammatory cytokines, IFN-γ, TNF-α and GM-CSF, to the antiinflammatory cytokines IL-4 and IL-10. By recognizing various distinctive ligands, it may be presumed that NT cells complement the MMRRs of macrophages in shaping antiinflammatory as well as proinflammatory responses, although the nature of the natural ligands await clarification [270]. One particular subset of NTs is NKT cells, with an invariant TCR recognizing glycolipid antigens restricted by CD1d presentation on APCs. NKT cells fulfill the criteria for innate immunity/tolerance because no prior antigen exposure or sensitization is required to achieve a rapid release of large amounts of either proinflammatory or antiinflammatory cytokines. The glycolipid, α-galactosylceramide (α-GalCer) can be presented by CD1d, and tends to predispose towards IL-4, but little is known about natural TCR ligands of NKT cells [271]. The difficulties in establishing causality of NKT phenotypes is illustrated by *Trypanosoma cruzi* infection, during which NKT cells limit parastaemia in the acute phase, presumably by an IFN-γ-dominated response, but in the chronic phase NKT cells secrete IFN-γ and IL-4 and augment the antibody (TH2-dependent) response. It is not certain whether binding of tyrpanosomal glycophosphoinositol to CD1d, or a surge of IL-12, or other proinflammatory cytokine(s), is responsible for the acute phase phenotype [272]. In filarial larvae infection of BALB/c or C57BL/6 mice, NKT cells are responsible for rapid IL-4 production, within 24 h of infection, and this innate response may contribute to early suppression of T cell proliferation along with initiation of a Th2 adaptive response [273]. The presence of anergy or tolerance (equivalent to innate tolerance) amongst some subsets of NT cells is already documented in HIV and TB [270] in xenograft or allograft acceptance [274, 275] and in advanced cancer [276]. In tolerance to rat pancreatic islet xenografting, NKT cells secrete combinations of IFN-γ and IL-4, but genetic deletion of either of these cytokines does not abrogate tolerance entirely, suggesting perhaps a suppressive role for another NKT-secreted cytokine, such as TGF-β, IL-10, IL-6 or TNF-α. However, lack of IFN-γ (from whatever source) does prolong graft survival somewhat [274]. On allografting of BALB/c tissues into C57BL/6 mice, graft acceptance is ascribed to blockade of T cell co-stimulatory pathways (LFA-1/ICAM-1, CD28/B7) and in this case is absolutely dependent on the presence of NKT cells. IFN-γ is also required for allograft tolerance. During the course of rejection, in the absence of immunosuppressive treatment, NKT cells do not seem to regulate alloreactive T cells. In advanced human prostate cancer, circulating NKT cells are diminished, proliferate less and show reduced production of IFN-γ but unchanged IL-4 compared to healthy controls. The assumption here is that loss of a proinflammatory profile in these NKT cells equates to lack of capability to destroy cancer cells [276]. The finding of NKT cells with antiinflammatory phenotypes in splanchnic nodes is consistent with an influx of antiinflammatory DCs from the gut [277]. The correlation of antiinflammatory NKT cells in the spleen with induction of anterior chamber-associated immune deviation (ACAID), resulting in systemic tolerance, suggests that these cells are involved in an antiinflammatory circuit, picking up their cytokine polarity in the eye or spleen from CD1d-mediated association with antigen carrying APCs (presumably DCs) of the same phenotype. IL-10 from NKT cells is essential for the differentiation of antigen-specific regulatory T cells (adaptive tolerance) in ACAID. Cells bearing the macrophage marker F4/80 in the anterior chamber also secrete IL-10. Thus, ACAID elegantly demonstrates the faithful carriage and amplification of a tolerogenic signal throughout the innate and adaptive immune system, as embodied in the principle of IFRL [278, 279]. In ACAID, B7 and CD40 are downregulated, possibly secondary to IL-10, so it is of interest that, in the absence of an intact CD40 costimulatory pathway, α-GalCer stimulation of NKT cells produces IL-4 but not IFN-γ [280]. In relation to differential cytokine secretion several lineages of human NKT cells have been defined, based on surface markers. Only $CD14^+$ NKT cells secrete IL-4 and IL-13 on primary stimulation while $CD4^-$ $CD8^-$ NKT cells secrete IFN-γ [281]. NKT cells can mediate antigen-specific oral tolerance [282], and can prevent EAE [283]. In contrast to the above litany of suppressive antiinflammatory responses, NKT cells can deliver strong proinflammatory activation on stimulation by IL-12/IL-15 from macrophages [284]. Indeed, IL-15, a major macrophage-secreted cytokine, is a survival factor for NKT cells [285].

The diverse phenotypes of NKT cells are another indication of the complex interdependence (usually cooperative) of cells involved in innate immunity, and demonstrate the crucial influence of sequential cytokine release on the balance between innate tolerance and innate immunity. Two hypotheses, which are not mutually exclusive, may be considered for the role of NKT cells in innate immunity. Firstly, they may be primarily ligand-driven by putative CD1d-restricted glycolipids so that, like γd T cells, they extend the range of innate immune recognition to include more non-peptide antigens. Secondly, they may be primarily driven by cytokines and other inflammatory mediators. In favor of the first suggestion, NKT cells on exposure to α-GalCer are involved at the earliest time points of immune reactions, a view based on the fact that they release IFN-γ and IL-4 before any other cells. They themselves appear to have no cytotoxic effector function in the setting of innate immunity but rapidly activate NK cells to secrete IFN-γ (within 6 h) and to become cytotoxic [286, 287]. NKT cells present α-GalCer to CD1d-expressing DCs during proinflammatory responses, inducing NKT cell secretion of IFN-γ but NKT cells additionally prime DCs to activate antigen-specific $CD4^+$ and $CD8^+$ T cells to IFN-γ-producing and cytotoxic phenotypes, consistent with an adaptive immune response [288]. Remarkably, NKT cells can also promote bystander activation and proliferation of memory $CD4^+$ and $CD8^+$ T cells, in an IFN-γ/IL-12-dependent fashion, a property they share with superantigen-activated T cells [289] and with IL-15-secreting macrophages [86]. NKT cells may themselves become effector cells later in a proinflammatory immune response, because in vitro they develop cytotoxicity when treated with combined IL-2 and α-GalCer. Activated NKT cells express CD40L, which engages CD40 on APCs and stimulates them to produce IL-12. There is a requirement for IL-12 to induce NKT cells to secrete IFN-γ [290, 291]. However, in vivo IL-12 is available rapidly from other sources in an innate immune response, and CD40 expression by DCs is an indication that they have already been activated, usually through TLR stimulation [292], negating any absolute requirement for NKT cells to trigger IL-12 release by APCs. Conversely, NKTs appear to require an external source of IL-12 to achieve a proinflammatory phenotype. Despite the specific and rapid ligand-driven proinflammatory burst, NKT cells are notorious for changing their phenotypes. Within hours or days of initial activation with α-GalCer the early IFN-γ-dominated output often gives way to a predominantly IL-4 secreting cell (unlike NK cells) and within 6 days of injection NKT cells polarize the adaptive response towards Th2 cells and IgE as an indicator of humoral immunity. Hence, α-GalCer may be used as an adjuvant to promote Th-2 polarization [293]. IL-4 secretion by NKT cells should encourage alternative macrophage activation. The second hypothesis suggests that NKT phenotypes, in vivo, may be more responsive to the existing cytokine milieu than to unknown glycolipid ligands. If this were the case, the complexion of the role of NKT cells would change quite significantly, because they would then be more likely to derive their activation cues from the phenotypes of other cells, such as TLR-stimulated macrophages and DCs, and behave as microenvironmental-responsive cells, rather than as microenvironment-directing cells. In the liver, for example, KCs are responsive to local conditions by reinforcing an antiinflammatory tolerant microenvironment. Polarization towards Th1 (increased IFN-γ, TNF-α) occurs if NKT cells are depleted, increasing susceptibility to LPS-induced hepatotoxicity. NKT cells are regulated by KC cytokines, dietary factors and norepinephrine, the latter promoting NKT cell survival and so helping to maintain the liver in a tolerogenic state [294]. In the steady state, with low dose continuing stimulation, NKTs may be primarily tolerogenic, and NKT deficiencies may predispose to organ-specific autoimmunity. In viral infection NKT cells can potentiate both the pro- and antiinflammatory activities of macrophages and DCs [295]. In *Salmonella* septicemia, NK, $\gamma\delta$T and NKT cells are upregulated and participate in a general proinflammatory reaction, but high numbers of activated natural T cells correlate with increased mortality and not survival [296]. The necessity to polarize NKT cells by cytokines to the required phenotype has become evident in their increasing use in cancer treatment. Although α-GalCer is used for initial stimulation it is necessary to combine activated NKT cells with IL-12 to prevent the drift towards Th2 polarization and to maintain the necessary enhanced IFN-γ levels during therapy [297–299]. The chameleon qualities of NKT cells suggest that, in the current paradigm, they are reinforcers and stabilizers of the activities of the dominant cells in an innate immune response rather than themselves being the dominant cell. Dominant polarizing cells are usually those responding directly to exogenous stimuli (or endogenous danger signals) and this, more often than not, coincides with APC TLR stimulation. Unless the elusive natural glycolipid ligand(s) of NKT cells suddenly declare themselves it is likely that the source of any primary stimulation of NKT cells will have to be sought elsewhere.

10.1.7
Definitions and Terminology

From the previous considerations, innate peripheral tolerance may be defined as any in vivo response to environmental, microenvironmental or intracellular stimuli that does not involve cells of the adaptive immune system or their products, which is antiinflammatory or non-inflammatory in nature, is not autoaggressive and may be associated with persistence of an infective agent and/or processing of self or foreign antigen in a non-proinflammatory manner. Defined in this way innate tolerance includes other immune cells making important and sometimes nonredundant contributions to innate tolerance, notably natural T cells, 'innate' B-1 cells and various other cells that can reinforce or initiate innate tolerance, such as epithelial and endothelial cells, masts cells and fibroblasts. Thus, 'non-professional' immunologically active cells can be agents of innate tolerance (as already recognized in innate immunity). Innate tolerance is not confined to cell-mediated changes and may include non-cellular processes that prevent pathogen-induced innate inflammation.

The use, in the paradigm, of the term 'antiinflammatory' as an alternative state of macrophage/DC activation is deliberate, though possibly contentious. It emphasizes that the tissue microenvironment is paramount in determining the direction and outcome of immune responses. Thus, 'Th2-mediated inflammation' is an aberrant/regulatory reaction to an ongoing (often chronic) proinflammatory state, or a specific response to an environmental (infective) agent. Adaptive tolerance, placed in context, is more than just a T cell-mediated 'state of nonresponsiveness to an antigen' [300] and may be defined as any antiinflammatory or non-inflammatory adaptive response to innate tolerance, which does not mediate immunity to micro-organisms. As alluded to earlier, the 'ideal' response to a helminth parasite might be an effectively targeted Th2/B cell/IgE-mediated attack accomplished entirely below the proinflammatory threshold, in which innate tolerance would instruct an adaptive response, tolerant to the parasite in every respect except for the specific and entirely focussed anti-parasitic IgE effector mechanism [301]. Although a build-up of effector cells would occur, failure to release any proinflammatory ('danger') signals, either exogenous or endogenous, would avert pro inflammatory tissue damage. The reality is somewhat different! Some phases of parasite life cycles, as well as microbial infections in general, almost invariably excite at least a degree of proinflammatory (macrophage) response, which often proves essential for successful elimination, but at the inevitable expense of tissue-damaging inflammation [302–308]. One particularly well-adapted 'silent invader', *Anaplasma phagoctophilum*, which may enter the macrophage cytosol without resort to TLRs, is possibly the ultimate expression of microbially-induced innate peripheral tolerance to date. Nevertheless, adaptive immunity or some other innate mechanism still manages to clear this particular infection by means as yet unknown [309].

Where macrophage innate overproduction of TGF-β occurs, or where innate proinflammatory responses are aberrantly linked to excessive T cell or eosinophil production of TGF-β, or associated molecules, exuberant fibrosis and scarring may result [310, 311]. It is debatable whether pathological exaggeration of normal healing processes (angiogenesis, fibrosis) can truly be described as 'inflammatory' [312, 313]. If the term inflammation is extended to accumulations of immune cells associated with Th-2 [314], or more strictly type-2 cytokines, a different, but perhaps no less valid, picture of inflammation emerges in which type 1 and type 2 inflammation [44] lie at either end of a continuum with a non-inflammatory region in the middle. The approach is very suited to describing responses to parasitic worms and *Leishmania*, and the microenvironment of tumors [315], but such a broad usage of the term 'inflammation' causes some difficulty interpreting the other much-used terms 'proinflammatory' and 'antiinflammatory'. These difficulties are seen in the literature where, for example, IL-12 may be described as antiinflammatory in conditions such as asthma and *Schistosoma* infection [316] and proinflammatory in conditions such as *Listeria* infection, even though in both cases the cytokine is polarizing immune responses away from Th2 and towards Th1. Granulomatous disease highlights the problems of terminology [317] because the histopathological appearance may belie the functional status of the immune cells, which could be producing predominantly type 1 or type 2 cytokines or a combination of the two [318, 319]. In *Schistosoma mansoni* infection, IL-13 is largely responsible for an eosinophilic granulomatous 'inflammation' that can lead on to hepatic fibrosis. Reduced granuloma formation, which follows treatment with IL-13Rα2, acting as a decoy for IL-13 may, therefore, be described as an antiinflammatory response [320]. Unfavorable outcomes of type 2 inflammation are over-exuberant restorative processes such fibrosis of lung or liver. Even though TGF-β may predispose to the proinflammatory condition of Alzheimer's disease, this is because the amyloid β protein activates proinflammatory pathways (NF-κB) and iNOS [321, 322].

The paradigm is readily cross-referenced to Matzinger's well-known 'danger' model [323] because a 'danger' signal is any endogenous signal, such as HSPs, HMGB1 or S100 proteins, that activates the innate immune system (neutrophils/macrophages/NK cells) to a proinflammatory phenotype as part of a local immune response. In the current macrophage-focussed paradigm, as in the danger model, inappropriate innate immune system responses can occur in autoimmunity, cancer and graft rejection. The paradigm presented here differs, though, from the danger model in some respects: firstly the danger model places the emphasis on endogenous alarm signals as the instigator of immune responses, rather than exogenous stimuli acting on macrophages/DCs. Secondly, in the present paradigm, macrophages can, additionally, be stimulated to an antiinflammatory phenotype, which actively offsets proinflammatory stimuli. Indeed, endogenous signals triggering this process are more intrinsically endogenous than signals triggering immune responses. Thirdly, the danger model is mainly concerned with macrophages as APCs whereas the current paradigm is concerned with all the integrated influences that affect macrophage effector function.

10.2
Ramifications of the Paradigm: Asthma

The paradigm implies that environmentally-triggered (innate) proinflammatory microenvironments, incorporating proinflammatory activated macrophages, could drive many subacute and chronic inflammatory conditions, regardless of whether there is a Th1 or Th2/Treg adaptive response. However, antiinflammatory microenvironments and macrophages would be expected to support only antiinflammatory (regulatory) T cell responses (or no response at all), and should not cause inflammation, even if eosinophils accumulate [324, 325], because the microenvironment (and eosinophil phenotype) should remain antiinflammatory. Survival of LPS-stimulated eosinophils is reduced by addition of IL-10 [325], whereas survival of eosinophils is prolonged by GM-CSF [326]. Eosinophils, themselves, can produce in excess of 20 cytokines and other bioactive molecules across the whole antiinflammatory/proinflammatory spectrum [327], ranging from TGF-β [328, 329], IL-4 [318], IL-5 [330], IL-10 [331] and IL-13 [332], to superoxide [333], iNOS [334], IDO [335] leukotriene C_4, cytotoxic granular proteins, GM-CSF [336, 337] and IFN-γ [338]. Eosinophils are expanded under the influence of circulating IL-5, a product of Th2 cells [339, 340] and recruited by eotaxins to immunologically-active sites as part of the adaptive immune response [336]. The eotaxin receptor CCR3 is induced in eosinophils by IL-5 [341]. Eotaxins are strongly induced in airway epithelial cells by the (predominantly) Th2 cytokine, IL-13 [342, 343], but also by proinflammatory stimuli – notably combinations of TNF-α, IL-1 and IFN-γ – and directly by helminth larvae in the lung [344]. According to the paradigm, if eosinophils retain plasticity, the final eosinophil phenotype should be determined by the tissue microenvironment. Eosinophils do indeed adopt either a non-degranulating cAMP-dominated, relatively inactive, phenotype [345] in antiinflammatory microenvironments or 'mature' to a proinflammatory, NFκ-B-activated phenotype, exhibiting enhanced survival in proinflammatory microenvironments [346, 347]. Unlike macrophages, eosinophils probably do not express TLRs or CD14 [348] and so would not initiate innate immune responses or survive in the absence of proinflammatory mediators, such as GM-CSF, inducible, for example, by lung exposure to LPS [326]. Like eosinophils, monocytes can express the eotaxin receptor, CCR3, and so both cells may be recruited together to proinflammatory microenvironments, in which eotaxin and other chemokines are upregulated [349–352]. Eotaxins are expressed by epithelium, fibroblasts, macrophages and eosinophils. Eotaxin 2 is upregulated in monocytes by LPS and IL-1β, but not TNF-α, and in macrophages by IL-4. Other eotaxins may have different kinetics and triggers [353]. Eotaxin(-1), eotaxin-2 and eotaxin-3 all directly stimulate a proinflammatory phenotype in eosinophils, including superoxide generation and degranulation [354]. In a proinflammatory microenvironment, eosinophils will reinforce inflammation [355, 356] as well as help to recruit more eosinophils and macrophages through chemokine expression. By contrast, during humoral (antiinflammatory) responses, increased IgE upregulates the high affinity IgE receptor on eosinophils [357] and on engagement of this receptor, eosinophils release IL-10

[358]. In the absence of proinflammatory reactions, eosinophils in peripheral tissues should therefore adopt an antiinflammatory phenotype, secreting IL-10. Indeed it may be conjectured that a combination of macrophages and eosinophils could mediate IgE-dependent cyotoxicity to helminth parasites, through upregulation of high affinity IgE receptors [359], at the same time releasing IL-10 but not NO, thus avoiding proinflammatory damage to the surrounding tissues [359–361]. This would be an example of innate tolerance driving adaptive (IgE-mediated) immunity, rather than the usual adaptive tolerance, and highlights the important distinction between inflammation and immunity, processes that are normally but not necessarily linked. In this situation the antiinflammatory cytokine IL-13 might contribute to parasite expulsion from the gut by increasing mucus production from goblet cells [362]. Although non-inflammatory control of infection is an attractive proposition, it is increasingly realized that many pathogens usually (perhaps invariably) need a proinflammatory reaction for their final elimination, to which both eosinophils and macrophages could contribute actively. In the gastrointestinal tract, where eotaxin is constitutively expressed, the presence of 'resting' eosinophils may enhance ongoing innate responses [363]. The phenotypic plasticity of eosinophils has even been thought to encompass a DC-like role since endobronchial eosinophils have been shown to capture particulate antigen and migrate to local lymph nodes [364]. Further investigation, however, has shown them to have low expression of MHC II and to be incapable of antigen presentation [365].

According to the paradigm, impairment of IL-12 [366–368], IL-18 [369], IL-18R [370–372] or other Th1-directing signal (such as Myd88) [373, 374], variations in inherent T cell-responsiveness [375], or aberrant DC responses to the prevailing microenvironment are all possible mechanisms through which proinflammatory macrophages could drive Th2/Treg responses [63, 376, 377]. Speculatively, for example, DCs could be disproportionately responsive to PGE_2 [378–380], a product of LPS-stimulated proinflammatory macrophages [381, 382], but with well-documented receptor-mediated antiinflammatory signaling properties. The partial breakdown of normal IFRL could account for the pathogenesis of some chronic inflammatory disorders, such as asthma [115, 383]. Under such circumstances, via DC signaling to draining lymph nodes, an increase in both Th1 and Th2, with the possibility of a Th2-preponderance, would be predicted by the paradigm (Figure 10.1). A mixed Th1/Th2 pattern has been verified in some animal models of asthma [384] and suggested in humans, where an increased ratio of IL-4/IFN-γ may predict an asthmatic tendency in infants [385]. Analysis of plasma and intracellular T cell cytokines in asthma shows elevation of both proinflammatory/Th1 (IL-18, IL-12 and IFN-γ) and antiinflammatory/Th2 (IL-13 and IL-10) cytokines, with a tendency to an overall increased Th2/Th1 ratio [386]. In status asthmaticus there is a marked increase in both proinflammatory (IL-1β and TNF-α) and antiinflammatory (IL-10, IL-1Ra and TGF-β1) mediators in BAL fluid, but here increases in Th1 may predominate over increases in Th2 [387]. In support of a dysregulation of PGE_2 in asthma, monocyte-derived DCs in asthma produce enhanced PGE_2 (COX-2 induction) and IL-10, but only equivalent increases in IL-12p70 (compared to normal controls) on LPS stimulation [388]. LPS-stimulated

AMs also produce more PGE_2, probably as a normal component of a proinflammatory phenotype (including IL-1β and IL-8) [389]. Whether PGE_2 from AMs in asthma could be elevated disproportionately, as found in BALB/c macrophages stimulated with LPS compared to other strains, and whether this runs in parallel with and/or determines in some way the asthmatic DC phenotype in vivo remains conjectural. In asthma, Th2-dependent eosinophil accumulation [390] and degranulation [391] would, therefore, be driven by an overriding innate proinflammatory response, reinforced by IFN-γ, in which AMs (AMs) would be prominent [392–394]. Thus, allergen challenge and clinical asthma are associated with release of proinflammatory cytokines such as IL-1β and TNF-α [395] that appear to drive an aberrant, but thwarted, downregulatory Th2 response, spearheaded by IL-13 and IL-4. In severe uncontrolled asthma the prototypical proinflammatory signal NF-κB is persistently activated, through increased phosphorylation of IκBα, and is associated with continuous output of inflammatory mediators [396]. Moreover, it has been suggested that a proinflammatory microenvironment, displaying typical upregulation of NF-κB activation [397], rather than antigen presentation, may even be essential for Th2 (and, therefore, eosinophil) recruitment in respiratory tract infection [398]. Exacerbation of asthma by microbial (proinflammatory) stimulation of TLRs expressed on AMs supports the paradigm [399], as does increased NF-κB activation in AMs in asthma [400, 401], reduced IL-10-positive cells in sputum after allergen challenge and decreased IL-10 after LPS challenge. Thus, anomalously, even though asthma is described as a 'Th2-mediated disease', LPS is a risk factor [402]. IL-15 from proinflammatory submucosal macrophages in asthma induces GM-CSF in eosinophils, both cells types being dependent on NF-κB for induction of inflammatory mediators and prolonged survival [403]. In addition, IL-15, like IL-12, enhances innate IFN-γ release from NK cells [404]. GM-CSF is well-recognized as a proinflammatory cytokine in the lung [405, 406] and is, for example, elevated (alongside NF-κB activation) in asthma-associated respiratory syncytial virus infection, which as the paradigm predicts, is characterized by macrophages and eosinophils with proinflammatory phenotypes [407]. Eotaxin mRNA is rapidly induced in AMs after allergen challenge [336]. PGE_2 at the inflammatory site may be instrumental in downregulating eosinophil [408] as well as macrophage and NK [264] proinflammatory activity.

It has already been suggested that the activation status of eosinophils in a local microenvironment will closely follow that of proinflammatory macrophages responsible for the initial response [409]. It is therefore of great interest that allergen challenge (assumed by the paradigm to induce proinflammatory macrophage activation) causes reduced expression of IL-5R and refractoriness to IL-5-provoked degranulation in recruited eosinophils recovered from BAL fluid. At the same time, eosinophil responsiveness to the proinflammatory cytokine GM-CSF is retained, giving a mechanism by which the eosinophil phenotype may be converted from anti- into proinflammatory on arrival in a proinflammatory lung microenvironment [410–412]. It has previously been observed that eosinophils can release IL-5, IL-13 and GM-CSF in blood, but only GM-CSF in tissues [336]. Monocytes release increased GM-CSF and reduced IL-10 on LPS stimulation

[413]. Lipid mediators produced by the 5-lipoxygensase pathway, and released from proinflammatory macrophages, are a strong endogenous chemoattractant for eosinophils, stimulating them to superoxide production and degranulation, as well as promoting GM-CSF secretion from cultured monocytes [414].

In the face of conspicuous evidence of proinflammatory lung changes in asthma, there are clearly problems interpreting asthma in terms of a simple Th1/Th2 polarization [415]. Of possible significance are Th2 cells retaining plasticity to revert to Th1/Th0, IL-4 increasing IL-12, and IL-12 promoting Th2 in the absence of IFN-γ [416]. In adoptive transfer studies in which human T cells and/or monocyte-derived DCs from asthmatic or normal subjects are transferred to SCID mice, DCs from allergic subjects migrate to lymph nodes more effectively and are more effective inducers of Th2 cells. Reasons offered for initial Th2 polarization are sustained T cell stimulation during specific antigen presentation and increased expression on DCs of CCR7 [417]. However, macrophages are also prominent in the lungs of DC donor mice, which, like the recipients, display combined increases in Th2 cells, eosinophils and IgE following antigen challenge. Additionally, constituents of the DC-priming microenvironment could easily include the proinflammatory cytokine TNF-α, as well as LPS, both of which have been implicated in the increased expression of CCR7 – results that are entirely consistent with a proinflammatory microenvironment and impaired IFRL leading to any observed aberrant Th2 responses. If (innate) proinflammatory responses are dampened, 'Th2-inflammation' should be reduced in parallel. In an in vivo murine model of asthma, injection of OVA-pulsed peritoneal macrophages, which produced high levels of IL-10 on interaction with antigen-specific T cells or LPS in vitro, inhibited lung Th1 (IFN-γ) and, more significantly, Th2 (IL-4 and IL-5) cytokines, eosinophil recruitment and asthma manifestations [418]. Again, both exogenous and endogenous IL-10 have been shown to suppress eosinophil infiltration [419, 420], and activation. Transgenic C57BL/6 mice lacking IL-10 exhibit more eosinophil infiltration on repeated OVA allergen challenge than wild-type mice but this does not affect IgE [419]. In terms of the paradigm, IL-10 would offset a proinflammatory macrophage phenotype, including iNOS induction [421] and would compete with innate proinflammatory stimuli, such as LPS [422].

There is much debate about the relationship between 'inflammation' and asthma. Asthma is undoubtedly a heterogeneous condition. The current paradigm may be useful in disentangling the contributions of three interrelated mechanisms: first, proinflammatory (innate) mechanisms, important in driving some, but probably not all, asthma manifestations (nor indeed all asthma); second, antiinflammatory mechanisms, mainly driven by the adaptive immune system and responsible for mucus overproduction and chronic aberrant fibrosis; third, mechanisms responsible for airway hyperresponsiveness (AHR). From this triple standpoint, IL-10 makes no proinflammatory contribution, but could arise from a partially compensated 'transitional' macrophage phenotype or from a pure antiinflammatory innate stimulus. In asthma arising after a proinflammatory trigger, adaptive (Th2) IL-10 could arise from aberrant DC transmission, but in pure asthma (as in a pure helminth reaction) it would arise from normal IFRL. In the lung microenvironment, IL-10 can increase AHR but reduces eosinophil activation and survival and reduces proinflammatory tendencies, including the innate proinflammatory drive to asthma, where relevant [423, 424]. TGF-β behaves similarly, but unlike IL-10 it might reduce AHR and is strongly fibrogenic. Although TGF-β1 has come under much scrutiny, not least as a possible treatment for asthma [425–427], TGF-β2 is in fact the most prominent isoform in severe asthma, when it is highly expressed by eosinophils. It may induce several mediators responsible for thickened sub-epithelial basement membrane [428]. IL-13, like IL-4, is classified as an antiinflammatory cytokine. Both can increase mucus and may cause AHR [332] and enhance IgE production by B cells. IL-13 is now recognized as perhaps the major effector molecule of asthma. In asthma, IL-13 increases adhesion molecule and chemokine expression in endothelium, recruits eosinophils, promotes collagen production, smooth muscle cholinergic contraction and proliferation, and stimulates matrix metalloproteinases and cathepsin proteases, all of which come under the umbrella of the term 'allergic inflammation' [427, 429]. However IL-13, is antiinflammatory in that it decreases release of TNF-α, IL-1β and proinflammatory chemokines from LPS-stimulated monocytes/macrophages and reduces iNOS. Application of the paradigm predicts two very different routes to IL-13-mediated asthma and allows interpretation of often conflicting experimental observations. In considering the antiinflammatory cytokines in asthma, DNA microarray analysis recently revealed that arginase (and the downstream product putrescine) is increased in murine models of asthma, triggered either by OVA or *Aspergillus fumigatus*, suggesting that alternatively activated macrophages generated by IL-4/IL-13 could play a part in pathogenesis [430, 431]. These findings do not sit well alongside a central role for proinflammatory mediators driving asthma [311], but cannot be ignored. Applying the 'triple analysis', such macrophages are unlikely to be proinflammatory because increased arginase reduces the metabolism of arginine through the iNOS pathway. However, both iNOS and arginase could be upregulated together [165] in a 'transitional' phenotype (Figure 9.3) and still compete for arginine or, assuming asthma is a segmental or patchy process, some macrophages could have fully proinflammatory phenotypes and some could be fully alternatively activated. Interstitial lung macrophages may be more accessible to alternative activation than AMs. Conceivably, in some asthma variants both innate triggering and effector macrophages could be antiinflammatory. Alternatively activated macrophages produce less NO, which would, hypothetically, increase AHR [431]. Turning to proinflammatory cytokines, IL-12 will increase any proinflammatory component in asthma by increasing the Th1 (IFN-γ) amplification of lung proinflammatory macrophage phenotypes, but at the same time will reduce all the asthma-promoting Th2 cytokines (IL-5, IL-4 and IL-10). Although the direct effect of this on AHR is unpredictable, the latter results are likely to be unfavorable, if asthma pathogenesis is fundamentally driven by an innate proinflammatory process [432]. In asthma generated in a relatively antiinflammatory setting, for example a grass allergy in mice, IFN-γ or IL-12 may be beneficial by reducing IgE and inhibiting Th2 cytokines, with a consequent reduction in AHR, IL-12, activated NK cells, reduced lung eosinophilia and increased IFN-γ [433]. Vaccination with a plasmid containing OVA cDNA fused to IL-18 cDNA polarizes towards Th1 and reverses AHR, associated with increased IFN-γ and reduced OVA-specific IgE [434]. Mast cell-derived histamine promotes IL-10 in DCs and programs a Th2 response with increased IgE. Allergens bind directly to IgE on mast cells, triggering several mediators, including histamine, creating an atopic rather than a proinflammatory feedback loop [435].

To summarize, in pure allergen driven asthma, AHR and mucus production can be induced in the absence of all proinflammatory cells and all proinflammatory mediators, requiring only IL-13. This cytokine can induce eotaxin and IL-5 in airway smooth muscle cells, can induce mucus through a STAT6 pathway, can upregulate profibrotic molecules, either by upregulating TGF-β or downregulating PGE_2, and can provoke AHR, partly perhaps by reducing NO through alternative macrophage activation. The sources of IL-13 are primarily T cells, but may be mast cells, basophils and eosinophils. The afferent route to IL-13 T cell output would be through antiinflammatory macrophage/DC (MDU) activation (Figure 10.1) [429]. The pure allergy route to asthma is reminiscent of the 'weep and sweep' reaction to helminths in which mucus and increased contractility help clear the gut [362]. At the other extreme, a purely pro-inflammatory innate response could drive an aberrant adaptive Th2 response with the consequences discussed above. In most cases the pathogenesis is likely to be a combination of both (Figure 10.1). AHR is not reliably related to either eosinophil influx or IL-4/IL-13-induced B cell isotype switching to IgE, and, in fact, may be more closely related to proinflammatory mediators such as TNF-α and GM CSF [367].

Finally, there are striking parallels between the pathogenesis of asthma and ulcerative colitis (UC) (Crohn's disease is clearly an IL-12-Th1-driven disease [436]). The innate stimulus in both ulcerative colitis and asthma support a mixed Th1 and Th2 response. In UC there are increases in IL-1β, IL-12p40, TNF-α and IFN-γ and in IL-4, IL-5, IL-10, IL-13 and TGF-β [437, 438]. Purer versions of each condition generating primarily Th2 may be compared. Asthma mediated primarily by induction of IL-13 corresponds well to oxazolone colitis in BALB/c mice in which the same cytokine is paramount in pathogenesis [439]. Characteristically, however, proinflammatory cytokines are prominent in both astma and UC and aberrant DC conduction may have a role in generating a mixed Th1/Th2 response in both. It is known, for example, that PGE_2 is increased in macrophages and eosinophils in UC [355, 356] and IDO is strongly overexpressed in proinflammatory macrophages and/or eosinophils in both conditions, constituting another putative bias towards Th2 [335, 440]. By contrast, another recent report suggests that induction of IDO in resident lung cells, rather than DCs, using a TLR9 ligand, could be beneficial to airway inflammation and AHR, perhaps because enhanced innate immunity somehow inhibits adaptive responses [441].

10.3
Ramifications of the Paradigm: Autoimmunity

The paradigm is readily applicable to autoimmunity. Organ-specific autoimmunity, as exemplified by multiple sclerosis (MS) or insulin-dependent diabetes mellitus (IDDM), could be initiated and sustained by environmentally driven [7, 442, 443], but misplaced, innate immune responses in which proinflammatory activated macrophages in the target organ overcome innate tolerance, cause 'bystander' tissue damage and release autoantigens to the immune system. Several mechanisms have been proposed but a common feature is that IFRL, and the DC link to lymphoid tissue, remains intact, accounting for specific Th1-mediated autoantigen-directed adaptive responses in genetically susceptible individuals [444–446]. In the paradigm, innate tolerance normally holds autoimmunity in check, preventing activation of self-reactive memory Th1 cells. In NOD mice, innate tolerance restrains IDDM, one indication of which is an antiinflammatory phenotype in NKT cells [447]. When 'inhibitory' NKT activation is promoted by multiple doses of α-GalCer in the NOD IDDM model, tolerogenic DCs are recruited to pancreatic lymph nodes and spleen [448]. Blockade of IL-10 activity, in vivo, abrogates protection from IDDM [447]. Likewise, reduction of antiinflammatory (inhibitory) NKT cells lowers the threshold to IDDM [448]. Thus, whilst stimulation of NKT cells to a proinflammatory phenotype exacerbates EAE or IDDM, stimulation of the same cells to an antiinflammatory phenotype, is protective against these conditions [283]. Nevertheless, experimental observations to date suggest NKT cells themselves do not initiate diabetes but, rather, modulate the threshold of proinflammatory macrophage activation in response to a putative environmental stimulus. There is good evidence that proinflammatory macrophages are responsible for 'single cell insulitis', the earliest known abnormality in 'spontaneous' IDDM development in NOD mice and diabetes-prone BioBreeding rats [446]. Interestingly, microarrays have revealed very high expression of OPN in active MS lesions. In murine EAE, OPN has a sustained proinflammatory exacerbating effect. Although not necessary for disease initiation, it appears to be able to tip the balance from a 'relapsing-remitting' to a progressive form. OPN-negative mice produce less IFN-γ and more IL-10. In the human disease, raised plasma OPN correlates with MS relapse or secondary progression [449]. In secondary progressive forms of MS, raised OPN is associated with reduced IL-10 [450], possibly signifying a low-grade chronic proinflammatory drive from the innate immune system. Notably, abnormalities of T cell responses and (cytokine) repertoire are not a necessary requirement for organ-specific autoimmunity since all the observed behavior of T cells, whether tolerogenic (disease-suppressing) or proinflammatory (disease-enhancing), can be explained by innate programming of the adaptive immune system; this includes the characteristic selection and spread of (autoimmune) T cell epitopes. In MS, one effect of IFN-β is to redress the proinflammatory bias by increasing IL-10 and reducing OPN [449, 451]. The assumption must be that restoration of innate peripheral tolerance, alone, will remove the drive to organ-specific autoimmunity.

Systemic autoimmunity, as exemplified by systemic lupus erythematosis (SLE) and its murine models, can (and perhaps must) also be linked to environmental triggers and abnormalities in innate tolerance. A salient feature is enhanced expression of innate tolerance, which, however, fails to counter an abnormality in the clearance of apoptotic cells [452]. Antiinflammatory clearance by macrophages is overwhelmed by an impaired capacity of the immune system to clear apoptotic cells and/or by a selective increase in the turnover of apoptotic cells (Figure 10.1) [452–454]. Although the substrate of autoimmunity in SLE is presumed to be the overloading of the combined 'antiinflammatory' mechanisms of the innate and adaptive immune systems, IFRL between innate and adaptive tolerance is preserved, as in organ-specific autoimmunity. Enhanced innate tolerance occurring in various lupus-prone murine strains is manifest, even in young pre-diseased animals, at the level of individual monocytes/macrophages, which, on LPS stimulation and, crucially, in the presence of apoptotic cells, produce increased IL-10 and reduced proinflammatory cytokines, including TNF-α, (IL-6), IL-1 and IL-12 [455–457]. Increased IL-10 from SLE monocytes reduces their ability to stimulate Th1 responses, a clear manifestation of innate tolerance [458]. Conversely, in the absence of apoptotic bodies or, on combined stimulation with LPS and IFN-γ, macrophages from lupus-prone strains show near-normal proinflammatory responses, including bioactive IL-12p70 generation, but no increase in IL-10 output [456, 457]. High innate production of IL-10 has been documented in presymptomatic SLE, in mild to moderately active SLE and in asymptomatic first-degree relatives of SLE patients [459]. Furthermore, increases in IL-10 production through FcγRIIb stimulation of monocytes/macrophages by (antigen-antibody) immune complexes (ICs) could occur, raising the possibility of a positive feedback loop involving ICs and antiinflammatory monocyte/macrophage phenotypes [460]. The potential protective value of an antiinflammatory innate tissue microenvironment might be partially offset by a tendency of SLE monocytes/macrophages themselves to undergo accelerated apoptosis [454, 461]. Uptake of apoptotic macrophages containing remnants of other apoptosed cells, or increased direct uptake of apoptosed cells by DCs programmed in an antiinflammatory microenvironment, would likely enhance 'antiinflammatory' antigen cross-presentation to CD4$^+$ and CD8$^+$ cells [454]. Increased stimulation of innate and adaptive tolerance, combined with only partially successful removal and immune processing of apoptotic cells/blebs displaying putative surface autoantigens (DNA, phosphatidylserine, cardiolipin) [452], would be predicted to promote the development of hypergammaglobulinaemia, IGg 1 antibodies (in mice), soluble ICs and increased IgE [455, 462]. Increased Th2 support to B cells may be inferred from the finding that in lupus patients CD4$^-$CD8$^-$ (double-negative) T cells interact with CD1c$^+$ B cells, including CD40–CD40L interaction, to produce high levels of IL-4, and mediate IgM to IgG isotype switching in a CD1c-restricted manner. Although this could be a mechanism promoting autoantibodies it is important to note that the ratio of IL-4/IFN-γ is highest in T cells from mildly active rather than very active disease (or normal controls) where the ratio is reversed. These findings are consistent with the operation of innate and adaptive tolerance [463, 464]. Before the onset of spontaneous lupus nephritis in NZB/W F1 mice, anti-IL-4 antibodies, which downregulate IgG1 and IgG3 anti-DNA autoantibodies, are effective in preventing disease onset. However, anti IL-12 antibodies, which reduce potentially nephrotoxic IgG2 antibodies, do not protect at this stage. These results suggest that the build up of autoantibodies via Th2 is more critical to disease initiation than an intact proinflammatory innate and adaptive immune apparatus. IL-12 may be suppressive to the 'priming' phase of lupus by inhibiting the number of DNA antibody-producing cells. Combined anti-IL-12 and anti-IL-4 antibodies reverse the protective effects of anti-IL-4 antibodies alone on disease initiation, perhaps because an increased ratio of IL-10/IL-12 will favor B cell proliferation, antibody secretion, MHC II expression and differentiation to plasma cells, at the same time downregulating the APC potential of tissue macrophages [458]. On this background, additional conditions that bring about class switching from (murine) IgG1 to IgG2 DNA autoantibodies could mark the transition ('checkpoint') to active disease.

One of the enigmas of SLE is the elevation of proinflammatory cytokines evident in peripheral blood, indicating the importance of inflammation in disease pathogenesis. Indeed combined elevations of Th1 and Th2 cytokines in peripheral blood are associated with the most severe disease. The present paradigm, however, predicts no tissue damage in SLE from the antiinflammatory reactions discussed above (much as would be expected in a pure antiinflammatory response to helminth infection). IL-12 is in abeyance and Th1 responses suppressed by IL-10 [458]. Conversely, the paradigm allows for the grafting on of an innate proinflammatory component, serving to alter the microenvironment, at least in the most vulnerable tissues, to a proinflammatory state. Because innate immunity programs adaptive immunity, disease severity should correlate most directly with markers of innate (proinflammatory) immunity. Loss of innate tolerance would be reflected in an increase in the IL-12/IL-10 ratio in the macrophage phenotype, although, according to the paradigm, absolute microenvironmental concentrations of each cytokine would be increased (Figure 10.1). There are, indeed, reports that IL-12 and Th1 cytokines are increased more than IL-10 in active disease whereas in 'early' SLE IL-12 is decreased [465, 466]. Serum IL-12p70/IL-10 ratio is increased in SLE and correlates with disease activity, whilst IL-10 levels correlate with anti-DNA antibody levels, as predicted [467]. IL-12p40 is elevated in peripheral blood in SLE [468, 469] but cannot be considered a reliable guide to bioactive IL-12p70 for several reasons. First, although essential for IL-12p70 production, IL-12p40 and IL-12p35 must be released from the same cell to combine into the active heterodimer [470]. Second, the binding properties of IL-12 make it likely that it is retained and exerts its action close to its site of release, stimulating neighboring cells in a paracrine manner [471]. For example, IL-12p70 is elevated in rheumatoid joints and MS brain, both conditions involving local proinflammatory activated macrophages and Th1 responses. Third, IL-12p40 may act as an antagonist to IL-12p70 at IL-12R, but whether this is relevant in the local tissue microenvironment, at a single cell level, is uncertain [472] since IL-12p40 is invariably secreted in excess of both IL-12p35 and active IL-12p70 [470]. As a further complication, it has recently been suggested that IL-12p40 monomer might behave as an agonist at IL-12R. Fourth, IL-12p40 may combine with other macrophage- and DC-generated molecules to create other 'IL-12-like' molecules, notably with p19, to form the novel cytokine IL-23. This may be induced by LPS or viral infection (along with IL-18) as part of an innate immune response [473], but unlike IL-12 preferentially activates memory T cells. Finally, the amount of IL-12p70 produced by DCs at the point of their interaction with T cells in lymphoid tissue will have most influence on T cell polarity and will usually correlate with a proinflammatory microenvironment in the tissue of origin (IFRL). The essential proinflammatory sequence in SLE (and many other conditions) is postulated to be primary stimulation in peripheral tissues of IL-12p70 production in both macrophages and DCs, leading to DC migration to lymphoid tissue, after which enhanced CD40 expression by DCs, also induced by IL-12p70, allows DC interaction with CD40L on T cells, amplifying IL12p70 production still further [292]. A pre-existing antiinflammatory microenvironment, in which IL-4 and/or IL-13 are present, may prime for IL-12 production when challenged with an appropriate proinflammatory environmental stimulus, increasing the instability of the SLE microenvironment.

Although there is uncertainty about the exact relevance of IL-12 to active lupus there is clear evidence in respect of a Th1 component and elevation of other proinflammatory monocyte/macrophage cytokines, including IL-18 and IL-15. It has been found, too, that proinflammatory macrophages secreting IFN-γ [474, 475] are required for the development of glomerulonephritis in the MRL-lpr/lpr murine model of SLE and are the primary cells initiating inflammation in kidney interstitium [476]. Deposition of anti-DNA antibody or ICs alone is not sufficient [477]. That plasma IL-18 is closely associated with disease severity [469, 478] is understandable when it is appreciated that IL-18 can enhance either Th1 or Th2 [479], depending critically on several other parameters, including relative availability of IL-12 [480], expression of T cell receptors for both IL18 [481, 482] and IL-12 and increased production of IL-4 from ligand-activated NKT cells under the influence of IL-18 [483]. Hypothetically, therefore, IL-18 could provide innate support both to Th2 [484], including IgE production, and to Th1 in the pathogenesis of SLE, depending on the prevailing microenvironmental IL-12/IL-18 ratio. In the early stages of a proinflammatory response to bacterial infection, IL-18 may even tend to limit IL-12 production [485]. In *Listeria* infection, IL-18 stimulates macrophages to secrete both TNF-α and NO [486]. Normally, IL-18 combines with IL-12 to give a strong drive to Th1 responses in T, NK and NKT cells. IL18 is elevated in MRL-lpr/lpr lupus-prone mice that overexpress IL-18R and show enhanced IFN-γ secretion [482]. IL-18 accelerates lupus nephritis and provokes lupus-like skin rashes in the same mouse model. Interestingly, in this model, the 'butterfly' rash is solely dependent on IL-18, does not occur with IL-18 and IL-12 together and is associated with an early systemic Th2 response [487]. Another proinflammatory cytokine of macrophage origin and increased in SLE, is IL-15 [404]. Like IL-18, IL-15 synergizes with IL-12 in innate immunity to induce NK cells to secrete IFN-γ, albeit much less potently than a combination of IL-18 and IL-12 [404, 488]. However, IL-15 and IL-12 is much more potent than combined IL-15 and IL-18 in inducing IL-10. The optimal combination for GM-CSF production is IL-15 and IL-18 [488]. Other strong candidate cytokines to drive SLE are type 1 interferons (IFNα/β), produced in abundance by plasmacytoid DCs but also by monocytoid DCs and macrophages. In the paradigm, any composite proinflammatory drive sufficient to counteract the default Th2 humoral bias will precipitate clinical disease. Transgenic lupus-prone mice engineered to lack either IFN-γ or IL-4 have demonstrated the non-redundant roles of Th2 and Th1 in lupus-associated tissue injury [489]. IgG class switching in lupus is driven firstly by antigenic determinants, notably nucleosomes of apoptotic cells, triggering interactions between T and B cells, and secondly by cytokines, with an increased IFN-γ/IL-4 ratio favoring pathogenic IgG2 over IgG1. Some T cell intrinsic abnormalities in lupus, particularly sustained induction of COX-2, increase T cell viability following activation. Of note, upregulated COX-2 occurs in proinflammatory macrophages, as found in tissues affected by active lupus. Antibody class switching from IgG1 to IgG2a in B cells is triggered by IFN-γ and is transduced through t-Bet.

What stimulates the proinflammatory innate response in SLE? An environmental link has long been suspected [490] and, in the current paradigm, additional environmental factors are involved in the loss of peripheral tolerance to lupus autoantigens. An environmental input initiates a swing towards a proinflammatory innate microenvironment that is superimposed on the pre-existing antiinflammatory microenvironment, thus pushing macrophages and DCs away from the (enhanced) expression of innate tolerance towards proinflammatory Th1-inducing phenotypes, considered responsible for the initiation, maintenance and relapse of clinical disease (Figure 10.1) [491]. Possible 'environmental' triggers that might promote a proinflammatory response in SLE include infection [486], UV light exposure, certain drugs (some of which may exacerbate apoptosis) and even 'therapeutic' IFN-γ itself [490], a situation reminiscent of early unsuccessful attempts to treat MS with IFN-γ. Both viral and some bacterial infections may be strongly influential in precipitating clinical disease by generating type-1 interferons in the microenvironment (from macrophages and DCs) and more specifically activating immunostimulatory DCs charged with self-antigens [492]. Experimentally, increased IFN-α exacerbates murine lupus and partial ablation of IFN-α reduces disease morbidity and prolongs survival. Expression of TLR7, TLR8 and TLR9 by (plasmacytoid) DCs and macrophages dictates preferential antigenic responses to DNA and RNA, which can be of viral, bacterial or self origin. Exogenous *E. coli* CpG-DNA causes progression of lupus nephritis in MRL-Fas lpr mice by activating a TLR9 pathway, leading to a significant elevation of IgG2a and a trend towards elevation of IgG1 serum DNA autoantibodies. Whilst B cell TLR9 could mediate this response directly, increased TLR9 expression occurs locally on macrophages/DCs in nephritic kidney, which are activated to drive Th1 responses. In this way stimulation by CpG-DNA of TLR 9 macrophages in kidney could exacerbate lupus nephritis, provoking cross reaction with self DNA and increased IGg2a autoantibodies [493]. In lupus, aberrant CD4$^+$ plasmacytoid DCs may be generated from monocytes, through exposure to IFN-α, and increase the transit of autoantigens to the T and B cell compartments. The combination of reduced plasmacytoid DCs in peripheral blood in active SLE and the presence of IFN-α-secreting plasmacytoid DCs in affected tissues, such as skin, suggests an unstable tissue microenvironment in which macrophages and DCs are secreting type 1 interferons [492]. The virtually universal finding of a signature of type-1 interferon-upregulated genes in peripheral blood mononuclear cells in active pediatric lupus has lead to the view that this cytokine could be the major driving force behind the disease, constituting a powerful mechanism to overcome innate and adaptive tolerance. The signature includes some hallmark lupus antigens (Ro52, Lamin 1b), the chemokine MCP-1, markers of transition to a mature DC phenotype (DC-LAMP) and several complement cascade and apoptosis-related genes [494]. The situation does not appear to be so clear-cut in adults, with only about half having an IFN signature, but this correlates well with clinical disease severity, particularly the involvement of kidneys and/or CNS [495]. Studying lupus glomeruli from renal biopsies by microarray analysis reveals that at least half show high expression of genes with type 1 IFN response elements in their promoters, but these tend to correlate inversely with fibrosis gene clusters, which are indicators of a worse prognosis. The cell type in the kidney displaying IFN I signatures could be recruited NK cells rather than locally-activated macrophages and whether these NK cells are protective or injurious is at present uncertain [496]. Just as TNF-α has been given prominence in the pathogenesis of rheumatoid arthritis, IL-4 in allergy and IFN-γ in uncomplicated proinflammatory responses, so too is IFN-α now receiving much attention as the driving force behind lupus. In one interpretation of the cytokine network, immune responses are portrayed as a dynamic system under the control of two opposing vectors, IFNα/β versus TNF-α and IFN-γ versus IL-4, thus incorporating the idea of regulatory balances between different proinflammatory cytokines, as well as the accepted balances between opposing pro- and antiinflammatory cytokines [492].

Another environmental agent, UV-B exposure, enhances innate tolerance at low exposure levels but overcomes it at higher levels. Single erythema-inducing doses of UV-B applied to human epidermis generate (innate) tolerogenic ('suppressor') macrophages secreting TGF-β and IL-10 and communicate a TGF-β-dependent anergic signal to T cells [497]. In this context, exposure of mice to low dose UV-B suppresses IL-12p70 (but not PGE_2, IL-1, IL-6 or TNF) in splenic APCs [82, 498], which in turn induces hapten-specific 'suppressor' T cells, and reduces Th1 responsiveness [499]. At higher doses, UVB is suggested to cause damage to keratinocytes sufficient to overcome the ability of innate tolerance to cope with increased apoptotic and/or necrotic products, thereby exposing cutaneous lymphocyte antigen and other potential autoantigens to the adaptive immune system. Thus, limited skin apoptosis may create an antiinflammatory microenvironment but high UVB exposure and increased apoptosis may trigger a proinflammatory tissue microenvironment. Other experimental techniques, which (like high dose UVB) create a chronic proinflammatory microenvironment in the skin, are IFN-γ overexpression linked to the keratinocyte involucrin promoter [490] and overexpression of CD40L linked to the keratin-14 promoter [500]. In both these models, systemic as well as localized autoimmune manifestations develop. In excess, apoptotic cells themselves may constitute a proinflammatory signal to DCs and also cause an increase in microenvironmental necrosis, another proinflammatory stimulus. Similarly, exposed putative autoantigens on apoptotic blebs can be opsonized by autoantibodies, such as anti-phospholipid antibodies, subsequently inducing proinflammatory rather than antiinflammatory responses on uptake by macrophages/DCs [501]. In the lupus-prone NZB mouse, pristane exposure, employed as a 'model environmental trigger', clearly exacerbates autoimmune manifestations through stimulating proinflammatory cytokines, notably IL-12 [502]. Conversely, IL-12p35-deficient mice treated with pristane develop harmless 'nephritogenic' antibodies, which deposit as ICs in the kidney, but, in the absence of proinflammatory IL-12, fail to induce nephritis [503]. Reducing IL-12 reduces nephritis and other organ damage in MRL-Fas mice [504]. It has been reported recently that the pristane-treated BALB/c mice, not housed in specific pathogen-free conditions, show a greater propensity to secrete IL-12, IL-6, TNF-α, IFN-γ and IL-4 and to develop greater increases in the IgG2a/IgG1 ratio than their pristane-treated germ-free littermates – evidence for two environmental agents acting synergistically [505].

Environmental triggers combine with less obvious internal triggers of 'innate' immunity in the pathogenesis of SLE, namely microenvironmental changes arising directly from the preexisting SLE-phenotype. For example, the association of complement factor deficiencies and SLE may be explained, in part, by a reduction of non-inflammatory clearance of complement-opsonized antigen–antibody ICs in the face of an increased IC load [506] and, in part, by impaired negative selection of autoantibody-producing B cells [507]. Complement could contribute to the balance of innate tolerance and innate immunity in various ways, not least through its effects on apoptosis, which range from pro- to anti-apoptotic and from opsonization to lysis [508]. The C1q and C4 components participate in clearance of ICs and apoptotic cells. Deficiency of C4 reduces the amount of C3b available to bind ICs and encourages precipitation of poorly opsonized ICs in tissues [506]. Deficiency of C1q impairs macrophage phagocytosis of apoptotic cells in a non-inflammatory microenvironment [509]. C1q binds to nuclear antigens/nucleosomes exposed on the surface of apoptotic cells and blebs [509, 510]. Impaired clearance by macrophages of apoptotic material may deliver autoantigens into the DC pathway, ultimately generating autoantibodies. This may be accomplished without inflammation but higher levels of apoptosis may trigger bystander proinflammatory changes in DC phenotypes [511]. Excess apoptotic bodies may provoke a proinflammatory tissue response, either directly or if they become secondarily necrotic, imparting a proinflammatory (Th1-generating) 'danger' signal to DCs and promoting an isotype switch to tissue-damaging antibodies. Apoptotic debris lodging in the glomeruli of C1q-deficient mice could constitute a proinflammatory stimulus, attracting proinflammatory macrophages and triggering damaging autoimmune glomerulonephritis. MBL, elevated in acute phase reactions, activates complement by the lectin pathway and so enhances opsonization. Genetically high MBL appears to protect against SLE. Another factor contributing to the balance between innate tolerance and innate immunity is the expression of immunoglobulin receptors. Uptake of ICs through one of three classes of Fcγ receptors on macrophages is either proinflammatory or antiinflammatory. The primary receptor for non-inflammatory clearance of ICs by macrophages is probably the FcγRIIb class of 'inhibitory' receptor, and its deficiency in mice predisposes to spontaneous lupus. The same receptor is downregulatory to B cell proliferation and induces B cell apoptosis. Balanced against these effects, activating FCγ receptors are proinflammatory and may enhance antibody responses. Some classes of proinflammatory FCγ receptors, in particular FcγRIII, are heavily implicated in lupus nephritis [512]. Sophisticated microarray analysis of lupus-affected renal biopsies indicates that fcRγc, a signaling γ-chain common to several FC receptors, is increased in the myeloid (macrophage) gene cluster, but other IC receptors may also be involved [496]. A common source of 'danger' signals may be autoimmunogenic fragments cleaved by granzyme from conventional autoantibodies on the surface of apoptotic blebs. Granzyme is an enzyme secreted by cytotoxic T cells and NK cells in a proinflammatory setting [513]. Some models of SLE invoke positive feedback loops between T and B cells, leading to epitope spread and autoantibody diversification, following an initial (B7 co-stimulation-independent) interaction between peripheral auto-reactive T cells and B cells, presenting autoimmunogenic peptides [514]. A specific hyposensitivity to certain (γ-chain) cytokines, notably IL-15 and IL-2, may indicate a specific T cell defect in SLE, leading to enhanced T cell apoptosis, particularly evident under proinflammatory conditions, adding to the apoptotic load [515]. In the presence of increased IL-10, combined with increased proinflammatory cytokines, as in the uncompensated active phase of SLE, activation induced cell death of lymphocytes may be accelerated by IL-10 and by upregulated expression of Fas receptors on T and B cells. Clinically, lymphopenia in active disease correlates with raised IL-10 [516].

Autoantibodies are considered to be leading players in systemic autoimmunity but in the present paradigm it is the change from non-damaging Th2-supported isotypes (IgG1 in mice) to Th1-supported isotypes (IgG2 in mice), that underlies the development of lupus pathogenesis and clinical disease [489]. In the pre-clinical phase, raised IL-10 may promote B cell differentiation and hyperactivity. Antibody production is critically dependent on $\alpha\beta$ T cells, first providing Th2 cell help to B cells to produce antibodies to exposed epitopes (IgG1 in mice) and second in response to a classic DTH reactions in which class I and class II presentation to T cells ultimately enhances tissue proinflammatory responses and B cells isotype switching to produce IgG2. Complement-activating IgG2a (and IgG3) antibodies dramatically increase IC and C3 deposition in glomerular capillaries and are associated with severe nephritis, in contrast to mice lacking the IFN-γ receptor that have raised levels of IgG1 anti-dsDNA antibodies but show no class switching to IFN-γ-dependent IgG2a and no GN [517]. It has recently been discovered that, in MRL-lpr/lpr mice, anti-C1q autoantibodies are present both in the circulation and in renal immune deposits early in life and prior to the development of nephritis, so the mere presence of autoantibodies (of any sort) may be insufficient to cause significant damage [518]. Imbalance towards Th1 (measured as an increased $CD4^+$ T cell IFN-γ/IL-4 ratio or as increased ratio of IgG2a/IgG1 or IgG3/IgG1) accelerates SLE in MRL-lpr/lpr mice [519]. One mechanism contributing to this may be the induction of a proinflammatory macrophage phenotype after phagocytosis of apoptotic cells opsonized with autoantibodies, such as antiphospholipid antibodies [520]. Uncoupling of IC deposition in the kidney from a proinflammatory microenvironment averts renal damage and can be achieved in MRL-lpr/lpr mice by high dose treatment with G-CSF, which downregulates (macrophage) FcγRIII. This fits with the observation that recruitment of proinflammatory macrophages, apparently secreting IFN-γ [474], and not just the presence of anti-DNA antibody or IC deposition [477], may be crucially required for the development of GN in murine models of SLE. Mice deficient in MCP-1 exhibit reduced macrophage recruitment and delayed onset of GN, despite IC deposition, indicating a significant pathogenic role for recruited proinflammatory macrophages [474], and suggesting parallels with organ-specific autoimmune disease. Comparing the significance of anti-C1q and anti-dsDNA antibodies in lupus nephritis, the absence of dsDNA antibodies excludes active nephritis, while both anti-C1q and anti-dsDNA antibodies have high specificity for active nephritis [521]. Inadequate induction of SOCS1 can lead to systemic autoimmunity (including anti-DNA antibodies) in SOCS+/- C57BL/6 mice, but not in the absence of CD4$^+$ T cells, again stressing the importance of the combination of IFN-γ and B cell-derived antibodies in pathogenesis [522]. Nevertheless, B7 co-stimulatory molecules must also be available for T cell activation and IgG autoantibody class-switching, without which MRL-lpr/lpr mice fail to develop autoimmunity [523].

All the multiple and sequential interactions suggested by the paradigm to culminate in full-blown nephrotoxic and tissue-damaging SLE are summarized in Figure 10.1. There are, however, several anomalies that appear to contradict the paradigm. The exact role of IL-10 in human SLE presents one such dilemma. It is well established that disease activity correlates positively with levels of endogenous serum IL-10 (whatever the source) and that genetically high IL-10 production is likely to predispose to increased incidence and severity of SLE [524]. These observations are well accommodated by the paradigm, but it is more difficult to explain why the administration of antibodies to IL-10, which reduce biologically-active IL-10, has been shown to decrease the clinical manifestations of the disease during, and for 6 months after, a three-week course of treatment [525]. This is in sharp contrast to IL-10-deficient lupus-prone mice, which, compared to their wild-type counterparts, show earlier onset and more severe clinical disease, together with increased IFN-γ and an earlier increase in potentially pathogenic IgG2a anti-dsDNA antibodies [526]. Peripheral blood cells from SLE patients can induce anti-dsDNA antibodies in SCID mice, an effect virtually abolished by treating the mice with anti-IL-10 antibody [525]. In the human study anti-dsDNA antibodies were not significantly decreased, except in one case, by anti-IL-10 antibody and endogenous IL-10 continued to be produced to varying degrees [525]. In direct contradistinction to the use of anti IL-10 antibody in the treatment of human SLE, other evidence from animal models might suggest that IL-10 administration itself could be considered a suitable treatment for human lupus, on the basis of its downregulation of IFN-γ, an accepted tissue and peripheral blood marker, for the most severe disease in both animals and humans [526, 527]. In line with this view, increased IL-10 is present in asymptomatic relatives, so that, if IL-10 exacerbates SLE rather than restraining it, some other suppressive mechanism would have to be postulated to hold the disease in check. According to the paradigm, high IL-10 primes for clinical disease by increasing antibody production and prolonging B cell survival. At the same time, IL-10 is an indicator of innate and adaptive tolerance that holds the disease in check. Therapeutic suppression of IL-10 in active SLE might reduce B cell activation and survival and, thereby, serve to reduce the pool of B cells available as a source of later pathogenic IC-forming opsonizing antibodies under the influence of proinflammatory cytokines during clinical exacerbations.

Another related difficulty concerns the sequencing of events. The paradigm suggests that the order of development of disease in humans is initially predominantly Th2, upon which Th1 is later superimposed [465, 526], and that the innate proinflammatory Th1-generating response that precipitates clinical SLE is a consequence of either an additional environmental insult or an overloading of non-inflammatory disposal of apoptotic material [465]. Remission is signaled by a return to predominantly Th2 adaptive responses. Against this view some animal experiments are interpreted as showing that a Th1 response precedes Th2 in the induction of lupus [528]. However, this may simply be an artefact of the induction process. Induction of murine lupus by injection of human anti-DNA causes an increase in proinflammatory peritoneal macrophage cytokines, leading to a Th1 response and an increased ratio of IgG2/IgG1 of autoantibodies, as would be anticipated, and later in the disease there is relative excess of Th2 cytokines. Such a sequence is entirely consistent with the paradigm except that in 'spontaneous' human or murine SLE there would be a preceding period of protective innate tolerance, associated with IgG1 antibodies and Th2 cytokines before the proinflammatory Th1 phase and later 'restorative' Th2 phase. Supporting the idea that enhanced innate tolerance eventually gives way to innate immunity, all murine lupus models show a macrophage defect in the generation of proinflammatory cytokines (IL-1, IL-6, IL-12, TNF-α and GM-CSF) when stimulated by apoptotic cells, from as early as 1 week of age [457]. Clinical disease development correlates (in the MRL-lpr/lpr mouse) with IgG2a antibodies (including anti DNA) and is driven by IL-12 and IFN-γ [529] as has been suggested for the human disease [465]. MRL-lpr/lpr and NZB/W macrophages are both poor IL-12 producers, a property that may allow the build-up of high levels of autoantibody B cells and plasma cells prior to eventual isotype switching [530]. Delivery of IL-12 plasmid into the MRL-lpr/lpr mouse 4-weekly from 4 weeks of age suppresses polyclonal B cell activation and autoantibody production; one interpretation of which is the necessity for a significant Th2 phase in the early evolution of lupus [531]. Similar reasoning can be applied to the delayed disease onset when young prediseased NZB/W F1 mice are treated with TNF-α or anti-IL-10 antibody. The immunomodulator, AS101, given regularly to NZB/W F1 mice from 4 weeks of age, reduces IL-10 and delays lupus onset despite increases in TNF-α, IFN-γ and IL-1 and is thought to act by removing an excessive and self-reinforcing drive towards B cell expansion and autoantibody production [532]. Once IgG2a antibodies have developed in the MRL-lpr/lpr mouse by 8 weeks, suppression of IL-12 effectively reduces toxic antibody production, IFN-γ and clinical disease parameters [533]. In terms of the current paradigm NZM 2328$^{+/+}$ mice lacking STAT6 are of great interest as they are defective in antiinflammatory (type 2) cytokines and produce a preponderance of IgG2a autoantibodies. However, little renal pathology is evident at 6 months, implying that no significant proinflammatory trigger is operating in the glomeruli. Further dissection of the mechanism of GN comes from NZ2410 mice, which overexpress IL-4 and develop severe glomerulosclerosis, which does not occur in STAT6 deficiency or on treatment with anti-IL-4 antibody. Another transgenic mouse, NZ 2328, which is STAT4 deficient, fails to secrete type 1 cytokines, and these animals, too, develop sclerotic renal disease [534]. It is concluded that type 2 cytokines (and STAT6) selectively promote sclerosis and that IFN-γ (and STAT4) has an important role in (IgG2a) autoantibody production. Various cytokine-modifying 'treatment' regimes have been investigated in murine models. Oligodeoxynucleotides, which suppress proinflammatory cytokines, delay the onset or progression of GN in NZB/W mice, accompanied by a significant reduction in anti-dsDNA antibodies, IFN-γ and IL-12 [535]. Histone deacetylase (HDAC) inhibitors downregulate both pro- and antiinflammatory cytokines (IL-12, IFN-γ, IL-6 and IL-10) and reduce GN in MRL-lpr/lpr mice, but there are no significant changes in IgG isotypes deposited in the glomeruli or changes in autoantibodies. A conclusion from this study is that renal disease cannot be directly linked to autoantibodies and that inflammatory mechanisms leading to tissue damage are distinct from those leading to autoantibody production. Global augmentation of cytokines in SLE is suggested to be a primary genetic defect, determined by a dysregulation of HDACs, rather than being a reactive process to other genetically-determined abnormalities, such as dysregulated apoptosis [536]. As in organ-specific autoimmunity there are reports of elevated OPN in active SLE, which is consistent with an ongoing proinflammatory response [537].

Therefore, according to the current paradigm, the pathogenesis of SLE commences with a perturbation of innate tolerance. In fact the microenvironmental input to the system initially provokes an exaggeration of innate (and adaptive) tolerance but this is eventually offset by an endogenous or exogenous input triggering proinflammatory macrophage phenotypes that initiate a switch towards adaptive (auto)immunity (Figure 10.1). The MDU at this stage has a phenotype supporting both Th1 and Th2, the ratio varying with disease severity. Disease resolution coincides with restoration of an antiinflammatory macrophage phenotype and innate tolerance. The puzzle of SLE is to decipher the inputs driving innate changes and relating the innate and adaptive effector mechanisms to the tissue damage caused. The paradigm can give a coherent explanation for most of the features of SLE, with the proviso that glomerosclerosis, a 'type 2 inflammation', equates to an antiinflammatory, reparative macrophage phenotype, even though severe kidney damage can ensue.

A comparison of organ-specific and systemic autoimmunity reveals a large degree of overlap in fundamental pathogenic mechanisms. In particular, a breakdown in innate and adaptive peripheral tolerance with the generation of proinflammatory macrophages and the involvement of Th1-orientated responses in the causation of major tissue damage would appear to be important, if not essential, features of both types of autoimmunity. Autoimmune diseases, both organ-specific and systemic, predominantly affect females, share some MHC susceptibility haplotypes and at least 18 non-random non-MHC linkage clusters, compatible with a common set of autoimmune susceptibility genes [538]. Autoantibodies are common in organ-specific as well as systemic autoimmunity and anti-nuclear antibodies are present in 16% of IDDM [539]. However, in early or pre-disease states, in both murine and human autoimmune diseases, antibodies are predominantly of non-opsonizing Th2-associated isotypes and are generally considered non-pathogenic. The NOD mouse, in which macrophages have a high potential to release IL-12 on LPS stimulation (even without IFN-γ co-stimulation) [540] has proved to be remarkably versatile and informative model of autoimmunity. Whereas female NOD mice develop IDDM 'spontaneously', the model can be diverted towards EAE by two injections of pertussis toxin (an environmental agent) intravenously, two days apart. Furthermore, NOD mice can be induced to develop an SLE phenotype, displaying IgG2a antibodies (including anti-dsDNA antibodies), two to three months after a single dose IV injection of *Mycobacterium* [521, 541]. In the NOD mouse, one gene (H2region) appears to govern both IDDM and SLE susceptibility [542]. Amongst non-MHC susceptibility genes, reduced expression of the antiinflammatory FCγR11b on macrophages is common to both NOD and SLE prone murine strains and is indicative of impaired innate tolerance whereas reduced expression of Fas in both NOD and MRL-lpr/lpr murine models would encourage persistence of autoreactive B cells [543]. Since T cells also have a central role in lupus, reduced Fas is a risk factor by allowing increased survival of autoimmune cells [544]. Another relevant non-MHC gene may be that of a macrophage-specific NO transporter, of importance both in the response to intracellular pathogens such a *Mycobacterium* and *Leishmania*, and for initiating Th1 responses, leading to autoimmunity [541]. Aged NOD mice tend to develop some features of systemic autoimmunity. Intriguingly, macrophages of prediseased (4–6 week) NOD mice exhibit a cytokine dysregulation akin to that of SLE-prone mice strain, elicited only in the presence of apoptotic cell constituents, and characterized by downregulation of IL-1β, IL-12p35, GM-CSF, TNF-α and MIP-1β but not of IL-12p40, TGF-β, M-CSF and RANTES. Although this unusual pattern of cytokine release is common to virtually all autoimmune strains it is conspicuously absent in macrophages from other strains not prone to autoimmunity and suggests that the genetic basis for the abnormality may comprise a shared background predisposing to, or permissive of, autoimmunity [545]. Even more remarkable, by 12 weeks, usually after the commencement of diabetes, there seems to be a dramatic reversal of the NOD phenotype to the excessive proinflammatory response to apototic material noted earlier. This clearly points to specific microenvironmental instruction. Environmental instruction, and hence microenvironmental instruction, is grist to the mill of the immune system and a powerful manipulator of macrophage (and other innate cell) phenotypes responsible for secondary adaptive change. For example, the Th2-orientated strain (BALB/c), sensitized either to myelin or testicular antigen, develops either EAE or experimental autoimmune orchitis (EAO) (both Th1 T cell mediated organ-specific autoimmune diseases) through a common disease susceptibility pathway modulated by different breeding environments [546]. Some breeding conditions enhance the expression of both EAE and EAO whilst others suppress them. In the paradigm, the most readily identified 'common pathway' consists of the integrated outcome of a balance between innate tolerance and innate immunity. As for disease associations, the paradigm predicts positive correlations between IL-12-driven organ-specific autoimmune diseases, notably MS and IDDM, and this appears to be the case epidemiologically. Prediabetics, as well as showing multiple T cell autoreactivities to islet antigens, commonly display autoimmunity to classical MS autoantigens [444]. However, although IDDM and SLE can be reciprocally induced in NOD mice, the coexistence of the two conditions is rare in humans. Associations have been documented between MS, IDDM, asthma, inflammatory bowel disease and autoimmune thyroid disease. The MS or IDDM and asthma associations, though not predictable within a polarized Th1/Th2 framework, would fit well with the current paradigm where both MS and asthma can be driven, essentially, by innate environmentally-triggered proinflammatory responses. SLE has shown both positive and negative associations with asthma, so is difficult to analyze. It should not be considered counter-intuitive that NOD mice with a strong genetic Th1 bias can be induced to an asthmatic phenotype, with lung expression of IL-4, IL-5, IL-13 and eotaxin, by a regime of allergen sensitization and intranasal challenge [547]. The opposite situation is found in the strongly Th2 biased BALB/c mice which can be stimulated to mount Th1 responses as, indeed, required for the successful elimination of *Leishmania* [548]. Pre-T cell regulation of autoimmunity is implied by the finding of inactive autoreactive T cells against both islet and CNS autoantigens in diabetics, relatives at risk of diabetes, MS patients and NOD mice alike. The interaction of infection and immunity has spawned vigorous debate. The hygiene hypothesis of asthma rests on the Th1/Th2 dichotomy, requiring eduction of the immune system towards Th1 by environmental instruction to prevent it, but unlike the current paradigm is difficult to reconcile with a concurrent marked increase in the prevalence of IDDM, which should be less prevalent in a cleaner society. One aspect of a cleaner society may be changed bowel biota, which maintains the immune system in a tolerogenic mode. The present paradigm predicts an increase in both organ-specific autoimmunity and asthma with loss of helminths, as from a declining prevalence of pinworms, estimated to have been present in 50% of American and North European children. Helminths inhibit the development of atopy by inducing regulatory T cells and IL-10 and by similar mechanisms inhibit IDDM onset in NOD mice. An opposing hypothesis, which must be tested against the epidemiological evidence, is that helminth infection increases the risk of asthma because the two conditions share many of the same 'effector' pathways (eosinophila, elevated IL-5 and IL-13 amongst others). Bearing in mind the lower prevalence of asthma in developing countries, the crux of the problem is to determine whether systemic, or even direct lung infection with helminths, or any similar infection that induces innate tolerance exacerbates or reduces the risk of asthma. Infection of BALB/c with Angiostrongylus suggests that in fact, the pulmonary inflammatory response to OVA is decreased, as measured by a reduction in IL-1β and a decreased total cell count in BAL fluid [549]. An important interaction on a global scale is the predicted reinforcement of innate tolerance by the interplay of two common pathogens, intestinal nematodes and pulmonary TB. The current paradigm predicts an increase in TB infection and progression, because innate tolerance equates to virulence, and this appears to be the case in practice [550]. Conversely, a reciprocal increase in nematode carriage might also be predicted as a result of this adverse synergism.

The current paradigm requires a proinflammatory innate stimulus as the starting point for most autoimmune reactions and this usually, but not necessarily, correlates with a Th1 and CD8$^+$ cytotoxic T cell response (IFRL). However, the ability to generate Th1 responses in the absence of the prototypical Th-1-directing cytokine IL-12p70 requires clarification. In fact IL-12p70 is not the only proinflammatory signaling molecule capable of carrying a Th1 polarizing signal to lymphoid tissue, either in animal models of MS (EAE) or SLE, and IL-23 is now recognized as an effective substitute, synergizing with IL-18 and IL-15. In exceptional cases the innate proinflammatory stimulus induces a Th2 response through aberrant innate-adaptive signaling in relation either to an organ specific or systemic autoimmune model. In a p35 knock-out model of EAE the Th2 cytokines IL-10 and IL-4 are upregulated [551], but significantly so too is IFN-γ (and TNF-α), indicating some stimulation of the Th1 pathway, or at least a proinflammatory component acting in the CNS. In addition, increased CNS NO [552] suggests that macrophage/microglial proinflammatory activation can be sustained in some way independently of IL-12p70, possibly through a partial Th1 response driven by IL-23 [553]. A primary abnormality of monocytes/macrophages is ideally placed to provide a mechanism, in line with the primary role of macrophage effector function, in shaping and sustaining unusual immune responses. Because non-specific tissue damage provides the initial stimulus to autoimmunity in vivo, more than one organ can be affected, so that MS and IDDM may coexist if there is coincidental delivery of an innate proinflammatory response to pancreas and brain and equal Th1-mediated autoantigen susceptibility. Multiple immunoregulatory defects, some proinflammatory and some antiinflammatory, are generated in autoimmunity and involve all immunologically active cells [554]. The concept of a balanced type 1 and 2 cytokine response has already gained credence as the basis of host defense during sepsis [119]. Adoptive transfer granulomatous experimental autoimmune thyroiditis (EAT) is a relatively benign self-limiting condition, ascribed to Th2, but addition of IL-12 converts it into severe necrotising form [555]. Common underlying mechanisms in autoimmune pathogenesis may extend to such overtly autoantibody-driven conditions as myasthenia gravis (MG). As in SLE, both Th1 and Th2 are prominent in its pathogenesis but damage is probably mediated by antibody switch to pathogenic IgG2 (in mice), occurring when B cells are driven by Th1 T cells [556]. Significantly, in adoptive transfer murine experimental autoimmune MG, disease progression is slowed [557] by anti-IL-18, which has been demonstrated to decrease Th1 and increase TGF-β. The increased production of B cells and autoantibodies in response to acetylcholine receptor injection is promoted in a murine model in which transgenic T cells secrete transient IL-10, but also IFN-γ. The antibodies are manifest mainly as IgG2, presumably signifying the influence of Th1 cells [556]. Extensive activity of allergic elements in the context of Th1-mediated autoimmunity has been described in MS [558].

In summary, the classic Th1-mediated diseases of IDDM and MS are caused by proinflammatory innate responses linked to genetically predetermined susceptibility to mount autoantigen-specific (tissue-specific) Th1 responses whereas systemic, autoantibody-mediated, autoimmunity is caused by adverse and interdependent alterations in the balance of innate tolerance and innate immunity (often changing over time), which drive combinations of polarized T and B cell responses and modulate apoptosis. The two are not mutually exclusive and can merge. An important consequence is that both organ-specific and systemic autoimmunity should be improved by suppressing innate proinflammatory responses and/or their consequences in order to re-establish effective innate tolerance. If innate immunity drives all autoimmunity it should come as no surprise that in some instances the adaptive immune system may make little or no contribution. At least some cases of macrophage activation syndrome may be included in this category. The term 'innate autoimmunity' is a recent addition to the classification of autoimmunity and is exemplified by reperfusion injury and fetal loss syndrome. Self-recognition is by components originally evolved through the adaptive immune system but now operating as specific pattern-recognition proteins/receptors of the innate immune system, such as specifically identifiable natural IgM antibodies, and the subsequent tissue damage is complement dependent [559]. This type of autoimmunity does not exclude secondary T cell responses, which may be marked, nor does it exclude other immune mediators and pathways triggering similar responses. A dramatic example is ischemia-reperfusion injury to the gut, which is totally abrogated in germ-free mice, under which circumstances IL-10 dominates a 'no inflammation' phenotype [560].

Finally, how does the concept of 'protective autoimmunity' fit into the framework of innate and adaptive tolerance and immunity? The observation that a Th1-mediated response to neuronal injury actually reduces the spread of secondary degeneration and improves the outcome for neuronal survival led to the hypothesis that a self-perpetuating destructive process following an acute focal insult can be offset by the contemporaneous generation of an alerting signal that channels healthy cells towards an alternative immune-mediated path to death. The concept can be formulated mathematically into a spatio-temporal model [561]. Experimental evidence to support it includes the observation that natural (CD4$^+$ CD25$^+$) regulatory T cells may exacerbate rather than ameliorate optic nerve injury [562], and that a vaccine that can stimulate an appropriate degree of (presumably) Th1 adaptive response to myelin epitopes, manifest as mild EAE, can improve the resolution of crush injury of the optic nerve, partially if the myelin peptide chosen is encephalitogenic, or more completely if non-encephalitogenic. The degree of neuroprotection achieved can be assessed by measuring the survival of retinal ganglion cells [563]. To explain these and similar findings, several pathophysiological mechanisms have been proposed, including removal of self antigen, reduction of neurotoxins (free radicals, eicosanoids, glutamate excess), increased uptake of glutamate by activated microglia in contact with IFN-γ-secreting T cells and secretion of neurotrophic substances by activated T cells, either directly or through their induction in microglia [561, 564, 565]. In terms of the paradigm presented here the immunological requirement for protective autoimmunity is for an innately-generated immune response to be switched off by an adaptive Th1 response. Since IFRL remains intact in this situation (as for a normal response to microbial infection) the mechanisms involved in delivering protective autoimmunity should be the same as those that would normally help to regulate and/or terminate a combined innate/adaptive proinflammatory immune response. A first autoprotective mechanism, on the evidence of a beneficial response of nerve crush injury to Th1 cell adoptive transfer EAE, must be a reduction in locally active proinflammatory macrophages (microglia) and or activated T cells, mediated by a Th1 mechanism. A well-recognized response to IFN-γ is apoptosis and this applies both to activated macrophages and T cells. Accordingly, Fas/FasL ligand interaction between T cells and macrophages is a major mechanism for deleting activated macrophages, which is important for clearing intracellular infection, but equally able to bring about a nonspecific reduction in inflammation [566]. Production of NO, from whatever source (activated macrophages or B cells), is enhanced by IFN-γ and induces T cell apoptosis [567]. Although this may seem an ideal way to downregulate inflammation (and would even mimic the 'transient amplified loss' [561] of the mathematical model before the apoptoic effect kicks-in), there is a fine balance between protective and destructive autoimmunity, influenced primarily by the extent of injury and genetic factors. In mycobacterial infection, for example, IFN-γ eliminates the responding T cells through induced apoptosis, an effect that is abolished in transgenic mice lacking IFN-γ. Macrophage depletion or iNOS inhibition also abrogates the apoptotic T cell response, indicating a crucial involvement of activated macrophages. Hence, during an immune response, or protective autoimmunity, a negative feedback loop maintains $CD4^+$ homeostasis because activated IFN-γ-secreting T cells indirectly trigger their own apoptosis by activating (or engaging already activated) macrophages [568]. Contrasting with a successful immune response to infection, EAE in IFN-γ-knock out mice demonstrate destructive autoimmunity, with $CD4^+$ cell accumulation in CNS and spleen, and ineffective elimination of self antigen [569]. As noted earlier, there is often a complex balance between factors supporting Th1 and Th2 responses in immune reactions, with differential effects on APC survival. IL-12 enhances survival of adherent monocytes but IL-10 induces apoptosis. Monocyte apoptosis is dose-dependent with IFN-γ, but this requires IL-10 secretion [570]. Another important activated macrophage and T cell pro-apoptoic cytokine is TNF-α (as discussed elsewhere). An intriguing candidate molecule for mediating protective autoimmunity is IDO, which is induced in proinflammatory activated macrophages and some DC sub-sets and a negative regulator of T cell clonal expansion [571]. Measures to enhance protective autoimmunity could be of use in treating low-grade inflammatory neurodegenerative conditions but may be a high-risk strategy because amplifying Th1 adaptive immunity could exceed a threshold where destructive autoimmunity supervenes (more damage caused than prevented). The applicably of protective autoimmunity to acute (innate) damage limitation is even less clear. For example, treatment with vaccinia virus complement control protein successfully reduces cord destruction by inhibiting immediate inflammation and macrophage recruitment [572]. Whether the resultant reduction of an adaptive Th1 response has a later adverse effect on protective autoimmunity can only be guessed at. Nevertheless, the concepts of protective autoimmunity and comprehensive immunity appear consistent with the paradigm presented here. Of note, apoptosis and corpse removal in *Drosophila*, and throughout mammalian ontogeny, might also be reasonably described as 'protective autoimmunity'. It is usually accomplished rapidly, discretely, ultra-efficiently and without inflammation.

10.4
Summary and Conclusions: Towards Immune System Modeling and Therapeutics

The related concepts of innate tolerance and macrophages as major transducers, integrators and effectors of microenvironmental stimuli have been developed as part of a general paradigm exploring the functional inter-relationships of immunologically active cells in peripheral tissues. In addition to the interactions of 'professional' immune cells, account must be taken of the complex immunomodulating effects of proinflammatory and antiinflammatory activation of epithelial cells, and of contributions from all other immunologically active cells that influence the microenvironment, e.g., mast cells, endothelium and fibroblasts. Activated neutrophils in early acute innate inflammation may strongly prime the microenvironment for a proinflammatory response, but even initial neutrophil recruitment and responsiveness could depend ultimately on the presence of proinflammatory, NFκ-B-activated [2], IL-8-secreting, monocytes/macrophages [573]. Equally, in the later stages of immune responses, phagocytosis of apoptotic neutrophils could tip the balance back again towards 'antiinflammation'. Thus, neutrophils when in 'proinflammatory mode' may be viewed as 'pre-amplifiers' of a proinflammatory input to the MDU, but when in 'apoptotic mode', may provide an input tending to polarize the MDU away from proinflammation and towards antiinflammation [243, 574]. Mechanisms of innate tolerance are of particular interest in tissues resistant to inflammation such as brain (where microglia constitute the resident macrophages) [575, 576] and the anterior chamber of the eye, where even minor inflammation may be catastrophic. Application of the general paradigm to the conditions discussed in detail is summarized in Figure 10.1. It may be seen that in asthma as in SLE the polarity and balance of the adaptive T cell response determines several significant feedback effects on macrophages, thereby, in essence, modulating macrophage (innate) effector function. Macrophages integrate innate and adaptive inputs, but when the final phenotype is proinflammatory, macrophages themselves become major effector cells, capable of killing microbes and of damaging normal tissue. It appears that eosinophils, although essentially deriving from secondary responses, may display a similar range of 'classical' and 'alternative' activated phenotypes as macrophages. The paradigm emphasizes sequential events in the orderly development of immune activation and rests on the fundamental tenet that, in vivo, changes in the innate immune system, whether pro- or antiinflammatory, both initiate and sustain changes in the adaptive immune system. An inescapable corollary of the paradigm is that the environment, as first filtered and modulated by epithelial barriers, must be the main determinant of innate macrophage and DC phenotypes, whilst macrophages themselves comprise critical signal integrators and effector cells for both innate and adaptive immunity, thereby effectively underpinning the whole gamut of subacute and chronic immune inflammatory responses. DCs transfer groupings of putative antigens to the adaptive immune system but for each individual DC the whole group of 'passenger' antigens is carried to the T cell compartment under either a proinflammatory or antiinflammatory 'banner'. Importantly, a trickle of semi-invariant antigen groupings, from normal cell turnover, carried under the antiinflammatory 'banner' will be translated into tolerogenic T cell responses, constituting an elegant system of continuous immune surveillance to distinguish 'self' from 'non-self'. The system characteristically requires an environmental input to bring about a change in polarity and this principle applies as much to autoimmunity as infection. Cancer would be viewed as an exceptional case in that a failure of immune surveillance allows a mismatch of antiinflammation (tumor tolerance) with dangerous 'altered self'. Apoptosis of all immunologically active cell types will affect the balance of innate tolerance and immunity. The treatment of many immunological diseases is simply to reduce proinflammation and/or increase antiinflammation in the relevant microenvironment. However, in cancer the best treatment may be to correct the mismatch and encourage tumor destructive proinflammatory responses.

Immunology has advanced largely on the basis of clinical observation in humans and experimental data in animal models. Animal models are versatile and can easily be manipulated both genetically and environmentally. As knowledge advances, qualitative functional paradigms of the mammalian immune system have become increasingly more sophisticated. Although new drugs can be designed on the basis of such paradigms and tested initially in animal models there may be distinct advantages to clinical immunology of attempting to develop mathematical models of the immune system. Qualitative models, which have become too complex to be expressed meaningfully in words or diagrams, could better be assembled into mathematical models and rigorously tested. Mathematical modeling is particularly useful to clarify the behavior of what is essentially a complex compartmentalized system in which inputs, outputs, and feedback loops change over time. Defining cell properties and processes mathematically excludes ambiguities and the model is constrained by the requirement for the properties of all cell types and the system as a whole to accord with observational data, even to the extent of explaining paradoxical data. All the major cell types and most influential known bioactive molecules engaged in immune reactions must be represented and must mimic faithfully the behavior of the immune system in health and diverse disease states, ranging from infection to autoimmunity and cancer. Equally, such a model could have a role in drug development. It is particularly exciting to contemplate that a mathematical model could evolve over time from simple beginnings, almost like the very system it is modeling. Even now, gene array studies are starting to describe the probability of individual gene expression mathematically [577]. Mathematical modeling is being applied to the interpretation of DNA microarrays, intracellular signaling pathways, specific diseases [578] and some aspects of immune cell function and interactions [579]. However, a 'working' computerized model of the whole mammalian immune system remains a dauntingly ambitious project. It necessarily depends on close collaboration between immunologist and mathematician and on the application of inductive reasoning to complex experimental data, as in the new specialities of proteomics and bioinformatics. Overall, mathematical modeling should facilitate a better understanding of the convoluted checks and balances of the mammalian immune system and its disruption in disease, a prerequisite for more focused and effective treatments of immune disorders. The paradigm presented here is offered as a small step towards that goal.

Acknowledgments

I thank the library staff of Epsom General Hospital, part of the Epsom and St Helier NHS Trust, for their unfailing support over the past seven years. This chapter is dedicated to my wife, Liz.

References

1 Fearon, D. T., Locksley, R. M. (1996): The instructive role of innate immunity in the acquired immune response. *Science* 272: 50–54.
2 Koay A, Gao X, Washington MK, Parman KS, Sadikot RT, Blackwell TS, Christman JW. (2002): Macrophages are necessary for maximal nuclear factor-kB activation in response to endotoxin. *Am. J. Respir. Cell. Mol. Biol.* 26: 572–578.
3 Rescigno, M. (2002): Dendritic cells and the complexity of microbial infection. *Trends Microbiol.* 10: 425–461.
4 S. Basu, S., R.J. Binder, R. J., R. Suto, R., K.M. Anderson, K. M., P.K. Srivastava, P. K. (2000): Necrotic but not apoptotic cell death releases heat shock proteins, which deliver a partial maturation signal to dendritic cells and activate the NF-kB pathway. *Int. Immunol.* 12: 1539–1546.
5 Sauter, B., Albert, M. L., Francisco, L., Larsson, M., Somersan, S., Bhardwaj, N. (2000): Consequences of cell death: exposure to necrotic tumour cells, but not primary tissue cells or apoptotic cells, induces the maturation of immunostimulatory dendritic cells. *J. Exp. Med.* 191: 423–233.
6 Reis e Sousa, C., S. Hieny, S., T. Scharton-Kersten, T., D. Jankovic, D., H Charest, H., R.N. Germain, R. N., A. Sher, A. (1997): In vivo microbial stimulation induces rapid CD40 ligand-independent production of interleukin 12 by dendritic cells and their redistribution to T cell areas. *J. Exp. Med.* 186: 1819–1829.

7 Vieira PL, de Jong EC, Wierenga EA, Kapsenberg ML, Kalinski P. (2000): Development of Th1-inducing capacity in myeloid dendritic cells requires environmental instruction. *J. Immunol.* 164: 4507–4512.

8 N. Stoy (2001): Macrophage biology and pathobiology in the evolution of immune responses: a functional analysis, *Pathobiology* 69: 179–211.

9 Desmedt, M., Rottiers, P., Dooms, H., Fiers, W., Grooten, J. (1998): Macrophages induce cellular immunity by activating Th1 cell responses and suppressing Th2 cell responses. *J. Immunol.* 160: 5300–5308.

10 Mills, C. D., Kincaid, K. Alt, J. M., Heilman, M. J., Hill, A. M. (2000): M-1/M-2 macrophages and the Th1/Th2 paradigm. *J. Immunol.* 164: 6166–6173.

11 Goerdt, S., Politz, O., Schledzewski, K., Birk, R., Gratchev, A., Guillot, P., Hakiy, N., Klemke, C.-D., Dippel, E., Kodelja, V., Orfanos, C. E. (1999): Alternative versus classical activation of macrophages. *Pathobiology* 67: 222–226.

12 Goerdt, S., Orfanos, C. E. (1999): Other functions, other genes: alternative activation of antigen-presenting cells. *Immunity* 10: 137–142.

13 Anderson, C. F., Mosser, D. M. (2002): A novel phenotype for an activated macrophage: the type 2 activated macrophage. *J. Leukoc. Biol.* 72: 101–106.

14 Welch, J. S., Escoubet-Lozach, L., Sykes, D. B., Liddiard, K., Greaves, D. R., Glass, C. K. (2002): Th2 cytokines and allergic challenge induce Ym1 expression in macrophages by a STAT6-dependent mechanism. *J. Biol. Chem.* 277: 42 821–42 829.

15 De Smedt, T., Van Mechelen, M., De Becker, G., Urbain, J., Leo, O., Moser, M. (1997): Effect of interleukin-10 on dendritic cell maturation and function. *Eur. J. Immunol.* 27: 1229–1235.

16 Shortman, K., Liu, Y.-J. (2002): Mouse and human dendritic cell subtypes. *Nat. Rev. Immunol.* 2: 151–161.

17 Kapsenberg, M. L., Kalinski, P. (1999): The concept of type 1 and type 2 antigen-presenting cells. *Immunol. Lett.* 69: 5–6.

18 P. Kalinski, C.M.U. Hilkens, E.A. Wierenga, M.L. Kapsenberg (1999): T-cell priming by type-1 and type-2 polarized dendritic cells: the concept of a third signal, *Immunol. Today*, 20: 561–567.

19 Belz, G. T., Heath, W. R., Carbone, F. R. (2002): The role of dendritic cell subsets in selection between tolerance and immunity. *Immunol. Cell Biol.*, 80: 463–468.

20 R.M. Steinman, D. Hawiger, M.C. Nussenzweig (2003): Tolerogenic dendritic cells, *Annu. Rev. Immunol.* 21: 685–711.

21 M.P. Everson, D.S. McDuffie, D.G. Lemak, W.J. Koopman, J.R. McGhee, K.W. Beagley (1996): Dendritic cells from different tissues induce production of different T cell cytokine profiles. *J. Leukoc. Biol.*, 59: 494–498.

22 Yamazaki, S., Iyoda, T., Tarbell, K., Olson, K., Velinzon, K., Inaba, K., Steinman, R.M. (2003): Direct expansion of functional $CD25^+$ $CD4^+$ regulatory T cells by antigen-processing dendritic cells. *J. Exp. Med.* 198: 235–247.

23 B.N. Lambrecht, M De Veerman, A.J. Coyle. J.-C. Gutierrez-Ramos, K. Thielemans, R.A. Pauwels (2000): Myeloid dendritic cells induce Th2 responses to inhaled antigen, leading to eosinophiloic airway inflammation. *J. Clin. Invest.* 106: 551–559.

24 W. Lauzon, I. Lemaire (1994): Alveolar macrophage inhibition of lung-associated NK activity: involvement of prostoglandins and transforming growth factor-β_1. *Exp. Lung Res.* 20: 331–349.

25 E. Fireman, A. Onn, Y. Levo, E. Bugolovov, S. Kivity (1999): Suppressive activity of bronchial macrophages recovered by induced sputum. *Allergy* 54: 111–118.

26 A. M. Mowat (2003): Anatomical basis of tolerance and immunity to intestinal antigens. *Nat. Rev. Immunol.* 3: 331–341.

27 Stumbles, P. A., Thomas, J. A., Pimm, C. L., Lee, P. T., Venaille, T. J., Proksch, S., Holt, P. G. (1998): Resting respiratory tract dendritic cells preferentially stimulate T helper cell type 2 (Th2) responses and require obligatory cytokine signals for induction of Th1 immunity. *J. Exp. Med.* 188: 2019–2031.

28 S. Gordon (2003): Alternative activation of macrophages. *Nat. Rev. Immunol.* 3: 23–35.
29 Yoshida, H., Hamano, S., Senaldi, G., Covey, T., Faggioni, R., Mu, S., Xia, M., Wakeham, A. C., Nishina, H., Potter, J., Saris, C. J. M., Mak, T. W. (2001): WSX-1 is required for the initiation of Th1 responses and resistance to L. major infection. *Immunity* 15: 569–578.
30 Y-J Liu (2001): Dendritic cell subsets and lineages, and their functions in innate and adaptive immunity. *Cell* 106: 259–262.
31 H. Hammad, B.N. Lambrecht, P. Pochard, P. Gosset, P. Marquillies, A-B. Tonnel, J. Pestel (2002): Monocyte-derived dendritic cells induce a house dust mite-specific Th2 allergic inflammation in the lung of humanized SCID mice: involvement of CCR7. *J. Immunol.* 169: 1524–1534.
32 R. Steinman, M.C. Nussenzweig (2002): Avoiding horror autotoxicus: the importance of dendritic cells in peripheral T cell tolerance. *Proc. Natl. Acad. Sci. U.S.A.*, 99: 351–358.
33 Walker, L. S., Abbas, A. K. (2002): The enemy within: keeping self-reactive T cells at bay in the periphery, *Nat. Rev. Immunol.*, 2: 11–19.
34 P. Pierre-Jacques, S. Emmanuelle, N. Olivier, D. Fradelizi (2002): Auto-protective redox buffering systems in stimulated macrophages. *BMC Immunol.* 3: 3.
35 I. Komuro, N. Keicho, A. Iwamoto, K.S. Akagawa (2001): Human alveolar macrophages and granulocyte-macrophage colony-stimulating factor-induced monocyte-derived macrophages are resistant to H2O2 via their high basal and inducible levels of catalase activity. *J. Biol. Chem..* 276: 24 360–24 364.
36 Lucas, S., Ghilardi, N., Li, J., de Sauvage, F. J. (2003): IL-27 regulates IL-12 responsiveness of naïve CD4 + T cells through Stat1-dependent and -independent mechanisms. *Proc. Natl. Acad. Sci. U.S.A* 100: 15 047–15 052.
37 Pulendran, B. (2004): Modulating TH1/TH2 responses with microbes, dendritic cells, and pathogen recognition receptors. *Immunol. Res.* 29: 187–196.

38 Tschoep, K., Manning, T.C., Harlin, H., George, C., Johnson, M., Gajewski, T.F. (2003): Disparate functions of immature and mature human myeloid dendritic cells: implications for dendritic cell-based vaccines. *J. Leukoc. Biol.* 74: 69–80.
39 P Loke, MG Nair, J Parkinson, D Guilliano, ML Blaxter, JE Allen (2002): IL-4 dependent alternatively-activated macrophages have a distinctive in vivo gene expression phenotype. *BMC Immunol.* 3: 7.
40 T. Fournier, N Bouach, C. Delafosse, B. Crestani, M. Aubier (1999): Inducible expression and regulation of the a_1-acid glycoprotein gene by alveolar macrophages: prostaglandin E_2 and cyclic AMP act as new positive stimuli, *J. Immunol.* 163: 2883–2890.
41 Hancock A, Armstrong L, Gama R, Millar A (1998): Production of interleukin 13 by alveolar macrophages from normal and fibrotic lung. *Am. J. Respir. Cell Mol. Biol.* 18: 60–65.
42 Lutz, M. B., Schnare, M., Menges, M., Rossner, S., Rollinghoff, M., Schuler, G., Gessner, A. (2002): Differential functions of IL-4 receptor types I and II for dendritic cell maturation and IL-12 production and their dependency on GM-CSF. *J. Immunol.* 169: 3574–3580.
43 V Kodelja, C Müller, O Politz, N Hakij, CE Orfanos, S Goerdt (1998): Alternative macrophage activation-associated CC-chemokine-1, a novel structural homologue of macrophage inflammatory protein-1a with a th-2-associated expression pattern. *J. Immunol.* 160: 1411–1418.
44 M.G. Nair, D.W. Cochrane, J.E. Allen (2003): Macrophages in chronic type 2 inflammation have a novel phenotype characterized by the abundant expression of Ym1 and Fizz1 that can be partly replicated in vitro. *Immunol. Lett.* 85: 173–180.
45 R.K. Coker (2001): Localisation of transforming growth factor b1 and b1 mRNA transcripts in normal and fibrotic lung. *Thorax* 56: 549–556.
46 A Gratchev, K Schledzewski, P Guillot, S Goerdt (2001): Alternatively activated

antigen-presenting cells: molecular repertoire, immune regulation, and healing. *Skin Pharmacol. Appl. Skin Physiol*. 14: 272–279.

47 A Gratchev, P Guillot, N Hakiy, O Politz, CE Orfanos, K Schledzewski, S Goerdt (2001): Alternatively activated macrophages differentially express fibronectin and its splice variants and the extracellular matrix protein βIG-H3. *Scand. J. Immunol*. 53: 386–392.

48 Kepra-Lenhart D, Mistry SK, Wu G, Morris, Jr, SM (2000): Arginase I: a limiting factor for nitric oxide and polyamine synthesis by activated macrophages? *Am. J. Physiol. Regulatory Integrative Comp. Physiol*. 279: R2237–R2242.

49 Mercado, R., Vijh, S., Allen, S.E., Kerksiek, K., Pilip, I. M., Pamer, E. G. (2000): Early programming of T cell populations responding to bacterial infection. *J. Immunol*. 165: 6833–6839.

50 Cohen IR (2000): Discrimination and dialogue in the immune system. *Semin. Immunol*. 12: 215–219.

51 C.A. Janeway Jr., R. Medzhitov (2002): Innate immune recognition, *Annu. Rev. Immunol*. 20: 197–216.

52 Wijburg OLC, Simmons CP, van Rooijen N, Strugnell RA (2000): Dual role for macrophages in vivo in pathogenesis and control of murine Salmonella enterica var. typhimurium infections. *Eur. J. Immunol*. 30: 944–953.

53 N. Seo, S. Hayakawa, Y. Tokura (2002): Mechanisms of immune privilege for tumour cells by regulatory cytokines produced by innate and acquired immune cells. *Cancer Biol*., 12: 291–300.

54 Mantovani, A., Allavena, P., Sica, A. (2004): Tumour-associated macrophages as a prototypic type II polarized phagocytic population: role in tumour progression. *Eur. J. Cancer* 40: 1660–1667.

55 Tzachanis, A., Berezovskaya, L.M., Nadler, V.A., Boussiotis, A. A. (2002): Blockade of B7/CD28 in mixed lymphocyte reaction cultures results in the generation of alternatively activated macrophages, which suppress T-cell responses *Blood* 99: 1465–1473.

56 Pritchard DI, Hewitt C, Moqbel R. (1997): The relationship between immunological responsiveness controlled by T-helper 2 lymphocytes and infections with parasitic helminths. *Parasitology* 115: S33–S44.

57 JT Attwood, DH Munn (1999): Macrophage suppression of T cell activation: a potential mechanism of peripheral tolerance. *Int. Rev. Immunol*. 18: 515–525.

58 K.W. Moore, R. de Waal Malefyt, RL Coffman, A. O'Garra (2001): Interleukin-10 and the interleukin-10 receptor. *Annu. Rev. Immunol*. 19: 683–765.

59 H.L. Weiner (2001): Induction and mechanism of action of transforming growth factor-β-secreting Th3 regulatory cells. *Immunol. Rev*. 182: 207–214.

60 H. Yssel, H. Groux (2000): Characterization of T cell subpopulations involved in the pathogenesis of asthma and allergic diseases. *Int. Arch. Allergy Immunol*. 121: 10–18.

61 G. Lezzi, E. Scotet, D. Scheidegger, A. Lanzavecchia (1999): The interplay between the duration of TCR and cytokine signaling determines the T cell polarization. *Eur. J. Immunol*. 29: 4092–4101.

62 Mc Guirk P, Mills KH. (2000): Direct anti-inflammatory effect of a bacterial virulence factor: IL-10-dependent suppression of IL-12 production by filamentous haemagglutinin from Bordetella pertussis. *Eur. J. Immunol*. 30: 415–422.

63 McGuirk P, McCann C, Mills KH. (2002): Pathogen-specific T regulatory 1 cells induced in the respiratory tract by a bacterial molecule that stimulates interleukin 10 production by dendritic cells: a novel strategy for evasion of protective T helper type 1 responses by Bordetella pertussis. *J. Exp. Med*. 195: 221–231.

64 Liu L, Rich BE, Inobe J-i Chen W, Weiner HL (1998): Induction of Th2 cell differentiation in the primary immune response: dendritic cells isolated from adherent cell culture treated with IL-10 prime naive CD4+ T cells to secrete IL-4. *Int. Immunol*. 10: 1017–1026.

65 H Groux, M. Bigler, J.E. de Vries, M-G. Roncarolo (1996): Interleukin-10

induces a long-term antigen-specific anergic state in human CD4+ T cells, *J. Exp. Med.*, 184: 19–29.
66. Zheng, Z., Narita, M., Takahashi, M., Liu, A., Furukawa, T., Toba, K., Aizawa, Y. (2004): Induction of T cell anergy by the treatment with IL-10-treated dendritic cells. *Comp. Immunol. Microbiol. Infect. Dis.* 27: 93–103.
67. Alard P., Clark, S. L., Kosiewicz, M. M. (2004): Mechanisms of tolerance induced by TGF-β-treated APC: CD4 regulatory T cells prevent the induction of the immune response possibly through a mechanism involving TGF-β. *Eur. J. Immunol.* 34: 1021–1030.
68. Alard P., Clark, S. L., Kosiewicz, M. M. (2003): Deletion, but not anergy, is involved in TGF-β-treated antigen-presenting cell-induced tolerance. *Int. Immunol.* 15: 945–953.
69. Kosiewicz MM, Alard P. (2004): Tolerogenic antigen-presenting cells: regulation of the immune response by TGF-b-treated antigen-presenting cells. *Immunol. Res.* 30: 155–170.
70. Camara, N. O., Sebille, F., Lechler, R. I. (2003): Human $CD4^+CD25^+$ regulatory cells have marked and sustained effects on $CD8^+$ T cell activation. *Eur. J. Immunol.* 33: 3473–3483.
71. I. Mellman, R. Steinman (2001): Dendritic cells: specialized and regulated antigen processing machines. *Cell* 106: 255–258.
72. K.M. Garza, S S Agersborg, E. Baker, K.S.K. Tung (2000): Persistence of physiological self antigen is required for the regulation of self tolerance. *J. Immunol.* 164: 3982–3989.
73. Lambolez F, Jooss K, Vasseur F, Sarukhan A. (2002): Tolerance induction of self antigens by peripheral dendritic cells. *Eur. J. Immunol.* 32: 5288–5297.
74. Steinman RM, Turley S, Mellman I, Inaba K (2000): The induction of tolerance by dendritic cells that have captured apoptotic cells. *J. Exp. Med.* 191: 411–416.
75. M.B. Lutz, G. Schuler (2002): Immature, semi-mature and fully mature dendritic cells: which signals induce tolerance or immunity? *Trends Immunol.* 23: 445–449.
76. Hawiger, D., Inaba, K., Dorsett, Y., Gou, M., Mahnke, K., Rivera, M., Ravetch, J. V., Steinman, R. M., Nussenzweig, M. C. (2001): Dendritic cells induce peripheral T cell unresponsiveness under steady state conditions in vivo. *J. Exp. Med.* 194: 769–779.
77. C. Nagler-Anderson (2001): Man the barrier! Strategic defences in the intestinal mucosa. *Nat. Rev. Immunol.* 1: 59–67.
78. Chitnis, T., Salama, A. D., Grusby, M. J., Sayegh, M. H., Khoury, S. J. (2004): Defining Th1 and Th2 immune responses in a reciprocal cytokine environment in vivo. *J. Immunol.* 172: 4260–4265.
79. U. Yrlid, M Svensson, C. Johansson, M. J. Wick (2000): Salmonella infection of bone marrow-derived macrophages and dendritic cells: influence on antigen presentation and initiating an immune response. *FEMS Immunol. Med. Biol.* 27: 313–320.
80. F-P Huang, N. Platt, M. Wykes, J.R. Major, T.J. Powell, C.D. Jenkins, G.G. MacPherson (2000): A discrete subpopulation of dendritic cells transports apoptotic intestinal epithelial cells to T cell areas of mesenteric lymph nodes. *J. Exp. Med.* 191: 435–443.
81. D. Grdic, R. Smith, A. Donachie, M. Kjerrulf, H. Hornquist, A. Mowat, N. Lycke (1999): The mucosal adjuvant effects of cholera toxin and immune-stimulating complexes differ in their requirement for IL-12, indicating different pathways of action. *Eur. J. Immunol.* 29: 1774–1784.
82. A. Boonstra, A. van Oudenaren, B. Barendregt, L. An, P.J.M. Leenen, H.F.J. Savelkoul (2000): UVB irradiation modulates systemic immune responses by affecting cytokine production of antigen-presenting cells. *Int. Immunol.* 12: 1531–1538.
83. Ito, T., Amakawa, R., Inaba, M., Ikehara, S., Inaba, K., Fukuhara, S. (2001): Differential regulation of human blood dendritic cell subsets by IFNs. *J. Immunol.* 166: 2961–2969.

84. Katz-Levy, Y., Neville, K. L., Girvin, A. M., Vanderlugt, C. L., Pope, J. G., Tan, L. J., Miller, S. D. (1999): Endogenous presentation of self myelin epitopes by CNS-resident APCs in Theiler's virus-infected mice. *Clin Invest.* 104: 599–610.
85. B.N. Lambrecht, H. Hammad (2002): Myeloid dendritic cells make it to the top. *Clin. Exp. Allergy* 32: 805–810.
86. Z. Liu, K. Geboes, S. Colpaert, G.R. D'Haens, P. Rutgeerts, J.L. Ceuppens (2000): IL-15 is highly expressed in inflammatory bowel disease and regulates local T cell-dependent cytokine production. *J. Immunol.* 164: 3608–3615.
87. Nishikomori R, Ehrhardt RO, Strober W (2000): T helper type 2 cell differentiation occurs in the presence of interleukin 12 receptor β2 chain expression and signalling. *J. Exp. Med*.191: 847–858.
88. K.J. Maloy, L. Salaun, R. Cahill, G. Dougan, N.J.Saunders, F. Powrie (2003): CD4+CD25+ Tr cells suppress innate immune patology through cytokine-dependent mechanisms. *J. Exp. Med.* 197: 111–119.
89. A. Shnyra, R. Brewington, A. Alipio, C. Amura, D.C. Morrison (1998): Reprogramming of lipopolysaccharide-primed macrophages is controlled by a counter-balanced production of IL-10 and IL-12. *J. Immunol.* 160: 3729–3736.
90. D.T. Denhardt, M Noda, A.W. O'Regan, D. Pavlin, J.S. Berman (2001): Osteopontin as a means to cope with environmental insults: regulation of inflammation, tissue remodeling, and cell survival. *J. Clin. Invest.* 107: 1055–1061.
91. S Ashkar, G.F. Weber, V. Panoutsakopoulou, ME Sanchirico, M. Jansson, S. Zawaideh, SR Rittling, D.T. Denhardt, M.J. Glimcher, H Cantor (2000): Eta-1 (osteopontin): an early component of type-1 (cell-mediated) immunity. *Science* 287: 860–864.
92. J.M. Weiss, A.C. Renkl, C.S. Mair, M. Kimmig, L.Liaw, T.Ahrens, S. Kon, M Maeda, H. Hotta, T. Uede, J.C. Simon (2001): Osteopontin is involve in the initiation of cutaneous contact hypersensitivity by inducing Langerhans and dendritic cell migration to lymph nodes. *J. Exp. Med.* 194: 1219–1229.
93. Nau, G. J., Liaw, L., Chupp, G. L., Berman, J. S., Hogan, B. L. M., Young, R. A. (1999): Attenuated host resistance against Mycobacterium bovis BCG infection in mice lacking osteopontin. *Infect. Immun.* 67: 4223–4230.
94. Nau, G. J., Chupp, G. L., Emile, J-F., Jouanguy, E., Berman, J. S., Casanova, J-L., Young, R. A. (2000): Osteopntin expression correlates with clinical outcome in patients with mycobacterial infection. *Am. J. Pathol.* 157: 37–42.
95. M. Mohamadzadeh, F. Berard, G Essert, C. Chalouni, B. Pulendran, J. Davoust, G. Bridges, A.K. Palucka, J. Banchereau (2001): Interleukin 15 skews monocyte differentiation into dendritic cells with features of Langerhans cells, *J. Exp. Med*. 194: 1013–1019.
96. Z. Su, M.M. Stevenson (2002): IL-12 is required for antibody-mediated protective immunity against blood-stage Plasmodium chabaudi AS malaria infection in mice. *J. Immunol.* 168: 1348–1355.
97. Streilein J W (1999): Immunoregulatory mechanisms of the eye. *Prog. Retinal Eye Res.* 18: 357–370.
98. G.F. Weber, S. Zawaideh, S. Hikita, V.A. Kumar, H. Cantor, S. Ashkar (2002): Phosphorylation-dependent interaction of osteopontin with its receptors regulates macrophage migration and activation. *J. Leukoc. Biol.* 72: 752–761.
99. van der Pouw Kraan TCTM, Boeije LCM, Smeenk RJT, Wijdenes J, Aarden LA (1995): Prostaglandin-E_2 is a potent inhibitor of human interleukin 12 production. *J. Exp. Med.* 181: 775–779.
100. T. Hanada, A. Yoshimura (2002): Regulation of cytokine signaling and inflammation. *Cytokine Growth Factor Rev.* 13: 413–421.
101. Janssen EM, Wauben MH, Nijkamp FP, van Eden W, van Oosterhout AJ. (2002): Immunomodulatory effects of antigen-pulsed macrophages in a murine model of allergic asthma. *Am. J. Respir. Cell Mol. Biol.* 27: 257–264.
102. Gerber JS, Mosser DM. (2001): Stimulatory and inhibitory signals originating

from the macrophage Fcγ receptors. *Microbes Infection* 3: 131–139.

103 Strassmann G, Patil-Koota V, Finkelman F, Fong M, Kambayashi T (1994): Evidence of the involvement of interleukin 10 in the differential deactivation of murine peritoneal macrophages by prostaglandin E_2. *J. Exp. Med.* 180: 2365–2370.

104 R.P. Phipps, S.H. Stein, R.L. Roper (1991): A new view of prostaglandin E regulation of the immune response. *Immunol. Today*, 12: 349–352.

105 Guo, C.Q. Cai, R.A. Schroeder, P.C. Kuo (2001): Osteopontin is a negative regulator of nitric oxide synthesis in murine macrophages. *J. Immunol.* 166: 1079–1086.

106 F. Takahashi, K. Takahashi, T. Okaxaki, K. Maeda, H. Ienaga, M. Maeda, S. Kon, T. Uede, Y. Fukuchi (2001): Role of osteopontin in the pathogenesis of bleomycin-induced pulmonary fibrosis. *Am. J. Respir. Cell Mol. Biol.*, 24: 264–271.

107 N. Hill, N. Sarvetnick (2002): Cytokines: promoters and dampeners of autoimmunity. *Curr. Opin. Immunol.* 14: 791–797.

108 P. Loke, J.P. Allison (2003): PD-L1 and PD-L2 are differentially regulated by Th1 and Th2 cells. *Proc. Natl. Acad. Sci. U.S.A.* 100: 5336–5341.

109 Lee, G. K., Park, H. J., Macleod, M., Chandler, P., Munn, D. H., Mellor, A. L. (2002): Tryptophan deprivation sensitizes activated T cells to apoptosis prior to cell division. *Immunology* 107: 452–460.

110 Mellor, A. L., Munn, D. H. (2003): Tryptophan catabolism and regulation of adaptive immunity. *J. Immunol.* 170: 5809–5813.

111 Grohman, U., Fallarino, F., Puccetti, P. (2003): Tolerance, DCs and tryptophan: much ado about IDO. *Trends Immunol.* 24: 242–248.

112 Mukhopadhyay, M. Mohanty, A. Mangla, A George, V. Bal, S. Rath, B. Ravindran (2002): Macrophage effector functions controlled by Bruton's tyrosine kinase are more crucial than the cytokine balance of T cell responses for microfilarial clearance. *J. Immunol.* 168: 2914–2921.

113 WR Heath, FR Carbone (2001): Cross-presentation, dendritic cells, tolerance and immunity. *Annu. Rev. Immunol.* 531.

114 S. Cobbold, H. Waldmann (1998): Infectious tolerance. *Curr. Opin. Immunol.* 10: 518–524.

115 Poulter, L. W., Burke, C. M. (1996): Macrophages and allergic lung disease. *Immunobiology* 195: 574–587.

116 Arras, M., Haux, F., Vink, A., Delos, M., Coutelier, J-P., Many, M.-C., Barbarin, V., Renauld, J.-C., Lison, D. (2001): Interleukin-9 reduces lung fibrosis and type 2 immune polarization induced by silica particles in a murine model. *Am. J. Respir. Cell Mol. Biol.* 24: 368–375.

117 Büttner, C., Skupin, A., Reimann, T., Rieber, E. P., Unteregger, G., Geyer, P., Frank, K.-H. (1997): Local production of interleukin-4 during radiation-induced pneumonitis and pulmonary fibrosis in rats: macrophages as a prominent source of IL-4. *Am. J. Respir. Cell Mol. Biol.* 17: 315–325.

118 Heath, V. L., Murphy, E. E., Crain, C., Tomlinson, M. G., O'Garra, A. (2000): TGF-β1 down-regulates Th2 development and results in decreased IL-4-induced STAT6 activation and GATA-3 expression. *Eur. J. Immunol.* 30: 2639–2649.

119 Matsukawa, A., Hogaboam, C. M., Lukacs, N. W., Lincoln, P. M., Evanoff, H.L., Strieter, R. M., Kunkel, S. L. (2000): Expression and contribution of endogenous IL-13 in an experimental model of sepsis. *J. Immunol.* 164: 2738–2744.

120 Drechsler, Y., Chavan, S., Catalano, D., Mandrekar, P., Szabo, G. (2002): FcγR cross-linking mediates NF-κB activation, reduced antigen-presentation capacity, and decreased IL-12 production in monocytes without modulation of myeloid dendritic cell development. *J. Leukoc. Biol.* 72: 657–667.

121 Sutterwala, F. S., Noel, G. J., Salgame, P., Mosser D. M. (1998): Reversal of proinflammatory responses by ligating

122 Sutterwala, F. S., Noel, G. J., Clynes, R., Mosser D. M. (1997): Selective suppression of interleukin-12 iduction after macrophage receptor ligation. *J. Exp. Med.* 185: 1977–1985.

the macrophage Fcγ receptor type 1. *J. Exp. Med.* 188: 217–222.

123 Anderson, C. F., Mosser D. M. (2002): Cutting edge: biasing immune responses by directing antigen to Fcγ receptors. *J. Immunol.* 168: 3697–3701.

124 Kane, M. M., Moser, D. M. (2000): *Leishmania* parasites and their ploys to disrupt macrophage activation. *Curr. Opin. Hematol.* 7: 26–31.

125 Hausmann, M., Kiessling, S., Mestermann, S., Webb, G., Spöttl, T., Andus, T., Schölmerich, J., Herfarth, H., Ray, K., Falk, W., Rogler, G. (2002): Toll-like receptors 2 and 4 are up-regulated during intestinal inflammation. *Gastroenterology* 122: 1987–2000.

126 Martinez-Pomares, L., Reid, D. M., Brown, G. D., Taylor, P. R., Stillion, R. J., Linehan, S.A., Zamze, S., Gordon, S., Wong, S. Y. C. (2003): Analysis of mannose receptor regulation by IL-4, IL-10, and proteolytic processing using novel monoclonal antibodies. *J. Leukoc. Biol.* 73: 604–613.

127 Aderem, A., Underhill, D. M. (1999): Mechanisms of phagocytosis in macrophages. *Annu. Rev. Immunol.* 17: 593–623.

128 Zamze, S., Martinez-Pomares, L., Jones, H., Taylor, P. R., Stillion, R. J., Gordon, S., Wong, S. Y. C. (2002): Recognition of bacterial capsular polysaccharides and lipopolysaccharides by the macrophage mannose receptor. *J. Biol. Chem.* 277: 41 613–41 623.

129 Marzolo, M. P., von Bernhardi, R., Inestrosa, N. C. (1999): Mannose receptor is present in a functional state in rat microglial cells. *J. Neurosci. Res.* 58: 387–395.

130 Gregory, C. D., Devitt, A. (2004): The macrophage and the apoptotic cell: an innate immune interaction viewed simplistically? *Immunology* 113: 1–14.

131 Burkart, V., Kim,Y.-E., Hartmann, B., Ghiea, I., Syldath, U., Kauer, M., Fingberg, W., Hanifi-Moghaddam, P., Müller, S., Kolb. H. E. (2002): Cholera toxin B pretreatment of macrophages and monocytes diminishes their proinflammatory responsiveness to lipopolysaccharide. *J. Immunol.* 168: 1730–1737.

132 Braun, M. C., He, J., Wu, C-Y., Kelsall, B. L. (1999): Cholera toxin suppresses interleukin (IL)-12 production and IL-12 receptor b1 and b2 chain expression. *J. Exp. Med.* 189: 541–552.

133 Bagley, K. C., Abdelwahab, S. F., Tuskan, R. G., Lewis, G. K. (2003): An enzymatically active A domain is required for cholera-like enterotoxins to induce a long-lived blockade on the induction of oral tolerance: new method for screening mucosal adjuvants. *Infect. Immun.* 71: 6850–6856.

134 Ryan, E. J., McNeela, E., Pizza, M., Rappuoli, R., O'Neill, L., Mills K. H. G. (2000): Modulation of innate and acquired immune responses by Escherichia coli heat-labile toxin: distinct pro- and anti-inflammatory effects of the nontoxic AB complex and the enzyme activity. *J. Immunol.* 165: 5750–5759.

135 Allen, J. E., Loke, P. (2001): Divergent roles for macrophages in lymphatic filariasis. *Parasite Immunol.* 23: 345–352.

136 Goodridge, H. S., Wilson, E. H., Harnett, W., Campbell, C.C., Harnett, M. M., Liew, F. Y. (2001): Modulation of macrophage cytokine production by ES-62, a secreted product of the filarial nematode Acanthocheilonema viteae. *J. Immunol.* 167: 940–945.

137 Whelan, M., Harnett, M. H., Houston, K. M., Patel, V., Harnett, W., Rigley, K. P. (2000): A filarial nematode-secreted product signals dendritic cells to acquire a phenotype that drives development of Th2 cells. *J. Immunol.* 164: 6453–6460.

138 Mizel, S. B., Snipes, J. A. (2002): Gram-negative flagellin-induced self-tolerance is associated with a block in interleukin-1 receptor-associated kinase release from Toll-like receptor 5. *J. Biol. Chem.* 277: 22 414–22 420.

139 Barral, A., Barral-Netto, M., Yong, E. C., Brownell, C. E., Twardzik, D. R., Reed, S. G. (1993): Transforming growth factor β as a virulence mechanism for

Leishmania braziliensis. *Proc. Natl. Acad. Sci. U.S.A.* 90: 3442–3446.

140 Melby, P. C., Tabares, A., Restrepo, B. I., Cardona, A. E., McGuff, E. S., Teale, J. M. (2001): *Leishmania donovani*: evolution and architecture of the splenic cellular immune response related to control of infection, *Exp. Parasitol.* 99: 17–25.

141 Reed, S. G. (1999): TGF-β in infections and infectious diseases. *Microbes Infect.* 1: 1313–1325.

142 Denkers, E. Y., Gazzinelli, R. T. (1998): Regulation and function of T-cell-mediated immunity during *Toxoplasma gondii* infection. *Clin. Microbiol. Rev.* 11: 569–588.

143 Letterio, J.J., Roberts, A.B. (1998): Regulation of immune responses by TGF-β. *Annu. Rev. Immunol.* 16: 137–161.

144 Shapira, S., Speirs, K., Gerstein, A., Caamano, J., Hunter, C. A. (2002): Suppression of NF-κB activation by infection with *Toxoplasma gondii*. *J. Infect. Dis.* 185 (suppl 1): S66–S72.

145 Butcher, B. A., Kim, L., Johnson, P. F., Denkers, E. Y. (2001): *Toxoplasma gondii* tachzoites inhibit proinflammatory cytokine induction in infected macrophages by preventing translocation of the transcription factor NF-κB. *J. Immunol.* 167: 2193–2201.

146 Rodriguez-Sosa, M., Satoskar, A. R., Calderón, R., Gomez-Garcia, L., Saavedra, R., Bojalil, R., Terrazas, L. I. (2002): Chronic helminth infection induces alternatively activated macrophages expressing high levels of CCR5 with low interleukin-12 production and TH2-biasing ability. *Infect. Immun.* 70: 3656–3664.

147 Balcewicz-Sablinska, M. K., Gan, H. X., Remold, H. G. (1999): Interleukin 10 produced by macrophages inoculated with Mycobacterium avium attenuates mycobacteria-induced apoptosis by reduction of TNF-a activity. *J. Infect. Dis.* 180: 1230–1237.

148 Loke, P., MacDonald, A. S., Allen, J. E. (2000): Antigen-presenting cells recruited by Brugia malayi induce Th2 differentiation of naive CD4[+] T cells. *Eur. J. Immunol.* 30: 1127–1135.

149 Falcone, F.H., Loke, P., Zang, X., MacDonald. A. S., Maizels, R. M., Allen, J. E. (2001): A Brugia malayi homolog of macrophage migration inhibitory factor reveals an important link between macrophages and eosinophil recruitment during nematode infection. *J. Immunol.* 167: 5348–5354.

150 Fadok, V. A., Bratton, D. L., Henson, P. M. (2001): Phagocyte receptors for apoptotic cells: recognition, uptake, and consequences. *J. Clin. Invest.* 108: 957–962.

151 Geske, J. F., Monks, J., Lehman, L., Fadok, V. A. (2002): The role of the macrophage in apoptosis: hunter gatherer and regulator. *Int. J. Hematol.* 76: 16–26.

152 Hu, B., Punturieri, A., Todt, J., Sonstein, J., Polak, T., Curtis, J. L. (2002): Recognition and phagocytosis of apoptotic T cells by resident tissue macrophages require multiple signal transduction events. *J. Leukoc. Biol.* 71: 881–889.

153 Hoffmann, P. R., deCathelineau, A. M., Ogden, C. A., Leverrier, Y., Bratton, D. L., Daleke, D. L., Ridley, A. J., Fadok, V. A., Henson, P. M. (2001): Phosphatidylserine (PS) induces PS receptor-mediated macropinocytosis and promotes clearance of apoptotic cells. *J. Cell Biol.* 155: 649–659.

154 Henson, P. M., Bratton, D. L., Fadok, V. A. (2001): The phosphatidyl receptor: a crucial molecular switch? *Nat. Rev. Mol. Cell. Biol.* 2: 627–633.

155 Böse, J., Gruber, A. D., Helming, L., Schiebe, S., Wegener, I., Hafner, M., Beales, M., Köntgen, F., Lengeling, A. (2004): The phosphatidylserine receptor has essential functions during embryogenesis but not in apoptotic cell removal. *J. Biol.* 3: 15.

156 Fadok, V. A., Bratton, D. L., Konowai, A., Freed, P. W., Westcott, J. Y., Henson P. M. (1998): Macrophages that have ingested apoptotic cells in vitro inhibit proinflammatory cytokine production through autocrine/paracrine mechanisms involving TGFβ, PGE$_2$, and PAF. *J. Clin. Invest.* 101: 890–898.

157 Voll, R. E., Hermann, M., Roth, E. A., Stach, C., Kalden, J. R., Girkontaite, I.

(1997): Immunosuppressive effects of apoptotic cells. *Nature* 390: 350–351.
158 McDonald, P. P., Fadok, V. A., Bratton, D., Henson, P.M. (1999): Transcriptional and translational regulation of inflammatory mediator production by endogenous TGF-β in macrophages that have ingested apoptotic cells. *J. Immunol.*163: 6164–6172.
159 Huynh, M.-L. N., Fadok, V. A., Henson, P. M. (2002): Phosphatidylserine-dependent ingestion of apoptotic cells promotes TGF-β1 secretion and the resolution of inflammation. *J. Clin. Invest.* 109: 41–50.
160 Sato, S., Takeuchi, O., Fujita, T., Tomizawa, H., Takeda, K., Akira, S. (2002): A variety of microbial components induce tolerance to lipopolysaccharide by differentially affecting MyD88-dependent and -independent pathways. *Int. Immunol.* 7: 783–791.
161 Varma, T. K., Toliver-Kinsky, T. E., Lin, C. Y., Koutrouvelis, A. P., Nichols, J. E., Sherwood, E. R. (2001): Cellular mechanisms that cause suppressed gamma interferon secretion in endotoxin-tolerant mice. *Infect. Immun.* 69: 5249–5263.
162 Karp, C. L., Wysocka, M., Ma, X., Marovich M., Factor, R.E., Nutman, T., Armant, M., Wahl, L., Cuomo, P., Trinchieri G. (1998): Potent suppression of IL-12 production from monocytes and dendritic cells during endotoxin tolerance. *Eur. J. Immunol.* 28: 3128–3136.
163 Wysocka, M., Robertson, S., Riemann, H., Caamano, J., Hunter, C., Mackiewicz, A., Montana, L. J., Trinchieri, G., Karp, C. L. (2001): IL-12 suppression during experimental endotoxin tolerance: dendritic cell loss and macrophage hyporesponsiveness. *J. Immunol.* 166: 7504–7513.
164 Corraliza, I. M., Soler, G., Eichmann, K., Modolell, M. (1995): Arginase induction by suppressors of nitric oxide synthesis (IL-4, IL-10, PGE$_2$) in murine bone-marrow-derived macrophages. *Biochem. Biophys. Res. Commun.* 206: 667–673.
165 Sonoki, T., Nagasaki, A., Gotoh, T., Takiguchi, M., Takeya, M., Matsuzaki, H., Mori, M. (1997): Coinduction of nitric oxide synthase and arginase I in cultured rat peritoneal macrophages and rat tissues in vivo by lipopolysaccharide. *J. Biol. Chem.* 272: 3689–3693.
166 Chang, C.-I., Zoghi, B., Liao, J. C., Kuo, L. (2000): The involvement of tyrosine kinases, cyclic AMP/protein kinase A, and p38 mitogen-activated protein kinase in IL-13-mediated arginase 1 induction in macrophages: its implications in IL-13-inhibited nitric oxide production. *J. Immunol.* 165: 2134–2141.
167 Vodovotz, Y., Chesler, L., Chong, H., Kim, S. J., Simpson, J. T., DeGraff, W., Cox, G. W., Roberts, A. B., Wink, D. A., Barcellos-Hoff. M. H. (1999): Regulation of transforming growth factor β1 by nitric oxide. *Cancer Res.* 59: 2142–2149.
168 Boutard, V., Havouis, R., Fouqueray, B., Philippe, C., Moulinoux, J.-P., Baud L. (1995): Transforming growth factor-β stimulates arginase activity in macrophages: implications for regulation of macrophage cytotoxicity. *J. Immunol.* 155: 2077–2084.
169 Sirois, J., Bissonnette, E. Y. (2001): Alveolar macrophages af allergic resistant and susceptible strains of rats show distinct cytokine profiles. *Clin. Exp. Immunol.* 126: 9–15.
170 Lentsch, A. B., Shanley, T.P., Sarma, V., Ward, P. A (1997): In vivo suppression of NF-κB and preservation of IκBa by interleukin-10 and interleukin-13. *J. Clin. Invest.* 100: 2443–2448.
171 Woods, J. M, Katschke, Jr., K. J., Tokuhira, M., Kurata, H., Arai, K.-I., Campbell, P. L., Koch, A.E. (2000): Reduction of inflammatory cytokines and prostaglandin E$_2$ by IL-13 gene therapy in rheumatoid arthritis synovium. *J. Immunol.* 165: 2755–2763.
172 Wang. P., Wu, P., Siegel, M. I., Egan, R. W., Billah, M. M. (1995): Interleukin (IL)-10 inhibits nuclear factor κB (NFκB) activation in human monocytes. IL-10 and IL-4 suppress cytokine synthesis by different mechanisms. *J. Biol. Chem.* 270: 9558–9563.
173 Munder, M., Eichmann, K., Modolell, M. (1998): Alternative metabolic states in murine macrophages reflected by the

173 ... nitric oxide synthase/arginase balance: competitive regulation by CD4$^+$ T cells correlates with Th1/Th2 phenotype. *J. Immunol.* 160: 5347–5354.

174 Wang, W. W., Jenkinson, C. P., Griscavage, J. M., Kern, R. M., Arabolos, N. S., Byrns, R. E., Cederbaum, S. D., Ignarro, L. J. (1995): Coinduction of arginase and nitric oxide synthase in murine macrophages activated by lipopolysaccharide. *Biochem. Biophys. Res. Commun.* 210: 1009–1016.

175 Stempin, C. C., Tanos, T. B., Coso, O. A., Cerban, F. M. (2004): Arginase induction promotes *Trypanosoma cruzi* intracellular replication in cruzipain-treated J774 cells through the activation of multiple signaling pathways. *Eur. J. Immunol.* 34: 200–209.

176 Barksdale, A. R., Bernard, A. C., Maley, M. E., Gellin, G. L., Kearney, P. A., Boulanger, B. R., Tsuei, B. J., Ochoa, J. B. (2003): Regulation of arginase expression by T-helper II cytokines and isoproterenol. *Surgery* 135: 527–535.

177 Lee, J., Ryu, H., Ferrante, R. J., Morris, S. M., Jr., Ratan, R. R. (2003): Translational control of inducible nitric oxide synthase expression by arginine can explain the arginine paradox. *Proc. Natl. Acad. Sci. U.S.A.* 100: 4843–4848.

178 El-Gayar, S., Thüring-Nahler, H., Pfeilschifter, J., Röllinghoff, M., Bogdan, C. (2003): Translational control of inducible nitric oxide synthase by IL-13 and arginine availability in inflammatory macrophages. *J. Immunol.* 171: 4561–4568.

179 Pauleau, A-L., Rutschman, R., Lang, R., Pernis, A., Watowich, S. S., Murray, P. J. (2004): Enhancer-mediated control of macrophage-specific arginase I expression. *J. Immunol.* 172: 7565–7573.

180 Khurana Hershey, G. K. (2003): IL-13 receptors and signaling pathways: an evolving web. *J. Allergy Clin. Immunol.* 111: 677–690.

181 Sinha, P., Clements, V. K., Ostrand-Rosenberg, S. (2005): Induction of myeloid-derived suppressor cells and induction of M1 macrophages facilitate the rejection of established metastatic disease. *J. Immunol.* 174: 636–645.

182 Matsukawa, A., Kaplan, M. H., Hogaboam, C. M., Lukacs, N. W., Kunkel, S. L. (2001): Pivotal role of signal transducer and activator of transcription (Stat)4 and Stat6 in the innate immune response during sepsis. *J. Exp. Med.* 193: 679–688.

183 Min, B., Legge, K. L., Bell, J. J., Gregg, R. K., Li, L., Caprio, J. C., Zaghouani, H. (2001): Neonatal exposure to antigen induces a defective CD40 ligand expression that undermines both IL-12 production by APC and IL-2 receptor upregulation on splenic T cells and perpetuates IFN-γ-dependent T cell anergy. *J. Immunol.* 166: 5594–5603.

184 Chang, H.-C., Zhang, S., Kaplan, M.H. (2002): Neonatal tolerance in the absence of Stat4- and Stat6-dependent Th cell differentiation. *J. Immunol.* 169: 4124–4128.

185 Shi, H. N., Grusby, M. J., Nagler-Anderson, C. (1999): Orally induced peripheral nonresponsiveness is maintained in the absence of functional Th1 or Th2 cells. *J. Immunol.* 162: 5143–5148.

186 Mowat, A. McI., Steel, M., Leishman, A. J., Garside, P. (1999): Normal induction of oral tolerance in the absence of a functional IL-12-dependent IFN-γ signaling pathway. *J. Immunol.* 163: 4728–4736.

187 Pack, C. D., Cestra, A. E., Min, B., Legge, K. L., Li, L., Caprio-Young, J. C., Bell, J. J., Gregg, R. K., Zaghouani, H. (2001): Neonatal exposure to antigen primes the immune system to develop responses to various lymphoid organs and promotes bystander regulation of diverse T cell specificities. *J. Immunol.* 167: 4187–4195.

188 Janeway, C. A. Jr. (2001): How the immune system works to protect the host from infection: a personal view. *Proc. Natl. Acad. Sci. U.S.A.* 98: 7461–7468.

189 Nio, J., Fujimoto, W., Konno, A., Kon, Y., Owhashi, M., Iwanaga, T. (2004): Cellular expression of murine Ym1 and Ym2, chitinase family proteins, as revealed by in situ hybridization and immunochemistry. *Histochem. Cell Biol.* 121: 473–482.

190 Chang, N.-Ca., Hung, S.-I., Hwa, K.-Y., Kato, I., Chen, J.-E., Liu, C.-H., Chang, A. C. (2001): A macrophage protein, Ym1, transiently expressed during inflammation is a novel mammalian lectin. *J. Biol. Chem.* 276: 17 497–17 506.

191 Noel, W., Hassanzadeh, G., Raes, G., Namangala, B., Daems I, Brys L, Brombacher, F., Baetselier, P. D., Beschin, A. (2002): Infection stage-dependent modulation of macrophage activation in Trypanosoma congolese-resistant and -susceptible mice. *Infect. Immun.* 70: 6180–6187.

192 Donnelly, S., O'Niell, S. M., Sekiya, M., Mulcahy, G., Dalton, J. P. (2005): Thioredoxin peroxidase secreted by Fasciola hepatica induces the alternative activation of macrophages. *Infect. Immun.* 73: 166–173.

193 Nair, M. G., Gallagher, I. J., Taylor, M. D., Loke, P. Coulson, P. S., Wilson, R. A., Maizels, R. M., Allen, J. E. (2005): Chitinase and Fizz family members are a generalized feature of nematode infection with selective upregulation of Ym1 and Fizz1 by antigen-presenting cells. *Infect. Immun.* 73: 385–394.

194 Pagliari, L. J., Perlman, H., Liu, H., Pope, R. M. (2000): Macrophages require constitutive NF-κB activation to maintain A1 expression and mitochondrial homeostasis. *Mol. Cell. Biol.* 20: 8855–8865.

195 Mitchell, R. A., Liao, H., Chesney, J., Fingerle-Rowson, G., Baugh, J., David, J., Bucala, R. (2002): Macrophage migration inhibitory factor (MIF) sustains macrophage proinflammatory function by inhibiting p53: regulatory role in the innate immune response. *Proc. Natl. Acad. Sci. U.S.A.* 99: 345–350.

196 Calandra, T., Spiegel, L. A., Metz, C. N., Bucala, R. (1998): Macrophage migratory inhibitory factor is a critical mediator of the activation of immune cells by exotoxins of Gram-positive bacteria. *Proc. Natl. Acad. Sci. U.S.A.* 95: 11 383–11 388.

197 Hsu, Y.-W., Chi, K.-H., Huang, W.-W., Lin, W-W. (2001): Ceramide inhibits lipopolysaccharide-mediated nitric oxide synthase and cyclooxygenase-2 induction in macrophages: effects on protein kinases and transcription factors. *J. Immunol.* 166: 5388–5397.

198 Lin, C-F., Chen, C-L., Chang, W-T., Jan, M-S., Hsu, L-J., Wu, R-H., Tang, M-J., Chang, W-C., Lin, Y-S. (2004): Sequential caspase-2 and caspase-8 activation upstream of mitochondria during ceramide- and etoposide-induced apoptosis. *J. Biol. Chem.* 279: 40 755–40 761.

199 Munn, D. H., Beall, A. C., Song, D., Wrenn, R. W., Throckmorton, D. C. (1995): Activation-induced apoptosis in human macrophages: developmental regulation of a novel cell death pathway by macrophage colony-stimulating factor and interferon γ. *J. Exp. Med.* 181: 127–136.

200 Gômez-Muñoz, A., Kong, J., Salh, B., Steinbrecher, U. P. (2003): Sphingosine-1-phosphate inhibits acid sphingomyelinase and blocks apoptosis in macrophages. *FEBS Lett.* 539: 56–60.

201 Gômez-Muñoz, A., Kong, J., Salh, B., Steinbrecher, U. P. (2004): Ceramide-1-phosphate blocks apoptosis through inhibition of acid sphingomyelinase. *J. Lipid Res.* 45: 99–105.

202 Monick, M. M., Mallampalli, R. K., Bradford, M., McCoy, D., Gross, T. J., Flaherty, D. M., Powers, L. S., Cameron, K., Kelly, S., Merrill, A. H., Jr., Hunninghake, G. W. (2004): Cooperative prosurvival activity by ERK and Akt in human alveolar macrophages is dependent on high levels of ceramidase activity. *J. Immunol.* 173: 123–135.

203 Kim, S. O., Ono, K., Tobias, P.S., Han, J. (2003): Orphan nuclear receptor Nur77 is involved in caspase-independent macrophage cell death. *J. Exp. Med.* 197: 1441–1452.

204 Preston, P. M., Dargbouth, M., Boulter, N. R., Hall, F. R., Tall, R., Kirvar, E., Brown, C. G. D. (2002): A dual role for immunosuppressor mechanisms in infection with Theileria annulata: well-regulated suppressor macrophages help in recovery from infection: profound immunosuppression promotes non-healing disease. *Parasitol. Res.* 88: 522–534.

205 Rhee, S. H., Hwang, D. (2000): Murine TOLL-like receptor 4 confers lipopolysaccharide responsiveness as determined by activation of NF-κB and expression of the inducible cyclooxygenase. *J. Biol. Chem.* 275: 34 035–34 040.

206 Moore, B. B., Coffey, M. J., Christensen, P., Sittering, S., Ngan, R., Wilke, C.A., McDonald, R., Phare, S. M., Peters-Golden, M., Paine III, R., Toews, G. B. (2000): GM-CSF regulates bleomycin-induced pulmonary fibrosis via a prostaglandin-dependent mechanism. *J. Immunol.* 165: 4032–4039.

207 Suzuki, T., Hashimoto, S.-I., Toyoda, N., Nagai, S., Yamazaki, N., Dong, H.-Y., Sakai, J., Yamashita, T., Nukiwa, T., Matsushima, K. (2000): Comprehensive gene expression profile of LPS-stimulated human monocytes by SAGE. *Blood* 96: 2584–2591.

208 J-J. Lee, M. Takei, S. Hori, Y. Inoue, Y. Harada, R. Tanosaki, Y Kanda, M. Kami. A. Makimoto, S.Mineishi, H. Kawai, A. Shimosaka, Y. Heike, Y. Ikarashi, H. Wakasugi, Y. Takaue, T-J. Hwang, H-J. Kim, T. Kakizoe (2002): The role of PGE_2 in the differentiation of dendritic cells: how do dendritic cells influence T-cell polarization and chemokine receptor expression? *Stem Cells* 20: 448–459.

209 Wu, D., Marko, M., Claycombe, K., Paulson, E., Meydani, S. N. (2003): Ceramide-induced and age-associated increase in macrophge COX-2 expression is mediated through upregulation of NF-κB activity. *J. Biol. Chem.* 278: 10 983–10 992.

210 Cho., Y. H., Lee, C. O., Kim, S. G. (2003): Potentiation of lipopolysaccharide-inducible cyclooxygenase 2 expression by C2-ceramide via c-Jun N-terminal kinase-mediated activation of CCAAT/enhancer binding protein β in macrophages. *Mol. Pharmacol.* 63: 512–523.

211 Eigler, A., Siegmund, B. Emmerich, U., Baumann, K. H., Hartmann, G., Endres, S. (1998): Anti-inflammatory activities of cAMP-elevating agents: enhancement of IL-10 synthesis and concurrent suppression of TNF production. *J. Leukoc. Biol.* 63: 101–107.

212 Zidek, Z. (1999): Adenosine-cyclic AMP pathways and cytokine expression. *Eur. Cytokine Netw.* 10: 319–328.

213 Lentsch, A. B., Ward, P. A. (2000): The NFκB/IκB system in acute inflammation. *Arch. Immunol. Therap. Exp. (Warsz)* 48: 59–63.

214 Liu, H., Sidiropoulos, P., Song, G., Pagliari, L. J., Birrer, M. J., Stein, B., Anrather, J., Pope, R. M. (2000): TNF-α gene expression in macrophages: regulation by NF-κB is independent of c-Jun or C/EBP$β^1$. *J. Immunol.* 164: 4277–4285.

215 G. Zhang, S. Ghosh (2001): Toll-like receptor-mediated NF-kB activation: a phylogenetically conserved paradigm in innate immunity, *J. Clin. Invest.* 107: 13–19.

216 Liu, Y., Cousin, J. M., Hughes, J., Van Damme, J., Secki, J. R., Haslett, C., Dransfield, I., Savill, J., Rossi, A. G. (1999): Glucocorticoids promote nonphlogistic phagocytosis of apoptotic leukocytes. *J. Immunol.* 162: 3639–3646.

217 John, M., Lim, S., Seybold, J., Jose, P., Robichaud, A., O'Connor, B., Barnes, P. J., Chung, K. F. (1988): Inhaled corticosteroids increase interleukin-10 but reduce macrophage inflammatory protein-1α, granulocyte-macrophage colony-stimulating factor, and interferon-γ release from alveolar macrophages in asthma. *Am. J. Respir. Crit. Care Med.* 157: 256–262.

218 Baugh, J. A., Bucala, R. (2002): Macrophage migratory inhibitory factor. *Crit. Care Med.* 30: S27–S35.

219 Roger T., Glauser M. P., Calandra, T. (2001): Macrophage migration inhibitory factor (MIF) modulates innate immune responses induced by endotoxin and Gram-negative bacteria. *J. Endotoxin Res.* 7: 456–460.

220 Laouar, A., Glesne, D., Huberman, E. (2001): Protein kinase C-β, fibronectin, α5β1-integrin, and tumor necrosis factor-α are required for phorbol diester-induced apoptosis in human myeloid leukaemia cells. *Mol. Carcinog.* 32: 195–205.

221 Oberholzer, C., Oberholzer, A., Clare-Salzler, M., Moldawer, L.L. (2001): Apoptosis in sepsis: a new target for therapeutic exploration. *FASEB J.* 15: 879–892.

222 Hermoso, M. A., Matsuguchi, T., Smoak, K., Cidlowski, J. A. (2004): Glucocorticoids and tumor necrosis factor alpha cooperatively regulate Toll-like receptor 2 gene expression. *Mol. Cell. Biol.* 24: 4743–4756.

223 Yoshida, Y., Kang, K., Berger, M., Chen, G., Gilliam, A. C., Moser, A., Wu, L., Hammerberg, C., Cooper, K. D. (1998): Monocyte induction of IL-10 and down-regulation of IL-12 by iC3b deposited in ultraviolet-exposed human skin. *J. Immunol.* 161: 5873–5879.

224 Misson, P. van den Brûle, S., Barbarin, V., Lison, D., Huaux, F. (2004): Markers of macrophage differentiation in experimental silicosis. *J. Leukoc. Biol.* 76: 926–932.

225 Barbarin, V., Arras, M., Misson, P., Delos, M., McGarry, B., Phan, S. H., Lison, D., Haux, F. (2004): Characterization of the effect of interleukin-10 on silica-induced lung fibrosis in mice. *Am. J. Respir. Cell Mol. Biol.* 31: 78–85.

226 Peterlin, B. M., Trono, D. (2003): Hide, shield and strike back: how HIV-infected cells avoid immune eradication. *Nat. Rev. Immunol.* 3: 97–107.

227 Burgert, H.G., Ruzsics, Z., Obermeier, S., Hilgendorf, A., Windheim, M., Elsing, A. (2002): Subversion of host defense mechanisms by adenoviruses. *Curr. Top. Microbiol. Immunol.* 269: 273–318.

228 Vossen, M. T. M., Westerhout, E. M., Söderberg-Nauclér, C., Wiertz, E. J. H. J. (2002): Viral immune evasion: a masterpiece of evolution. *Immunogenetics* 54: 527–542.

229 Smith, S. A., Kotwal, G. J. (2002): Immune response to poxvirus infections in various animals. *Crit. Rev. Microbiol.* 28: 149–185.

230 Biron, C. A., Nguyen, K. B., Pien, G. C., Cousens, L. P, Salazar-Mather, T. P. (1999): Natural killer cells in antiviral defense: function and regulation by innate cytokines. *Annu. Rev. Immunol.* 17: 189–220.

231 Satoh, M., Shaheen, V. M., Kao, P. N., Okano, T., Shaw, M., Yoshida, H., Richards, H. R., Reeves, W. H. (1999): Autoantibodies define a family of proteins with conserved double-stranded RNA-binding domains as well as DNA binding activity. *J. Biol. Chem.* 274: 34 598–34 604.

232 Pannetier, D., Faure, C., Georges-Courbot, M-C., Deubel, V., Baize, S. (2004): Human macrophages, but not dendritic cells, are activated and produce alpha/beta interferons in response to Mopeia virus infection. *J. Virol.* 78: 10 516–10 524.

233 Hay, S., Kannourakis, G. (2002): A time to kill: viral manipulation of the cell death program. *J. Gen. Virol.* 83: 1547–1564.

234 Derfuss, T., Meinl, E. (2002): Herpesviral proteins regulating apoptosis. *Curr. Top. Microbiol. Immunol.* 269: 257–272.

235 Benedict, C. A., Norris, P. S., Ware, C. F. (2002): To kill or be killed: viral evasion of apoptosis. *Nat. Immunol.* 3: 1013–1018.

236 C.L. Bevins, C. L. (1999): Scratching the surface; inroads to a better understanding of airway host defense. *Am. J. Respir. Cell Mol. Biol.* 20: 861–863.

237 Gallo, R. L., Murakami, M., Ohtake, T., Zaiou, M. (2002): Biology and clinical relevance of naturally occurring antimicrobial peptides. *J. Allergy Clin. Immunol.* 110: 823–831.

238 Thielens, N. M., Tacnet-Delorme, P., Arlaud, G. J. (2002): Interaction of C1q and mannan-binding lectin with viruses. *Immunobiology* 205: 563–574.

239 Neth, O., Jack, D. L., Dodds, A. W., Holzel, H., Klein, N. J., Turner, M. W. (2000): Mannose-binding lectin binds to a range of clinically relevant microorganisms and promotes complement deposition. *Infect. Immun.* 68: 688–693.

240 Ogden, C. A., deCathelineau, A., Hoffmann, P. R., Bratton, D., Ghebrehiwet, B., Fadok, V., Henson, P. M. (2001): C1q and mannose binding lectin engagement of cell surface calreticulin and CD91 initiates macropinocytosis and

241 Nauta, A. J., Castellano, G., Xu, W., Woltman, A. M., Borrias, M. C., Daha, M. R., van Kooten, C., Roos, A. (2004): Opsonization with C1q and mannose-binding lectin targets apoptotic cells to dendritic cells. *J. Immunol.* 173: 3044–3050.

242 Schagat, T. L., Wofford, J. A., Wright, J. R. (2001): Surfactant protein A enhances alveolar macrophage phagocytosis of apoptotic neutrophils. *J. Immunol.* 166: 2727–2733.

243 Watford, W. T., Smithers, M. B., Frank, M. M., Wright, J. R. (2002): Surfactant protein A enhances the phagocytosis of C1q-coated particles by alveolar macrophages. *Am. J. Physiol. Lung Cell Mol. Physiol.* 283: L1011–L1022.

244 Palaniyar, N., Nadeaalingham, J., Reid, K. B. M. (2003): Pulmonary innate immune proteins and receptors that interact with gram-positive bacterial ligands. *Immunobiology* 205: 575–594.

245 Okano, M., Satoskar, A. R., Nishizaki, K., Abe, M., Harn, D. A. Jr. (1999): Induction of Th2 responses and IgE is largely due to carbohydrates functioning as adjuvants on Schistosoma mansoni egg antigens. *J. Immunol.* 163: 6712–6717.

246 Carroll, M. C. (2000): The role of complement in B cell activation and tolerance. *Adv. Immunol.* 74: 61–88.

247 Quartier, P., Potter, P. K., Ehrenstein, M. R., Walport, M. J., Botto, M. (2005): Predominant role of IgM-dependent activation of the classical pathway in the clearance of dying cells by murine bone marrow-derived macrophages in vitro. *Eur. J. Immunol.* 35: 252–260.

248 Borrego, F., Kabat, J., Kim, D-K., Lieto, L., Maasho, K., Pena, J., Solana, R., Coligan, J.E. (2001): Structure and function of major histocompatability complex (MHC) class 1 specific receptors expressed on human natural killer (NK) cells. *Mol. Immunol.* 38: 637–660.

249 Lanier, L. L. (2000): Turning on natural killer cells. *J. Exp. Med.* 191: 1259–1262.

250 Dunne, J., Lynch, S., O'Farrelly, C., Todryk, S., Hegarty, J. E., Feighery, C., Doherty, D. G. (2001): Selective expansion and partial activation of human NK cells and NK receptor-positive T cells by IL-2 and IL-15. *J. Immunol.* 167: 3129–3138.

251 Ohteki, T. (2002): Critical role for IL-15 in innate immunity. *Curr. Mol. Med.* 2: 371–380.

252 Biber, J. L., Jabbour, S., Parihar, R., Dierksheide, J., Hu, Y., Baumann, H., Bouchard, P., Caligiuri, M. A., Carson, W. (2002): Administration of two macrophage-derived interferon-γ-inducing factors (IL-12 and IL-15) induces a lethal systemic inflammatory response in mice that is dependent on natural killer cells but does not require interferon-γ, *Cell. Immunol.* 216: 31–42.

253 Granucci, F., Zanoni, I., Pavelka, N., van Dommelen, S. L. H., Andoniou, C. E., Belardelli, F., Degli Eposti, M. A., Ricciardi-Castagnoli, P. (2004): A contribution of mouse dendritic cell-derived IL-2 for NK cell activation. *J. Exp. Med..* 200: 287–295.

254 J.R. Ortaldo, H.A. Young (2003): Expression of IFN-γ upon triggering of activating Ly49D NK receptors in vitro and in vivo: costimulation with IL-12 or IL-18 overrides inhibitory receptors. *J. Immunol.* 170: 1763–1769.

255 Yamamoto, K., Shibata, F., Miyasaka, N., Miura, O. (2002): The human perforin gene is a direct target of STAT4 activated by IL-12 in NK cells. *Biochem. Biophys. Res. Commun.* 297: 1245–1252.

256 Kinoshita, N., Hiroi, H., Ohta, N., Fukuyama, S., Park. E. J., Kiyono, H. (2002): Autocrine IL-15 mediates intestinal epithelial cell death via the activation of neighbouring intraepithelial NK cells. *J. Immunol.* 169: 6187–6192.

257 Wu, C. Y., Gadina, M., Wang, K., O'Shea, J., Seder, R. A. (2000): Cytokine regulation of IL-12 receptor $\beta 2$ expression: differential effects on human T and NK cells. *Eur. J. Immunol.* 30: 1364–1374.

258 Nguyen, K. B., Salazar-Mather, T. P., Dalod, M. Y., Van Deusen, J. B., Wei, X-q, Liew, F. Y., Caligiuri, M. A., Durbin, J. E., Biron, C. A. (2002): Coordinated and distinct roles for IFN-alpha beta,

IL-12, and IL-15 regulation of NK cell responses to viral infection. *J. Immunol.* 169: 4279–4287.
259. Sivori, S., Falco, M., Chiesa, M. D., Carlomagno, S., Vitale, M., Moretta, L., Moretta, A. (2004): CpG and double-stranded RNA trigger human NK cells by Toll-like receptors: induction of cytokine release and cytotoxicity against tumors and dendritic cells. *Proc. Natl. Acad. Sci. USA*. 101: 10116–10121.
260. Vilches, C., Parham, P. (2002): KIR: diverse, rapidly evolving receptors of innate and adaptive immunity, *Annu. Rev. Immunol.* 20: 217–251.
261. Warren, H. S., Kinnear, B. F., Phillips, J. H., Lanier, L. L. (1995): Production of IL-5 by human NK cells and regulation of IL-5 secretion by IL-4, IL-10, and IL-12. *J. Immunol.* 154: 5144–5152.
262. Peritt, S., Robertson, G., Gri, L., Showe, M., Aste-Amezaga, G., Trinchieri, G. (1998): Cutting edge: differentiation of human NK cells into NK1 and NK2 subsets, *J. Immunol.* 161: 5821–5824.
263. Todd, D. J., Greiner, D. L., Rossini, A. A., Mordes, J. P., Bortell, R. (2001): An atypical population of NK cells that spontaneously secrete IFN-γ and IL-4 is present in the intraepithelial lymphoid compartment of the rat. *J. Immunol.* 167: 3600–3609.
264. Joshi, P. C., Zhou, X., Cuchens, M., Jones, Q. (2001): Prostaglandin E_2 suppressed IL-15-mediated human NK cell function through down-regulation of common γ-chain. *J. Immunol.* 166: 885–891.
265. Ferlazzo, G., Morandi, B., D'agostino, A., Meazza, R., Melioli, G., Moretta, A., Moretta, L. (2003): The interaction between NK cells and dendritic cells in bacterial infections results in rapid induction of NK cell activation and in the lysis of uninfected dendritic cells. *Eur. J. Immunol.* 33: 306–313.
266. Orange, J. S., Fassett, M. S., Koopman, L. A., Boyson, J. E., Strominger, J. L. (2002): Viral evasion of natural killer cells. *Nat. Immunol.* 3: 1006–1012.
267. R. Parihar, R., J. Dierksheide, J., Y Hu, Y., W.E. Carson, W. E. (2002): IL-12 enhances the natural killer cell cytokine response to Ab-coated tumor cells. *J. Clin. Invest.* 110: 983–992.
268. Nylén, S., Maasho, K., Söderström, K., Ilg, T., Akuffo, H. (2003): Live *Leishmania* promastigotes can directly activate primary human natural killer cells to produce interferon-γ. *Clin. Exp. Immunol.* 131: 457–467.
269. Demarco, R. A., Fink, M. P., Lotze, M. T. (2005): Monocytes promote natural killer cell interferon γ production in response to the endogenous danger signal HMGB1. *Mol. Immunol.* 42: 433–444.
270. Poccia, F., Agrati, C., Ippolito, G., Colizzi, V., Malkovsky, M. (2001): Natural T cell immunity to intracellular pathogens and nonpeptidic immunoregulatory drugs. *Curr. Mol. Med.* 1: 137–151.
271. Godfrey, D. I., Hammond, K. J. L., Poulton, L. D., Smyth, M. J., Baxter, A. G. (2000): NKT cells: facts, functions and fallacies, *Immunol. Today* 21: 573–576.
272. Duthie, M. S., Wleklinski-Lee, M., Smith, S., Nakayama, T., Taniguchi, M., Kahn, S. J. (2002): During *Trypanosoma cruzi* infection CD1d-restricted NK T cells limit parasitemia and augment the antibody response to a glycophosphinositol-modified surface protein. *Infect. Immun.* 70: 36–48.
273. Balmer. P., Devaney, E. (2002): NK T cells are a source of early interleukin-4 following infection with third-stage larvae of the filarial nematode Brugia pahangi. *Infect. Immun.* 70: 2215–2219.
274. Ikehara, Y., Yasunami, Y., Kodama, S., Maki, T., Nakano, M., Nakayama, T., Taniguchi, M., Ikeda, S. (2000): $CD4^+Va14$ natural killer cells are essential for acceptance of rat islet xenografts in mice. *J. Clin. Invest.* 105: 1761–1767.
275. Seino, K., Fukao, K., Muramoto, K., Yanagisawa, K., Takeda, Y., Kakuta, S., Iwakura, Y., Van Kaer, L., Takeda, K., Nakayama, T., Kakuta, S., Bashuda, H., Yagita, H., Okumura, K. (2001): Requirement for natural killer T (NKT) cells in the induction of allograft tolerance. *Proc. Natl. Acad. Sci. U.S.A.*, 98: 2577–2581.

276 Tahir, S. M. A., Cheng, O., Shaulov, A., Koezuka, Y., Bubley, G. J., Wilson, S. B., Balk, S. P., Exley, M. A. (2001): Loss of IFN-γ production by invariant NK T cells in advanced cancer. *J. Immunol.* 167: 4046–4050.

277 Laloux, V., Beaudoin, L., Ronet, C., Lehuen, A. (2002): Phenotypic and functional differences between NKT cells colonizing splanchic and peripheral lymph nodes. *J. Immunol.* 168: 3251–3258.

278 Sonada, K-H, Exley, M., Snapper, S., Balk, S. P., Stein-Streilein, J. (1999): CD1-reactive natural killer T cells are required for development of systemic tolerance through an immune-privileged site. *J. Exp. Med.* 190: 1215–1225.

279 Sonada, K-H., Faunce, D. E., Taniguchi, M., Exley, M., Balk, S., Stein-Streilein, J. (2001): NK T cell-derived IL-10 is essential for the differentiation of antigen-specific T regulatory cells in systemic tolerance. *J. Immunol.* 166: 42–50.

280 Hayakawa, Y., Takeda, K., Yagita, H., Van Kaer, L., Saiki, I., Okumura, K. (2001): Differential regulation of Th1 and Th2 functions of NKT cells by CD28 and CD40 costimulatory pathways. *J. Immunol.* 166: 6012–6018.

281 Lee, P. T., Benlagha, K., Teyton, L., Bendelac, A. (2002): Distinct functional lineages of human Vα24 natural killer T cells. *J. Exp. Med.* 195: 637–641.

282 Margenthaler, J. A., Landeros, K., Kataoka, M., Flye, M. W. (2002): CD1-dependent natural killer (NK1.1$^+$) T cells are required for oral and portal venous tolerance induction. *J. Surg. Res.* 104: 29–35.

283 Singh, A. K., Wilson, M. T., Hong, S., Olivares-Villagomez, D., Du, C., Stanic, A. K., Joyce, S., Sriram, S., Koezuka, Y., Kaer, L. V. (2001): Natural killer T cell activation protects mice against experimental autoimmune encephalomyelitis. *J. Exp. Med.* 194: 1801–1811.

284 Loza, M. J., Metelitsa, L. S., Perussia, B. (2001): NKT and T cells: coordinate regulation of NK-like phenotype and cytokine production. *Eur. J. Immunol.* 32: 3453–3462.

285 Ranson, T., Vosshenrich, C. A. J., Corcuff, E., Richard, O., Laloux, V., Lehuen, A. (2003): IL-15 availability conditions homeostasis of peripheral natural killer cells. *Proc. Natl. Acad. Sci. U.S.A.* 100: 2663–2668.

286 Chamoto, K., Takeshima, T., Kosaka, A., Tsuji, T., Matsuzaki, J., Togashi, Y., Ikeda, H., Nishimura, T. (2004): NKT cells act as regulatory cells rather than killer cells during activation of NK cell-mediated cytotoxicity by alpha-galactosylceramide in vivo. *Immunol. Lett.* 95: 5–11.

287 Carnaud, C., Lee, D., Donnars, O., Park, S.-H., Beavis, A., Kpoezuka, Y., Bendelac, A. (1999): Cutting edge: cross-talk between cells of the innate immune system: NKT cells rapidly activate NK cells. *J. Immunol.* 163: 4647–4650.

288 Nishimura, T., Kitamura, H., Iwakabe, K., Yahata, T., Ohta, A., Sato, M., Takeda, K., Okumura, K., Van Kaer, L., Kawano, T., Taniguchi, M., Nakui, M., Sekimoto, M., Koda, T. (2000): The interface between innate and acquired immunity: glycolipid antigen presentation by CD1d-expressing dendritic cells to NKT cells induces the differentiation of antigen-specific cytotoxic T lymphocytes. *Int. Immunol.* 12: 987–994.

289 Eberl, G., Brawand, P., MacDonald, H. R. (2000): Selective bystander proliferation of memory CD4$^+$ and CD8$^+$ T cells upon NK T or T cell activation. *J. Immunol.* 165: 4305–4311.

290 Tomura, M., Yu, W.-G., Ahn, H.-J., Yamashita, M., Yang, Y.-F., Uno, S., Hamaoka, T., Kawano, T., Taniguchi, M., Koezuka, Y., Fujiwara, H. (1999): A novel function of Vα14$^+$CD4$^+$NKT cells: stimulation of IL-12 production by antigen-presenting cells in the innate immune system. *J. Immunol.* 163: 93–101.

291 Kitamura, H., Iwakabi, K., Yahata, T., Nishimura, S.-i., Ohta, A., Ohmi, Y., Sato, M., Takeda, K., Okumura, K., Van Kaer, L., Kawano, T., Taniguchi, M., Nishimura, T. (1999): The natural killer T (NKT) cell ligand a-galactosylceramide demonstrates its immunopotentiating effect by inducing interleukin (IL)-12

291 production by dendritic cells and IL-12 receptor expression on NKT cells. *J. Exp. Med.*, 189: 1121–1127.
292 Schulz, O., Edwards, A. D., Schito, M., Aliberti, J., Manickasingham, S., Sher, A., Reis e Sousa, C. (2000): CD40 triggering of heterodimeric IL-12p70 production by dendritic cells in vivo requires a microbial priming signal. *Immunity* 13: 453–462.
293 Singh, N., S. Hong, S., Scherer, D. C., Serizawa, I., Burdin, N., Kronenberg, M., Koezuka, Y., Van Kaer, L. (1999): Cutting edge: activation of NK T cells by CD1d and a-galactosylceramide directs conventional T cells to the acquisition of a Th2 phenotype. *J. Immunol.* 163: 2373–2377.
294 Li, Z., Diehl, A. (2003): Innate immunity in the liver. *Curr. Opin. Gastroenterol.* 19: 565–571.
295 Wilson, S. B., Delovitch, T. L. (2003): Janus-like role of regulatory iNKT cells in autoimmune disease and tumour immunity. *Nat. Rev. Immunol.* 3: 211–222.
296 J. Jason, I. Buchanan, L.K. Archibald, O.C. Nwanyanwu, M. Bell, T.A. Green, A. Eick, A.Han, D. Razsi, P.N. Kazembe, H. Dobbie, M. Midathada, W.R. Jarvis (2000): Natural T, $\gamma\delta$ and NK cells in mycobacterial, Salmonella and human immunodeficiency virus infection. *J. Infect. Dis.* 182: 474–481.
297 Smyth, M. J., Crowe, N. Y., Pellici, D. G., Kyparissoudis, K., Kelly, J. M., Takeda, K., Yagita, H., Godfrey, D. I. (2002): Sequential production of interferon-γ by NK1.1$^+$ T cells and natural killer cells is essential for the antimetastatic effect of a-galactosylceramide. *Blood* 99: 1259–1266.
298 Nakui, M., Ohta, A., Sekimoto, M., Sato, M., Iwakabe, K., Yahata, T., Kitamura, H., Kawano, T., Makuuchi, H., Taniguchi, M., Nishimura, T. (2000): Potentiation of antitumor effect of NKT cell ligand, a-galactosylceramide by combination with IL-12 on lung metastases of malignant melanoma cells. *Clin. Exp. Metastasis* 18: 147–153.
299 Godfrey, D. I., Kronenberg, M. (2004): Going both ways: immune regulation via CD1d-dependent NKT cells. *J. Clin. Invest.* 114: 1379–1388.
300 Cohn L. (2001): Food for thought. Can immunological tolerance be induced to treat asthma? *Am. J. Respir. Cell Mol. Biol.* 24: 509–512.
301 Spencer, L. A., Porte, P., Zetoff, C., Rajan, T. V. (2003): Mice genetically deficient in immunoglobulin E are more permissive hosts than wild-type mice to a primary, but not secondary, infection with the filarial nematode Brugia malayi. *Infect. Immun.* 71: 2462–2467.
302 Su, Z., Stevenson, M. M. (2002): IL-12 is required for antibody-mediated protective immunity against blood-stage Plasmodium chabaudi AS malaria infection in mice. *J. Immunol.* 168: 1348–1355.
303 M. Rodriguez-Sosa, J.R. David, R. Bojalil, A.R. Satoskar, L.I. Terrazas (2002): Cutting edge: susceptibility to the larval stage of the helminth parasite Taenia crassiceps is mediated by a Th2 response induced via STAT6 signaling. *J. Immunol.* 168: 3135–3139.
304 Pérez-Santos, J. L. M., Talamás-Rohana, P. (2001): In vitro indimetacin administration upregulates interleukin-12 production and polarizes the immune response towards a Th1 type in susceptible BALB/c mice infected with Leishmania mexicana. *Parasite Immunol.* 23: 599–606.
305 Scanga, C. A., Alberti, J., Jankovic, D., Tilloy, F., Bennouna, S., Denkers, E. Y., Medzhitov, R., Sher, A. (2002): Cutting edge: MyD88 is required for resistance to *Toxoplasma gondii* infection and regulates parasite-induced IL-12 production by dendritic cells. *J. Immunol.* 168: 5997–6001.
306 Nickdel, M. B., Roberts, F., Brombacher, F., Alexander, J., Rogers, C. W. (2001): Counter-protective role for interleukin-5 during acute *Toxoplasma gondii* infection. *Infect. Immun.* 69: 1044–1052.
307 Wynn, T. A. (1997): The debate over the effector function of eosinophils in helminth infection: new evidence from studies on the regulation of vaccine immunity by IL-12. *Mem. Inst. Oswaldo Cruz* 92 (suppl. 2): 105–108.

308 Hoffmann, K. F., Wynn, T. A., Dunne, D. W. (2002): Cytokine-mediated host responses during Schistosome infections; walking the fine line between immunological control and immunopathology. *Adv. Parasitol.* 52: 265–307

309 Ehlers, S. (2004): Commentary: adaptive immunity in the absence of innate immune responses? The un-Tolled truth of the silent invaders. *Eur. J. Immunol.* 34: 1783–1788.

310 Ohno, I., Nitta,Y., Yamauchi, K., Hoshi, H., Honma, M., Woolley, K., O'Bryne, P., Tamura, G., Jordana, M., Shirato, K. (1996): Transforming growth factor $\beta 1$ (TGF$\beta 1$) gene expression by eosinophils in asthmatic airways inflammation. *Am. J. Respir. Cell Mol. Biol.* 15: 404–409.

311 Vignola, A. M., La Grutta, S., Chiappara, G., Benkeder, A., Bellia, V., Bonsignore, G. (2002): Cellular network in airways inflammation and remodelling. *Paed. Respir. Rev.* 3: 41–46.

312 Hansen, G., McIntire. J. J., Yeung, P., Berry, G., Thorbecke, G. J., Chen, L., DeKruyff, R. H., Umetsu, D. T. (2000): CD4$^+$ T helper cells engineered to produce latent TGF-$\beta 1$ reverses allergen-induced airway hyperreactivity and inflammation. *J. Clin. Invest.* 105: 61–70.

313 Sheppard, D. (2001): Pulmonary fibrosis: a cellular overreaction or a failure of communication? *J. Clin. Invest.* 107: 1501–1502.

314 Chiaramonte, J. M. G., Hesse, M., Cheever, A. W., Wynn, T. A. (2000): CpG oligonucleotides can prophylactically immunize against Th2-mediated Schistosoma egg-induced pathology by an IL-12-independent mechanism. *J. Immunol.* 164: 973–985.

315 Balkwill, F., Mantovani, A. (2001): Inflammation and cancer: back to Virchow? *Lancet* 357: 539–545.

316 M. Hesse, M. Modolell, A. C. La Flamme, M. Schito, J. M. Fuentes, A.W. Cheever, E. J. Pearce, T.A. Wynn (2001): Differential regulation of nitric oxide synthase-2 and arginase-1 by type 1/type 2 cytokines in vivo: granulomatous pathology is shaped by the pattern of L-arginine metabolism. *J. Immunol.* 167: 6533–6544.

317 M. Hesse, A.W. Cheever, D. Jankovic, T.A. Wynn (2000): NOS-2 mediates the protective anti-inflammatory and antifibrotic effects of the Th1-inducing adjuvant, IL-12, in a Th2 model of granulomatous disease. *Am. J. Pathol.* 157: 945–955.

318 Rumbley, C. A., Sugaya, H., Zekavat, S. A., El Refaei. M., Perrin, P. J., Phillips, S. M. (1999): Activated eosinophils are the major source of Th2-associated cytokines in the schistosome granule. *J. Immunol.* 162: 1003–1009.

319 Chen, K., Wei, Y., Sharp, G. C., Braley-Mullen, H. (2002): Inhibition of TGF$\beta 1$ by TGF$\beta 1$ antibody or lisinopril reduces thyroid fibrosis in granulomatous experimental autoimmune thyroiditis. *J. Immunol.* 169: 6530–6538.

320 Mentink-Kane, M. M., Wynn, T. A. (2004): Opposing roles for IL-13 and IL-13 receptor α in health and disease. *Immunol. Rev.* 202: 191–202.

321 Buckwalter, M. S., Wyss-Coray, T. (2004): Modelling neuroinflammatory phenotypes in vivo. *Neuroinflammation* 1: 10.

322 Akama, K. T., Albanese, C., Pestell, R. G., Van Eldik, L. J. (1998): Amyloid β-peptide stimulates nitric oxide production in astrocytes through an NF-κB-dependent mechanism. *Proc. Natl. Acad. Sci. U.S.A.* 95: 5795–5800.

323 Matzinger, P. (2002): The danger model: a renewed sense of self. *Science* 296: 301–305.

324 Kita, H., Jorgensen, R. K., Reed C. E., Dunnette, S. L., Swanson, M. C., Bartemes, K. R., Squillace, D., Blomgren, J., Bachman, K., Gleich, G. J. (2000): Mechanism of topical glucocorticoid treatment of hay fever: IL-5 and eosinophil activation during natural allergen exposure are suppressed but IL-4, IL-6, and IgE antibody production are unaffected. *J. Allergy Clin. Immunol.* 106: 521–529.

325 Takanaski, S., Nonaka, R., Xing, Z., O'Byrne, P., Dolovich, J., Jordana, M. (1994): Interleukin-10 inhibits lipopolysaccharide-induced survival and cyto-

kine production by human peripheral blood eosinophils. *J. Exp. Med*. 180: 711–715.
326 Meerschaert, J., Busse, W. W., Bertics, P. J., Mosher, D. F. (2000): CD14$^+$ cells are necessary for increased survival of eosinophils in response to lipopolysaccharide. *Am. J. Respir. Cell Mol. Biol*. 23: 780–787.
327 Mahmudi-Azer, S., Velazquez, J. R., Lacy, P., Denburg, J. A., Moqbet, R. (2000): Immunoflorescence analysis of cytokine and granule protein expression during eosinophil maturation from cord blood-derived CD34$^+$ progenitors. *J. Allergy Clin. Immunol*. 105: 1178–1184.
328 Minshall, E. M., Leung, D. Y. M., Martin, R. J., Song, Y. L., Cameron, L., Ernst, P., Hamid, Q. (1997): Eosinophil-associated TGF-β_1 mRNA expression and airways fibrosis in bronchial asthma. *Am. J. Respir. Cell Mol. Biol*. 17: 326–333.
329 Ohkawara, G., Tamura, T., Iwasaki, A., Tanaka, T., Kikuchi, Shirato, K. (2000): Activation and transforming growth factor-β production in eosinophils by hyaluronan. *Am. J. Respir. Cell Mol. Biol*. 23: 444–451.
330 Adachi, T., Alam, R. (1998): The mechanism of IL-5 signal transduction. *Am. J. Physiol*. 275: C623–C633.
331 Nakajima, H., Gleich, G. J., Kita, H. (1996): Constitutive production of IL-4 and IL-10 and stimulated production of IL-8 by normal peripheral blood eosinophils. *J. Immunol*. 156: 4859–4866.
332 Schmid-Grendelmeier, P., Altznauer, F., Fischer, B., Bizer, C., Straumann, A., Menz, G., Blaser, K., Wuthrich, B., Simon, H.-U. (2002): Eosinophils express functional IL-13 in eosinophilic inflammatory diseases. *J. Immunol*. 169: 1021–1027.
333 Lacy, P., Moqbel, R. (2001): Immune effector functions of eosinophils in allergic airway inflammation. *Curr. Opin. Allergy Clin. Immunol*. 1: 79–84.
334 del Pozo, V., de Arruda-Chaves, E., de Andres, B., Carbaba, B., Lopez-Farre, A., Gallardo, S., Corageno, I., Vidarte, L., Jurado, A., Sastra, J., Palomino, P., Laboz, C. (1997): Eosinophils transcribe and translate messenger RNA for inducible nitric oxide synthase. *J. Immunol*. 158: 859–864.
335 Odemuyiwa, S. O., Ghahary, A., Li, Y., Puttagunta, L., Lee, J. E., Musat-Marcu, S., Ghahary, A., Moqbel, R. (2004): Cutting edge: human eosinophils regulate T cell subset selection through indoleamine 2,3-dioxygenase. *J. Immunol*. 173: 5909–5913.
336 Gauvreau, G. M., O'Byrne, P. M., Moqbel, R., Velazquez, J., Watson, R. M., Howie, K. J., Denburg, J. A. (1998): Enhanced expression of GM-CSF in differentiating eosinophils of atopic and atopic asthmatic subjects. *Am. J. Respir. Cell Mol. Biol*. 19: 55–62.
337 Bhattacharya, S., Stout, B. A., Bates, M. E., Bertics, P. J., Malter, J. S. (2001): Granulocyte macrophage colony-stimulating factor and interleukin-5 activate STAT5 and induce CIS1 mRNA in human peripheral blood eosinophils. *Am. J. Respir. Cell Mol. Biol*. 24: 312–316.
338 Woerly, G., Roger, N., Loiseau, S., Dombrowicz, D., Capron, A., Capron, M. (1999): Expression of CD28 and CD86 by human eosinophils and role in the secretion of type 1 cytokines (interleukin 2 and interferon γ): inhibition by immunoglobulin A complexes. *J. Exp. Med*. 190: 487–495.
339 Wang, J., Palmer, K., Lotvall, J., Milan, S., Lei, X.-F., Matthael, K. I., Gauldie, J., Inman, M.D., Jordana, M., Xing, Z. (1998): Circulating, but not local lung, IL-5 is required for the development of antigen-induced airways eosinophilia. *J. Clin. Invest*. 102: 1132–1141.
340 Cieslewicz, G., Tomkinson, A., Adler, A., Duez, C., Schwarze, J., Takeda, K., Larson, K.A., Lee, J. J., Irvin, C. G., Gelfand, E. W. (1999): The late, but not early, asthmatic response is dependent on IL-5 and correlates with eosinophil infiltration. *J. Clin. Invest*. 104: 301–308.
341 Stirling, R. G., Van Rensen, E. L. J., Barnes, P J., Chung, K. F. (2001): Interleukin-5 induces CD34$^+$ eosinophil progenitor mobilization and eosinophil CCR3 expression in asthma. *Am. J. Respir. Crit. Care Med*. 164: 1403–1409.

342 Zhu, Z., Homer, R. J., Wang, Z., Chen, Q., Geba, G. P., Wang, J., Zhang, Y., Elias, J. A. (1999): Pulmonary expression of interleukin-13 causes inflammation, mucus hypersecretion, subepithelial fibrosis, physiologic abnormalities, and eotaxin production. *J. Clin. Invest.* 103: 779–788.

343 Li, L., Xia, Y., Nguyen, A., Hon Lai, Y., Feng, L., Mosmann, T. R., Lo, D. (1999): Effects of Th2 cytokines on chemokine expression in the lung: IL-13 potently induces eotaxin expression by airway epithelial cells. *J. Immunol.* 162: 2477–2487.

344 Culley, F. J., Brown, A., Girod, N., Pritchard, D. I., Williams, T. J. (2002): Innate and cognate mechanisms of pulmonary eosinophilia in helminth infection. *Eur. J. Immunol.* 32: 1376–1385.

345 Kita, H., Abu-Ghazaleh, R. I., Gleich, G. J., Abraham, R.T. (1991): Regulation of Ig-induced eosinophil degranulation by adenosine 3′,5′-cyclic monophosphate. *J. Immunol.* 146: 2712–2718.

346 Fujihara, S., Ward, C., Dransfield, I., Hay, R. T., Uings, I. J., Hayes, B., Farrow, S. N., Haslett, C., Rossi, A. G. (2002): Inhibition of nuclear factor-κ B activation un-masks the ability of TNF-α to induce human eosinophil apoptosis. *Eur. J. Immunol.* 32: 457–466.

347 Yamashita, N., Koizumi, L. H., Murata, M., Mano, K., Ohta, K. (1999): Nuclear factor-κ B mediates interleukin-8 production in eosinophils. *Int. Arch. Allergy Immunol.* 120: 230–236.

348 Sabroe, I., Jones, E. C., Usher, L. R., Whyte, M. K., Dower, S. K. (2002): Toll-like receptor (TLR)2 and TLR4 in human peripheral blood granulocytes: a critical role for monocytes in leukocyte lipopolysaccharide responses. *J. Immunol.* 168: 4701–4710.

349 Nagase, H., Miyamasu, M., Yamaguchi, M., Fujisawa, T., Kawasaki, H., Ohta, K., Yamamoto, K., Morita, Y., Hirai, K. (2001): Regulation of chemokine receptor expression in eosinophils. *Int. Arch. Allergy Immunol.* 125 (suppl 1): 29–32.

350 Kudlacz, E., Whitney, C., Andresen, C., Conklyn, M. (2002): Functional effects of eotaxin are selectively upregulated on IL-5 transgenic mouse eosinophils. *Inflammation* 26: 111–119.

351 Kampen, G. T., Stafford, S., Adachi, T., Jinquan, T., Quan, S., Grant, J. A., Skov, P. S., Poulsen, L. K., Alam, R. (2000): Eotaxin induces degranulation and chemotaxis of eosinophils through the activation of ERK2 and p38 mitogen-activated protein kinases. *Blood* 95: 1911–1917.

352 Menzies-Gow, A., Ying, S., Sabroe, I., Stubbs, V. L., Soler, D., Williams, T. J., Kay, A. B. (2002): Eotaxin (CCL11) and eotaxin-2 (CCL24) induce recruitment of eosinophils, basophils, neutrophils, and macrophages as well as features of early- and late-phase allergic deactions following cutaneous injection in human atopic and nonatopic volunteers. *J. Immunol.* 169: 2712–2717.

353 Watanabe, K., Jose, P. J., Rankin, S. M. (2002): Eotaxin-2 generation is differentially regulated by lipopolysaccharide and IL-4 in monocytes and macrophages. *J. Immunol.* 168: 1911–1918.

354 Badewa, A. P., Hudson, C. E., Heiman, A. S. (2002): Regulatory effects of eotaxin, eotaxin-2, and eotaxin-3 on eosinophil degranulation and superoxide anion generation. *Exp. Biol. Med.* 227: 645–651.

355 Raab, Y., Sundberg, C., Hallgren, R., Knutson, L., Gerdin, B. (1995): Mucosal synthesis and release of prostaglandin E_2 from activated eosinophils and macrophages in ulcerative colitis. *Am. J. Gastroenterol.* 90: 614–620.

356 Raab, Y., Fredens, K., Gerdin, B., Hallgren, R. (1998): Eosinophil activation in ulcerative colitis: studies on mucosal release and localization of eosinophil granule constituents. *Dig. Dis. Sci.* 43: 1061–1070.

357 Kinet, J-P. (1999): The high affinity IgE receptor (FcεRI): from physiology to pathology. *Annu. Rev. Immunol.* 17: 931–972.

358 Kayaba, H., Dombrowicz, D., Woerly, G., Papin, J-P., Loiseau, S., Capron, M. (2001): Human eosinophils and human high affinity IgE receptor transgenic mouse eosinophils express low levels of high affinity IgE receptor, but release

IL-10 upon receptor actrivation. *J. Immunol.* 167: 995–1003.
359 Dombrowicz, D., Quatannens, B., Papin, J.-P., Capron, A., Capron, M. (2000): Expression and functional FceRI on rat eosinophils and macrophages. *J. Immunol.* 165: 1266–1271.
360 Capron, M., Woerly, G., Kayaba, H., Loiseau, S., Roger, N., Dombrowicz, D. (2001): Invited lecture: role of membrane receptors in the release of T helper 1 and 2 cytokines by eosinophils. *Int. Arch. Allergy Immunol.* 124: 223–226.
361 Novak, N., Bieber, T., Katoh, N. (2001): Engagement of FcεR1 on human monocytes induces the production of IL10 and prevents their differentiation in dendritic cells. *J. Immunol.* 167: 797–804.
362 Brombacher, F. (2000): The role of interleukin-13 in infectious diseases and allergy. *Bioessays* 22: 646–656.
363 Rothenberg, M. E., Mishra, A., Brandt, E. B., Hogan, S. P. (2001): Gastrointestinal eosinophils. *Immunol. Rev.* 179: 139–155.
364 Shi, H-Z., Humbles, A., Gerard, C., Jin, Z., Weller, P. F. (2000): Lymph node trafficking and antigen presentation by endobronchial eosinophils. *J. Clin. Invest.* 105: 945–953.
365 van Rijt, L. S., Vos, N., Hijdra, D., de Vries, V. C., Hoogsteden, H. C., Lambrecht, B. N. (2003): Airway eosinophils accumulate in the mediastinal lymph nodes but lack antigen-presenting potential for naive T cells. *J. Immunol.* 171: 3372–3378.
366 van der Pouw Kraan, T. C. M., Boeije, L. C. M., de Groot, E. R., Stapel, S. O., Snijders, A., Kapsenberg, M. L., van der Zee, J. S., Aarden, L. A. (1997): Reduced production of IL-12 and IL-12-dependent IFN-γ release in patients with allergic asthma. *J. Immunol.* 158: 5560–5565.
367 Kips, J. C. (2001): Cytokines in asthma. *Eur. Respir. J.* 18 (suppl 34): 24S–33S.
368 Plummeridge, M.J., Armstrong, L., Birchall, M. A., Millar, A. B. (2000): Reduced production of interleukin 12 by interferon γ primed alveolar macrophages from atopic asthmatic subjects. *Thorax* 55: 842–847.
369 Ho, L.-P., Davis, M., Denison, A., Wood, F. T., Greening, A. P. (2002): Reduced interleukin-18 levels in BAL specimens from patients with asthma compared to patients with sarcoidosis and healthy control subjects. *Chest* 121: 1421–1426.
370 Kuribayashi, K., Kodama, T., Okamura, H., Sugita, M., Matsuyama, T. (2002): Effects of post-inhalation treatment with interleukin-12 on airway hyperreactivity, eosinophilia and interleukin-18 receptor expression in a mouse model of asthma. *Clin. Exp. Allergy* 32: 641–649.
371 Xu, D., Chan, W. L., Leung, B. P., Hunter, D., Schulz, K., Carter, R. W., McInnes, L. B., Robinson, J. H., Liew, F. Y. (1998): Selective expression and function of interleukin-18 receptor on T helper (Th) type 1 but not Th2 cells. *J. Exp. Med.* 188: 1485–1492.
372 Yoshimoto, T., Takeda, K., Tanaka, T., Ohkusu, K., Kashiwamura, S.-i., Okamura, H., Akira, S., Nakanishi, K. (1998): IL-12 upregulates IL-18 receptor expression on T cells, Th1 cells, and B cells: synergism with IL-18 for IFN-γ production. *J. Immunol.* 161: 3400–3407.
373 Jankovic, D., Kullberg, M. C., Hieny, S., Caspar, P., Collazo, C, M., Sher, A. (2002): In the absence of IL-12, $CD4^+$ T cell responses to intracellular pathogens fail to default to a Th2 pattern and are host protective in an $IL\text{-}10^{-/-}$ setting. *Immunity* 16: 429–439.
374 Schnare, M., Barton, G. M., Holt, A. C., Takeda, K., Akira, S., Medzhitov, R. (2001): Toll-like receptors control activation of adaptive immune responses. *Nat. Immunol.* 2: 947–950.
375 Bellinghausen, U. Brand, J. Knop, J. Saloga (2000): Comparison of allergen-stimulated dendritic cells from atopic and nonatopic donors dissecting their effect on autologous naive and memory T helper cells of such donors. *J. Allergy Clin. Immunol.* 105: 988–996.
376 Hayashi, T., Gong, X., Rossetto, C., Shen, C., Takabayashi, K., Redecke, V., Spiegelberg, H., Broide, D., Raz, E.

(2005): Induction and inhibition of the Th2 phenotype spread: implications for childhood asthma. *J. Immunol.* 174: 5864–5873.

377 Papadopoulos, N. G., Stanciu, L. A., Papi A, Holgate ST, Johnston SL. (2002): A defective type 1 response to rhinovirus in atopic asthma. *Thorax* 57: 328–32.

378 Kapsenberg, M. L., Hilkens, C. M. U., van der Pouw Kraan, T. C. M. T., Wierenga, E. A., Kalinski, P. (2000): Atopic allergy: a failure of antigen-presenting cells to properly polarize helper T cells? *Am. J. Respir. Crit. Care Med.* 162: S76–S880.

379 Scandella, E., Men, Y., Legler, D.F., Gillessen, S., Prickler, L., Ludewig, B., Groettrup, M. (2004): CCL9/CCL21-triggered signal transduction and migration of dendritic cells requires prostaglandin E_2. *Blood* 103: 1596–1601.

380 Kuroda, E., Yamashita, U. (2003): Mechanisms of enhanced macrophage-mediated prostaglandin E_2 production and its suppressive role in Th1 activation in Th2-dominant BALB/c mice. *J. Immunol.* 170: 757–764.

381 Uematsu, S., Matsumoto, M., Takeda, K., Akira, S. (2002): Lipopolysaccharide-dependent prostaglandin E_2 production is regulated by the glutathione-dependent prostaglandin E_2 synthase gene induced by the Toll-Like receptor 4/MyD88/NF-IL6 pathway. *J. Immunol.* 168: 5811–5816.

382 Kuipers, H., Heirman, C., Hijdra, D., Muskens, F., Willart, M., van Meirvenne, S., Thielemans, K., Hoogsteden, H. C., Lambrecht, B. N. (2004): Dendritic cells retrovirally overexpressing IL-12 induce strong Th1 responses to inhaled antigen in the lung but fail to revert established Th2 sensitization. *J. Leukoc. Biol.* 76: 1028–1038.

383 Van Rijt, L. S., Lambrecht, B. N. (2001): Role of dendritic cells and Th2 lymphocytes in asthma: lessons from eosinophilic airway inflammation in the mouse. *Microsc. Res. Technol.* 53: 256–272.

384 Sung, S. J., Rose, C. E., Jr., Man Fu, S. (2001): Intratracheal priming with ovalbumin- and ovalbumin 323-339 peptide-pulsed dendritic cells induces airways hyperresponsiveness, lung eosinophilia, goblet cell hyperplasia, and inflammation. *J. Immunol.* 166: 1261–1271.

385 Roman, M., Calhoun, W. J., Hinton, K. L., Avendano, L. F., Simon, V., Escobar, A. M., Gaggero, A., Diaz, P. V. (1997): Respiratory syncytial virus infection in infants is associated with predominant Th-2-like response. *Am. J. Respir. Crit. Care Med.* 156: 190–195.

386 Wong, C. K., Ho, C. Y., Ko, F. W., Chan, C. H., Ho, A. S., Hui, D. S., Lam, C. W. (2001): Proinflammatory cytokines (IL-17, IL-6, IL-18, and IL-12) and Th cytokines (IFN-γ):, IL-4, IL-10, and IL-13) in patients with allergic asthma. *Clin. Exp. Immunol.* 125: 177–183.

387 Tillie-Leblond, I., Pugin, J., Marquette, C.-H., Lamblin, C., Saulnier, F., Brichet, A., Wallaert, B., Tonnel, A.-B., Gosset, P. (1999): Balance between proinflammatory cytokines and their inhibitors in bronchial lavage from patients with status asthmaticus. *Am. J. Respir. Crit. Care Med.* 159: 487–494.

388 Long, J. A., Fogel-Petrovic, M., Knight, D. A., Thompson, P. J., Upham, J. W. (2004): Higher prostaglandin E_2 production by dendritic cells from subjects with asthma compared with normal controls. *Am. J. Respir. Crit. Care Med.* 170: 485–491.

389 Catena, E., Mazzarella, G., Peluso, G. F., Micheli, P., Cammarata, A., Marsico, S. A. (1993): Phenotypic features and secretory pattern of alveolar macrophages in atopic asthmatic patients. *Monaldi Arch. Chest Dis.* 48: 6–15.

390 Hamelmann, E., Takeda, K., Schwarze, J., Vella, A. T., Irvin, C. G., Gelfand, E. W. (1999): Development of eosinophilic airway inflammation and airway hyperresponsiveness requires interleukin-5 but not immunoglobulin E or B lymphocytes. *Am. J. Respir. Cell Mol. Biol.* 21: 480–489.

391 Erjefält, J. S., Greiff, L., Andersson, M., Adelroth, E., Jeffery, P. K., Persson, C. G. A. (2001): Degranulation patterns of eosinophil granulocytes as determi-

nants of eosinophil driven diseases. *Thorax* 56: 341–344.

392 Pujol, J.-L., Cosso, B., Daurès, J.-P., Clot, J., Michel, F.-B., Godard, P. (1990): Interleukin-1 release by alveolar macrophages in asthmatic patients and healthy subjects. *Int. Arch. Allergy Appl. Immunol.* 91: 207–210.

393 Hart, L. A., Krishnan, V. L., Adcock, I. M., Barnes, P. J., Chung, K. F. (1998): Activation and localization of transcription factor, nuclear factor-κB, in asthma. *Am. J. Respir. Crit. Care Med.* 158: 1585–1592.

394 Smart, J. M., Horak, E., Kemp, A. S., Robertson, C. F., Tang, M. L. (2002): Polyclonal and allergen-induced cytokine responses in adults with asthma: resolution of asthma is associated with normalization of IFN-γ responses. *J. Allergy Clin. Immunol.* 110: 450–456.

395 Townley, R. G., Horiba, M. (2003): Airway hyperresponsiveness. A story of mice and men and cytokines. *Clin. Rev. Allerg. Immunol.* 24: 85–109.

396 Gagliardo, R., Chanez, P., Mathieu, M., Bruno, A., Costanzo, G., Gougat, C., Vachier, I., Bosquet, J., Bonsignore, G., Vignola, A. M. (2003): Persistent activation of nuclear factor-κB signaling pathway in severe uncontrolled asthma. *Am. J. Respir. Crit. Care Med.* 168: 1190–1198.

397 Yang, L., Cohn, L., Zhang, D.-H., Homer, R., Ray, A., Ray, P. (1998): Essential role of nuclear factor κB in the induction of eosinophilia in allergic airway inflammation. *J. Exp. Med.* 188: 1739–1750.

398 Chaplin, D. D. (2002): Cell cooperation in development of eosinophil-predominant inflammation in airways. *Immunol. Res.* 26: 55–62.

399 Becker, S., Fenton, M. J., Soukup, J. M. (2002): Involvement of microbial components and toll-like receptors 2 and 4 in cytokine responses to air pollution particles. *Am. J. Respir. Cell Mol. Biol.* 27: 611–618.

400 Poynter, M. E., Cloots, R., van Woerkom, T., Butnor, K. J., Vacek, P., Taatjes, D. J., Irvin, C. G., Janssen-Heininger, Y. M. (2004): NF-κB activation in airways modulates allergic inflammation but not hyperresponsiveness. *J. Immunol.* 173: 7003–7009.

401 Choi, I. W., Kim, D. K., Ko, H. M., Lee, H. K. (2004): Administration of antisense phosphorothioate oligonucleotide to the p65 subunit of NF-κB inhibits established asthmatic reaction in mice. *Int. Pharmacol.* 4: 1817–1828.

402 Salez, L., Singer, M., Balloy, V., Crémi-non, C., Chignard, M. (2000): Lack of IL-10 synthesis by murine alveolar macrophages upon lipopolysaccharide exposure. Comparison with peritoneal macrophages. *J. Leukoc. Biol.* 67: 545–552.

403 Hoontrakoon, R., Chu, H. W., Gardai, S. J., Wenzel, S. E., McDonald, P., Fadok, V. A., Henson, P. M., Bratton, D. L. (2002): Interleukin-15 inhibits spontaneous apoptosis in human eosinophils via autocrine production of granulocyte macrophage-colony stimulating factor and nuclear factor-κB activation. *Am. J. Respir. Cell Mol. Biol.* 26: 404–412.

404 Fehniger, T. A., Yu, H., Cooper, M. A., Suzuki, K., Shah, M. H., Caligiuri, M. A. (2000): Cutting edge: IL-15 costimulates the generalized Shwartzman reaction and innate immune IFN-γ production in vivo. *J. Immunol.* 164: 1643–1647.

405 Trapnell, B. C., Whitsett, J. A. (2002): GM-CSF regulates pulmonary surfactant homeostasis and alveolar macrophage-mediated innate host defense. *Annu. Rev. Physiol.* 64: 775–802.

406 Paine III, P. R., Preston, A. M., Wilcoxen, S., Jin, H., Siu, B. B., Morris, S. B., Reed, J. A., Ross, G., Whitsett, J. A., Beck, J. M. (2000): Granulocyte-macrophage colony-stimulating factor in the innate response to Pneumocystis carinii pneumonia in mice. *J. Immunol.* 164: 2602–2609.

407 Garofalo, R. P., Haeberle, H. (2000): Epithelial regulation of innate immunity to respiratory syncytial virus. *Am. J. Respir. Cell Mol. Biol.* 23: 581–585.

408 Gauvreau, G. M., Watson, R. M., O'Byrne, P. M. (1999): Protective effects of inhaled PGE$_2$ on allergen-induce airway responses and airway inflamma-

tion. *Am. J. Respir. Crit. Care Med.* 159: 31–36.

409 Ruckert, R., Brandt, K., Braun, A., Hoymann, H. G., Herz, U., Budagian, V., Durkop, H., Renz, H., Bulfone-Paus, S. (2005): Blocking IL-15 prevents the induction of allergen-specific T cells and allergic inflammation in vivo. *J. Immunol.* 174: 5507–5515.

410 Liu, L. Y., Sedgwick, J. B., Bates, M. A., Vrtis, R. F., Gern, J.E., Kita, H., Jarjour, N. N., Busse, W. W., Kelly, E. A. B. (2002): Decreased expression of membrane IL-5 receptor α on human eosinophils: I. Loss of membrane IL-5 receptor α on airway eosinophils and increased soluble IL-5 receptor α in the airway after allergen challenge. *J. Immunol.* 169: 6452–6459.

411 Liu, L. Y., Sedgwick, J. B., Bates, M. A., Vrtis, R. F., Gern, J.E., Kita, H., Jarjour, N. N., Busse, W. W., Kelly, E. A. B. (2002): Decreased expression of membrane IL-5 receptor α on human eosinophils: II. IL-5 down-modulates its receptor via a proteinase-mediated process. *J. Immunol.* 169: 6459–6466.

412 Julius, P., Hochheim, D., Boser, K., Schmidt, S., Myrtek, D., Bachert, C., Luttmann, W., Virchow, J. C. (2004): Interleukin-5 receptors on human lung eosinophils after segmental allergen challenge. *Clin. Exp. Allergy* 34: 1064–1070.

413 Seldon, P. M., Giembycz, M. A. (2001): Suppression of granulocyte/macrophage colony-stimulating factor release from human monocytes by cyclic AMP-elevating drugs: role of interleukin-10. *Br. J. Pharmacol.* 134: 58–67.

414 Stamatiou, P., Chan, C.-C., Monneret, G., Ethier, D., Rokach, J., Powell, W. S. (2004): 5-Oxo-6,8,11,14-eicosatetraenoic acid stimulates the release of the eosinophil survival factor granulocyte/macrophage colony-stimulating factor from monocytes. *J. Biol. Chem.* 279: 28 159–28 164.

415 Yssel, H., Groux, H. (2000): Characterization of T cell subpopulations involved in the pathogenesis of asthma and allergic diseases. *Int. Arch. Allergy Immunol.* 121: 10–18.

416 Brusselle, G. G., Kips, J. C., Peleman, R. A., Joos, G. F., Devos, R. R., Tavernier, J. H., Pauwels, R. A. (1997): Role of IFN-γ in the inhibition of the allergic airway inflammation caused by IL-12. *Am. J. Respir. Cell Mol. Biol.* 17: 767–771.

417 Hammad, H., Duez, C., Fahy, O., Tsicopoulos, A., André, C., Wallaert, B., Lebecque, S., Tonnel, A.-B., Pestel, J. (2000): Human dendritic cells in the severe combined immunodeficiency mouse model: their potentiating role in the allergic reaction. *Lab. Invest.* 80: 605–614.

418 Janssen, E. M., Wauben, M. H., Nijkamp, F. P., van Eden, W., van Oosterhout, A. J. (2002): Immunomodulatory effects of antigen-pulsed macrophages in a murine model of allergic asthma. *Am. J. Respir. Cell Mol. Biol.* 27: 257–264.

419 Tournoy, K. G., Kips, J.C., Pauwels, R. A. (2000): Endogenous interleukin-10 suppresses allergen-induced airway inflammation and nonspecific airway responsiveness. *Clin. Exp. Allergy* 30: 775–783.

420 Tournoy, K. G., Kips, J. C., Pauwels, R. A. (2001): Counterbalancing of TH2-driven allergic airway inflammation by IL-12 does not require IL-10. *J. Allergy Clin. Immunol.* 107: 483–491.

421 B.T. Ameredes, R. Zamora, K.F. Gibson, T.R. Billiar, B.Dixon-McCarthy, S. Watkins, W.J. Calhoun (2001): Increased nitric oxide production by airway cells of sensitized and challenged IL-10 knockout mice, *J. Leukoc. Biol.* 70: 730–736.

422 Quinn, T. J., Taylor, S., Wohlford-Lenane, C. L., Schwartz, D. A. (2000): IL-10 reduces grain dust-induced airway inflammation and airway hyperreactivity. *J. Appl. Physiol.* 88: 173–179.

423 Van Scott, M. R. Justice, J P., Bradfield, J. F., Enright, E., Sigounas, A., Sur, S. (2000): IL-10 reduces Th2 cytokine production and eosinophilia but augments airway reactivity in allergic mice. *Am. J. Physiol. Am. J. Physiol. Lung Cell Mol. Physiol.* 278: L667–L674.

424 Mäkelä, M. J., Kanehiro, A., Borish, L. Dakhama, A., Loader, J., Joetham, A., Xing, Z., Jordana, M., Larsen, G. L., Gelfand, E. W. (2000): IL-10 is necessary for the expression of airway hyperrespon-

siveness but not pulmonary inflammation after allergic sensitization, *Proc. Natl. Acad. Sci. U.S.A.* 97: 6007–6012.

425 Nakao, A. (2001): Is TGF-β1 the key to suppression of human asthma? *Trends Immunol.* 22: 115–118.

426 Magnan, A., Retornaz, F., Tsicopoulos, A., Brisse, J., Van Pee, D., Gosset, P., Chamlian, A., Tonnel, A. B., Vervloet, D. (1997): Altered compartmentalization of transforming growth factor-β in asthmatic airways. *Clin. Exp. Allergy* 27: 389–395.

427 Hansen, G., Berry, G., DeKruyff, R. H., Umetsu, D. T. (1999): Allergen-specific Th1 cells fail to counterbalance Th2 cell-induced airway hyperreactivity but cause severe airway inflammation. *J. Clin. Invest.* 103: 175–183.

428 Balzar, S., Chu, H. W., Silkoff, P., Cundall, M., Trudeau, J. B., Strand, M., Wenzel, S. (2005): Increased TGF-β2 in severe asthma with eosinophilia. *J. Allergy Clin. Immunol.* 115: 110–117.

429 Wills-Karp, M., Chiaramonte, M. (2003): Interleukin-13 in asthma. *Curr. Opin. Pulm. Med.* 9: 21–27.

430 Zimmermann, N., King, N. E., Laporte, J., Yang, M., Mishra, A., Pope, S. M., Muntel, E. E., Witte, D. P., Pegg, A. A., Foster, P. S., Hamid, Q., Rothenberg, M. E. (2003): Dissection of experimental asthma with DNA microarray analysis identifies arginase in asthma pathogenesis. *J. Clin. Invest.* 111: 1863–1874.

431 Vercelli, D. (2003): Arginase: marker, effector, or candidate gene for asthma? *J. Clin. Invest.* 111: 1815–1817.

432 Castro, M., Chaplin, D. D., Walter, M. J., Holtzman, M. J. (2000): Could asthma be worsened by stimulating the T-helper type 1 immune response? *Am. J. Respir. Cell Mol. Biol.* 22: 143–146.

433 Hussell T (2000): IL-12-activated NK cells reduce lung eosinophilia to the attachment protein of respiratory syncytial virus but do not enhance the severity of the illness in CD8 T cell-immunodeficient conditions. *J. Immunol.* 165: 7109–7115.

434 Maecker, H. T., Hansen, G., Walter, D. M., DeKruyff, R. H., Levy, S., Umetsu, D. T. (2001): Vaccination with allergen-IL-18 fusion DNA protects against, and reverses established, airway hyperreactivity in a murine asthma model. *J. Immunol.* 166: 959–965.

435 Mazzoni, A., Young, H. A., Spitzer, J. H., Visintin, A., Segal, D. M. (2001): Histamine regulates cytokine production in maturing dendritic cells, resulting in altered T cell polarization. *J. Clin. Invest.* 108: 1865–1873.

436 Okazawa, A., Kanai, T., Watanabe, M., Yamazaki, M., Inoue, N., Ikeda, M., Kurimoto, M., Ishii, H., Hibi, T. (2002): Th-1 mediated intestinal inflammation in Crohn's disease may be induced by activation of lamina propria lymphocytes through synergistic stimulation of interleukin-12 and interleukin-18 without T receptor engagement. *Am. J. Gastroenterol.* 97: 3108–3117.

437 Sawa, Y., Oshitani, N., Adachi, K., Higuchi, K., Matsumoto, T., Arakawa, T. (2003): Comprehensive analysis of intestinal messenger RNA profile by real-time quantitative polymerase chain reaction in patients with inflammatory bowel disease. *Int. J. Mol. Med.* 11: 175–179.

438 Kadivar, K., Ruchelli, E. D., Marcowitz, J. E., Defelice, M. L., Strogatz, M. L., Kanzaria, M. M., Reddy, K. P., Baldassano, R. N., von Allmen, D., Brown, K. A. (2004): Intestinal interleukin-13 in pediatric inflammatory bowel disease patients. *Inflamm. Bowel Dis.* 10: 593–598.

439 Kojima, R., Kuroda, S., Ohkishi, T., Nakamaru, K., Hatakeyama, S. (2004): Oxazolone-induced colitis in BALB/c mice: a new method to evaluate the efficacy of therapeutic agents for ulcerative colitis. *J. Pharmacol. Sci.* 96: 307–313.

440 Wolf, A. M., Wolf, D., Rumpold, H., Moschen, A. R., Kraser, A., Obrist, P., Fuchs, D., Brandacher, G., Winkler, C., Geboes, K., Rutgeerts, P., Tilg, H. (2004): Overexpression of indoleamine 2,3-dioxygenase in human inflammatory bowel disease. *Clin. Immunol.* 113: 47–55.

441 Hayashi, T., Beck, L., Rossetto, C., Gong, X., Takihawa, O., Takabayashi, K., Broide, D. H., Carson, D. A., Raz, E.

(2004): Inhibition of experimental asthma by indoleamine 2,3-dioxygenase. *J. Clin. Invest.* 114: 270–279.

442 Segal, B. M., Klinman, D. M., Shevach, E. M. (1997): Microbial products induce autoimmune disease by an IL-12-dependent pathway. *J. Immunol.* 158: 5087–5090.

443 Rothe, H., Kolb, H. (1998): The APC1 concept of type 1 diabetes. *Autoimmunity* 27: 179–184.

444 Winer, S., Astsaturov, I., Cheung, R. K., Gunaratnam, L., Kubiak, V., Cortez, M. A., Moscarello, M., O'Connor, P. W., McKerlie, C., Becker, D. J., Dosch, H.-M. (2001): Type-1 diabetes and multiple sclerosis patients target islet plus central nervous system autoantigens: nonimmunized nonobese diabetic mice can develop autoimmune encephalitis. *J. Immunol.* 166: 2831–2841.

445 Beyan, H., Buckley, L. R., Yousaf, N., Londei, M., Leslie, R. D. G. (2003): A role for innate immunity in type 1 diabetes? *Diabetes Metab. Res. Rev.* 19: 89–100.

446 Stoy, N. (2002): Monocyte/macrophage initiation of organ-specific autoimmunity: the ultimate 'bystander' hypothesis? *Med. Hypoth.* 58: 312–326.

447 Sharif, S., Arreaza, G. A., Zucker, P., Delovitch, T. L. (2002): Regulatory natural killer T cells protect against spontaneous and recurrent type 1 diabetes. *Ann. New York Acad. Sci.* 958: 77–88.

448 Naumov, Y. N., Bahjat, K. S., Gausling, R., Abraham, R., Exley, M. A., Koezuka, Y., Balk, S. B., Strominger, J. L., Clare-Salzer, M., Wilson S. B. (2001): Activation of CD1d-restricted T cells protects NOD mice from developing diabetes by regulating dendritic cell subsets. *Proc. Natl. Acad. Sci. U.S.A.* 98: 13838–13843.

449 Comabella, M., Pericot, I., Goertsches, R., Nos, C., Castillo, M., Blas Navarro, J., Rio, J., Montalban, X. (2005): Plasma osteopontin levels in multiple sclerosis. *J. Neuroimmunol.* 158: 231–239.

450 Soldan, S. S., Alvarez Retuerto, A. I., Sicotte, N. L., Voskuhl, R. R. (2004): Dysregulation of IL-10 and IL-12p40 in secondary progressive multiple sclerosis. *J. Neuroimmunol.* 146: 209–215.

451 Ersoy, E., Kus, C. N., Sener, U., Coker, I., Zorlu, Y. (2005): The effects of interferon-β on interleukin-10 in multiple sclerosis patients. *Eur. J. Neurol.* 12: 208–211.

452 Andrade, F., Casciola-Rosen, L., Rosen, A. (2000): Apoptosis in systemic lupus erythematosus. *Rheum. Dis. Clinics N. Am.* 26: 215–227.

453 Charles, P. J. (2003): Defective waste disposal: does it induce autoantibodies in SLE? *Ann. Rheum. Dis.* 62: 1–3.

454 Mevorach, D. (2003): Systemic lupus erythematosis. A question of balance. *Clin. Rev. All. Clin. Immunol.* 25: 49–59.

455 Alleva, D. G., Kaser, S. B., Beller, D. I. (1998): Intrinsic defects in macrophage IL-12 production associated with immune dysfunction in the MRL/++ New Zealand black/white F_1 lupus prone mice and the Leishmania major-susceptible BALB/c strain. *J. Immunol.* 161: 6878–6884.

456 Alleva, D. G., Pavlovich, R. P., Grant, C., Kaser, S. B., Beller, D. I. (2000): Aberrant macrophage cytokine production is a conserved feature among autoimmune mouse strains. Elevated interleukin (IL)-12 and an imbalance in tumor necrosis factor-α and IL-10 define a unique cytokine profile in macrophages from young nonobese diabetic mice. *Diabetes* 49: 1106–1115.

457 Koh, J. S., Wang, J. Z., Levine, J. S. (2000): Cytokine dysregulation induced by apoptotic cells is a shared characteristic of murine lupus. *J. Immunol.* 165: 4190–4201.

458 Lauwerys, B. R., Houssiau F. A. (2003): Involvement of cytokines in the pathogenesis of systemic lupus erythematosis. *Adv. Exp. Med. Biol.* 520: 237–251.

459 van der Linden, M. W., Westendorp, R. G., Sturk, A., Bergman, W., Huizinga, T. W. (2000): High interleukin-10 production in first-degree relatives of patients with generalized but not cutaneous lupus erythematosus. *J. Invest. Med.* 48: 327–334.

460 Rönnelid, J., Tejde, A., Mathsson, L., Nilsson-Ekdahl, K., Nilsson, B. (2003):

Immune complexes from SLE sera induce IL 10 production from normal peripheral blood mononuclear cells by an FcγRII dependent mechanism: implications for a possible vicious cycle maintaining B cell hyperactivity in SLE. *Ann. Rheum. Dis.* 62: 37–42.

461 Shoshan, Y., Shapira, I., Toubi, E., Ftrolkis, I., Yaron, M., Mevorach, D. (2001): Accelerated Fas-mediated apoptosis of monocytes and maturing macrophages from patients with systemic lupus erythematosus: relevance to in vitro impairment of interaction with iC3b-opsonized cells. *J. Immunol.* 167: 5963–5969.

462 Beebe, A. M., Cua, D. J., de Waal Malefyt, R. (2002): The role of interleukin-10 in autoimmune disease:systemic lupus erythematosus (SLE) and multiple sclerosis (MS). *Cytokine Growth Factor Rev.* 13: 403–412.

463 Sieling, P. A., Porcelli, S. A., Duong, B. T., Spada, F., Bloom, B. R., Diamond, B., Hahn, B. H. (2000): Human double-negative T cells in systemic lupus erythematosus provide help for IgG and are restricted by CD1c. *J. Immunol.* 165: 5338–5344.

464 Nakajima, A., Hirose, S., Yagita, S., Okumura, K. (1997): Roles of IL-4 and IL-12 in the development of lupus inNZB/W F1 mice. *J. Immunol.* 158: 1466–1472.

465 Viallard, J. F., Pellegrin, J. L., Ranchin, V., Schaeverbeke, T., Dehais, J., Longy-Boursier, M., Ragnaud, J. M., Leng, B., Moreau, J. F. (1999): Th1 (IL-2), interferon-gamma (IFN-γ) and Th2 (IL-10, IL-4) cytokine production by peripheral blood mononuclear cells (PBMC) from patients with systemic lupus erythematosus (SLE). *Clin. Exp. Immunol.* 115: 189–195.

466 Horwitz, D. A., Gray, J. D., Behrendsen, S. C., Kubin, M., Rengaraju, M. Ohtsuka, K., Trinchieri, G. (1998): Decreased production of interleukin-12 and other Th1-type cytokines in patients with recent-onset systemic lupus erythematosus. *Arthr. Rheum.* 41: 838–844.

467 Gomez, D., Correa, P. A., Gomez, L. M., Cadena, J., Molina, J. F., Anaya, J. M. (2004): Th1/Th2 cytokines in patients with systemic lupus erythematosus; is tumour necrosis factor α protective? *Semin. Arthritis Rheum.* 33: 404–413.

468 Robak, E., Robak, T., Wozniacka, A., Zak-Prelich, M., Sysa-Jedrzejowska, A., Stepien, H. (2002): Proinflammatory interferon-g-inducing monokines (interleukin-12, interleukin-18, interleukin-15) – serum profile in patients with systemic lupus erythematosus, *Eur. Cytokine Netw.* 13: 364–368.

469 Wong, C. K., Li, E. K., Ho, C. Y., Lam, C. W. K. (2000): Elevation of plasma interleukin-18 concentration is correlated with disease activity in systemic lupus erythematosus. *Rheumatology* 39: 1078–1081.

470 Ma, X., Trinchieri, G. (2001): Regulation of interleukin-12 production in antigen-presenting cells. *Adv. Immunol.* 79: 55–92.

471 Hasan, M., Najjam. S., Gordon, M. Y., Gibbs, R. V., Rider, C. C. (1999): IL-12 is a heparin-binding cytokine. *J. Immunol.* 162: 1064–1070.

472 Walter, M. J., Kajiwara, N., Karanja, P., Castro, M., Holtzman, M. J. (2001): Interleukin 12 p40 production by barrier epithelial cells during airway inflammation. *J. Exp. Med.* 193: 339–351.

473 Pirhonen, J., Matikainen, S., Julkunen, I. (2002): Regulation of virus-induced IL-12 and IL-23 expression in human macrophages. *J. Immunol.* 169: 5673–5678.

474 Carvalho-Pinto, C. E., García, M. I., Mellado, M., Rodríguez-Frade, J. M., Martin-Caballero, J., Flores, J., Martínez-A, C., Balomenos, D. (2002): Autocrine production of IFN-γ by macrophages controls their recruitment to kidney and development of glomerulonephritis in MRL/lpr mice. *J. Immunol.* 169: 1058–1067.

475 Oates, J. C., Gilkeson, G. S. (2002): Mediators of injury in lupus nephritis. *Curr. Opin. Rheumatol.* 14: 498–503.

476 Tarzi, R. M., Davies, K. A., Robson, M. G., Fossati-Jimack, L., Saito, T., Walport, M. J., Cook, H. T. (2002): Nephrotoxic nephritis is mediated by Fcγ recep-

477 Suen, J.-L., Chuang, Y.-H., Chiang, B.-L. (2002): In vivo tolerance breakdown with dendritic cells pulsed with U1A protein in non-autoimmune mice: the induction of a high level of autoantibodies but not renal pathological changes. *Immunology* 106: 326–335.

478 Calvani, N., Richards, H. B., Tucci, M., Pannarale, G., Silvestris, F. (2004): Upregulation of IL-18 and predominance of a Th1 immune response is the hallmark of lupus nephritis. *Clin. Exp. Immunol.* 138: 171–178.

479 Hoshino, T., Kawase, Y., Okamoto, M., Yokota, K., Yoshino, K., Yamamura, K.-I., Miyaxaki, J.-I., Young, H. A., Oizumi. K. (2001): Cutting edge: IL-18-transgenic mice: in vivo evidence of a broad role for IL-18 in modulating immune function. *J. Immunol.* 166: 7014–7018.

480 Veenstra, K. G., Jonak, Z. L., Trulli, S., Gollob, J. A. (2002): IL-12 induces monocyte IL-18 binding protein expression via IFN-γ. *J. Immunol.* 168: 2282–2287.

481 Smeltz, R. B., Chen, J., Hu-Li, J., Shevach, E. M. (2001): Regulation of interleukin (IL)-18 receptor alpha chain expression on $CD4^+$ T cells during T helper (Th)1/Th2 differentiation. Critical downregulatory role of IL-4. *J. Exp. Med.* 194: 143–153.

482 Neumann, D., Del Giudice, E., Ciaramella, A., Boraschi, D., Bossù, P. (2001): Lymphocytes from autoimmune MRL lpr/lpr mice are hyperresponsive to IL-18 and overexpress the IL-18 receptor accessory chain. *J. Immunol.* 166: 3757–3762.

483 Leit-de-Moraes, M. C., Hameg, A., Pacilio, M., Koezuka, Y., Taniguchi, M., Van Kaer, L., Schneider, E., Dy, M., Herbelin, A. (2001): IL-18 enhances IL-4 production by ligand-activated NKT lymphocytes: a pro-Th2 effect of IL-18 exerted through NKT cells. *J. Immunol.* 166: 945–951.

484 Kashiwamura, S-i., Ueda, H., Okamura, H. (2002): Roles of interleukin-18 in tissue destruction and compensatory reactions. *J. Immunother.* 25 (suppl 1): S4–S11.

485 Bohn, E., Sing, A., Zumbihl, R., Bielfeldt, C., Okamura, H., Kurimoto, M., Heesemann, J., Autenrieth, I. B. (1998): IL-18 (IFN-γ-inducing factor) regulates early cytokine production in, and promotes resolution of, bacterial infection in mice. *J. Immunol.* 160: 299–307.

486 Neighbors, M., Xu, X., Barrat, F. J., Ruuls, S. R., Churakova, T., Debets, R., Bazan, J. F., Kastelein, R. A., Abrams, J. S., O'Garra, A. (2001): A critical role for interleukin 18 in primary and memory effector responses in Listeria monocytogenes that extends beyond its effect on interferon γ production. *J. Exp. Med.* 194: 343–354.

487 Esfandiari, E., McInnes, I. B., Lindop, G., Huang, F.-P., Field, M., Komai-Koma, M., Wei, X.-q., Liew, F. Y. (2001): A proinflammatory role of IL-18 in the development of spontaneous autoimmune disease. *J. Immunol.* 167: 5338–5347.

488 Fehniger, T. A., Shah, M. H., Turner, M. J., Van Deusen. J. B., Whitman, S. P., Cooper, M. A., Suzuki, K., Wechser, M., Goodsaid, F., Caliguri, M. A. (1999): Differential cytokine and chemokine expression by human NK cells following activation with IL-18 or IL-15 in combination with IL-12: implications for the innate immune response. *J. Immunol.* 162: 4511–4520.

489 Peng, S. L., Moslehi, J., Craft, J. (1997): Roles of interferon-γ and interleukin-4 in murine lupus. *J. Clin. Invest.* 99: 1936–1946.

490 Seery, J. P. (2000): IFN-γ transgenic mice: clues to the pathogenesis of systemic lupus erythematosus? *Arthritis Res.* 2: 437–440.

491 Shlomchik, M. J., Craft, J. E., Mamula, M. J. (2001): From T to B and back again: positive feedback in systemic autoimmune disease. *Nat. Rev. Immunol.* 1: 147–153.

492 Bancereau, J., Pascual, V., Palucka, A. K. (2004): Autoimmunity through cytokine-induced dendritic cell activation. *Immunity* 20: 539–550.

493 Anders, H. J., Banas, B., Schlondorff, D. (2004): Signaling danger: toll-like receptors and their potential roles in kidney disease. *J. Am. Soc. Nephrol.* 15: 854–867.

494 Bennett, L., Palucka, A. K., Arce, E., Cantrell, V., Borvak, J., Banchereau, J., Pascual, V. (2003): Interferon and granulopoiesis signatures in systemic lupus erythematosus blood. *J. Exp. Med.* 197: 681–685.

495 Baechler, E. C., Gregersen, P. K., Behrens, T. W. (2004): The emerging role of interferon in human systemic lupus erythematosus. *Curr. Opin. Immunol.* 16: 801–807.

496 Peterson, K. S., Huang, J. F., Zhu, J., D'Agati, V., Liu, X., Miller, N., Erlander, M. G., Jackson, M. R., Winchester, R. J. (2004): Characterization of heterogeneity in the molecular pathogenesis of lupus nephritis from transcriptional profiles of laser-captured glomeruli. *J. Clin. Invest.* 113: 1722–1733.

497 Stevens, S. R., Shibaki, A., Meunier, L., Cooper, K. D. (1995): Suppressor T cell-activating macrophages in ultraviolet irradiated skin induce a novel, TGF-β dependent form of T cell activation characterized by deficient IL-2Rα expression. *J. Immunol.* 155: 5601–5607.

498 Kasahara, S., Wago, H., Cooper E. L. (2002): Dissociation of innate and adaptive immunity by UVB irradiation. *Int. J. Immunopathol. Pharmacol.* 15: 1–11.

499 Shreedhar, V. K., Pride, M. W., Sun, Y., Kripe, M. L., Strickland, F. M. (1998): Origin and characteristics of ultraviolet-B radiation-induced suppressor T lymphocytes. *J. Immunol.* 161: 1327–1335.

500 Mehling, A., Loser, K., Varga, G., Metze, D., Luger T. A., Schwarz, T., Grabbe, S., Beissert, S. (2001): Overexpression of CD40 ligand in murine epidermis results in chronic skin inflammation and systemic autoimmunity. *J. Exp. Med.* 194: 615–628.

501 McHugh, N. J. (2002): Systemic lupus erythematosus and dysregulated apoptosis – what is the evidence? *Rheumatology* 41: 242–245.

502 Yoshida, H., Satoh, M., Behney, K. M., Lee, C.-G., Richards, H. B., Shaheen, V. M., Yang, J.-Q., Singh, R. R., Reeves, W. H. (2002): Effect of an exogenous trigger on the pathogenesis of lupus in (NZB X NZW)F$_1$ mice. *Arthr. Rheum.* 46: 2235–2244.

503 Calvani, N., Satoh, M., Croker, B. P., Reeves, W. H., Richards, H.B. (2003): Nephritogenic autoantibodies but absence of nephritis in IL-12p35-deficient mice with pristane-induced lupus. *Kidney Int.* 64: 897–905.

504 Kitawada, E., Lenda, D. M., Kelley, V. R. (2003): IL-12 deficiency in MRL-Fas(lpr) mice delays nephritis and intrarenal IFN-γ expression, and diminishes systemic pathology. *J. Immunol.* 170: 3915–3925.

505 Mizutani, A., Shaheen, V. M., Yoshida, H., Akoagi, J., Kuroda, Y., Nacionales, D. C., Yamasaki, Y., Hirakata, M., Ono, N., Reeves, W. H., Satoh, M. (2005): Pristane induced autoimmunity in germ-free mice. *Clin. Immunol.* 114: 110–118.

506 Traustadottir, K. H., Sigfusson, A., Steinsson, K., Erlendsson, K. (2002): C4A deficiency and elevated levels of immune complexes: the mechanism behind increased susceptibility to systemic lupus erythematosus. *J. Rheumatol.* 29: 2359–2366.

507 Carroll, M. (2001): Innate immunity in the etiopathology of autoimmunity. *Nat. Immunol.* 2: 1089–1090.

508 Fishelson, Z., Attali, G., Mevorach, D. (2001): Complement and apoptosis. *Mol. Immunol.* 38: 207–219.

509 Taylor, P. R., Carugati, A., Fadok, V. A., Cook, H. T., Andrews, M., Carroll, M. C., Savill, J S., Henson, P. M., Botto, M., Walport, M. J. (2000): A hierarchical role for the clasical pathway complement proteins in the clearance of apoptotic cells in vivo. *J. Exp. Med.* 192: 359–366.

510 Molina, H. (2002): Update on complement in the pathogenesis of systemic lupus erythematosus. *Curr. Opin. Rheumatol.* 14: 492–497.

511 Rovere, P., Vallinoto, C., Bondanza, A., Crosti, M. C. Rescigno, M., Ricciardi-Castagnoli, P., Rugarli, C., Manfredi, A. A. (1998): Cutting edge: bystander apoptosis triggers dendritic cell matura-

512 Dijstelbloem, H. M., van de Winkel, J. G. J., Kallenberg, C. G. M. (2001): Inflammation in autoimmunity: receptors for IgG revisited, *Trends Immunol.* 22: 510–516.

513 Casciola-Rosen, L., Andrade, F., Ulanet, D., Wong, W. B., Rosen, A. (1999): Cleavage by granzyme B is strongly predictive of autoantigen status: implications for autoimmunity. *J. Exp. Med.* 190: 815–825.

514 Doyle, H. A., Yan, J., Liang, B., Mamula, M. J. (2001): Lupus autoantigens: origins, forms, and presentation. *Immunol. Res.* 24: 131–147.

515 Lorenz, H-M., Grünke, M., Hieronymus, T., Winkler, S., Blank, N., Rascu, A., Wendler, J., Geiler, T., Kalden, J. R. (2002): Hyporesponsiveness to γc-chain cytokines in activated lymphocytes from patients with systemic lupus erythematosus leads to accelerated apoptosis. *Eur. J. Immunol.* 32: 1253–1263.

516 Georgescu, L., Vakkalanka, R. K., Elkon, K. B., Crow, M. K. (1997): Interleukin-10 promotes activation-induced cell death of SLE lymphocytes mediated by Fas ligand. *J. Clin. Invest.* 100: 2622–2633.

517 Haas, C., Ryffel, B., Le Hir, M. (1997): IFN-γ is essential for the development of autoimmune glomerulonephritis in MRL/lpr mice. *J. Immunol.* 158: 5484–5491.

518 Trouw, L. A., Groeneveld, T. W. L., Seelen, M. A., Duijs, M. G. J., Bajema, I. M., Prins, F. A., Kishore, U., Salant, D. J., Verbeek, J. S., van Kooten, C., Daha, M. R. (2004): Anti-C1q autoantibodies deposit in glomeruli but are only pathogenic in combination with glomerular C1q-containing immune complexes, *J. Clin. Invest.* 114: 679–688.

519 S. Takahashi, L. Fossati, M. Iwamoto, R. Motta, T. Kobayakawa, S. Izui (1996): Imbalance towards Th1 predominance is associated with acceleration of lupus-like autoimmune syndrome in MRL mice. *J. Clin. Invest.* 97: 1597–1604.

520 Zavala, F., Masson, A., Hadaya, K., Ezine, S., Schneider, E., Babin, O., Bach, J.-F. (1999): Granulocyte-colony stimulating factor treatment of lupus autoimmune disease in MRL-lpr/lpr mice. *J. Immunol.* 163: 5125–5132.

521 Oelzner, P., Deliyska, B., Funstuck, R., Hein, G., Herrmann, D., Stein, G. (2003): Anti-C1q antibodies and antiendothelial cell antibodies in systemic lupus erythematosus – relationship with disease activity. *Clin. Rheumatol.* 22: 271–278.

522 Fujimoto, M., Tsutsui, H., Xinshou, O., Tokumoto, M., Watanabe, D., Shima, Y., Yoshimoto, T., Hirakata, H., Kawase, I., Nakanishi, K., Kishimoto, T., Naka, T. (2004): Inadequate induction of suppressor of cytokine signaling-1 causes systemic autoimmune diseases. *Int. Immunol.* 16: 303–314.

523 Kinoshita, K., Tesch, G., Schwarting, A., Maron, R., Sharpe, A. H., Kelley, V. R. (2000): Costimulation by B7-1 and B7-2 is required for autoimmune disease in MRL-Fas^{lpr} mice. *J. Immunol.* 164: 6046–6056.

524 A.W. Gibson, J.C. Edberg, J. Wu, R.G.J. Westendorp, T.W.J. Huizinga, R.P. Kimberley, (2001): Novel single nucleotide polymorphisms in the distal IL-10 promoter affect IL-10 production and enhance the risk of systemic lupus erythematosus. *J. Immunol.* 166: 3915–3922.

525 Llorente, L., Richaud-Patin, Y., García-Padilla, C., Claret, E., Jakez-Ocampo, J., Cardiel, M. H., Alcocer-Varela, J., Grangeot-Keros, L., Alarcón-Segovia, D., Wijdenes, J., Galanaud, P., Emilie, D. (2000): Clinical and biological effects of anti-interleukin-10 monoclonal antibody administration in systemic lupus erythematosus. *Arthritis Rheum.* 43: 1790–1800.

526 Yin, Z., Bahtiyar, G., Zhang, N., Liu, L., Zhu, P., Robert, M. E., McNiff, J., Madaio, M. P., Craft, J. (2002): IL-10 regulates murine lupus. *J. Immunol.* 169: 2148–2155.

527 Theofilopoulos, A. N., Koundouris, S., Kono, D. H., Lawson, B R. (2001): The role of IFN-γ in systemic lupus erythematosus: a challenge to the TH1/TH2 paradigm in autoimmunity. *Arthritis Res.* 3: 136–141.

528 Segal, R., Bermas, B. L., Dayan, M., Kalush, F., Shearer, G. M., Mozes, E. (1997): Kinetics of cytokine production in experimental systemic lupus erythematosus. Involvement of T helper cell 1/T helper cell 2-type cytokines in disease. *J. Immunol.* 158: 3009–3016.

529 Schwarting, A., Tesch, G., Kinoshita, K., Maron, R., Eweiner, H. L., Kelly, V. R. (1999): IL-12 drives IFN-γ-dependent autoimmune kidney disease in MRL-fas^{lpr} mice. *J. Immunol.* 163: 6884–6891.

530 Liu, J., Beller, D. I. (2003): Distinct pathways of NF-κB regulation are associated with aberrant macrophage IL-12 production in lupus- and diabetes-prone mouse strains. *J. Immunol.* 170: 4489–4496.

531 Hagiwara, E., Okubo, T., Aoki, I., Ohno, S., Tsuji, T., Ihata, A., Ueda, A., Shirai, A., Okuda, K., Miyazaki, J., Ishigatsubo, Y. (2000): IL-12-encoding plasmid has a beneficial effect on spontaneous autoimmune disease in MRL/MRL lpr/lpr mice. *Cytokine* 12: 1035–1041.

532 Kalechman, Y., Gafter, U., Ping Da, J., Albeck, M., Alarcon-Segovia, D., Sredni, B. (1997): Delay in the onset of systemic lupus erythematosus following treatment with the immunomodulator AS101. Association with IL-10 inhibition and increase in TNF-a levels. *J. Immunol.* 159: 2658–2667.

533 Furuya, Y., Kawakita, T., Nomoto, K. (2001): Immunomodulating effect of a traditional Japanese medicine, hachimijiogan (ba-wei-di-huang-wan), on Th1 predominance in autoimmune MRL/MP lpr/lpr mice. *Int. Immunopharmacol.* 1: 551–559.

534 Jacob, C. O., Zang, S., Li, L., Ciobanu, Y., Quismorio, F., Mizutani, A., Satoh, M., Koss, M. (2003): Pivotal role of Stat 4 and Stat6 in the pathogenesis of the lupus-like disease in the New Zealand mixed 2328 mice. *J. Immunol.* 171: 1564–1571.

535 Dong, L., Ito, S., Ishii, K. J., Klinman, D. M. (2005): Suppressive oligonucleotides delay the onset of glomerulonephritis and prolong survival in lupus-prone NZB x NZW mice. *Arthr. Rheum.* 52: 651–658.

536 Mishra, N., Reilly, C. M., Brown, D. R., Ruiz, P., Gilkeson, G. S. (2003): Histone deacetylase inhibitors modulate renal disease in the MRL-lpr/lpr mouse. *J. Clin. Invest.* 111: 539–552.

537 Wong, C. K., Lit, L. C., Tam, L. S., Li, E. K., Lam, C. W. (2005): Elevation of plasma osteopontin concentration is correlated with disease activity in patients with systemic lupus erythematosus. *Rheumatol. Oxford* 44: 602–606.

538 Becker, K. G., Simon, R. M., Bailey-Wilson, J. E., Freidlin, B., Biddison, W. E., McFarland, H. F., Trent, J. M. (1998): Clustering of non-major histocompatability complex susceptibility candidate loci in human autoimmune diseases. *Proc. Natl. Acad. Sci. U.S.A.* 95: 9979–9984.

539 Silveira, P. A., Baxter, A. G. (2001): The NOD mouse as a model of SLE. *Autoimmunity* 34: 35–64.

540 Stoffels, K., Overbergh, L., Guilietti, A., Kasran, A., Bouillon, R., Gysemans, C., Mathieu, C. (2002): NOD macrophages produce high levels of inflammatory cytokines upon encounter of apoptotic or necrotic cells. *J. Autoimmunity* 23: 9–15.

541 Baxter, A. G., Horsfall, A. C., Healey, D., Ozegbe, P., Day, S., Williams, D. G. (1994): Mycobacteria precipitate an SLE-like syndrome in diabetes-prone NOD mice. *Immunology* 83: 227–231

542 Jordan, M. A., Silveira, P. A., Shepherd, D. P., Chu, C., Kinder, S. J., Chen, J., Palmisano, L. J., Poulton, L. D., Baxter, A. G. (2000): Linkage analysis of systemic lupus erythematosus induced in diabetes-prone nonobese diabetic mice by Mycobacterium bovius. *J. Immunol.* 165: 1673–1684.

543 Seery, J. P., Wang, E., Cattell, V., Carroll, J. M., Owen, M. J., Watt, F. M. (1999): A central role for $\alpha\beta$ T cells in the pathogenesis of murine lupus. *J. Immunol.* 162: 7241–7248.

544 Drappa, J., Brot, N., Elkon, K. B. (1993): The Fas protein is expressed at high levels on CD4$^+$CD8$^+$ thymocytes and activated mature lymphocytes in normal mice but not in the lupus-prone strain,

MRL lpr/lpr. *Proc. Natl. Acad. Sci. U.S.A.* 90: 10 340–10 344.

545 Fan, H., Longacre, A., Meng, F., Patel, V., Hsiao, K., Koh, J. S., Levine, J. S. (2004): Cytokine dysregulation induced by apoptotic cells is a shared characteristic of macrophages from non obese diabetic and systemic lupus erythematosus-prone mice. *J. Immunol.* 172: 4834–4843.

546 Teuscher, C., Hickey, W. F., Grafer, C. M., Tung, K. S. K. (1998): A common immunoregulatory locus controls susceptibility to actively induced experimental allergic encephalomyelitis and experimental allergic orchitis in BALB/c mice. *J. Immunol.* 160: 2751–2756.

547 Araujo, L. M., Lefort, J., Nahori, M. A., Diem, S., Zhu, R., Dy, M., Leit-de-Moraes, M. C., Bach, J. F., Vargaftig, B. B., Herbelin, A. (2004): Exacerbated Th2-mediated airway inflammation and hyperresponsiveness in autoimmune diabetes-prone NOD mice: a critical role for CD1d-dependent NKT cells. *Eur. J. Immunol.* 34: 327–325.

548 Misslitz, A. C., Bonhagen, K., Harbecke, D., Lippuner, C., Kamradt, T., Aebischer, T. (2004): Two waves of antigen-containing dendritic cells in vivo in experimental Leishmania major infection. *Eur. J. Immunol.* 34: 715–725.

549 Pinto, L. A., Pitrez, P. M., Fontoura, G. R., Machado, D. C., Jones, M. H., Graeff-Tiexeira, C., Stein, R.T. (2004): Infection of BALB/c mice with Angiostrongylus costaricensis decreases pulmonary inflammation response to ovalbumin. *Parasite Immunol.* 26: 151–155.

550 Tristao-sa, R., Ribeiro-Rodrigues, R., Johnson, L. T., Pereira, F. E., Dietze, R. (2002): Intestinal nematodes and pulmonary uberculosis. *Rev. Soc. Bras. Med. Trop.* 35: 533–535.

551 Becher, B., Durell, B. G., Noelle, R. J. (2002): Experimental autoimmune encephalitis and inflammation in the absence of interleukin-12. *J. Clin. Invest.* 110: 493–497.

552 Gran, B., Zhang, G. X., Yu, S., Li, J., Chen, X. H., Ventura, E. S., Kamoun, M., Rostami, A. (2002): IL12p35-deficient mice are susceptible to experimental autoimmune encephalomyelitis: evidence for redundancy in the IL-12 system in the induction of central nervous system autoimmune demyelination. *J. Immunol.* 169: 7104–7110.

553 Andersson, A., Kokkolaa, R., Wefer, J., Erlandsson-Harris, H., Harris, R. A. (2004): Differential macrophage expression of IL-12 and IL-23 upon innate immune activation defines rat autoimmune susceptibility. *J. Leukoc. Biol.* 76: 1118–1124.

554 A. Kukreja, G. Cost, J. Marker, C. Zhang, Z Sun, K. Lin-Su, S. Ten, M. Sanz, M. Exley, B. Wilson, S. Porcelli, N. Maclaren (2002): Multiple immunoregulatory defects in type-1 diabetes. *J. Clin. Invest.* 109: 131–140.

555 Braley-Mullen, H., Sharp, G. C., Tang, H., Chen, K., Kyriakos, M., Bickel, J. T. (1998): Interleukin-12 promotes activation of effector cells that induce a severe destructive granulomatous form of murine experimental autoimmune thyroiditis. *Am. J. Pathol.* 152: 1347–1358.

556 Ostlie, N. S., Karachunski, P. I., Wang, W., Monfardini, C., Kronenberg, M., Conti-Fine, B. M. (2001): Transgenic expression of IL-10 in T cells facilitates development of experimental myasthenia gravis. *J. Immunol.* 166: 4853–4862.

557 Im, S. H., Barchan, D., Maiti, P. K., Raveh, L., Souroujon, M. C., Fuchs, S. (2001): Supresion of experimental myasthenia gravis, a B cell-mediated disease, by blockade of IL-18. *FASEB J.* 15: 2140–2148.

558 Pedotti,R., DeVoss, J. J., Youssef, S., Mitchell, D., Wedemeyer, J., Madanat, R., Garren, H., Fontoura, P., Tsai, M., Galli, S. J., Sobel, R. A., Steinman, L. (2003): Multiple elements of the allergic arm of the immune system modulate autoimmune demyelination. *Proc. Natl. Acad. Sci. U.S.A.* 100: 1867–1872.

559 Carroll, M. C., Holers, V. M. (2005): Innate autoimmunity. 86: 137–157.

560 Souza, D. G., Vieira, A. T., Soares, A. C., Pinho, V., Nicoli, J. R., Vieire L. Q., Teixeira, M. M. (2004): The essential role of the intestinal microbiota in facilitating acute inflammatory responses. *J. Immunol.* 173: 4137–4146.

561 U. Nevo, I. Golding, A. U. Neumann, M. Schwartz, S. Akselrod (2004): Autoimmunity as an immune defense against degenerative processes: a primary mathematical model illustrating the bright side of autoimmunity. *J. Theor. Biol.* 227: 583–592.

562 J. Kipnis, T. Mizrahi, E. Hauben, I. Shaked, E. Shevach, M. Schwartz (2002): Neuroprotective autoimmunity: naturally occurring $CD4^+CD25^+$ regulatory T cells suppress the ability to withstand injury to the central nervous system. *Proc. Natl. Acad. Sci. U.S.A.* 99: 15620–15625.

563 Fisher, J., Levkovitch-Verbin, H., Schori, H., Yoles, E., Butovsky, Kaye, J. F., Ben-Nun. A., Schwartx, M. (2001): Vaccination for neuroprotection in the mouse optic nerve: implication for optic neuropaties. *J. Neurosci.* 21: 136–142.

564 Schwartz, M. (2001): Protective autoimmunity as a T cell response to central nervous system trauma: prospects for therapeutic vaccines. *Prog. Neurobiol.* 65: 489–496.

565 Schwartz, M., Shaked, I., Fisher, J., Mizrahi, T., Schori, H. (2003): Protective autoimmunity against the enemy within: fighting glutamate toxicity. *Trends Neurosci.* 26: 297–302.

566 Koide, N., Sugiyama, T., Mu, M. M., Mori, I., Yoshida, T., Hamano, T., Yokochi, T. (2003): Gamma-interferon-induced nitric oxide production in mouse CD5+ B1-like cell line and its association with apoptotic death. *Microbiol. Immunol.* 47: 669–679.

567 D. K. Dalton, L. Haynes, C-Q Chu, S. L. Swain, S. Wittmer (2000): Interferon γ eliminates responding CD4 T cells during Mycobacterial infection by inducing apoptosis of activated CD4 T cells. *J. Exp. Med.* 192: 117–122.

568 C-Q Chu, S. Wittmer, D. K. Dalton (2000): Failure to suppress the expansion of the activated CD4 T cell population in interferon-γ deficient mice leads to exacerbation of experimental autoimmune encephalomyelitis. *J. Exp. Med.* 192: 123–128.

569 J. Estaquier, J. C. Ameisen (1997): A role for T-helper type-1 and type-2 cytokines in the regulation of human monocyte apoptosis. *Blood* 90: 1618–1625.

570 M. J. Kaplan, D. Ray, R-R Mo, R. L. Yung, B.C. Richardson (2000): TRAIL (Apo2 ligand) and TWEAK (Apo3 ligand) mediate CD4+ T cell killing of antigen-presenting macrophages. *J. Immunol.* 164: 2897–2904.

571 Mellor, A., Baban, B., Chandler, P., Marshall, B., Jhaver, K., Hansen, A., Koni, P. A., Iwashima, M., Munn, D. H. (2003): Cutting edge: Induced indoleamine 2,3-dioxygenase expression in dendritic cell subsets suppresses T cell clonal expansion. *J. Immunol.* 171: 1652–1655.

572 Reynolds, D. N., Smith, S. A., Zhang, Y. P., Mengsheng, Q., Lahiri, D. K., Morassutti, D. J., Shields, C. B., Kotwal, G. J. (2004): Vaccinia virus complement control protein reduces inflammation and improves spinal cord integrity following spinal cord injury. *Ann. New York Acad. Sci.* 1035: 147–164.

573 Kooguchi, K., Hashimoto, S., Kobayashi, A., Kitamura, Y., Kudoh, I., Wiener-Kronish, J., Sawa, T. (1998): Role of alveolar macrophages in initiation and regulation of inflammation in Pseudomonas aeruginosa pneumonia. *Infect. Immun.* 66: 3164–3169.

574 C. Godson, S. Mitchell, K. Harvey, N.A. Petasis, N. Hogg, H.R. Brady (2000): Cutting edge: Lipoxins rapidly stimulate nonphlogistic phagocytosis of apoptotic neutrophils by monocyte-derived macrophages. *J. Immunol.* 164: 1663–1667.

575 H. Wenkel, J.W. Streilein, M.J. Young (2000): Sytemic immune deviation in the brain that does not depend on the integrity of the blood-brain barrier. *J. Immunol.* 164: 5125–5131.

576 Aloisi F, De Simone R, Columba-Cabezas S, Levi G. (1999): Opposite effects of interferon-g and prostaglandin E_2 on tumour necrosis factor and interleukin-10 production in microglia: a regulatory loop controlling microglia pro- and anti-inflammatory activities. *J. Neurosci. Res.* 56: 571–580.

577 T. Ravasi, C. Wells, A. Forrest, D.M. Underhill, B.J. Wainwright, A. Aderem, S. Grimmond, D.A. Hume (2002):

Generation of diversity in the innate immune system: macrophage heterogeneity arises from gene-autonomous transcriptional probability of individual inducible genes. *J. Immunol.* 168: 44–50.

578 B.F. De Blasio, P. Bak, F. Pociot, A.E. Karlsen, J. Nerup (1999): Onset of type i diabetes. A dynamic instability. *Diabetes* 48: 1677–1685.

579 Byrne, H. M., Owen, M. R. (2002): Use of mathematical models to simulate and predict macrophage activity in diseased tissues, in Burke, B., Lewis, C. E. (Eds.) *The Macrophage*: Second edition, Oxford University Press, Oxford, UK. 599–633.

11
Polymorphonuclear Neutrophils as Antigen-presenting Cells
Amit R. Ashtekar and Bhaskar Saha

11.1
Introduction

Polymorphonuclear neutrophils (PMN) play a fundamental role in the innate arm of the immune response. Their role as effective phagocytic scavengers is well documented and constitutes an important mechanism of primary host defense. They make up about two-thirds of peripheral blood leukocytes, with approximately 1.0×10^{11} cells being produced daily by normal adults [1]. PMN are absent from uninfected, uninjured, 'healthy' tissues and are the first cells to migrate and massively infiltrate the injured or pathogen-invaded tissue [2]; their migration is guided by chemotactic stimuli, which marks the initiation of inflammation. PMN entry into the tissue is controlled by dynamic interaction between adhesion molecules like integrins, selectins and immunoglobulins, expressed by PMN and the endothelium [3]. PMN maintain a formidable defense system against bacterial invaders and their principal role in an inflammatory immune response is the acute phagocytic clearance of foreign pathogens, through the generation of reactive oxygen intermediates and the release of pre-formed lytic enzymes and inflammatory mediators [4]. In keeping with this view PMN were generally believed to be terminally differentiated cells lacking the ability to synthesize proteins and playing only a passive role through phagocytic clearance of pathogens (Figure 11.1). Only in recent years is the role of PMN in antigen processing and presentation being given a shape.

Since research on the molecular nature of antigen processing and presentation during 1970s and 1980s was carried out primarily on B cells, macrophages and, later, on dendritic cells, the role of B cells, dendritic cells (DCs) and macrophages as professional antigen presenting cells (APCs) – capable of inducing efficient primary T cell-mediated immune response to foreign antigens – is well documented and accepted. DCs are the most specialized APCs, inducing T cell activation after antigen capture and migration in the regional lymph nodes. Thus, they are proposed to be the initiators of immune responses to foreign antigens [5]. They are also considered to be key players in the induction and maintenance of autoimmune reactions [6]. Similarly, macrophages are designed to function as potential

---

Antigen Presenting Cells: From Mechanisms to Drug Development. Edited by H. Kropshofer and A. B. Vogt
Copyright © 2005 WILEY-VCH Verlag GmbH & Co. KGaA, Weinheim
ISBN: 3-527-31108-4

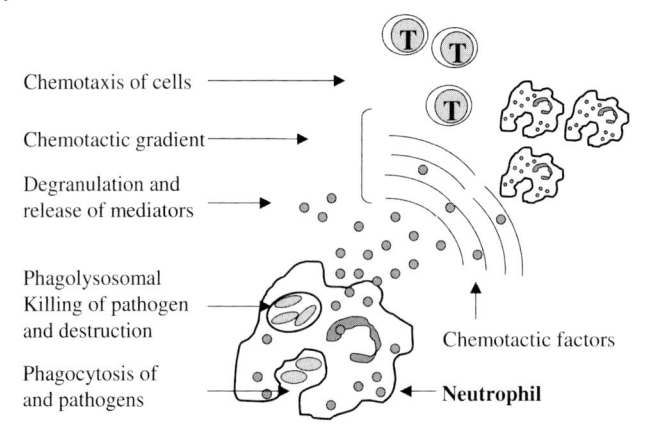

Figure 11.1 Earlier concepts about the functions performed by polymorphonuclear neutrophils (PMN). According to this early concept, PMN are terminally differentiated, transcriptionally inactive cells. The primary functions are phagocytic clearance of the antigens, foreign pathogens and tissue debris. The preformed mediators in the neutrophilic granules are released upon activation and kill pathogens extracellularly, induce chemotaxis of different blood cells, and host-tissue destruction in the case of exaggerated or prolonged activation.

APCs as well as efficient phagocytic killers [7]. Also, B cells bearing the antigen-specific receptors in the form of immunoglobulins can capture antigens directly and present to T cells. In this scenario neutrophils have received much less attention and have never been viewed as inducers of T cell response. Basically, the need to change the view towards neutrophils never become evident till some discrete and scattered reports, though not directly, pointed towards the prominent role that neutrophils can play in the afferent limb of the immune response.

Interestingly, research accumulated over the past several years have revealed a novel role of PMN in various infectious and autoimmune diseases, thus raising their caliber from a mere servile phagocyte to dominant inducers of a T cell dependant immune response [8]. PMN exhibit most of the parameters that categorize a particular cell as a prominent contributor towards the adaptive immune response. PMN contribute significantly to the inductive limb of the immune response by modulating both cellular and humoral immunity, particularly by the production of various immunoregulatory cytokines [9]. The ability to respond to and secrete various cytokines endows PMN with the ability to dictate a specific T_h subset promotion or modulate the cell pool at the site of infection and thereby the disease outcome. Additionally, the potential role of PMN to directly influence the inductive phase of an immune response was established when it was shown that granulocyte-macrophage colony stimulating factor (GM-CSF), IFN-γ and IL3 induced the expression of major histocompatibility complex (MHC) class II molecules on PMN [10]. Corroborating these findings were reports showing the expression of co-stimulatory molecules and the capacity to process and present antigens to T cells in MHC class II restricted manner. Though the ability of PMN to act as an

APC might not be comparable with professional APCs, their existence in large numbers can contribute significantly towards the total antigen presentation. Moreover, the findings that immediate precursors of end stage neutrophilic PMN can be reverted in their functional maturation program and driven to acquire the features of DC, which are more potent APCs [11, 12], define a new aspect of how PMN can modulate antigen presentation. Compared with the well-acclaimed heterogeneity of macrophages and dendritic cells that contribute to the differential regulation of immune responses, the existence and purpose of PMN heterogeneity [13, 14] remains largely unknown, but has the potential to shed light on the variable role of neutrophil subsets in various diseases. PMN can also interact with other tissue histiocytes [15, 16], thus altering their effector functions and contributing towards the final T cell response. Another fundamental aspect of neutrophils is their capacity to migrate, allowing them to exert a continuous surveillance for incoming antigens in almost all body tissues and a prompt report to T cells in secondary lymphoid organs.

Despite the controversy that surrounds the role of PMN as inducers of T cell response, enough evidence exists to positively propose a novel role for them. This chapter tackles a still debatable but definitively maturing concept and is intended to review the latest findings that support the concerned notion and provide explanations to certain limitations, highlighting work that remains to be done. The chapter is thus intended to help the reader to de-establish long held concepts about the functions of neutrophils as mere scavengers and to develop a different, novel perspective on them while appreciating their contributions as the 'first line of defense against invading pathogens'.

11.2
PMN as Antigen-presenting Cells

11.2.1
Basic Criteria of an APC for T Cells

To function as an APC, a cell needs to satisfy several criteria, including the capacity to collect and cleave antigens, generate antigenic determinants, form stable complexes of the peptides with MHC class I/ II molecules and to express (constitutive or inducible) co-stimulatory molecules (Figure 11.2). To guide T-cell differentiation to a particular subset or phenotype, they also need to secrete cytokines, creating a milieu conducive for T-cell differentiation. It is stated here that neutrophils satisfy all these criteria sufficiently to be regarded as APCs and inducers of T cell dependent immune response.

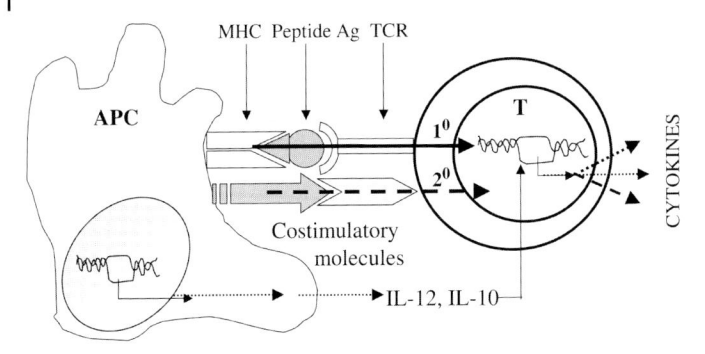

Figure 11.2 An antigen-presenting cell (APC) has to satisfy two minimum basic pre-requisites. There are three basic requirements that a cell needs to satisfy to be considered as an APC. Besides expression of MHC class-I expression, professional antigen-presenting cells express MHC class-II molecules. Besides MHC molecule expression, the APC is also required to express co-stimulatory molecule and cytokines. The MHC molecules bind the antigenic peptide, excepting superantigens that do not need to be processed and are presented intact, and present the antigen to T cell receptor (TCR), forming a ternary complex. The ternary complex formation results in the first signal that initiates T cell activation and gene transcription. The co-stimulatory receptors on APC bind their respective ligands on T cells to deliver the second signal to T cells that results in either potentiation or inhibition of the effects of the first signal. Cytokines secreted by the APC are often crucially important as they may modulate the T cell differentiation further.

11.2.2
Acquisition of Antigens

The efficiency of antigen capture is one of the major determinants of the level of expression of antigenic peptides on class II MHC molecules [17] and is determined by (a) the presence of cell surface receptors, (b) clustering of those receptors in invaginating endocytic regions, (c) the overall endocytic activity of cells, and (d) the physical form of antigen, i.e., soluble versus particulate [18]. Although particulate antigens are more pliable than soluble antigens when intracellular processing is concerned, the overall recognition of exogenous antigens by class II restricted T cells follow more complex rules. In addition, the concentration of the MHC class II-antigen complexes that accumulate on the APC exposed to an antigen depends on several factors: (a) the concentration of free antigen, (b) the efficiency of antigen capture by APC, (c) the efficiency of processing, (d) the concentration of class II binding molecules with available binding sites, and (e) the stability of the complex. Although a defined comparison exists between the established APCs and PMN in these regards, neutrophils can process and present soluble as well as particulate antigens in a MHC class II restricted manner to T cells [19, 20].

Unassisted phagocytosis, macro-pinocytosis, and receptor-mediated antigen uptake that also includes uptake of opsonized particles, all contribute to antigen uptake by neutrophils (Figure 11.3). The phagocytic recognition of particles is mediated in part by complement receptor 1 (CR1) (CD35) [21] and CR3 (CD11b/CD18) [22], which recognize C3b and iC3b, respectively, and Fc-gamma receptor I (FcγRI) (CD64) and FcγRIII (CD16), which bind the Fc portion of IgG molecules [23]. PMN can modulate the expression of these molecules and can rapidly increase, within 15 min, the expression of CR1 and CR3, by up to 10-fold upon exposure to chemotactic peptides [24, 25]. Additionally, human neutrophils express a receptor that binds the C3d region of iC3b and C3dg and is distinct from CR1, CR2, and CR3 [26]. Indeed, neutrophils may have a distinct and more efficient antigen uptake mechanism than other non-professional APCs like eosinophils or epithelial cells, which can confer more antigen presentation functions upon them.

Under variable in vivo conditions any of these factors can become limiting and can be compensated by increasing any of the others. Thus, a decrease in antigen concentration can be compensated by increasing the concentration of class II molecules [27] or increasing the efficiency of antigen uptake or possibly by improving the subsequent antigen processing. Indeed, antigen-specific B cells present antigen to T cells at 10^3–10^4 times lower antigen concentration than do nonspecific B cells [28]. As neutrophils can transcribe, translate and express MHC II molecules on the surface only after appropriate stimulation [10] it seems that this con-

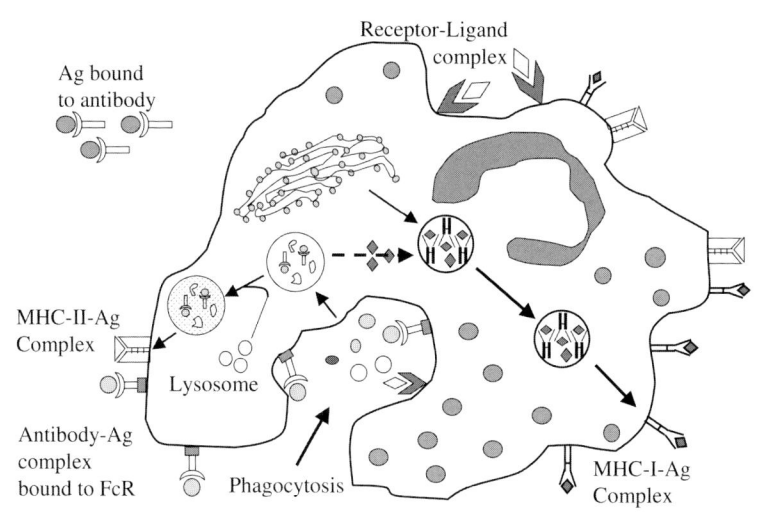

Figure 11.3 Neutrophils can acquire antigens by several different mechanisms. PMN are highly active phagocytes. As a result, the antigens are often phagocytosed unassisted and, sometimes, as an opsonized complex. In the latter mode, the antigens are complexed with the antibody or a soluble protein like C3b, and the complex is recognized by respective receptors on the PMN surface. Once the complex binds to the receptor in sufficient number, they are internalized and degraded. The other pathway is direct receptor mediated endocytosis, where an antigen binds directly to its receptor on the PMN surface and is endocytosed. Another characteristic feature of PMN antigen acquisition is extracellular proteolysis. Like immature dendritic cells, PMN secretes metalloproteinases that cleave extracellular proteins and the resulting peptides are often internalized and presented.

dition may be limiting. Therefore, an immense capacity of neutrophils to collect antigens by various means may be viewed as compensatory mechanisms for their low MHC class II expression.

The effectiveness of MHC class II antigen presentation is extremely dependent on the quantity of antigen delivered to the processing compartment [17, 18]. Only small amounts of antigen will be acquired by fluid phase endocytosis unless the local concentration of antigen is high. While phagocytosis provides a large source for particulate materials, receptor-mediated uptake is critical for non-particulate soluble antigens. The ability of cells to use antibodies to facilitate antigen acquisition is most relevant. By constructing a fusion protein that targets prostate specific antigen (PSA) to FcγRI on antigen presenting cells it has been shown that PSA was internalized and processed by the human myeloid THP-1 cell line, resulting in presentation of MHC class I-associated PSA peptides and lysis of THP-1 by PSA-specific human CTL [29].

Because PMN are the prototypical innate immune cell, and toll-like receptors (TLRs) are the archetypal innate immune pattern recognition receptors, many researchers have reported an important role of TLRs in modulating the neutrophil function [30–32]. Human neutrophils express most of the TLRs so far described: TLRs 1, 2, and 4–10 [33]. Upon binding to respective ligands, these receptors on PMN signal to regulate the rate of phagocytosis, IL-8 production, chemotaxis in response to IL-8, shedding of L-selectin from their surface and priming for N-formylated-methionine-leucine-phenylalanine peptide (fMLP)-mediated super oxide production [33]. Interestingly, TLR-2 internalizes antigens for presentation to T cells [34] and this describes how TLRs can potentiate a cell like neutrophil, so far conceived to have functions only in innate immunity, to act as an inducer of T cell immunity. This also describes a very novel and direct mechanism whereby TLRs can link innate to adaptive immunity. Though TLRs can modulate various functions in different cells, they themselves are under the control of different modulators [35], implying the existence of a well-coordinated functional loop whereby appropriate immune homeostasis is preserved even in the face of disturbing antigenic challenges. Therefore, at the site of infection, TLRs help prepare neutrophils to better combat the intruders in balanced fashion.

11.2.3
Antigen Processing

Antigen processing refers to the multiple biochemical and cellular events that occur within APCs that result in proteolytic degradation of antigens, association of the fragments with MHC molecules, and expression of the peptide-MHC molecules at the cell surface where they can be recognized by the T cell receptor on a T cell. The generation of appropriate epitopes is crucial and requires limited proteolytic processing to generate appropriate peptides for binding to MHC molecules [36]. This process requires intricate orchestration and interplay among MHC class II molecules, the associated chaperon termed Ii (invariant chain), antigenic peptides and acidic proteases [cathepsins (Cat)] residing within endosomal compartments. Various proteolytic enzymes – including cysteine, aspartic and serine proteases, and metalloproteases – reside in the endocytic pathway of most cell types. Those prevalent in APCs are the aspartic protease Cat D and the cysteine proteases Cat B, F, L, S and Z, and asparaginyl endopeptidase (AEP) [37].

Identification of the proteases involved in degradation of endocytosed antigens and Ii has been a matter of intense investigation. Proteolytic enzymes secreted by neutrophils include cathepsins B, D, S, G, etc. [38, 39]. But it seems that there is much redundancy in the functions of cathepsins. Indeed, even in the absence of CatB and CatD, MHC class II processing can function normally [40]. This points to the fact that neutrophils and other APCs may have yet undiscovered enzymes that function actively in antigen processing to generate immunogenic (and/or tolerogenic) epitopes. Conversely, the regulation of endosomal protease activity through the activity of inhibitors also plays an important role in modulating the endosomal environment. The modulation of these parameters in neutrophils may explain the variation among the capacity of individual subsets to process and present antigens in different experimental situations. While GM-CSF and IFN-γ induce the expression of MHC class II molecules in PMN, IFN-γ up-regulates CatS [37], which is highly expressed in APCs and plays a key role in processing Ii in human B cells [41]. This proves that cytokines have a multi-dimensional role to play in modulating the APC functions. As TLRs have been increasingly implicated as active linkers of innate and adaptive immunity [34], signaling through them may also be linked to the expression of proteases, whereby TLRs may up- or down-regulate specific proteases or inhibitors of proteases, depending upon the ligand detected.

Current knowledge on antigen processing is based on (a) the use of inhibitors of different proteases, (b) localization of enzymes to particular cells and sub-cellular compartments, and (c) attempts to reconstruct antigen (and invariant chain) processing in vitro with defined proteases [18]. Depending upon the presence and localization of appropriate proteolytic enzymes in a particular APC, an entirely different degradative scheme of proteolytic cleavages could occur, thus generating either a variety or, more appropriately, unique antigenic determinants. Among various factors that regulate the generation of antigenic determinants, proteases play the most important roles. In certain cases, a crucial first step may involve a single cathepsin and thus allow the processing of diverse determinants on a molecule. As shown for AEP, it carries out the rate-limiting cleavage during the processing of microbial tetanus toxoid C fragments [42]. This variable presence of degradative enzymes thus explains the heterogeneity in antigen processing systems. Indeed, neutrophil derived cathepsins and gelatinases are reportedly involved in generating neo-epitopes that can be candidates for presentation to T cells [43, 44].

The processing of soluble antigens, like tetanus toxoid (TT), needs the presence of specific proteases and precise access to generate efficient antigenic determinants. In fact, neutrophils are capable of inducing proliferation of TT-specific T cells [19]. Proliferation of only $CD4^+$ (not $CD8^+$) T cells was induced, and only with autologous but not heterologous PMN. Three hours of contact with exogenous TT was sufficient to induce proliferation of TT-specific T cells provided that the PMN had been pre-incubated with IFN-γ and GM-CSF. Therefore, with the use of TT it was demonstrated that PMN can process protein and present it as peptides. To the contrary, in a different study, activation of human T cells by MHC class II expressing neutrophils was not observed with TT [20]. However, the negative finding might be due to any one or all of the following factors. As it has been shown that the expression pattern of MHC class II molecules is donor and subset dependent [10], the antigen processing cascades may also vary accordingly. Additionally, the generation of determinants depends not only on protease specificity but also by protease accessibility and effective presentation requires a balance between generation and destruction of T cell determinants. Sometimes certain potential determinants fail to gain the opportunity to address T cells. In this case, the concept of dominant and cryptic determinants can be called into question. A dominant T cell determinant induces a strong T cell response and, conversely, cryptic determinant makes little impact on the immune response, either to induce anergy or tolerance, unless its display is up-regulated. Though neutrophils may be able to process TT, they may not be able to generate dominant determinants, due to which T cell proliferation is not initiated. Though this could rely on many factors, in the case of neutrophils with extreme degradative capabilities, it seems that excessive processing of antigens may lead to enzymatic destruction of determinants; in other words, in certain conditions, the half-life of the antigenic determinants generated in PMN can be too short to form the complex with MHC-II and subsequent presentation to T cells.

During such situations where certain peptides are not presented by neutrophils another route of antigen presentation can become pre-eminent. DCs have proteases and peptide-receptive MHC molecules – H-2M or HLA-DM – all together at the cell surface, suggesting that these cells are able to present antigens by direct binding at the cell surface [45]. The fact that neutrophils exhibit intracellular as well as surface expression of MHC class II molecules after culture with IFN-γ GM-CSF and IL-3 [10] and also posses a wide array of extracellular as well as intracellular proteases [18, 37, 46] suggests that neutrophils might also be able to present antigens by direct binding at the cell surface without the requirement for internalization and endosomal processing. In fact, recombinant expression of HLA-DR and HLA-DM at the cell surface is sufficient for efficient surface binding and presentation of peptides in transfected COS-7 cells [47]. Additionally, neutrophils may partly contribute or reciprocate to DC or B cells by either processing antigens extracellularly or presenting the antigens processed by other cells. This mode of antigen processing and presentation may assume a crucial importance in different disease conditions like autoimmune diseases.

Extracellular proteolytic enzymes play a major role in pathogenesis of most autoimmune diseases [44, 48–50] and various physiological and pathologic phenomena, such as leukocyte migration, bone resorption and cancer cell metastasis [51–56]. Matrix metalloproteinases (MMPs) are a host cell-derived proteolytic enzyme family that degrades the extracellular matrix. The proteolytic activities of matrix metalloproteinases as well as their inhibitors are important in maintaining the integrity of the extracellular and MMPs are increasingly being implicated in the pathogenesis of several autoimmune diseases [57]. Immune mechanisms differ from one autoimmune disease to another. In many, including multiple sclerosis, it appears that the specific immune system (T-cells and B-cells) are targeting antigens derived from the body's own tissue. Neutrophil-derived MMPs have a great impact on the generation of unique extracellular antigenic determinants in various autoimmune diseases. It is produced mainly by neutrophils but is also produced by various other blood-derived cell types, e.g., monocytes, macrophages, eosinophils, and leukemic cells [58, 59]. Of those enzymes, MMP-2 (gelatinase A) and MMP-9 (gelatinase B), also called type IV collagenases or gelatinases, are related enzymes that break down type IV collagen. Increased levels of expression of MMP-9 and MMP-2 have been observed in Alzheimer's disease, stroke, multiple sclerosis, acute respiratory distress syndrome and amyotrophic lateral sclerosis [60–63]. Demyelination is the major underlying factor responsible for the symptoms of multiple sclerosis [56], which causes the destructive removal of myelin, an insulating and protective protein that covers neurons, and MMPs could be responsible for the influx of inflammatory mononuclear cells into the central nervous system, contribute to myelin destruction and disrupt the integrity of the blood–brain barrier.

Immature DC have a more potent capacity to generate MMPs and have been implicated in many autoimmune and demyelinating diseases [45]. As neutrophils trans-differentiate to and acquire DC like characteristics [11, 12], in this way they may be more potent in synthesizing MMPs and this could explain another mechanism by which neutrophils can add to extracellular antigen processing. Indeed, ultraviolet (UV) irradiation increases expression of MMP-8 in human skin in vivo, an increment associated with infiltration into the skin of neutrophils, which are the major cell type that expresses MMP-8 [64, 65]. Additionally, UV preferentially induces the recruitment of memory $CD4^+$ T cells in normal human skin [66], which can now interact with neutrophils and initiate an immune response. Apart from the UV-induced effects of the APC on T_h cells, it was shown that specific macrophage functions, like phagocytosis and killing of intracellular organisms, were impaired [67]. Thus, in these situations where localized APC are impaired of the basic functions of antigen presentation, neutrophils can compensate the loss in immune functions and initiate a T cell response [8].

Despite the apparent segregation of the class I and class II pathways, antigens from intracellular pathogens, including *Mycobacterium* spp., *Escherichia coli*, *Salmonella typhimurium*, *Brucella abortus* and *Leishmania* spp., elicit an MHC class-I-dependent $CD8^+$ T-cell response, a process referred to as cross-presentation [68]. Neutrophils process phagocytosed bacteria via an alternative MHC-I antigen-processing pathway that allows MHC-I presentation of peptides derived from the bacteria to $CD8^+$ T cells [68, 69]. This alternative antigen-processing pathway starts with phagocytosis of exogenous antigens because cytochalasin D, an inhibitor of phagocytosis, inhibits the processing. Unlike the classical MHC I antigen-processing pathway, lactacystin, a proteasome inhibitor, or brefeldin A, which blocks anterograde transport from the endoplasmic reticulum through the Golgi apparatus, do not affect this alternative antigen processing. Therefore, it was proposed that neutrophils phagocytose bacteria and either process them wholly in the vacuolar compartments or deliver the exogenous antigens from vacuolar compartment to the cytosol for further processing. But the continued processing of bacterial antigens by neutrophils in conditions that block cytosolic processing indicates that alternative MHC-I antigen processing by neutrophils occurs by vacuolar mechanisms. Additionally, neutrophils and macrophages are both capable of alternative MHC-I processing and presentation, but neutrophils are more efficient than macrophages at the same titer [68]. This can be explained by the extensive phagocytosis, microbicidal functions and high catabolic capabilities of neutrophils over macrophages.

Another way by which neutrophils can aid antigen presentation was observed when they were shown to regurgitate peptides after initial processing, and can thus directly or indirectly boost antigen presentation [68, 69]. Regurgitated peptides can contribute to direct antigen presentation by binding to MHC-I molecules on neutrophils themselves or in an indirect manner by delivering peptides onto MHC class I or even class II molecules of other APCs, thus affecting different stages of T cell responses like amplification of CD8 responses or affecting the longevity of memory T cells.

11.2.4
Expression of MHC Class I/II and Co-stimulatory Molecules

Being educated how to distinguish between self and non-self, T cells leave thymus and detect antigens displayed in context with MHC molecules on APCs in the periphery. Class I MHC molecules are expressed constitutively on almost all nucleated cells of the body, while constitutive expression of class II molecules is restricted to certain cells of the immune system – B cells, macrophages and DC – although expression of class II MHC may be induced on other cell types at sites of inflammation. According to the binary logic of antigen presentation, the path leading to the association of protein fragments differs for MHC class I and class II molecules [70]. Though alternatives exist, conventionally, MHC class I molecules present degradation products derived from intracellular (endogenous) proteins in the cytosol and MHC class II molecules present fragments derived from extracellular (exogenous) proteins that are located in an intracellular compartment.

Neutrophils express MHC class I molecules and present endogenous antigens to $CD8^+$ T cells. Additionally, as stated earlier, they can also present exogenous antigens via alternative routes of presentation [68, 69]. Constitutive expression of MHC class II antigens and of the co-stimulatory receptors CD80 and CD86 is restricted to professional antigen presenting cells. Neutrophils of healthy donors are reported to be negative for those antigens. The appropriate constitutive and inducible expression of class II MHC antigens is essential for normal immune function; thus, unsurprisingly, an aberrant expression on cell types normally class II MHC negative has been correlated with various autoimmune disorders, and lack of expression results in a severe combined immunodeficiency disorder called bare lymphocyte syndrome (BLS) [71]. In a related study, however, in addition to the other factors contributed by neutrophils, PMN of patients with active Wegener's granulomatosis acquired MHC class II antigens, thus implying their role in autoimmune diseases [72]. Likewise, in rheumatoid arthritis synovial fluid neutrophils synthesize and express large amounts of class II MHC but not co-stimulatory molecules [73]. Therefore, in entirely contrasting situations where PMN express only MHC class II molecules or both MHC class II molecules and co-stimulatory molecules, the T cells may undergo anergy or be activated optimally, respectively. This might underlie a novel interaction with T cells that is important in terms of disease pathology. Indeed it has been shown that, during Ag presentation by monocytes to T cells, bystander PMN are induced to express class II molecules. Recent studies have shown that neutrophils, when activated by the correct combination of cytokines, can be induced to express cell surface MHC class II (DR) antigen, CD80 (B7.1) and CD86 (B7.2) molecules, which are required for antigen presentation and subsequent T-cell activation. Corroborating this fact are results that demonstrated the expression of MHC class II molecules in PMN in response to GM-CSF, IFN-γ and IL-3 in order of descending efficiency [10].

Some possible explanations for the apparent lack of MHC class II expression on neutrophils, as reported by some researchers, can be rendered. Primarily, PMN expressing class II molecules ranged from as low as 13% to a high of 72% [10]. Similarly, the number of MHC class II molecules per cell on PMN ranged from 3313 to 15 713. These findings suggest that MHC class II distribution varies in different individuals and the failure in obtaining class II expression in some experiments may partially be attributed to the sensitivity of the detection system used. A different study, examining the effect of neutrophil isolation on MHC class II expression, found that cell surface expression of MHC class I and class II molecules is dramatically reduced on neutrophils after purification [74]. The level of expression of MHC class I antigens is seasonal and donor-dependent and rapidly decreases after in vitro culture despite negligible necrosis and apoptosis of neutrophils. Although treatment with IFN-γ partially prevents the loss of MHC class I molecules on neutrophils, it is relatively inefficient to induce MHC class II antigens. The unique in vivo conditions at the site of infection may be more conducive for MHC expression on neutrophils and that the expression seems to be more stringent and regulated.

Another way by which PMN can gain MHC class II molecules and thus participate in direct antigen presentation in a class II-restricted manner is through trans-differention to either DC or macrophages [11, 12, 75]. Trans-differentiation involves cells of a specific lineage and committed to a phenotype, altering their functional as well as phenotypic traits and finally leading to a different cell type. Various reports confirm the capacity of pre-end stage neutrophils to trans-differentiate. Highly purified lactoferrin-positive, immediate precursors of end-stage neutrophilic PMN (PMNp), can be reverted in their functional maturation program and driven to acquire characteristic DC features (Figure 11.4). Upon culture with the cytokine combination GM-CSF, IL4 and TNF-α, they develop DC morphology and acquire molecular features characteristic of DCs. These molecular changes include neo-expression of the DC-associated surface molecules – CD1a, CD1b, CD1c, human leukocyte antigen (HLA)-DR, HLA-DQ, CD80, CD86, CD40, CD54, and CD5 – but down-regulation of the markers of other lineages, e.g., CD15 and CD65s. Additional stimulation with CD40 ligand also induces expression of CD83 and augments CD80, CD86, and HLA-DR expression. Neutrophil-derived DCs are potent T cell stimulators in allogeneic, as well as autologous, mixed lymphocyte reactions, whereas freshly isolated neutrophils are unable to do so. In addition, neutrophil-derived DCs are at least 10 000 times more efficient than freshly isolated monocytes in presenting soluble antigen to autologous T cells. Also, in functional terms, these neutrophil-derived DCs thus closely resemble "classical" DC populations.

Neutrophils can also be trans-differentiated into monocytes [75]. It is generally recognized that post-mitotic neutrophils give rise to polymorphonuclear neutrophils alone. Nonetheless, a lineage switch of human post-mitotic neutrophils into macrophages in culture has been demonstrated. When the $CD15^+CD14^-$ cell population, which predominantly consists of band neutrophils, was cultured with GM-CSF, TNF-α, IFN-γ IL4 and subsequently with macrophage colony-stimulating factor (M-CSF) alone, the resultant cells had morphologic, cytochemical, and phenotypic features of macrophages. In contrast to the starting population, these cells are negative for myeloperoxidase, neutrophil-specific esterase, and lactoferrin, and they up-regulated non-specific esterase activity and the expression of the receptor for M-CSF, mannose receptor, and HLA-DR. $CD15^+CD14^-$ cells pro-

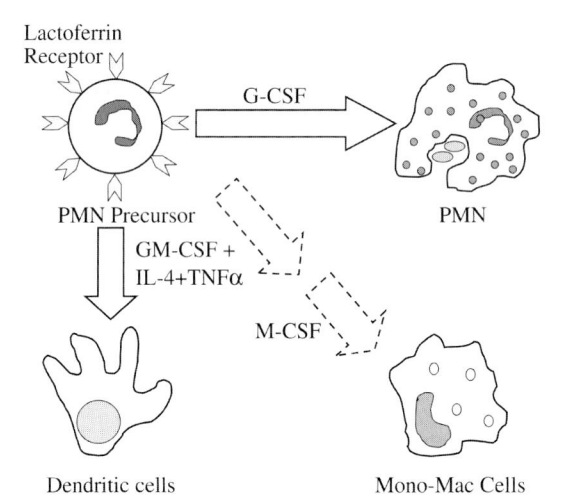

Figure 11.4 Neutrophils can be trans-differentiated to macrophages (mono-mac cells) and dendritic cells. The lactoferrin receptor-positive PMN precursors that belong to "granulocyte lineage" can be differentiated to cells of "myeloid lineage" such as monocyte-macrophages and dendritic cells. Since this phenomenon, termed trans-differentiation, results in cells that are much more potent as antigen presenting cells, the phenomenon is of crucial significance in the elicitation of T cell responses.

ceeded to macrophages through the CD15⁻CD14⁻ cell population. Macrophages derived from CD15⁺CD14⁻ neutrophils had phagocytic function. Therefore, in response to cytokines, post mitotic neutrophils can become macrophages and this may serve as an alternate route for neutrophils trans-differentiation.

The factors that regulate neutrophil activation and trans-differentiation to DCs or mono-mac cells can also regulate their life span. PMN are considered to be short-lived cells that undergo spontaneous apoptosis if not appropriately stimulated. Granulocyte apoptosis is now acknowledged to have a critical role in the progression of inflammatory responses. Granulocytes are preprogrammed to die with important physiological mechanisms for non-inflammatory clearance. Shutdown of secretory capacity represents an important aspect of the program of biochemical events that accompany neutrophil apoptosis together with surface molecular changes that serve to identify apoptotic cells as targets for phagocytic removal. In contrast to necrosis, apoptosis provides a granulocyte clearance mechanism that would tend to limit tissue injury and promote resolution, rather than persistence, of inflammation [76]. The control of neutrophil turnover in the circulation is a key event in homeostasis and inflammation. Various cytokines modulate neutrophil apoptosis and can be viewed as a strategy to modulate the antigen presenting function of neutrophils. In order that neutrophils can present antigens it becomes necessary that they be long lived so that they can acquire necessary surface molecules, process and present antigens. Cytokines detected in the circulation during sepsis like TNF-α, IFN-γ, granulocyte colony stimulating factor (G-CSF), GM-CSF inhibit neutrophil apoptosis while IL-10 counteracts inhibition of neutrophil apoptosis induced by LPS, TNF-α, IFN-γ, G-CSF and GM-CSF, whereas IL-4 or IL-13 are ineffective [77–79], thus associating the prolonged life span of neutrophils with the inflammatory conditions, which are counteractively regulated by the anti-inflammatory cytokines like IL-10 and TGF-β. The prolonged life span is important in that it can give neutrophils a better chance to interact and present antigens to T cells.

11.2.5
Delivery of Second Signal

Although antigen-receptors are able to recognize a wide array of antigens, this alone provides very little information about the context in which these antigens exist – whether they arise from pathogens or whether they are self-antigens. Therefore, although antigen-specific signals are required, a second co-stimulatory, which adds an additional layer of regulation, is necessary (Figure 11.5). Thus, signal 1 is generated by interaction of an MHC-antigenic peptide complex with the TCR-CD3 complex. The interaction of CD28 with one of the B7 molecules (CD80 [B7.1] and CD86 [B7.2]) on professional APCs is generally considered as the most important co-stimulatory signal for T cell activation. APC in a resting condition express either no or only low levels of B7 molecules and are up-regulated as a result of interactions with microbial products or inflammation. These cellular interactions, which occur in the specialized microenvironment of secondary lym-

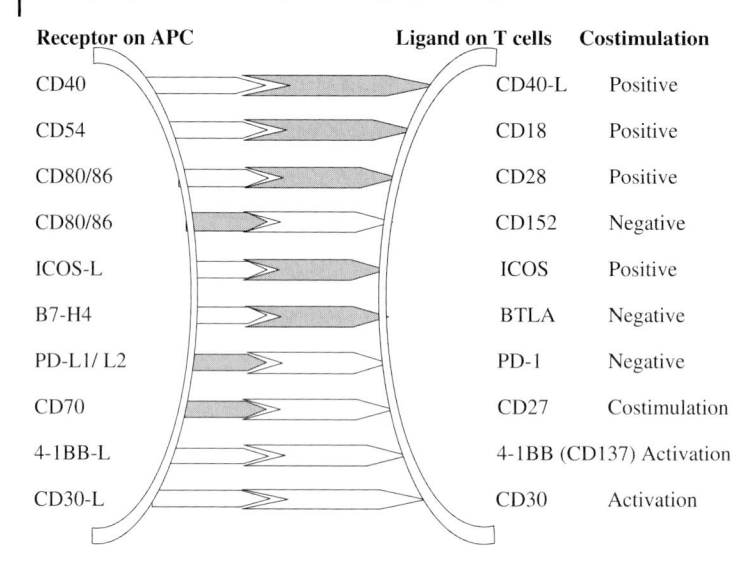

Receptor on APC	Ligand on T cells	Costimulation
CD40	CD40-L	Positive
CD54	CD18	Positive
CD80/86	CD28	Positive
CD80/86	CD152	Negative
ICOS-L	ICOS	Positive
B7-H4	BTLA	Negative
PD-L1/L2	PD-1	Negative
CD70	CD27	Costimulation
4-1BB-L	4-1BB (CD137)	Activation
CD30-L	CD30	Activation

Figure 11.5 Several co-stimulatory molecules are reported to deliver the second signal to T cells. Receptors on the APC and their respective ligands on T cells are listed here. Of all these ligands on T cells, some deliver positive co-stimulation (dark shading) while some deliver negative co-stimulation (no shading). While positive co-stimulation potentiates the TCR-driven activation of T cells, negative co-stimulation suppresses the first signal. Therefore, the observed T cell activation is often a resultant effect of both signals. Details of how these three signals are quantitatively and qualitatively interrelated to induce the observed T cell function are yet to be deciphered.

phoid tissue, are critical for IL-2 production and activation of naïve T cells. Fundamentally, these co stimulatory interactions serve the following functions: they strengthen TCR signals, promoting cell survival, proliferation and cytokine production. Newer families of B-7 molecules are differentially expressed on various APCs and can differentially modulate the initiation or perpetuation of the ongoing immune response [80].

Surprisingly, in view of their important co-stimulatory activity, B7 molecules are either not expressed or have only low expression levels on professional APC. DCs in peripheral blood are $CD80^-$ and only weakly express CD86, but both molecules are rapidly induced during culture [81]. Conversely, DC generated in vitro from $CD34^+$ precursors expresses CD80 earlier than CD86 in culture [82]. Human peripheral blood monocytes constitutively express low levels of CD86 [83], while CD80 is expressed on monocytes only after activation with IFN-γ or GM-CSF, and cross-linking of FcR on monocytes strongly inhibits the up-regulation of CD80 and CD86 with, as a functional consequence, severe impairment of the capacity to function as antigen presenting cells (APC) and to stimulate T cell activation [84, 85]. Resting human B cells are $CD80^-$ and, upon activation, CD86 appears more rapidly than CD80 [86]. An important signal for induction or up-regulation of B7 molecules on all three types of professional APC is the interaction of CD40L on activated T cells with CD40 on the APC [87]. Thus, in infection where these cells limit the expression of co-stimulatory molecules, neutrophils have a fair chance to act as APCs and influence the immune response.

Therefore, to be a complete APC, neutrophils will have to express B7 or similar molecules, serving the same function. As studies have shown that B7-CD28 signaling is not as effective at regulating effector and memory T cell responses [88], the newer B7 molecules can act as additive regulators of T cell response. Indeed a study by Windhagen et al. has demonstrated that peripheral blood PMN express a CD80-like molecule that interacts with CD28, regulating T-cell function [89]. CD80 mRNA is expressed in bone marrow cells and lipopolysaccharide (LPS)-stimulated but not in unstimulated PMN. The CD80-like molecule is localized to the cytoplasmic granules and translocated to the cell surface after stimulation with LPS or IL-12 in some donors. CD80-like molecules can interact with functional B7 ligand and might be important in the immunobiology of PMN. The expression of CD80 on APCs requires external stimuli and is tightly regulated. CD80 appears on B cells and monocytes about 2 days after stimulation and is, therefore, thought to influence the later stages of the immune response. Consequently, surface expression of CD80 on PMN may be of biological significance when under optimal in vivo conditions survival is prolonged, particularly at the site of inflammation. Because surface expression of the CD80-like molecule was only detected in the sub-acute phase of bacterial meningitis it can be hypothesized that other cells, possibly in combination with other factors like LPS, are important to induce the translocation of CD80 molecules to the cell surface [89]. Furthermore, the surface expression of CD80-like molecules might reflect the age or maturation phases of PMNs during chronic inflammation.

The co stimulatory signal can be delivered by interaction of either CD80 or CD86 expressed by antigen-presenting cells with CD28 on the T cells. Comparison of the function of CD80 and CD86 in different experimental animal systems generated conflicting data on the roles of the co-stimulatory molecules. Therefore, it was investigated whether there are differences between CD80 and CD86-mediated co-stimulation in an alloantigen-specific primary T cell response induced by B7-transfected human cell lines of epithelial origin. Both transfected keratinocyte cell lines efficiently induce T cell proliferation and the ratios of stimulator versus responder cells are similar. The kinetics of proliferation and IL-2, IL-4 and IFN-γ production are also comparable between both transfectant lines. However, despite equal B7 expression levels, the CD80-induced T cell proliferation was consistently higher in magnitude than that of CD86 [90]. Comparison of precursor frequencies of helper T lymphocytes responsive with either CD80 or CD86 revealed that the frequency of CD80-responsive T cells was higher than that of CD86, and that the frequency of cells activated by a combination of CD80 and CD86 did not differ significantly from that of CD80 alone. Therefore, the constitutive expression of CD80 by neutrophils arguably has a more potent co-stimulatory function. Similarly, CD80 but not CD86 expression was differentially modulated on interaction with *Candida* yeasts or hyphae in both murine and human PMNs in vitro and in *Candida*-infected mice [91]. This suggests that parasites may have evolved a functional strategy whereby it only modulates the host mechanism that may help its prolonged survival in the host.

In vitro, PMNs inhibit the activation of IFN-γ-producing CD4$^+$ T cells and induce apoptosis through a CD80–CD28-dependent mechanism, which is against the conventional idea of anti-apoptotic function of CD28 in T cells. Corroborating these in vitro data, expansion of CD80 PMNs was observed in disseminated candidiasis in mice and their depletion increased the IFN-γ-mediated anti-fungal resistance [91]. By contrast, human neutrophils augment IFN-γ secretion from T cells [92]. Therefore, further studies are required to decipher the role of neutrophils in delivering CD80$^-$ or CD86$^-$-mediated co-stimulation because they might differentially engage CD28 or cytotoxic T-lymphocyte-associated antigen-4 (CTLA-4), imparting different effector functions to T cells. Possibly, these phenotypically different subsets of neutrophils display characteristic profiles of co stimulatory molecule expression and lead to differential Th-subset differentiation.

Modulation or intervention of T cell activation has been an attractive target for the modulation of immune response, especially in the treatment of autoimmune diseases and other immunological disorders. It is suggested that multiple mechanisms contribute to CD28/B7-mediated T cell co-stimulation in disease settings that include expansion of activated pathogenic T cells, differentiation of Th1/Th2 cells, and the migration of T cells into target tissues [93]. As neutrophils express and up-regulate certain B7 molecules, they can act as potential targets in altering the immune mechanisms.

11.2.6
Alteration in Cytokine Milieu

The cytokine milieu is of prime importance and has a multifactorial impact on various cells, thus deciding the onset and course of the immune response. During both acute and chronic inflammatory processes, various soluble factors are involved in leukocyte recruitment through increased expression of cellular adhesion molecules and chemo-attraction. Many of these soluble mediators regulate the activation of the resident cells (such as fibroblasts, endothelial cells, tissue macrophages, and mast cells) and the newly recruited inflammatory cells (such as monocytes, lymphocytes, neutrophils, and eosinophils). Some of these mediators resulting in the systemic responses to the inflammatory process fall into four main categories: (1) inflammatory lipid metabolites such as platelet-activating factor (PAF) and the numerous derivatives of arachidonic acid (prostaglandins, leukotrienes, lipoxins), which are generated from cellular phospholipids; (2) three cascades of soluble proteases/substrates (clotting, complement, and kinins), which generate numerous pro-inflammatory peptides; (3) nitric oxide, a potent endogenous vasodilator, whose role in the inflammatory process has only recently begun to be explored; and (4) a group of cell-derived polypeptides, cytokines, which to a large extent orchestrate the inflammatory response, i.e. they are major determinants of the make-up of the cellular infiltrate, the state of cellular activation, and the systemic responses to inflammation [94]. Most cytokines are multifunctional. They are pleiotropic molecules that elicit their effects locally or systemically in an autocrine or paracrine manner. Cytokines are involved in extensive networks that involve synergistic as well as antagonistic interactions and exhibit both negative and positive regulatory effects on various target cells. Several cytokines play key roles in mediating acute inflammatory reactions, namely IL-1, TNF-α, IL-6, IL-11, IL-8 and other chemokines, G-CSF, and GM-CSF. Of these, IL-1 (α and β) and TNF are extremely potent inflammatory molecules: they are the primary cytokines that mediate acute inflammation induced in animals by intra-dermal injection of bacterial lipopolysaccharide and two of the primary mediators of septic shock. Chronic inflammation may develop following acute inflammation and may last for weeks or months, and in some instances for years. During this phase of inflammation, cytokine interactions result in monocyte chemotaxis to the site of inflammation where macrophage-activating factors (MAF), such as IFN-γ, MCP-1, and other molecules then activate the macrophages while migration inhibition factors (MIF), such as GM-CSF and IFN-γ, retain them at the inflammatory site. Cytokines known to mediate chronic inflammatory processes can be divided into those participating in humoral inflammation, such as IL-3–7, IL-9, IL-10, IL-13, and TGF-β and those contributing to cellular inflammation such as IL-1–4, IL-7, IL-9, IL-10, IL-12, IFNs, IFN-γ-inducing factor (IGIF), TGF-β, and TNF-α and -β.

Being the first candidates to appear and actively phagocytose intruders, neutrophils surely are the first line of defense against invaders. In addition, it has been investigated whether these cells can activate both the humoral and cell mediated branches of the immune system. It would also be very profitable to the overall immune system if neutrophils can modulate functions of different immune cell types, viz. T cells, NK cells, macrophages, etc., by the release of different cytokines and, therefore, effectively modulate the subsequent adaptive immune responses. The ability of neutrophils to serve as a cytokine source, in combination with their large numbers in peripheral blood and the ability to rapidly migrate to a focus of infection, suggest that PMN may be a key cell type in cytokine-initiated immune system triggering. The immunoregulatory role of neutrophils in modulating T cell subset selection has been proved in many experimental models. Neutrophils have a profound impact on the outcome of the infection and help in mounting an appropriate adaptive immune response. Indeed, PMN-rich cell populations of different types of activity are recruited in the mouse peritoneum after 2–3 h by glycogen and thioglycolate and these cells produce factors capable of potentiating, enhancing, or suppressing responses to T- or B-cell mitogens by normal syngeneic lymphocytes [95].

After phagocytosis, but not after exposure to LPS, the PMN progressively release considerable amounts of IL-8 into the culture medium (18.6–50 ng ml^{-1} in 18 h) [96]. This could be the very first task of neutrophils, as IL-8 acts in an autocrine manner and activate neutrophils, chemotactically attracting neutrophils and T cells. Moreover, human neutrophils stimulated in vitro with IL-8 were found to release granule-derived factor(s) that induce in vitro T cell and monocyte chemotaxis and chemokinesis. Together, the existing information now suggests that neutrophils store and release, upon stimulation with IL-8 or other neutrophil activators, chemoattractants that mediate T cell and monocyte accumulation at sites of inflammation.

The contribution of neutrophils to the early cytokine balance governing Th1 and Th2 cell development was examined in mice with candidiasis [97]. Neutrophils secreted IL-12 and IL-10, correlating with the respective development of self-limiting (Th1-associated) and progressive (Th2-associated) disease. Exogenous IL-12 was effective in protecting neutropenic hosts susceptible to infection, suggesting that neutrophils, via their ability to release cytokines, play an active role in determining the qualitative development of the T cell response and their early role in anti-candidal immunity can be replaced by exogenous IL-12.

PMN recruited by the bacterium *Actinomyces viscosus* (AV) show their exclusive effects on the T cell lymphocyte population, which indicate that PMN act on a helper T cell population to enhance both proliferation and differentiation in lymphocyte populations [98]. A study by Appelberg et al investigated the participation of neutrophils in the mechanisms of resistance during the immune phase of the antimicrobial response to *Listeria monocytogenes* infection. BALB/c mice were unable to express T-cell-mediated immunity to this pathogen in the absence of granulocytes [99]. Again, this supports the view that neutrophils should be included in the concept of cell-mediated immunity.

Various studies point towards the involvement of neutrophils in perpetuating certain autoimmune diseases, predominantly by secreting cytokines and other chemo attractants. Blood neutrophils can be stimulated to express and rapidly release large quantities of oncostatin M [100]. This important cytokine is, proposedly, released from neutrophils as they infiltrate rheumatoid joints and, thus, contributes to the complex cytokine network that characterizes rheumatoid arthritis. Additionally, synovial fluid neutrophils in rheumatoid arthritis express macrophage inflammatory protein 1 alpha [101]. Thus, through their production of various cytokines and constitutive chemokines they directly alter the behavior of lymphocytes that accumulate within chronically inflamed joints, leading to their inappropriate survival and retention.

In another study, peritoneal exudate polymorphonuclear neutrophils (PEC-PMN) and mononuclear cells (PEC-MNC) were obtained from normal BALB/c and from autoimmune MRL-lpr/lpr mice (lpr) with different disease severities. The spontaneous and mitogen-stimulated expression of T-helper lymphocyte type-1 (Th1) – represented by IFN-γ and IL-2 – and T-helper lymphocyte type-2 (Th2) – represented by IL-4 and IL-10 cytokine mRNA in these cells was detected [102]. The spontaneous expression of Th1/Th2 cytokine mRNA in PEC-PMN from autoimmune mice was progressively increased in parallel with disease severity but was not changed by LPS stimulation. By contrast, spontaneous expression of Th1/Th2 cytokine mRNA in PEC-MNC from these mice was progressively decreased in parallel with disease severity but retained the responsiveness to phytohemagglutinin stimulation. PEC-PMN from the autoimmune mice progressively suppressed the production of IL-4, IL-10 and IFN-γ whereas the production of IL-2 was enhanced by autologous MNC in parallel with disease severity. These results suggest that a reciprocal relationship exists in the expression of Th1/Th2 cytokine mRNA between PEC-PMN and PEC-MNC in lpr mice in parallel with disease severity. Autoimmune PEC-PMN can exert significant modulatory effects on Th1/Th2 cytokine production by autologous MNC in stimulation. This study thus points towards the differential role that neutrophils may play in various autoimmune diseases.

Resistance and susceptibility to infection with the protozoan parasite *Leishmania major* have been correlated with the sustained activation of parasite-reactive CD4$^+$ Th1 or Th2 cells, respectively. Early PMN infiltration at the site of infection with *L. major* has been reported, with qualitative and quantitative differences between susceptible BALB/c and resistant C57BL/6 mice. The possible immunomodulatory role of PMN in CD4$^+$ T cell differentiation in mice was examined by studying the effect of transient depletion of PMN during the early phase after *L. major* delivery [103, 104]. A single injection of the PMN-depleting NIMP-R14 mAb 6 h before infection with *L. major* prevented the early burst of IL-4 mRNA transcription otherwise occurring in the draining lymph node of susceptible BALB/c mice. Since this early burst of IL-4 mRNA transcripts had previously been shown to instruct Th2 differentiation in mice from this strain, the effect of PMN depletion on Th subset differentiation at later time points after infection was examined. The transient depletion of PMN in BALB/c mice was sufficient to inhibit Th2 cell development otherwise occurring after *L. major* infection. Decreased Th2 responses were paralleled with partial resolution of the footpad lesions induced by *L. major*. The draining lymph node-derived CD4$^+$ T cells from PMN-depleted mice remained responsive to IL-12 after *L. major* infection, unlike those of infected BALB/c mice receiving control Ab. The protective effect of PMN depletion was shown to be IL-12 dependent, as concomitant neutralization of IL-12 reversed the protective effect of PMN depletion. These results suggest a role for an early wave of PMN in the development of the Th2 response characteristic of mice susceptible to infection with *L. major*. Strange it is that depletion of PMN in BALB/c mice infected with *L. major* hampered the development of a polarized Th2 response but the same treatment did not significantly affect the development of a Th1 response in resistant (C57BL/6) mice. These findings corroborate the idea that PMN could play an early role in the induction of the Th2 response that develops in BALB/c mice following infection with *L. major*. Two cytokines, IL-10 and TGF-β, which are secreted by murine neutrophils, could potentially play such an inhibitory role. Both cytokines have been reported to counteract IL-12-mediated effects in Th differentiation.

A possible immunomodulatory role for neutrophils in the protective mechanisms against *Mycobacterium tuberculosis* is supported by the fact that mouse neutrophils secrete several cytokines, such as IL-1α/β, IL-6, IL-10, IL-12, and TNF-α [106, 107]. Several of these cytokines are known to modulate the expression of IFN-γ, which is a key cytokine in the host defense mechanisms against *M. tuberculosis* in the mouse [108]. A second major component of the protective response is the macrophage product, nitric oxide. This molecule is required to create a toxic environment within the infected macrophage and depends on the enzyme iNOS for its production. In turn, the expression of iNOS is highly dependent upon IFN-γ expression [109]. Interestingly, the neutrophil-depleted mice expressed less mRNA for IFN-γ early during depletion (day 3), and the loss of this IFN-γ resulted in reduced expression of iNOS mRNA later (day 7). These observations clearly suggest that the presence of neutrophils in the infection foci is important for the early production of IFN-γ and that the decreased production of IFN-γ in neutropenic mice affects the nitric oxide-mediated anti-mycobacterial activity of infected macrophages.

Thus, neutrophils can make unique contributions to modulation of T cell proliferation and differentiation. Direct or indirect contributions by neutrophils could allow enhanced presentation of antigen and thus subsequent activation of T cells. In view of the above findings, PMN should be considered not only as active and central elements of the inflammatory response, but also as cells that, through cytokine secretion, may significantly influence the direction and evolution of the immune processes.

11.3
Evolution of Newer Thoughts as PMN March to a Newer Horizon

The concept of neutrophil as a phagocytic cell long remained stagnant in the framework of immune response. Suddenly, assisted by several key observations on its ability to express MHC class-II, co-stimulatory molecules and myriad cytokines, neutrophils' role as a modulator of T cell function is fast becoming a part in integrated immune homeostasis. The new trend in neutrophil research, so conspicuously visible, is analysis of phenotypic and functional varieties of neutrophils, obviating the need for their further classification as it is done for other immunocompetent cells such as B cells, T cells, dendritic cells and so on. Logically, when neutrophils are so versatile in their functions, it would be a natural waste of their potential for finer tuning the immune responses if they were all uniformly bestowed with all the capacities and if their origin is also regulated similarly. Therefore, it is no wonder that neutrophil subsets with distinguishable phenotypic features with functional correlates are being described, albeit scant and scattered. Thus on the new gleaming horizon where immune regulation is fast becoming a dominant theme, neutrophils are going to take a major part, showing themselves in different forms and performing many different functions with unique regulatory capacities.

Acknowledgment

The work is supported by the Department of Biotechnology, Government of India.

References

1. Cline, M. J. (1975). *The White Cell*. Harvard University Press, Cambridge MA, p. 28.
2. Woodman, D. M. et al. (1998). The functional paradox of CD43 in leukocyte recruitment: a study using CD43-deficient mice. *J. Exp. Med*. 188(11):2181–2186.
3. Hogg, N. (1992). Roll, roll, roll your leukocyte gently down the vein. *Immunol. Today*. 13(4): 113–115.
4. Abramson, S. and Weissmann, G. (1981). The release of inflammatory mediators from neutrophils. *Ric. Clin. Lab*. 11, 91–99
5. Banchereau, J. et al. (2000). Immunobiology of dendritic cells. *Annu. Rev. Immunol*. 18: 767–811
6. Burkhard Ludewig et al. (2001). Dendritic cells in autoimmune diseases. *Curr. Opin. Immunol*. 13: 657–662
7. Unanue, E. R. (1984). Anitgen presenting function of macrophages. *Annu. Rev. Immunol*. 2: 395–428.
8. Ashtekar, A. R., Saha, B. (2003). Poly's plea: membership to the club of APCs. *Trends Immunol*. 24 (9): 485–490.
9. Lloyd A. R., Oppenheim, J. J. (1992). Poly's lament: the neglected role of the polymorphonuclear neutrophil in the afferent limb of the immune response. *Immunol. Today*. 13(5): 169–172.
10. Gosselin, E. J. et al. (1993). Induction of MHC class II on human polymorphonuclear neutrophils by granulocyte/macrophage colony-stimulating factor, IFN-γ, and IL-3. *J. Immunol*. 151: 1482–1490.
11. Oehler, L. et al. (1998). Neutrophil granulocyte-committed cells can be driven to acquire dendritic cell characteristics. *J. Exp. Med*. 187: 1019–1028.
12. Iking-Konert C. et al. (2001). Transdifferentiation of polymorphonuclear neutrophils: acquisition of CD83 and other functional characteristics of dendritic cells. *J. Mol. Med*. 79(8): 464–474.
13. Venuprasad, K. et al. (2001). Immunobiology of CD28 expression on human peripheral blood neutrophils. I. CD28 plays important role in regulating neutrophil migration by modulation of CXCR-1 expression. *Eur. J. Immunol*. 31: 1536–1543.
14. Westwood, N.B. et al. (1995). Activated phenotype in neutrophils and monocytes from patients with primary proliferative polycythaemia. *J. Clin. Pathol*. 48(6): 525–530.
15. Lefkowitz DL et al. (1995). Neutrophil-macrophage interaction: a paradigm for chronic inflammation. *Med. Hypotheses* 44(1): 58–62.
16. Venuprasad, K. et al. (2002). Human neutrophil expressed CD28 interacts with macrophage expressed B7 to induce IFN-γ and restrict Leishmania growth. *J. Immunol*. 169, 920–928.
17. Lanzavecchia A. (1990). Receptor-mediated antigen uptake and its effect on antigen presentation to class II-restricted T lymphocytes. *Annu. Rev. Immunol*. 8: 773–793.
18. Colin Watts. (1997). Capture and processing of exogenous antigens for presentation on MHC molecules. *Annu. Rev. Immunol*. 15: 821–850.
19. Radsak M. et al. (2000). Polymorphonuclear neutrophils as accessory cells for T-cell activation: major histocompatibility complex class II restricted antigen-dependent induction of T-cell proliferation. *Immunology* 101: 521–530.
20. Fanger, N. A. et al. (1997). Activation of human T cells by major histocompatability complex class II expressing neutrophils: proliferation in the presence of superantigen, but not tetanus toxoid. *Blood* 89: 4128–4135
21. Ahearn, J. M., Fearon, D. T. (1989). Structure and function of complement receptors, CR1 (CD35) and CR2 (CD21). *Adv. Immunol*. 46: 183–219.
22. Springer, T. A. (1990). Adhesion receptors of the immune system. *Nature*. 346: 425–434.
23. Ravetch, J. V., Kinet, J. P. (1991). Fc receptors. *Annu. Rev. Immunol*. 9: 457.
24. Fearon, D. T., Collins, L. A. (1983). Increased expression of C3b receptors on polymorphonuclear leukocytes induced by chemotactic factors and by

purification procedures. *J. Immunol.* 130: 170–175.

25 O'Shea, J. J. et al. (1985). Evidence for distinct intracellular pools of receptors for C3b and C3bi in human neutrophils. *J. Immunol.* 134: 2580–2587.

26 Vik, D. P and Fearon, D. T. (1985). Neutrophils express a receptor for iC3b, C3dg, and C3d that is distinct from CR1, CR2, and CR3. *J Immunol.* 134(4): 2571–2579.

27 Matis, L. A. et al. (1983). Magnitude of response of histocompatibility-restricted T cell clones is a function of the products of the concentration of antigen and Ia molecules. *PNAS* 80(19): 6019–6023.

28 Rock, K. L. et al. (1984). Antigen presentation by hapten specific B-lymphocytes. I: Role of surface immunoglobulin receptors. *J. Exp. Med.* 160: 1102–1125.

29 Wallace, P. K. et al. (2001). Exogenous antigen targeted to FcγRI on myeloid cells is presented in association with MHC class I. *J. Immunol. Methods.* 248(1-2): 183–194.

30 Sabroe, I. et al. (2003). Selective roles for Toll-like receptor (TLR) 2 and TLR4 in the regulation of neutrophil activation and life span. *J. Immunol.* 170(10): 5268–5275.

31 Fan, J. and Malik, A. B. (2003). Toll-like receptor 4 (TLR4) signaling augments chemokines induced neutrophil migration by modulating cell surface expression of chemokines receptors. *Nat. Med.* 9: 315–321.

32 Remer, K. A. et al. (2003). Toll-like receptor-4 is involved in eliciting an LPS-induced oxidative burst in neutrophils. *Immunol. Lett.* 85(1): 75–80.

33 Fumitaka H. et al. (2003). Toll-like receptors stimulate human neutrophil function. *Blood* 102(7): 2660–2669.

34 Schjetne, K. W. et al. (2003). Cutting edge: link between innate and adaptive immunity: Toll-like receptor 2 internalizes antigen for presentation to CD4+ T cells and could be an efficient vaccine target. *J. Immunol.* 171(1): 32–36.

35 Kurt-Jones, E. A. et al. (2002). Role of toll-like receptor 2 (TLR2) in neutrophil activation: GM-CSF enhances TLR2 expression and TLR2-mediated interleukin 8 responses in neutrophils. *Blood* 100(5): 1860–1868.

36 Blum, J. S., Cresswell, P. (1998). Role for intracellular proteases in the processing and transport of class II HLA antigens. *PNAS* 85(11): 3975–3979.

37 Lennon-Dumenil, A.-M. et al. (2002). A closer look at proteolysis and MHC-class-II-restricted antigen presentation. *Curr. Opin. Immunol.* 14: 15–21.

38 Campbell, E.J. (1982). Human leukocyte elastase, cathepsin G and lactoferrin: family of neutrophil granule glycoproteins that bind to an alveolar macrophage receptor. *PNAS* 79: 6941–6945.

39 Kimura Y and Yokoi-Hayashi K. (1996). Polymorphonuclear leukocyte lysosomal proteases, cathepsins B and D affect the fibrinolytic system in human umbilical vein endothelial cells. *Biochim. Biophys. Acta* 1310(1): 1–4.

40 Riese, R. J., Chapman, H. A. (2000). Cathepsins and compartmentalization in antigen presentation. *Curr. Opin. Immunol.* 12: 107–113.

41 Riese, R. J. et al. (1996). Essential role for cathepsin S in MHC class II-associated invariant chain processing and peptide loading. *Immunity* 4: 357–366.

42 Antoniou, A. N. et al. (2000). Control of antigen presentation by a single protease cleavage site. *Immunity* 12: 391–398.

43 Agostini, A. de et al. (1988). A common neoepitope is created when the reactive center of C1 inhibitor is cleaved by plasma Kallikrein, activated factor XII fragment, C1 esterase, or neutrophil elastase. *J. Clin. Invest.* 82: 700–705.

44 P. E. Van Den Steen, et al. (2002). Cleavage of denatured natural collagen type II by neutrophil gelatinase B reveals enzyme specificity, post-translational modifications in the substrate, and the formation of remnant epitopes in rheumatoid arthritis. *FASEB J.* 16(3): 379–389.

45 Laura Santambrogio et al. (1999). Extracellular antigen processing and presentation by immature dendritic cells. *PNAS* 96(26): 15 056–15 061.

46 Marina Molino et al. (1995). Proteolysis of the human platelet and endothelial

cell thrombin receptor by neutrophil derived cathepsin G. *J. Biol. Chem.* 270(19): 11 168–11 175.

47 Sherman, M. A., Weber, D. A., Spotts, E. A., Moore, J. C. and Jensen, P. E. (1997). Inefficient peptide binding by cell-surface class II MHC molecules. *Cell Immunol.* 182(1): 1–11.

48 Romanic, A. M., Madri, J. A. (1994). Extracellular matrix-degrading proteinases in the nervous system. *Brain Pathol.* 4(2): 145–156.

49 Momohara, S. et al. (1997). Elastase from polymorphonuclear leukocyte in articular cartilage and synovial fluids of patients with rheumatoid arthritis. *Clin Rheumatol.* 16(2): 133–140.

50 Descamps, F. J. et al. (2003). Remnant epitopes generate autoimmunity: from rheumatoid arthritis and multiple sclerosis to diabetes. *Adv. Exp. Med. Biol.* 535: 69–77.

51 Sitia, G. et al. (2002). Depletion of neutrophils blocks the recruitment of antigen-nonspecific cells into the liver without affecting the antiviral activity of hepatitis B virus-specific cytotoxic T lymphocytes. *PNAS* 99(21): 13 717–13 722.

52 Sitia G. et al. (2004). MMPs are required for recruitment of antigen-nonspecific mononuclear cells into the liver by CTLs. *J. Clin. Invest.* 113(8): 1158–1167.

53 Harlan, J.M. (1987). Consequences of leukocyte-vessel wall interactions in inflammatory and immune reactions. *Semin. Thromb. Hemost.* 13: 434–444.

54 Hou, P. et al. (2004). Matrix metalloproteinase-12 (MMP-12) in osteoclasts: new lesson on the involvement of MMPs in bone resorption. *Bone* 34(1): 37–47.

55 Tazawa, H. et al. (2003). Infiltration of neutrophils is required for acquisition of metastatic phenotype of benign murine fibrosarcoma cells: implication of inflammation-associated carcinogenesis and tumor progression. *Am. J. Pathol.* 163(6): 2221–2232.

56 Martin, R. et al. (1992). Immunological aspects of demyelinating diseases. *Annu. Rev. Immunol.* 10: 153–187.

57 Forsyth, P.A. et al. (2001). Matrix metalloproteinases and diseases of the CNS. *Trends Neurosci.* 24(1): 8–9.

58 Masure, S, Proost, P, Van Damme, J., Opdenakker, G. (1991). Purification and identification of 91-kDa neutrophil gelatinase: release by the activating peptide interleukin-8. *Eur. J. Biochem.* 198: 391–398.

59 Vu TH, Werb Z. (1998). Gelatinase B: structure, regulation, and function. In: Parks, W.C., Mecham, R.P., eds. *Matrix Metalloproteinases*. St. Louis: Academic, p. 115–148.

60 Avolio, C. et al. (2003). Serum MMP-2 and MMP-9 are elevated in different multiple sclerosis subtypes. *J. Neur. Immunol.* 136(1-2): 46–53.

61 Beuche, W. et al. (2000). Matrix metalloproteinase-9 is elevated in serum of patients with amyotrophic lateral sclerosis. *Neuroreport.* 11(16): 3419–3422.

62 Lanchou, J. et al. (2003). Imbalance between matrix metalloproteinases (MMP-9 and MMP-2) and tissue inhibitors of metalloproteinases (TIMP-1 and TIMP-2) in acute respiratory distress syndrome patients. *Crit. Care Med.* 31(2): 536–542.

63 Lorenzl, S. et al. (2003). Increased plasma levels of matrix metalloproteinase-9 in patients with Alzheimer's disease. *Neurochem. Int.* 43(3): 191–196.

64 Teunissen, M.B. et al. (2002). Ultraviolet B radiation induces a transient appearance of IL-4[+] neutrophils, which support the development of Th2 responses. *J. Immunol.* 168(8): 3732–3739.

65 Fisher, G.J. et al. (2001). Ultraviolet irradiation increases matrix metalloproteinase-8 protein in human skin in vivo. *J. Invest. Dermatol.* 117(2): 219–226.

66 Di Nuzzo, S. et al. (1998). UVB preferentially induces the recruitment of memory CD4+ T cells in normal human skin: long-term effect after a single exposure. *J. Invest. Dermatol.* 110(6): 978–981.

67 Jeevan, A. et al. (1995). Ultraviolet radiation reduces phagocytosis and intracellular killing of mycobacteria and inhibits nitric oxide production by macro-

phages in mice. *J. Leuk. Biol.* 57(6): 883–890.
68. Potter, N. S and Harding, C. V. (2001). Neutrophils process exogenous bacteria via an alternate class I MHC processing pathway for presentation of peptides to T lymphocytes. *J. Immunol.* 167: 2538–2546.
69. Tvinnereim, A.R., Hamilton, S.E., Harty, J.T. (2004). Neutrophil involvement in cross-priming CD8+ T cell responses to bacterial antigens. *J. Immunol.* 173(3): 1994–2002.
70. Yewdell, J. W., Bennick, J. R. (1990). The binary logic of antigen processing and presentation to T cells. *Cell* 62: 203–206.
71. Douhan, J. 3rd, Hauber, I., Eibl, M. M., Glimcher, L. H. (1996). Genetic evidence for a new type of major histocompatibility complex class II combined immunodeficiency characterized by a dyscoordinate regulation of HLA-D alpha and beta chains. *J. Exp. Med.* 183(3): 1063–1069.
72. Iking-Konert, C., Vogt, S., Radsak, M., Wagner, C., Hansch, G.M., Andrassy, K. (2001). Polymorphonuclear neutrophils in Wegener's granulomatosis acquire characteristics of antigen presenting cells. *Kidney Int.* 60(6): 2247–2262.
73. Cross, A., Bucknall, R.C., Cassatella, M.A., Edwards, S.W., Moots, R.J. (2003). Synovial fluid neutrophils transcribe and express class II major histocompatibility complex molecules in rheumatoid arthritis. *Arthritis Rheum.* 48(10): 2796–806.
74. Vachiery, N., Totte, P., Balcer, V., Martinez, D., Bensaid, A. (1999). Effect of isolation techniques, in vitro culture and IFN gamma treatment on the constitutive expression of MHC class I and class II molecules on goat neutrophils. *Vet. Immunol. Immunopathol.* 70(1-2): 19–32.
75. Araki, H. et al. (2004). Reprogramming of human postmitotic neutrophils into macrophages by growth factors. *Blood* 103(8): 2973–2980.
76. Haslett, C. (1999). Granulocyte apoptosis and its role in the resolution and control of lung inflammation. *Am. J. Respir. Crit. Care Med.* 160(5): S5–S11.

77. Wagner, C. et al. (2000). Differentiation of polymorphonuclear neutrophils in patients with systemic infections and chronic inflammatory diseases: evidence of prolonged life span and de novo synthesis of fibronectin. *J. Mol. Med.* 78(6): 337–345.
78. McLoughlin, R. M. et al. (2003). Interplay between IFN-γ and IL-6 signaling governs neutrophil trafficking and apoptosis during acute inflammation. *J. Clin. Invest.* 112(4): 598–607.
79. Keel, M. et al. (1997). Interleukin-10 counterregulates proinflammatory cytokine-induced inhibition of neutrophil apoptosis during severe sepsis. *Blood* 90(9): 3356–3363.
80. Liang, L., Sha, W C. (2002). The right place at the right time: novel B7 family members regulate effector T cell responses. *Curr. Opin. Immunol.* 14: 384–390.
81. McLellan, A. D. et al. (1995). Activation of human peripheral blood dendritic cells induces the CD86 co-stimulatory molecule. *Eur. J. Immunol.* 25(7): 2064–2068.
82. Caux, C. et al. (1994). B70/B7-2 is identical to CD86 and is the major functional ligand for CD28 expressed on human dendritic cells. *J. Exp. Med.* 180(5): 1841–1847.
83. Azuma, M. et al. (1993). B70 antigen is a second ligand for CTLA-4 and CD28. *Nature* 366(6450): 76–79.
84. Freedman, A. S. et al. (1991). Selective induction of B7/BB1 on interferon-γ stimulated monocytes: a potential mechanism for amplification of T cell activation through the CD28 pathway. *Cell Immunol.* 137(2): 429–437.
85. Barcy, S. et al. (1995). FcR cross-linking on monocytes results in impaired T cell stimulatory capacity. *Int. Immunol.* 7(2): 179–189.
86. Hathcock, K. S. et al. (1994). Comparative analysis of B7-1 and B7-2 co-stimulatory ligands: expression and function. *J. Exp. Med.* 180(2): 631–640.
87. Foy, T. M. et al. (1996). Immune regulation by CD40 and its ligand gp39. *Annu. Rev. Immunol.* 14: 591–617.

88. London, C. A. et al. (2000). Functional responses and costimulator dependence of memory CD4+ T cells. *J. Immunol.* 164(1): 265–272.
89. Windhagen, A. et al. (1999). Human polymorphonuclear neutrophils express a B7-1-like molecule. *J. Leukoc. Biol.* 66: 945–952.
90. van Dijk, A. M. et al. (1996). Human B7-1 is more efficient than B7-2 in providing co-stimulation for alloantigen-specific T cells. *Eur J Immunol.* 26(9): 2275–2278.
91. Mencacci, A. et al. (2002). CD80 Gr-1 myeloid cells inhibit development of antifungal Th1 immunity in mice with candidiasis. *J. Immunol.* 15, 3180–3190.
92. Venuprasad, K. et al. (2003). CD28 signaling in neutrophil induces Tcell chemotactic factor(s) modulating T cell response. *Hum. Immunol.* 46: 38–43.
93. Salomon, B., Bluestone, J. A. (2001). Complexities of CD28/B7: CTLA-4 costimulatory pathways in autoimmunity and transplantation. *Annu. Rev. Immunol.* 19: 225–252.
94. Carol, A. et al. (1997). Cytokines in acute and chronic inflammation. *Frontiers Biosci.* 2: d12–26.
95. Rodrick, M. L. et al. (1982). Effects of supernatants of polymorphonuclear neutrophils recruited by different inflammatory substances on mitogen responses of lymphocytes. *Inflammation* 6(1): 1–11.
96. Bazzoni, F. et al. (1991). Phagocytosing neutrophils produce and release high amounts of the neutrophil-activating peptide 1/ interleukin 8. *J. Exp. Med.* 173(3): 771–774.
97. Luigina, R. et al. (1997). An immunoregulatory role for neutrophils in CD4 T helper subset selection in mice with candidiasis. *J. Immunol.* 158: 2356–2362.
98. Fitzgerald, J. et al. (1983). Polymorphonuclear leukocyte regulation of lymphocyte proliferation and differentiation. *Immunobiology* 165(5): 421–431.
99. Appelberg, R. et al. (1994). Neutrophils as effector cells of T-cell-mediated, acquired immunity in murine listeriosis. *Immunology* 83(2): 302–307.
100. Cross, A. et al. (2004). Secretion of oncostatin M by neutrophils in rheumatoid arthritis. *Arthritis Rheum.* 50(5): 1430–1436.
101. Hurst, S. M. et al. (2002). Secretion of Oncostatin M by infiltrating neutrophils: regulation of IL6 and chemokines expression in human mesothelial cells. *J. Immunol.* 169: 5244–5251.
102. Yu, C. L. et al. (1998). Expression of Th1/Th2 cytokine mRNA I peritoneal exudative polymorphonuclear neutrophils and their effects on mononuclear cell Th1/Th2 cytokine production in MLR-lpr/lpr mice. *Immunology* 95: 480–487.
103. Tacchini-Cottier, et al. (2000). An immunomodulatory function for neutrophils during the induction of a CD4+ Th2 response in BALB/c mice infected with *Leishmania major*. *J. Immunol.* 165(5): 2628–2636.
104. Lima, G. M. et al. (1998). The role of polymorphonuclear leukocytes in the resistance to cutaneous Leishmaniasis. *Immunol. Lett.* 64(2-3): 145–151.
105. Beil, W. J. et al. (1992). Differences in the onset of the inflammatory response to cutaneous leishmaniasis in resistant and susceptible mice. *J. Leukocyte Biol.* 52(2): 135–142.
106. Scapini, P. et al. (2000). The neutrophil as a cellular source of chemokines. *Immunol. Rev.* 177: 195–203.
107. Senaldi, G. et al. (1994). Role of polymorphonuclear neutrophil leukocytes and their integrin CD11a (LFA-1) in the pathogenesis of severe murine malaria. *Infect. Immunol.* 62(4): 1144–1149.
108. Ellis, T. N., Beaman, B. L. (2004). Interferon-γ activation of polymorphonuclear neutrophils function. *Immunology* 112(1): 2–12.
109. Hertz, C.J., Mansfield, J. M. (1999). IFN-gamma-dependent nitric oxide production is not linked to resistance in experimental African trypanosomiasis. *Cell Immunol.* 192(1): 24–32.

12
Microglia – The Professional Antigen-presenting Cells of the CNS?

Monica J. Carson

12.1
Introduction: Microglia and CNS Immune Privilege

12.1.1
What are Microglia?

All healthy tissues in the body maintain populations of macrophages and/or immature dendritic cells that act as sentinels and first responders to tissue damage and pathogenic insults [1]. Microglia are the resident tissue macrophage of the central nervous system (CNS) (Figure 12.1). Found throughout the CNS and making up 5–15% of all CNS cells, microglia are strategically placed to control the onset, progression and resolution of inflammatory responses occurring within the CNS [2, 3]. Under non-pathological conditions, microglia in the CNS of healthy

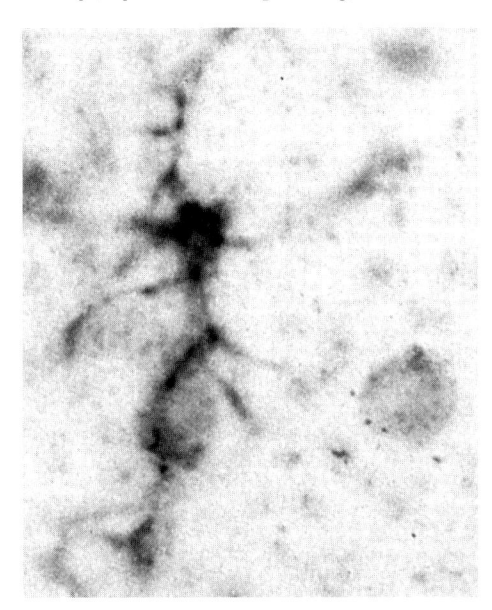

Figure 12.1 Unactivated microglia in the healthy adult murine CNS. Note the processes extending out from the microglia to all features in its local environment.

Antigen Presenting Cells: From Mechanisms to Drug Development. Edited by H. Kropshofer and A. B. Vogt
Copyright © 2005 WILEY-VCH Verlag GmbH & Co. KGaA, Weinheim
ISBN: 3-527-31108-4

rodents and humans express low to negligible levels of co-stimulatory molecules B7.2 and CD40 and of MHC class I and II. Consequently, in the absence of other signals, microglia are incapable of acting as antigen-presenting cells and, thus, of retaining or locally activating T cells within the CNS.

Microglia do appear poised to transform rapidly into antigen-presenting cells. Among the most rapid and universal responses of microglia to nearly any activating stimulus is an enhanced expression of molecules required to present antigen to T cells (such as MHC class I and II). At first glance, this universal response of microglia to injury and pathogens appears counter-intuitive to the well-described immune privileged status of the CNS. Indeed, literature analyses of microglial function have often characterized microglia as "enemies within" that aid and abet an invading immune system in destroying the CNS. However, since the CNS of all mammals contains microglia, and since induction of MHC expression is seen in all species examined, it is likely that microglial activation and MHC expression is not in of itself maladaptive for CNS function. Rather, it is highly likely that dysregulated microglial responses, caused by either primary microglial dysfunction or dysregulated environmental signals do substantially contribute to ongoing CNS pathogenesis. Thus, before we can discuss whether and in what context antigen-presentation by microglia may be beneficial for CNS function, we first need to define the current concept of CNS immune privilege.

12.1.2
Is Immune Privilege Equivalent to Immune Isolation?

The CNS is considered an immune privileged site based on the following seminal observation. Allografts (tissue grafts from the same species but expressing a different major histocompatibility complex [MHC] haplotype) placed into most non-CNS tissue sites are rapidly destroyed by the immune system. By contrast, allografts placed within the parenchyma of the CNS are not immediately rejected and survive for substantially longer periods. Experimental data from numerous groups exploring immune privilege have confirmed the two key tenets that define the current concept of CNS immune privilege:

1. The threshold for initiating T cell responses against antigens found primarily within the CNS is much higher than for antigens found primarily outside the CNS [4, 5].
2. The types of immune cells recruited to a tissue by inflammatory insults (LPS, viral infection, mechanical damage to stroma) and the kinetics of recruitment differ when the insult occurs in the CNS versus in other peripheral tissue sites. Within the CNS, the kinetics of inflammation are delayed and there is a tendency for fewer granulocytes to be recruited [5, 6].

For decades, dogma suggested that three elements were essential for the maintenance of CNS immune privilege:

1. An intact blood–brain barrier (BBB) to prevent immune cell infiltration into the healthy CNS.

2. The absence of a lymphatic system to drain soluble antigens to the cervical lymph nodes.
3. The absence of a resident population of macrophages/dendritic cell population capable of acting as an effective antigen-presenting cell able to initiate T cell mediated responses.

In essence, immune privilege was felt to rely on isolating the CNS from the immune system. Failure to isolate the CNS was thought to trigger an inevitable cascade of destructive CNS inflammation. In support of this view, in vitro assays clearly documented the potential of activated macrophages and microglia to produce vast quantities of molecules with demonstrated neurotoxic potential (cytokines, proteases, free radicals) (reviewed in Refs. 2, 5, 7, 8). Experimental rodent models of CNS demyelinating autoimmunity were associated with increased microglial expression of MHC, and a substantial influx of T cells and hematogenously-derived macrophages. Furthermore, drug therapies for various neuropathologies aimed at suppressing both microglia and macrophage function have provided limited (but rarely complete) amelioration of symptoms/pathology.

In direct contrast to this viewpoint, data gathered over the last 15 years conclusively demonstrates that immune privilege is neither simply based on isolating the CNS from the immune system nor on simply preventing T cells from encountering CNS antigens. Numerous studies have revealed that

1. Immune cells are readily able to transmigrate across an intact BBB.
2. Several brain areas have incomplete or leaky BBB and are not inflamed in the healthy individuals.
3. Soluble antigens do drain to the cervical lymph nodes.
4. The CNS does possess several populations able to present antigen; within the CNS parenchyma, that cell is the microglia.
5. The CNS is able to successfully down-regulate and terminate robust T cell responses occurring within the CNS and associated with clearing several forms of viruses and bacteria from the CNS.

While several factors are likely to play key roles in maintaining CNS immune privilege, here we explore the role of one cell type: the microglia. Specifically, we discuss to what extent microglia differ from other tissue/inflammatory macrophages in their potential to present antigen and regulate T cell responses. In brief, we suggest that microglia are a unique CNS-specific type of antigen presenting cell and that regulated antigen-presentation by microglia can play measurable beneficial roles in maintaining CNS function. In part this beneficial antigen-presentation is due to eliciting neuroprotective T cell responses and to promoting regulatory T cell effector function.

12.2
Do Microglia Differ from Other Macrophage Populations?

12.2.1
Microglia are Likely of Mesodermal Origin

In Section 12.1 we defined microglia as the tissue macrophage of the CNS. This definition is based on microglial expression of many common macrophage markers, including iba-1, Fc receptor (FcR), CD11b (also known as mac-1) and F4/80 [2, 7, 8]. There have been some contentions that some or all microglia, like neurons and macroglia (oligodendrocytes and astrocytes), are of neuroectodermal origin. However, attempts to induce microglial differentiation from neuroectodermal stem cells have consistently failed while these same studies were able to convincingly demonstrate the common neuroectodermal stem cell origins of CNS neurons and macroglia. Therefore, it is currently presumed that microglia are of mesodermal origin (like all other tissue macrophages). While it is still debated when microglia first appear in the CNS, histological studies clearly show that microglia colonize the CNS early in embryonic development. Microglia are easily detected within CNS tissue by embryonic day 15 in rodents and by gestational week 11 in humans [8].

12.2.2
Parenchymal Microglia are not the only Myeloid Cells in the CNS

Two distinct CNS macrophage populations are readily apparent on histological examination of CNS tissue sections: a parenchymal population, referred to as parenchymal microglia and a perivascular population referred to as either perivascular macrophages or perivascular microglia. Morphologically parenchymal microglia tend to have a stellate ramified morphology with processes extending to all features of their environment. Figure 12.1 depicts a microglia within the CNS of a healthy adult mouse. Note the extremely small size of the microglial cell body with processes extending out to all cells in its vicinity. In contrast to parenchymal microglia, perivascular macrophages have cell bodies and processes aligned along the vasculature. In addition to these two populations, myeloid populations are also found within the meninges that surround the CNS and in the choroid plexus located within the ventricles [2, 4].

12.2.3
In Contrast to other Macrophages, Parenchymal Microglia are not Readily Replaced by Bone Marrow Stem Cells

The first clear indications that parenchymal microglia differed from nearly all other macrophage populations came from studies examining the kinetics by which stem cells replaced differentiated macrophages within adult tissues. Specifically, several groups using irradiation bone marrow chimeric rats and mice have illustrated that while most tissue macrophage populations are replaced relatively rapidly by bone marrow derived cells within a few weeks, parenchymal microglia are not (Figure 12.2) [9, 10]. Bone marrow chimeric rodents are generated by killing the bone marrow, the stem cell source for most immune cell populations, with a lethal dose of whole body irradiation.

Immediately, following irradiation, animals receive replacement bone marrow from a genetically distinct donor. The turnover rate of macrophages within each tissue is then determined by monitoring the time required to replace macrophages displaying the recipient genotype with macrophages displaying the donor genotype.

A few key caveats for these types of studies do need to be mentioned. The irradiation dose used in these studies must be carefully titered to avoid giving a "super-lethal" dose that leads to permanent radiation-induced damage to the intestine, tissue scarring and non-specific innate inflammatory responses within internal organs. In addition, several studies have indicated that irradiation affects more than just the bone marrow. Specifically, irradiation leads to widespread if transient activation of glia and endothelial cells throughout the CNS. For this reason, some investigators choose to shield the head during the irradiation procedure.

Despite these caveats, Vallieries and Sawchenko have recently confirmed and extended the original irradiation chimera studies using bone marrow isolated from green fluorescent (GFP) mice (actin promoter driving GFP expression in every cell) grafted into irradiated congenic mice [11]. Due to the high degree of vascularization in the CNS, the authors state that cells located within a blood vessel could be mistaken for being within the parenchyma if the location of the vasculature is not carefully documented. Consequently, in their studies, the investigators carefully defined the location of GFP+ cells expressing macrophages markers (iba1) as: (1) clearly within the CNS parenchyma, (2) closely associated with cere-

How can we separate antigen-presentation by microglia from that by peripheral immune cells?

Using bone marrow chimeric mice, we generated two types of mice
in which antigen-presentation would occur primarily *in the CNS*
In which antigen-presentation would occur primarily *in the peripheral immune system*

CNS Microglia as APCs (peripheral immune system MHC II KO) → ← (CNS Microglia MHC II KO) Peripheral immune system as APCs

APCs=antigen-presenting cells

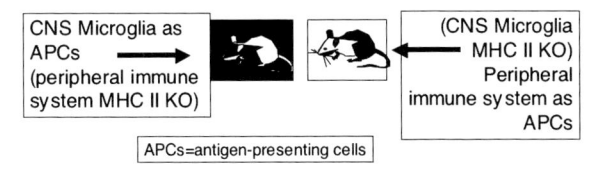

Figure 12.2 Irradiation chimeras can be used to separately assay in vivo antigen presentation by microglia and peripheral immune cells. APC: antigen-presenting cell.

bral vasculature (either within or aligned alongside), or (3) within the leptomeninges. In brief, they found that GFP+ donor cells replaced macrophages located in perivascular regions and in the leptomeninges at a rate that would ensure complete replacement of these populations by the GFP+ donor cells within 1 year.

The studies by Vallieries and Sawchenko also confirmed previous observations that most parenchymal microglia were a relatively stable self-renewing population [9, 10]. Furthermore, their analysis indicated that morphology alone was an inaccurate predictor of whether a myeloid cell within the CNS parenchyma was blood-derived. Hematogenous (bone marrow derived) macrophages with morphologies similar to other non-bone marrow derived parenchymal microglia were rarely but selectively found in specific brain regions 6 months post-bone marrow transfer [11]. In some brain regions such as the cerebral cortex, caudoputamen and hippocampal formation, donor derived GFP+ parenchymal microglia were never found. In some brain regions with an incomplete blood–brain barrier (BBB) such as the arcuate nucleus, or adjacent to the ventricular system, a rare donor-derived GFP+ parenchymal microglia was observed. The only brain region where donor-derived GFP+ parenchymal microglia were reliably found was the cerebellum. Within the cerebellum, donor derived microglia were always located in the molecular layer and most abundantly the paraflocculus. Considered together with several other similar studies, these data indicate that the CNS is populated by two different types of macrophage populations: a population of parenchymal microglia that is only rarely (if ever) replenished by bone marrow stem cells, and a population of macrophages (perivascular and perhaps the occasional parenchymal cell) that is rapidly replenished by bone marrow stem cells.

12.2.4
Microglia Display Stable Differences in Gene Expression that Distinguish them from Other Macrophage Populations

As early as 1991, Sedgwick and colleagues had made the seminal observation that microglia differed from other macrophage populations by their differential expression of CD45 [13]. Like all other nucleated cells of hematopoietic lineage, microglia express CD45, a transmembrane domain protein tyrosine phosphatase, also known as leukocyte common antigen [12–14]. However, unlike all other differentiated cells of hematopoietic lineage that constitutively express high levels of CD45, parenchymal microglia express uniformly low levels of CD45 [12–14]. Interestingly, peripheral immune cells display this same low level of CD45 expression early in development (embryonic day 14 in the mouse), but by birth only microglia maintain this low level of CD45 expression [12–14]. Recently, CD45 has been identified as a negative regulator of cytokine receptor signaling in a wide variety of immune cells [15–17]. With respect to microglia, antibody mediated crosslinking of CD45 inhibits β-amyloid or CD154 (the ligand for CD40) induced microglial activation and thus inhibited microglial production of nitric oxide and TNFα [18, 19]. Therefore, the lower level of CD45 expression stably maintained by parenchymal microglia as compared to all other macrophage populations is not only a useful marker to discriminate the two types of cells but is an indicator of stable differences in cellular physiology [17].

12.2.5
Morphology is not a Reliable Parameter to Differentiate Microglia from Other Macrophage Populations

When analyzed by flow cytometry, the differential expression of CD45 can reliably distinguish between microglia and other macrophage populations [12–14]. However, this requires dissociation of the tissue and loss of the spatial localization of each cell type with respect to pathology and/or brain structures. When analyzed histologically, correct determination of relative levels of CD45 expression is less precise. Both microglia and macrophages increase their relative expression of CD45 upon activation [12–14]. Activated microglia express CD45 levels intermediate between unactivated microglia and unactivated macrophages and thus may be misidentified as CNS-infiltrating macrophages if compared to unactivated microglial populations [12–14]. Frequently, morphological criteria are used to differentiate between parenchymal microglia and inflammatory macrophages that acutely infiltrate the CNS in responses to various pathological signals. Cells with ramified morphologies are generally labeled as microglia, while cells with amoeboid morphologies are labeled as macrophages. The imprecision of morphological based criteria to determine cell type has been experimentally demonstrated. When fluorescently-labeled rodent microglia were placed on cultured rodent brain slices they developed both ramified and amoeboid morphologies that appeared dependent on brain region and type of tissue damage [20]. Conversely, the previously discussed studies of Vallieries and Sawchenko clearly demonstrate that ramified morphologies do not necessarily represent microglia. Consistent with their observations, fluorescently labeled myeloid dendritic cells retained a ramified morphology when injected into the CNS parenchyma, even after migration into CNS parenchyma [21]. Similarly, in autoimmune responses outside of the CNS, both macrophages and dendritic cells can display ramified rather than amoeboid morphologies [22].

These data raise the following questions: are the rare donor-derived GFP+ stellate cells in the parenchyma of the irradiation bone marrow chimeric CNS indicative of a very slow rate of hematogenous replenishment of parenchymal microglia? Or are the rare donor derived GFP+ stellate macrophages performing a necessary immune surveillance function? More precisely, in addition to acquiring a stellate morphology, do the donor-derived GFP+ stellate parenchymal cells acquire other features indicative of a microglia phenotype? Do these GFP+ stellate cells display the characteristically low level of CD45 expression?

To move away from an imprecise morphological discrimination between microglia and macrophages, and to define the potential functions of microglia within the CNS, several groups have performed comprehensive screens of microglial gene expression using several different methods (microarray chips, TOGA, candidate gene testing) [23–27]. In brief, these studies have failed to identify a single marker that unambiguously identifies and distinguishes microglia from other antigenically related macrophages. However, a few themes are beginning to emerge from careful examination of the gene expression profiles of microglia [23–27]. First, nearly all stimuli induce microglial expression of at least a subset of the molecules required to present antigen and thus to activate T cells. Strikingly, many neurodegenerative signals stimulate an incomplete induction of the antigen-presenting cell program. Myeloid cells that express an incomplete antigen-presenting cell program (i.e. deficient induction of co-stimulatory molecules CD40, B7) are unable to effectively stimulate proinflammatory T cell response. Rather, myeloid cells with incomplete antigen-presenting cell phenotypes are more effective at promoting either "immunosuppressive" Th2 T cell responses or antigen-specific inactivation of T cells.

12.3
To What Extent is Microglial Phenotype Determined by the CNS Microenvironment?

The bone marrow and molecular studies discussed in the previous sections dramatically demonstrate that microglia are phenotypically distinct from other macrophage populations. Until recently, most research focused on how activated microglia negatively affect neuronal function. Now several studies have dramatically demonstrated that neurons are not helpless victims. Rather, several new studies reveal that CNS neurons actively control and define microglial function [28]. For example, microglial expression of molecules required for antigen-presentation is inhibited by electrically active neurons, but is induced when electrical activity is suppressed [28]. Neuronal production of neurotrophins also suppresses the IFNγ-induced microglial expression of MHC class II [29]. Neuropeptides such as α-MSH and VIP inhibit pro-inflammatory cytokine and NO production by LPS-activated microglia in in vitro assays [30, 31]. In vivo studies using knock-out mice illustrate that CD200 expression by neurons actively prevents activation of myeloid cells expressing the CD200 receptor (i.e. microglia, macrophages, dendritic cells) [32]. Even in the CNS of healthy CD200 deficient mice, microglia display an activated phenotype (elevated expression of CD45, MHC class II, complement receptor 3). Furthermore, in CD200 deficient mice, microglial activation was accelerated in responses to axonal degeneration, and EAE occurred with a more rapid onset. CNS neurons also express CD22, an endogenous ligand for CD45 [33]. As discussed in an earlier section of this article, CD45 is negative regulator of microglial and macrophage pro-inflammatory cytokine production. Based on their relative levels of CD45 expression, neurons would be predicted to more efficiently inhibit macrophage activation as compared to microglial activation! Notably, not all neuronal products inhibit microglia. Fractalkine released from injured neurons and the neuropeptide substance P augment microglial production of pro-inflammatory factors [34, 35].

In aggregate, these studies raise the following question: when microglial cell lines or primary microglia are cultured in the absence of neurons and glia, is their phenotype representative of normal microglia in either the developing or adult CNS? Consistent with the described neuronal regulation of microglial phenotype and function, several studies confirm that cultured microglia display a quasi-activated phenotype (elevated expression of CD45 and MHC) and are semi-primed to respond aggressively to pathogenic signals [14, 36, 37].

12.4 Microglia versus Macrophages/Dendritic Cells as Professional Antigen-presenting Cells

12.4.1 In vitro and Ex Vivo Assays of Antigen-presentation

To what extent microglia can act as antigen-presenting cells, and when, is much debated. The difficulty in resolving this debate is partly due to the different types of microglia (cultured, neonatal, or adult) and T cells being assayed (naïve, memory or T cell lines). As discussed in the previous section, the very phenotype and function of the assayed microglia is likely to be strongly altered by extensive culture or by exposure to in vivo pathology.

Microglia isolated from healthy adult rodent CNS were relatively ineffective as compared to splenic or CNS-infiltrating macrophages at promoting CD4+ T cell proliferation, when tested ex vivo, even when MHC expression was induced by IFNg treatment [12, 14, 38]. However, microglia activated by in vivo pathology (during EAE, or transgenic CNS overexpression of IL-3), to become not only MHC class II positive but also $CD45^{intermediate}$, gave a slightly different picture [13, 36, 39]. These microglia were still relatively poor at stimulating T-cell proliferation, due in part to their stimulated production of prostaglandins and nitric oxide, but were found to be very potent at promoting T-cell production of pro-inflammatory Th1 cytokines [36, 39]. Indeed, microglia were much more potent at driving Th1 T-cell effector function than the other macrophage populations examined in these studies because in the absence of antigen-induced T-cell proliferation fewer T cells were producing greater levels of IFNγ [36].

Studies by Miller and colleagues indicate that the relative potential of microglia and hematogenously derived macrophages to act as antigen-presenting cells does not necessarily remain constant throughout at least some forms of CNS inflammation. For example, Theiler's virus infection of the murine CNS results in a demyelinating encephalomyelitis (TMEV-IDD) due to the activation of myelin-specific autoreactive CD4+ T cells. Early in disease, microglia and macrophages were found to be equally effective at stimulating T cell proliferation and IFNγ production by myelin-specific T cell lines [38]. However, as the disease progressed, hematogenous macrophages became much more effective, presumably due to their much higher levels of MHC class II and co-stimulatory molecules. In TMEV-IDD, microglia and macrophages are among the cells infected by the virus [40]. Therefore, it still remains unknown whether the observed early increase in the antigen-presenting function of microglia was a consequence of direct viral infection or of signal provided by virally infected cells in the vicinity of the activated microglia.

Most of these studies examined the ability of microglia isolated from healthy and inflamed rodent CNS to present pre-processed peptide antigens to T cells in vitro. However, in vivo, it is more likely that antigen would be processed after phagocytosis of cellular debris. While unactivated cultured microglia cannot efficiently process antigen, upon activation by IFNγ or by viral infection, microglia can be stimulated to efficiently process and present endogenous (viral antigens) and exogenous myelin antigens [41]. Several studies also suggest that myelin phagocytosis itself can prime microglia to become antigen-presenting cells [42, 43]. Having the ability to process antigen indicates that microglia have the potential to participate in the epitope switching that has been observed in viral models of CNS demyelination and that is presumed to occur during MS [40]. Here, an initial immune response is generated against viral antigens. However, as myelin damaged occurs (as a consequence of oligodendrocyte infection and/or death or inflammatory products), macrophages and microglia phagocytose the debris. Presentation of the myelin epitopes causes the immune response to shift from a viral specific response to a myelin-specific response.

12.4.2
Culture Conditions can have Profound Effects on Microglia Effector Functions as Assayed In Vitro

The microglia used in these studies (described in Section 12.4.1) were isolated from adult rodent CNS and assayed for their antigen-presenting cell function immediately after purification. However, analysis of cultured microglia illustrates how strongly environmental factors can modify microglial effector function and therefore confound our in vitro analyses. For instance, cultured microglia, even if maintained in the presence of other CNS cells and in the absence of known stimulatory factors display a quasi-activated phenotype [13, 36, 39]. They express intermediate levels of CD45, and dependent on culture conditions may be weakly positive for co-stimulatory factors and MHC class I and II. Using these cells, one set of studies has suggested that weakly activated microglia induce T cell anergy or unresponsiveness, due in part to their low level expression of co-stimulatory molecules, CD40 and B7.2 [44]. In this same set of studies, treating microglia with GM-CSF and IFNγ, molecules found in abundance in inflammatory infiltrates, rapidly transformed microglia into potent antigen-presenting cells [44]. Similar culture conditions also induce microglia to express high levels of the dendritic cell markers, CD11c and Dec-205 [45]. Even in the absence of stimulatory factors a small percentage of cultured microglia spontaneously express the dendritic cell marker, Dec-205 [36]. Considered together these experiments provide a cautionary tale about the potential relevance of using cultured microglia and in vitro assay systems.

12.4.3
In Vivo Assays of Antigen-presentation

To examine the antigen-presenting function of microglia in vivo, some groups have selectively depleted the peripheral macrophage populations by treating animals with mannosylated liposome-encapsulated dichloromethylene diphosphonate (Cl_2MDP) [46, 47]. Following depletion of the peripheral myeloid population, rodents were immunized with myelin proteins in complete Freund's adjuvant to induce experimentally-induced encephalomyelitis (EAE), an animal model for multiple sclerosis [47]. In EAE, T cell mediated anti-myelin responses result in CNS inflammation and decreased motor function associated with CNS demyelination. While untreated rodents developed EAE, the Cl_2MDP-treated animals did not. Strikingly, Cl_2MDP-treatment did not prevent T-cell extravasation or Th1 cytokine production following adoptive transfer of myelin-specific T cells, but Cl_2MDP-treatment did inhibit CNS demyelination, $TNF\alpha$ production, T-cell infiltration into the CNS parenchyma and induction of clinical EAE seen in the non-Cl_2MDP-treated rodents. These studies suggest the ability of microglia to present antigen in vivo is neither identical to nor as potent as other antigen-presenting cells. These studies also show that parenchymal infiltration of T cells and/or activated macrophages are required for full-blown EAE. However, these studies do not distinguish between the roles of activated macrophages as initiators vs. effectors of disease (antigen-presenting cells vs. phagocytes/producers of neurotoxic molecules). An additional complication is that the effects of Cl_2MDP are not limited to peripheral macrophages and may alter microglial function.

A different approach was taken to directly compare in vivo the relative abilities of microglia and peripherally derived antigen-presenting cells to recruit and retain T cells within the CNS. In these studies, antigen-pulsed dendritic cells and microglia were injected into the striatum of healthy adult mice in which all the T cells were specific for the pulsed antigen. In these in vivo assays, dendritic cells were dramatically more effective than microglia in recruiting and retaining antigen-specific T cells in the CNS.

These data might be interpreted as demonstrations that, when considered as antigen-presenting cells, microglia are relatively inefficient and perhaps functionally redundant with peripheral antigen-presenting cell populations. In stark contrast, recent studies using irradiation bone marrow chimeric mice have dramatically demonstrated that microglia and peripheral antigen-presenting cells have distinct and non-redundant functions in regulating T cell effector function in both beneficial and maladaptive directions.

12.4.4
Antigen-presentation by Microglia is Necessary to Evoke or Sustain Neuroprotective T Cell Effector Function

The first set of studies examined the relative roles of microglia and macrophages in eliciting and driving neuroprotective CD4+T cell responses following facial axotomy [48] (Figure 12.3). In this model, motoneuron degeneration following facial axotomy is associated with microglial activation and CNS-infiltration of macrophages and T cells [49]. This inflammation was initially believed to be maladaptive and a contributing cause of motoneuron degeneration. However, in the absence of CD4+ T cells, motoneuron degeneration was much more rapid and severe [50]. Using bone marrow chimeric mice, animals were generated in which either only microglia or only hematogenous immune cells could act as antigen-presenting cells [48]. Strikingly, microglia were not able to initiate the protective

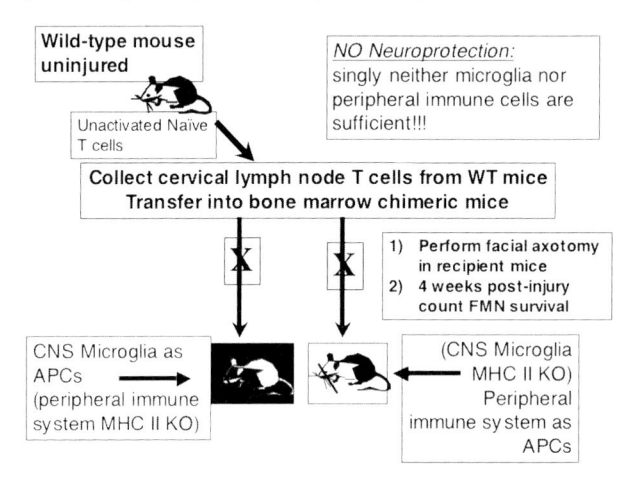

Figure 12.3 Neither antigen-presentation by microglia nor peripheral immune cells is sufficient to drive neuroprotective T cell responses. APC: antigen-presenting cell, WT: wild-type, FMN: facial motoneuron.

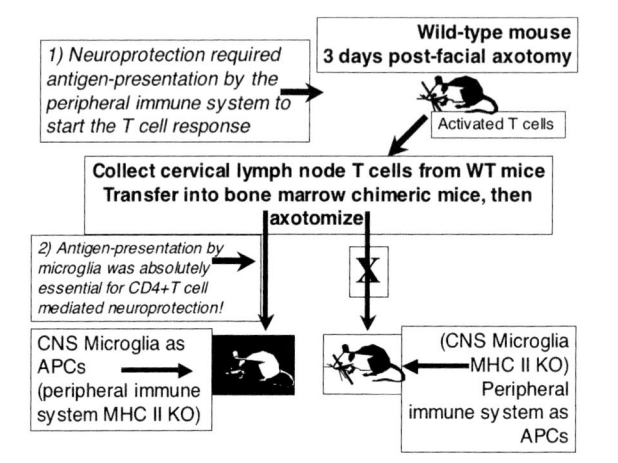

Figure 12.4 Peripheral immune cells are required to initiate the neuroprotective immune response, while microglia are required to evoke or sustain neuroprotective T cell responses. APC: antigen-presenting cell, WT: wild-type.

CD4+T cell response. Initiation of the T cell response was entirely dependent on hematogenous macrophages. While macrophages could initiate the response and did infiltrate the CNS, antigen-presentation solely by hematogenous macrophages failed to support CD4+T cell mediated neuroprotection of motoneurons. However, once T cell activation was initiated by peripheral macrophages, protection of motoneurons following facial axotomy was absolutely dependent on antigen-presentation by microglia activated by local neurodegenerative signals (Figure 12.4) [48].

12.4.5
Why were Microglia Unable to Initiate Protective T Cell Responses?

In this facial axotomy model, why couldn't microglia initiate the protective CD4+ T cell responses? In most tissues, resident populations of tissue macrophages or immature dendritic cells activate T cells responses as follows [1, 51]. In response to pathogenic signals (tissue damage, pathogens), the tissue macrophage captures antigens from cellular debris. The tissue macrophage then is activated to emigrate out of the tissue and migrate to the draining lymph node. Within the draining lymph node the activated tissue macrophage/dendritic cell presents antigen to the unactivated T cells in the lymph node. Upon activation, the T cells leave the lymph node in search of their target antigen. Within the original damaged/diseased tissue, tissue macrophages now present their captured antigens to the infiltrating T cells [1, 51]. This second presentation of antigen serves to tether the T cells to the tissue and to elicit T cell effector function.

To initiate T cell responses, microglia must be able to migrate out of the CNS and into the draining cervical lymph nodes. Bone marrow chimeric mice demonstrated that the parenchymal microglia population is only rarely replenished from the bone marrow, but do not address the reverse issue of whether microglia leave the CNS. To answer this question, fluorescently labeled dendritic cells and cultured primary microglia were injected into the CNS [21]. While fluorescently labeled dendritic cells could be detected within the T dependent regions of the cervical lymph nodes within 24 h, fluorescently labeled microglia were never detected in the cervical lymph nodes. These data indicate that either microglia are incapable of migrating out of the CNS into the cervical lymph nodes or that they migrate at a substantially lower rate than dendritic cells (below the limits of detection of the histological assay used in this study). Thus, the role of antigen-capture and migration into the cervical lymph nodes may reside in either the perivascular macrophage populations or the occasional hematogenous-derived macrophage with stellate morphology found within the CNS parenchyma.

These studies suggest that microglia play essential roles in minimizing neurodegeneration and maintaining neuronal integrity. However, several other studies illustrate the potential for microglia. In contrast, two additional studies using bone marrow chimeric mice also demonstrate unique microglial contributions to T cell mediated neuropathology. In both these studies, Becher and colleagues examined the relative roles of microglia and peripheral immune cells in the onset and progression of MOG EAE. MOG EAE is induced by immunizing C57Bl/6 mice with the myelin protein MOG. The resulting inflammation and demyelination occurs primarily within the cerebellum and spinal cord.

First, Becher et al. generated bone marrow chimeric mice in which only parenchymal microglia or hematogenous macrophages express the pro-inflammatory cytokine IL-23 [52]. Although IL-23 has multiple effects, when expressed by an antigen-presenting cell IL-23 promotes promotes T cells to develop pro-inflammatory Th1 effector functions. Becher et al. showed that when microglia could not express IL-23, the severity of clinical symptoms of MOG-EAE was drastically reduced. Additionally, while there was little impact on the degree of inflammation occurring within the CNS, the T cell cytokine profile shifted from a proinflammatory TH1 response to a protective TH2 response.

Second, Becher et al. also generated bone marrow chimeric mice in which microglia could not express CD40, the receptor for CD154 [53]. Activated T cells express CD154 and can trigger antigen-presenting cells (microglia, macrophages and dendritic cells) to produce IL-12 and IL-23 via CD40. Microglia deficient in CD40 expression will not produce these cytokines when presenting antigen to T cells. As with the previous study, the severity, duration and degree of T cell infiltration during MOG-EAE were all reduced when microglia could not express CD40. Thus, even in the presence of CNS-infiltrating macrophages, interrupting the ability of microglia to support pro-inflammatory T cell responses is sufficient to alter the course of the disease.

Why did activated microglia play key functions in promoting beneficial T cell responses in the facial axotomy yet in the EAE model play a key function in promoting destructive T cell responses? Although the answer to this question is not yet defined it is likely to be the consequence of the differential activations states of microglia in the two models. Due to the use of strong adjuvants (factors which amplify and promote immune responses), microglia express dramatically higher levels of MHC II, CD40 and other co-stimulatory molecules than microglia activated by wallerian degeneration of a small number of axons.

12.5
TREM-2 Positive Microglia may Represent Subsets Predisposed to Differentiate into Effective Antigen-presenting Cells

For most of this chapter, microglial heterogeneity has been largely ignored. However, regional differences in cell morphology, antigenic markers, response to cytokines, constitutive and inducible MHC expression have long been realized to be indicators of microglial heterogeneity in vivo [35, 54, 55]. Recent characterizations of microglial gene expression have revealed that a subset of microglia may also have a higher potential to differentiate into effective antigen-presenting cells and thus a higher potential to activate T cells [26]. In the TOGA® based screens of microglial gene expression, the orphan receptor, triggering receptor expressed on myeloid cells-2 (TREM-2) was identified as being expressed by microglia [26]. When expression was examined in the adult mouse brain by in situ hybridization, a subset of microglia were observed to constitutively express TREM-2 [26]. Interestingly, the highest percentage of microglia expressing TREM-2 was located in regions predisposed to develop Alzheimer's disease pathology (entorhinal cortex, hippocampus). Conversely, the highest percentage of TREM-2 negative microglia was in regions with an incomplete or leaky blood–brain barrier (BBB), such as in the hypothalamic and circumventricular regions. Although TREM-2 expression was strongly down-regulated by bacterial signals such as LPS, microglial expression of TREM-2 was dramatically upregulated in pathologies with abundant neurodegeneration and necrosis [26]. Most notably, TREM-2 expression was dramatically induced in microglial bordering the outer edges of amyloid plaques in transgenic mice overexpressing mutant forms of amyloid precursor protein.

As yet, TREM-2 function remains ill-defined; however, Colonna and colleagues have implicated TREM-2 in mediating the differentiation of human monocyte derived dendritic cells into antigen-present cells [56]. Recently, Daws and colleagues have detected a TREM-2 binding activity in astrocyte cell lines, implying that at least some forms of astrocytes may express the endogenous TREM-2 ligand and thus could trigger microglial differentiation into more potent antigen-presenting cells [57]. Strikingly, humans lacking TREM-2 develop early onset cognitive dementia, white matter deficits and die by their early 40s [58]. Considered with data presented in earlier sections, it is tempting to speculate that TREM-2 mediated activation of microglia, and the subsequent development of the ability to present antigen to T cells, is a necessary and ongoing mechanism underlying the normal maintenance of the human CNS. If true, the induction of TREM-2 in neurodegenerative diseases may not be in of itself maladaptive, but an attempt to recruit T cell mediated neuroprotection.

However, the induction of microglial expression of TREM-2 near amyloid plaques may also have consequences for vaccine based therapies for Alzheimer's disease. Recently, a novel vaccine based therapy for Alzheimer's disease was tested in phase I and phase II clinical trials [59–61]. Based on results in animal models, individuals diagnosed with Alzheimer's disease were immunized with amyloid and adjuvant to initiate an effective immune attack on the amyloid plaques within the CNS of these patients. Unfortunately, a small percentage (reportedly less than 10%) of treated individuals developed severe clinical signs of encephalitis, necessitating the abrupt halt of the phase II trials. It is presently unknown whether human amyloid pathology induced activation of microglia altered their antigen-presenting cell functions. For example, would TREM-2 triggered microglia have contributed to the disastrous response to amyloid vaccination observed in 10% of the patients or to the tolerization of this protocol in the other 90% of patients?

12.6
Are Microglia the "Professional Antigen-presenting Cell of the CNS?"

The term professional antigen-presenting cell is usually reserved to describe mature dendritic cells able to efficiently activate and drive naïve T cell differentiation. Here we have discussed the unique phenotype of the tissue macrophage of the CNS: the microglia. Like all cells of myeloid lineage, microglia participate in both the adaptive (T cell-mediated) and innate arms of the immune response to injury and pathogens. The studies discussed in this chapter indicate the potential for microglia to either limit and amplify CNS pathology by acting as effective antigen-presenting cells.

Thus, when we consider whether microglia are antigen-presenting cells, we have to consider what is the ultimate outcome of their interactions with antigen-inexperienced (but chemokine recruited) as well as with T cells previously activated by other peripheral antigen-presenting cells. Rather than driving pro-inflammatory T cell responses, under healthy or well-regulated conditions, the role of the microglia may be to redirect T cell effector function toward neuroprotective outcomes. Recent studies clearly demonstrate that neurons can actively regulate microglial function (i.e. via CD200 and CD22 mediated mechanisms). These data imply that CNS inflammatory disease may not always be a primary disease of microglia and/or the immune system, but may result from primary neuronal dysfunction!

Acknowledgments

NIH grants NS045735 and NS39508 (M.J.C.).

References

1 Lo D, Feng LL, Li L, Carson MJ, Crowley M, Pauza M, et al. Integrating innate and adaptive immunity in the whole animal. *Immunol. Rev.* (1999) 169: 225–239.
2 Carson MJ. Microglia as liaisons between the immune and central nervous systems: functional implications for multiple sclerosis. *Glia* (2002) 40(2): 218–231.
3 Streit WJ. Microglial response to brain injury: a brief synopsis. *Toxicol. Pathol.* (2000) 28(1): 28–30.
4 Perry VH. A revised view of the central nervous system microenvironment and major histocompatibility complex class II antigen presentation. *J. Neuroimmonol.* (1998) 90 (2): 113–121
5 Matyszak MK. Inflammation in the CNS: balance between immunological privilege and immune responses. *Prog. Neurobiol.* (1998) 56(1): 19–35.
6 Carson MJ, Sutcliffe JG. *The role of microglia in CNS inflammatory disease: Friend or foe?* in: Bondy SC, Campbell A, eds. Inflammatory events in neurodegeneration. Scottsdale, Prominent Press (2001), p. 1–14.
7 Aloisi F. Immune function of microglia. *Glia* (2001) 36(2): 165–179.

8 Kreutzberg GW. Microglia: a sensor for pathological events in the CNS. *Trends Neurosci.* (1996) 19(8): 312–318.

9 Hickey WF, Kimura H. Perivascular microglial cells of the CNS are bone marrow-derived and present antigen in vivo. *Science* (1988) 239(4837): 290–292.

10 Matsumoto Y, Fujiwara M. Absence of donor-type major histocompatibility complex class I antigen-bearing microglia in the rat central nervous system of radiation bone marrow chimeras. *J. Neuroimmunol.* (1987) 17(1): 71–82.

11 Vallieres L, Sawchenko PE. Bone marrow-derived cells that populate the adult mouse brain preserve their hematopoietic identity. *J. Neurosci.* (2003) 23(12): 5197–5207.

12 Sedgwick JD, Schwender S, Imrich H, Dorries R, Butcher GW, ter Meulen V. Isolation and direct characterization of resident microglial cells from the normal and inflamed central nervous system. *Proc. Natl. Acad. Sci. U.S.A.* (1991) 88(16): 7438–7442.

13 Renno T, Krakowski M, Piccirillo C, Lin JY, Owens T. TNF-alpha expression by resident microglia and infiltrating leukocytes in the central nervous system of mice with experimental allergic encephalomyelitis. Regulation by Th1 cytokines. *J. Immunol.* (1995) 154(2): 944–953.

14 Carson MJ, Reilly CR, Sutcliffe JG, Lo D. Mature microglia resemble immature antigen-presenting cells. *Glia* (1998) 22(1): 72–85.

15 Irie-Sasaki J, Sasaki T, Penninger JM. CD45 regulated signaling pathways. *Curr. Top. Med. Chem.* (2003) 3(7): 783–796.

16 Hermiston ML, Xu Z, Weiss A. CD45: a critical regulator of signaling thresholds in immune cells. *Annu. Rev. Immunol.* (2003) 21: 107–137.

17 Penninger JM, Irie-Sasaki J, Sasaki T, Oliveira-dos-Santos AJ. CD45: new jobs for an old acquaintance. *Nat. Immunol.* (2001) 2(5): 389–396.

18 Tan J, Town T, Mori T, Wu Y, Saxe M, Crawford F, et al. CD45 opposes beta-amyloid peptide-induced microglial activation via inhibition of p44/42 mitogen-activated protein kinase. *J. Neurosci.* (2000) 20(20): 7587–7594.

19 Tan J, Town T, Mullan M. CD45 inhibits CD40L-induced microglial activation via negative regulation of the Src/p44/42 MAPK pathway. *J. Biol. Chem.* (2000) 275(47): 37 224–37 231.

20 Hailer NP, Heppner FL, Haas D, Nitsch R. Fluorescent dye prelabelled microglial cells migrate into organotypic hippocampal slice cultures and ramify. *Eur. J. Neurosci.* (1997) 9(4): 863–866.

21 Carson MJ, Reilly CR, Sutcliffe JG, Lo D. Disproportionate recruitment of CD8+ T cells into the central nervous system by professional antigen-presenting cells. *Am. J. Pathol.* (1999) 154(2): 481–494.

22 Ploix C, Lo D, Carson MJ. A ligand for the chemokine receptor CCR7 can influence the homeostatic proliferation of CD4 T cells and progression of autoimmunity. *J. Immunol.* (2001) 167(12): 6724–6730.

23 Baker CA, Manuelidis L. Unique inflammatory RNA profiles of microglia in Creutzfeldt-Jakob disease. *Proc. Natl. Acad. Sci. U.S.A.* (2003) 100(2): 675–679.

24 Baker CA, Martin D, Manuelidis L. Microglia from Creutzfeldt-Jakob disease-infected brains are infectious and show specific mRNA activation profiles. *J. Virol.* (2002) 76(21): 10 905–10 913.

25 Moran LB, Duke DC, Turkheimer FE, Banati RB, Graeber MB. Towards a transcriptome definition of microglial cells. *Neurogenetics* (2004) 5(2): 95–108.

26 Schmid CD, Sautkulis LN, Danielson PE, Cooper J, Hasel KW, Hilbush BS, et al. Heterogeneous expression of the triggering receptor expressed on myeloid cells-2 on adult murine microglia. *J. Neurochem.* (2002) 83(6): 1309–1320.

27 Inoue H, Sawada M, Ryo A, Tanahashi H, Wakatsuki T, Hada A, et al. Serial analysis of gene expression in a microglial cell line. *Glia* (1999) 28(3): 265–271.

28 Neumann H. Control of glial immune function by neurons. *Glia* (2001) 36(2): 191–199.

29 Neumann H, Misgeld T, Matsumuro K, Wekerle H. Neurotrophins inhibit major histocompatibility class II inducibility of microglia: involvement of the p75 neurotrophin receptor. *Proc. Natl. Acad. Sci. U.S.A.* (1998) 95(10): 5779–5784.

30 Delgado R, Carlin A, Airaghi L, Demitri MT, Meda L, Galimberti D, et al. Melanocortin peptides inhibit production of proinflammatory cytokines and nitric oxide by activated microglia. *J. Leukoc. Biol.* (1998) 63(6): 740–745.

31 Kim WK, Kan Y, Ganea D, Hart RP, Gozes I, Jonakait GM. Vasoactive intestinal peptide and pituitary adenylyl cyclase-activating polypeptide inhibit tumor necrosis factor-alpha production in injured spinal cord and in activated microglia via a cAMP-dependent pathway. *J. Neurosci.* (2000) 20(10): 3622–3630.

32 Hoek RM, Ruuls SR, Murphy CA, Wright GJ, Goddard R, Zurawski SM, et al. Down-regulation of the macrophage lineage through interaction with OX2 (CD200). *Science* (2000) 290(5497): 1768–1771.

33 Mott RT, Ait-Ghezala G, Town T, Mori T, Vendrame M, Zeng J, et al. Neuronal expression of CD22: Novel mechanism for inhibiting microglial proinflammatory cytokine production. *Glia* (2004) 46(4): 369–379.

34 Sawynok J, Liu XJ. Adenosine in the spinal cord and periphery: release and regulation of pain. *Prog. Neurobiol.* (2003) 69(5): 313–340.

35 McCluskey LP, Lampson LA. Local immune regulation in the central nervous system by substance P vs. glutamate. *J. Neuroimmunol.* (2001) 116(2): 136–146.

36 Carson MJ, Sutcliffe JG, Campbell IL. Microglia stimulate naive T-cell differentiation without stimulating T-cell proliferation. *J. Neurosci. Res.* (1999) 55(1): 127–134.

37 Becher B, Antel JP. Comparison of phenotypic and functional properties of immediately ex vivo and cultured human adult microglia. *Glia* (1996) 18(1): 1–10.

38 Mack CL, Vanderlugt-Castaneda CL, Neville KL, Miller SD. Microglia are activated to become competent antigen presenting and effector cells in the inflammatory environment of the Theiler's virus model of multiple sclerosis. *J. Neuroimmunol.* (2003) 144(1–2): 68–79.

39 Juedes AE, Ruddle NH. Resident and infiltrating central nervous system APCs regulate the emergence and resolution of experimental autoimmune encephalomyelitis. *J. Immunol.* (2001) 166(8): 5168–5175.

40 Miller SD, Olson JK, Croxford JL. Multiple pathways to induction of virus-induced autoimmune demyelination: lessons from Theiler's virus infection. *J. Autoimmun.* (2001) 16(3): 219–227.

41 Olson JK, Girvin AM, Miller SD. Direct activation of innate and antigen-presenting functions of microglia following infection with Theiler's virus. *J. Virol.* (2001) 75(20): 9780–9789.

42 Cash E, Zhang Y, Rott O. Microglia present myelin antigens to T cells after phagocytosis of oligodendrocytes. *Cell Immunol.* (1993) 147(1):129–138.

43 Cash E, Rott O. Microglial cells qualify as the stimulators of unprimed CD4+ and CD8+ T lymphocytes in the central nervous system. *Clin. Exp. Immunol.* (1994) 98(2): 313–318.

44 Matysak MK, Denis-Donini S, Citterio S, Longhi R, Granucci F, Ricciardi-Castagnoli P. Microglia induce myelin basic protein-specific T cell anergy or T cell activation, according to their state of activation. *Eur. J. Immunol.* (1999)29 (10): 3063–76.

45 Santambrogio L, Belyanskaya SL, Fischer FR, Cipriani B, Brosnan CF, Ricciardi-Castagnoli P, et al. Developmental plasticity of CNS microglia. *Proc. Natl. Acad. Sci. USA* (2001) 98(11): 6295–6300.

46 Tran EH, Hoekstra K, vanRooijen N, Dijkstra CD, Owens T. Immune invasion of the central nervous system parenchyma and experimental allergic encephalomyelitis, but not leukocyte extravasation from blood, are prevented

in macrophage-depleted mice. *J. Immunol.* (1998) 161(7): 3767–3775.

47 Bauer J, Huitinga I, Zhao W, Lassmann H, Hickey WF, Dijkstra CD. The role of macrophages, perivascular cells, and microglial cells in the pathogenesis of experimental autoimmune encephalomyelitis. *Glia* (1995) 15(4): 437–446.

48 Byram SC, Carson MJ, Deboy CA, Serpe CJ, Sanders VM, Jones KJ. CD4+T cell-mediated neuroprotection requires dual compartment antigen presentation. *J. Neurosci.* (2004) 24(18): 4333–4339.

49 Raivich G, Jones LL, Kloss CU, Werner A, Neumann H, Kreutzberg GW. Immune surveillance in the injured nervous system: T-lymphocytes invade the axotomized mouse facial motor nucleus and aggregate around sites of neuronal degeneration. *J. Neurosci.* (1998) 18(15): 5804–5816.

50 Serpe CJ, Kohm AP, Huppenbauer CB, Sanders VM, Jones KJ. Exacerbation of facial motoneuron loss after facial nerve transection in severe combined immunodeficient (scid) mice. *J. Neurosci.* (1999) 19(11).

51 Medzhitov R, Janeway CA, Jr. Innate immune recognition and control of adaptive immune responses. *Semin. Immunol.* (1998) 10(5):351–353.

52 Becher B, Durell BG, Noelle RJ. IL-23 produced by CNS-resident cells controls T cell encephalitogenicity during the effector phase of experimental autoimmune encephalomyelitis. *J. Clin. Invest.* (2003) 112(8):1186–1191.

53 Becher B, Durell BG, Miga AV, Hickey WF, Noelle RJ. The clinical course of experimental autoimmune encephalomyelitis and inflammation is controlled by the expression of CD40 within the central nervous system. *J. Exp. Med.* (2001) 193(8): 967–974.

54 Pedersen EB, McNulty JA, Castro AJ, Fox LM, Zimmer J, Finsen B. Enriched immune-environment of blood-brain barrier deficient areas of normal adult rats. *J. Neuroimmunol.* (1997) 76(1-2): 117–131.

55 Flaris NA, Densmore TL, Molleston MC, Hickey WF. Characterization of microglia and macrophages in the central nervous system of rats: definition of the differential expression of molecules using standard and novel monoclonal antibodies in normal CNS and in four models of parenchymal reaction. *Glia* (1993) 7(1): 34–40.

56 Bouchon A, Hernandez-Munain C, Cella M, Colonna M. A dap12-mediated pathway regulates expression of cc chemokine receptor 7 and maturation of human dendritic cells. *J. Exp. Med.* (2001) 194(8): 1111–1122.

57 Daws MR, Sullam PM, Niemi EC, Chen TT, Tchao NK, Seaman WE. Pattern recognition by TREM-2: binding of anionic ligands. *J. Immunol.* (2003) 171(2): 594–599.

58 Paloneva J, Autti T, Raininko R, Partanen J, Salonen O, Puranen M, et al. CNS manifestations of Nasu-Hakola disease: a frontal dementia with bone cysts. *Neurology* (2001) 56(11): 1552–1558.

59 Brower V. Harnessing the immune system to battle Alzheimer's: some of the most promising approaches to fight Alzheimer's disease aim to develop vaccines. *EMBO Rep.* (2002) 3(3): 207–209.

60 Senior K. Dosing in phase II trial of Alzheimer's vaccine suspended. *Lancet Neurol.* (2002) 1(1): 3.

61 Birmingham K, Frantz S. Set back to Alzheimer vaccine studies. *Nat. Med.* (2002) 8(3): 199–200.

13
Contribution of B Cells to Autoimmune Pathogenesis
Thomas Dörner and Peter E. Lipsky

13.1
Introduction

Autoantibodies are features of most systemic autoimmune diseases. The discovery of characteristic autoantibodies mostly recognizing specific subcellular structures and cellular products, respectively, served to classify many systemic inflammatory diseases, such as rheumatoid arthritis (RA), systemic lupus erythematosus (SLE), Sjögren's syndrome (SS), mixed connective tissue disease, anticardiolipin syndrome, ANCA positive vasculitides, as well as organ-specific autoimmune diseases, such as Grave's disease, myasthenia gravis etc. In RA, Ig itself, targeted by rheumatoid factor, is a major autoantigen.

The generation of autoantibodies in these conditions is thought to occur because of immune dysregulation with an apparent break in tolerance. However, autoreactivity also occurs under non-autoimmune circumstances, such as during infections, immunizations and certain traumatic accidents, and then is presumed to be beneficial. Only a small fraction of autoantibodies, such as anti-double-stranded DNA, anti-Ro and anti-cardiolipin antibodies have been shown to be of pathogenic importance, whereas others are associated with autoimmune diseases, but not necessarily pathogenic. Importantly, persistent autoantibodies in systemic autoimmune diseases represent a break in tolerance regardless of whether they are pathogenic.

The cellular and molecular bases of autoantibody production remain largely unknown. Generation of some autoantibodies manifests a strong association with certain specific MHC class-II molecules [1, 2] as well as the usage of particular Ig variable region genes. In this regard, anti-CCP antibodies are very specific and occur early in RA [3–5]. Recent studies have contributed significantly to our understanding of pathogenic mechanisms inducing autoimmunity in RA [6–8]. Hill et al. [9] identified the specific role of MHC class II molecules in presenting citrullinated peptides to the immune system. Since the nature of the arthritogenic antigen is unknown, they studied T cell responses to citrulline-containing peptides in HLA-DRB1*0401 transgenic (DR4-IE tg) mice. Most interestingly, they could demonstrate that the conversion of arginine into citrulline at the peptide side-

Antigen Presenting Cells: From Mechanisms to Drug Development. Edited by H. Kropshofer and A. B. Vogt
Copyright © 2005 WILEY-VCH Verlag GmbH & Co. KGaA, Weinheim
ISBN: 3-527-31108-4

chain position interacting with the shared epitope significantly increased peptide-MHC affinity and led to the activation of CD4+ T cells. These results suggest a mechanism whereby DRB1 alleles with the shared epitope could initiate an autoimmune response to citrullinated peptides in RA patients. Another study [10] analyzed the involvement of peptidylarginine deiminase citrullinating enzymes (encoded by PADI genes) in the generation of citrullinated peptides in RA. In this case-control linkage disequilibrium study, PADI type 4 was identified as a susceptibility locus for RA and was found to be expressed in RA synovial tissues. Interestingly, the haplotype of PADI4 associated with susceptibility to RA was identified as having a higher stability of mRNA transcripts and was associated with antibody levels to citrullinated peptides in sera from individuals with RA. However, a recent study of RA patients in the UK could not confirm the association between the PADI4 polymorphism and RA [11]. Expression of citrullinated proteins in affected joints of humans and mice has been confirmed by another recent study [12], whereas no anti-CCP antibodies could be generated in mice. This may confirm that genetic predisposition plays a significant role in generating these autoantibodies and provides an interesting perspective into the interplay of genotype and phenotype leading to disease manifestations depending on the break of tolerance in the B cell system reflected by the production of autoantibodies.

Whereas intrinsic perturbances of B cell function are considered to be essential features of autoimmune diseases, several influences extrinsic to B cells can also contribute to the emergence of autoimmunity (Figure 13.1).

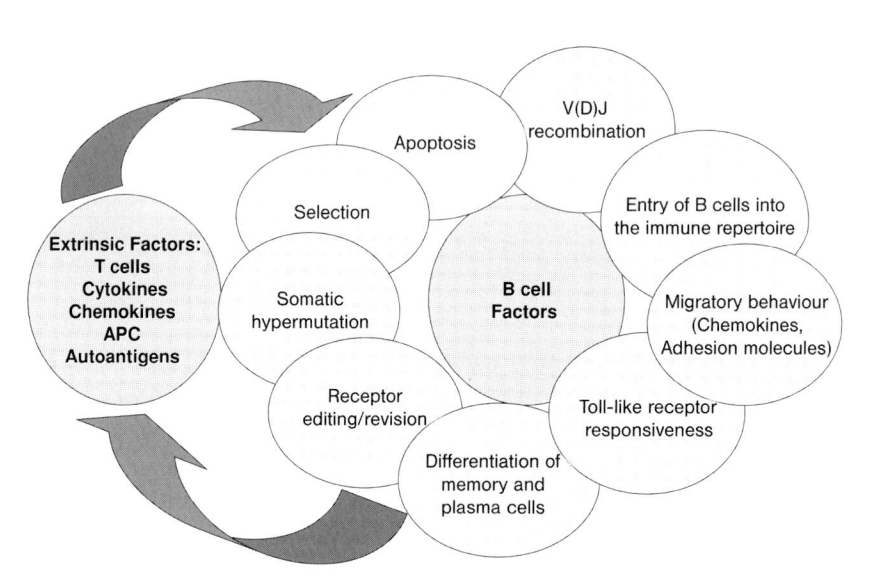

Figure 13.1 Extrinsic as well as intrinsic B cell factors involved in normal regulation of B cell development as well as potential factors disturbed under the conditions of autoimmunity.

13.2
Autoimmunity and Immune Deficiency

Notably, the humoral immune system, especially B cells and their descendants, plasma cells producing protective antibodies maintain the aspect of immunological memory that cannot be carried out by other cellular components of the innate or adaptive immune system. It is apparent that disturbances in the self-regulating circuits of these cellular components with their products can lead to clinically important disorders. Human primary immunodeficiency and autoimmunity appear to be more closely related than previously thought. In this context, hypergammaglobulinemia but also hypogammaglobulinemia can be associated with autoimmunity, indicating an essential fine tuning of this arm of humoral immunity by relevant qualitative and quantitative differences in production of Ig. It is also well accepted that inherited deficiencies of the complement system are associated with an increased incidence of SLE, glomerulonephritis, and vasculitis. Several antibody deficiencies are associated with autoimmune disease. Autoimmune cytopenias are commonly observed in individuals with selective IgA deficiency and common variable immune deficiency. Polyarticular arthritis can be seen in children with X-linked agammaglobulinemia. Combined cellular and antibody deficiencies, such as Wiskott-Aldrich syndrome, also carry an increased risk for juvenile rheumatoid arthritis and autoimmune hemolytic anemia. Recent advances in subcellular regulation of immune activation may allow further insights into the mechanisms involved [13]. Since engagement of the BCR plays a decisive role for the subsequent outcome of each B cell (Figure 13.2), we will initiate this discussion by analyzing IgV gene usage in health and disease.

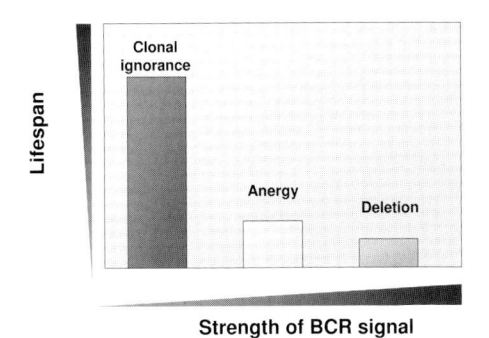

Figure 13.2 Basic mechanisms known to be involved in the prevention of B cell autoreactivity.

13.2.1
Basic Mechanisms Providing Diversity to the B Cell Receptor

The specificity of an antibody for an antigen is largely determined by three complementary-determining regions (CDRs) in the variable (V) region of both heavy (H) and light (L) chains (Figure 13.3). Each Ig molecule contains a unique set of CDRs that is interposed in the tertiary structure of the protein to form the classical antigenic binding site [14, 15]. The frame work regions (FRs) are thought to provide a scaffold to support the antigen binding site [16]. At the molecular level, a high grade of diversity of the Ig variable region (IgV) is generated during early B cell development (Figure 13.3) by the somatic recombination process that assembles functional genes by successive rearrangements of one of a number of joining (J), diversity (D) and finally, variable (V) minigene elements of the heavy chain, followed by V-J rearrangement of the light chain. During V(D)J recombination, further diversity can be introduced by exonuclease activity as well as the addition of nontemplated (N) or templated (P) nucleotides at the joining sites that particularly influence the amino acids encoded by the CDR3, the major site of antigen binding. Of importance, the pairing of heavy and light chain genes is apparently random [17], generating an additional degree of diversity.

In contrast to the T cell receptor repertoire, in which 10^{14} to 10^{16} diverse receptors can potentially be generated using the various germline genes, it is estimated that IgV genes may only be able to generate 10^8 different B cell receptors. As a result, the T cell receptor repertoire may be sufficiently large to recognize all linear peptide epitopes, whereas the B cell repertoire may be insufficient to recognize all possible pathogens and conformational epitopes that can only be recognized by B cells. Additional mechanisms, however, are used by B cells to increase diversity further after initial V(D)J recombination. In this regard, somatic hypermutation

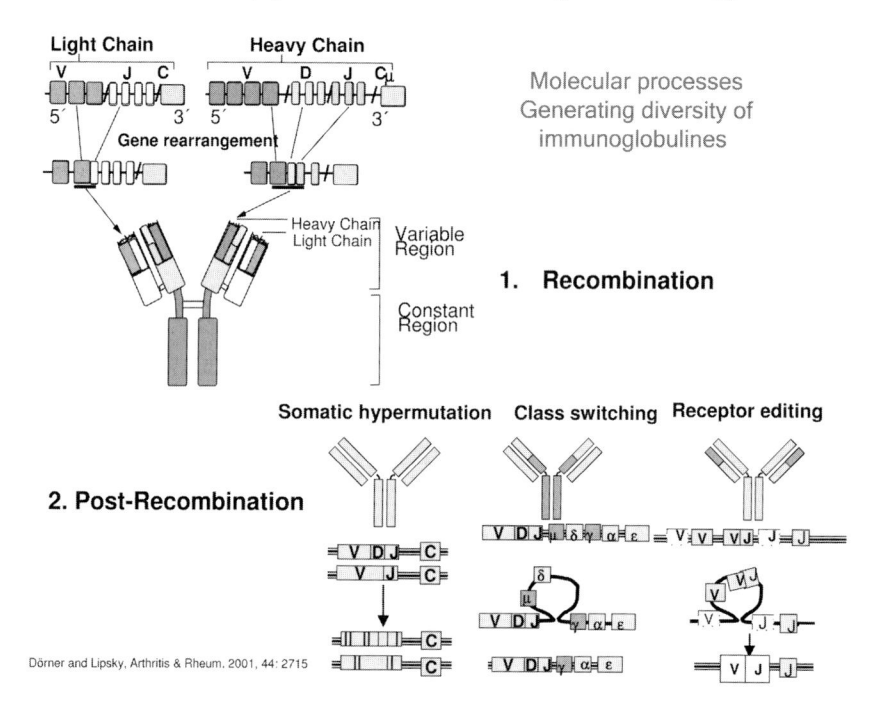

Figure 13.3 Molecular mechanisms involved in the generation of IgV variability leading to the diversity of the BCR as well as secreted Ig. (This figure also appears with the color plates.)

and secondary rearrangement of upstream V gene segments (receptor replacement or editing) may further diversify the IgV gene repertoire following antigen or autoantigen exposure [18–20]. In addition, Ig class switching serves to refine the functional capabilities of the Ig molecule. These various mechanisms contribute to the generation of a highly diversified array of IgV gene products that conceivably could recognize all potential antigens.

13.2.2
Ig V Gene Usage by B Cells of Healthy Individuals

The distribution of V, D, and J gene usage and the imprints of the recombinational machinery and the frequency and nature of somatic mutations have been delineated for the heavy and light chains in normals [21–27]. Whereas several specific V_H and $V\kappa$ genes were noted to be positively or negatively selected in the normal repertoire [21–24], $V\lambda$ gene rearrangements were not positively selected in normals [23]. Studies of V_H gene distribution in normals [21, 22] found that the frequency of usage in general was similar to the germline complexity. Notably, previous studies have demonstrated that most human anti-dsDNA antibodies are encoded by V_H3 family members and to a lesser degree by other V_H families [28–30]. It is likely that this distribution reflects the fact that this V_H family is the largest family of V_H genes and is the most frequently expressed one in normals [31–33].

At the level of individual genes, ten V_H genes (V_H3-23, 4-59, 4-39, 3-07, 3-30, 1-18, 3-30.3, DP-58, 4-34 and 3-09) were found to be employed by approximately 60% of normal peripheral blood B cells [21, 22]. One particular V_H3 family member, 3-23/DP-47/V_H26 was used by about 13% of B cells. Previous studies [33, 34] had suggested that over-representation of V_H3-23 occurred at the pre-B cell stage of development.

13.2.3
Potential Abnormalities in Molecular Mechanisms Underlying IgV Gene Usage in Systemic Autoimmune Diseases

There are several reasons to consider the possibility that fundamental abnormalities in the molecular mechanisms generating IgV genes may play a role in producing autoantibodies. First, biases in recombination resulting in the preferential usage of certain V, D, and J genes might lead to the generation of a B cell repertoire enriched in Ig receptors potentially binding autoantigens. Moreover, alterations in the recombination enzymes (RAG1 and 2, Ku60 and Ku80, DNA phosphokinase, terminal deoxynucleotidyl transferase, exonuclease activity) may generate Ig rearrangements with D inversions and D-D fusions or other CDR3 abnormalities that could increase the likelihood of autoantigen reactivity. The biased pairing of heavy and light chains might also contribute to autoantigen binding. Subsequently, the frequency and pattern of somatic hypermutation and, potentially, abnormalities in the enzymatic machinery (activation induced deaminase AID

[35]) could also contribute to the production of autoreactive B cells [29, 36–38]. Finally, deficient receptor editing/revision could permit the persistence of autoantibody encoding B cells. Thus, several molecular mechanisms could potentially play a role in the emergence of autoimmune B cells.

Importantly, each of these putative intrinsic defects of B cells, as well as the various extrinsic influences, including abnormalities of ligand interactions (Figure 13.1), could contribute to the production of autoantibodies, either alone or in association. This is of particular interest since at least two types of autoantibody-producing populations in SLE seem to exist. The first is exemplified by anti-DNA producing cells. The anti-DNA titer exhibits a striking correlation with disease activity and usually decreases at the end of a disease flare and in response to conventional immunosuppressive therapies. These results imply that these autoantibodies are produced by a subset of cells that is rapidly and transiently stimulated and actively proliferating and is susceptible to antiproliferative therapy. Secondly, titers of antibodies to Ro, La, Sm, RNP, and cardiolipin are persistent, lack a predictable change with therapy and do not correlate with disease activity. These results suggest that these antibodies may be produced by persistent long-lived plasma cells [39] that may not be influenced by conventional immunosuppressive or anti-proliferative therapy. Whether differences in the degree of T cell help (TI versus TD) account for these distinct types of autoantibodies needs to be delineated.

13.2.4
Lack of Molecular Differences in V(D)J Recombination in Patients with Systemic Autoimmune Diseases

Previous analyses of autoimmune B cells focused preferentially on productive V gene rearrangements. Therefore, these studies were not able to discern the impact of molecular and selective influences. This was possible by analyses of the nonproductive repertoire that is not expressed but reflects the immediate impact of molecular processes, such as recombination and somatic hypermutation [17, 21–24, 40–44] and is not influenced by selection. In contrast, the distribution of B cells and their productive IgV gene rearrangements can be influenced by various selective events during development and subsequent antigenic stimulation because of the nature of the expressed heavy and/or light chain [18, 19]. Analyses of the nonproductive IgV gene repertoire in SLE patients and patients with SS documented minimal abnormalities in the nonproductive V_H, $V\kappa$ and $V\lambda$ gene repertoires [41–47]. The overall data indicate that IgV gene usage in the nonproductive repertoire is not significantly different from normal, suggesting that the basic process of IgV gene recombination is largely normal in patients with autoimmune diseases, but strikingly influenced by post-recombinational processes.

13.2.5
Receptor Editing/Revision and Autoimmunity

Formerly, it was thought that expression of membrane-bound Ig B cell receptor (BCR) extinguished subsequent Ig rearrangements by down-regulating the expression of recombination activating gene (RAG) 1 and RAG2 enzymes in the bone marrow. However, several studies provide evidence that immature B cells outside the bone marrow [48–56] retain RAGs activity and therefore can replace their receptors by secondary Ig gene recombination (receptor editing/revision) (Figure 13.3). This is noted with increased frequency in secondary lymphoid organs [49, 50, 55, 56] and in the fetus [57]. The question of whether RAGs are reexpressed in mature B cells after antigen exposure is a more complex issue. RAGs are expressed by germinal center cells and by post-switch memory cells that also express IgG in man [55].

Receptor revision in B cells, along with the process of clonal selection, contributes to the somatic evolution of appropriate immune responses under normal circumstances [19, 54].

There is controversy over whether receptor editing/revision or secondary rearrangements or defects of this mechanism are involved in shaping the B cell repertoire in autoimmunity. Deficiencies in central or peripheral receptor editing might play a role in generating autoimmunity [51–53, 58–63]. In addition, analysis of autoreactive hybridomas [61] generated from SLE patients demonstrated an enhanced usage of J proximal Vκ1 genes and preferential use of J elements proximal to Vκ, suggesting that receptor editing in SLE might be defective, since skewing towards the usage of Jκ distal Vκ genes and Jκ5 expressing V gene products [42, 51, 52, 56, 62, 63] has been taken as an indication of active receptor editing. As receptor editing at the V_L loci is thought to play a major role in rescuing autoreactive B cells from deletion [51, 52, 62, 63], defects in receptor editing may play a role in the etiology of SLE [58–63].

Defects in receptor editing were also suggested by the finding that the proximally located Vκ gene A30 known to encode anti-dsDNA antibodies when combined with Jκ2 was not deleted in SLE patients with nephritis [59, 60]. By contrast, the use of Jκ distal V genes in some autoantibodies of SLE patients [64, 65] has suggested that receptor editing might be intact in these patients and, in fact, that receptor editing might perversely serve to introduce the usage of V genes that encode autoantibodies. Additional analyses of VκJκ rearrangements have supported the possibility of increased receptor editing in SLE [42]. The B3 gene, the single member of the Vκ4 family, was overrepresented in the productive repertoire of an SLE patient [42] with clear evidence of enhanced receptor editing. B3 is the most Jκ proximal Vκ gene but needs to be rearranged by inversion and, therefore, may not be used frequently in primary rearrangements. However, when the initial Vκ rearrangement occurs to a 5′ Vκ gene, it occurs by inversion because of the orientation of the Jκ distal Vκ cassette of genes. This retains B3 in the locus in an opposite orientation that now can be secondarily rearranged by deletion. In SLE, the B3 gene is frequently rearranged to Jκ5 with fewer mutations than when it was rearranged to Jκ1-4, indicating that Jκ5-employing rearrangements had been introduced after the mutational process had been initiated, which is consistent with peripheral receptor revision [42]. This gene also encodes for autoantibodies, including the anti-dsDNA specific idiotype, F4+ [66], and was detected frequently in B cells infiltrating the rheumatoid synovial membrane [56], suggesting that receptor editing may contribute to autoantibody formation in RA as well as SLE.

A Vλ repertoire analysis of an SLE patient also revealed a preferential usage of the Jλ distal V genes of cluster C and the most distal Jλ7 element [23], which was significantly different than their usage in normals [43]. Analysis of mutations indicated that receptor editing of Vλ genes occurred centrally, before mutational activity was activated.

The extent of receptor editing/revision in normals has not been fully elucidated. Whereas some studies in normals [21–24] did not show indications of active receptor editing/revision, de Wildt and colleagues [67] identified B cell clones among 365 human IgG$^+$ V$_H$/V$_L$ pairs with the same heavy chain rearrangement but different light chains (VκL1 and Vλ1c) and concluded that receptor editing had been active in these B cells. Receptor editing was also found in an analysis of fetal spleen rearrangements [57].

Studies in patients with RA [56, 68] provided evidence that receptor editing/revision may also be more active in the synovium of patients than in normals. In contrast to these patients, individuals with SS appear to have decreased receptor editing/revision, as identified by enhanced usage of V proximal located J$_L$ segments, likely reflecting a defect or infrequent usage of receptor editing [44, 45]. In this regard, another analysis of six monoclonal antibodies with RF activity obtained from the peripheral blood of SS patients showed that all used Vλ proximal Jλ2/3 gene segments [69], which is consistent with the conclusion that receptor editing/revision may be defective or less frequently employed in SS. These data are consistent with the conclusion that the use of receptor editing is different

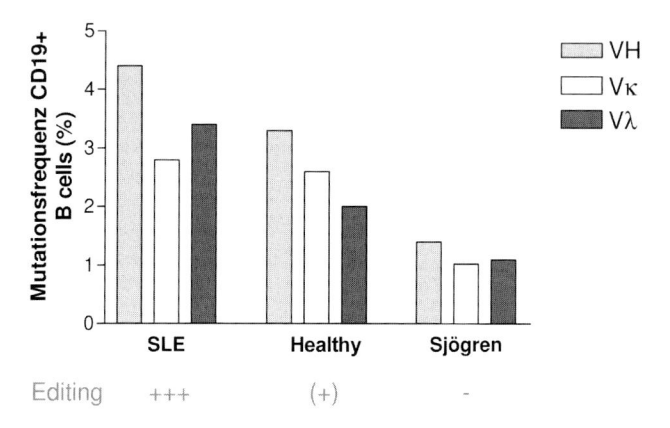

Figure 13.4 Comparison of the mutational frequencies of individual B cells obtained from patients with SLE, Sjögren's and normal controls compared to indications of receptor editing/revision.

in different autoimmune diseases (Figure 13.4). However, the possibility must be considered that enhanced receptor editing in SLE patients reflects increased B cell activity but is insufficient to delete autoantibody encoding B cells.

13.2.6
Selective Influences Shaping the Ig V Gene Repertoire in Autoimmune Diseases

An early hypothesis was that a fundamental abnormality in autoimmunity might bias IgV gene usage, altering the entire repertoire and leading to a greater tendency of B cells to bind autoantigen. The bulk of available data, however, do not support this hypothesis, but rather suggested the absence of genetic abnormalities in IgV gene usage as a mechanism causing B cells to generate an autoreactive repertoire in patients with systemic autoimmune diseases [42–47, 70, 71]. Thus, no genetic polymorphisms of the heavy and light chain IgV gene loci and no fundamental abnormalities in the V(D)J recombination process appear to underlie the generation of autoantibodies. However, there are major distortions of the IgV gene repertoire that derive from the impact of post-recombinational processes, such as selection and somatic hypermutation [42-46].

13.2.6.1 IgV Gene Usage by Autoantibodies

There are very few examples of an association of particular V_H genes and autoimmune disorders. The clearest example is the use of V_H4-34 by cold agglutinins [72–74]. V_H4-34, expressing the idiotype 9G4, has been shown to encode cold agglutinins. Antibodies employing the V_H4-34 gene also recognize different autoantigens, as is the case of some rheumatoid factors [75], anti-DNA antibodies [76], and the anti-D antibodies [77]. Among anti-DNA antibodies, V_H 4-34 is employed by both the T14 and 9G4 idiotypes. Most of the anti-I cold agglutinins V_L chains are derived from Vκ3 and, to a lesser degree, Vκ1 or Vκ2 germline genes, whereas cold agglutinins with anti-i activity are not restricted to certain L chains, suggesting that associated light chains can refine the binding capacities.

V_H4-34 has been reported in 5/10 (50%) of V_H4 employing anti-DNA antibodies [78], but in only 19.2% of CD5$^+$ B cells and 15.6% of CD5$^-$ B cells expressing productively rearranged V_H4 genes from normals [22].

One study [79] surveyed a broad variety of human anti-dsDNA binding hybridomas to define sequence characteristics that may be related to pathogenicity. These human anti-dsDNA monoclonal antibodies did not exhibit a preferential use of particular V_H and V_L genes. However, certain features of the CDR3, such as usage of an uncommon reading frame of the D segment, D-D fusions or the frequent presence of arginine were found to be characteristic of anti-dsDNA antibodies [36–38]. Analyses of IgV gene usage in autoreactive monoclonal autoantibodies and IgG Fab libraries confirm the conclusion that there is no preferential usage of particular V_H and V_L genes. In general, the Ig V gene usage of monoclonal antibodies recapitulates the frequency of gene usage in normals. However, there are some possible exceptions. Thus, V_H3-23 that has been reported to encode the 16/6 idiotype was detected in 18% of IgM anti-DNA human autoantibodies [78], exceeding the frequency of this gene in normals (13%) [22]. From all analyses, autoantibodies clearly frequently employ V_H genes that are frequent in the normal repertoire, such as V_H3-23, V_H1-69, V_H4-34, V_H4-39 and V_H3-07. However, anti-DNA antibodies also use uncommon V_H genes, such as V_H1-46, V_H3-64, V_H3-74, and V_H4-61 [78, 79]. Thus, the detection of a broad variety of V_H genes used by these autoantibodies is consistent with the conclusion that there is no typical pattern of "autoimmune V genes" and, therefore, there is no molecular abnormality or specific defect in selection in IgV gene usage causing the production of autoantibodies.

Moreover, there is a broad V_L gene usage by human monoclonal autoantibodies. Vκ1-3 family members are used somewhat more frequently than Vλ genes, making up 40% of the overall repertoire in anti-DNA antibodies and are characteristic of cold agglutinins [80]. However, this distribution is also similar to the normal repertoire, supporting the contention that the choice of IgV genes encoding autoantibodies follows similar rules as seen in antibodies to exogenous antigens.

13.2.7
Role of Somatic Hypermutation in Generating Autoantibodies

Somatic hypermutation resulting in the accumulation of basic residues in the CDRs can contribute to the generation of autoantibodies. Since the underlying mechanisms and enzymes involved in somatic hypermutation have not been completely delineated, most studies have examined the overall frequency and pattern of mutations, rather than the activity of the enzymes involved in generating the mutations, such as AID [35].

There is some evidence that potentially pathogenic autoantibodies may arise by somatic hypermutation during the immune response to foreign antigens from antibodies that have no reactivity to autoantigens in their germline configuration. An abnormality in the unknown "mutator" mechanism is suggested to predispose to increased somatic mutations, since most pathogenic anti-DNA antibodies are heavily mutated. Although the pattern of mutations of anti-dsDNA antibodies present in SLE is suggestive of an antigen-driven response with an increased frequency of replacement mutations in CDRs compared to FRs, the antigen(s) driving this process has/have not been clearly identified. Immunization with pneumococcus antigen [81] can stimulate the production of anti-DNA antibodies that cross-reacted with both dsDNA and phosphorylcholine, the dominant hapten on the pneumococcal cell wall. Binding analysis of a panel of anti-DNA encoding heavy chain sequences with single amino acid substitutions documented that the mutations altered antigen binding, and the capacity to induce renal damage. These data suggested that exogenous antigen can generate self reactive antibodies during a germinal center reaction by inducing somatic hypermutation of IgV genes [82]. Although the role of somatic hypermutation in generating autoantibodies has only been directly addressed in a limited fashion, one study reported six human anti-DNA monoclonal antibodies that contained replacement mutations in the CDRs which introduced basic residues apparently involved in DNA binding [29]. These mutations were essential for autoantibody activity, as demonstrated by the finding that back mutation to the germline configuration resulted in a loss of DNA binding capacity. These data indicate that mutational activity can lead to the development of autoantibodies from V(D)J rearrangements that had no intrinsic autoreactivity.

The importance of somatic mutations and the subsequent introduction of basic residues in the CDRs of autoantibodies have also been reported [83–86]. Various studies [79] have considered enhanced R/S ratios in the CDRs of gene rearrangements encoding autoantibodies as an indication for antigen-mediated selection [86]. This assumption has been questioned, however, since nonproductive Ig V_H gene rearrangements without selective influences exhibited an intrinsically high R/S ratio in the CDRs [40, 41]. Thus, increased R/S ratios in the CDRs of V_H genes reflect an intrinsic feature of the sequences [87] and the characteristic action of the mutator operating on V_H genes independent of antigen. By contrast, reduced R/S ratios have been repeatedly reported for V_H gene rearrangements obtained from the salivary glands of SS patients [88–90]. Notably, this was not found in V_L rearrangements from the peripheral blood of these patients [44, 46], suggesting that there is a selection process confined to the salivary glands in these patients.

One of the most striking features of SLE patients is the marked and generalized increase in mutational frequency. The increased impact of somatic hypermutation was a generalized abnormality in SLE and not limited to B cells expressing specific B cell receptors, but rather diffusely noted in memory $CD27^+$ B cells and plasma cells. These findings are consistent with other results suggesting an increased mutational activity in SLE [66]. Since mutational activity in general is thought to be induced in response to T cell dependent antigens, the B cells of SLE patients appeared to have been stimulated in a T cell dependent manner more intensively or more persistently than in normals [91]. Not only was the mutational frequency in memory cells and plasma cells increased in SLE [45], but the frequency of plasma cells was also increased in patients with active SLE [70, 92]. Neither the frequency of mutations nor the percentage of circulating memory B cells or plasma cells was increased in SS patients [93]. The results suggest that mutational activity is differentially upregulated in SLE but not in SS. Whether the increased mutational frequency in SLE reflects a primary abnormality in the mutational machinery, the intensity or persistence of stimulation, intensive germinal center reaction with CD40/154 involvement [94] and/or ICOS/ICOS-L engagement [91, 95], a defect in apoptosis of B cells expressing mutated receptors or an exaggerated role of T cells involved in B cell activation remains to be determined.

13.3
Disturbed Homeostasis of Peripheral B Cells in Autoimmune Diseases

Recent studies [45, 93, 96] examined the phenotype and Ig heavy chain gene usage of peripheral B cell subpopulations of patients with autoimmune diseases. Since B cell development can be followed using several surface phenotypic markers (Figure 13.5), analysis of this expression profile permits the detection of B cell differentiation status in health and disease. In the peripheral blood of healthy controls, ca. 60% of B cells have the phenotype of naïve cells, and 40% have that of memory cells. Typically, in such healthy donors, less than 2% of the peripheral B cells are plasma cells ($CD19^{dim}$, $CD20^-$, $CD38^{++}$, and $CD27^{high}$). Analyses of SLE and SS patients identified a marked reduction of $CD19^+/CD27^-$ naïve B cells, retention of $CD19^+/CD27^+$ memory B cells and increased numbers of $CD19^{dim}/CD27^{high}$ plasma cells in SLE and a predominance of naïve B cells in SS [92, 93, 96]. Molecular analysis of IgV genes in one SLE patient showed that these B cell subpopulations differed in their V_H gene usage. With regard to individual genes, the V_H3-23 gene was found most often in $CD27^-/IgD^+$ naïve (6/14) and $CD27^+/IgD^+$ memory B cells (4/15), whereas V_H4-34 and V_H4-59 were frequently found in $CD27^{high}/IgD^-$ plasma cells [92]. In the latter population, a heavily mutated clone using V_H4-61 was identified. Preferential usage of V_H4 genes by post-switch cells has been reported by others in patients with RA [97, 98]. By contrast, V_H3 was most frequently found in naïve B cells or unfractionated peripheral B cells of

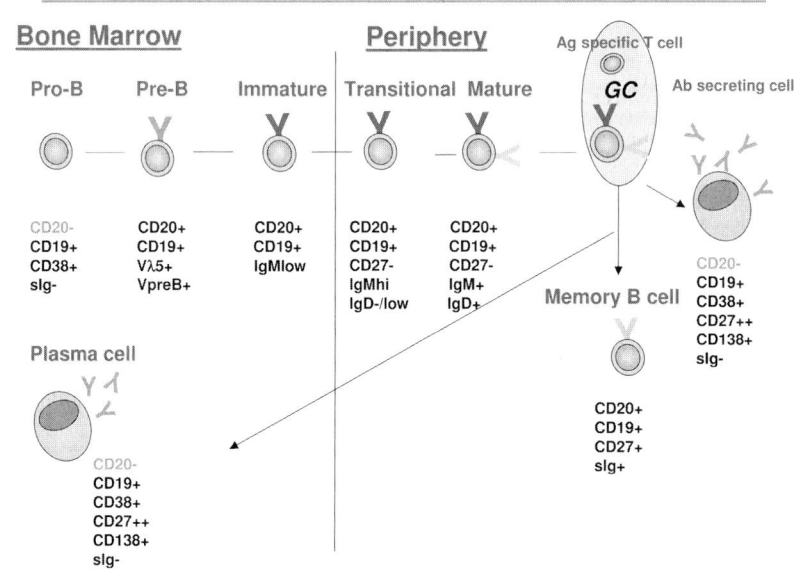

Figure 13.5 Schematic development of B cells in the bone marrow as well as in the periphery based on the expression of several activation/differentiation markers. (Modified according to Weiss and Silverman, 2003.)

normals [21, 22, 31–33]. Moreover, the gene V_H4-34 frequently used in clonally unrelated CD27^{high} plasma cells of the SLE patient has been reported to be involved in the formation of anti-dsDNA antibodies in these patients [99–101] and to be expanded in patients with active disease [78, 92]. In normals, this particular gene occurred at a frequency of 3.5% among peripheral CD5$^+$ and 3.9% among CD5$^-$ B cells [22] was found to be negatively selected in an analysis of CD19$^+$ peripheral cells in normals [21] and was found to be excluded from post-switch tonsil plasma cells [102]. Further studies confirmed abnormalities in peripheral B cell subsets in SLE, SS and RA [103–105]. The overall data indicate that there are differences of peripheral B cell composition related to distinctive V gene usage in specific B cell subpopulations that likely result from antigen-specific selection and, therefore, the result of post-recombinational processes. Analysis of blood samples from patients with RA showed a clear change in the B cell populations, with a shift toward cells expressing an activated and differentiated phenotype [106].

13.4
Signal Transduction Pathways in B Cells

In general, the immune system is maintained by a fine balance between activation and inhibition (Figure 13.6). On one hand, it must possess adequate reactivity to generate an effective immune response to target non-self molecules while, on the

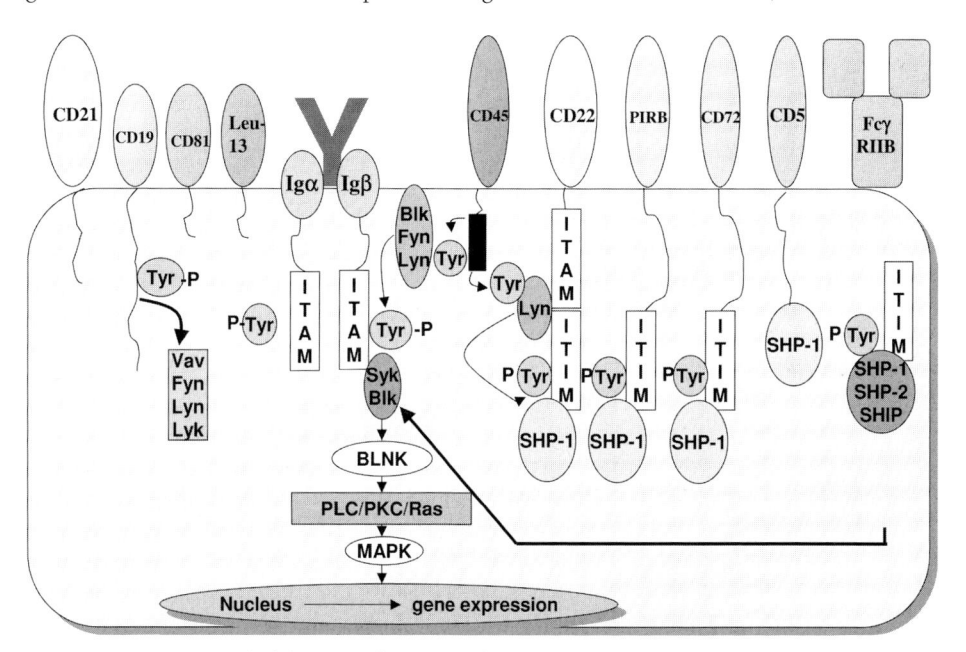

Figure 13.6 Activation and inhibitory markers on B cells and their signaling pathways involved (for further details see text). (This figure also appears with the color plates.)

other hand, avoiding the emergence of autoimmunity. Essential to this process is the ability to control the timing and place of activation and to limit the extent of activation. This is regulated by several extrinsic and intrinsic mechanisms. Failure to maintain this balance could result in either immunodeficiency or autoimmunity. Inhibitory receptors are involved in this regulation.

Autoimmune diseases are complex disorders characterized by adaptive immune responses supported by innate mechanisms (i.e. TNF in RA) that are inappropriately directed against self tissues. Previous research has primarily focused on the contributions of dysfunctional T cells, whereas recent studies have demonstrated that B cells are important contributors in the pathogenesis of autoimmunity, besides their ability to produce autoantibodies. Our understanding of the important cellular and signaling components involved in B cell development and the maintenance of normal humoral immune responses has expanded greatly in the last decade, especially by multiple genetic studies in mice. This has provided new insight into the role of B cells in several autoimmune diseases.

13.4.1
B Cell Function Results from Balanced Agonistic and Antagonistic Signals

B cells undergo a tightly regulated developmental pathway from early progenitors to terminally differentiated plasma cells (Figure 13.5). Many of these developmental steps depend on signals mediated through soluble factors and receptor–ligand interactions. Multiple checkpoints permit both positive and negative selection of B cells, both centrally in the bone marrow and in the peripheral lymphoid tissues, such as the spleen and lymph nodes. These checkpoints are necessary to ultimately produce a diverse population of B cells capable of generating high affinity effector antibodies in the absence of pathologic autoreactivity. Transgenic mice with perturbations in selective regulatory pathways that affect B cell development often develop autoimmune disease. By altering gene expression in mice, two major categories of defects that lead to autoantibody production have been identified: one set that alters B cell longevity and others that alter thresholds for immune cellular activation.

13.4.1.1 Altered B Cell Longevity can Lead to Autoimmunity
After their generation in the bone marrow, the process of negative selection eliminates most immature B cells before entry into peripheral lymphoid tissues. Resident B cells in the periphery undergo a second screening process for reactivity with peripheral self antigens that results in apoptosis, receptor editing or anergy depending on the strength of BCR signaling (Figure 13.2). Autoreactive B cells that have survived this screening process have received various growth and anti-apoptotic signals. Thus, alterations in the expression of genes that regulate B cell survival can lead to the development of autoimmunity. A classical example of dysregulated apoptotic regulatory genes leading to autoimmunity was found in bcl-2 transgenic mice. Enhanced bcl-2 expression allows inappropriate survival of auto-

reactive B cell clones [107]. Bcl-2 transgenic mice develop anti-nuclear antibodies and have glomerulonephritis caused by immune complex deposition. A second example is provided by MRL mice homozygous for mutations in the Fas gene, a death-inducing receptor required for normal regulation of B cell and T cell lifespans. MRL$^{lpr/lpr}$ mice develop a spectrum of autoreactivity resembling that found in human SLE and other autoimmune diseases.

In addition to intrinsic defects that can lead to increased B cell longevity, external signals permit autoreactive B cells to escape deletion. One such signal that is particularly important in B cell growth, differentiation, and survival is BAFF (also known as BlyS, TALL-1, THANK and zTNF4). It is a member of the TNF family of cytokines that is produced by dendritic cells, monocytes and macrophages and induces immature B cell survival and growth of mature B cells within peripheral lymphoid tissues. BAFF binds three receptors: BCMA (B-cell maturation antigen), TACI (transmembrane activator and calcium-modulator and cyclophilin ligand interactor) and BAFF-Receptor. Through these receptors, BAFF acts as a potent co-stimulator for B cell survival when coupled with B cell antigen receptor ligation. BAFF ligation increased bcl-2 expression and increased activation of NF-κb, both of which increase B cell survival [108]. Mice transgenic for either BCMA or BAFF display mature B-cell hyperplasia and develop an SLE-like disease, with anti-DNA antibodies, elevated serum IgM, vasculitis and glomerulonephritis [109]. Moreover, BAFF expression is elevated in MRL$^{lpr/lpr}$ mice and (NZW×NZB)F1 hybrid mice and correlates with disease progression [108]. Conversely, BAFF-deficient mice show a complete loss of follicular and marginal-zone B lymphocytes [110a,b]. Attempts to determine the role of BAFF in autoimmune disease development and progression have used BCMA-immunoglobulin (Ig) and TACI-Ig fusion proteins as decoy BAFF receptors. Administration of BMCA-Ig or TACI-Ig to (NZW×NZB)F1 mice leads to increased survival, decreased proteinuria and delayed disease progression [108, 109]. Elevated BAFF levels are also found in some SLE and SS patients [109].

Studies of knockout mice have shown that BCMA, TACI and BAFF-R are not directly equivalent in function [111, 112]. Mice lacking BCMA show normal B-cell development and antibody responses [113, 114], whereas TACI-deficient mice are deficient only in T-cell-independent antibody responses [115, 116]. Very recently, BCMA has been identified to be involved in the generation of long-lived plasma cells [117]. Paradoxically, mice lacking TACI show increased B-cell proliferation and accumulation, suggesting an inhibitory role for TACI in B-cell homeostasis. Gene-targeted mice lacking BAFF-R have yet to be reported, but the natural mouse mutant A/WySnJ has a disruption of the intracellular domain of BAFF-R. A/WySnJ mice display a phenotype that is similar to BAFF$^{-/-}$ mice, although follicular and marginal-zone B cells are not completely abolished [118]. In addition, A/WySnJ mice are impaired only in T-cell-dependent antibody responses, unlike the more comprehensive defect observed in BAFF-deficient mice. These results suggest that, while BAFF-R may be the major receptor relaying BAFF-mediated signals for B-cell survival, redundancy in function may be provided by the other two receptors, especially by TACI.

13.4.1.2 Altered B Cell Activation can Lead to Autoimmunity

As noted above, signals generated through the B cell antigen receptor (BCR) are critical for the development and B cell responses to antigen. The BCR is non-covalently associated with the signal transduction elements, Igα (CD79a) and Igβ (CD79b) (Figure 13.6). The cytoplasmic domains of Igα and β contain highly conserved motifs that are the sites of Src family kinase docking and tyrosine phosphorylation, termed the immunoreceptor tyrosine-based activation motifs (ITAM). Phosphorylation of tyrosines within these motifs is mediated by Src family kinases, including Lyn, Fyn or Blk. These phosphorylation events promote BCR recruitment of another tyrosine kinase, Syk, which facilitates receptor phosphorylation and initiates downstream signaling cascades that promote B cell activation [119].

The generation and maintenance of self-reactive B cells is regulated by autoantigen signaling through the BCR complex. These responses are further influenced by other cell surface signal transduction molecules, including CD19, CD21 and CD22, that function as response regulators to amplify or inhibit BCR signaling. CD19, CD21 and CD22 modulate BCR-mediated signals by altering intrinsic intracellular signal transduction thresholds and thereby adjusting the strength of signal needed to initiate BCR-mediated activation [120]. Intracellular regulatory molecules that also control BCR signaling intensity include Lyn, Btk, Vav and the SHP1 protein tyrosine phosphatase [121, 122]. Notably, CD19, CD21, CD22, Lyn, Vav and SHP1 are functionally linked in a common signaling pathway (Figure 13.6).

CD19

Mice with altered CD19, CD21, CD22, Lyn or SHP1 expression produce autoantibodies and develop a spectrum of autoimmunity. Peripheral tolerance is disrupted in mice that overexpress CD19, which results in hyperactive B cells and the spontaneous production of IgG subclass autoantibodies [123]. Mice that have only a 15–30% increase in CD19 expression have a distinct phenotype from normal controls, developing SLE-like manifestations [124].

Although multiple molecules involved in a common CD19 signal transduction pathway influence autoimmunity in mice, similar examples in humans are very limited. In systemic sclerosis, autoantibodies are detected in more than 90% of SSc patients and are considered to play a critical role in the pathogenesis of SSc. Surprisingly, CD19 and CD21 expression levels are 20% higher on B cells from SSc patients compared with healthy individuals, whereas the expression of other cell surface markers such as CD20, CD22, and CD40 is normal. As in SSc patients, CD19 overexpression by 20% induced autoantibody production in a non-autoimmune strain of mice [125]. Antinuclear antibodies and rheumatoid factor were induced in these mice, but not wild-type controls. Like mice that overexpress CD19, the tight-skin mouse, a genetic model for human SSc, also contains spontaneously activated B cells and autoantibodies against SSc-specific target autoantigens [125]. Tight-skin mice also develop cutaneous fibrosis. In contrast to mice that overexpress CD19, mice that are CD19$^{-/-}$ have a markedly elevated BCR signaling threshold as compared with wild-type mice. CD19-deficiency in tight-skin mice results in quiescent B cells, with significantly reduced autoantibody production and skin fibrosis. Overall, modest alterations in CD19 expression could contribute to the development of autoantibodies in humans, as reported for SSc [126]. Graded alterations in expression or function in these "response-regulators" may play a role in autoimmune diseases.

CD21

CD19 physically interacts with CD21, a complement receptor on the surface of B cells. Thereby, CD19 transduces signals generated by CD21 binding complement fragment C3d, which may amplify signals generated by simultaneous BCR ligation. Multiple studies have suggested that altered CD21 function correlates with autoimmunity in animal models. In addition, self-reactive B cells with 60% reduced CD21 expression are not anergized by soluble self-antigen in mouse models of tolerance [127]. Although these studies suggest a direct role for CD21 in regulating B cell function and autoantibody production, this may actually reflect a role for CD21 in regulating cell surface CD19 expression.

13.4.1.3 Inhibitory Receptors of B Cells

There are two broad classes of inhibitory receptors that share several structural and functional similarities. Each inhibitory receptor contains one or more immunoreceptor tyrosine-based inhibitory motifs (ITIMs) within its cytoplasmic domain that are essential for generation and transduction of inhibitory signals. Ligation of the inhibitory receptor to an immunoreceptor tyrosine-based activatory motif (ITAM)-containing activatory molecule results in tyrosine kinase phosphorylation of the tyrosine residue within the ITIM [128] by lyn [129] (Figure 13.6). Tyrosine phosphorylation of the ITIM allows it to bind and activate phosphatases containing an src homology 2 (SH2) domain. Two classes of SH2-containing inhibitory phosphatases have been identified: the protein tyrosine phosphatases SHP-1 and SHP-2, and the phosphoinositol phosphatases SHIP and SHIP2. These classes have separate downstream signaling pathways through which they modulate cellular inhibition. In general, each class of phosphatases interacts with the ITIMs of different inhibitory receptors but each inhibitory receptor appears to act predominantly through only one class of phosphatase [130].

Of note, FcγRII, CD22, PD-1, Lyn, SHP-1 and SHIP are crucial elements in the signaling pathways of the inhibitory receptors. Experimental evidence suggests that defective regulation by B cell inhibitory receptors may be of importance in autoimmunity.

FcγRIIb

Three classes of FcγR have been described in humans, FcγRI, FcγRII and FcγRIII. FcγRII and III are further expressed in a and b forms. FcγRI, IIa and IIIa are activating receptors, whereas FcγRIIb is an inhibiting receptor. The function of FcγRIIIb, which lacks an intracellular domain, is unknown. Coordinate expression of FcγR has been implicated in various diseases involving immune complexes, such as diabetes, SLE, RA, multiple sclerosis and autoimmune anemia. FcγRIIb is a member of the Ig superfamily and represents a single-chain, low-affinity receptor for the Fc portion of IgG [131]. It is a 40 kDa protein that consists of two extracellular Ig-like domains, a transmembrane domain and an intracytoplasmic domain that contains a single ITIM. It binds IgG either complexed to multivalent soluble antigens as immune complexes or bound to cell membranes [132]. The isoform on B cells is unique in containing an intracytoplasmic motif that prevents its internalisation [133, 134].

In B cells, which do not express any other Fc receptors, it acts to inhibit signaling through the B cell receptor (BCR), whereas in myeloid cells FcγRIIb inhibits activation through activatory Fc receptors. It is cross-linked to the BCR by immune complexes containing IgG and antigen recognized by the BCR. Coligation of FcγRIIb to the BCR leads to tyrosine phosphorylation of the ITIM by the tyrosine kinase lyn, recruitment of SHIP and inhibition of Ca^{2+} flux and proliferation. The precise mechanism by which SHIP prevents B cell proliferation is uncertain [135]. FcγRIIb also induces apoptosis on aggregation of the receptor in the absence of BCR signaling. In this circumstance an apoptotic signal is generated through Btk and Jnk independent of the ITIM, which is abrogated when FcγRIIb is cross-linked with the BCR. Coligation of FcγRIIb is thought to provide feedback control of the B cell immune response, shutting off or preventing a response if sufficient antigen-specific IgG is present.

Evidence of a role for defective FcγRIIb inhibition in the pathogenesis of autoimmunity is provided by studies of FcγRII-deficient mice, mouse models of autoimmune disease and human SLE as well as RA. FcγRIIb deficiency renders normally resistant strains of mice susceptible to collagen-induced arthritis and Goodpasture's syndrome [136].

FcγRIIb$^{-/-}$ mice derived on a C57BL/6 but not Balb/c background produce autoantibodies and develop immune complex-mediated autoimmune disease resembling SLE [137], including an immune complex-mediated glomerulonephritis and renal failure.

Genetic studies of polygenic murine models of human autoimmune diseases implicate FcγRIIb in pathogenesis. Several independent linkage studies in murine models of SLE and RA have identified disease susceptibility loci that contain *fcgr2*. Notably, the region on chromosome 1 containing *fcgr2* also contains numerous other candidate genes, e.g. complement receptor 2 [138]. Genetic studies of the human autoimmune diseases SLE and insulin dependent diabetes mellitus (IDDM) have shown significant linkage to the region of chromosome 1 (1q23) containing the low-affinity Fc receptors (both activatory and inhibitory) [139–142].

Several studies have also found a correlation between specific polymorphisms in FcγRIIA, FcγRIIIA and FcγRIIIB and the development of several different autoimmune diseases [143, 144], although this has not been a consistent finding [145]. One of the studies above has implicated FcγRIIB directly, all three of the Fc receptor genes are clustered very tightly at 1q23 and are thus in linkage dysequilibrium with each other [144].

Genetic studies have linked polymorphisms in FcγRIIB to disease pathogenesis in humans. A recent study has identified a single nucleotide polymorphism in the *Fcgr2b* gene that results in an Ile232Thr substitution [146]. The 232T/T genotype was found at a significantly higher frequency in Japanese SLE patients compared with controls. The precise effect of this mutation is unknown, but it lies within the trans-membrane region of the molecule and it is known that an intact TM region is required for induction of apoptotic signals through FcγRIIB in the mouse. Thus, *Fcgr2b* is clearly a candidate gene for human autoimmune disease.

CD22

CD22 is a B cell-specific glycoprotein that first appears intracellularly during the late pro-B cell stage of ontogeny, with expression shifting to the plasma membrane with B cell maturation until plasma cell differentiation. CD22, with seven extracytoplasmic Ig-like domains, belongs to the Ig superfamily that serves as receptors for carbohydrate determinants on a wide variety of cell surface and soluble molecules in vivo. In contrast to CD19, CD22 can act as an antagonist to B cell activation, most likely by enhancing the threshold of BCR-induced signals. Following BCR engagement, CD22 is predominantly phosphorylated within immunoreceptor tyrosine-based inhibitory motifs (ITIM) present in its cytoplasmic domain. Phosphorylation is predominantly mediated by Lyn, downstream of the CD19-dependent Lyn kinase amplification loop. If phosphorylated by Lyn, CD22 recruits the SHP-1 and SHIP phosphatases, leading to activation of a CD22/SHP-1/SHIP regulatory pathway that down-regulates CD19 phosphorylation and BCR-mediated signal transduction. Thus, CD19 and CD22 function as general "rheostats" that define signaling thresholds critical for expansion of the peripheral B cell pool [147]. Ligation of CD22 to the BCR, and subsequent SHP-1 activation inhibits B cell activation by inhibiting the MAP kinases ERK2, JNK and p38 and dephosphorylating molecules involved in the early events of BCR mediated activation. These include the BCR itself, tyrosine kinases activated by phosphorylation of Igαβ (such as syk) and the targets of these kinases (including the adaptor protein BLNK and PLCγ). Coligation of CD22 to the BCR reduces B cell activation. Thus, the interaction of CD22 with its ligand may promote B cell activation [148]. Alternatively, increased levels of ligand on inflamed endothelium would recruit CD22 [149].

CD22-deficient mice have an expanded B1 cell population and increased serum IgM, and their B cells are hyper-responsive to stimulation through the BCR [150]. With age they develop high-affinity isotype-switched autoantibodies to dsDNA, myeloperoxidase and cardiolipin, although not overt autoimmune disease [151].

CD22 has been linked genetically to disease in both mice and humans. All these point to a possible role for defects in CD22 contributing to the development of autoimmunity.

That autoimmunity and immune responses are regulated or "fine-tuned" by a CD19-dependent signal transduction pathway explains why subtle increases in CD19 expression or decreased CD22 function leads to autoantibody production [147].

PD-1

The PD-1 receptor is a 55 kDa inhibitory receptor of the Ig superfamily that is highly conserved between humans and mice [148, 149]. It is expressed on resting B cells, T cells and macrophages and is induced strongly on activation [149]. It is composed of a single extracellular Ig-like domain, a transmembrane region and has two tyrosine residues in the cytoplasmic tail, one of which forms part of an ITIM. Two PD-1 ligands (PD-Ls) have been identified and are constitutively expressed on dendritic cells and on heart, lung, thymus and kidney and also on monocytes after IFN-γ stimulation [150, 151]. In vitro studies on a B cell lymphoma line using a chimæric molecule with the FcγRII extracellular domain and the PD-1 cytoplasmic domain have shown that ligation of the PD-1 cytoplasmic domain to the BCR can inhibit signaling through it. This inhibition prevented BCR-mediated proliferation, Ca^{2+} mobilization and tyrosine phosphorylation of molecules, including CD79beta, syk, PLCγ2 and ERK1/2. It is mediated by recruitment of SHP-2 to a non-ITIM cytoplasmic tyrosine residue [152]. The physiological role of PD-1 in B cells is unclear, but it may play a role in maintaining peripheral tolerance by limiting activation of autoreactive B cells by cross-linking PD-1 during interactions with PD-L expressing cells [153].

Linkage studies in human autoimmune diseases have identified susceptibility loci for both SLE [154] and IDDM [155] that lie close to the gene PD-1, but no direct evidence for abnormal PD-1 function in human disease has been identified. Nonetheless, PD-1 knockout mice develop autoantibodies and autoimmune disease.

13.4.1.4 Inhibitory Receptor Pathways and Autoimmunity

Inhibitory receptors are subserved by remarkably similar signaling pathways. To date, lyn is the only tyrosine kinase that has been identified as phosphorylating ITIMs on the B cell inhibitory receptors, and most of these ITIMs then associate with SHP-1 or SHIP (Figure 13.6).

SHP-1

SHP-1 is a protein tyrosine phosphatase and is similar in structure to SHP-2. SHP-1 is the phosphatase that is utilized most widely in the inhibitory receptor signaling pathways. SHP-1 plays the predominant role in regulating through

ITIMs, whereas increasing evidence suggests that SHP-2 may well have an additional activating role [156–158]. Clearly, these molecules have an important role in regulation of a normal immune system, which is related, at least in part, to their recruitment by inhibitory receptors.

Consistent with its role in mediating inhibitory receptor function, SHP-1 deficiency results in the development of spontaneous autoimmune disease. SHP-1 also associates with BCR, FcR, growth factor, complement and cytokine receptors [159]. The respective knockout phenotype is consistent with SHP-1 having a predominant role in the inhibitory receptor pathways [160, 161]. These mice have B cells that are hyper-responsive to BCR stimulation [162], raised levels of serum immunoglobulin [162], develop autoantibodies [160] and severe autoimmune disease with immune complex deposition in skin, lung and kidney [161], patchy alopecia [160], splenomegaly and inflamed paws.

There are no clear data that show linkage between SHP-1 and the development of autoimmune disease in humans. However, defects in SHP-1 expression have been associated with SLE in humans; reduced levels of SHP-1 (and lyn) are seen in the lymphocytes of patients with SLE during inactive phases of the disease [163], suggesting a potential role in pathogenesis.

SHIP

SHIP is an SH2-containing inositol phosphatase related to SHIP-2 [164], and they share a conserved N-terminal SH2 catalytic domain. SHIP acts predominantly on the FcγRIIb signaling pathway. The molecule is highly conserved and is expressed widely in myeloid and lymphoid lineages, including B cells [165].

The pattern of B cell abnormalities seen in the SHIP-deficient mouse is consistent with an inhibitory role in B cell signaling. Splenic B cells have an activated phenotype with lower surface levels of IgM and higher levels of IgD and are hyper-responsive to BCR-mediated stimulation measured by expression of the activation markers CD69 and CD86 [166].

Genetic studies in humans have identified susceptibility loci for both diabetes and SLE [167] mapping to the region of the genome containing SHIP, but no direct evidence exists for abnormal SHIP function in human disease.

Inhibitory receptors control the activation threshold of many immune cells, including B cells. There are many similarities in the signaling pathways of these inhibitory receptors. Consistent with this is the fact that B cells from inhibitory receptor-deficient mice have similarities in phenotype and lowered thresholds for activation as has been reported for SLE. Patients with SLE have hyperactive B cells, hypergammaglobulinaemia and develop autoantibodies. These are deposited as immune complexes, initiating an inflammatory reaction, resulting in end organ damage and the clinical features of the disease. Inhibitory receptors also have specific effects, as they bind different ligands, and signal through different phosphatases.

The phenotype of most SLE mouse models suggests impaired inhibitory receptor function, with hyperactive B cells and a similar pattern of autoantibody production and glomerular disease to that seen in inhibitory receptor knockout mice. Nonetheless, at almost every genetic susceptibility locus containing an inhibitory receptor implicated in SLE there are large numbers of genes encoding other immunologically relevant molecules that may also play a role in disease pathogenesis. The role of inhibitory receptors in spontaneous disease is, therefore, yet to be established firmly; nonetheless, the evidence favors contributions by defective inhibitory receptor function to the pathogenesis of B cell-mediated autoimmune diseases.

13.5
B Cell Abnormalities Leading to Rheumatoid Arthritis

In RA, considerable interest has focused on abnormalities in fibroblasts, macrophages, T cell responses or production of T cell cytokines and/or defective control by regulatory T cells in RA, which led to one of the early biological treatment of RA, i.e. the use of anti-CD4 antibodies [166]. However, recent evaluation of the role of B cells – the evolutionarily most recent immune cells – in the immune system has indicated that they are more than just the precursors of antibody-secreting cells (Table 13.1, Figure 13.7). B cells have more essential functions in regulating immune responses than had previously been appreciated. Therefore, it is possible that exaggeration of any of these B cell activities could contribute to the development of autoimmune disease. In this regard, B cells can function as antigen-presenting cells (Table 13.1), as has been shown in animal models [167–170] where mice that had B cells but lacked secreted Ig nevertheless developed autoimmunity [167]. Lymphotoxin produced by B cells is essential for the differentiation of follicular dendritic cells into secondary lymphoid organs and the organization of effective lymphoid architecture [171]. In addition, antigen-presenting M cells do not develop in the gastrointestinal mucosa in the absence of B cells [170]. Similarly, activated B cells express co-stimulatory molecules that can be essential for the evolution of T effector cells [171–173]. B cells themselves may differentiate into polarized cytokine-producing effector cells that can influence the differentiation of T effector cells [173]. Most recently, IL-10 positive B cells with immunoregulatory functions have been identified [173]. The cytokines produced by activated B cells may influence the function of antigen-presenting dendritic cells. Taken together, these results indicate that B cells play an essential role in lymphoid organogenesis as well as in the initiation and regulation of T and B cell responses. Therefore, these cells represent a central component of the immune system, but have only recently emerged as an important therapeutic target.

Table 13.1 Immune functions of B cells.

1. Precursors of (auto)antibody-secreting plasma cells
2. Essential functions of B cells in regulating immune responses:
   (i) Antigen-presenting cells
   (ii) Differentiation of follicular dendritic cells in secondary lymphoid organs
   (iii) Essential role in lymphoid organogenesis as well as in the initiation and regulation of T and B cell responses
   (iv) Development of effective lymphoid architecture (antigen-presenting M cells)
   (v) Activated B cells express co-stimulatory molecules and may differentiate into polarized cytokine-producing effector cells that can be essential for the evolution of T effector cells
   (vi) Differentiation of T effector cells
   (vii) Immunoregulatory functions by IL-10 positive B cells
   (viii) Cytokine production by activated B cells may influence the function of antigen-presenting dendritic cells

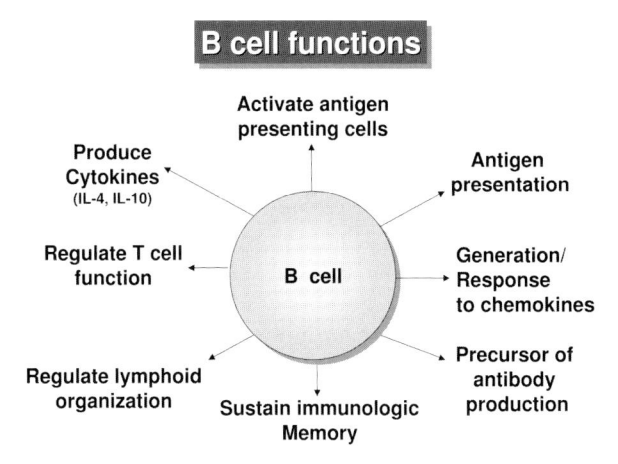

Figure 13.7 Summary of B cell functions under normal conditions.

13.5.1
Activated B Cells may Bridge the Innate and Adaptive Immune System

Conceptually, the relationship between the immunopathogenic role of the innate and adaptive immune system in RA has regained significant attention [174–177]. There is a general perception of three stages in the course of RA – disease initiation, perpetuation and a terminal destruction process [174–179]. However, the distinct role of antigen-specific lymphocytes remains a matter of debate since, i.e., neutrophils dominate synovial effusions. Recent concepts [179] repostulated that the initiation of RA may be antigen-independent by involving joint constituents. Secondly, the inflammatory phase appears to be driven by specific antigens – either foreign or native and either integral to the joint or presented in the periphery. The third stage – destruction of the synovium – seems to be again antigen-independent. Although it is unclear to what extent B cells are involved in certain stages of the disease, their role appears to be significant, either as a link to other immune cells potentially bridging innate and the adaptive immune system or as directing cellular components in inflammation. Nonetheless, B cells can be considered as an "enhancing element" of RA severity.

A recent model, the K/BxN mouse, has raised particular interest. In this model, spontaneous arthritis occurs in mice that express both the transgene encoded KRN T cell receptor and the IAg^7 MHC class II allele [178–182]. The transgenic T cells have a specificity for glucose-6-phosphate isomerase (G6PI) and can break tolerance in the B cell compartment, resulting in the production of autoantibodies to G6PI. Affinity-purified anti-G6PI Ig from these mice can transfer joint specific inflammation to healthy recipients [182]. A mechanism for joint-specific disease arising from autoimmunity to G6PI has been suggested recently. G6PI bound to the surface of cartilage serves as the target for anti-G6PI binding and subsequent complement-mediated damage. In this model, the inciting event is the expression of an autoreactive T cell receptor in the periphery. However, joint destruction is delegated by the adaptive response to innate immune mechanisms and can be transferred to animals that lack B and T cells [182, 183]. Whereas these animal studies are very compelling and intriguing, analyses of anti-G6PI antibodies in the serum of RA patients indicate that these autoantibodies apparently do not play a frequent or significant role [183, 184].

13.5.2
"Humoral Imprinting" in Rheumatoid Arthritis

Autoantibodies are the serologic hallmark of autoimmune diseases and serve as indicators of the break in self tolerance. In this context, rheumatoid factor (RF) is considered as a serologic marker in RA. Several other autoantibodies are also present in RA (reviewed in Ref. 185), whereas the role of precursor B cells in this disease is less clear. Nevertheless, patients with RA who produce RF and other autoantibodies have a more severe course of the disease.

Several new antigen systems, such as antibodies against citrullinated proteins [186], with a striking specificity for RA have been identified, demonstrating that such humoral immune disturbances are associated with a more aggressive form of RA and may precede the disease by several years [4, 5]. Baeten et al. [187] demonstrated recently that intracellular citrullinated proteins, as well as binding of an antifillagrin antibody, was confined to RA synovium, indicating a specificity and role of these autoantigenic structures in the disease. In general, demonstration of autoantibodies against several structures in RA indicates a more active disease and radiologic progression, respectively – with the major exception of antibodies against IL-1 [188]. The bulk of the available data indicate that RA patients producing autoantibodies and, therefore, having a disturbed humoral immune response suffer from a more severe type of the disease.

13.5.3
Indications of Enhanced B Cell Activity in RA

Indicators of enhanced B cell activation in RA include migration into the synovium via chemokine/chemokine receptors [189] with subsequent formation of T/B cell aggregates and the development of tertiary follicular structures [190], expression of co-stimulatory molecules, such as the ligand for CD40 (CD154) and ICOS/ICOS-L, enhanced production of several cytokines that further stimulate B cell function, including interleukin 6 (IL-6) and IL-10, markedly enhanced mutational activity in the synovium and abnormalities in positive and negative selection of B cells in RA.

B cells have been characterized in the inflamed synovium. One study identified CD20+/CD38– B cells from RA synovium with a dramatically reduced proliferative capacity [191, 192] when cultured with IL-2 and IL-10 in combination with CD40L-expressing fibroblasts. The authors suggest that these cells represent a subset of effector B cells. However, there is a need to demonstrate further their impact in the pathogenesis of RA, especially in man.

The group of Weyand and Goronzy has demonstrated the distinct role of B cells in the rheumatoid synovium [190, 193, 194]. In one study [193], $CD4^+$ T cells in synovial GCs were isolated by microdissection and the activation requirements for these follicle-derived $CD4^+$ T cells were analyzed in adoptive transfer experiments. Distinct GCs from the same patient contained identical $CD4^+$ T cell clones that, upon transfer into heterologous synovial tissues, were able to increase the production of proinflammatory mediators. Two factors were identified to be critical in developing arthritis, matching with the MHC class II polymorphism of the implanted synovium and the presence of B cells in the tissue. The data suggest a critical role of B cells in regulating the activation of tissue-invading $CD4^+$ T cells. Treatment of chimeric mice with anti-CD20 mAb inhibited the production of IFN-γ and IL-1β, indicating that APCs other than B cells could not substitute in maintaining T cell activation. Given the ability of B cells to capture antigen specifically with their receptors and present it to T cells, B cells may be uniquely situated to stimulate proinflammatory T cells in rheumatoid synovitis. It is widely accepted that a specific role of MHC class II molecules confers susceptibility to the disease through involvement in critical antigen presentation events, either during thymic selection or in the periphery.

Although identification of the central arthritogenic antigen in RA is still lacking, formation of T cell/B cell aggregates and the formation of GC-like structures can be correlated with increasing disease activity, including the production of autoantibodies, such as RF [195]. It needs to be shown whether there are apparent differences in the underlying pathologic process, with the disease activity likely escalating from (1) non-organized infiltrate to (2) T/B aggregates through (3) ectopic germinal center formation. Such distinctions might indicate differences in the course of the disease correlating with the degree of B cell activity and may provide the benefit of fine-tuning therapeutic strategies, such as the decision of using conventional DMARDs, TNF blockers vs. B cell depletion.

13.5.4
T Cell Independent B Cell Activation

Recent evidence indicates that B cells play an interactive role between the innate and adaptive immune system. In detail, Marshak-Rothstein et al. [195] have shown that effective activation of RF+ B cells can be mediated by IgG2a-chromatin immune complexes requiring synergistic engagement of the B cell receptor and a member of the MyD88-dependent toll like receptor (TLR) family, likely TLR9. Bacterial and vertebrate DNAs differ by the absence of CpG methylation in bacteria. The immune system uses TLR9 to detect the presence of unmethylated CpG dinucleotides as a signal of infection; these motifs are rare in vertebrate DNA. In humans, the expression of TLR9 appears to be relatively restricted to B cells and $CD123^+$ dendritic cells. Upon the detection of CpG motifs, B cells are induced to proliferate and secrete Ig, and DCs secrete a wide array of cytokines, interferons and chemokines that activate T_H1 cells. Bacterial DNA or CpG motifs co-stimulate B cell activation through cell membrane Ig, thereby promoting the development of antigen-specific responses. This study found that immune complexes containing self-DNA activate RF-specific B cells as a result of two distinct signals, (1) engagement of the B cell antigen receptor (BCR) and (2) activation of TLR9 through the histone/DNA portion of the immune complex. Although the implications of these findings for certain autoimmune diseases need to be delineated, the evidence that TLR9 activation co-stimulates autoreactive B cells provides a mechanism of action for an established therapy for systemic autoimmune diseases and suggests new therapeutic approaches. Decades ago, it was found that chloroquine is an effective therapy for systemic autoimmunity, but the mechanism of its activity was not identified. Chloroquine and other compounds that interfere with endosomal acidification and maturation specifically block all CpG-mediated signals [196]. The established efficacy of chloroquine and related compounds in treating autoimmune diseases could be related to a requirement for continuous co-stimulation of the BCR and TLR9 pathways in sustaining disease activity. Moreover, sulfasalazine also has effects on reducing B cell activity, further indicating that known DMARDs may have the capacity to influence the biologic activity of these cells [198]. Moreover, other TLRs can also be considered as candidates in RA pathogenesis, i.e. TLR 4 recognizing LPS, but need further investigation.

An important implication of this work is that beyond overall B cell depletion that decreases both T cell-dependent and -independent responses, there needs to be more precise targeting directed towards the category of B cell response to be specifically eliminated (Figure 13.8).

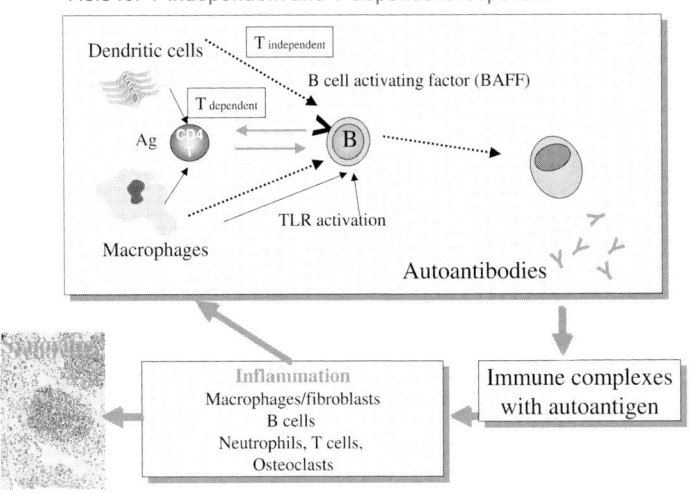

Figure 13.8 Central role of B cells in T cell dependent and T cell independent immune responses with an important involvement of BAFF.

13.6
Depleting anti-B Cell Therapy as a Novel Therapeutic Strategy

In general, therapy for severe autoimmune disease has primarily relied on broadly immunosuppressive agents such as cyclophosphamide, methotrexate, cyclosporine, leflunomide, mycophenolate mofetil and corticosteroids [197]. Although survival rates have improved dramatically, none of these therapies offers a cure and most have significant toxicity. With the advent of monoclonal antibody and specific small molecule based therapies, more specific and effective therapies are possible. Therapy directed at specifically reducing B cell numbers has recently gained attention and enthusiasm [198]. Based on the ability of the chimerized anti-CD20 monoclonal (rituximab) to reduce B cell numbers without significant toxicity, it is also being evaluated in human clinical trials for patients with autoimmune diseases [198, 199]. Rituximab functions by binding the membrane-embedded CD20 surface molecule on B cells, leading to B cell elimination by host immune effector mechanisms such as ADCC. Considering our current understanding of the role of B cells in the pathogenesis of autoimmune disease, the potential specificity of rituximab with minimal toxicity, and the encouraging preliminary results in human clinical trials, one can expect to see a significant expansion in the use of rituximab in human clinical trials with patients with various autoimmune disorders. Notably, however, most autoimmune patients that have benefited from rituximab therapy have not manifested remarkable decreases in measurable Ig levels, but have had some decreases in autoantibody titers, such as RF [199]. This suggests that the therapeutic effect does not simply rely on deleting precursors of ab producing cells, but likely by deleting antigen-presenting B cells.

Rituximab (RTX), initially developed as a therapy for B cell malignancies, is a humanized anti-CD20 monoclonal antibody (IgG1kappa) and depletes B cells by complement-dependent cytotoxicity and antibody-dependent cell-mediated cytotoxicity. Several studies have demonstrated the beneficial effect of anti-CD20 in RA, SLE, ITP, autoimmune hemolytic anemia, cold agglutinin disease, myasthenia gravis, and Wegener's granulomatosis as well as in a growing number of other diseases [200–210].

In RA, Edwards and Cambridge [207] initiated a first open trial of B cell depletion in 5 patients with refractory disease by using the humanized anti-CD20 monoclonal antibody in RA and could show improvement in 4/5 patients. Five patients with refractory RA were given a regimen of anti-CD20 with intravenous cyclophosphamide and high-dose corticosteroids. All of the patients achieved responses, meeting the American College of Rheumatology 70% improvement criteria (ACR70) after the first or second treatment cycle. The use of relatively high-level immunosuppression with cyclophosphamide and corticosteroids in the first trial made it difficult to discern the specific contribution of B cell depletion. Subsequently, this group and other investigators pioneered further studies on the effect of anti-CD20 therapy in RA.

In a more recent study, the same group [208] reported 22 patients with RA with five different combinations of RTX, cyclophosphamide and/or high dose prednisolone on an open trial basis. Although the study did not expand knowledge beyond the previous study and again did not allow a distinction of therapeutic effects by RTX vs. cyclophosphamide vs. prednisolone, the safety and major clinical improvements of this therapy were documented.

A further, independent study by De Vita and colleagues [209] reported results on anti-CD20 treatment of 5 patients with RA that had not been responsive to combination therapy with methotrexate and cyclosporine, with or without anti-TNF therapy. After a washout period of 1 month, these five patients received anti-CD20 therapy alone. Four of the 5 patients (80%) achieved ACR20 responses; in 1 of the 4 (20%), the response was at the ACR50 level, and 1 (20%) had an ACR70 response. At this point, the study by De Vita et al. [209] confirmed the previous experience and provided more direct evidence that B cell-ablative therapy may have a role in patients with RA.

Recently, Edwards and colleagues [210] reported data on the efficacy and safety of the first randomized placebo-controlled trial in 161 patients with RA. This study contained 4 treatment arms with MTX alone ($n = 40$), RTX alone ($n = 40$), RTX+cyclophosphamide ($n = 41$) and RTX+MTX ($n = 40$). Interim analysis at 24 weeks demonstrated a significant ACR20 (65–76% vs. 38%) and ACR50 (33–43% vs. 13%) response of RTX protocols versus MTX alone, whereas significant ACR70 responses were only evident in the RTX+MTX group (23%). At 48 weeks, RTX+MTX showed the highest ACR responses (ACR20 65%; ACR50 35%; ACR70 15%) which is remarkably since these patients continued taking MTX alone after they received only two infusions 17 days apart at entry of the study after they failed to respond to MTX. In this study, the monoclonal anti-CD20 antibody was reported to be well tolerated in all three treatment arms, although one patient died in the RTX+MTX group, reportedly not related to the study drug. So far no enhanced rates of infections have been reported in autoimmune patients under RTX treatment, in particular no tuberculosis in the limited number of patients to date. It is emphasized that the ACR20, 50 and 70 responses were apparently very similar to those reported in the infliximab and etanercept trials (reviewed in Ref. 211). A surprising result of the last study [211] is that a combination with MTX ($\leq 10\,mg\,wk^{-1}$) was superior to a combination with cyclophosphamide. These stimulating data in RA have generated an expanding interest to study the value of RTX in a number of other diseases, such as SLE, Wegener's granulomatosis, IGM associated polyneuropathy, multiple sclerosis etc. where a number of studies is underway.

Remarkably, patients who received RTX have B cell depletion for 6–14 months but do not develop a decrease in their Ig level, in contrast to a reduced RF level, and apparently do not develop infectious complications.

Although the role of B cells in RA has been discussed and several questions remain, there is now increasing evidence for a pathogenic role of B cells in RA. Based on these observations, B cell ablation appears to be a further candidate as another biologic therapy in RA.

Other potential targets for treating B cell-mediated human autoimmune diseases include BAFF antagonists and decoy receptors utilizing BMCA or TACI. Recent trials of CTLA-4-lg fusion proteins that disrupt T-B cell interactions [212] and T cell activation have been promising in the treatment of RA, showing moderate efficacy with no evidence of significant toxicity. Considering the important roles of CD22 and CD19 in the regulation of B cell function, it is conceivable that modulators or antagonists of CD22 and/or CD19 function will enter human clinical trials. The development of therapeutic MoAbs that block certain ligand engagement or intracellular pathways may have considerable benefit for the treatment of autoimmunity, without the risk of eliminating bulk B cell populations as with anti-CD20-directed therapies. Since CD19 deficiency ameliorates autoimmunity in mice, a further understanding of the molecular aspects of CD19/Src-family kinase interactions may lead to the identification of target molecules for therapeutic intervention during human autoimmunity.

Although CD40–CD40 ligand (CD40L) interactions are critical for normal B and T cell interactions and monoclonals against these therapeutic targets have demonstrated efficacy in animal experiments, human clinical trials targeting this pathway were disappointing, either because of a lack of efficacy or unexpected toxicity, especially thrombembolism [213, 214]. Rapidly advancing molecular understanding of regular and disturbed immune responses will provide abundant targets appropriate for drug development. It is expected and likely that many of these drugs will target B cell function directly since it is becoming more obvious that B cells contribute substantially to multiple human autoimmune diseases.

Acknowledgments

Supported by the Sonderforschungsbereiche 421 and 650, C7 and DFG grants 491/4-7, 5-3, 4.

References

1 Bigazzi PE und Reichlin M. Systemic autoimmunity. *Immunological Series*, Vol. 54, M. Dekker Inc., 1991, New York.

2 Tan EM. Antinuclear antibodies: diagnostic markers for autoimmune diseases and probes for cell biology. *Adv. Immunol.* (1989) 44: 93–151.

3 Reparon-Schuijt CC, van Esch WJ, van Kooten C, Ezendam NP, Levarht EW, Breedveld FC, Verweij CL. Presence of a population of CD20+, CD38- B lymphocytes with defective proliferative responsiveness in the synovial compartment of patients with rheumatoid arthritis. *Arthritis Rheum.* (2001) 44: 2029–2037.

4 van Gaalen FA, Linn-Rasker SP, van Venrooij WJ, de Jong BA, Breedveld FC, Verweij CL, Toes RE, Huizinga TW. Autoantibodies to cyclic citrullinated peptides predict progression to rheumatoid arthritis in patients with undifferentiated arthritis: a prospective cohort study. *Arthritis Rheum.* (2004) 50: 709–715.

5 Nielen MM, van Schaardenburg D, Reesink HW, van de Stadt RJ, van der Horst-Bruinsma IE, de Koning MH, Habibuw MR, Vandenbroucke JP, Dijkmans BA. Specific autoantibodies precede the symptoms of rheumatoid arthritis: a study of serial measurements in blood donors. *Arthritis Rheum.* (2004) 50: 380–386.

6 Edwards JC, Cambridge G. Sustained improvement in rheumatoid arthritis following a protocol designed to deplete B lymphocytes. *Rheumatology* (2001) 40: 205–211.

7 Leandro MJ, Edwards JC, Cambridge G. Clinical outcome in 22 patients with rheumatoid arthritis treated with B lymphocyte depletion. *Ann. Rheum. Dis.* (2002) 61: 883–888.

8 De Vita S, Zaja F, Sacco S, De Candia A, Fanin R, Ferraccioli G. Efficacy of selective B cell blockade in the treatment of rheumatoid arthritis: evidence for a pathogenetic role of B cells. *Arthritis Rheum.* (2002) 46: 2029–2033.

9 Hill JA, Southwood S, Sette A, Jevnikar AM, Bell DA, Cairns E. Cutting edge: the conversion of arginine to citrulline allows for a high-affinity peptide interaction with the rheumatoid arthritis-associated HLA-DRB1*0401 MHC class II molecule. *J. Immunol.* (2003) 171: 538–541.

10 Suzuki A, Yamada R, Chang X, Tokuhiro S, Sawada T, Suzuki M, Nagasaki M, Nakayama-Hamada M, Kawaida R, Ono M, Ohtsuki M, Furukawa H, Yoshino S, Yukioka M, Tohma S, Matsubara T, Wakitani S, Teshima R, Nishioka Y, Sekine A, Iida A, Takahashi A, Tsunoda T, Nakamura Y, Yamamoto K. Functional haplotypes of PADI4, encoding citrullinating enzyme peptidylarginine deiminase 4, are associated with rheumatoid arthritis. *Nat. Genet.* (2003) 34: 395–402.

11 Barton A, Bowes J, Eyre S, Spreckley K, Hinks A, John S, Worthington J. A functional haplotype of the PADI4 gene associated with rheumatoid arthritis in a Japanese population is not associated in a United Kingdom population. *Arthritis Rheum.* (2004) 50: 1117–1121.

12 Vossenaar ER, Nijenhuis S, Helsen MM, van der Heijden A, Senshu Z, van den Berg WB, van Venrooij WJ, Joosten LA. Citrullination of synovial proteins in murine models of rheumatoid arthritis. *Arthritis Rheum.* (2003) 48: 2489–2500.

13 Fischer A. Human primary immunodeficiency diseases: a perspective. *Nat. Immunol.* (2004) 5: 23–30.

14 Kabat EA, Wu TT, Perry HM, Gottesmann KS, Foeller C. *Sequences of Proteins of Immunological Interest* (National Institutes of Health, Bethesda, MD), 5th edition, 91–3242, NIH Publ., 1991.

15 Tonegawa S. Somatic generation of antibody diversity. *Nature* (1983) 302: 575–581.

16 Pospisil R, Young-Cooper GO, Mage RG. Preferential expansion and survival of B lymphocytes based on VH framework 1 and framework 3 expression: "Positive" selection in appendix of normal and VH-mutant rabbits. *Proc. Natl. Acad. Sci. U.S.A.* (1995) 92: 6961–6965.

17 Brezinschek HP, Foster SJ, Dörner T, Brezinschek RI, Lipsky PE: Pairing of variable heavy and variable kappa chains in individual naive and memory B cells. *J. Immunol.* (1998) 160: 4762–4767.

18 Rolink A, Melchers F: Molecular and cellular origins of B lymphocyte diversity. *Curr. Opin. Immunol.* (1993) 5: 207–217.

19 Rajewsky K. Burnet's unhappy hybrid. *Nature* (1998) 394: 624–625.

20 Itoh K, Meffre E, Albesiano E, Farber A, Dines D, Stein P, Asnis SE, Furie RA, Jain RI, Chiorazzi N. Immunoglobulin heavy chain variable region gene replacement as a mechanism for receptor revision in rheumatoid arthritis synovial tissue B lymphocytes. *J. Exp. Med.* (2000) 192: 1151–1164.

21 Brezinschek HP, Brezinschek RI, Lipsky PE: Analysis of the heavy chain repertoire of human peripheral blood B cells using single-cell polymerase chain reaction. *J. Immunol.* (1995) 155: 190–202.

22 Brezinschek HP, Foster SJ, Brezinschek RI, Dörner T, Domiati-Saad R, Lipsky PE: Analysis of the human VH gene repertoire. Differential effects of selection and somatic hypermutation on peripheral CD5+/IgM+ and CD5-/IgM+ B cells. *J. Clin. Invest.* (1997) 99: 2488–2501.

23 Farner NL, Dörner T, Lipsky PE: Molecular mechanisms and selection influence the generation of the human VlambdaJlambda repertoire. *J. Immunol.* (1999) 162: 2137–2145.

24 Foster SJ, Brezinschek HP, Brezinschek RI, Lipsky PE: Molecular mechanisms and selective influences that shape the kappa gene repertoire of IgM+ B cells. *J. Clin. Invest.* (1997) 99: 1614–1627.

25 Barbie V, Lefranc M-P. The human immunoglobulin kappa variable (IGKV) genes and joining (IGKJ) segments. *Exp. Clin. Immunogenet.* (1998) 15: 171–183.

26 Pallares N, Frippiat JP, Giudicelli V, Lefranc MP. The human immunoglobulin lambda variable (IGKL) genes and joining (IGLJ) segments. *Exp. Clin. Immunogenet.* (1998) 15: 8–18.

27 Pallares N, Lefebvre S, Contet V, Matsuda F, Lefranc MP. The human immunoglobulin heavy variable genes. *Exp. Clin. Immunogenet.* (1999) 16: 36–60.

28 Stewart AK, Huang C, Stollar B, Schwartz RS: High-frequency representation of a single VH gene in the expressed human B cell repertoire. *J. Exp. Med.* (1993) 177: 409.

29 Winkler TH, Fehr H, Kalden JR: Analysis of immunoglobulin variable region genes from human IgG anti-DNA hybridomas. *Eur. J. Immunol.* (1992) 22: 1719–1728.

30 Harada T, Suzuki N, Mizushima Y, Sakane T: Usage of a novel class of germline immunoglobulin variable region genes for cationic anti-DNA autoantibodies in human lupus nephritis and its role for the development of the disease. *J. Immunol.* (1994) 1153: 4806–4815.

31 Huang SC, Jiang R, Glas AM, Milner ECB. Nonstochastic utilization of Ig V region genes in unselected human peripheral B cells. *Mol. Immunol.* (1996) 33: 553–560.

32 Rao SP, Huang SC, Milner ECB. Analysis of the VH3 repertoire among genetically disparate individuals. *Exp. Clin. Immunogenet.* (1996) 13: 131–138.

33 Kraj P, Rao SP, Glas AM, Hardy RR, Milner EC, Silberstein LE. The human heavy chain Ig V region gene repertoire is biased at all stages of B cell ontogeny, including early pre-B cells. *J. Immunol.* (1997) 158: 5824–5832.

34 Schwartz RS, Stollar BD. Heavy-chain directed B-cell maturation: continous clonal selection beginning at the pre-B cell stage. *Immunol. Today* (1994) 15: 27–32.

35 Muramatsu M, Kinoshita K, Fagarasan S, Yamada S, Shinkai Y, Honjo T. Class switch recombination and hypermutation require activation-induced cytidine deaminase (AID), a potential RNA editing enzyme. *Cell* (2000) 102: 553–563.

36 van Es JH, Gmelig-Meyling FHJ, van de Akker WRM, Aanstoot H, Derksen RHWM, Logtenberg T. Somatic mutations in the variable regions of human IgG anti-double stranded DNA antibodies suggest a role for antigen in the induction of SLE. *J. Exp. Med.* (1991) 173: 461–470.

37 Shlomchik M, Mascelli M, Shan H, Radic MZ, Pisetsky D, Marshak-Rothstein A, Weigert M. Anti-DNA antibodies from autoimmune mice arise by clonal expansion and somatic mutation. *J. Exp. Med.* (1990) 171: 265–292.

38 Wloch MK, Alexander AL, Pippen AM, Pisetsky DS, Gilkeson GS. Molecular properties of anti-DNA induced in preautoimmune NZB/W mice by immunization with bacterial DNA. *J. Immunol.* (1997) 158: 4500–4506.

39 Manz RA, Thiel A, Radbruch A. Lifetime of plasma cells in the bone marrow. *Nature* (1997) 388: 133–134.

40 Dörner T, Brezinschek HP, Foster SJ, Brezinschek RI, Farner NL, Lipsky PE. Comparable impact of mutational and selective influences in shaping the expressed repertoire of peripheral IgM+/CD5- and IgM+/CD5+ B cells. *Eur. J. Immunol.* (1998) 28: 657–668.

41 Dörner T, Brezinschek HP, Brezinschek RI, Foster SJ, Domiati-Saad R, Lipsky PE. Analysis of the frequency and pattern of somatic mutations within nonproductively rearranged human V_H genes. *J. Immunol.* (1997) 158: 2779–2789.

42 Dörner T, Foster SJ, Farner NL, Lipsky PE. Immunoglobulin kappa chain receptor editing in systemic lupus erythematosus. *J. Clin. Invest.* (1998) 102: 688–694.

43 Dörner T, Farner NL, Lipsky PE. Immunoglobulin lambda and heavy chain gene usage in early untreated systemic lupus erythematosus suggests intensive B cell stimulation. *J. Immunol.* (1999) 163: 1027–1036.

44 Heimbächer C, Hansen A, Pruss A, Jacobi A, Reiter K, Lipsky PE, Dörner T. Immunoglobulin Vκ light chain analysis in patients with Sjögren's syndrome. *Arthritis Rheum.* (2001) 44: 626–637.

45 Odendahl M, Jacobi A, Hansen A, Feist E, Hiepe F, Burmester GR, Lipsky PE, Radbruch A, Dörner T. Disturbed peripheral B lymphocyte homeostasis. *J. Immunol.* (2000) 165: 5970–5979.

46 Kaschner S, Hansen A, Jacobi A, Reiter K, Monson NL, Odendahl M, Burmester GR, Lipsky PE, Dörner T. Immunoglobulin Vλ light chain gene usage in patients with Sjögren's syndrome. *Arthritis Rheum.* (2001) 44: 2620–2632.

47 Hansen A, Farner NL, Dörner T, Lipsky PE (2000). Use of immunoglobulin variable genes in normals and patients with systemic lupus erythematosus. *Int. Arch. Allergy Immunol.* 123: 36–45.

48 Kelsoe G. Life and death in germinal centers (Redux). *Immunity* (1996) 4: 107–111.

49 Han S, Dillon SR, Zheng B, Shimoda M, Schlissel MS, Kelsoe G. V(D)J recombinase activity in a subset of germinal center B lymphocytes. *Science* (1997) 278: 301–305.

50 Papavasiliou F, Casellas R, Suh H, Qin XF, Besmer E, Pelanda R, Nemazee D, Rajewsky K, Nussenzweig MC. V(D)J recombination in mature B cells: a mechanism for altering antibody responses. *Science* (1997) 278: 298–302.

51 Luning Prak E, Trounstine M, Huszar D, Weigert M. Light chain editing in -deficient animals: a potential mechanism of B cell tolerance. *J. Exp. Med.* (1994) 180: 1805–1815.

52 Luning Prak E, Weigert M. Light chain replacement: a new model for antibody gene rearrangement. *J. Exp. Med.* (1995) 182: 541–548.

53 Tiegs SL, Russell DM, Nemazee D. Receptor editing in self-reactive bone

marrow B cells. *J. Exp. Med.* (1993) 177: 1009–1020.
54 Nemazee D, Weigert M. Revising B cell receptors. *J. Exp. Med.* (2000) 191: 1813–7.
55 Girschick HJ, Grammer AC, Nanki T, Mayo M, Lipsky PE. RAG1 and RAG2 expression by B cell subsets from human tonsil and peripheral blood. *J. Immunol.* (2001) 166: 377–386.
56 Meffre E, Davis E, Schiff C, Cunningham-Rundles C, Ivashkiv LB, Staudt LM, Young JW, Nussenzweig MC. Circulating human B cells that express surrogate light chains and edited receptors. *Nat. Immunol.* (2000) 1: 207–213.
57 Lee J, Monson NL, Lipsky PE. The VλJλ repertoire in human fetal spleen: evidence for positive selection and extensive receptor editing. *J. Immunol.* (2000) 165: 6322–6333.
58 Radic MZ, Zouali M. Receptor editing, immune diversification and self-tolerance. *Immunity* (1996) 5: 505–511.
59 Suzuki N, Harada T, Mihara S, Sakane T. Characterization of a germline encoding cationic anti-DNA antibody and role of receptor editing for development of the autoantibody in patients with systemic lupus erythematosus. *J. Clin. Invest.* (1996) 98: 1843–1850.
60 Suzuki N, Mihara S, Sakane T. Development of pathogenic anti-DNA antibodies in patients with systemic lupus erythematosus. *FASEB J.* (1997) 11: 1033–1038.
61 Bensimon C, Chastagner P, Zouali M. Human lupus anti-DNA autoantibodies undergo essentially primary V kappa gene rearrangements. *EMBO J.* (1994) 13: 2951–2962.
62 Chen C, Luning-Prak E, Weigert M. Editing disease-associated autoantibodies. *Immunity* (1997) 6: 97–105.
63 Gay D, Saunders T, Camper S, Weigert M. Receptor editing: an approach by autoreactive B cells to escape tolerance. *J. Exp. Med.* (1993) 177: 999–1008.
64 Manheimer-Lory AJ, Irignoyen M, Gaynor B, Monhian R, Splaver A, Diamond B. Analysis of V kappa I and V lambda II light chain genes in the expressed B cell repertoire. *Ann. New York Acad. Sci.* (1995) 764: 301–311.
65 Manheimer-Lory AJ, Monhian R, Splaver A, Gaynor B, Diamond B. Analysis of the V kappa I family: germline genes from an SLE patient and expressed autoantibodies. *Autoimmunity* (1995) 20: 259–265.
66 Manheimer-Lory AJ, Zandman-Goddard G, Davidson A, Aranow C, Diamond B. Lupus-specific antibodies reveal an altered pattern of somatic mutation. *J. Clin. Invest.* (1997) 100: 2538–2546.
67 de Wildt RM, Hoet RM, van Venrooij WJ, Tomlinson IM, Winter G. Analysis of heavy and light chain pairings indicates that receptor editing shapes the human antibody repertoire. *Mol. Biol.* (1999) 285: 895–901.
68 Itoh K, Meffre E, Albesiano E, Farber A, Dines D, Stein P, Asnis SE, Furie RA, Jain RI, Chiorazzi N. Immunoglobulin heavy chain variable region gene replacement As a mechanism for receptor revision in rheumatoid arthritis synovial tissue B lymphocytes. *J. Exp. Med.* (2000) 192: 1151–1164.
69 Elagib KE, Borretzen M, Thompson KM, Natvig JB. Light chain variable (VL) sequences of rheumatoid factors (RF) in patients with primary Sjogren's syndrome (pSS): moderate contribution of somatic hypermutation. *Scand. J. Immunol.* (1999) 50: 492–498.
70 Dörner T, Heimbächer C, Farner NL, Lipsky PE. Enhanced mutational activity of Vκ gene rearrangements in systemic lupus erythematosus. Clin. Immunol. (1999) 92: 188–196.
71 de Wildt RM, Tomlinson IM, van Venrooij WJ, Winter G, Hoet RM. Comparable heavy and light chain pairings in normal and systemic lupus erythematosus IgG(+) B cells. *Eur. J. Immunol.* (2000) 30: 254–261.
72 Pascual V, Victor K, Lelsz D, Spellerberg MB, Hamblin TJ, Thompson KM, Randen I, Natvig J, Capra JD, Stevenson FK. Nucleotide sequence analysis of the V regions of two IgM cold agglutinins: evidence that the VH4-21 gene segment is responsible for the major cross reactive

idiotype. *J. Immunol.* (1991) 146: 4385–4391.

73 Silberstein LE, Jefferies LC, Goldman J, Friedman D, Moore JS, Nowell PC, Roelcke D, Pruzanski W, Roudier J, Silverman GJ. Variable region gene analysis of pathologic human autoantibodies to the related i and I red blood cell antigens. *Blood* (1991) 78: 2372–2386.

74 Ruzickova S, Pruss A, Odendahl M, Wolbart K, Burmester GR, Scholze J, Dörner T, Hansen A. Chronic lymphocytic leukemia preceded by cold agglutinin disease: intraclonal immunoglobulin light-chain diversity in V(H)4-34 expressing single leukemic B cells. *Blood* (2002) 100: 3419–3422.

75 Kraj P, Friedman DF, Stevenson F, Silberstein LE. Evidence for the overexpression of the VH4-34 (VH4.21) Ig gene segment in the normal adult human peripheral blood B cell repertoire. *J. Immunol.* (1995) 154: 6406–6420.

76 Stevenson FK, Longhurst C, Chapman CJ, Ehrenstein M, Spellerberg MB, Hamblin TJ, Ravirajan CT, Latchman D, Isenberg D. Utilization of the VH4-21 gene segment by anti-DNA antibodies from patients with systemic lupus erythematosus. *J. Autoimmun.* (1993) 6: 809–825.

77 Borretzen M, Chapman C, Stevenson FK, Natvig JB, Thompson KM. Structural analysis of VH4-21 encoded human IgM allo- and autoantibodies against red blood cells. *Scand. J. Immunol.* (1995) 42: 90–97.

78 Grammer A, Dörner T, Lipsky PE. Immunglobulin variable gene usage in systemic autoimmune diseases. In Theofilopoulos A. and Fathman G. (Eds.) *Current Directions in Autoimmunity*. Karger Publishers, 2001, Basel, Switzerland.

79 Rahman A, Latchman DS, Isenberg DA. Immunoglobulin variable region sequences of human monoclonal anti-DNA antibodies. *Semin. Arthritis Rheum.* (1998) 28: 141–154.

80 Cauerhff A, Braden BC, Carvalho JG, Aparicio R, Polikarpov I, Leoni J, Goldbaum FA. Three-dimensional structure of the Fab from a human IgM cold agglutinin. *J. Immunol.* (2000) 165: 6422–6428.

81 Putterman C, Limpanasithkul W, Edelman M, Diamond B. The double edge sword of the immune response: mutational analysis of a murine anti-pneumococcal anti-DNA antibody. *J. Clin. Invest.* (1996) 97: 2251–2259.

82 Ray SK, Putterman C, Diamond B. Pathogenic autoantibodies are routinely generated during the response to foreign antigen: a paradigm for autoimmune disease. *Proc. Natl. Acad. Sci. U.S.A.* (1996) 93: 2019–2024.

83 Ehrenstein MR, Hartley B, Wilkinson LS, Isenberg DA. Comparison of a monoclonal and polyclonal anti-idiotype against a human IgG anti-DNA antibody. *J. Autoimmun.* (1994) 7: 349–367.

84 Stewart AK, Huang C, Long AA, Stollar BD, Schwartz RS. VH-gene representation in autoantibodies reflects the normal human B-cell repertoire. *Immunol. Rev.* (1992) 128: 101–122.

85 Diamond B, Katz JB, Paul E, Aranow C, Lustgarten D, Scharff MD. The role of somatic mutation in the pathogenic anti-DNA response. *Annu. Rev. Immunol.* (1992) 10: 731–757.

86 Shlomchik M, Mascelli M, Shan H, Radic MZ, Pisetsky D, Marshak-Rothstein A, Weigert M. Anti-DNA antibodies from autoimmune mice arise by clonal expansion and somatic mutation. *J. Exp. Med.* (1990) 171: 265–292.

87 Chang B, Casali P. The CDR1 sequences of a major proportion of human germline Ig VH genes are inherently susceptible to amino acid replacement. *Immunol. Today* (1994) 15: 367–373.

88 Wallace DJ, Linker-Israeli M. It's not the same old lupus or Sjogren's any more: one hundred new insights, approaches, and options since 1990. *Curr. Opin. Rheumatol.* (1999) 11: 321–329.

89 Gellrich S, Rutz S, Borkowski A, Golembowski S, Gromnica-Ihle E, Sterry W, Jahn S. Analysis of V(H)-D-J(H) gene transcripts in B cells infiltrating the salivary glands and lymph node tissues of patients with Sjögren's syn-

drome. *Arthritis Rheum.* (1999) 42: 240–247.

90 Miklos JA, Swerdlow SH, Bahler DW. Salivary gland mucosa-associated lymphoid tissue lymphoma immunoglobulin VH genes show frequent use of V1-69 with distinctive CDR3 features. *Blood* (1995) 2000: 3878–3884.

91 Hutloff A, Büchner K, Reiter K, Odendahl M, Baelde HJ, Jacobi A, Dörner T, Kroczek RA. Involvement of ICOS in the exaggerated memory B-cell and plasma cell generation in systemic lupus erythematosus. *Arthritis Rheum.* (2004) 50: 3211–3220.

92 Jacobi A, Hansen A, Burmester GR, Dörner T, Lipsky PE (2001). Enhanced mutational activity and disturbed selection of mutations in V_H gene rearrangements in systemic lupus erythematosus. *Autoimmunity* (2001) 33: 61–76.

93 Hansen A, Odendahl M, Reiter K, Jacobi AM, Feist E, Scholze J, Burmester GR, Lipsky PE, Dörner T. Diminished peripheral blood memory B cells and accumulation of memory B cells in the salivary glands of patients with Sjogren's syndrome. *Arthritis Rheum.* (2002) 46: 2160–2171.

94 Grammer AC, Slota R, Fischer R, Gur H, Girschick H, Yarboro C, Illei GG, Lipsky PE. Abnormal germinal center reactions in systemic lupus erythematosus demonstrated by blockade of CD154-CD40 interactions. *J. Clin. Invest.* (2003) 112: 1506–1520.

95 Iwai H, Abe M, Hirose S, Tsushima F, Tezuka K, Akiba H, Yagita H, Okumura K, Kohsaka H, Miyasaka N, Azuma M. Involvement of inducible costimulator-B7 homologous protein costimulatory pathway in murine lupus nephritis. *J. Immunol.* (2003) 171: 2848–2854.

96 Jacobi AM, Odendahl M, Reiter K, Bruns A, Burmester GR, Radbruch A, Valet G, Lipsky PE, Dörner T. Correlation between circulating $CD27^{high}$ plasma cells and disease activity in patients with systemic lupus erythematosus. *Arthritis Rheum.* (2003) 48: 1332–1342.

97 Kim HJ, Krenn V, Steinhauser G, Berek C. Plasma cell development in synovial germinal centers in patients with rheumatoid and reactive arthritis. *J. Immunol.* (1999) 162: 3053–3062.

98 Voswinkel J, Pfreundschuh M, Gause A. Evidence for a selected humoral immune response encoded by VH4 family genes in the synovial membrane of a patient with RA. *Ann. New York Acad. Sci.* (1997) 815: 312–315.

99 Mockridge CI, Chapman CJ, Spellerberg MB, Sheth B, Fleming TP, Isenberg DA, Stevenson FK. Sequence analysis of V(4-34)-encoded antibodies from single B cells of two patients with systemic lupus erythematosus (SLE). *Clin. Exp. Immunol.* (1998) 114: 129–136.

100 van Vollenhoven RF, Bieber MM, Powell MJ, Gupta PK, Bhat NM, Richards KL, Albano SA, Teng NN. VH4-34 encoded antibodies in systemic lupus erythematosus: a specific diagnostic marker that correlates with clinical disease characteristics. *J. Rheumatol.* (1999) 26: 1727–1733.

101 Isenberg DA, McClure C, Farewell V, Spellerberg M, Williams W, Cambridge G, Stevenson F. Correlation of 9G4 idiotope with disease activity in patients with systemic lupus erythematosus. *Ann. Rheum. Dis.* (1998) 57: 566–570.

102 Yavuz AS, Monson NL, Yavuz S, Grammer AC, Longo N, Girschick HJ, Lipsky PE. Different patterns of bcl-6 and p53 gene mutations in tonsillar B cells indicate separate mutational mechanisms. *Mol. Immunol.* (2002) 39: 485–493.

103 Potter KN, Mockridge CI, Rahman A, Buchan S, Hamblin T, Davidson B, Isenberg DA, Stevenson FK. Disturbances in peripheral blood B cell subpopulations in autoimmune patients. *Lupus* (2002) 11: 872–827.

104 Bohnhorst JO, Bjorgan MB, Thoen JE, Natvig JB, Thompson KM. Bm1-Bm5 classification of peripheral blood B cells reveals circulating germinal center founder cells in healthy individuals and disturbance in the B cell subpopulations in patients with primary Sjogren's syndrome. *J. Immunol.* (2001) 167: 3610–3618.

105 Bohnhorst JO, Thoen JE, Natvig JB, Thompson KM. Significantly depressed

percentage of CD27+ (memory) B cells among peripheral blood B cells in patients with primary Sjogren's syndrome. *Scand. J. Immunol.* (2001) 54: 421–427.

106 Lindenau S, Scholze S, Odendahl M, Dörner T, Radbruch A, Burmester GR, Berek C. Aberrant activation of B cells in patients with rheumatoid arthritis. *Ann. New York Acad. Sci.* (2003) 987: 246–248.

107 Strasser A, Whittingham S, Vaux DL, Bath ML, Adams JM, Cory S, Harris AW. Enforced BCL2 expression in B-lymphoid cells prolongs antibody responses and elicits autoimmune disease. *Proc. Natl. Acad. Sci. U.S.A.* (1991) 88: 8661–8665.

108 Mackay F, Browning JL. BAFF: a fundamental survival factor for B cells. *Nat. Rev. Immunol.* (2002) 2: 465–475.

109 Mackay F, Mackay CR. The role of BAFF in B-cell maturation, T-cell activation and autoimmunity. *Trends Immunol.* (2002) 23: 113–115.

110 (a) Schiemann B, Gommerman JL, Vora K, Cachero TG, Shulga-Morskaya S, Dobles M, Frew E, Scott ML. An essential role for BAFF in the normal development of B cells through a BCMA-independent pathway. *Science* (2001) 293: 2111–2114. (b) Gross JA, Johnston J, Mudri S, Enselman R, Dillon SR, Madden K, Xu W, Parrish-Novak J, Foster D, Lofton-Day C, Moore M, Littau A, Grossman A, Haugen H, Foley K, Blumberg H, Harrison K, Kindsvogel W, Clegg CH. TACI and BCMA are receptors for a TNF homologue implicated in B-cell autoimmune disease. *Nature* (2000) 404: 995–999.

111 Thompson JS, Bixler SA, Qian F, Vora K, Scott ML, Cachero TG, Hession C, Schneider P, Sizing ID, Mullen C, Strauch K, Zafari M, Benjamin CD, Tschopp J, Browning JL, Ambrose C. BAFF-R, a newly identified TNF receptor that specifically interacts with BAFF. *Science* (2001) 293: 2108–2111.

112 Xu S, Lam KP. B-cell maturation protein, which binds the tumor necrosis factor family members BAFF and APRIL, is dispensable for humoral immune responses. *Mol. Cell Biol.* (2001) 21: 4067–4074.

113 Yan M, Wang H, Chan B, Roose-Girma M, Erickson S, Baker T, Tumas D, Grewal IS, Dixit VM. Activation and accumulation of B cells in TACI-deficient mice. *Nat. Immunol.* (2001) 2: 638–643.

114 von Bulow GU, van Deursen JM, Bram RJ. Regulation of the T-independent humoral response by TACI. *Immunity* (2001) 14: 573–582.

115 Carsetti R, Kohler G, Lamers MC. Transitional B cells are the target of negative selection in the B cell compartment. *J. Exp. Med.* (1995) 181: 2129–2140.

116 O'Connor BP, Raman VS, Erickson LD, Cook WJ, Weaver LK, Ahonen C, Lin LL, Mantchev GT, Bram RJ, Noelle RJ. BCMA is essential for the survival of long-lived bone marrow plasma cells. *J. Exp. Med.* (2004) 199: 91–98.

117 Schiemann B, Gommerman JL, Vora K, Cachero TG, Shulga-Morskaya S, Dobles M, Frew E, Scott ML. An essential role for BAFF in the normal development of B cells through a BCMA-independent pathway. *Science* (2001) 293: 2111–2114.

118 Benschop RJ, Cambier JC. B cell development: signal transduction by antigen receptors and their surrogates. *Curr. Opin. Immunol.* (1999) 11: 143–151.

119 Tedder TF, Tuscano J, Sato S, Kehrl JH. CD22, a B lymphocyte-specific adhesion molecule that regulates antigen receptor signaling. *Annu. Rev. Immunol.* (1997) 15: 481–504.

120 Cornall RJ, Cyster JG, Hibbs ML, Dunn AR, Otipoby KL, Dark EA, Goodnow CC. Polygenic autoimmune traits; Lyn, CD22, and SHP-1 are limiting elements of a biochemical pathway regulating BCR signaling and selection. *Immunity* (1998) 8: 497–508.

121 Tedder TF. Response-regulators of B lymphocyte signaling thresholds provide a context for antigen receptor signal transduction. *Semin. Immunol.* (1998) 10: 259–265.

122 Inaoki M, Sato S, Weintraub BC, Goodnow CC, Tedder TF. CD19-regulated signaling thresholds control peripheral tolerance and autoantibody production

in B lymphocytes. *J. Exp. Med.* (1997) 186: 1923–1931.
123 Chan OT, Madaio MP, Shiomchik MJ. The central and multiple roles of B cells in lupus pathogenesis. *Immunol. Rev.* (1999) 169: 107–121.
124 Saito E, Fujimoto M, Hasegawa M, Komura K, Hamaguchi Y, Kaburagi Y, Nagaoka T, Takehara K, Tedder TF, Sato S. CD19-dependent B lymphocyte signaling thresholds influence skin fibrosis and autoimmunity in the tight-skin mouse. *J. Clin. Invest.* (2002) 109: 1453–1462.
125 Prodeus AP, Goerg S, Shen LM, Pozdnyakova OO, Chu L, Alicot EM, Goodnow CC, Carroll MC. A critical role for complement in maintenance of self-tolerance. *Immunity* (1998) 9: 721–731.
126 Doody GM, Justement LB, Delibrias CC, Matthews RJ, Lin J, Thomas ML, Fearon DT. A role in B cell activation for CD22 and the protein tyrosine phosphatase SHP. *Science* (1995) 269: 242–244.
127 Smith KGC, Tarlinton DM, Doody GM, Hibbs ML, Fearon DT. Inhibition of the B cell by CD22: a requirement for Lyn. *J. Exp. Med.* (1998) 187: 807–811.
128 Tamir I, Dal Porto JM, Cambier JC. Cytoplasmic protein tyrosine phosphatases SHP-1 and SHP-2: regulators of B cell signal transduction. *Curr. Opin. Immunol.* (2000) 12: 307–315.
129 Daeron M. Fc receptor biology. *Annu. Rev. Immunol.* (1997) 15: 203–234.
130 Malbec O, Fridman WH, Daeron M. Negative regulation of hematopoietic cell activation and proliferation by Fc gamma RIIB. *Curr. Top. Microbiol. Immunol.* (1999) 244: 13–27.
131 Brooks DG, Qiu WQ, Luster AD, Ravetch JV. Structure and expression of human IgG FcRII (CD32). Functional heterogeneity is encoded by the alternatively spliced products of multiple genes. *J. Exp. Med.* (1989) 170: 1369–1385.
132 Latour S, Fridman WH, Daeron M. Identification, molecular cloning, biologic properties, and tissue distribution of a novel isoform of murine low-affinity IgG receptor homologous to human Fc gamma RIIB1. *J. Immunol.* (1996) 157: 189–197.
133 Pritchard NR, Smith KG. B cell inhibitory receptors and autoimmunity. *Immunology* (2003) 108: 263–273.
134 Nakamura A, Yuasa T, Ujike A, Ono M, Nukiwa T, Ravetch JV, Takai T. Fcgamma receptor IIB-deficient mice develop Goodpasture's syndrome upon immunization with type IV collagen: a novel murine model for autoimmune glomerular basement membrane disease. *J. Exp. Med.* (2000) 191: 899–906.
135 Bolland S, Ravetch JV. Spontaneous autoimmune disease in Fc (gamma) RIIB-deficient mice results from strain-specific epistasis. *Immunity* (2000) 13: 277–285.
136 Boackle SA, Holers VM, Chen X, Szakonyi G, Karp DR, Wakeland EK, Morel L. Cr2, a candidate gene in the murine Sle1c lupus susceptibility locus, encodes a dysfunctional protein. *Immunity* (2001) 15: 775–785.
137 Hashimoto L, Habita C, Beressi JP, Delepine M, Besse C, Cambon-Thomsen A, Deschamps I, Rotter JI, Djoulah S, James MR, et al. Genetic mapping of a susceptibility locus for insulin-dependent diabetes mellitus on chromosome 11q. *Nature* (1994) 371: 161–164.
138 Shai R, Quismorio FP Jr, Li L, Kwon OJ, Morrison J, Wallace DJ, Neuwelt CM, Brautbar C, Gauderman WJ, Jacob CO. Genome-wide screen for systemic lupus erythematosus susceptibility genes in multiplex families. *Hum. Mol. Genet.* (1999) 8: 639–644.
139 Gaffney PM, Kearns GM, Shark KB, Ortmann WA, Selby SA, Malmgren ML, Rohlf KE, Ockenden TC, Messner RP, King RA, Rich SS, Behrens TW. A genome-wide search for susceptibility genes in human systemic lupus erythematosus sib-pair families. *Proc. Natl. Acad. Sci. U.S.A.* (1998) 95: 14875–14879.
140 Gaffney PM, Ortmann WA, Selby SA, Shark KB, Ockenden TC, Rohlf KE, Walgrave NL, Boyum WP, Malmgren ML, Miller ME, Kearns GM, Messner RP, King RA, Rich SS, Behrens TW.

Genome screening in human systemic lupus erythematosus results from a second Minnesota cohort and combined analyses of 187 sib-pair families. *Am. J. Hum. Genet.* (2000) 66: 547–556.

141 Harley JB, Moser KL, Gaffney PM, Behrens TW. The genetics of human systemic lupus erythematosus. *Curr. Opin. Immunol.* (1998) 10: 690–696.

142 Wakeland EK, Liu K, Graham RR, Behrens TW. Delineating the genetic basis of systemic lupus erythematosus. *Immunity* (2001) 15: 397–408.

143 Botto M, Theodoridis E, Thompson EM, Beynon HL, Briggs D, Isenberg DA, Walport MJ, Davies KA. Fc gamma RIIa polymorphism in systemic lupus erythematosus (SLE): no association with disease. *Clin. Exp. Immunol.* (1996) 104: 264–268.

144 Kyogoku C, Dijstelbloem H, Tsuchiya N, Hatta Y, Kato H, Yamaguchi A, Fukazawa T, Jansen MD, Hashimoto H, van de Winkel JG, Kallenberg CG, Tokunaga K. Fcγ receptor gene polymorphisms in Japanese patients with systemic lupus erythematosus. *Arthritis Rheum.* (2002) 46: 1242–1254.

145 Fujimoto M, Bradney AP, Poe JC, Steeber DA, Tedder TF. Modulation of B lymphocyte antigen receptor signal transduction by a CD19yCD22 regulatory loop. *Immunity* (1999) 11: 191–200.

146 Smith KGC, Fearon DT. Receptor modulators of BCR signalling –CD19/22. *Curr. Top. Microbiol. Immunol.* (1999) 245: 195–212.

147 Neuberger MS, Lanoue A, Ehrenstein MR, Batista FD, Sale JE, Williams GT. Antibody diversification and selection in the mature B-cell compartment. *Cold Spring Harb. Symp. Quant. Biol.* (1999) 64: 211–216.

148 O'Keefe TL, Williams GT, Davies SL, Neuberger MS. Hyperresponsive B cells in CD22-deficient mice. *Science* (1996) 274: 798–801.

149 O'Keefe TL, Williams GT, Batista FD, Neuberger MS. Deficiency in CD22, a B cell-specific inhibitory receptor, is sufficient to predispose to development of high affinity autoantibodies. *J. Exp. Med.* (1999) 189: 1307–1313.

150 Okazaki T, Maeda A, Nishimura H, Kurosaki T, Honjo T. PD-1 immunoreceptor inhibits B cell receptor-mediated signalling by recruiting src homology 2-domain-containing tyrosine phosphatase 2 to phosphotyrosine. *Proc. Natl. Acad. Sci. U.S.A.* (2001) 98: 13 866–13 867.

151 Nishimura H, Honjo T. PD-1: an inhibitory immunoreceptor involved in peripheral tolerance. *Trends Immunol.* (2001) 22: 265–268.

152 Moser KL, Neas BR, Salmon JE, Yu H, Gray-McGuire C, Asundi N, Bruner GR, Fox J, Kelly J, Henshall S, Bacino D, Dietz M, Hogue R, Koelsch G, Nightingale L, Shaver T, Abdou NI, Albert DA, Carson C, Petri M, Treadwell EL, James JA, Harley JB. Genome scan of human systemic lupus erythematosus: evidence for linkage on chromosome 1q in African-American pedigrees. *Proc. Natl. Acad. Sci. U.S.A.* (1998) 95: 14 869–14 874.

153 Davies JL, Kawaguchi Y, Bennett ST, Copeman JB, Cordell HJ, Pritchard LE, Reed PW, Gough SC, Jenkins SC, Palmer SM, et al. A genome-wide search for human type 1 diabetes susceptibility genes. *Nature* (1994) 371: 130–136.

154 Ishida Y, Agata Y, Shibahara K, Honjo T. Induced expression of PD-1, a novel member of the immunoglobulin gene superfamily, upon programmed cell death. *EMBO J.* (1992) 11: 3887–3895.

155 Agata Y, Kawasaki A, Nishimura H, Ishida Y, Tsubata T, Yagita H, Honjo T. Expression of the PD-1 antigen on the surface of stimulated mouse T and B lymphocytes. *Int. Immunol.* (1996) 8: 765–772.

156 Huyer G, Alexander DR. Immune signalling: SHP-2 docks at multiple ports. *Curr. Biol.* (1999) 9: R129–R132.

157 Bolland S, Ravetch JV. Inhibitory pathways triggered by ITIM-containing receptors. *Adv. Immunol.* (1999) 72: 149–177.

158 Green MC, Shultz LD. Motheaten, an immunodeficient mutant of the mouse. I. Genetics and pathology. *J. Hered.* (1975) 66: 250–258.

159 Shultz LD, Coman DR, Bailey CL, Beamer WG, Sidman CL. 'Viable motheaten'; a new allele at the motheaten locus. I. Pathology. *Am. J. Pathol.* (1984) 116: 179–192.

160 Sidman CL, Shultz LD, Hardy RR, Hayakawa K, Herzenberg LA. Production of immunoglobulin isotypes by Ly-1$^+$ B cells in viable motheaten and normal mice. *Science* (1986) 232: 1423–1425.

161 Yu CC, Tsui HW, Ngan BY, Shulman MJ, Wu GE, Tsui FW. B and T cells are not required for the viable motheaten phenotype. *J. Exp. Med.* (1996) 183: 371–380.

162 Lioubin MN, Algate PA, Tsai S, Carlberg K, Aebersold A, Rohrschneider LR. p150Ship, a signal transduction molecule with inositol polyphosphate-5-phosphatase activity. *Genes Dev.* (1996) 10: 1084–1095.

163 Huber M, Helgason C, Damen J, Scheid M, Duronio V, Liu L, Ware MD, Humphries RK, Krystal G. The role of SHIP in growth factor induced signalling. *Prog. Biophys. Mol. Biol.* (1999) 71: 423–424.

164 Helgason CD, Damen JE, Rosten P, Grewal R, Sorensen P, Chappel SM, Borowski A, Jirik F, Krystal G, Humphries RK. Targeted disruption of SHIP leads to hemopoietic perturbations, lung pathology, and a shortened life span. *Genes Dev.* (1998) 12: 1610–1620.

165 Luo D, Maclaren N, Huang H, Muir A, She J. Intrafamilial and case-control association analysis of D2S152 in insulin-dependent diabetes. *Autoimmunity* (1995) 2: 143–147.

166 Horneff G, Burmester GR, Emmrich F, Kalden JR. Treatment of rheumatoid arthritis with an anti-CD4 monoclonal antibody. *Arthritis Rheum.* (1991) 34: 129–140.

167 Chan OT, Hannum LG, Haberman AM, Madaio MP, Shlomchik MJ. A novel mouse with B cells but lacking serum antibody reveals an antibody-independent role for B cells in murine lupus. *J. Exp. Med.* (1999) 189: 1639–1648.

168 Shlomchik MJ, Craft JE, Mamula MJ. From T to B and back again: positive feedback in systemic autoimmune disease. *Nat. Rev. Immunol.* (2001) 1: 147–153.

169 Gonzalez, M., Mackay, F., Browning, J. L., Kosco-Vilbois MH, Noelle RJ. The sequential role of lymphotoxin and B cells in the development of splenic follicles. *J. Exp. Med.* (1998) 187: 997–1007.

170 Golovkina TV, Shlomchik M, Hannum L, Chervonsky A. Organogenic role of B lymphocytes in mucosal immunity. *Science* (1999) 286: 1965–1968.

171 Flynn S, Toellner KM, Raykundalin C, Goodall M, Lane P. CD4 T cell cytokine differentiation: the B cell activation molecule, OX40 ligand, instructs CD4 T cells to express interleukin 4 and upregulates expression of the chemokine receptor, Blr-1. *J. Exp. Med.* (1998) 188: 297–304.

172 Harris DP, Haynes L, Sayles PC, Duso DK, Eaton SM, Lepak NM, Johnson LL, Swain SL, Swain L, Swain FE. Reciprocal regulation of polarized cytokine production by effector B and T cells. *Nat. Immunol.* (2000) 1: 475–481.

173 Fillatreau S, Sweenie CH, McGeachy MJ, Gray D, Anderton SM. B cells regulate autoimmunity by provision of IL-10. *Nat. Immunol.* (2002) 3: 944–950.

174 Corr M, Firestein GS. Innate immunity as a hired gun: but is it rheumatoid arthritis? *J. Exp. Med.* (2002) 195: 33–35

175 Arend WP. The innate immune system in rheumatoid arthritis. *Arthritis Rheum.* (2001) 44: 2224–2234.

176 Thomas R, Lipsky PE. Presentation of self peptides by dendritic cells: possible implications for the pathogenesis of rheumatoid arthritis. *Arthritis Rheum.* (1996) 39: 151–160.

177 Firestein GS, Zvaifler NJ. How important are T cells in chronic rheumatoid synovitis?: II. T cell- independent mechanisms from beginning to end. *Arthritis Rheum.* (2002) 46: 298–308.

178 Kouskoff V, Korganow AS, Duchatelle V, Degott C, Benoist C, Mathis D. Organ-specific disease provoked by systemic autoimmunity. *Cell* (1996) 87: 811–822.

179 Mangialaio S, Ji H, Korganow AS, Kouskoff V, Benoist C, Mathis D. The

arthritogenic T cell receptor and its ligand in a model of spontaneous arthritis. *Arthritis Rheum.* (1999) 42: 2517–2523.

180 Matsumoto I, Staub A, Benoist C, Mathis D. Arthritis provoked by linked T and B cell recognition of a glycolytic enzyme. *Science* (1999) 286: 1732–1735.

181 Maccioni M, Zeder-Lutz G, Huang H, Ebel C, Gerber P, Hergueux J, Marchal P, Duchatelle V, Degott C, van Regenmortel M, Benoist C, Mathis D. Arthritogenic monoclonal antibodies from K/BxN mice. *J. Exp. Med.* (2002) 195: 1071–1077.

182 Korganow AS, Ji H, Mangialaio S, Duchatelle V, Pelanda R, Martin T, Degott C, Kikutani H, Rajewsky K, Pasquali JL, Benoist C, Mathis D. From systemic T cell self-reactivity to organ-specific autoimmune disease via immunoglobulins. *Immunity* (1999) 10: 451–461.

183 Kassahn D, Kolb C, Solomon S, Bochtler P, Illges H. Few human autoimmune sera detect GPI. *Nat. Immunol.* (2002) 3: 411–2.

184 Schubert D, Schmidt M, Zaiss D, Jungblut PR, Kamradt T. Autoantibodies to GPI and creatine kinase in RA. *Nat. Immunol.* (2002) 3: 411.

185 Smolen JS, Steiner G. Are autoantibodies active players or epiphenomena? *Curr. Opin. Rheumatol.* (1998) 10: 201–206.

186 Schellekens GA, de Jong BA, van den Hoogen FH, van de Putte LB, van Venrooij WJ. Citrulline is an essential constituent of antigenic determinants recognized by rheumatoid arthritis-specific autoantibodies. *J. Clin. Invest.* (1998) 101: 273–281.

187 Baeten D, Peene I, Union A, Meheus L, Sebbag M, Serre G, Veys EM, De Keyser F. Specific presence of intracellular citrullinated proteins in rheumatoid arthritis synovium: relevance to antifilaggrin autoantibodies. *Arthritis Rheum.* (2001) 44: 2255–2262.

188 Miossec P. Anti-interleukin 1alpha autoantibodies. *Ann. Rheum. Dis.* (2002) 61: 577–579.

189 Ruth JH, Rottman JB, Katschke KJ Jr, Quin S, Wu L, LaRosa G, Ponath P, Pope RM, Koch AE. Selective lymphocyte chemokine receptor expression in the rheumatoid joint. *Arthritis Rheum.* (2001) 44: 2750–2760.

190 Weyand CW, Kurtin PJ, Goronzy JJ. Ectopic lymphoid organogenesis. A fast track for autoimmunity. *Am. J. Pathol.* (2001) 159: 787–793.

191 Webster EA, Khakoo AY, Mackus WJ, Karpusas M, Thomas DW, Davidson A, Christian CL, Lederman S. An aggressive form of polyarticular arthritis in a man with CD154 mutation (X-linked hyper-IgM syndrome). *Arthritis Rheum.* (1999) 42: 1291–1296.

192 Reparon-Schuijt CC, van Esch WJ, van Kooten C, Ezendam NP, Levarht EW, Breedveld FC, Verweij CL. Presence of a population of CD20+, CD38- B lymphocytes with defective proliferative responsiveness in the synovial compartment of patients with rheumatoid arthritis. *Arthritis Rheum.* (2001) 44: 2029–2037.

193 Takemura S, Braun A, Crowson C, Kurtin PJ, Cofield RH, O'Fallon WM, Goronzy JJ, Weyand CM. Lymphoid neogenesis in rheumatoid synovitis. *J. Immunol.* (2001) 167: 1072–1080.

194 Weyand CM, Kurtin PJ, Goronzy JJ. Ectopic lymphoid organogenesis: a fast track for autoimmunity. *Am. J. Pathol.* (2001) 159: 787–793.

195 Leadbetter EA, Rifkin IR, Hohlbaum AM, Beaudette BC, Shlomchik MJ, Marshak-Rothstein A. Chromatin-IgG complexes activate B cells by dual engagement of IgM and Toll-like receptors. *Nature* (2002) 416: 603–607.

196 Yi AK, Tuetken R, Redford T, Waldschmidt M, Kirsch J, Krieg AM. CpG motifs in bacterial DNA activate leukocytes through the pH-dependent generation of reactive oxygen species. *J. Immunol.* (1998) 160: 4755–4761.

197 Hirohata S, Ohshima N, Yanagida T, Aramaki K. Regulation of human B cell function by sulfasalazine and its metabolites. *Int. Immunopharmacol.* (2002) 2: 631–640.

198 Cambridge G, Leandro MJ, Edwards JC, Ehrenstein MR, Salden M, Bodman-Smith M, Webster AD. Serologic changes following B lymphocyte deple-

tion therapy for rheumatoid arthritis. *Arthritis Rheum.* (2003) 48: 2146–2154.

199 Saleh MN, Gutheil J, Moore M, Bunch PW, Butler J, Kunkel L, Grillo-Lopez AJ, LoBuglio AF. A pilot study of the anti-CD20 monoclonal antibody rituximab in patients with refractory immune thrombocytopenia. *Semin. Oncol.* (2000) 27: 99–103.

200 Zecca M, De Stefano P, Nobili B, Locatelli F. Anti-CD20 monoclonoal antibody for the treatment of severe, immune-mediated, pure red cell aplasia and hemolytic anemia. *Blood* (2001) 97: 3995–3997.

201 Ratanatharathorn V, Carson E, Reynolds C, Ayash LJ, Levine J, Yanik G, Silver SM, Ferrara JL, Uberti JP. Anti-CD20 chimeric monoclonal antibody treatment of refractory immune-mediated thrombocytopenia in a patient with chronic graft-versus-host disease. *Ann. Intern. Med.* (2000) 133: 275–279.

202 Ahrens N, Kingreen D, Seltsam A, Salama A. Treatment of refractory autoimmune haemolytic anaemia with anti-CD20 (rituximab). *Br. J. Haematol.* (2001) 114: 244–245.

203 Berentsen S, Tjonnfjord GE, Brudevold R, Gjertsen BT, Langholm R, Lokkevik E, Sorbo JH, Ulvestad E. Favourable response to therapy with the anti-CD20 monoclonal antibody rituximab in primary chronic cold agglutinin disease. *Br. J. Haematol.* (2001) 115: 79–83.

204 Layios N, van Den Neste E, Jost E, Deneys V, Scheiff JM, Ferrant A. Remission of severe cold agglutinin disease after rituximab therapy. *Leukemia* (2001) 15: 187–188.

205 Stasi R, Pagano A, Stipa E, Amadori S. Rituximab chimeric anti-CD20 monoclonal antibody treatment for adults with chronic idiopathic thrombocytopenic purpura. *Blood* (2001) 98: 952–957.

206 Specks U, Fervenza FC, McDonald TJ, Hogan MC. Response of Wegener's granulomatosis to anti-CD20 chimeric monoclonal antibody therapy. *Arthritis Rheum.* (2001) 44: 2836–2840.

207 Edwards JC, Cambridge G. Sustained improvement in rheumatoid arthritis following a protocol designed to deplete B-lymphocytes. *Rheumatology* (2001) 40: 205–211.

208 Leandro MJ, Edwards JC, Cambridge G. Clinical outcome in 22 patients with rheumatoid arthritis treated with B-lymphocyte depletion. *Ann. Rheum. Dis.* (2002) 61: 883–888.

209 De Vita S, Zaja F, Sacco S, De Candia A, Fanin R, Ferraccioli G. Efficacy of selective B cell blockade in the treatment of rheumatoid arthritis: evidence for a pathogenetic role of B cells. *Arthritis Rheum.* (2002) 46: 2029–2033.

210 Edwards JC, Szczepanski L, Szechinski J, Filipowicz-Sosnowska A, Emery P, Close DR, Stevens RM, Shaw T. Efficacy of B-cell targeted therapy with vituximas in patients with rheumatoid arthritis. *N. Engl. J. Med.* (2004) 350: 2572–2581.

211 Pisetsky DS, St Clair EW. Progress in the treatment of rheumatoid arthritis. *JAMA* (2001) 286: 2787–2790.

212 Kremer JM, Westhovens R, Leon M, Di Giorgio E, Alten R, Steinfeld S, Russell A, Dougados M, Emery P, Nuamah IF, Williams GR, Becker JC, Hagerty DT, Moreland LW. Treatment of rheumatoid arthritis by selective inhibition of T-cell activation with fusion protein CTLA4Ig. *New Engl. J. Med.* (2003) 349: 1907–1915.

213 Davis JC Jr, Totoritis MC, Rosenberg J, Sklenar TA, Wofsy D. Phase I clinical trial of a monoclonal antibody against CD40-ligand (IDEC-131) in patients with systemic lupus erythematosus. *J. Rheumatol.* (2001) 28: 95–101.

214 Huang W, Sinha J, Newman J, Reddy B, Budhai L, Furie R, Vaishnaw A, Davidson A. The effect of anti-CD40 ligand antibody on B cells in human systemic lupus erythematosus. *Arthritis Rheum.* (2002) 46: 1554–1562.

14
Dendritic Cells (DCs) in Immunity and Maintenance of Tolerance

Magali de Heusch*, Guillaume Oldenhove* and Muriel Moser

14.1
Introduction

There is increasing evidence that cells of the dendritic family display opposite functions. They have the unique capacity to induce optimal activation of T lymphocytes, and in particular to sensitize naive T cells that have never encountered the antigen they are specific for. Several recent reports have demonstrated that dendritic cells (DCs) also play a major role in peripheral tolerance. These two complementary functions would ensure the integrity of the organism in an environment full of pathogens.

It is unclear whether autocrine or paracrine factors determine the stimulatory versus suppressive function of DCs. Several reports suggest indeed that the nature of the DC itself, its activation status and/or the microenvironment may determine the fate of the T lymphocytes.

14.2
Dendritic Family

In the 1960s, several reports suggested that antigen-specific activation of T and B cells required the presence of "accessory" cells [1]. The authors postulated that a "transfer of information" was required between two cell types. More than three decades later, there is strong evidence that antigen-presenting-cells, which play much more than a "accessory" role, transmit multiple signals to cells in the lymphoid organs, related to the nature and the amplitude of the infection, the character of the infected tissue, etc.

DCs were discovered in 1973 by Steinman and Cohn [2] during observations on the cells of mouse spleen that adhered to surfaces. In addition to mononuclear phagocytes, granulocytes and lymphocytes, they found various large, nucleated cells, whose cytoplasm was arranged in pseudopods of varying forms and lengths.

* Both authors share first authorship.

They further reported that these "dendritic" cells undergo characteristic movements and do not appear to engage in active endocytosis. Since then, numerous studies have described the hallmarks of this family and, in particular, their capacity to sensitize naive T cells that have not encountered the antigen they are specific for.

Among antigen-presenting-cells, which include B cells, DCs and macrophages, DCs display unique properties: (1) at some stage, they have a dendritic morphology (from δενδρεον, tree) which favors interaction with numerous lymphocytes; (2) they express various receptors that specifically recognize microbial products and enable them to discriminate between self and non self; (3) they are motile and, in particular, migrate from the periphery to the lymphoid organs; (4) they have a specialization of function over time: they shift from an antigen-capturing mode to a T cell sensitizing mode during a phenomenon called maturation; (5) they express various co-stimulatory molecules and produce cytokines with stimulatory and inhibitory function.

14.3
DCs at Various Stages of Maturation

Cells of the dendritic family display a specialization of function over time. This was discovered in 1985, when Schuler and Steinman [3] reported that freshly isolated Langerhans cells were weak stimulators of T cell proliferation but underwent a progressive increase in stimulatory capacity in culture. Steinman and his colleagues subsequently showed that the capacity of DC to present protein varied inversely with stimulatory activity in MLR, suggesting that DC function involves two components that develop in sequence: a presentation step and a sensitization step [4]. It is generally admitted that DCs at the immature state are posted outside the lymphoid organs where they act as standing guards for pathogen invasion. In case of infection and/or inflammation, they move to the draining lymph nodes, thereby forming a link between the periphery and the secondary lymphoid tissues, where immune responses are initiated. Interestingly, the maturation confers to DCs the capacity to delay antigen presentation of antigenic/MHC complexes, i.e., to present in the lymphoid organs antigens encountered earlier in peripheral tissues [5]. The coordinated migration and maturation of DCs are considered critical for T cell priming, as mature DCs express membrane molecules and secrete cytokines that are required for optimal T cell sensitization.

Various stimuli have been shown to induce maturation of DCs and include microbial products and inflammatory cytokines. In particular, injection of lipopolysaccharide induces DC migration in situ, as assessed by upregulation of immunostimulatory properties and downmodulation of processing capacity [6]. These functional changes correlate with a rapid migration of DC to the T cell area, resulting in co-localization of T lymphocytes and mature DCs that have encountered microbial antigens. Interestingly, DCs express pattern-recognition receptors, such as Toll-like receptors, that distinguish infectious non-self from infectious self, and engagement of TLRs by various ligands leads to their maturation, migration and production of cytokines, thereby favoring the induction of immunity to infectious (rather than self) antigens [7, 8].

14.4
Immature DCs

In addition to their role of sentinels in most tissues, there is some evidence that immature DCs may induce a state of non-responsiveness in naive T cells and may, therefore, contribute to peripheral tolerance to antigens that do not induce their maturation.

Several groups have reported that immature DCs can induce the differentiation and expansion of type 1-regulatory T cells (Tr1). Roncarolo and colleagues [9] developed a protocol to differentiate Tr cells by repetitive exposure of naive peripheral blood $CD4^+$ T cells to allogeneic immature DCs. The resulting Tr1 cells produce IL-10, TGF-β and suppress proliferation and cytokine production by autologous $CD4^+$ T cells via an IL-10- and TGF-β-dependent mechanism. These Tr1 cells do not express high levels of FoxP3 and appear distinct from $CD4^+$ $CD25^+$ regulatory T cells. Other groups have reported that immature DCs induce a state of tolerance by triggering the IL-10-dependent differentiation of Tr1 cells in mice and humans [10–12]. Of note, a natural DC subset that induces the development of Tr1 cells has been recently identified in mice by Groux and colleagues [13].

However, other reports suggest that immature or semi-mature DCs may induce the clonal expansion of $CD4^+$ T cells and their differentiation into a memory pool that displays effector function only upon a secondary antigenic stimulation. Thus, adoptive transfer of freshly isolated (immature) splenic DCs pulsed extracorporeally with OVA peptide induced the proliferation of antigen-specific $CD4^+$ T lymphocytes that do not produce IFN-γ in vitro upon antigenic challenge. By contrast, injection of antigen-pulsed mature DCs resulted in proliferation of naive T cells and their rapid differentiation into Th1 effector cells. Mice injected with antigen-pulsed immature DCs did not become tolerant but displayed a memory response upon antigen boost [14]. Consistent with these observations, Lanzavecchia and colleagues have shown that T cells progressively differentiate, depending on the level of signal that accumulates during T cell–DC interaction [15]. Thus, a pool of central memory cells may develop at an intermediate stage and further differentiate into effector cells upon secondary encounter with the antigen.

14.5
Homing of DCs into Secondary Lymphoid Organs

While examining the behavior of Langerhans cells after skin transplantation, Larsen and colleagues found that these cells in epidermal sheets from grafts and explants increased in size and expression of MHC molecules, and then decreased in numbers. By 24 h, cells resembling LC were found close to the epidermal–dermal junction, and by 3 days they formed cords in dermal lymphatics before leaving the skin [16]. The same group showed that donor-derived MHC class II-positive DCs migrated rapidly out of the mouse cardiac allografts into the recipient spleens where they homed to the peripheral white pulp and associated predominantly with CD4$^+$ T lymphocytes [17]. The impact of DC migration to T cell priming was further demonstrated by Martin-Fontecha et al. [18] who showed that CCR7$^{-/-}$ DCs, injected subcutaneously, failed to migrate to the draining lymph nodes and did not induce immunity. The same authors further demonstrated that the magnitude and quality of CD4$^+$ T cell response were proportional to the number of antigen-carrying DCs that reached the lymph node, and that the migration is regulated at the level of entry into lymphatic vessels by inflammatory cytokines through upregulation of CCL21.

Interestingly, DCs appear to migrate not only in inflammatory conditions but also in the steady state. Vermaelen and colleagues have studied the antigen transport from the airway mucosa to the thoracic lymph nodes following intratracheal instillation of fluorescein isothiocyanate (FITC)-conjugated macromolecules. They found FITC$^+$ cells with stellate morphology in the T cell area of lymph nodes and identified these cells as migratory airway-derived lymph node DCs [19]. Similarly, the antigen-presenting activities of DCs have been shown to shift from lung to lymph node after an airway challenge with soluble antigen. MacPherson's group has also reported that a discrete DC subset constitutively endocytosed and transported apoptotic cells to T cell areas [20]. Co-localization of immature DCs with T lymphocytes is likely to be involved in the induction of peripheral tolerance (see Section 14.9).

Several chemokines and receptors appear to regulate DC trafficking at different stages of maturation in vivo [21]. In particular, Caux and colleagues have demonstrated that immature and mature DCs were selectively recruited by distinct chemokines expressed in different anatomic sites [22, 23]. Their observations suggest a role for MIP-3α/CCR6 in recruitment of immature DCs at site of injury and for MIP-3β/CCR7 in accumulation of antigen-loaded mature DCs in T cell rich areas. Upon maturation, DCs lose their expression of CCR6 and responsiveness to MIP-3α, to become sensitive to MIP-3β through induction of CCR7. A recent report [24] demonstrates that CCR7 is also required for the steady-state migration of skin-derived DC in the absence of inflammatory signals. In particular, CCR7-deficient mice lack a subset of skin-derived LC that display a CD11c$^+$ MHCII^{hi} phenotype in draining lymph nodes, and this semi-mature population of DC may be involved in induction of tolerance in vivo. In addition, CCR2-deficient mice have impaired Langerhans cell migration, reduced numbers of CD11c$^+$ splenic DCs and a block in the *Leishmania major* infection-induced relocalization of splenic DCs from the marginal zone to the T cell areas, suggesting that CCR2 is an important determinant of DC migration and localization as well as the development of protective immunity to *Leishmania major* [25].

14.6
DCs as Adjuvants

In 1978, Steinman and Witmer provided the first evidence that DCs purified from mouse spleens were potent stimulators of the primary mixed leukocyte reaction [26]. The authors used a dose–response assay to compare the potency of purified DCs with that of other spleen cell populations. They found that the MLR-stimulating capacity of fractionated spleen cells correlated with the DC numbers and that macrophages, T and B cells stimulated weakly. Since then, numerous reports have shown that DCs can activate both $CD4^+$ and $CD8^+$ T lymphocytes in various models.

In vivo, DCs are powerful stimulators of immune responses when adoptively transferred into recipients. Injection of epidermal Langerhans cells [27] or splenic DCs [28], coupled to trinitrophenyl, was shown to activate effector cells in mice. The in vivo priming capacity of DCs has been largely documented in several $CD4^+$ [29, 30] and $CD8^+$ T cell responses in rodents and in humans (see Section 14.15). In addition to T cell responses, humoral responses are induced following transfer of DCs, which are characterized by the secretion of IgG1 and IgG2a antibodies [31].

Based on these observations, it was admitted that DCs that have undergone maturation and presented high levels of MHC/antigen complexes and co-stimulatory molecules were capable of inducing optimal immune responses. However, two recent reports challenge this view. Pasare and Medzhitov have shown that DC maturation and migration to lymph nodes are not sufficient for T cell activation in the absence of TLR-induced inflammatory cytokines. Of note, the requirement for TLR-induced DC-derived cytokines can be bypassed by depletion of Treg, suggesting that TLR engagement counteracts the suppressive activity of Treg [32]. These observations are reminiscent of very old studies by Gershon, Steinman and colleagues, who showed that splenic DCs and peritoneal exudate cells induced by mycobacteria activate effector cells that are resistant to suppression [28]. They found that intravenous injection of TNP-DC and TNP-coupled PEC that had been induced by the complete Freund's adjuvant produced an immune response that could not be suppressed by coadministration of an intravenous injection of TNBS known to induce TNP-specific suppressor cells.

A second report by Sporri and Reis e Sousa [33] demonstrates that the nature of the signal that leads to DC activation determines the outcome of DC/T cell interaction. They show that DCs activated indirectly by inflammatory mediators are able to upregulate MHC molecules and co-stimulation and to drive T cell clonal expansion. However, such DCs do not produce IL-12 and are unable to drive differentiation of $CD4^+$ T cells into Th1 effectors in vivo. In contrast, exposure to pathogen components resulted in fully activated DCs that promoted T helper cell responses. Interestingly, directly and indirectly activated DCs may have complementary roles in the induction and maintenance of immunity: DCs activated by engagement of pathogen recognition receptors (PRR) may prime for Th1 whereas DCs activated in trans by inflammatory mediators may induce clonal expansion and/or trigger memory cells.

14.7
DC Subsets

The dendritic family appears to be heterogeneous and includes various DC subclasses [34].

14.7.1
Classical DCs

Phenotypic analysis of DC from different tissues has revealed some heterogeneity in lymphoid organs. In the spleen, three DC types have been identified that differ by their expression of CD4 and CD8α/α molecules [35]. The CD4$^-$8α^+, CD4$^-$8$^-$, CD4$^+$8α^- subpopulations share many common markers, including MHC class II and CD11c, and express similar levels of the co-stimulatory molecules CD40, CD80 and CD86. The three splenic subtypes behave as rapidly-turning-over products of three independent developmental lineages [36].

DC populations of the lymph nodes appear even more complex: in addition to the three cell types present in the spleen, two DC populations have been described that may be dermal and epidermal-derived DC [37]. Mouse Peyer's patches contain two distinct populations of DCs: one that resides underneath the follicle-associated epithelium and one present in the interfollicular T cell regions [38].

14.7.2
Plasmacytoid DCs

In 2004, O'Doherty and colleagues identified in human peripheral blood a subpopulation of CD11c$^-$ immature DCs with low MHC class II expression and poor T cell stimulatory capacity [39]. Similar cells were subsequently purified from tonsils and were shown to be identical to the cells producing interferon in peripheral blood in response to viruses [40]. Human pDCs are CD4$^+$IL-3RA (CD123)$^+$ CD11c$^-$, whereas murine pDCs are CD123$^-$ and express the surface antigens CD11c, B220 and Ly6C. In addition, murine pDC upregulate CD8α expression during their maturation [41, 42].

Although both pDCs and classical DCs can secrete type I interferon in response to viruses through TLR-dependent and TLR-independent pathways, pDCs produce larger amounts. Of note, type I interferons stimulate the function of various cell types: they activate cytolytic activity and IFN-γ production by NK cells, promote survival of T cells and Th1 differentiation, induce the maturation and stimulatory properties of DCs [43] and induce B cells to produce immunoglobulins.

Freshly isolated human and mouse pDC are poorly stimulatory, as they do not endocytose antigens efficiently, have minimal expression of co-stimulatory molecules and low MHC class II expression, as compared to classical DCs. In contrast, activated pDCs can present antigens and induce expansion and activation of CD4$^+$ and CD8$^+$ T lymphocytes.

The contribution of pDCs to immune responses is unclear [44]. In particular, it is still unknown whether they initiate immunity or are mainly involved in the induction of innate responses and the regulation of adaptive responses. In addition, a few recent reports suggest that pDCs may be involved in the negative regulation of some immune responses (see Section 14.9).

Although the DC subsets were thought to be committed to the DC lineage, a recent report does not support this view. Zuniga et al. [45] have indeed shown that virus infection can reprogram bone marrow plasmacytoid precursor DCs to convert into cells with phenotypic and functional characteristics of $CD11c^+$ conventional DCs through the production of type 1 interferon. The cell transformation is accompanied by the acquisition of potent antigen-presenting-capacity and the ability to secrete IL-12 in response to a TLR4 ligand. Whether this unexpected plasticity of bone marrow pDCs plays a major role for resolution of viral infections remains to be determined.

14.8
DCs in T Cell Polarization

DCs are more than a simple "on/off" switch of the immune response, but in addition seem to contribute significantly to the polarization of immune responses. The plasticity of the DCs allows them to modulate their function according to the nature of the infection and the tissue damaged, thereby providing this useful information to T cells confined in the lymphoid organs. Various mechanisms have been reported by which DCs may regulate the Th1/Th2 balance in vivo, including the intensity and the duration of the (antigenic and co-stimulatory) signal [46], the DC subclass [47–49], the maturation stage [14], the nature and the concentration of the antigen [50], the cytokine microenvironment, etc. (for a review see Ref. 51).

There is recent evidence that other cell types, in particular effector cells, participate in T cell priming and affect the quality of the developing T cell response induced by DCs in vivo (for a review, see Kalinski and Moser, Nat. Rev. Immunol. 2005. 5, 251–260). B cells regulate the capacity of DCs to promote IL-4 secretion, possibly by downregulating their secretion of IL-12. Indeed, administration of DCs from B cell deficient mice induced the development of cells secreting IL-4 only (no IFN-γ), whereas injection of DCs from wild-type animals primes for Th1- and Th2-type cells [52]. Natural killer cells, which are normally excluded from lymph nodes, have been shown to be recruited to lymph nodes on stimulation by the injection of mature DCs and this movement provides a potent boost for Th1 responses. NK cell depletion and reconstitution experiments demonstrated indeed that NK cells provide an early source of IFN-γ that is necessary for Th1 polarization [53].

Steinman and colleagues have reported that α-galactoceramide acts as a stimulus in vivo for the full maturation of DCs in mice. The phenotypic and functional changes associated with the maturation were not induced directly by this glycolipid but required natural killer T cells. Of note, DCs from mice given α-galactoceramide with antigen were able to induce antigen specific, IFN-γ producing T cells upon transfer into naive animals, suggesting that DC maturation accounts for the induction of Th1 immunity [54]. The authors further showed that inflammatory cytokines and a distinct CD40/CD40L signal were both required for the Th1 adjuvant properties of this glycolipid [55].

14.9
Tolerogenic DC

Although ignored for several decades, the tolerogenic properties of DCs are now gaining increasing attention from immunologists. An essential question was how autoimmunity was avoided when fully mature DCs were presenting autoantigens together with non-self infectious antigens. This question long remained unresolved, until several reports suggested the existence of active tolerogenic mechanisms that continuously silenced autoreactive cells in the periphery [56].

Deletion has been reported as a mechanism of extrathymic tolerance to various antigens (including self antigens) expressed in tissues outside the normal recirculation pathway for naive T cells. Kurts and colleagues [57] have shown that OVA-specific transgenic T cells, injected into mice expressing OVA in the pancreatic β cells and proximal kidney, proliferated in the lymph nodes draining the pancreas and the kidney and were subsequently deleted. An important step in the understanding of this process was the demonstration that a bone marrow-derived antigen-presenting-cell was capable of processing and presenting antigens expressed by peripheral tissues. Using mice in which only CD11c$^+$ DC can activate OT-I cells, the authors further demonstrated that DC were sufficient to cross-present exogenous self-antigens in vivo and are likely involved in the deletion process [58].

Other studies have used direct DC targeting to antigens to show that DCs can be tolerogenic antigen presenting cells in vivo. Targeting antigens in situ through DEC-205 resulted in deletional tolerance of CD4$^+$ T cells, whereas combined injection of anti-CD40 antibody resulted in sustained T cell activation [59]. Using an inducible expression and presentation of lymphocytic choriomeningitis virus-derived CTL epitopes, thereby avoiding any cell transfer, Probst et al. demonstrated that DCs induce immunity or tolerance depending solely on their activation status [60]. In humans, Dhodapkar et al. reported a decline in antigen-specific IFN-γ-producing T cells and the onset of IL-10-secreting cells upon injection of immature DCs [61].

The nature of the tolerogenic DCs is still elusive. Likely candidates include:
1. DCs at the immature stage, which have been shown to induce a state of unresponsiveness in T lymphocytes (see Section 14.4).
2. DCs made tolerogenic by various treatments, such as IL-10 alone or in combination with TGF-β, immunosuppressive agents such as vitamin D3.
3. Some specialized DC subsets, which have been shown strictly dedicated to tolerance induction, even in a mature state.

In particular, CD8α^+ DCs have been shown to display tolerogenic properties in vitro [62] and in vivo using the tumor/self-peptide P815 AB [63]. The same subset was involved in the phenomenon of cross-tolerance, i.e., the induction of peripheral tolerance to tissue-associated antigens [64, 65]. The tolerogenic role of CD8α^+ DCs is, however, still elusive, as other studies have demonstrated that CD8α^+ DCs were the major producers of IL-12 and efficiently primed for Th1 in vivo [66, 67].

A population of CD11c^{low}CD45Rb^{high} DCs was isolated from spleen and lymph nodes from mice, which display an immature phenotype even after in vitro activation and induce tolerance and Tr1 cell differentiation in vivo [13].

Special attention has been focused lately on plasmacytoid DCs. Several recent studies have demonstrated that pDCs display tolerogenic potential in some conditions. pDCs have been shown to prevent disease in a mouse model of asthma: Lambrecht and colleagues addressed the role of pDCs in a mouse model of asthma by depleting this subset using anti-Gr1 antibodies, and showed that pDCs depleted mice developed increased features of asthma and Th2 cell cytokine production in mesenteric lymph nodes [68]. Although both classical and plasmacytoid DCs can transport FITC-labeled OVA to the draining lymph nodes, only classical DCs primed naive T cells in vivo. By contrast, pDCs induced the generation of suppressive T cells in vitro and, upon transfer, prevented the development of asthma in vivo. In a bone marrow transplantation model, plasmacytoid precursor DCs facilitate hematopoietic stem cell engraftment in mismatched recipients and induce durable tolerance to transplanted grafts [69].

Two recent reports suggest that stroma may drive mature DCs to differentiate into regulatory DCs. Zhang et al. [70] have reported that endothelial-like splenic stromal cells promoted the proliferation of mature DCs and their development into a unique type of DCs that display decreased levels of CD11c, MHC class II, CD86, and secreted more IL-10 and NO than mature DCs. The authors further showed that these DCs exerted inhibitory effects on CD4 T cells in vitro and in vivo, and this suppressive function appears mediated by nitric oxide. An in vivo counterpart of this subset was identified in spleen, and these cells undergo proliferation after immunization. These data support the possibility that maturing DCs migrate to secondary lymphoid organs, activate naive T cells and then further differentiate into regulatory DCs under the influence of the local microenvironment. This view challenges previous observations that suggested that fully mature DCs undergo spontaneous apoptosis in vivo [6, 71]. Rather, mature DCs may not be terminally differentiated cells but may revert to a form of regulatory cell that would play a role in the termination of the immune response and the homeostasis of the immune system. A second report shows that spleen-derived stromal cells promote selective development of CD11c^{lo} CD45RB$^+$ IL-10-producing regulatory DC from lineage-negative c-kit$^+$ progenitor cells. These DC have the capacity to suppress T cell responses and induce IL-10-producing regulatory T cells in vitro and to induce antigen-specific tolerance in vivo [72].

14.10
Mechanisms of Tolerance

The molecular mechanisms underlying the induction of peripheral tolerance by DCs are still poorly understood. Several mechanisms have been identified, which include antigenic presentation in the absence of co-stimulation, deletion of antigen-reactive T cells by tryptophan metabolites, regulation of T cell/DC contacts or Fas/Fas-ligand interaction, and the induction of regulatory T cells.

14.10.1
Lack of Co-stimulation

In vitro studies have shown that T cell clones recognizing antigen in the absence of co-stimulatory signals show impaired proliferation and may become anergic to further immunogenic stimulation. Whether a state of systemic tolerance can be reached under conditions of suboptimal co-stimulation remains an open question but has been challenged by the observation that experienced rather than naive T cells seem to be sensitive to anergy by signal alone [73, 74]. In addition, virtually all antigen-specific T cells would have to interact with DCs expressing antigen in the absence of co-stimulation in order to induce effective tolerance in the whole animal.

14.10.2
Peripheral Deletion of Autoreactive T Cells

There is some evidence that tryptophan metabolites, or signaling through CD95, may induce T cell deletion. Populations of DC producing indoleamin 2,3-dioxygenase (IDO) have been described by several groups. Of particular interest is the finding that human DCs that express IDO constitutively have been identified in normal lymphoid tissues and in larger numbers in tumor-draining lymph nodes [75]. Notably, the presence of IL-10 during DC maturation has been shown to potentiate the function of IDO, suggesting a new mechanism of action for IL-10. In addition, engagement of B7 molecules on DC with CTLA-4 appears to activate the immunosuppressive pathway of tryptophan metabolism (see Section 14.11).

Signaling through CD95 seems to be involved in tolerance induction, as CD95-deficient OT-I cells are not susceptible to deletion by cross-presentation of tissue-derived ovalbumin in the renal and pancreatic lymph nodes [76].

14.10.3
Dynamics of Cellular Contacts

Nussenzweig and colleagues reported recently on the dynamics of T/DC interactions and showed that the DCs resident in the lymph nodes form an extensive cellular network. Interestingly, DCs injected intradermally initially moved faster than resident DC and then became progressively dispersed into the resident network, losing motility. Probably, the DCs maintained in a fixed position in the network scan the passing T cells by probing with the dendrites. In the steady state, the DC network would present T cells with an enormous surface area rich in the self-MHC class II-peptide complexes required to maintain self tolerance. Mature DCs carrying antigen into the lymph nodes from the periphery disperse through the T cell zones, thereby facilitating their interaction with migrating T cells specific for the antigen they carry. Of note, the mature DCs would be ideally positioned to relay antigens to other DCs in the network. These observations are in line with previous reports showing that antigen/MHC complexes expressed on donor DC are detected on recipient DCs in the draining lymph nodes [77].

Recent evidence suggests distinct dynamics of T cell interactions with antigen-presenting DCs in tolerogenic versus immunogenic conditions. Amigorena and colleagues have analyzed the dynamics of the interactions between DCs and T cells during priming in explanted lymph nodes using two-photon laser scanning microscopy [78]. CMTMR-labeled OT-I cells were adoptively transferred into mice that were injected 4 h later with anti-DEC205-OVA plus a maturation stimulus (LPS or anti-CD40). They found that, during T cell priming, DC–T cell dynamics followed a sequence of three phases of interactions: transient, stable and transient. A stop signal was delivered to T cells 15–20 h after immunization and most of the DC–T cell conjugates remained stable for more than 30 min. T cells began secreting IFN-γ at that time and began dividing after 30–48 h of interaction. Interestingly, the stable phase is dictated by the kinetics of DC maturation s rather than by T cells. During the induction of tolerance, i.e., in the absence of adjuvant, T cells established brief and unstable contacts with DCs throughout the whole observation period. The authors conclude that DCs can deliver brief or prolonged stimulation to $CD8^+$ T cells that will determine their differentiation. In particular, stable T cell contacts are required for priming, whereas brief DC–T cell interactions result in induction of T cell tolerance. These data are in keeping with previous observations showing that the duration of T cell receptor stimulation is a critical parameter in T cell fate determination [46].

14.10.4
Induction of Regulatory T Cells

There is evidence that DCs may promote tolerance by inducing and/or sustaining the differentiation of regulatory T cells. This active mechanism would lead to a memory of tolerance and ensure long-term, antigen-specific hyporesponsiveness in animals even in stimulatory conditions.

Three main populations of cells with suppressive activity have been described and renamed regulatory T cells: Tr1 and Th3 cells, which are induced by IL-10 and/or TGF-β, and $CD4^+$ $CD25^+$ cells, which are present naturally in the periphery. Several reports have suggested that DCs may trigger Tr1 cells mainly at the immature stage (see Section 14.4). However, mature DCs may induce Tr1 cells in some conditions, as Akbari et al. have reported that phenotypically mature pulmonary DCs produced IL-10 when exposed to respiratory antigens and stimulate the development of Tr1 cells [79].

Although it was originally thought that DCs that have not undergone maturation would sustain/activate the population of naturally occurring regulatory T cells, there is evidence that these $CD4^+ CD25^+$ control ongoing immune responses, induced by mature DCs or adjuvant [80–82]. In particular, these cells seem to exert a negative feedback mechanism on Th1 responses [82].

Interestingly, recent data have demonstrated that normal $CD4^+ CD25^-$ T cells have the potential to convert spontaneously in the periphery into a $CD4^+ CD25^+$ regulatory T cells which are physically and functionally indistinguishable from naturally occurring $CD4^+ CD25^+$ regulatory T cells [83]. This conversion does not require a thymus. When $CD4^+ CD25^-$ cells were injected into $B7-1/B7-2^{-/-}$ mice, very few $CD4^+ CD25^+$ were recovered that were unable to suppress proliferation of responder cells in vitro or in vivo. The requirement for B7 suggests that antigen-presenting-cells may be involved, in particular DCs that have been shown to express B7 in vivo in the absence of any intentional stimulation [84].

These observations suggest that immature and mature DCs may act through distinct regulatory T cell populations. Immature DCs would present mainly tissue self antigens and trigger Tr1 cells, thereby limiting autoimmune reactions. By contrast, fully competent mature DCs would sustain the survival/function of $CD4^+ CD25^+$ Treg and prevent excessive (Th1?) immune responses directed to self and (mainly?) non-self antigens.

14.11
CD28-B7 Bidirectional Signaling

Recent reports have shown that CD28 and CTLA-4 as ligands of B7 molecules can signal DCs to initiate distinct effector responses. The first evidence was provided by Grohmann et al., who showed that a soluble form of CTLA-4 [85] and membrane anchored CTLA-4 on the surface of regulatory T cells [86] activate the immunosuppressive pathway of tryptophan catabolism in DCs, leading to the induction of antigen-specific tolerance. In contrast to results with CTLA4-Ig treatment, DCs conditioned with CD28-Ig display increased adjuvant capacities [87]. The authors report that DCs treated for 24 h with various concentrations of either CD28-Ig or CTLA-4-Ig produce IFN-γ, but that only CD28-Ig-conditioned DCs produce IL-6. Interestingly, blocking the effects of IL-6 resulted in enhanced tryptophan catabolism in DCs exposed to CD28-Ig, suggesting that IL-6 effects are dominant, a hypothesis consistent with previous studies by Medzhitov and colleagues [88]. The authors further demonstrated an adjuvant activity of soluble CD28 in vivo, as assessed by enhanced T cell-mediated immunity to tumor and self peptides, fungal clearance and tumor growth inhibition. Collectively these observations suggest that T cells can instruct DCs to manifest tolerogenic properties after CTLA-4 engagement of B7 or Th1-prone properties after CD28-B7 interaction.

14.12
Crosspriming

It still unclear whether new migrant DCs prime T cells following direct interaction. There is evidence that transferred DCs establish stable interaction with T cells in the lymph nodes draining the site of injection [89, 90] and that this stability is crucial for T cell activation (see Section 14.10.3). However, some observations suggest that exchanges of peptide–MHC complexes occur between different DC populations in vitro and in vivo and may play a major role in the amplification of antigen presentation and, thereby, T cell responses. In vitro studies support the concept that a major component of the MLR is the secondary presentation of alloantigens acquired from stimulator DC by DC of responder type [91]. Kleindienst and Brocker used transgenic mouse models with targeted MHC class II expression specifically on DCs to test whether DCs loaded in vitro with antigen and used as cellular vaccines directly stimulate antigen-specific T cells [92]. Their results suggest that participation of resident DCs, but not endogenous B cells, is crucial to obtain optimal expansion and effector function of T cells in vivo. The transfer of antigen to endogenous DCs occurs in the lymph nodes and requires the migration of viable antigen-loaded DCs. Although the mechanism of antigen-transfer is unknown, vesicles of endosomal origin are likely candidates. Amigorena and colleagues have demonstrated [93] that, in addition to carrying antigen, exosomes promote the exchange of functional peptide–MHC complexes between DCs. Of note, naive T cell stimulation by exosomes in vitro requires two distinct DC populations, one that produces exosomes and a second that acquires peptides–MHC complexes and stimulates T lymphocytes. These observations are consistent with a role for exosomes in spreading antigen-specific signals by exchanging antigens and/or MHC complexes between transferred DCs and host resident DCs.

14.13
Cross-presentation and Cross-tolerization

Among the antigen-presenting-cells, DCs have the unique capacity of processing soluble exogenous proteins to MHC class I restricted, $CD8^+$ T lymphocytes. In particular, Albert and colleagues have found that human DCs can efficiently present influenza antigens derived from apoptotic, influenza-infected cells and stimulate class I-restricted $CD8^+$ cytotoxic T lymphocytes [94].

Notably, the cross-presentation confers to DCs the capacity to generate cytotoxic responses to viruses that do not infect them and to react to dangers associated with altered protein composition, whatever the affected cells. This pathway may also account for the phenomenon of cross-priming whereby antigens derived from tumor cells or transplants are presented by host APCs. The dominant processing pathway of cross-presented antigens appears to implicate the TAP transporters and proteasome complexes. It was discovered that ER membranes deliver all elements required for MHC class I antigen processing to early phagosomes, which become autonomous cross-presentation compartments. After antigen export to the cytosol and degradation by the proteasome, peptides are translocated by TAP into the lumen of the same phagosomes and loaded on phagosomal MHC class I molecules [95]. In addition to this ER-like phagosomes, a recent report provided evidence that pinocytosed soluble proteins can gain access to the lumen of the perinuclear ER after internalization by DCs. Thus, distinct mechanisms appear to be responsible for cross-presentation by the ER-associated machinery of antigens taken up by phagocytosis and pinocytosis [96].

Notably, macrophages form ER-like phagosomes that are competent for cross-presentation [97] but there is no evidence that exogenous proteins can access the ER of macrophages, an observation that could result from the rapid degradation of proteins in these cells, as compared to DCs.

The authors propose that the separation of internalized antigens into two distinct pathways may properly regulate immunity and tolerance to MHC class I-restricted antigens [96]. During infection, microbial antigens would be captured (from whole pathogens or fragments of infected cells) by phagosomes generating peptide–MHC complexes, and engagement of Toll-like receptors would induce DC maturation. In the steady state, pinocytosis of self antigens present in the microenvironment would allow their low constitutive presentation, leading to peripheral tolerance to self.

A key issue is the identification of the signal that determines whether such presentation ultimately results in a cytotoxic T cell (CTL) response (cross-priming) or in $CD8^+$ T cell inactivation (cross-tolerance). Reis e Sousa and colleagues have recently described a mechanism that promotes cross-priming during viral infections [98]. They show that murine $CD8alpha^+$ DCs are activated by double-stranded RNA present in virally infected cells but absent from uninfected cells. DC activation requires phagocytosis of infected material, followed by signaling through the dsRNA receptor, toll-like receptor 3 (TLR3).

14.14
DC as Regulators of T Cell Recirculation

In addition to their major role in the initiation of immunity, DCs appear to direct cell traffic. A recent report suggests indeed that skin- and gut-associated DCs can alter the homing properties of effector T cells [99]. Thus, DC from gut-associated lymphoid tissues induce on activated CD8 T cells the expression of the gut-homing molecules $\alpha 4\beta 7$ and CCR9, as well as the capacity to migrate to the small bowel. By contrast, naive T cells activated by DC from peripheral lymph nodes express more selectin ligands and mRNA for CCR4, and home more efficiently to inflamed skin. Of note, reactivation of tissue committed memory cells modified their tissue tropism according to the last activating DC's origin, a property that would allow the immune system to fight pathogens that colonize discrete sites.

14.15
DC-based Immunotherapy of Cancer

The unique properties of DCs as adjuvant, in particular for Th1 priming, suggested their use for inducing tumor-specific immunity in cancer patients. This approach led to tumor rejection of established tumors in several murine models and gave limited but encouraging results in melanoma patients [100–102].

Interestingly, recent observations suggest that anti-vaccine CTL may not be the effector cells that kill the bulk of tumor cells, but rather that T cells directed against tumor antigens other than the vaccine antigen may be involved in tumor rejection. Boon, Coulie and colleagues have reported that new antitumor clonotypes arise after vaccination and can be detected in the blood and in high numbers in regressing metastases [103, 104].

There is strong evidence that a negative environment is produced in the tumor that would lead to local paralysis of the T cells specific to the tumor. Several mechanism of inhibition of T cells have been described and include tumor-specific human $CD4^+ CD25^+$ [105], TGF-b and IL-10 production, IDO-producing DCs possibly under the influence of Treg, etc.

Collectively, these observations suggest that critical parameters need to be better defined to optimize DC-based immunotherapy [Figdor, *Nat. Med.* 2004, 10(5), 475–480] – in particular the choice of tumor antigen, the nature of the DC vaccine and the strategy to counteract suppressive mechanisms.

14.16
Conclusion

DCs can perform various – even opposite – functions with great efficiency. Their capacity to induce immunity versus tolerance and their specialization of function over time render them ideally suited to organize the defense against pathogens while avoiding autoimmunity. Our understanding of DC function in the pathogenesis of human diseases suggests that these cells may be potential targets for therapeutic intervention.

References

1. Mosier, D.E., A requirement for two cell types for antibody formation in vitro. *Science*, 1967. **158**(808), 1573–5.
2. Steinman, R.M. and Z.A. Cohn, Identification of a novel cell type in peripheral lymphoid organs of mice. I. Morphology, quantitation, tissue distribution. *J. Exp. Med.*, 1973. **137**(5), 1142–62.
3. Schuler, G. and R.M. Steinman, Murine epidermal Langerhans cells mature into potent immunostimulatory dendritic cells in vitro. *J. Exp. Med.*, 1985. **161**(3), 526–46.
4. Romani, N., et al., Presentation of exogenous protein antigens by dendritic cells to T cell clones. Intact protein is presented best by immature, epidermal Langerhans cells. *J. Exp. Med.*, 1989. **169**(3), 1169–78.
5. Mellman, I. and R.M. Steinman, Dendritic cells: specialized and regulated antigen processing machines. *Cell*, 2001. **106**(3), 255–8.
6. De Smedt, T., et al., Regulation of dendritic cell numbers and maturation by lipopolysaccharide in vivo. *J. Exp. Med.*, 1996. **184**(4), 1413–24.
7. Medzhitov, R. and C.A. Janeway, Jr., Innate immunity: the virtues of a nonclonal system of recognition. *Cell*, 1997. **91**(3), 295–8.
8. Medzhitov, R. and C.A. Janeway Jr, Decoding the patterns of self and non self by the innate immune system. *Science*, 2002. **296**, 298–300.
9. Roncarolo, M.G., M.K. Levings, and C. Traversari, Differentiation of T regulatory cells by immature dendritic cells. *J. Exp. Med.*, 2001. **193**(2), F5–9.
10. Jonuleit, H., et al., Induction of interleukin 10-producing, nonproliferating CD4(+) T cells with regulatory properties by repetitive stimulation with allogeneic immature human dendritic cells. *J. Exp. Med.*, 2000. **192**(9), 1213–22.
11. Menges, M., et al., Repetitive injections of dendritic cells matured with tumor necrosis factor alpha induce antigen-specific protection of mice from autoimmunity. *J. Exp. Med.*, 2002. **195**(1), 15–21.
12. Hugues, S., et al., Tolerance to islet antigens and prevention from diabetes induced by limited apoptosis of pancreatic beta cells. *Immunity*, 2002. **16**(2), 169–81.
13. Wakkach, A., et al., Characterization of dendritic cells that induce tolerance and T regulatory 1 cell differentiation in vivo. *Immunity*, 2003. **18**(5), 605–17.
14. de Heusch, M., et al., Depending on their maturation state, splenic dendritic cells induce the differentiation of CD4(+) T lymphocytes into memory and/or effector cells in vivo. *Eur. J. Immunol.*, 2004. **34**(7), 1861–9.
15. Lanzavecchia, A. and F. Sallusto, Dynamics of T lymphocyte responses: intermediates, effectors, and memory cells. *Science*, 2000. **290**(5489), 92–7.
16. Larsen, C.P., et al., Migration and maturation of Langerhans cells in skin transplants and explants. *J. Exp. Med.*, 1990. **172**(5), 1483–93.
17. Larsen, C.P., P.J. Morris, and J.M. Austyn, Migration of dendritic leukocytes from cardiac allografts into host spleens. A novel pathway for initiation of rejection. *J. Exp. Med.*, 1990. **171**(1), 307–14.
18. Martin-Fontecha, A., et al., Regulation of dendritic cell migration to the draining lymph node: impact on T lymphocyte traffic and priming. *J. Exp. Med.*, 2003. **198**(4), 615–21.
19. Vermaelen, K.Y., et al., Specific migratory dendritic cells rapidly transport antigen from the airways to the thoracic lymph nodes. *J. Exp. Med.*, 2001. **193**(1), 51–60.
20. Huang, F.P., et al., A discrete subpopulation of dendritic cells transports apoptotic intestinal epithelial cells to T cell areas of mesenteric lymph nodes. *J. Exp. Med.*, 2000. **191**(3), 435–44.
21. McColl, S.R., Chemokines and dendritic cells: a crucial alliance. *Immunol. Cell Biol.*, 2002. **80**(5), 489–96.
22. Caux, C., et al., Dendritic cell biology and regulation of dendritic cell trafficking by chemokines. *Springer Semin Immunopathol*, 2000. **22**(4), 345–69.

23. Dieu, M.C., et al., Selective recruitment of immature and mature dendritic cells by distinct chemokines expressed in different anatomic sites. *J. Exp. Med.*, 1998. **188**(2), 373–86.
24. Ohl, L., et al., CCR7 governs skin dendritic cell migration under inflammatory and steady-state conditions. *Immunity*, 2004. **21**(2), 279–88.
25. Sato, N., et al., CC chemokine receptor (CCR)2 is required for langerhans cell migration and localization of T helper cell type 1 (Th1)-inducing dendritic cells. Absence of CCR2 shifts the Leishmania major-resistant phenotype to a susceptible state dominated by Th2 cytokines, b cell outgrowth, and sustained neutrophilic inflammation. *J. Exp. Med.*, 2000. **192**(2), 205–18.
26. Steinman, R.M. and M.D. Witmer, Lymphoid dendritic cells are potent stimulators of the primary mixed leukocyte reaction in mice. *Proc. Natl. Acad. Sci. U.S.A.*, 1978. **75**(10), 5132–6.
27. Ptak, W., et al., Role of antigen-presenting cells in the development and persistence of contact hypersensitivity. *J. Exp. Med.*, 1980. **151**(2), 362–75.
28. Britz, J.S., et al., Specialized antigen-presenting cells. Splenic dendritic cells and peritoneal-exudate cells induced by mycobacteria activate effector T cells that are resistant to suppression. *J. Exp. Med.*, 1982. **155**(5), 1344–56.
29. Inaba, K., et al., Dendritic cells pulsed with protein antigens in vitro can prime antigen- specific, MHC-restricted T cells in situ. *J. Exp. Med.*, 1990. **172**(2), 631–40.
30. De Becker, G., et al., Regulation of T helper cell differentiation in vivo by soluble and membrane proteins provided by antigen-presenting cells. *Eur. J. Immunol.*, 1998. **28**(10), 3161–71.
31. Sornasse, T., et al., Antigen-pulsed dendritic cells can efficiently induce an antibody response in vivo. *J. Exp. Med.*, 1992. **175**(1), 15–21.
32. Pasare, C. and R. Medzhitov, Toll-dependent control mechanisms of CD4 T cell activation. *Immunity*, 2004. **21**(5), 733–41.
33. Sporri, R. and C. Reis e Sousa, Inflammatory mediators are insufficient for full dendritic cell activation and promote expansion of CD4+ T cell populations lacking helper function. *Nat. Immunol.*, 2005. **6**(2), 163–70. Epub 2005 Jan 16.
34. Shortman, K. and Y.J. Liu, Mouse and human dendritic cell subtypes. *Nat. Rev. Immunol.*, 2002. **2**(3), 151–61.
35. Vremec, D. and K. Shortman, Dendritic cell subtypes in mouse lymphoid organs: cross-correlation of surface markers, changes with incubation, and differences among thymus, spleen, and lymph nodes. *J. Immunol.*, 1997. **159**(2), 565–73.
36. Kamath, A.T., et al., The development, maturation, and turnover rate of mouse spleen dendritic cell populations. *J. Immunol.*, 2000. **165**(12), 6762–70.
37. Henri, S., et al., The dendritic cell populations of mouse lymph nodes. *J. Immunol.*, 2001. **167**(2), 741–8.
38. Iwasaki, A. and B.L. Kelsall, Freshly isolated Peyer's patch, but not spleen, dendritic cells produce interleukin 10 and induce the differentiation of T helper type 2 cells. *J. Exp. Med.*, 1999. **190**(2), 229–39.
39. O'Doherty, U., et al., Human blood contains two subsets of dendritic cells, one immunologically mature and the other immature. *Immunology*, 1994. **82**(3), 487–93.
40. Kadowaki, N., et al., Natural interferon alpha/beta-producing cells link innate and adaptive immunity. *J. Exp. Med.*, 2000. **192**(2), 219–26.
41. Asselin-Paturel, C., et al., Mouse type I IFN-producing cells are immature APCs with plasmacytoid morphology. *Nat. Immunol.*, 2001. **2**(12), 1144–50.
42. Nakano, H., M. Yanagita, and M.D. Gunn, Cd11c(+)b220(+)gr-1(+) cells in mouse lymph nodes and spleen display characteristics of plasmacytoid dendritic cells. *J. Exp. Med.*, 2001. **194**(8), 1171–8.
43. Santini, S.M., et al., Type I interferon as a powerful adjuvant for monocyte-derived dendritic cell development and activity in vitro and in Hu-PBL-SCID

44 Colonna, M., G. Trinchieri, and Y.J. Liu, Plasmacytoid dendritic cells in immunity. *Nat. Immunol.*, 2004. **5**(12), 1219–26.

45 Zuniga, E.I., et al., Bone marrow plasmacytoid dendritic cells can differentiate into myeloid dendritic cells upon virus infection. *Nat. Immunol.*, 2004. **5**(12), 1227–34. Epub 2004 Nov 07.

46 Langenkamp, A., et al., Kinetics of dendritic cell activation: impact on priming of TH1, TH2 and nonpolarized T cells. *Nat. Immunol.*, 2000. **1**(4), 311–6.

47 Rissoan, M.C., et al., Reciprocal control of T helper cell and dendritic cell differentiation. *Science*, 1999. **283**(5405), 1183–6.

48 Maldonado-Lopez, R., et al., Role of CD8alpha+ and CD8alpha- dendritic cells in the induction of primary immune responses in vivo. *J. Leukoc. Biol*, 1999. **66**(2), 242–6.

49 Pulendran, B., et al., Distinct dendritic cell subsets differentially regulate the class of immune response in vivo. *Proc. Natl. Acad. Sci. U.S.A.*, 1999. **96**(3), 1036–41.

50 Boonstra, A., et al., Flexibility of mouse classical and plasmacytoid-derived dendritic cells in directing T helper type 1 and 2 cell development: dependency on antigen dose and differential toll-like receptor ligation. *J. Exp. Med.*, 2003. **197**(1), 101–9.

51 Moser, M. and K.M. Murphy, Dendritic cell regulation of TH1-TH2 development. *Nat. Immunol.*, 2000. **1**(3), 199–205.

52 Moulin, V., et al., B lymphocytes regulate dendritic cell (DC) function in vivo: increased interleukin 12 production by DCs from B cell-deficient mice results in T helper cell type 1 deviation. *J. Exp. Med.*, 2000. **192**(4), 475–82.

53 Martin-Fontecha, A., et al., Induced recruitment of NK cells to lymph nodes provides IFN-gamma for T(H)1 priming. *Nat. Immunol.*, 2004. **5**(12), 1260–5. Epub 2004 Nov 07.

54 Fujii, S., et al., Activation of natural killer T cells by alpha-galactosylceramide rapidly induces the full maturation of dendritic cells in vivo and thereby acts as an adjuvant for combined CD4 and CD8 T cell immunity to a coadministered protein. *J. Exp. Med.*, 2003. **198**(2), 267–79.

55 Fujii, S., et al., The linkage of innate to adaptive immunity via maturing dendritic cells in vivo requires CD40 ligation in addition to antigen presentation and CD80/86 costimulation. *J. Exp. Med.*, 2004. **199**(12), 1607–18. Epub 2004 Jun 14.

56 Moser, M., Dendritic cells in immunity and tolerance – do they display opposite functions? *Immunity*, 2003. **19**(1), 5–8.

57 Kurts, C., et al., Class I-restricted cross-presentation of exogenous self-antigens leads to deletion of autoreactive CD8(+) T cells. *J. Exp. Med.*, 1997. **186**(2), 239–45.

58 Kurts, C., et al., Cutting edge: dendritic cells are sufficient to cross-present self-antigens to CD8 T cells in vivo. *J. Immunol.*, 2001. **166**, 1439–1442.

59 Hawiger, D., et al., Dendritic cells induce peripheral T cell unresponsiveness under steady state conditions in vivo. *J. Exp. Med.*, 2001. **194**(6), 769–79.

60 Probst, H.C., et al., Inducible transgenic mice reveal resting dendritic cells as potent inducers of CD8(+) T cell tolerance. *Immunity*, 2003. **18**(5), 713–20.

61 Dhodapkar, M.V., et al., Antigen-specific inhibition of effector T cell function in humans after injection of immature dendritic cells. *J. Exp. Med.*, 2001. **193**(2), 233–8.

62 Suss, G. and K. Shortman, A subclass of dendritic cells kills CD4 T cells via Fas/Fas-ligand-induced apoptosis. *J. Exp. Med.*, 1996. **183**(4), 1789–96.

63 Grohmann, U., et al., IFN-gamma inhibits presentation of a tumor/self peptide by CD8 alpha- dendritic cells via potentiation of the CD8 alpha+ subset. *J. Immunol.*, 2000. **165**(3), 1357–63.

64 Belz, G.T., et al., The CD8alpha(+) dendritic cell is responsible for inducing peripheral self-tolerance to tissue-associated antigens. *J. Exp. Med.*, 2002. **196**(8), 1099–104.

65. Liu, K., et al., Immune tolerance after delivery of dying cells to dendritic cells in situ. *J. Exp. Med.*, 2002. **196**(8), 1091–7.
66. Maldonado-Lopez, R., et al., CD8alpha+ and CD8alpha− subclasses of dendritic cells direct the development of distinct T helper cells in vivo. *J. Exp. Med.*, 1999. **189**(3), 587–92.
67. Reis e Sousa, C., et al., In vivo microbial stimulation induces rapid CD40 ligand-independent production of interleukin 12 by dendritic cells and their redistribution to T cell areas. *J. Exp. Med.*, 1997. **186**(11), 1819–29.
68. de Heer, H.J., et al., Essential role of lung plasmacytoid dendritic cells in preventing asthmatic reactions to harmless inhaled antigen. *J. Exp. Med.*, 2004. **200**(1), 89–98.
69. Fugier-Vivier, I.J., et al., Plasmacytoid precursor dendritic cells facilitate allogeneic hematopoietic stem cell engraftment. *J. Exp. Med.*, 2005. **201**(3), 373–83.
70. Zhang, M., et al., Splenic stroma drives mature dendritic cells to differentiate into regulatory dendritic cells. *Nat. Immunol.*, 2004. **5**(11), 1124–33. Epub 2004 Oct 10.
71. Hou, W.S. and L. Van Parijs, A Bcl-2-dependent molecular timer regulates the lifespan and immunogenicity of dendritic cells. *Nat. Immunol.*, 2004. **5**(6), 583–9. Epub 2004 May 9.
72. Svensson, M., et al., Stromal cells direct local differentiation of regulatory dendritic cells. *Immunity*, 2004. **21**(6), 805–16.
73. Andris, F., et al., Naive T cells are resistant to anergy induction by anti-CD3 antibodies. *J. Immunol.*, 2004. **173**(5), 3201–8.
74. Schwartz, R.H., T cell anergy. *Annu. Rev. Immunol.*, 2003. **21**, 305–34. Epub 2001 Dec 19.
75. Munn, D.H., et al., Potential regulatory function of human dendritic cells expressing indoleamine 2,3-dioxygenase. *Science*, 2002. **297**(5588), 1867–70.
76. Kurts, C., et al., The peripheral deletion of autoreactive CD8+ T cells induced by cross-presentation of self-antigens involves signaling through CD95 (Fas, Apo-1). *J. Exp. Med.*, 1998. **188**(2), 415–20.
77. Lindquist, R.L., et al., Visualizing dendritic cell networks in vivo. *Nat. Immunol.*, 2004. **5**(12), 1243–50. Epub 2004 Nov 14.
78. Hugues, S., et al., Distinct T cell dynamics in lymph nodes during the induction of tolerance and immunity. *Nat. Immunol.*, 2004. **5**(12), 1235–42. Epub 2004 Oct 31.
79. Akbari, O., R.H. DeKruyff, and D.T. Umetsu, Pulmonary dendritic cells producing IL-10 mediate tolerance induced by respiratory exposure to antigen. *Nat. Immunol.*, 2001. **2**(8), 725–31.
80. Yamazaki, S., et al., Direct expansion of functional CD25+ CD4+ regulatory T cells by antigen-processing dendritic cells. *J. Exp. Med.*, 2003. **198**(2), 235–47.
81. Lohr, J., et al., The inhibitory function of B7 costimulators in T cell responses to foreign and self antigens. *Nat. Immunol.*, 2003. 4, 664–669.
82. Oldenhove, G., et al., CD4+ CD25+ regulatory T cells control T helper cell type 1 responses to foreign antigens induced by mature dendritic cells in vivo. *J. Exp. Med.*, 2003. **198**(2), 259–66.
83. Liang, S., et al., Conversion of CD4+ CD25- cells into CD4+ CD25+ regulatory T cells in vivo requires B7 costimulation, but not the thymus. *J. Exp. Med.*, 2005. **201**(1), 127–37.
84. Inaba, K., et al., The tissue distribution of the B7-2 costimulator in mice: abundant expression on dendritic cells in situ and during maturation in vitro. *J. Exp. Med.*, 1994. **180**(5), 1849–60.
85. Grohmann, U., et al., CTLA-4-Ig regulates tryptophan catabolism in vivo. *Nat. Immunol.*, 2002. **3**(11), 1097–101.
86. Fallarino, F., et al., Modulation of tryptophan catabolism by regulatory T cells. *Nat. Immunol.*, 2003. **4**(12), 1206–12. Epub 2003 Oct 26.
87. Orabona, C., et al., CD28 induces immunostimulatory signals in dendritic cells via CD80 and CD86. *Nat. Immunol.*, 2004. **5**(11), 1134–42. Epub 2004 Oct 03.

88 Pasare, C. and R. Medzhitov, Toll pathway-dependent blockade of CD4+CD25+ T cell-mediated suppression by dendritic cells. *Science*, 2003. **299**(5609), 1033–6.

89 Ingulli, E., et al., In vivo detection of dendritic cell antigen presentation to CD4(+) T cells. *J. Exp. Med.*, 1997. **185**(12), 2133–41.

90 Stoll, S., et al., Dynamic imaging of T cell-dendritic cell interactions in lymph nodes. *Science*, 2002. **296**, 1873–1876.

91 Bedford, P., K. Garner, and S.C. Knight, MHC class II molecules transferred between allogeneic dendritic cells stimulate primary mixed leukocyte reactions. *Int. Immunol.*, 1999. **11**(11), 1739–44.

92 Kleindienst, P. and T. Brocker, Endogenous dendritic cells are required for amplification of T cell responses induced by dendritic cell vaccines in vivo. *J. Immunol.*, 2003. **170**(6), 2817–23.

93 Thery, C., et al., Indirect activation of naive CD4+ T cells by dendritic cell-derived exosomes. *Nat. Immunol.*, 2002. **3**(12), 1156–62. Epub 2002 Nov 11.

94 Albert, M.L., B. Sauter, and N. Bhardwaj, Dendritic cells acquire antigen from apoptotic cells and induce class I-restricted CTLs. *Nature*, 1998. **392**(6671), 86–9.

95 Guermonprez, P., et al., ER-phagosome fusion defines an MHC class I cross-presentation compartment in dendritic cells. *Nature*, 2003. **425**(6956), 397–402.

96 Ackerman, A.L., et al., Access of soluble antigens to the endoplasmic reticulum can explain cross-presentation by dendritic cells. *Nat. Immunol.*, 2005. **6**(1), 107–13. Epub 2004 Dec 12.

97 Houde, M., et al., Phagosomes are competent organelles for antigen cross-presentation. *Nature*, 2003. **425**(6956), 402–6.

98 Schulz, O., et al., Toll-like receptor 3 promotes cross-priming to virus-infected cells. *Nature*, 2005. **13**, 13.

99 Mora, J.R., et al., Reciprocal and dynamic control of CD8 T cell homing by dendritic cells from skin- and gut-associated lymphoid tissues. *J. Exp. Med.*, 2005. **201**(2), 303–16. Epub 2005 Jan 10.

100 Schuler-Thurner, B., et al., Rapid induction of tumor-specific type 1 T helper cells in metastatic melanoma patients by vaccination with mature, cryopreserved, peptide-loaded monocyte-derived dendritic cells. *J. Exp. Med.*, 2002. **195**(10), 1279–88.

101 Schultz, E.S., et al., Functional analysis of tumor-specific Th cell responses detected in melanoma patients after dendritic cell-based immunotherapy. *J. Immunol.*, 2004. **172**(2), 1304–10.

102 Paczesny, S., et al., Expansion of melanoma-specific cytolytic CD8+ T cell precursors in patients with metastatic melanoma vaccinated with CD34+ progenitor-derived dendritic cells. *J. Exp. Med.*, 2004. **199**(11), 1503–11. Epub 2004 Jun 01.

103 Lurquin, C., et al., Contrasting frequencies of antitumor and anti-vaccine T cells in metastases of a melanoma patient vaccinated with a MAGE tumor antigen. *J. Exp. Med.*, 2005. **201**(2), 249–57.

104 Germeau, C., et al., High frequency of antitumor T cells in the blood of melanoma patients before and after vaccination with tumor antigens. *J. Exp. Med.*, 2005. **201**(2), 241–8.

105 Wang, H.Y., et al., Tumor-specific human CD4+ regulatory T cells and their ligands: implications for immunotherapy. *Immunity*, 2004. **20**(1), 107–18.

15
Thymic Dendritic Cells

Ken Shortman and Li Wu

15.1
Thymic Dendritic Cells

The major if not sole function of the thymus is the production and selection of T lymphocytes [1, 2]. Accordingly, the function of thymus cells not of the T lineage is interpreted in terms of their role in inducing or supporting the complex series of developmental steps involved in the production of T cells. For thymic dendritic cells (DC) there is little evidence of any role in the early steps of T cell development from bone-marrow derived precursors. Rather, thymic DC, as specialised antigen presenting cells, have been implicated in the later steps of T-cell receptor repertoire selection. The finding that thymic DC appear during ontogeny about the same time as $CD4^+8^+$ thymocytes supports a role in repertoire selection [3]. Negative selection, the induction of apoptotic death in potentially self-reactive developing T cells, is the usual role ascribed to thymic DC [4–6]. However, the relative importance of negative selection in preventing autoimmune responses, compared to the generation of regulatory T cells or deflection of cortical thymocytes to alternative T cell developmental states, is currently being reassessed. Thymic DC are likely to be involved in these additional tolerogenic mechanisms.

In this chapter we will first consider the subtypes of DC found within the thymus and will compare them to the DC in peripheral lymphoid tissues. Next, we will consider the origin of thymic DC and the developmental steps that produce them. Finally, we will discuss the role of thymic DC in T cell repertoire selection and tolerance. Most data will be derived from mice, but we will include, where possible, work on human thymus.

15.2
Localisation and Isolation of Thymic DC

The bone-marrow derived component of the thymus, although dominated by T-lineage cells, also includes a low level of B cells, macrophages and DC. Most thymic DC are localised in the medulla or at the cortico-medullary junction in the mouse [4, 6], as in the rat [7, 8]. Although the proportion of DC in the thymus overall is only around 0.5%, lower than in other lymphoid organs, within the medulla itself the incidence of DC would be comparable to that in the spleen or lymph nodes.

The reduced incidence of DC in thymus suspensions makes DC isolation more demanding than from other tissues. A further problem is that, on preparing a thymus cell suspension by mechanical means or by collagenase digestion, the majority of the thymic DC remain associated with T-lineage thymocytes in the form of "rosettes", consisting of a central DC with multiple attached thymocytes [9, 10]. These rosettes must be disassociated by EDTA treatment (to chelate Ca^{2+} and Mg^{2+}) and mild shear force, to ensure segregation of DC and T-lineage cells. Centrifugation in an iso-osmotic density medium to select the 3–5% lightest cells is then an effective first step in enriching the DC component [11]. Immunomagnetic depletion of T-lineage and other non-DC lineage cells is a useful second enrichment step; however, care must be taken not to deplete some DC because of Fc-receptor binding of depleting antibodies, or because of DC pickup of T-lineage thymocyte antigens [11]. If a completely pure preparation of thymic DC is required, there is at present no substitute for fluorescence-activated cell sorting using at least two DC-selective markers [11].

15.3
Pickup of Antigens by Thymic DC

The isolation of thymic DC and the separation into subtypes is usually based on fluorescent staining or particular surface components. The marked tendency of thymic DC to pick-up surface antigens from T-lineage thymocytes is a major hazard in subsequent flow-cytometric analyses or sorting. These surface components, probably in the form of fragments of thymocyte membranes, appear to remain associated with the DC when the DC-thymocyte rosette complexes are dissociated. CD4, $CD8\alpha\beta$ and Thy-1 are all picked up by the DC from thymocytes [11, 12]. Although the level of staining for such acquired antigens is generally lower than for authentic DC components, it is sufficient to muddy the analyses [11]. In some cases an incubation in culture is sufficient to remove much of the acquired antigen. However, full verification that the marker is a product of the DC has involved two sorts of tests. The first is to determine whether the DC contain mRNA coding for the markers stained on the DC surface. The second is to check that a given allotype of the marker is found only on the DC of the right genetic origin, when chimeric thymuses are constructed by transferring into irradiated mice bone marrow from two strains of mice differing in the marker allotype [11, 12].

15.4
Subtypes of Thymic DC

We can define a distinct DC subtype as a DC with a unique pattern of surface markers and which is not the immediate precursor or product of another DC subtype. In this definition we exclude separate activation states as representing separate DC subtypes. However, the DC subtypes so defined may still be closely related, perhaps having branched off during development from a common precursor. Using such a definition we can segregate at least five subtypes of conventional DC (cDC) in peripheral lymphoid tissues of the mouse [11, 13, 14], as well as distinguishing the class 1 interferon producing plasmacytoid cell (pDC) which can function as a DC precursor [15–17]. In many cases there are clear functional differences associated with this division based on DC surface markers [18]. The same approach leads to three subtypes of DC-lineage cells within the thymus, and each has a close equivalent within the peripheral DC subtypes.

15.5
Major Thymic cDC Population

The total cDC of mouse thymus, representing about 0.3% of thymocytes, can be isolated as $CD11c^{hi}$, $MHC\ II^{hi}$, $CD45RA^-$ or $CD45R^-$ cells (Figure 15.1A). Most of these are $CD8^+$, the proportion varying between around 75% in C57BL/6 mice to around 90% in Balb c mice [11, 19–21]. The CD8 is in the form of the $CD8\alpha\alpha$ homodimer and the DC express the mRNA for $CD8\alpha$ [11, 19]. However, the DC may also have on their surface a smaller amount of $CD8\alpha\beta$, picked up from associated thymocytes [11]. The staining for $CD8\alpha$ may be as high as on T cells, although there are lower staining cells, which include some less developed $CD8\alpha\alpha^+$ DC as well as DC that lack $CD8\alpha\alpha$ production but pick up surface $CD8\alpha\beta$ from thymocytes (Figure 15.1B). This $CD8^+$ population does not express CD4, although there may be significant pickup of CD4 from thymocytes [11]. This majority population of thymic DC resembles in many respects the minority (20%) subset of $CD8^+$ spleen and lymph node DC, in being $CD205^+$ and $CD11b^-$. However, one marked difference is that a proportion of thymic $CD8^+$ DC, but not of peripheral $CD8^+DC$, stain for BP-1 (6C3) and express mRNA for BP-1 [21, 22]. To date no other differences have been noted between $BP-1^+$ and $BP-1^-$ thymic $CD8^+DC$.

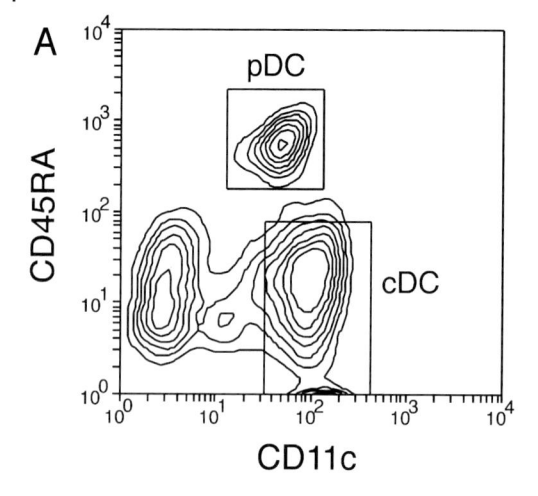

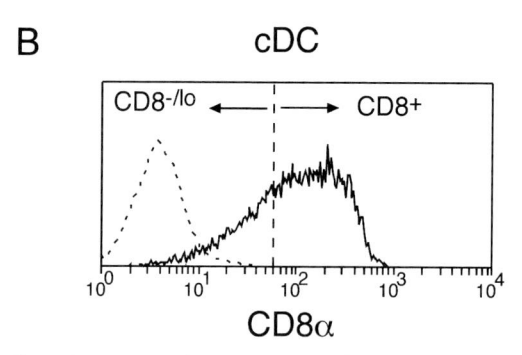

Figure 15.1 Mouse thymic conventional DC (cDC) and plasmacytoid DC (pDC) populations. Thymic DC were separated as light density cells, then non-DC cells were depleted using immunomagnetic beads. The enriched DC preparation was fluorescence stained for CD45RA, CD11c and CD8α. cDC were selected as CD11c^{hi} CD45RA$^-$, and pDC selected as CD11c^{int} CD45RA$^+$, using the gates shown in part (A). (B) Staining of cDC for CD8α. Most bright-staining cells are CD8αα$^+$. The tail of low staining cells (CD8$^{-/lo}$) includes some cells with low CD8αα expression and many cells that are CD8αα$^-$ but have picked up CD8αβ from thymocytes.

15.6
Minor Thymic cDC Population

Mouse thymus contains a minority population of cDC (around 25% in C57BL/6 mice, less in Balb c mice) that does not synthesise and express CD8αα on its cell surface [11, 20]. Recognising and separating this subtype is particularly difficult because of the pick up of surface CD8αβ from thymocytes, so they appear as a lower staining shoulder on a CD8α fluorescence histogram [11] (Figure 15.1B). However, better segregation can be obtained by two-parameter staining for CD8α and Sirp α, where the minority CD8α^{low} Sirp α^+ population stands out (Wu and Lahoud, unpublished). Despite these problems in separation, gene expression profiles indicate clear differences between this minor subset and the majority CD8$^+$ thymic cDC (Lahoud and Shortman unpublished).

Because of all the difficulties in defining and analyzing this minor thymic cDC subset, it is not yet clear whether it represents a single CD8$^-$ DC subtype with a special intrathymic role, or whether it is simply an indicator of the entry into the thymus from the blood stream of a number of peripheral DC subtypes.

15.7
Thymic pDC

About 35% of the DC lineage cells in the thymus are the natural interferon producing plasmacytoid pre-dendritic cells, pDC, which closely resemble the pDC found in peripheral lymphoid organs. These pDC can be separated from a DC-enriched preparation as CD11c^{medium} MHC II^{low}, CD45RA^{high}, CD45R^{high} cells [22]. They are round cells with plasmacytoid rather than dendritic morphology, but transform into dendritic morphology on activation in culture [6]. Like their counterparts in the periphery, thymic pDC are mainly Ly6C$^+$; it remains to be established whether the small Ly6C$^-$ component is either an earlier developmental stage of pDC or is a different lineage entirely. A proportion of thymic pDC express CD4 and/or CD8α; in this respect they resemble spleen or lymph node pDC rather than bone-marrow or blood pDC, which are CD4$^-$ CD8$^-$ [16, 17]. However, as with splenic pDC, the expression of CD4 and CD8α appears to represent different activation or developmental states of the one pDC lineage, rather than marking discrete subtypes as for cDC [16]. Like the pDC in the periphery, and in contrast to cDC, thymic pDC express high levels of toll-like receptors (TLR) 7 and 9, but only low levels of TLR 2–4 [22]. Accordingly, thymic pDC respond to the TLR9 ligand CpG, but not to *Escherichia coli* lipopolysaccharide [16, 22].

15.8
Maturation State and Antigen Processing Capacity of Thymic DC

The cDC of the thymus are in a non-activated or "immature" state, similar to those in the spleen of a steady-sate mouse [23]. Thus, although they express moderate levels of surface MHC II, most of the MHC II is within the cell in endosomes and only shifts to the surface after deliberate activation ([23], Wilson, Villadangos and Shortman, unpublished). Although thymic cDC already show some surface expression of co-stimulatory molecules, this is low compared to the high levels achieved on deliberate activation [11, 23]. However, the major CD8$^+$ thymic cDC population generally appears slightly more activated than their peripheral CD8$^+$ cDC counterparts. The modal levels of the co-stimulatory molecules CD80 and CD86 are a little higher and a few thymic DC express relatively high levels.

However, despite this apparent small shift towards a more "activated" state in the terms of surface markers, thymic CD8$^+$ DC, like their counterparts in the periphery in steady-state mice, have the antigen uptake and processing properties considered characteristic of "immature" DC. They are capable of phagocytic or endocytic uptake of antigens, and of processing antigens for presentation on MHC II. The CD8$^+$ thymic DC, like their peripheral counterparts, are also efficient at cross-presenting antigens on MHC I ([23]; Wilson, Villadangos and Shortman, unpublished).

The thymic pDC are also in an "immature" maturation state, very similar to their peripheral pDC counterparts. They have only low levels of surface MHC II and very low surface expression of co-stimulatory molecules [16, 22]. They have little phagocytic ability but are capable of endocytosis. Although they can process for MHC presentation antigens that enter the cell by endocytosis, and can process endogenous proteins, in their non-activated state they are very poor stimulators of T cells (Wilson, Villadangos and Shortman, unpublished). However, once activated by microbial stimuli, such as oligonucleotides containing bacterial DNA CpG motifs, they can activate T cells into proliferation [16, 22].

15.9
Cytokine Production by Thymic DC

A special capacity of thymic DC to make particular cytokines would be a pointer to an intrathymic role beyond antigen presentation. However, results with the few cytokines surveyed so far suggest there is little cytokine production in situ and that the capacity of thymic DC to produce cytokines on appropriate activation is very similar to their peripheral DC counterparts. Thus thymic CD8$^+$ cDC, if stimulated with CpG oligonucleotides or polyI-C or CD40-ligand, together with appropriate cytokines (GM-CSF, IFN-γ and IL-4), will produce very high levels of bioactive IL-12p70, much more than CD8$^-$ cDC although somewhat less than splenic CD8$^+$ cDC [24, 25]. Thymic CD8$^-$ DC, in the presence of IL-12 and IL-18, will produce IFN-γ [24]. Thymic pDC, if stimulated with CpG oligonucleotides, will produce the extremely high levels of type 1 interferons characteristic of splenic pDC [16, 22]. They will also produce IL-6 and moderate levels of IL-12p70 ([16], O'Keefe and Shortman unpublished). In all these cases the cytokines are produced only after experimental activation and the production by freshly isolated thymic DC in culture without the stimuli is very low. Even when the potentially more activated small subgroup of CD8$^+$ thymic cDC expressing relatively high levels of MHC II was isolated, it produced little IL-12p70 without additional microbial stimulus (Hochrein, O'Keefe and Shortman, unpublished). Thus to date there is no evidence for a significant cytokine production by freshly isolated thymic DC, although they have a potential for cytokine production if appropriately stimulated.

15.10
DC of the Human Thymus

The DC of human thymus, like those of rodent thymus, are localised mainly in the medulla and at the cortico-medullary junction [8, 26–28]. They may be isolated by procedures basically similar to those used for mouse thymus, but with changes in the depleting and positive sorting antibody specificities, in accordance with the differences in surface antigen expression between mouse and human DC [27]. A major difference is that human thymic DC do not express CD8, but do express CD4 [8, 27, 29, 30]. Despite these differences, a range of other markers and functional properties suggest human thymus contains three distinct DC populations [8, 27] and that these correspond closely to the three populations in the mouse thymus.

The cDC may be sorted from enriched human thymus DC preparations on the basis of some combinations of the following markers: HLA-DR$^+$, CD11c$^+$, CD4$^+$, CD45RA^{low} [8, 27]. These cDC may then be divided into two subgroups, most readily on the basis of CD11b expression. The major subgroup (CD11b$^-$) expresses higher levels of co-stimulatory molecules, notably CD86, and expresses several markers characteristic of interdigitating DC, such as DC-LAMP [8, 27]. This subgroup has many of the functional markers of the major mouse thymus CD8α^+ DC population, including expression of the chemokine TECK and the production of large amounts of IL-12p70 when appropriately stimulated [27]. In contrast, the more myeloid-like, minor CD11b$^+$ subgroup of cDC expresses lower levels of co-stimulatory molecules, has several myeloid features also found in monocyte-derived DC and germinal centre DC, and produces little if any TECK or IL-12p70 [8, 27]. Both thymic cDC populations are effective stimulators of T cells in an allogenic mixed leucocyte reaction [27].

Human thymus also contains pDC resembling closely the pDC in human peripheral lymphoid tissues and the pDC in the mouse thymus [8, 27, 28, 31]. They are found along with cDC in the medulla and the cortico-medullary junction, but some are also present in the cortex [8, 27–29]. Human thymic pDC may be sorted from an enriched thymic DC preparation and distinguished from cDC as being HLA-DR^{low}, CD45RA$^+$, CD11c$^-$ and IL-3Rα^+ (CD123$^+$). The IL-3Rα level is higher than on mouse pDC and sufficiently high to serve as an immunofluorescent marker. Human thymic pDC also express a number of lymphoid-related RNA transcripts, including pre-Tα and Spi-B [8, 28]. On activation in culture with IL-3 and CD40-ligand, the pDC develop into mature DC, as do their peripheral pDC equivalents [8, 27, 28]. On activation in culture with influenza virus, these pDC produce large amounts of IFN-α; under these stimulation conditions human thymic cDC do not produce IFN-α [8]. The capacity on activation to transform into DC and also to produce type 1 interferons indicates that these human thymic pDC are the functional equivalents of the pDC in mouse thymus and of the pDC in peripheral lymphoid tissues in both species.

15.11
Turnover Rate and Lifespan of the Thymic DC

Studies on the regeneration of an irradiated thymus following transfer of precursor cells suggested that, once generated, thymic DC survive around 15 days [21]. More accurate estimates of thymic cDC turnover in steady-state adult mice have been obtained based on the kinetics of bromodeoxyuridine uptake into DC DNA. Thymic cDC are not themselves dividing, but the DNA label first enters a dividing precursor and thence the DC product pool. In such studies all unlabeled thymic DC were replaced by labeled cells after 10 days of continuous bromodeoxyuridine administration [32]. This 10 day lifespan for the bulk of the thymic cDC is roughly in line with the average turnover rate of the T-lineage cells in the thymus, but is slower than the turnover of the cDC in the spleen. However, the detailed labeling kinetics also revealed a subgroup of around 20% of thymic cDC that showed a more rapid turnover of around 3 days, which is very similar to spleen cDC [32]. The turnover of pDC in the thymus has not been investigated; however, the similar pDC in the spleen are relatively long lived with a lifespan over 14 days [16].

15.12
Endogenous versus Exogenous Sources of Thymic DC

Thymic DC have their ultimate origin in haematopoietic stem cells; purified bone-marrow multipotent stem cells have been shown to reconstitute the cDC of an irradiated thymus [33]. However, the immediate origin of thymic DC is an important issue in relating them to the rest of the DC network. Are thymic DC renewed by continuous input of fully developed or near-fully developed DC from the bloodstream, or are thymic DC generated within the thymus itself, from an early precursor cell, along with the T-lineage thymocytes? Current evidence from studies on mice, discussed below, suggests that the major thymic cDC population is generated within the thymus, whereas the minor thymic cDC population derives from the bloodstream. The immediate origin of the thymic pDC is still uncertain, but human thymus contains both the precursor cells and an appropriate environment for pDC generation, suggesting an endogenous origin [34].

The concept that thymic DC might be generated from an endogenous precursor derived from the then surprising finding that the earliest recognisable T-precursor population purified from the adult mouse thymus, the so-called "low CD4 precursor" [35], was also capable of producing DC [33]. On transfer into the thymus of irradiated recipients this precursor formed both T-lineage thymocytes and $CD8^+$ cDC [17, 21, 33, 36]; the ratio of these products suggested the precursors were able to account for the production of most thymic DC. These T-precursors were also able to generate cDC in culture [37].

This picture of the generation of thymic $CD8^+$ cDC from an early thymocyte precursor cell, and so linked to the generation of the T-lineage, has been reinforced by a careful study of thymus precursor cell seeding and thymus population dynamics, using normal steady-state mice, parabiotic mice with linked bloodstreams, and irradiation depleted mice [38]. Most of the CD8$^+$ thymic cDC population showed developmental dynamics indicating an early pro-thymocyte origin, including cycles of gated precursor seeding into the thymus, competition for developmental niches and lack of exchange between parabiotic partners.

However, this same study [38] revealed a second developmental stream leading to thymic DC, a stream with totally different developmental dynamics and fitting the model of continuous importation of more developed DC from the bloodstream. This developmental stream displayed extensive exchange between parabiotic partners and did not follow the gated kinetics of development characteristic of T-lineage cells. The products of this stream included CD8$^-$ cDC. Such a developmental stream of blood origin could be the source of the minor thymic cDC population, although it was estimated to generate 50% rather than 20% of thymic DC. It might also correspond to the subgroup of thymic cDC showing a fast, spleen DC-like, turnover rate in the labeling kinetic studies described above.

15.13
Lineage Relationship and Differentiation Pathways of Thymic cDC

The finding of an early thymus cDC precursor with the characteristics of a lymphoid-restricted precursor first raised the possibility that some DC could be of lymphoid rather than myeloid origin [33]. The further observation that generation of DC in culture from this precursor required a set of cytokines that did not include the "myeloid hormone" GM-CSF added weight to this view [37]. However, the generalisation that all CD8 bearing murine DC, whether thymic or peripheral, were of lymphoid-precursor origin proved to be incorrect. It is now clear that all subtypes of DC can be generated from either a myeloid or a lymphoid precursor [39–41], particularly from those that express Flt-3 [41], and that these "restricted" precursor populations have a degree of developmental flexibility. In the periphery, the predominance of myeloid-restricted precursors makes these the likely source of most DC. In the thymus, the reverse is likely to apply, at least for the CD8$^+$ cDC that derive from endogenous precursors.

How closely is thymic cDC development linked to T-cell development? The thymic "low CD4 precursor", which has T-cell, B-cell and NK-cell, but little myeloid potential, is effectively a lymphoid-restricted precursor population with a capacity to form cDC [42]. This population has not yet rearranged T-cell receptor (TCR) genes. The next downstream T-precursor retains some capacity to produce cDC, but this capacity is lost once TCR gene rearrangement commences [36]. This suggests that thymic cDC development branches from T-development just prior to TCR-gene rearrangement. However, although these highly purified thymic T-precursor populations appeared homogeneous, it remained possible that they consisted of a mix of separate but phenotypically similar precursors, each committed to a separate lineage. The very high (70%) cloning efficiency of DC generation from single precursor cells in culture argued against this [37]. However, in some mice in which T-lineage development is blocked, thymic cDC development is maintained, indicating that T cell development and thymic cDC development can be "unlinked". This was shown for mice lacking c-kit and the cytokine receptor common γ chain [43] and for a conditional knockout of the Notch 1 gene [44]. Although these results argue against a common DC/T precursor, they could also be explained by a blockage of only the T-cell lineage branch, or by the existence of alternative lymphoid or myeloid routes to DC formation. This uncertainty prompted a search for markers of a lymphoid or myeloid origin that could be applied to normal, steady-state mice.

The cDC of the normal adult mouse thymus do show low expression at the RNA level of several genes characteristic of T-lineage cells, including CD3 chains and pre-Tα [45]. These may be relics of a lymphoid past. More convincing is the finding that many thymic $CD8^+$ cDC have, like T-lineage cells, D-J but not V-D-J rearrangements of their IgH genes [45]. This would require not only mRNA transcription but also protein translation for the RAG-1, RAG-2 and other genes, then the functional activity of this recombinase system, leaving an indelible marker in the DNA of all subsequent progeny. It is a strong argument that these thymic DC derive from a pathway common to B and T cells. At the very least, these thymic cDC have had in their past a series of developmental events entirely characteristic of lymphoid cells. However, it is important to note that not all thymic $CD8^+$ cDC displayed these IgH D-J rearrangements, that the minor thymic $CD8^-$ cDC had much reduced in incidence, and that few if any peripheral cDC displayed such rearrangements [45]. Those DC lacking IgH gene rearrangements could well be of myeloid origin, although we lack a positive marker of a myeloid past.

In human thymus studies of thymic cDC development have, of necessity, been mainly based on culture systems allowing the development of cDC from early $CD34^+$ precursors from fetal or post-natal thymus [46–48]. These studies have demonstrated the potential for cDC to be generated from an early intrathymic precursor, believed to also generate the T lineage cells. The $CD34^+$ precursor population produced both T-cell and cDC progeny when transferred into human thymus grafts in lymphoid deficient mice [49]. The cloning efficiencies from single precursor cell cultures demonstrated common NK/T precursors and NK/DC precursors [46–48] and common T/NK/DC precursors have been proposed. Overall, this suggests that in the human thymus, as in the mouse, many thymic cDC derive from a lymphoid-restricted developmental pathway.

15.14
Lineage Relationships and Developmental Pathways of Thymic pDC

The developmental pathway of thymic pDC, as of pDC in peripheral tissues, presents even more of a challenge to traditional views of haemopoiesis [50]. Thymic and peripheral pDC, both human and mouse, express a range of lymphoid and particular B-cell related gene products [8, 28, 45, 50, 51]. In the mouse, they also carry D-J but not V-D-J rearrangements of IgH genes [45, 51]. This applies regardless of whether pDC derive from lymphoid or myeloid restricted precursors. Whether this reflects a remarkable developmental plasticity in the activation of lymphoid-related genes, or too rigid an initial view of the developmental potential of the isolated precursor populations, remain to be settled.

The developmental relationship between the pDC and the cDC of the thymus is not clear. They appear to be products of separate sub-lineages, but the possibility that one is the product of the other has not been excluded. The mouse thymus "low CD4" precursor, which produces both T cells and cDC, is a very poor producer of pDC [41]. This suggests either that thymic pDC branch off earlier than thymic cDC from a common T-lymphocyte pathway or that they are of entirely separate origin. However, in a system where human $CD34^+$ thymus precursor cells were transferred into human thymus grafts in a lymphoid-deficient mouse host, the precursor cells produced, as well as T cells, both pDC and cDC [49]. This strongly supports the view that human thymic pDC develop within the thymus from an intrathymic precursor, and suggests a common lymphoid origin for both the pDC and the cDC.

15.15
Thymic cDC do not Mediate Positive Selection

Radiation-resistant thymic stromal cells, rather than bone-marrow derived elements, have generally been considered to dictate positive selection. Numerous in vivo studies point to the thymic cortical epithelium as the site [52]. Nevertheless, as efficient antigen-presenting cells, thymic cDC have the potential to contribute to positive selection on self-MHC-peptide complexes. A strong argument against such a role is location: most cDC are in the medullary zone, while the $CD4^+8^+$ thymocytes being selected are in the cortex. There is also direct experimental evidence that thymic cDC do not cause positive selection. Embryonic thymus re-aggregation cultures, where individual cell types can be added and the effects on T-cell development monitored, indicate that epithelial cells, but not cDC, can induce positive selection [53, 54]. Finally, selective expression of MHC class II IE molecules on thymic DC, under the control of the CD11c promoter, did not produce positive selection for responses to IE presentation [55].

15.16
Thymic cDC and Negative Selection

A current model of thymic T-cell receptor repertoire selection by self-MHC-peptide presentation proposes that a low-affinity interaction initiates positive selection and survival, a high-affinity interaction initiates negative selection and death, and no interaction at all leads to programmed apoptosis, death by neglect. There are, however, several problems in monitoring the negative selection component and determining the role played by thymic DC. The first is the existence of a series of developmental options for self-reactive CD4$^+$8$^+$ thymocytes, which may deflect cortical thymocytes from the classical circulating $\alpha\beta$T cell pathway, or which may lead to $\alpha\beta$T cells with blocked immune responses, without inducing intrathymic death. These include the production of regulatory (suppressor) T cells, the induction of anergic T cells, and the deflection of CD4$^+$8$^+$ thymocytes to intestinal CD8$\alpha\alpha$ T cells [56, 57]. Some of these alternative developmental fates are probably induced by interaction with epithelial cells rather than DC. The second problem is that the level of apoptotic death due to negative selection is hard to estimate [58]. Apoptotic thymocytes do not stay around to be counted and are rapidly eliminated by thymic macrophages. In addition, it is likely that most apoptotic death in the thymus is due to neglect, so the component caused by negative selection is hard to estimate against this large background. Thirdly, many other cell types in the thymus have the potential to present self-MHC and induce negative selection, including cortical and medullary epithelium, B cells, macrophages and even CD4$^+$8$^+$ thymocytes [59–64]. The final problem is the anatomical separation of the cells proposed to play a role in the repertoire selection process. The epithelial cells mediating the initial selective events are in the cortex, interacting with CD4$^+$8$^+$ thymocytes, whereas the cDC are in the medullary zone where they will encounter thymocytes that have already undergone positive selection and are in a different physiological state.

Despite these complexities, it does some likely that thymic cDC do play an important role in eliminating self-reactive T cells. They are certainly efficient at negative selection in systems when they are the only cell presenting the model self-antigen or self MHC. Thus, in foetal thymus reaggregation cultures, added DC, but not cortical epithelium, induced negative selection [54]. Selective expression of MHC class II I-E molecules in DC resulted in the deletion of I-E reactive CD4 T cells from the repertoire [55]. Experiments involving mixed bone marrow chimeras have demonstrated that around 30% of the bone-marrow derived (presumed cDC) cells that present the selecting I-E molecules are required to ensure complete negative selection [65]. Presumably, this incidence is required to ensure all developing thymocytes have an effective encounter with the selecting cells.

However, in the thymus of a normal steady state mouse, where many antigen-presenting cell types in the thymus present self-antigens, it is unclear what component of the overall negative selection events are then carried out by cDC. Some negative selection will presumably occur in the cortex, on interaction with cortical epithelium. At this stage the TCR levels on the CD4$^+$8$^+$ thymocytes are low. After positive selection, when TCR levels are elevated, potential reactivity with self MHC-self peptides will be increased. It therefore seems logical to have a second network of negatively selecting elements in the medulla and at the cortico-medullary junction, to cope with a potentially enhanced self-reactivity. Thymic cDC are well positioned for this role. The recently positively selected, "semi-mature" subset of medullary T cells may be particularly susceptible to negative selection signals [66]. However, in the medulla another cell type, the medullary epithelial cell, which also expresses MHC II, appears to share the overall negative selection load. Medullary epithelium presents a mosaic of peripheral-tissue specific proteins, including several hormones, associated with expression of the AIRE gene [62, 63, 67–70]. Although cDC do express low levels of AIRE, the medullary epithelium expresses higher levels and appears specialised for presentation of this subset of peripheral self antigens. The major thymus subtype, the $CD8^+$ cDC in the mouse, being generated within the thymus rather than migrating in from the periphery, should present mainly endogenous self antigens for negative selection. However, being a $CD8^+$ cDC, it also has a capacity to take up and cross present antigens derived from other thymus cells undergoing apoptosis.

The cDC that enter the thymus via the blood stream, which probably include the minor $CD8^-$ thymic cDC, may have a different role. Such DC have the potential to carry into the thymus both self and foreign antigens. cDC carrying antigens introduced into a peripheral site (the eye) have been found to migrate to the thymus [71]. These immigrant DC have been proposed to generate a lineage of regulatory T cells on interaction with developing thymocytes. This would enable the thymus to play a dynamic role in the regulation of peripheral immune responses to both self and foreign antigens.

15.17
Role of pDC in the Thymus

Although the details are unclear, there are obvious potential roles for thymic cDC in central tolerance mechanisms, as components of a negative selection network and as inducers of T cell anergy and of regulatory T cells. The function of thymic pDC is, by contrast, a complete mystery. They too could be involved in the negative selection. However, they are very poor presenters of antigens unless subject to microbial stimulation, and so this seems unlikely. They might produce cytokines such as IFN-α, and so influence the development and function of cDC. As the production of cytokines, though, requires microbial stimulation, this also seems unlikely. The function of pDC within the thymus, as in other tissues, may simply be to protect the tissue from viral infections.

References

1 J. F. A. P. Miller, D. Osoba, *Physiol. Rev.* **1967**, *47*, 437–520.
2 C. R. M. J. Janeway, *Immunity* **1994**, *1*, 3–6.
3 A. Dakic, Q. X. Shao, A. D'Amico, M. O'Keeffe, W. F. Chen, K. Shortman, L. Wu, *J. Immunol.* **2004**, *172*, 1018–1027.
4 C. Ardavin, *Immunol. Today* **1997**, *18*, 350–361.
5 K. Shortman, L. Wu, in *Dendritic Cells: Biology and Clinical Applications* (Eds.: M. T. Lotze, A. W. Thomson), Academic Press Ltd., **1998**, pp. 15–28.
6 J. Sprent, S. Webb, *Curr. Opin. Immunol.* **1995**, *7*, 196–205.
7 A. M. Duijvestijn, A. N. Barclay, *J. Leukocyte Biol.* **1984**, *36*, 561–568.
8 N. Bendriss-Vermare, C. Barthelemy, I. Durand, C. Bruand, C. Dezutter-Dambuyant, N. Moulian, S. Berrih-Aknin, C. Caux, G. Trinchieri, F. Briere, *J. Clin. Invest.* **2001**, *108*, 1237.
9 B. A. Kyewski, R. V. Rouse, H. S. Kaplan, *Proc. Natl. Acad. Sci. U.S.A.* **1982**, *79*, 5646–5650.
10 K. Shortman, D. Vremec, *Dev. Immunol.* **1991**, *1*, 225–235.
11 D. Vremec, J. Pooley, H. Hochrein, L. Wu, K. Shortman, *J. Immunol.* **2000**, *164*, 2978–2986.
12 K. Shortman, L. Wu, D. Ardavin, D. Vremec, F. Stozik, K. Winkel, G. Suss, in *Dendritic cells in Fundamental and Clinical Immunology* (Eds.: J. Banchereau, D. Schmitt), Plenum Publishing Corp., New York, **1995**, pp. 21–29.
13 S. Henri, D. Vremec, A. Kamath, J. Waithman, S. Williams, C. Benoist, K. Burnham, S. Saeland, E. Handman, K. Shortman, *J. Immunol.* **2001**, *167*, 741–748.
14 G. T. Belz, C. M. Smith, L. Kleinert, P. Reading, A. Brooks, K. Shortman, F. R. Carbone, W. R. Heath, *Proc. Natl. Acad. Sci. U.S.A.* **2004**, *101*, 8670–8675.
15 C. Asselin-Paturel, A. Boonstra, M. Dalod, I. Durand, N. Yessaad, C. Dezutter-Dambuyant, A. Vicari, O. G. A., C. Biron, F. Briere, G. Trinchieri, *Nat. Immunol.* **2001**, *2*, 1144–1150.
16 M. O'Keeffe, H. Hochrein, D. Vremec, I. Caminschi, J. L. Miller, E. M. Anders, L. Wu, M. Lahoud, S. Henri, B. Scott, P. Hertzog, L. Tatarczuch, K. Shortman, *J. Exp. Med.* **2002**, *196*, 1307–1319.
17 M. O'Keeffe, H. Hochrein, D. Vremec, B. Scott, P. Hertzog, L. Tatarczuch, K. Shortman, *Blood* **2003**, *101*, 1453–1459.
18 K. Shortman, Y.-J. Liu, *Nat. Rev. Immunol.* **2002**, *2*, 153–163.
19 D. Vremec, M. Zorbas, R. Scollay, D. J. Saunders, C. F. Ardavin, L. Wu, K. Shortman, *J. Exp. Med.* **1992**, *176*, 47–58.
20 D. Vremec, K. Shortman, *J. Immunol.* **1997**, *159*, 565–573.
21 L. Wu, D. Vremec, C. Ardavin, K. Winkel, G. Suss, H. Georgiou, E. Maraskovsky, W. Cook, K. Shortman, *Eur. J. Immunol.* **1995**, *25*, 418–425.
22 T. Okada, Z.-X. Lian, M. Naiki, A. A. Ansari, S. Ikehara, M. E. Gershwin, *Eur. J. Immunol.* **2003**, *33*, 1012–1019.
23 N. S. Wilson, D. El-Sukkari, G. T. Belz, C. M. Smith, R. J. Steptoe, W. R. Heath, K. Shortman, J. A. Villadangos, *Blood* **2003**, *102*, 2187–2194.
24 H. Hochrein, K. Shortman, D. Vremec, B. Scott, P. Hertzog, M. O'Keeffe, *J. Immunol.* **2001**, *166*, 5448–5455.
25 H. Hochrein, M. O'Keeffe, T. Luft, S. Vandenabeele, Grumont R. J, E. Maraskovsky, K. Shortman, *J. Exp. Med.* **2000**, *192*, 823–833.
26 M. Pelletier, C. Tautu, D. Landry, S. Montplaisir, C. Chardrand, C. Perreault, *Immunology* **1986**, *58*, 263–270.
27 S. Vandenabeele, H. Hochrein, N. Mavaddat, K. Winkel, K. Shortman, *Blood* **2001**, *97*, 1733–1741.
28 P. C. Res, F. Couwenberg, F. A. Vyth-Dreese, H. Spits, *Blood* **1999**, *94*, 2647–2657.
29 F. Sotzik, Y. Rosenberg, A. W. Boyd, M. Honeyman, D. Metcalf, R. Scollay, L. Wu, K. Shortman, *J. Immunol.* **1994**, *152*, 3370–3377.
30 K. Winkel, F. Sotzik, D. Vremec, P. U. Cameron, K. Shortman, *Immunol. Lett.* **1994**, *40*, 93–99.

31 C. Schmitt, H. Fohrer, S. Beaudet, P. Palmer, M. J. Alpha, B. Canque, J. C. Gluckman, A. H. Dalloul, *J. Leukoc. Biol.* **2000**, *668*, 836–844.

32 A. T. Kamath, S. Henri, F. Battye, D. F. Tough, K. Shortman, *Blood* **2002**, *100*, 1734–1741.

33 C. Ardavin, L. Wu, C. L. Li, K. Shortman, *Nature* **1993**, *362*, 761–763.

34 H. Spits, F. Couwenberg, Q. Bakker A., K. Weijer, C. H. Uittenbogaart, *J. Exp. Med.* **2000**, *192*, 1775–1783.

35 L. Wu, R. Scollay, M. Egerton, M. Pearse, G. J. Spangrude, K. Shortman, *Nature* **1991**, *349*, 71–74.

36 L. Wu, C. L. Li, K. Shortman, *J. Exp. Med.* **1996**, *184*, 903–911.

37 D. Saunders, K. Lucas, J. Ismaili, L. Wu, E. Maraskovsky, A. Dunn, D. Metcalf, K. Shortman, *J. Exp. Med.* **1996**, *184*, 2185–2196.

38 I. Goldschneider, E. Donskoy, *J. Immunol.* **2003**, *170*, 3514–3521.

39 M. G. Manz, D. Traver, T. Miyamato, I. L. Weissman, K. Akashi, *Blood* **2001**, *97*, 3333–3341.

40 L. Wu, A. D'Amico, H. Hochrein, K. Shortman, K. Lucas, *Blood* **2001**, *98*, 3376–3382.

41 A. D'Amico, L. Wu, *J. Exp. Med.* **2003**, *198*, 293–303.

42 L. Wu, M. Antica, G. R. Johnson, R. Scollay, K. Shortman, *J. Exp. Med.* **1991**, *174*, 1617–1627.

43 H. R. Rodewald, T. Brocker, C. Haller, *Proc. Natl. Acad. Sci. U.S.A.* **1999**, *96*, 15 068–15 073.

44 F. Radtke, I. Ferrero, A. Wilson, R. Lees, M. Aguet, H. R. MacDonald, *J. Exp. Med.* **2000**, *191*, 1085–1094.

45 L. Corcoran, I. Ferrero, D. Vremec, K. Lucas, J. Waithman, M. O'Keeffe, L. Wu, A. Wilson, K. Shortman, *J. Immunol.* **2003**, *170*, 4926–4932.

46 P. Res, E. Martinez-Cáceras, A. Christina Jaleco, F. Staal, E. Noteboom, K. Weijer, H. Spits, *Blood* **1996**, *87*, 5196–5206.

47 M. J. Sanchez, M. O. Muench, M. G. Roncarolo, L. L. Lanier, J. H. Phillips, *J. Exp. Med.* **1994**, *180*, 569–576.

48 C. Márquez, C. Trigueros, J. M. Franco, A. R. Ramiro, Y. R. Carrasco, M. López-Botet, M. L. Toribio, *Blood* **1998**, *91*, 2760–2771.

49 K. Weijer, C. H. Uittenbogaart, A. Voordouw, F. Couwenberg, J. Seppen, B. Blom, F. A. Vyth-Dreese, H. Spits, *Blood* **2002**, *99*, 2752–2759.

50 Y. H. Wang, Y. J. Liu, *Immunity* **2004**, *21*, 1–5.

51 H. Shigematsu, B. Reizis, H. Iwasaki, S. I. Mizuno, D. Hu, D. Traver, P. Leder, N. Sakaguchi, K. Akashi, *Imunity* **2004**, *21*, 43–53.

52 C. Benoist, D. Mathis, *Cell* **1989**, *58*, 1027–1033.

53 G. Anderson, J. J. T. Owen, N. C. Moore, E. J. Jenkinson, *J. Exp. Med.* **1994**, *179*, 2027–2031.

54 G. Anderson, K. M. Partington, E. J. Jenkinson, *J. Immunol.* **1998**, *161*, 6599–6603.

55 T. Brocker, M. Riedinger, K. Karjalainen, *J. Exp. Med.* **1997**, *185*, 541–550.

56 F. Ramsdell, B. J. Fowlkes, *Science* **1990**, *248*, 1342–1348.

57 A. Bendalac, *Nat. Immunol.* **2004**, *5*, 557–558.

58 C. D. Surh, J. Sprent, *Nature* **1994**, *372*, 100–103.

59 H. Pircher, K. Brduscha, U. Steinhoff, M. Kasai, R. M. Zinkernagel, H. Hengartner, B. A. Kyewski, K. P. Muller, *Eur. J. Immunol.* **1993**, *23*, 669–674.

60 S. R. Webb, J. Sprent, *Eur. J. Immunol.* **1990**, *20*, 2525–2528.

61 A. G. Farr, A. Y. Rudensky, *J. Exp. Med.* **1998**, *188*, 1–4.

62 L. Klein, T. Klein, U. Ruther, B. A. Kyewski, *J. Exp. Med.* **1998**, *188*, 5–16.

63 P. Kleindienst, I. Chretien, T. Winkler, T. Brocker, *Blood* **2000**, *95*, 2610–2616.

64 H. Pircher, K. P. Muller, B. A. Kyewski, H. Hengartner, *Int. Immunol.* **1992**, *4*, 1065–1069.

65 H. Taniguchi, M. Abe, T. Shirai, K. Fukao, H. Nakauchi, *J. Immunol.* **1995**, *155*, 5631–5636.

66 H. Kishimoto, J. Sprent, *J. Exp. Med.* **1997**, *185*, 263–271.

67 J. Derbinski, A. Schulte, B. A. Kyewski, L. Klein, *Nat. Immunol.* **2001**, *2*, 1032–1039.
68 J. Pitkanen, P. Peterson, *Genes Immun.* **2003**, *4*, 12–21.
69 M. S. Anderson, E. S. Venanzi, L. Klein, Z. Chen, S. Berzins, S. Turley, H. von Boehmer, R. Bronson, A. Dierich, C. Benoist, D. Mathis, *Science* **2002**, *298*, 1395–1401.
70 J. Sprent, C. D. Surh, *Nat. Immunol.* **2003**, *4*, 303–304.
71 I. Goldschneider, R. E. Cone, *Trends Immunol.* **2003**, *24*, 77–81.

Part V
Antigen Presenting Cell-based Drug Development

16
Antigen Presenting Cells as Drug Targets

Siquan Sun, Robin L. Thurmond and Lars Karlsson

16.1
Introduction

Dendritic cells (DC), macrophages, and B cells are considered to be the major antigen presenting cells (APC). DC are highly specialized APC whose major function is to activate T cells. Macrophages also have important functions as phagocytes that serve as the first line of defense to clear invading pathogens along with their role as APC. While mainly providing function as antibody producing cells, B cells can also take up antigens through surface antigen-specific receptor and then present these antigens to antigen-specific $CD4^+$ helper T cells. Even though the APC function of both macrophages and B cells is believed to be involved in pathogenesis of diseases and may offer exciting drug discovery opportunities, most of discussion in this chapter will focus on DC.

APC are now clearly critically important in the immune response by bridging innate and adaptive immunity, as well as for maintenance of self-tolerance. When exposed to an infection or local tissue injury, DC acquire, process and present antigens to T lymphocytes. Using pathogen-pattern recognition receptors such as toll-like receptors (TLR), DC also sense and relay information on the nature of invading pathogens by producing cytokines that bias $CD4^+$ T cell differentiation and promote effector functions towards the specific pathogens. In addition, DC also receive cues from local damaged tissues or previously activated $CD4^+$ T cells to direct naïve $CD4^+$ T cell differentiation. Certain subtypes of DC, particularly immature DC, are essential for the maintenance of self tolerance by the induction of regulatory T cells or by deleting autoreactive T cells. Thus, as a central player of immune responses DC serve as ideal targets for drug discovery, providing both opportunities to potentiate the immune response against pathogens or tumors and to inhibit unwanted immune responses in immune-mediated inflammatory diseases (IMID) like autoimmune diseases, allergy, asthma, and transplantation.

Many facets of APC function can be viewed as possible targets for therapeutic intervention, including APC generation and differentiation; antigen capture, processing, and presentation; maturation, activation, migration and survival. In this chapter, we first review the roles of DC in autoimmune disease, allergy/asthma, transplantation and cancer. Second, we provide an overview of how currently marketed immunomodulating and immunosuppressive therapies may affect APC function. Third, we will discuss some potential drug targets specifically targeting APC. Lastly, we will discuss the progress of using APC themselves as therapeutic agents, i.e. DC-based cancer vaccine as well as ex vivo generated tolerogenic DC for transplantation and autoimmune diseases.

16.2
Roles of DC in disease

16.2.1
Transplantation

Recognition and rejection of organ allografts are mediated by interstitial "passenger" donor DC as well as host DC that are recruited to the allografts (Gould and Auchincloss, 1999). Donor DC originating in the allograft migrate to host secondary lymphoid tissues and present highly immunogenic donor antigenic peptides in the context of donor major histocompatibility complexes (MHC) for direct allorecognition by naïve host T cells. At the same time, host DC are able to invade the allograft, where they can capture donor antigens in the context of host MHC for indirect allorecognition by T cells. Both mechanisms generate T cell responses directed against the allograft and can lead to rejection. Direct allorecognition plays a major role in the rejection of allografts since most host T cells (>90%) reactive against allograft recognize alloantigens in the context of donor MHC molecules (Benichou et al., 1999; Illigens et al., 2002). Therefore, it may be necessary to inhibit both donor DC and host DC for the treatment of allograft rejection (Gould and Auchincloss, 1999).

16.2.2
Autoimmune Diseases

Activation of autoreactive T cells is the underlying mechanism for initiation and exacerbation of various autoimmune diseases. This activation is driven by APC function and, therefore, inhibition of APC is a potentially viable therapy for the treatment of autoimmune diseases. Phenotypical and functional abnormalities of DC have been reported in several autoimmune diseases, e.g. systemic lupus erythematosus (SLE), rheumatoid arthritis (RA), diabetes, and Sjögrens syndrome (Morel et al., 2003). In SLE patients, there is an overt activation of plasmacytoid DC (pDC), the so-called professional IFN-α producers, which results in high levels of IFN-α in sera of these patients. The number of circulating pDC in the blood is reduced and the few pDC found mostly show an active phenotype and reside in peripheral tissues (Blanco et al., 2001; Scheinecker et al., 2001) such as cutaneous lupus erythematosus lesions (Farkas et al., 2001). Mostly due to the elevated level of IFN-α, myeloid DC were found to be activated and are very potent in their ability to activate T cells in vitro (Blanco et al., 2001). Increased numbers of DC were also found in the affected joints of RA patients (Summers et al., 1995; Pettit and Thomas, 1999). This observation is also supported by the finding that cell-free RA synovial fluid facilitated DC differentiation and maturation from myeloid DC progenitors as well as CD14-derived DC (Santiago-Schwarz et al., 2001). Moreover, the function of synovial DC from RA appeared to be resistant to IL-10-induced down-modulation (MacDonald et al., 1999). Normally IL-10 down-regulates the expression of MHC class II (MHCII) and B7 on DC and dampens their ability to stimulate T cells. However, IL-10 had no such effect on RA synovial DC. Multiple sclerosis (MS) was also found to be associated with high levels of circulating DC secreting proinflammatory cytokines, such as IFN-α, TNF-γ, and IL-6 (Huang et al., 1999) and reduced IL-10 production (Ozenci et al., 1999). In addition, myeloid and plasmacytoid DC are present in cerebrospinal fluid in elevated numbers in MS, which may be due to enhanced DC expression of CCR5, a chemotactic receptor for chemokines RANTES and MIP-1α/β (Pashenkov et al., 2002).

16.2.3
Allergy/Asthma

DC are crucial in driving activation of allergen-specific CD4$^+$ T cells as well as for biasing the differentiation of these T cells to a Th2-type phenotype, thought to be responsible for the development of allergy and asthma. Increased DC density was observed in nasal or bronchial mucosa in allergic rhinitis and asthma patients (Fokkens et al., 1989; Godthelp et al., 1996; Moller et al., 1996; Tunon-De-Lara et al., 1996) and, upon exposure to allergen, DC were quickly recruited to bronchial mucosa (Jahnsen et al., 2001). Inhaled corticosteroids down-regulate the number of DC in bronchial mucosa of atopic asthmatic patients with a reduction of airway inflammation (Moller et al., 1996). An added suggestion for the importance of DC in pathogenesis of allergic asthma comes from mice, where systemic administration of antigen-pulsed DC induces Th2-dependent experimental allergic asthma (Graffi et al., 2002).

16.2.4
Cancer

Various abnormalities of DC are associated with cancer, most of which lead to a reduction in APC function (Turtle and Hart, 2004). This is one of the major underlying mechanisms for decreased cellular immunity against tumors. DC counts fluctuate following hematopoietic stem cells transplantation and are reduced following chemotherapy or post-surgery (Savary et al., 1998; Fearnley et al., 1999; Ho et al., 2001). Systemically released factors produced by tumors, including IL-10, TGF-β, vascular endothelial growth factor (VEGF) and cyclooxygenase, have a generalized suppressive effect on the activation and differentiation of DC, which may lead to induction of regulatory T cells or a skewing away from Th1 anti-tumor responses (Morel et al., 2003; Cerundolo et al., 2004; Turtle and Hart, 2004). More-

16.3
Marketed Drugs Affecting APC function

Not surprisingly, due to the central role played by APC in immune responses, many drugs with immunomodulatory properties may attribute their therapeutic efficacy, at least partially, to their effects on modulating APC function. These drugs include adjuvants, immunosuppressives, anti-inflammatory small molecules, as well as new biologic agents (Figure 16.1).

Among currently used anti-inflammatory and immunosuppressive drugs, it is now well established that many interfere with the functions of APC such as DC. For example, corticosteroids (CS) modulate gene expression by inhibition of transcription factors such as NFκB and AP-1. This results in down-regulation of DC differentiation and maturation (Piemonti et al., 1999a; Piemonti et al., 1999b; Woltman et al., 2000; Woltman et al., 2002) as well as inhibiting activation (Moynagh, 2003) leading to reduced production of cytokines, such as IL-12 (Blotta et al., 1997), down-regulation of MHC molecules and co-stimulatory molecules on

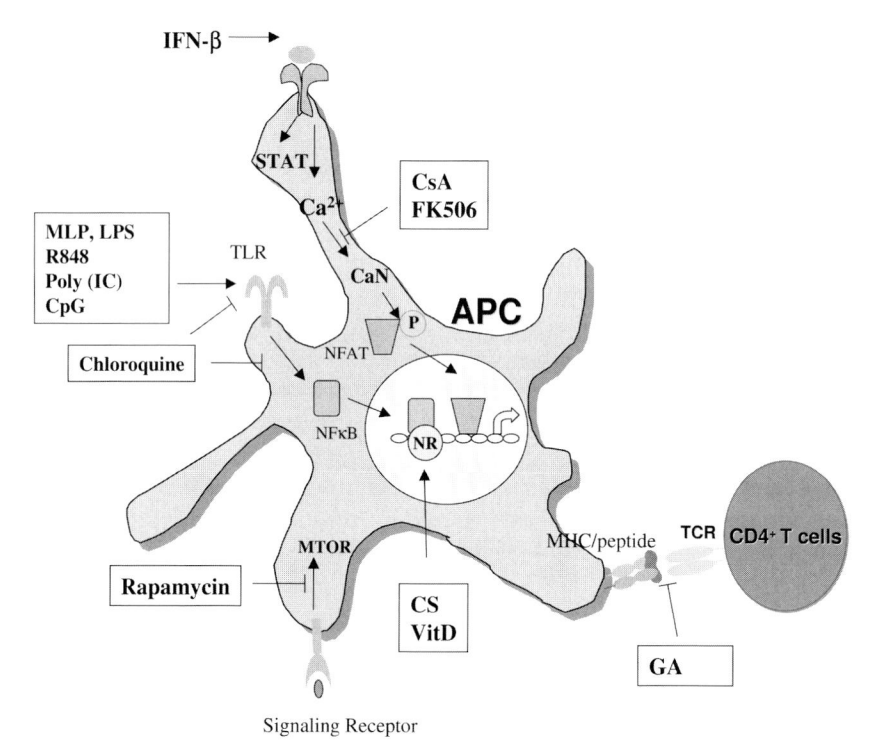

Figure 16.1 Mechanisms of action of marketed drugs affecting functions of APC. (This figure also appears with the color plates.)

the APC surface (Moser et al., 1995; Moller et al., 1996). CS also affect migration and recruitment of APC to peripheral lymphoid organs (Moser et al., 1995; Cumberbatch et al., 1999; Koopman et al., 2001) as well as to other tissues (Moller et al., 1996; Till et al., 2001). In addition, CS also induce apoptosis of DC (Woltman et al., 2002) in a caspase-independent manner (Kim, K.D. et al., 2001).

FK506 (tacrolimus), cyclosporin A (CysA) and rapamycin (sirolimus) represent a class of immunosuppressives used successfully in transplantations, markedly reducing the rate of acute rejection and increasing short-term graft survival rates. In addition, CysA and FK506 also show effects in IMID such as RA, nephrotic syndrome, psoriasis and atopic dermatitis (AD). By complexing with different members of immunophilin family proteins, FK506 or CysA inhibits the activity of calcineurin, a phosphatase that is critical for the activation of transcription factor NFAT. Rapamycin inhibits MTOR, a kinase involved in cellular activation and proliferation. While the effect of these agents on T cells has been well documented (Bishop and Li, 1992; Almawi et al., 2001), they also exert effects on DC generation, maturation, activation, and apoptosis (Woltman et al., 2000; Woltman et al., 2001; Castedo et al., 2002; Woltman et al., 2003; Adorini et al., 2004). Treatment of DC with these agents leads to reduced endocytosis (Monti et al., 2003; Tajima et al., 2003; Hackstein and Thomson, 2004), reduced cytokine production and down-regulation of MHC molecules and co-stimulatory molecules such as B7.1 and CD40 (Matsue et al., 2002; Monti et al., 2003; Tajima et al., 2003).

A steroid hormone derived from vitamin D_3, $1\alpha,25(OH)_2D_3$, shows marked immunomodulatory effects by affecting APC (Hackstein and Thomson, 2004). It inhibits the differentiation, as well as phenotypic and functional activation of human DC and may also induce DC apoptosis (Piemonti et al., 2000). Thus, $1\alpha,25(OH)_2D_3$ inhibits human DC maturation and activation induced by LPS, TNF-α or CD40L and reduces the capacity of DC to stimulate T cell responses. In addition, $1\alpha,25(OH)_2D_3$-treatment seemed to favor generation of self-tolerance by promoting the development of regulatory or anergic T cells. Consistent with this notion, $1\alpha,25(OH)_2D_3$-treated DC produced increased concentration of IL-10 and CCL18 chemokine. CCL18 recruits immature DC (Vulcano et al., 2003), which are more tolerogenic than immunostimulatory. Moreover, pretreatment of female mice with vitamin D_3 analogue-conditioned DC from male mice prolonged survival of male skin grafts (Griffin et al., 2001).

Although most adjuvants presently used in humans were discovered empirically, it is now clear that they affect the functions of APC, such as activation, cytokine production and antigen capture. Although the underlying mechanism for alum salts (the main adjuvant licensed for human use) has not been completely elucidated, protein aggregates complexed to the salt are more immunogenic than soluble proteins. This may be due to enhanced antigen uptake by APC. In addition, adjuvant systems that function by creating particles, e.g. liposomes, virosomes, aliginate microspheres, nonionic block copolymers, also enhance antigen capture by APC. Certain adjuvants contain constituents that activate APC via toll-like receptors (TLR) or other pathogen pattern-recognition receptors (Burdin et al., 2002; Rhodes, 2002). For example, monophosphoryl lipid A (MPL) used in several vaccines activates TLR4, which leads to a Th1 response and induction of cellular immunity. Therefore, the discovery of TLR and their cognate agonistic ligands presents new opportunities in adjuvant research, where it may be possible to move into a more rationally designed approach to the discovery of adjuvants with better efficacy and specificity towards targeted pathogens or tumors.

Choloroquine, hydroxychloroquine, and quinacrine are best known for the treatment of malaria, but these drugs, especially hydroxychloroquine (most frequently prescribed due to its lower toxicity), are also classified as slow-onset disease-modifying antirheumatic drugs (DMARD) for IMID, including SLE, RA, and sarcoidosis (Tett et al., 1990) and graft versus host disease (GVHD)(Schultz et al., 1995). The underlying mechanism of action of chloroquine and its analogs in IMID is still under investigation. Chloroquine is a weak base, which may be preferably enriched in lysosomes, thereby inhibiting lysosomal acidification and antigen processing by APC (Streicher et al., 1984; Schultz et al., 1995). However, these drugs are also functional antagonists of certain TLR, especially for TLR9 (Hacker et al., 1998), and possibly for TLR7 (Lee et al., 2003) and TLR4 (Weber and Levitz, 2000). The antagonistic effect of chloroquine on TLR9 was initially perceived as evidence of a requirement for lysosomal acidification for the binding of TLR9 ligand, the immunostimulatory CpG DNA. However, recent evidence suggests that chloroquine might directly compete with CpG DNA for binding to TLR9 (Rutz et al., 2004) or, in the case of TLR4, inhibit activation of MAP kinases (Weber et al., 2002). For TLR9, these drugs inhibit production of cytokines and chemokines, such as IFN-α or IP-10, as well as inhibiting the upregulation of MHC molecules and co-stimulatory molecules on APC. It has been postulated that inhibition by these drugs of TLR9-mediated activation of plasmacytoid DC and autoreactive B cells is the underlying mechanism of their clinical efficacy in SLE (Anders et al., 2004b; Boule et al., 2004). In fact, activation of APC through various TLR may be involved in pathogenesis of various IMID such SLE, and RA as well as sepsis (Cook et al., 2004). Therefore, discovery of new TLR antagonists represents an exciting opportunity for drug discovery as well.

Statins, inhibitors of 3-hydroxy-3-methylglutaryl coenzymes A reductase, inhibit the biosynthesis of mevalonate, which results in reduction of cholesterol synthesis. Statins have been commercially successful drugs, showing therapeutic benefits in reducing cardiovascular events as well as atherosclerosis. The beneficial effects of statins are well accepted as being at least partially due to their immunomodulatory functions (Steinman, 2004). Statins strongly inhibit the activation and antigen presentation functions of APC (Youssef et al., 2002; Yilmaz et al., 2004). In a mouse model of experimental autoimmune encephalomyelitis (EAE), an animal model of human MS, oral atorvastatin prevents or reverses paralysis (Youssef et al., 2002). Atorvastatin inhibits the IFN-γ-induced upregulation of MHCII molecules and co-stimulatory molecules such as CD40, CD80, and CD86 in cultures of microglia, the APC of the central nervous system. As a result, atorvastatin treatment decreases activation of antigen-specific T cells. In addition, atorvastatin also shifts the balance of cytokine production by T cells from Th1-type cytokines, such as IL-2, IL-12, IFN-γ to Th2-type-cytokines such as IL-4, IL-5 and IL-10. Similarly, incubation of isolated human DC with simvastatin or atorvastatin decreased expression of HLA-DR as well as co-stimulatory or homing molecules such as CD83, CD40, CD86 and CCR7, resulting in a reduced ability to stimulate T cells (Yilmaz et al., 2004). Statins have shown promising efficacy in reducing inflammatory lesions in pilot clinical trials to treat MS (Vollmer et al., 2004).

Two other agents currently indicated for treatment of relapsing-remitting MS, recombinant IFN-β and synthetic glatiramer acetate (GA), may also exert their immunomodulatory activity by affecting APC functions. IFN-β down-regulates expression of MHCII molecules on microglial cells (Hall et al., 1997). GA is the acetate salt of a mixture of synthetic polypeptides of four amino acids [L-glutamic acid, L-lysine, L-alanine, and L-tyrosine (thus GLATiramer)]. It was initially synthesized in an attempt to mimic the physicochemical properties of myelin basic protein (MBP), one of the known inducers of EAE in rodent models. Surprisingly, GA failed to induce EAE in animals. On the contrary, it showed remarkable efficacy in suppressing EAE in animals as well as MS in human. The underlying mechanisms of GA's therapeutic effects are multiple. Beside its possible neuroprotective effect by inducing neurotropic factors, the main mechanism is its effect on the function of APC. GA binds strongly to MHCII molecules, and may thus effectively compete with MBP for MHC binding, which would result in reduced antigen presentation by APC and T cell activation (Dhib-Jalbut et al., 2003). Moreover, it appears that GA skewed CD4$^+$ T cells to a Th2 response by inhibiting IL-12 production by APC (Dhib-Jalbut et al., 2002; Losy et al., 2002).

16.4
New Potential APC Drug Targets

We now discuss some current drug discovery efforts specifically targeting APC functions, including APC activation, antigen presentation, co-stimulation, cell adhesion and migration as well as APC survival. Figure 16.2 illustrates several of these APC drug targets.

16.4.1
APC Activation

Before APC can present antigens to T cells they need to be activated. Activation also provides a crucial link between the innate and adaptive immune responses, since APC can be activated directly by pathogen molecules or indirectly via a host response. In autoimmune diseases, inappropriate activation of host responses may lead to the initiation of the disease. Regardless of the sources, activation of DC leads to the upregulation of the machinery necessary for antigen presentation, upregulation of co-stimulatory molecules, production of cytokines and chemokines, and chemotaxis to secondary lymphoid organs. Therefore, interfering with APC activation would obviously be a desirable target for immunomodulation. In fact the activation of APC by adjuvants has long been used to ensure successful

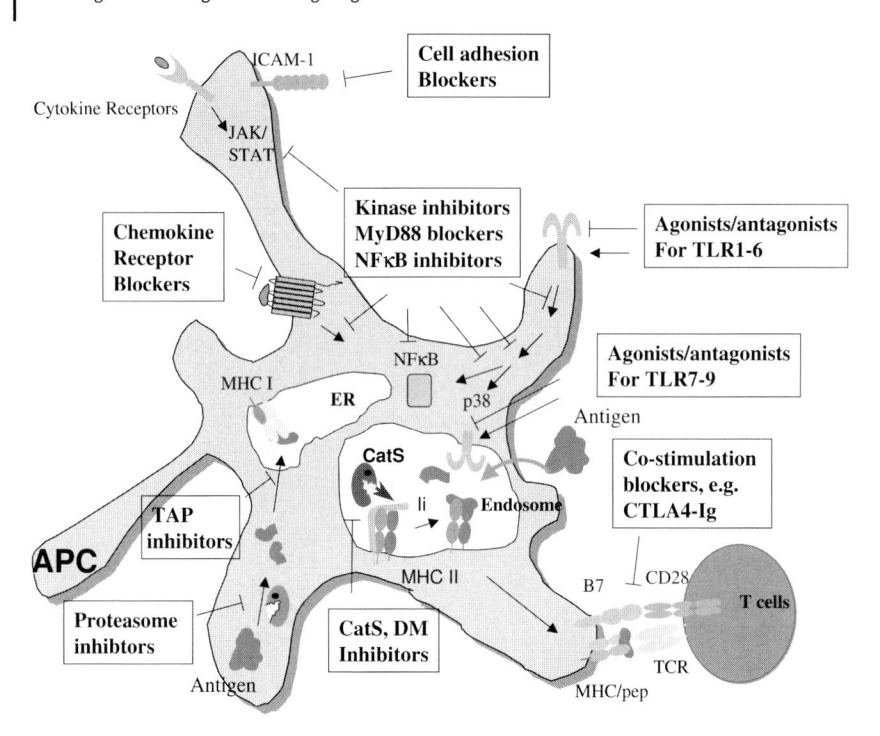

Figure 16.2 New potential APC drug targets. (This figure also appears with the color plates.)

vaccinations. DC can be activated by various cytokines, such as TNFα, IFNγ and IL-1, and antibody therapeutics directed at these cytokines should in theory inhibit DC activation (see Lanzavecchia and Sallusto, 2001 for a review).

Recently, the role of pattern recognition receptors like TLR has emerged as important regulators of the activation of APC. In fact, activation of APC through various TLR may be involved in pathogenesis of various IMID (Cook et al., 2004). Thus, TLR9-mediated activation of plasmacytoid DC and autoreactive B cells may play a role in the pathogenesis of SLE or RA (Anders et al., 2004a; Anders et al., 2004b; Boule et al., 2004). TLR4 may contribute to atherosclerosis and play a role in sepsis since patients bearing a mutated TLR4 (D299G) show attenuated response to endotoxin and also show reduced risk of arthrosclerosis (Kiechl et al., 2002). As described earlier, various adjuvants such as a TLR4 ligand MLP, a functional antagonist of TLR9 chloroquine, as well as agonists of TLR7/8 such as imiquimod (Hemmi et al., 2002; Jurk et al., 2002) are several examples of known agents/drugs that have shown therapeutic benefits in humans. Therefore, the TLR, their cognate ligands and the TLR signal transduction pathways provide exciting opportunities for drug discovery (Quesniaux and Ryffel, 2004; Ulevitch, 2004).

With identification of cognate ligands for various TLR, it may be possible to develop small molecule agonists/antagonists of TLR or antagonistic antibodies. For example, a synthetic TLR4 antagonist, E5564, is efficacious in inhibiting LPS-mediated activation of immune cells both in vitro and in vivo in mice and humans (Chilman-Blair et al., 2003; Hawkins et al., 2004; Lynn et al., 2004). In mice an antibody that blocks ligand binding to TLR2 suppressed lethality in a septic shock model induced by either a TLR2 ligand, the synthetic lipopeptide P_3CSK_4, or heat-inactivated Gram-positive bacteria *B. subtilis* (Meng et al., 2004). In addition, immunostimulatory CpG oligodeoxynucleotides (ODN) have shown efficacies in animal models of allergy/asthma, cancer, as well as vaccines, and are currently being tested in clinical trials (Klinman, 2004; Krieg, 2004). However, it has been recently shown in mice that chronic administration of CpG ODN may cause considerable pathological side effects (Heikenwalder et al., 2004).

Another approach would be to target the signaling mechanisms that lead to APC activation. Regarding TLR signaling, there are MyD88 dependent and independent pathways. The MyD88 dependent pathway is mediated by the association of the cytosolic Toll/IL-1 receptor (TIR) domain with MyD88 and signaling via several kinases, including IRAK1 and 4. In fact, a small molecule MyD88 mimic, hydrocinnamoyl-l-valyl-pyrrolodine, specifically designed to disrupt the interaction between TIR and MyD88, showed efficacy in both in vitro and in vivo systems (Bartfai et al., 2003). Other kinases downstream of IRAK4, such as TAK1 and the mitogen activated kinases such as p38 and JNK, have also been implicated in these pathways and represent interesting targets. Inhibitors of some of these kinases would be expected to significantly affect TLR-dependent APC activation although their effects would not be specific to these pathways (Boehm and Adams, 2000; Foster et al., 2000; Harper and LoGrasso, 2001; Ninomiya-Tsuji et al., 2003).

The main downstream target of the MyD88 pathway is NFκB. Therefore, NFκB inhibitors (Umezawa and Chaicharoenpong, 2002) are also promising candidates for blocking TLR-mediated APC activation. Activation of NFκB requires the phosphorylation of IκB by the IKK complex, which can be targeted with kinase inhibitors (Castro et al., 2003). This phosphorylation eventually leads to the degradation of IκB by the proteasome and, therefore, proteasome inhibitors may also be useful in blocking APC activation. Indeed many of the in vivo immunosuppressive properties of proteasome inhibitors may be due to this mechanism rather than blocking the generation of antigenic peptides as described below. The MyD88 independent pathway can also lead to NFκB activation and this appears to be mediated via RIP1 activation. Finally, IRF3 is also involved in TLR signaling and it is thought to be activated by TBK1, representing an additional pathway. There are also several areas where disruption of protein–protein interactions could inhibit activation, but the targeting of such interactions with small molecules has met with limited success.

16.4.2
Antigen Presentation

One of the key functions of APC, of course, is antigen presentation to T cells. Professional APC are responsible for the stimulation of naïve and memory $CD4^+$ and $CD8^+$ T cells. One advantage of this approach over other immunosuppressive targets is the ability to selectively target either $CD8^+$ or $CD4^+$ T cell activation since the presentation pathways are distinct.

APC, like all cells, can present endogenous antigens to $CD8^+$ T cells via the MHC class I (MHCI) pathway. However, only professional APC express the proper co-stimulatory molecules to activate naïve $CD8^+$ T cells and thus are crucial for mounting cellular responses to infection (Kurts et al., 2001; Jung et al., 2002; Norbury et al., 2002). The T cell epitopes are generated by the normal protein degradation machinery in the cytosol, namely the proteasome, although other proteases like tripeptidyl peptidase II (TPPII) have also been implicated. Peptides generated by these proteases are transported into the ER by the transporter associated with antigen presentation (TAP) where they are trimmed to the required 8-11 amino acid size by aminopeptidases like ERAP1. These peptides are then loaded onto MHCI and transported to the cell surface where they assist in activating $CD8^+$ T cells. Nature provides validation for targeting the MHCI pathway as immunomodulatory therapy since many viruses encode proteins that disrupt this pathway to avoid immune detection (see Yewdell and Hill, 2002, for a review). Not only do viruses use this mechanism to prevent the killing of the host cell but many infect DC and therefore could disrupt activation of naïve $CD8^+$ T cells (Engelmayer et al., 1999; Norbury et al., 2002).

Naïve $CD8^+$ T cells can only be activated by professional APC. Therefore, to generate a $CD8^+$ T cell response to antigens that are not expressed in professional APC, like some viral or tumor antigens, these antigens must first be taken up by the APC. This is especially important for antigens that generate autoimmune responses that should not normally be recognized and antigens that mediate transplant rejection since they only exist in the target cells. In these cases cross presentation of exogenous antigens to $CD8^+$ T cells by professional APC is likely to be the most important pathway. APC take up these exogenous antigens and then present them on MHCI as well as MHCII. The presentation on MHCI can occur via TAP-dependent or TAP-independent pathways (Belz et al., 2002; Ramirez and Sigal, 2004). The machinery for TAP-dependent cross presentation is similar to that for normal MHCI presentation except that the antigen does not originate in the cytosol but arrives there via the endosomal system.

Within the MHCI pathway the proteasome has emerged as a viable drug target. Several viruses also use this approach to evade immune detection by encoding proteins that inhibit proteasome activity (Levitskaya et al., 1997; Apcher et al., 2003; Zhang and Coffino, 2004). In cell culture, proteasome inhibitors can block MHCI presentation but, despite many reports of efficacy of proteasome inhibitors in vivo, there has been no direct proof that this is due to effects on antigen presentation (Rock et al., 1994; Bai and Forman, 1997; Cerundolo et al., 1997; Craiu et al., 1997; Lopez and Del Val, 1997). There are two proteasome inhibitors in clinical use. The first is ritonavir, which is marketed as a HIV protease inhibitor, but has also been shown to be a proteasome inhibitor at therapeutically relevant concentrations (Andre et al., 1998; Schmidtke et al., 1999; Groettrup et al., 2004). There is only circumstantial evidence that the activity has any benefit in humans but ritonavir treatment in mice decreases the generation of CTL against LCMV (Andre et al., 1998). In 2003 the first inhibitor specifically targeting the proteasome, bortezomib, was approved for the treatment of multiple myeloma. It is yet to be seen whether this inhibitor has any immunomodulatory properties and there have been no reports of inhibition of antigen presentation using this inhibitor. One significant disadvantage of proteasome inhibitors is the ubiquitous role of the proteasome in general protein catabolism, as well as its specific functions in the degradation of regulatory proteins such as IκB and several proteins involved in cell cycle control. The approval of bortezomib implies that it is possible to separate therapeutic from unwanted effects, although this remains to be proven for the treatment of autoimmune diseases where the therapeutic window may need to be larger than for oncology indications. Nevertheless, there are hints in vitro that antigen presentation activity can be separated from other essential functions (Andre et al., 1998).

TPPII has also been implicated in the generation of T cell epitopes and, therefore, represents another possible target. Recent work has shown that TPPII is solely responsible for the degradation of some antigens and can assist the proteasome in the degradation of others (Samady et al., 2003; Seifert et al., 2003; Reits et al., 2004). Therefore, as for inhibition of the proteasome, inhibition of TPPII should reduce the presentation of antigens. Inhibitors of TPPII have already been developed for the treatment of obesity due to its role in degrading cholecystokinin-8 (Ganellin et al., 2000; Breslin et al., 2002).

Another point of intervention is to block peptide entry into the ER by TAP. One human immunodeficiency that interferes with $CD8^+$ T cell function is due to absence of functional TAP (Gadola et al., 2000). These individuals are susceptible to bacterial, fungal and parasitic infections, underscoring the importance of this molecule in the immune response. Furthermore, several viruses encode proteins that inhibit TAP function, including the HCMV-encoded US6 gene product that inhibits the ATPase activity of TAP (Hewitt et al., 2001; Momburg and Hengel, 2002; Reits et al., 2002). The advantage of such an approach is that it should be fairly selective for MHCI presentation since this is the only known function of TAP.

Once the peptides enter the ER they are subjected to several aminopeptidases that are required to trim them to the proper size before they are loaded onto MHCI molecules. One of the key enzymes for this is ERAP1. Using either non-selective inhibitors or siRNA, inhibition of ERAP1 reduces the presentation of many antigens (Saric et al., 2002; York et al., 2002). However, ERAP1 is not necessary for the processing of antigens for all MHCI haplotypes since some of them require longer peptides. For these haplotypes an increased expression of ERAP1 can actually inhibit presentation (York et al., 2002).

Perhaps the most important function of APC is the presentation of exogenous antigens to $CD4^+$ T cells, a function that is almost exclusively the domain of professional APC (DC, B cells and macrophages). Once again viruses yield clues as to the feasibility of targeting $CD4^+$ T cell activation since several viral proteins interfere with this process to dampen immune responses (Hegde et al., 2003). All APC internalize exogenous antigens either through endocytosis or receptor-mediated internalization and deliver them to endosomal compartments where they are subjected to acidification and the action of various proteases. Therefore, one way to inhibit antigen presentation is to inhibit the proteases involved in generating antigenic peptides. One potential problem with this approach is that many of the proteases implicated in this process, such as cathepsins, are ubiquitously expressed and are thought to play a role in general protein catabolism. Therefore, inhibition of these proteases may lead to serious side effects. Indeed many lysosomal storage diseases are caused by inactivation of lysosomal proteases. There are, however, a few proteases that seem to be preferentially expressed in APC and may have more specific roles in antigen processing. One example is the asparaginyl endopeptidase (AEP, also known as legumain). AEP is a cysteine protease that cleaves specifically after asparagine residues and is found in the lysosomes of APC. AEP is involved in the cleavage of certain antigens like tetanus toxoid and its inhibiting proteolytic activity inhibits the presentation of the antigen (Manoury et al., 1998; Loak et al., 2003). However, proteases can have roles both in producing and destroying antigenic peptides. This is true for AEP, which generates tetanus T cell epitopes but degrades a key epitope of myelin basic protein and prevents its presentation (Manoury et al., 2002; Loak et al., 2003). Therefore, inhibition of AEP would block the activation of tetanus specific T cells but may allow for the generation of myelin basic protein reactive T cells that may be damaging. AEP also highlights the general toxicity problems of inhibiting proteases that have more than one function. Mice deficient in AEP show defects in the processing of several other proteases in kidney proximal tubule cells, leading to enlargement of the lysosomes and accumulation of macromolecules very similar to that is seen with lysosomal storage diseases in humans (Shirahama-Noda et al., 2003).

In contrast to inhibiting the proteolysis of antigens, a more specific inhibition may be carried out by interfering with the loading of the antigenic peptides onto the MHCII molecules. MHCII molecules are synthesized in the ER as heterodimers, which contain the α and β peptide binding subunits. Three of these dimers are bound to a trimeric invariant chain (Ii) to form a nonameric complex. Ii acts as a chaperone that stabilizes the MHCII, targets it to the endosomal system and prevents peptide loading. Once the MHCII is delivered to the lysosome the invariant chain must be removed before antigenic peptides can be loaded. This is carried out by a series of proteolytic steps that results in the final degradation product, the class II-associated invariant chain peptide (CLIP) remaining in the MHCII peptide-binding groove. The exchange of CLIP for high-affinity antigenic peptides is mediated via the interaction of MHCII with a nonclassical MHC molecule known as HLA-DM. The MHCII-peptide complexes can then be transported to the cell surface where they are able to interact with T cell receptors. There are several possible points of intervention in this process. First, if the proteolysis of the invariant chain is inhibited then antigen presentation should be blocked. Many of the proteases responsible for the early proteolytic steps have not been identified, although AEP has been implicated as performing the initiating cleavage (Manoury et al., 2003). In contrast, the final cleavage step that generates CLIP in DC and B cells is clearly due to cathepsin S activity. Cathepsin S is a cysteine protease that is expressed mainly in the lysosome of DC, B cells and macrophages. Disruption of cathepsin S activity should block antigen loading onto MHCII and attenuate $CD4^+$ T cell activity. This has been validated with cathepsin S deficient mice, which show a decrease in invariant chain degradation in DC and B cells, but only a moderate effect in macrophages (Nakagawa et al., 1999; Shi et al., 1999). These mice also display a diminished capacity to present antigens and show a reduced susceptibility to collagen-induced arthritis (CIA). Inhibitors of cathepsin S block both invariant chain processing and antigen presentation in cells and in vivo (Riese et al., 1996; Villadangos et al., 1997; Allen et al., 2001; Fiebiger et al., 2001; Podolin et al., 2001; Thurmond et al., 2004). Furthermore, in vivo treatment with cathepsin S inhibitors leads to the attenuation of the antibody responses in mice immunized with ovalbumin, as well as smaller increases in IgE titer and less lung eosinophil infiltration in a mouse model of pulmonary hypersensitivity (Riese et al., 1998). Inhibitors of CatS also decrease the degree of inflammation in both the rat adjuvant-induced arthritis (AIA) model (Biroc et al., 2001) and the CIA model in mice (Podolin et al., 2001) and are effective in treating a murine model of Sjögren syndrome (Saegusa et al., 2002). In addition to effects on antigen presentation to $CD4^+$ T cell, cathepsin S has also been implicated in TAP-independent cross presentation to $CD8^+$ T cells and, therefore, inhibitors may have the benefit of affecting both $CD4^+$ and $CD8^+$ T cells (Shen et al., 2004).

Another potential target in the MHCII presentation pathway is HLA-DM. As described earlier, DM in necessary for the removal of CLIP and the loading of antigenic peptides. As for inhibition of cathepsin S, blocking the activity of DM will also reduce the presentation of antigens on the surface of APC and block $CD4^+$ T cell activation. Indeed in H2-M (the mouse homologue of DM) deficient mice there is a profound decrease in the ability of APC to present antigens (Fung-Leung et al., 1996). In human B lymphoblasts, presentation of the immunodominant insulin-dependent diabetes mellitus autoantigen, glutamine decarboxylase, was regulated by the levels of HLA-DM (Lich et al., 2003). However, disrupting DM/MHC interactions may be problematic since targeting intracellular protein–protein interactions has proven to be difficult, although the crystal structure does reveal a potential small molecule binding site (Mosyak et al., 1998).

16.4.3
Co-stimulation

An antigenic peptide bound to either MHCII or MHCI is not sufficient to activate T cells. There also has to be other interactions mediated by co-stimulatory molecules, and this offers other targets for the disruption APC/T cell interactions. One of the most import interactions is between CD40 on the APC and CD40L on the T cell, which leads to the activation of both APC and T cells. In DC this interaction is necessary for the upregulation of B7 molecules and in B cells CD40 ligation triggers activation, proliferation, Ig production and isotype switching. The CD40/CD40L interaction is also important in cytokine production in macrophages. The most advanced attempts to inhibit this interaction have involved antibodies directed against CD40L. This disrupts T cell activation, as well as the activation of APC. So far the results from two studies of anti-CD40L in SLE patients have been mixed. One phase II study with 85 SLE patients failed to demonstrate efficacy (Kalunian et al., 2002). However, another small study with a different antibody showed reduction in activated B cells in the periphery, decreases in serum anti-dsDNA titers and overall disease activity (Huang et al., 2002). A larger study with this antibody also appeared to be effective in patients with proliferative lupus nephritis but the antibody has been discontinued due to toxicity issues (Boumpas et al., 2003; Sidiropoulos and Boumpas, 2004). Efficacy has also been seen in patients with refractory immune thrombocytopenic purpura (Kuwana et al., 2004). The opposite approach would be to direct antibodies against CD40 on the APC. This has some drawbacks because many antibodies that bind CD40 activate it and lead to activation of B cells and macrophages. One antibody has been described that does not activate CD40 but still blocks CD40L interaction (Kwekkeboom et al., 1993). This antibody is effective in blocking EAE in marmosets and kidney allograft rejection in monkeys (Boon et al., 2001; Laman et al., 2002; Haanstra et al., 2003).

B7 molecules (CD80 and CD86) interact with CD28 on the surface of T cells and provide co-stimulation to drive the clonal expansion of naïve T cells. One approach to block the B7/CD28 interaction that has had success in clinical trials is a CTLA4-Ig fusion protein. CTLA-4, which is another ligand for B7, is upregulated by T cells after activation. CTLA4-Ig binds to both CD80 and CD86 on the surface of APC and has a higher affinity for the B7 molecules than does CD28 and, therefore, blocks CD28 mediated activation of T cells. The use of CTLA4-Ig has been effective in several animal models of autoimmune disease (Finck et al., 1994; Knoerzer et al., 1995; Webb et al., 1996; Takiguchi et al., 1999). One CTLA4-Ig (abatacept) tested in a trial with psoriasis patients afforded a significant improvement in clinical disease activity (Abrams et al., 1999; Abrams et al., 2000). In addition, antibody responses to T cell-dependent neoantigens were altered, supporting the hypothesis that the drug interferes with naïve T cell activation mediated by APCs (Abrams et al., 1999). Abatacept has been shown to be effective in two human RA trials (Moreland et al., 2002; Kremer et al., 2003). In the largest study, with 339 patients, the highest dose of CTLA4-Ig (10 mg kg^{-1}) showed a significant increase in the number of subjects with an ACR20 response relative to the placebo, starting at two months and continuing until the end of the study at six months (Kremer et al., 2003). The rates of ACR50 and ACR70 responses were also higher. Abatacept is currently in phase III trials for RA and phase II for MS.

Clinical data has also been reported on an antibody to CD80, galiximab, which appeared to show efficacy in a phase I trial in patients with plaque psoriasis (Gottlieb et al., 2004). Antibodies directed against CD86 have not reached human clinical trials but have been shown to have activity in a murine lung inflammation model and in a model of Sjögren syndrome (Mathur et al., 1999; Saegusa et al., 2000). Success with this approach has prompted the search for small molecules that antagonize the CD28–B7 interaction. Several small molecules have been identified that bind to CD80 and block the binding of CD28 with nanomolar IC_{50}s (Erbe et al., 2002; Green et al., 2003). However, these compounds had no effect on the binding of CD28 to CD86 and were not effective in blocking CD28 co-stimulation-dependent cell-based assays. It is unclear as to whether more potent inhibitors will yield the desired results.

Another interaction involves CD137L on APC and CD137 (4-1BB) on T cells. CD137 is an important co-stimulation signal for T cells, with $CD8^+$ cells being more sensitive than $CD4^+$ cells (Shuford et al., 1997; Melero et al., 1998; Mittler et al., 1999; Takahashi et al., 1999; Tan et al., 2000; Lee et al., 2004). This interaction appears to be very important for the function of $CD8^+$ T cells since mice deficient in either CD137 or CD137L have defects in the generation of cytotoxic T lymphocyte (CTL) responses (DeBenedette et al., 1999; Kwon et al., 2002). Anti-CD137L antibodies block the development of GVHD and prolong allograft survival in mice (Nozawa et al., 2001; Cho et al., 2004). This suggests that disrupting CD137/CD137L interactions may be a plausible mechanism for treating $CD8^+$ mediated conditions. In addition to the effects on $CD8^+$ cells, these antibodies are also able to suppress $CD4^+$ T cell help during T cell dependent humoral responses in vivo, possibly due to the expression of CD137 on DC (Mittler et al., 1999). This has been proven to translate to $CD4^+$ T cell mediated disease models where agonistic anti-CD137 antibodies have positive effects in two different models of SLE (Sun et al., 2002; Foell et al., 2003) as well as murine CIA (Foell et al., 2004). However, CD137 deficient mice have reduced Th2 responses (Vinay et al., 2004) and signaling through CD137 inhibits the suppressive function of regulatory T cells (Choi et al., 2004). Therefore, it is currently inconclusive as to which disease indications warrant activation of CD137 and which would be better served by inhibiting the interaction of CD137 with its ligand.

Another potential target yet to be fully explored is the interaction between the inducible co-stimulator (ICOS), which is expressed on activated T cells, and the B7 homologous protein (B7h) on APC. Ligation of ICOS induces the production of several different cytokines from activated T cells and may be a potent stimulatory of effector T cells (Hutloff et al., 1999). These features differentiate it from CD28 stimulation, which mainly targets naïve cell activation, making it an attractive therapeutic target. An antibody directed against B7h that blocks the interaction with ICOS does seem to have positive effects on CIA in mice (Iwai et al., 2002).

16.4.4
Cell Adhesion

The initial interaction between T cells and APC is thought to occur through cell adhesion molecules. APC express LFA-3, ICAM-1, ICAM-2 and DC-SIGN, all of which can interact with molecules such as CD2, LFA-1 and ICAM-3 on T cells. These are probably transient interactions that allow the T cells to sample many different APC expressing many different MHC-peptide complexes. Adhesion molecules are also important for interacting with the endothelium and facilitating migration of APC. One of the key interactions is between LFA-1 and ICAM-1. The importance of this interaction for the proper functioning of immune responses is evidenced by the reoccurring bacterial infection of patients with leukocyte adhesion deficiency (LAD) syndrome who lack LFA-1 (Arnaout, 1990). Dendritic cells appear to express both LFA-1 and ICAM-1. During DC migration LFA-1 can interact with ICAM-1 expressed on endothelium, facilitating the migration of DC to sites of inflammation. However, the adhesion of DC to T cells is thought to be mediated by the interaction of ICAM-1 on DC with LFA-1 on T cells and in turn helps to stabilize immune synapse formation (Grakoui et al., 1999). This interaction requires the prior stimulation of the T cell receptor, leading to a conformation change in LFA-1 that increases the affinity for ICAM-1 (Dustin and Springer, 1989). Therefore, inhibition of the LFA-1/ICAM interaction should inhibit DC migration and the activation of T cells. An antibody to LFA-1, efalizumab, that blocks its interaction with ICAM-1 has been approved for the treatment of psoriasis (Doggrell, 2004; Marecki and Kirkpatrick, 2004). The immunosuppressive effects of this antibody are thought to be mainly mediated by disruption of leukocyte trafficking. It may also have effects on T cell activation although it is unknown if this contributes to its efficacy in humans. However, the importance of the LFA-1/ICAM-1 interaction in activating T cells has been shown by the ability of anti-LFA-1 or anti-ICAM-1 antibodies to inhibit mixed-lymphocyte reactions (Werther et al., 1996). In addition, splenocytes from ICAM-1 deficient mice are unable to stimulate mixed-lymphocyte reactions although they can be stimulated themselves, implying that it is ICAM-1 on APC that is important for activating T cells (Sligh et al., 1993). Antibodies to ICAM-1 and ICAM-2 can partially block antigen-specific proliferation of T cells but a mixture of antibodies to all three ICAM isoforms completely inhibits proliferation (de Fougerolles et al., 1994). This interaction may also help in the polarization of T cells. Blocking LFA-1/ICAM-1 interactions with antibodies to either molecule can alter DC-induced polarization of naïve T cells to the Th1 phenotype (Smits et al., 2002). In humans an antibody to ICAM-1, enlinmomab, has had success in trials of RA (Kavanaugh et al., 1994; Kavanaugh et al., 1996). In this case the efficacy appeared to not to correlate to changes in lymphocyte migration but instead to the induction of anergy in naïve $CD4^+$ T cells. This may occur if the T cells are exposed to antigen in the absence of co-stimulation, which may be expected to happen if the interactions between LFA-1 and ICAM are blocked (Davis et al., 1995). This indicates that the mechanism of action may be due to blocking APC–T cell interactions instead of disruptions in leukocyte migration (Davis et al., 1995). However, longer-term studies with this particular antibody did lead to immunogenicity, which limited its therapeutic utility (Kavanaugh et al., 1997). Enlimomab has also been tested in the clinic for the treatment of acute clinical stroke with a negative outcome (Furuya et al., 2001) and did not reduce the rate of acute rejection after renal transplantation (Salmela et al., 1999). Another approach using an antisense ODN, alicaforsen, to reduce the expression of ICAM-1 has reached human trials (Cullell-Young et al., 2002). This treatment has shown some promise in the treatment Crohn's disease but failed to show efficacy in a small trail in RA (Yacyshyn et al., 1998; Maksymowych et al., 2002; Yacyshyn et al., 2002a; Yacyshyn et al., 2002b). Small molecule inhibitors of this interaction have also been described that do appear to affect APC–T cell interaction, although most appear to bind to LFA-1 (Liu, 2001).

DC-SIGN has also been shown to interact with ICAM molecules, where it facilitates both migration of DC and interaction with T cells. In DC it can bind to ICAM-2 on endothelium and can mediate DC tethering, rolling and transmigration (Geijtenbeek et al., 2000a). DC-SIGN can also interact with ICAM-3 expressed on resting T cells. Here it plays a role in initiating the contact between DC and T cells and is involved in the formation of the immune synapse (Geijtenbeek et al., 2000c). Antibodies to DC-SIGN can block DC-induced T cell proliferation of resting but not activated T cells (Geijtenbeek et al., 2000c). Therefore, inhibiting the interaction of DC-SIGN with ICAM-3 should interfere with T cell activation and blocking the interactions with ICAM-2 may affect DC trafficking. In addition to its role in cell–cell interactions, DC-SIGN has also been implicated as an antigen receptor for mannose- and galactose-containing structures. Antigens captured by DC-SIGN are rapidly internalized to lysosomal compartments and can be presented to $CD4^+$ T cells (Engering et al., 2002). DC-SIGN also facilitates the uptake of viruses like HIV into DC (Geijtenbeek et al., 2000b; Geijtenbeek and Van Kooyk, 2003).

Another important interaction is that between LFA-3 on APC and CD2 on T cells. Antibodies to CD2 are being tested in clinical trials for the treatment of psoriasis and for use in transplantation (Sorbera et al., 2002). Interaction of LFA-3 and CD2 enhances the responses of T cells to antigens (Moingeon et al., 1989). Antibodies to LFA-3 can inhibit the activation of T cells by APC (Bierer et al., 1988; Koyasu et al., 1990; Wingren et al., 1993; Teunissen et al., 1994). T cells from CD2 deficient mice require higher levels of antigen for stimulation than T cells from wild-type (WT) mice although it was shown that CD48 on APC may be the preferred ligand in mice (Bachmann et al., 1999). Antibodies to CD48 can block in vivo priming and MLR (Chavin et al., 1994). There is also evidence that CD2 is expressed on DC and that it plays a role in activation (Crawford et al., 2003).

16.4.5
APC Chemotaxis

Immature DC need to migrate to sites of inflammation where they encounter antigens and become activated. After activation mature DC then migrate to the lymph node where they can activate T cells. These steps are accomplished via the expression of chemokines receptors. Immature DC express receptors like CCR2, CCR5, CCR6, CXCR1 and CXCR2, which respond to pro-inflammatory chemokines like MIP-1, RANTES and MCP-1 or to tissue-homing chemokines like

CCL20. Many of these receptors are not specific for DC and also affect the recruitment of other leukocytes, such as neutrophils and monocytes, to the sites of inflammation. These chemokine receptors are G-protein coupled receptors and are amenable to antagonism by small molecules. Several molecules are in clinical development for various indications, including autoimmune diseases (Onuffer and Horuk, 2002; Schwarz and Wells, 2002; Gao and Metz, 2003). Most of the anti-inflammatory properties of these molecules are thought to be mediated through their interaction with neutrophils and monocytes. However, a protein antagonist of both CCR1 and CCR5 can reduce the baseline numbers of tracheal intraepithelial DC in mice and also inhibit the increase seen upon exposure to bacteria (Stumbles et al., 2001). One chemokine receptor of particular interest for DC migration is CCR6, which is thought to be crucial for mucosal immunity. Mice deficient in CCR6 have defects in DC localization in Peyer's patches, impaired humoral immune responses to orally administered antigen and have diminished DTH responses (Cook et al., 2000; Varona et al., 2001). In addition these mice have reduced airway hyperresponsiveness and airway eosinophilia in a cockroach antigen model of pulmonary inflammation (Lukacs et al., 2001). Both eosinophilia and airway hyperresponsiveness are reduced in the absence of CCR6, implying that it plays a role in the activation of T cells.

Maturation of DC leads to the down-regulation of the expression of chemokines receptors that respond to inflammatory stimuli and an upregulation of CCR7. CCR7 binds to CCL21 on endothelial cells and to CCL19 expressed in the T cell area of the lymph node. CCR7 is express by T cells and DC and plays a role in the trafficking of these cells to lymph nodes. Mice that lack CCR7 have impaired migration of activated DC to the lymph node and lack contact and delayed type hypersensitivity reactions (Forster et al., 1999). The same findings are also observed in ptl mutant mice, which lack both ligands for CCR7, CCL21 and CCL19 (Gunn et al., 1999; Mori et al., 2001). Antibodies to CCL21 block DC migration in vivo and can inhibit contact hypersensitivity reactions (Saeki et al., 1999; Engeman et al., 2000). There are no reported small molecule antagonists of CCR7 but protein antagonists can be generated by truncating either of the naturally occurring ligands. These antagonists can inhibit CTL responses and block the generation of GVHD (Sasaki et al., 2003; Pilkington et al., 2004). Therefore, it appears that interfering with chemokine binding to DC could lead to immunomodulation by blocking migration either to or from sites of inflammation.

Migration of B cells can also be specifically inhibited by antagonizing CXCR5 (BLR1), which binds to CXCL13 (BCA-1) (Gunn et al., 1998; Legler et al., 1998). Mice deficient in CXCR5 have defects in Peyer's patches and in the migration of activated B cells to the proper location in the spleen (Forster et al., 1996).

16.4.6
APC Survival

Each APC has a limited lifetime and requires signals for survival. The different signals necessary for B cell survival are the best understood. The most important seems to come from the B lymphocyte stimulator (BAFF, also known as BlyS). BAFF binds to three known receptors, BCMA, BAFF-R and TACI, on B cells and is thought to be active in its soluble form. BAFF is absolutely required for B cell maturation and survival. In BAFF-deficient mice B cell development is disrupted and virtually no mature B cells are present (Gross et al., 2001; Schiemann et al., 2001). This appears to be mediated through interactions with BAFF-R since only BAFF-R deficient mice share the same phenotype (Schiemann et al., 2001; Thompson et al., 2001). BAFF may also be involved in autoimmune disorders. Mice that overexpress BAFF have increases in the number of B cells and exhibit autoimmune symptoms (Mackay et al., 1999; Groom et al., 2002). Furthermore, high levels of BAFF have been detected in patients with rheumatic autoimmune diseases (Cheema et al., 2001; Zhang et al., 2001; Groom et al., 2002). An Ig-fusion protein with TACI, one of the BAFF receptors, has proved effective in blocking the progression of arthritis in the mouse CIA model (Gross et al., 2001) and in a murine model of SLE (Gross et al., 2000; Ramanujam et al., 2004). An Ig-fusion protein of BAFF-R has also been reported to be efficacious in a mouse SLE model (Kayagaki et al., 2002). An antibody to BAFF (belimumab) blocks the interaction with all three receptors and reduces B cell population in several animal models (Baker et al., 2003). A phase I study in SLE patients has been completed and the expected reduction in peripheral B cells was found (Stohl, 2004). Recently, BAFF has also been implicated in providing co-stimulation to B and T cells (Huard et al., 2001; Huard et al., 2004; Ng et al., 2004). This appears to be a property of the membrane-bound form rather than the soluble molecule and is also thought to occur through interaction with BAFF-R. Antibodies to BAFF-R that block interaction with BAFF also block co-stimulation (Ng et al., 2004). Thus, inhibition of BAFF function may lead to beneficial effects by reducing B cell survival and by blocking co-stimulation, thereby reducing T cell activation.

16.4.7
Intracellular Signaling

APC can respond to many different inflammatory stimuli such as cytokines, chemokines and other small molecule mediators. Many of these responses have been described in detail above. Inhibiting the intracellular signaling pathways can inhibit the action of these stimuli. However, these pathways are numerous and in almost all cases they are not specific for APC function. For example, inhibition of targets like p38 kinase or NFκB activation will inhibit APC functions, as well as having effects in other leukocytes. This very broad field is beyond the scope of this chapter; several reviews discuss the targeting of kinases inhibitors and other signaling pathway for immunosuppression (Roshak et al., 2002; Blease and Raymon, 2003; Kumar et al., 2003; Adcock and Caramori, 2004; O'Shea et al., 2004; Sweeney and Firestein, 2004; Uckun and Mao, 2004; Yamamoto and Gaynor, 2004). However, one kinase worth mentioning is Bruton's Tyrosine Kinase, BTK, which is involved in B cell receptor (BCR) signaling (Khan, 2001; Maas and Hendriks, 2001). This kinase is mutated in human X-linked agammaglobulemia, which leads to a lack of B cells in the periphery due to a block in B cell maturation (Conley et al., 2000; Vihinen et al., 2000). Furthermore, X-linked immunodeficiency in mice is also linked to mutations in this kinase and is characterized by abnormal activation of B cells (Rawlings et al., 1993). There are several inhibitors of this kinase reported, including one with efficacy in a mouse model of GVHD (Mahajan et al., 1999; Cetkovic-Cvrlje and Uckun, 2004).

16.4.8
APC Depletion

Another approach to inhibit APC function is to just remove the APC. The best example of an APC depleting therapy is rituximab, which is a human/mouse chimeric monoclonal antibody directed against CD20 (Kazkaz and Isenberg, 2004; Rastetter et al., 2004). CD20 is expressed on pre and mature B cells but not on plasma cells. Its exact function is unknown although is it thought to be a calcium channel subunit and have a variety of effects on various B cell activities (Golay et al., 1985; Bubien et al., 1993; Tedder and Engel, 1994). In vivo the binding of rituximab to CD20 depletes B cells by initiating complement or antibody-mediated cytotoxicity and apoptosis. Indeed, treatment with the antibody leads to a dramatic reduction in the number of peripheral blood B cells that is long lasting (Edwards et al., 2004). This drug was approved for the treatment of B cell non-Hodgkin's lymphoma in the USA in 1997. Due to its effect on B cell population, several clinical studies have been carried out on various autoimmune diseases where B cells have been implicated, including SLE, autoimmune thrombocytopenia, IgM mediated neuropathies, cold agglutinin disease, hemolytic anemia, myasthenia gravis, Wegener's granuloma, and pure red cell aplasia (Edwards et al., 2002; Silverman and Weisman, 2003; Kazkaz and Isenberg, 2004). There have been several reports of clinical trials of rituximab in RA (Edwards and Cambridge, 2001; De Vita et al., 2002; Leandro et al., 2002; Edwards et al., 2004). Data from a large randomized trial in RA has been published (Edwards et al., 2004). Rituximab, in combination with methotrexate, showed a statistically significant increase in the percentage of patients with ACR20, ACR50 or ACR70 responses at 24 weeks that was maintained up to 48 weeks. In all patients B cell counts and rheumatoid factor levels were reduced but serum IgG levels remained in the normal range. However, it remains unclear whether the effect of B cell depletion on RA is due to effects on antigen presentation, antibody production or some other B cell activity. Other strategies for depleting B cells are also undergoing clinical evaluation, including epratuzumab, which targets another marker expressed on all pro to mature B cells, CD22, and leads to B cell depletion (Cesano and Gayko, 2003; Poe et al., 2004). In a small 9 patient study epratuzumab was associated with clinical benefits for at least one month in patients with moderate SLE (Kaufmann et al., 2004).

16.5
APC per se as Drugs – DC-based Immunotherapy Therapy

Recent developments in the understanding of DC biology both in immunity and tolerance as well as technologies for the purification and manipulation of DC ex vivo have brought about opportunities for employing DC themselves as therapeutic agents. Thus, autologous DC can be purified and conditioned ex vivo by exposure to antigens and/or differentiation factors to render them either immunostimulatory or tolerogenic. The cells can then be re-infused back into patients to modulate immune responses.

16.5.1
DC-based Cancer Vaccines

Taking advantage of DC as a strong "natural adjuvant", DC-based cancer vaccines have shown promise in eliciting anti-tumor responses in clinical trials (Zhang et al., 2002; Schuler et al., 2003; Blattman and Greenberg, 2004; Brody and Engleman, 2004; Cerundolo et al., 2004; Turtle and Hart, 2004). The first DC-based cancer vaccine study was published in 1996 (Hsu et al., 1996). In this study, a group of four patients with non-Hodgkin's lymphoma (NHL) were treated with purified autologous DC loaded ex vivo with specific recombinant idiotype proteins. Although the number of patients in this study was rather small, the positive clinic responses (two complete responses and one stabilization of the disease) and anti-tumor immune responses observed provided a valuable proof of concept. Since then, over 30 trials have been carried out, targeting various cancers. Although DC-based vaccines are still considered at an early stage (Schuler et al., 2003), significant progress has been made with vaccines tested in several different cancers, including melanoma, prostate carcinoma, renal cell carcinoma, multiple myeloma and non-Hodgkin's lymphoma, breast carcinoma, colorectal carcinoma and non-small cell lung cancer (Brody and Engleman, 2004; Turtle and Hart, 2004). Encouragingly, in all clinical trials to date, positive clinical results were observed, such as partial or complete regressions of tumors and cancer-specific humoral or cellular responses. Although concerns remain about possible complications such as the development of autoimmune disease, few side effects have been noted, except for vitiligo in melanoma trials (Nestle et al., 2001), and DC-based vaccines have generally been considered safe and well tolerated (Gilboa, 2001).

The findings from these trials, while promising, have pointed out several key issues or variables for further investigation, including types of DC used, antigen loading, DC activation, and routes of vaccine administration. It has not yet been conclusively determined which subtypes of DC are the best suited for cancer vaccines, and DC prepared in different ways differ in their ability to stimulate tumor-specific T cell responses (Osugi et al., 2002). The loading of DC with tumor-associated antigens has been explored in several ways (Zhang et al., 2002; Turtle and Hart, 2004), including "pulsing" DC with peptides, whole proteins, unseparated tumor lysates or necrotic tumor cells (O'Rourke et al., 2003), DC-tumor fusions

(Kugler et al., 2000; Trevor et al., 2004), transduction with recombinant viral expression vectors (Trevor et al., 2001), or transfection with TAA-coding naked DNA (Van Tendeloo et al., 1998; Van Tendeloo et al., 2001) or mRNA (Nair and Boczkowski, 2002). One of most critical variables for DC vaccines development may be the protocol used to optimally activate immature DC after antigen loading since immature DC are poorly immunogenic and may, in fact, be tolerogenic (Hackstein and Thomson, 2004). Finally, the mode in delivery of DC may be important since the numbers of DC found to reach secondary lymphoid tissues such as the lymph node varies, depending on the route of administration (Lappin et al., 1999). In a pilot clinical trial (Fong et al., 2001a), antigen-specific T cell responses were observed in all three cohorts of patient with metastatic prostate cancer immunized with antigen-pulsed DC delivered by intravenous (i.v.), intradermal (i.d.), or intralymphatic (i.l.) injection. However, a distinct Th1 response, i.e., production of IFN-γ by primed T cells, was only obvious with i.d. and i.l. routes of administration. In contrast, production of antigen-specific antibodies was more evident with the i.v. route (5/9 patients) compared with i.d. (1/6 patients) and for i.l. (2/6 patients).

16.5.2
Targeting and Activating DC In Vivo

Although ex vivo generated DC-based tumor vaccines have shown promise, the procedures are time-consuming, expensive, and difficult to apply to large-scale vaccination in a human population. In addition, the culture conditions for ex vivo manipulation of DC may not be ideal for harnessing the full potential of DC. Therefore, activating and delivering specific antigens to DC in vivo may prove to be a more efficient strategy for induction of anti-tumor immune responses (Foged et al., 2002).

Several strategies have been used to target DC in vivo to enhance immune responses, which exploit various aspect of DC biology such as differentiation, antigen capture, activation, maturation and migration. Administration of the DC growth factor Flt3 ligand (Flt3L) systematically increases the number of DC by 20-fold in vivo (Fong et al., 2001b), which increases the number of available DC to capture vaccine antigen and be activated. However, notably, most DC mobilized by Flt3L appear to have an immature phenotype and may be tolerogenic rather than immunostimulatory if not properly activated by other constituents of a vaccine (Viney et al., 1998).

Specific antigens can be delivered directly to DC by targeting surface receptors on DC or through enhancing antigen uptake via receptor-mediated phagocytosis, macropinocytosis, and endocytosis. In addition, it appears that different DC surface receptors may deliver antigens into different subcellular compartments of DC for processing, and may thus induce different immune responses to targeted antigens. Several DC receptors have been exploited, mostly with promising findings, including Fc receptors (Akiyama et al., 2003; Yada et al., 2003), DEC-205 (Bonifaz et al., 2002; Bonifaz et al., 2004; Geijtenbeek et al., 2004), chemokine receptors (Biragyn et al., 1999), Flt3 (Hung et al., 2001), B7 (Deliyannis et al., 2000), macrophage mannose receptor (Engering et al., 1997) and glycolipid globotriacylceramide (Gb30), which is a receptor for Shiga B toxin (Haicheur et al., 2000; Haicheur et al., 2003). Targeted delivery of antigens can also be enhanced by exploiting different antigen delivery systems such as lipid particles, polymer particles, or viral vectors (Foged et al., 2002).

16.5.3
DC-based Immunotherapy for Transplantation and Autoimmune Diseases

Opposite to immunostimulation, the other crucial function of DC is to induce tolerance of T cells to self antigens both in the thymus and in peripheral tissues. Thus, under normal conditions, DC in peripheral tissues constantly capture and present self-antigens to auto-reactive T cells. However, in the absence of co-stimulation this may result in tolerance of these T cells by one of several processes, including deletion, anergy, Th2 skewing or the induction of regulatory T cells. For example, targeted delivery of antigens via DEC-205 receptor into DC in the absence of immunostimulation in mice induced antigen-specific T cell tolerance (Hawiger et al., 2001; Bonifaz et al., 2002), which was refractory to antigen rechallenge in the presence of adjuvant.

There is now ample evidence that tolerogenic DC generated ex vivo can be efficacious in suppressing immune responses in vivo after re-infusion. Mouse bone marrow (BM)-derived DC (BM-DC), pulsed with a mixture of islet antigen-derived peptides, prevent the development of diabetes in non-obese diabetic (NOD) mice following i.v. injection (Feili-Hariri et al., 1999). Rat immature BM-DC cultured with GM-CSF and IL-4 induced tolerance to EAE when injected subcutaneously (s.c.) (Huang et al., 2000; Xiao et al., 2001). Interestingly, immature DC isolated from EAE rats, after ex vivo culture with GM-CSF and IL-4, effectively transferred tolerance to naïve rats. In this experiment there was no need for antigen loading ex vivo (Xiao et al., 2001), which is significant since it suggests that ex vivo antigen loading may be omitted when autologous DC isolated from patients with ongoing autoimmune diseases are manipulated in vitro.

Tolerogenic DC can also be generated by culturing DC with selected cytokines (Simon et al., 1991; Lu et al., 1999; Shinomiya et al., 1999; Griffin et al., 2001; Menges et al., 2002; Muller et al., 2002; Yarilin et al., 2002; Sato et al., 2003a; Sato et al., 2003b; Duan et al., 2004; Zhang et al., 2004). For example, DC cultured with GM-CSF and TGF-β ex vivo and then administered to mice prolonged survival of allografts in vivo (Lu et al., 1999). In another study, transfer of IFN-γ-treated splenic DC suppressed diabetes in NOD mice following intraperitoneal (i.p.) injection (Shinomiya et al., 1999). However, as is the case with tumor vaccines, the routes of administration are important in determining the final functional outcome of injected tolerogenic DC. For example, s.c. – but not i.v. – injected IFN-γ-treated splenic DC effectively suppressed EAE in Lewis rats. Genetic engineering can be used to construct tolerogenic DC that express certain cytokines or proteins. For example, DC infected with an adenoviral vector expressing IL-4 could effectively suppress established murine CIA (Kim, S.H. et al., 2001) or prevent diabetes in NOD mice (Feili-Hariri et al., 2003). A mixture of BM-DC infected with an adenoviral vector expressing IL-10 or TGF-β prolonged survival of renal grafts in a mouse model of transplantation (Gorczynski et al., 2000), which was correlated with both inhibition of the induction of $CD8^+$ CTL and enhancement of a polarization to produce Th2 cytokines (IL-4, IL-10, and TGF-β) on antigen-specific restimulation in vitro. DC expressing IL-10 (Takayama et al., 1998), the co-stimulation-blocking agent CTLA4-Ig (Lu et al., 1999; Takayama et al., 2000) or apoptosis inducer FasL (Matsue et al., 1999; Min et al., 2000) have also been used.

Encouragingly, human immature DC generated from $CD34^+$ progenitors or $CD14^+$ monocyte-derived in vitro cultures can be induced to become tolerogenic DC (Jonuleit et al., 2000; Dhodapkar et al., 2001; Roncarolo et al., 2001). As observed with mouse DC, IL-10 treated human DC can inhibit alloactivation of human $CD4^+$ T cells or antigen-specific $CD4^+$ and $CD8^+$ T cells (Steinbrink et al., 1997; Steinbrink et al., 2002). This may be due to the effect of IL-10 inhibiting both IL-12 production and the upregulation of co-stimulatory molecules on the DC. Certain DC subtypes, such as pDC, may also function as tolerogenic DC in the absence of adequate stimulation to induce CD4 tolerance or activated by CD40L to induce T-regulatory (Tr) cells (Gilliet and Liu, 2002; Kuwana, 2002). In a small human study, administration of antigen-pulsed immature DC in vivo resulted in induction of regulatory IL-10 producing T cells and reduction of IFN-γ producing T cells as well as $CD8^+$ CTL activity (Dhodapkar et al., 2001).

16.6
Conclusion

As a central player in immune responses, APC serve as ideal targets for drug discovery, providing opportunities both to potentiate the immune response against pathogens or tumors and to inhibit unwanted immune responses in immune-mediated inflammatory diseases (IMID) like autoimmune diseases, allergy/asthma, and transplantation. This notion is validated by the fact that several marketed immunomodulatory drugs may attribute some of their therapeutic efficacy to their effects on APC function. In addition, ex vivo manipulated DC have shown promising results in induction of tumor immunity as well as tolerance in man. The ever-expanding understanding of the biology of APC has brought about exciting opportunities for the discovery of drugs that specifically target APC function, including APC activation, antigen presentation, and migration. These efforts hold the promise to deliver therapeutics that may offer better medicines for many indications.

References

Abrams, J.R., Kelley, S.L., Hayes, E., Kikuchi, T., Brown, M.J., Kang, S., Lebwohl, M.G., Guzzo, C.A., Jegasothy, B.V., Linsley, P.S. and Krueger, J.G. (2000): Blockade of T lymphocyte costimulation with cytotoxic T lymphocyte-associated antigen 4-immunoglobulin (CTLA4Ig) reverses the cellular pathology of psoriatic plaques, including the activation of keratinocytes, dendritic cells, and endothelial cells. *J. Exp. Med.* 192: 681–693.

Abrams, J.R., Lebwohl, M.G., Guzzo, C.A., Jegasothy, B.V., Goldfarb, M.T., Goffe, B.S., Menter, A., Lowe, N.J., Krueger, G., Brown, M.J., Weiner, R.S., Birkhofer, M.J., Warner, G.L., Berry, K.K., Linsley, P.S., Krueger, J.G., Ochs, H.D., Kelley, S.L. and Kang, S. (1999): CTLA4Ig-mediated blockade of T-cell costimulation in patients with psoriasis vulgaris. *J. Clin. Invest.* 103: 1243–1252.

Adcock, I.M. and Caramori, G. (2004): Chemokine receptor inhibitors as a novel option in treatment of asthma. *Current Drug Targets: Inflamm. Allergy* 3: 257–261.

Adorini, L., Giarratana, N. and Penna, G. (2004): Pharmacological induction of tolerogenic dendritic cells and regulatory T cells. *Semin. Immunol.* 16: 127–134.

Akiyama, K., Ebihara, S., Yada, A., Matsumura, K., Aiba, S., Nukiwa, T. and Takai, T. (2003): Targeting apoptotic tumor cells to Fc gamma R provides efficient and versatile vaccination against tumors by dendritic cells. *J. Immunol.* (Baltimore, Md.: 1950) 170: 1641–1648.

Allen, E.M., Vitali, N., Underwood, S., Sweeney, D., Wheeler, D., Lawrence, C., Patterson, J., Graupe, M., Wong, D., Palmer, J. and Aldous, D. (2001): Reversible cathepsin S (CATS) inhibitors block invariant chain degradation both *in vitro* and *in vivo*. *Inflam. Res.* 50: S159.

Almawi, W.Y., Assi, J.W., Chudzik, D.M., Jaoude, M.M. and Rieder, M.J. (2001): Inhibition of cytokine production and cytokine-stimulated T-cell activation by FK506 (tacrolimus)1. *Cell Transplant.* 10: 615–623.

Anders, H.-J., Banas, B. and Schloendorff, D. (2004a): Signaling Danger: Toll-Like Receptors and their Potential Roles in Kidney Disease. *J. Am. Soc. Nephrol.* 15: 854–867.

Anders, H.-J., Vielhauer, V., Eis, V., Linde, Y., Kretzler, M., Perez de Lema, G., Strutz, F., Bauer, S., Rutz, M., Wagner, H., Grone, H.-J. and Schlondorff, D. (2004b): Activation of Toll-like receptor-9 induces progression of renal disease in MRL-Fas(lpr) mice. *FASEB J.* 18: 534–536, 510.1096/fj.1003-0646fje.

Andre, P., Groettrup, M., Klenerman, P., De Giuli, R., Booth, B.L., Jr., Cerundolo, V., Bonneville, M., Jotereau, F., Zinkernagel, R.M. and Lotteau, V. (1998): An inhibitor of HIV-1 protease modulates proteasome activity, antigen presentation, and T cell responses. *Proc. Natl. Acad. Sci. U.S.A.* 95: 13 120–13 124.

Apcher, G.S., Heink, S., Zantopf, D., Kloetzel, P.-M., Schmid, H.-P., Mayer, R.J. and Kruger, E. (2003): Human immunodeficiency virus-1 Tat protein interacts with distinct proteasomal a and b subunits. *FEBS Lett.* 553: 200–204.

Arnaout, M.A. (1990): Leukocyte adhesion molecules deficiency: its structural basis, pathophysiology and implications for modulating the inflammatory response. *Immunol. Rev.* 114: 145–180.

Bachmann, M.F., Barner, M. and Kopf, M. (1999): CD2 sets quantitative thresholds in T cell activation. *J. Exp. Med.* 190: 1383–1392.

Bai, A. and Forman, J. (1997): The effect of the proteasome inhibitor lactacystin on the presentation of transporter associated with antigen processing (TAP)-dependent and TAP-independent peptide epitopes by class I molecules. *J. Immunol.* 159: 2139–2146.

Baker, K.P., Edwards, B.M., Main, S.H., Choi, G.H., Wager, R.E., Halpern, W.G., Lappin, P.B., Riccobene, T., Abramian, D., Sekut, L., Sturm, B., Poortman, C., Minter, R.R., Dobson, C.L., Williams, E., Carmen, S., Smith, R., Roschke, V., Hilbert, D.M., Vaughan, T.J. and Albert, V.R. (2003): Generation and characterization of LymphoStat-B, a human monoclonal antibody that antagonizes the bioactivities of

B lymphocyte stimulator. *Arthritis Rheum.* 48: 3253–3265.

Bartfai, T., Behrens, M.M., Gaidarova, S., Pemberton, J., Shivanyuk, A. and Rebek, J., Jr. (2003): A low molecular weight mimic of the Toll/IL-1 receptor/resistance domain inhibits IL-1 receptor-mediated responses. *Proc. Natl. Acad. Sci. U.S.A.* 100: 7971–7976.

Belz, G.T., Carbone, F.R. and Heath, W.R. (2002): Cross-presentation of antigens by dendritic cells. *Crit. Rev. Immunol.* 22: 439–448.

Benichou, G., Valujskikh, A. and Heeger, P.S. (1999): Contributions of direct and indirect T cell alloreactivity during allograft rejection in mice. *J. Immunol.* 162: 352–358.

Bierer, B.E., Barbosa, J., Herrmann, S. and Burakoff, S.J. (1988): Interaction of CD2 with its ligand, LFA-3, in human T cell proliferation. *J. Immunol.* 140: 3358–3363.

Biragyn, A., Tani, K., Grimm, M.C., Weeks, S. and Kwak, L.W. (1999): Genetic fusion of chemokines to a self tumor antigen induces protective, T-cell dependent antitumor immunity. *Nat. Biotechnol.* 17: 253–258.

Biroc, S.L., Gay, S., Hummel, K., Magill, C., Palmer, J.T., Spencer, D.R., Sa, S., Klaus, J.L., Michel, B.A., Rasnick, D. and Gay, R.E. (2001): Cysteine protease activity is up-regulated in inflamed ankle joints of rats with adjuvant-induced arthritis and decreases with in vivo administration of a vinyl sulfone cysteine protease inhibitor. *Arthritis Rheum.* 44: 703–711.

Bishop, D.K. and Li, W. (1992): Cyclosporin A and FK506 mediate differential effects on T cell activation in vivo. *J. Immunol.* 148: 1049–1054.

Blanco, P., Palucka, A.K., Gill, M., Pascual, V. and Banchereau, J. (2001): Induction of dendritic cell differentiation by IFN-alpha in systemic lupus erythematosus. *Science* 294: 1540–1543.

Blattman, J.N. and Greenberg, P.D. (2004): Cancer immunotherapy: A treatment for the masses. *Science* 305: 200–205.

Blease, K. and Raymon, H.K. (2003): Small molecule inhibitors of cell signaling: novel future therapeutics for asthma and chronic obstructive pulmonary diseases. *Curr. Opin. Invest. Drugs* 4: 544–551.

Blotta, M.H., DeKruyff, R.H. and Umetsu, D.T. (1997): Corticosteroids inhibit IL-12 production in human monocytes and enhance their capacity to induce IL-4 synthesis in CD4+ lymphocytes. *J. Immunol. (Baltimore, Md.: 1950)* 158: 5589–5595.

Boehm, J.C. and Adams, J.L. (2000): New inhibitors of p38 kinase. *Expert Opin. Therapeutic Pat.* 10: 25–37.

Bonifaz, L., Bonnyay, D., Mahnke, K., Rivera, M., Nussenzweig, M.C. and Steinman, R.M. (2002): Efficient targeting of protein antigen to the dendritic cell receptor DEC-205 in the steady state leads to antigen presentation on major histocompatibility complex class I products and peripheral CD8+ T cell tolerance. *J. Exp. Med.* 196: 1627–1638.

Bonifaz, L.C., Bonnyay, D.P., Charalambous, A., Darguste, D.I., Fujii, S.-I., Soares, H., Brimnes, M.K., Moltedo, B., Moran, T.M. and Steinman, R.M. (2004): In vivo targeting of antigens to maturing dendritic cells via the DEC-205 receptor improves T cell vaccination. *J. Exp. Med.* 199: 815–824.

Boon, L., Brok, H.P.M., Bauer, J., Ortiz-Buijsse, A., Schellekens, M.M., Ramdien-Murli, S., Blezer, E., Van Meurs, M., Ceuppens, J., De Boer, M., Hart, B.A. and Laman, J.D. (2001): Prevention of experimental autoimmune encephalomyelitis in the common marmoset (Callithrix jacchus) using a chimeric antagonist monoclonal antibody against human CD40 is associated with altered B cell responses. *J. Immunol.* 167: 2942–2949.

Boule, M.W., Broughton, C., Mackay, F., Akira, S., Marshak-Rothstein, A. and Rifkin, I.R. (2004): Toll-like receptor 9-dependent and -independent dendritic cell activation by chromatin-immunoglobulin G complexes. *J. Exp. Med.* 199: 1631–1640.

Boumpas, D.T., Furie, R., Manzi, S., Illei, G.G., Wallace, D.J., Balow, J.E. and Vaishnaw, A. (2003): A short course of BG9588 (anti-CD40 ligand antibody) improves serologic activity and decreases hematuria in patients with proliferative lupus glomerulonephritis. *Arthritis Rheum.* 48: 719–727.

Breslin, H.J., Miskowski, T.A., Kukla, M.J., Leister, W.H., De Winter, H.L., Gauthier,

D.A., Somers, M.V.F., Peeters, D.C.G. and Roevens, P.W.M. (2002): Design, synthesis, and tripeptidyl peptidase II inhibitory activity of a novel series of (S)-2,3-dihydro-2-(4-alkyl-1H-imidazol-2-yl)-1H-indoles. *J. Med. Chem.* 45: 5303–5310.

Brody, J.D. and Engleman, E.G. (2004): DC-based cancer vaccines: lessons from clinical trials. *Cytotherapy* 6: 122–127.

Bubien, J.K., Zhou, L.J., Bell, P.D., Frizzell, R.A. and Tedder, T.F. (1993): Transfection of the CD20 cell surface molecule into ectopic cell types generates a calcium conductance found constitutively in B lymphocytes. *J. Cell Biol.* 121: 1121–1132.

Burdin, N., Guy, B., and Moingeon, P. (2002): Immunological foundations to the quest for new vaccine adjuvants. *Biodrugs* 18: 79–93.

Castedo, M., Ferri, K.F. and Kroemer, G. (2002): Mammalian target of rapamycin (mTOR): pro- and anti-apoptotic. *Cell Death Different.* 9: 99–100.

Castro, A.C., Dang, L.C., Soucy, F., Grenier, L., Mazdiyasni, H., Hottelet, M., Parent, L., Pien, C., Palombella, V. and Adams, J. (2003): Novel IKK inhibitors: b-carbolines. *Bioorg. Med. Chem. Lett.* 13: 2419–2422.

Cerundolo, V., Benham, A., Braud, V., Mukherjee, S., Gould, K., Macino, B., Neefjes, J. and Townsend, A. (1997): The proteasome-specific inhibitor lactacystin blocks presentation of cytotoxic T lymphocyte epitopes in human and murine cells. *Eur. J. Immunol.* 27: 336–341.

Cerundolo, V., Hermans, I.F. and Salio, M. (2004): Dendritic cells: a journey from laboratory to clinic. *Nat. Immunol.* 5: 7–10.

Cesano, A. and Gayko, U. (2003): CD22 as a target of passive immunotherapy. *Semin. Oncol.* 30: 253–257.

Cetkovic-Cvrlje, M. and Uckun, F.M. (2004): Dual targeting of Bruton's tyrosine kinase and Janus kinase 3 with rationally designed inhibitors prevents graft-versus-host disease (GVHD) in a murine allogeneic bone marrow transplantation model. *Br. J. Haematol.* 126: 821–827.

Chavin, K.D., Qin, L., Lin, J., Woodward, J., Baliga, P., Kato, K., Yagita, H. and Bromberg, J.S. (1994): Anti-CD48 (murine CD2 ligand) mAbs suppress cell mediated immunity in vivo. *Int. Immunol.* 6: 701–709.

Cheema, G.S., Roschke, V., Hilbert, D.M. and Stohl, W. (2001): Elevated serum B lymphocyte stimulator levels in patients with systemic immune-based rheumatic diseases. *Arthritis Rheum.* 44: 1313–1319.

Chilman-Blair, K., Leeson, P.A. and Bayes, M. (2003): E-5564: treatment of septic shock TLR4 (LPS) receptor antagonist. *Drugs Future* 28: 633–639.

Cho, H.R., Kwon, B., Yagita, H., La, S., Lee, E.A., Kim, J.E., Akiba, H., Kim, J., Suh, J.H., Vinay, D.S., Ju, S.A., Kim, B.S., Mittler, R.S., Okumura, K. and Kwon, B.S. (2004): Blockade of 4-1BB (CD137)/4-1BB ligand interactions increases allograft survival. *Transplant Int.: Off. J. Eur. Soc. Organ Transplant.* 17: 351–361.

Choi, B.K., Bae, J.S., Choi, E.M., Kang, W.J., Sakaguchi, S., Vinay, D.S. and Kwon, B.S. (2004): 4-1BB-dependent inhibition of immunosuppression by activated CD4+CD25+ T cells. *J. Leukocyte Biol.* 75: 785–791.

Conley, M.E., Rohrer, J. and Minegishi, Y. (2000): X-linked agammaglobulinemia. *Clin. Rev. Allergy Immunol.* 19: 183–204.

Cook, D.N., Pisetsky, D.S. and Schwartz, D.A. (2004): Toll-like receptors in the pathogenesis of human disease. *Nat. Immunol.* 5: 975–979.

Cook, D.N., Prosser, D.M., Forster, R., Zhang, J., Kuklin, N.A., Abbondanzo, S.J., Niu, X.D., Chen, S.C., Manfra, D.J., Wiekowski, M.T., Sullivan, L.M., Smith, S.R., Greenberg, H.B., Narula, S.K., Lipp, M. and Lira, S.A. (2000): CCR6 mediates dendritic cell localization, lymphocyte homeostasis, and immune responses in mucosal tissue. *Immunity* 12: 495–503.

Craiu, A., Gaczynska, M., Akopian, T., Gramm, C.F., Fenteany, G., Goldberg, A.L. and Rock, K.L. (1997): Lactacystin and clasto-lactacystin b-lactone modify multiple proteasome b-subunits and inhibit intracellular protein degradation and major histocompatibility complex class I antigen presentation. *J. Biol. Chem.* 272: 13 437–13 445.

Crawford, K., Stark, A., Kitchens, B., Sternheim, K., Pantazopoulos, V., Triantafellow, E., Wang, Z., Vasir, B., Larsen, C.E.,

Gabuzda, D., Reinherz, E. and Alper, C.A. (2003): CD2 engagement induces dendritic cell activation: Implications for immune surveillance and T-cell activation. *Blood* 102: 1745–1752.

Cullell-Young, M., Del Fresno, M., Leeson, P.A. and Bayes, M. (2002): Alicaforsen sodium: treatment of IBD antipsoriatic. *Drugs Future* 27: 439–445.

Cumberbatch, M., Dearman, R.J. and Kimber, I. (1999): Inhibition by dexamethasone of Langerhans cell migration: influence of epidermal cytokine signals. *Immunopharmacology* 41: 235–243.

Davis, L.S., Kavanaugh, A.F., Nichols, L.A. and Lipsky, P.E. (1995): Induction of persistent T cell hyporesponsiveness in vivo by monoclonal antibody to ICAM-1 in patients with rheumatoid arthritis. *J. Immunol.* 154: 3525–3537.

de Fougerolles, A.R., Qin, X. and Springer, T.A. (1994): Characterization of the function of intercellular adhesion molecule (ICAM)-3 and comparison with ICAM-1 and ICAM-2 in immune responses. *J. Exp. Med.* 179: 619–629.

De Vita, S., Zaja, F., Sacco, S., De Candia, A., Fanin, R. and Ferraccioli, G. (2002): Efficacy of selective B cell blockade in the treatment of rheumatoid arthritis: evidence for a pathogenetic role of B cells. *Arthritis Rheum.* 46: 2029–2033.

DeBenedette, M.A., Wen, T., Bachmann, M.F., Ohashi, P.S., Barber, B.H., Stocking, K.L., Peschon, J.J. and Watts, T.H. (1999): Analysis of 4-1BB ligand (4-1BBL)-deficient mice and of mice lacking both 4-1BBL and CD28 reveals a role for 4-1BBL in skin allograft rejection and in the cytotoxic T cell response to influenza virus. *J. Immunol.* 163: 4833–4841.

Deliyannis, G., Boyle, J.S., Brady, J.L., Brown, L.E. and Lew, A.M. (2000): A fusion DNA vaccine that targets antigen-presenting cells increases protection from viral challenge. *Proc. Natl. Acad. Sci. U.S.A.* 97: 6676–6680.

Dhib-Jalbut, S., Chen, M., Henschel, K., Ford, D., Costello, K. and Panitch, H. (2002): Effect of combined IFNb-I a and glatiramer acetate therapy on GA-specific T-cell responses in multiple sclerosis. *Multiple Sclerosis* 8: 485–491.

Dhib-Jalbut, S., Chen, M., Said, A., Zhan, M., Johnson, K.P. and Martin, R. (2003): Glatiramer acetate-reactive peripheral blood mononuclear cells respond to multiple myelin antigens with a Th2-biased phenotype. *J. Neuroimmunol.* 140: 163–171.

Dhodapkar, M.V., Steinman, R.M., Krasovsky, J., Munz, C. and Bhardwaj, N. (2001): Antigen-specific inhibition of effector T cell function in humans after injection of immature dendritic cells. *J. Exp. Med.* 193: 233–238.

Doggrell, S.A. (2004): Efalizumab for psoriasis? *Expert Opin. Investigational Drugs* 13: 551–554.

Duan, R.-S., Bandara Adikari, S., Huang, Y.-M., Link, H. and Xiao, B.-G. (2004): Protective potential of experimental autoimmune myasthenia gravis in Lewis rats by IL-10-modified dendritic cells. *Neurobiol. Disease* 16: 461–467.

Dustin, M.L. and Springer, T.A. (1989): T-cell receptor cross-linking transiently stimulates adhesiveness through LFA-1. *Nature* 341: 619–624.

Edwards, J.C.W. and Cambridge, G. (2001): Sustained improvement in rheumatoid arthritis following a protocol designed to deplete B lymphocytes. *Rheumatology* 40: 205–211.

Edwards, J.C.W., Leandro, M.J. and Cambridge, G. (2002): B-lymphocyte depletion therapy in rheumatoid arthritis and other autoimmune disorders. *Biochem. Soc. Trans.* 30: 824–828.

Edwards, J.C.W., Szczepanski, L., Szechinski, J., Filipowicz-Sosnowska, A., Emery, P., Close, D.R., Stevens, R.M. and Shaw, T. (2004): Efficacy of B-cell-targeted therapy with rituximab in patients with rheumatoid arthritis. *New England J. Med.* 350: 2572–2581.

Engelmayer, J., Larsson, M., Subklewe, M., Chahroudi, A., Cox, W.I., Steinman, R.M. and Bhardwaj, N. (1999): Vaccinia virus inhibits the maturation of human dendritic cells: a novel mechanism of immune evasion. *J. Immunol.* 163: 6762–6768.

Engeman, T.M., Gorbachev, A.V., Gladue, R.P., Heeger, P.S. and Fairchild, R.L. (2000): Inhibition of functional T cell priming and contact hypersensitivity responses by treatment with anti-secondary lymphoid chemokine antibody during hapten sensitization. *J. Immunol.* 164: 5207–5214.

Engering, A., Geijtenbeek, T.B.H., Van Vliet, S.J., Wijers, M., Van Liempt, E., Demaurex, N., Lanzavecchia, A., Fransen, J., Figdor, C.G., Piguet, V. and Van Kooyk, Y. (2002): The dendritic cell-specific adhesion receptor DC-SIGN internalizes antigen for presentation to T cells. *J. Immunol.* 168: 2118–2126.

Engering, A.J., Cella, M., Fluitsma, D., Brockhaus, M., Hoefsmit, E.C.M., Lanzavecchia, A. and Pieters, J. (1997): The mannose receptor functions as a high capacity and broad specificity antigen receptor in human dendritic cells. *Eur. J. Immunol.* 27: 2417–2425.

Erbe, D.V., Wang, S., Xing, Y. and Tobin, J.F. (2002): Small molecule ligands define a binding site on the immune regulatory protein B7.1. *J. Biol. Chem.* 277: 7363–7368.

Farkas, L., Beiske, K., Lund-Johansen, F., Brandtzaeg, P. and Jahnsen, F.L. (2001): Plasmacytoid dendritic cells (natural interferon- alpha/beta-producing cells) accumulate in cutaneous lupus erythematosus lesions. *Am. J. Pathol.* 159: 237–243.

Fearnley, D.B., Whyte, L.F., Carnoutsos, S.A., Cook, A.H. and Hart, D.N.J. (1999): Monitoring human blood dendritic cell numbers in normal individuals and in stem cell transplantation. *Blood* 93: 728–736.

Feili-Hariri, M., Dong, X., Alber, S.M., Watkins, S.C., Salter, R.D. and Morel, P.A. (1999): Immunotherapy of NOD mice with bone marrow-derived dendritic cells. *Diabetes* 48: 2300–2308.

Feili-Hariri, M., Falkner, D.H., Gambotto, A., Papworth, G.D., Watkins, S.C., Robbins, P.D. and Morel, P.A. (2003): Dendritic cells transduced to express interleukin-4 prevent diabetes in nonobese diabetic mice with advanced insulitis. *Human Gene Therapy* 14: 13–23.

Fiebiger, E., Meraner, P., Weber, E., Fang, I.F., Stingl, G., Ploegh, H. and Maurer, D. (2001): Cytokines regulate proteolysis in major histocompatibility complex class II-dependent antigen presentation by dendritic cells. *J. Exp. Med.* 193: 881–892.

Finck, B.K., Linsley, P.S. and Wofsy, D. (1994): Treatment of murine lupus with CTLA4Ig. *Science* 265: 1225–1227.

Foell, J., Strahotin, S., O'Neil, S.P., McCausland, M.M., Suwyn, C., Haber, M., Chander, P.N., Bapat, A.S., Yan, X.-J., Chiorazzi, N., Hoffmann, M.K. and Mittler, R.S. (2003): CD137 costimulatory T cell receptor engagement reverses acute disease in lupus-prone NZB * NZW F1 mice. *J. Clin. Invest.* 111: 1505–1518.

Foell, J.L., Diez-Mendiondo, B.I., Diez, O.H., Holzer, U., Ruck, P., Bapat, A.S., Hoffmann, M.K., Mittler, R.S. and Dannecker, G.E. (2004): Engagement of the CD137 (4-1BB) costimulatory molecule inhibits and reverses the autoimmune process in collagen-induced arthritis and establishes lasting disease resistance. *Immunology* 113: 89–98.

Foged, C., Sundblad, A. and Hovgaard, L. (2002): Targeting vaccines to dendritic cells. *Pharm. Res.* 19: 229–238.

Fokkens, W.J., Vroom, T.M., Rijntjes, E. and Mulder, P.G. (1989): Fluctuation of the number of CD-1(T6)-positive dendritic cells, presumably Langerhans cells, in the nasal mucosa of patients with an isolated grass-pollen allergy before, during, and after the grass-pollen season. *J. Allergy Clin. Immunol.* 84: 39–43.

Fong, L., Brockstedt, D., Benike, C., Wu, L. and Engleman, E.G. (2001a): Dendritic cells injected via different routes induce immunity in cancer patients. *J. Immunol.* 166: 4254–4259.

Fong, L., Hou, Y., Rivas, A., Benike, C., Yuen, A., Fisher, G.A., Davis, M.M. and Engleman, E.G. (2001b): Altered peptide ligand vaccination with Flt3 ligand expanded dendritic cells for tumor immunotherapy. *Proc. Natl. Acad. Sci. U.S.A.* 98: 8809–8814.

Forster, R., Mattis, A.E., Kremmer, E., Wolf, E., Brem, G. and Lipp, M. (1996): A putative chemokine receptor, BLR1, directs B cell migration to defined lymphoid organs and specific anatomic compartments of the spleen. *Cell* 87: 1037–1047.

Forster, R., Schubel, A., Breitfeld, D., Kremmer, E., Renner-Muller, I., Wolf, E. and

Lipp, M. (1999): CCR7 coordinates the primary immune response by establishing functional microenvironments in secondary lymphoid organs. *Cell* 99: 23–33.

Foster, M.L., Halley, F. and Souness, J.E. (2000): Potential of p38 inhibitors in the treatment of rheumatoid arthritis. *Drug News Perspect.* 13: 488–497.

Fung-Leung, W.-P., Surh, C.D., Liljedahl, M., Pang, J., Leturcq, D., Peterson, P.A., Webb, S.R. and Karlsson, L. (1996): Antigen presentation and T cell development in H2-M-deficient mice. *Science* 271: 1278–1281.

Furuya, K., Takeda, H., Azhar, S., McCarron, R.M., Chen, Y., Ruetzler, C.A., Wolcott, K.M., DeGraba, T.J., Rothlein, R., Hugli, T.E., Del Zoppo, G.J. and Hallenbeck, J.M. (2001): Examination of several potential mechanisms for the negative outcome in a clinical stroke trial of enlimomab, a murine anti-human intercellular adhesion molecule-1 antibody: A bedside-to-bench study. *Stroke* 32: 2665–2674.

Gadola, S.D., Moins-Teisserenc, H.T., Trowsdale, J., Gross, W.L. and Cerundolo, V. (2000): TAP deficiency syndrome. *Clin. Exp. Immunology* 121: 173–178.

Ganellin, C.R., Bishop, P.B., Bambal, R.B., Chan, S.M.T., Law, J.K., Marabout, B., Luthra, P.M., Moore, A.N.J., Peschard, O., Bourgeat, P., Rose, C., Vargas, F. and Schwartz, J.-C. (2000): Inhibitors of tripeptidyl peptidase II. 2. Generation of the first novel lead inhibitor of cholecystokinin-8-inactivating peptidase: A strategy for the design of peptidase inhibitors. *J. Med. Chem.* 43: 664–674.

Gao, Z. and Metz, W.A. (2003): Unraveling the chemistry of chemokine receptor ligands. *Chem. Rev.* 103: 3733–3752.

Geijtenbeek, T.B.H., Krooshoop, D.J.E.B., Bleijs, D.A., Van Vliet, S.J., Van Duijnhoven, G.C.F., Grabovsky, V., Alon, R., Figdor, C.G. and Van Kooyk, Y. (2000a): DC-SIGN-ICAM-2 interaction mediates dendritic cell trafficking. *Nat. Immunol.* 1: 353–357.

Geijtenbeek, T.B.H., Kwon, D.S., Torensma, R., Van Vliet, S.J., Van Duijnhoven, G.C.F., Middel, J., Cornelissen, I.L.M.H.A., Nottet, H.S.L.M., KewalRamani, V.N., Littman, D.R., Figdor, C.G. and Van Kooyk, Y. (2000b): DC-SIGN, a dendritic cell-specific HIV-1-binding protein that enhances trans-infection of T cells. *Cell* 100: 587–597.

Geijtenbeek, T.B.H., Torensma, R., Van Vliet, S.J., Van Duijnhoven, G.C.F., Adema, G.J., Van Kooyk, Y. and Figdor, C.G. (2000c): Identification of DC-SIGN, a novel dendritic cell-specific ICAM-3 receptor that supports primary immune responses. *Cell* 100: 575–585.

Geijtenbeek, T.B.H. and Van Kooyk, Y. (2003): DC-SIGN: a novel HIV receptor on DCs that mediates HIV-1 transmission. *Curr. Top. Microbiol. Immunol.* 276: 31–54.

Geijtenbeek, T.B.H., van Vliet, S.J., Engering, A., t Hart, B.A. and van Kooyk, Y. (2004): Self- and nonself-recognition by C-type lectins on dendritic cells. *Annu. Rev. Immunol.* 22: 33–54.

Gilboa, E. (2001): The risk of autoimmunity associated with tumor immunotherapy. *Nat. Immunol.* 2: 789–792.

Gilliet, M. and Liu, Y.-J. (2002): Human plasmacytoid-derived dendritic cells and the induction of T-regulatory cells. *Human Immunol.* 63: 1149–1155.

Godthelp, T., Fokkens, W.J., Kleinjan, A., Holm, A.F., Mulder, P.G., Prens, E.P. and Rijntes, E. (1996): Antigen presenting cells in the nasal mucosa of patients with allergic rhinitis during allergen provocation. *Clin. Exp. Allergy: J. Br. Soc. Allergy Clin. Immunol.* 26: 677–688.

Golay, J.T., Clark, E.A. and Beverley, P.C.L. (1985): The CD20 (Bp35) antigen is involved in activation of B cells from the G0 to the G1 phase of the cell cycle. *J. Immunol.* 135: 3795–3801.

Gorczynski, R.M., Bransom, J., Cattral, M., Huang, X., Lei, J., Xiaorong, L., Min, W.P., Wan, Y. and Gauldie, J. (2000): Synergy in induction of increased renal allograft survival after portal vein infusion of dendritic cells transduced to express TGFb and IL-10, along with administration of CHO cells expressing the regulatory molecule OX-2. *Clin. Immunol.* (Orlando, FL) 95: 182–189.

Gottlieb, A.B., Kang, S., Linden, K.G., Lebwohl, M., Menter, A., Abdulghani, A.A., Goldfarb, M., Chieffo, N. and Totoritis, M.C. (2004): Evaluation of safety and clinical activity of multiple doses of the anti-CD80 monoclonal antibody, galiximab, in patients with moderate to severe plaque psoriasis. *Clin. Immunol.* (San Diego, CA) 111: 28–37.

Gould, D.S. and Auchincloss, H., Jr. (1999): Direct and indirect recognition: the role of MHC antigens in graft rejection. *Immunol. Today* 20: 77–82.

Graffi, S.J., Dekan, G., Stingl, G. and Epstein, M.M. (2002): Systemic administration of antigen-pulsed dendritic cells induces experimental allergic asthma in mice upon aerosol antigen rechallenge. *Clinical Immunol.* (San Diego, CA) 103: 176–184.

Grakoui, A., Bromley, S.K., Sumen, C., Davis, M.M., Shaw, A.S., Allen, P.M. and Dustin, M.L. (1999): The immunological synapse: A molecular machine controlling T cell activation. *Science* 285: 221–227.

Green, N.J., Xiang, J., Chen, J., Chen, L., Davies, A.M., Erbe, D., Tam, S. and Tobin, J.F. (2003): Structure-activity studies of a series of dipyrazolo[3,4-b: 3′,4′-d]pyridin-3-ones binding to the immune regulatory protein B7.1. *Bioorg. Med. Chem.* 11: 2991–3013.

Griffin, M.D., Lutz, W., Phan, V.A., Bachman, L.A., McKean, D.J. and Kumar, R. (2001): Dendritic cell modulation by 1alpha,25 dihydroxyvitamin D3 and its analogs: a vitamin D receptor-dependent pathway that promotes a persistent state of immaturity in vitro and in vivo. *Proc. Natl. Acad. Sci. U.S.A.* 98: 6800–6805.

Groettrup, M., de Giuli, R. and Schmidtke, G. (2004): Effects of the HIV-1 protease inhibitor ritonavir on proteasome activity and antigen presentation. *Proteasome Inhib. Cancer Therapy.* 207–216.

Groom, J., Kalled, S.L., Cutler, A.H., Olson, C., Woodcock, S.A., Schneider, P., Tschopp, J., Cachero, T.G., Batten, M., Wheway, J., Mauri, D., Cavill, D., Gordon, T.P., Mackay, C.R. and Mackay, F. (2002): Association of BAFF/BLyS overexpression and altered B cell differentiation with Sjogren's syndrome. *J. Clin. Invest.* 109: 59–68.

Gross, J.A., Dillon, S.R., Mudri, S., Johnston, J., Littau, A., Roque, R., Rixon, M., Schou, O., Foley, K.P., Haugen, H., McMillen, S., Waggie, K., Schreckhise, R.W., Shoemaker, K., Vu, T., Moore, M., Grossman, A. and Clegg, C.H. (2001): TACI-Ig neutralizes molecules critical for B cell development and autoimmune disease: impaired B cell maturation in mice lacking BLyS. *Immunity* 15: 289–302.

Gross, J.A., Johnston, J., Mudri, S., Enselman, R., Dillon, S.R., Madden, K., Xu, W., Parrish-Novak, J., Foster, D., Lofton-Day, C., Moore, M., Littau, A., Grossman, A., Haugen, H., Foley, K., Blumberg, H., Harrison, K., Kindsvogel, W. and Clegg, C.H. (2000): TACI and BCMA are receptors for a TNF homologue implicated in B-cell autoimmune disease. *Nature* 404: 995–999.

Gunn, M.D., Kyuwa, S., Tam, C., Kakiuchi, T., Matsuzawa, A., Williams, L.T. and Nakano, H. (1999): Mice lacking expression of secondary lymphoid organ chemokine have defects in lymphocyte homing and dendritic cell localization. *J. Exp. Med.* 189: 451–460.

Gunn, M.D., Ngo, V.N., Ansel, K.M., Ekland, E.H., Cyster, J.G. and Williams, L.T. (1998): A B-cell-homing chemokine made in lymphoid follicles activates Burkitt's lymphoma receptor-1. *Nature* 391: 799–803.

Haanstra, K.G., Ringers, J., Sick, E.A., Ramdien-Murli, S., Kuhn, E.-M., Boon, L. and Jonker, M. (2003): Prevention of kidney allograft rejection using anti-CD40 and anti-CD86 in primates. *Transplantation* 75: 637–643.

Hacker, H., Mischak, H., Miethke, T., Liptay, S., Schmid, R., Sparwasser, T., Heeg, K., Lipford, G.B. and Wagner, H. (1998): CpG-DNA-specific activation of antigen-presenting cells requires stress kinase activity and is preceded by non-specific endocytosis and endosomal maturation. *EMBO J.* 17: 6230–6240.

Hackstein, H. and Thomson, A.W. (2004): Dendritic cells; Emerging pharmacological targets of immunosuppressive drugs. *Nat. Rev. Immunol.* 4: 24–34.

Haicheur, N., Benchetrit, F., Amessou, M., Leclerc, C., Falguieres, T., Fayolle, C., Bismuth, E., Fridman, W.H., Johannes, L. and Tartour, E. (2003): The B subunit of Shiga toxin coupled to full-size antigenic protein elicits humoral and cell-mediated immune responses associated with a Th1-dominant polarization. *Int. Immunol.* 15: 1161–1171.

Haicheur, N., Bismuth, E., Bosset, S., Adotevi, O., Warnier, G., Lacabanne, V., Regnault, A., Desaymard, C., Amigorena, S., Ricciardi-Castagnoli, P., Goud, B., Fridman, W.H., Johannes, L. and Tartour, E. (2000): The B subunit of shiga toxin fused to a tumor antigen elicits CTL and targets dendritic cells to allow MHC class I-restricted presentation of peptides derived from exogenous antigens. *J. Immunol.* 165: 3301–3308.

Hall, G.L., Wing, M.G., Compston, D.A.S. and Scolding, N.J. (1997): b-Interferon regulates the immunomodulatory activity of neonatal rodent microglia. *J. Neuroimmunol.* 72: 11–19.

Harper, S.J. and LoGrasso, P. (2001): Inhibitors of the JNK signaling pathway. *Drugs Future* 26: 957–973.

Hawiger, D., Inaba, K., Dorsett, Y., Guo, M., Mahnke, K., Rivera, M., Ravetch, J.V., Steinman, R.M. and Nussenzweig, M.C. (2001): Dendritic cells induce peripheral T cell unresponsiveness under steady state conditions in vivo. *J. Exp. Med.* 194: 769–779.

Hawkins, L.D., Christ, W.J. and Rossignol, D.P. (2004): Inhibition of endotoxin response by synthetic TLR4 antagonists. *Curr. Top. Med. Chem.* 4: 1147–1171.

Hegde, N.R., Chevalier, M.S. and Johnson, D.C. (2003): Viral inhibition of MHC class II antigen presentation. *Trends Immunol.* 24: 278–285.

Heikenwalder, M., Polymenidou, M., Junt, T., Sigurdson, C., Wagner, H., Akira, S., Zinkernagel, R. and Aguzzi, A. (2004): Lymphoid follicle destruction and immunosuppression after repeated CpG oligodeoxynucleotide administration. *Nat. Med.* 10: 187–192.

Hemmi, H., Kaisho, T., Takeuchi, O., Sato, S., Sanjo, H., Hoshino, K., Horiuchi, T., Tomizawa, H., Takeda, K. and Akira, S. (2002): Small anti-viral compounds activate immune cells via the TLR7 MyD88-dependent signaling pathway. *Nat. Immunol.* 3: 196–200.

Hewitt, E.W., Gupta, S.S. and Lehner, P.J. (2001): The human cytomegalovirus gene product US6 inhibits ATP binding by TAP. *EMBO J.* 20: 387–396.

Ho, C.S., Lopez, J.A., Vuckovic, S., Pyke, C.M., Hockey, R.L. and Hart, D.N. (2001): Surgical and physical stress increases circulating blood dendritic cell counts independently of monocyte counts. *Blood* 98: 140–145.

Hsu, F.J., Benike, C., Fagnoni, F., Liles, T.M., Czerwinski, D., Taidi, B., Engleman, E.G. and Levy, R. (1996): Vaccination of patients with B-cell lymphoma using autologous antigen-pulsed dendritic cells. *Nat. Med.* 2: 52–58.

Huang, W., Sinha, J., Newman, J., Reddy, B., Budhai, L., Furie, R., Vaishnaw, A. and Davidson, A. (2002): The effect of anti-CD40 ligand antibody on B cells in human systemic lupus erythematosus. *Arthritis Rheum.* 46: 1554–1562.

Huang, Y.M., Xiao, B.G., Ozenci, V., Kouwenhoven, M., Teleshova, N., Fredrikson, S. and Link, H. (1999): Multiple sclerosis is associated with high levels of circulating dendritic cells secreting pro-inflammatory cytokines. *J. Neuroimmunol.* 99: 82–90.

Huang, Y.M., Yang, J.S., Xu, L.Y., Link, H. and Xiao, B.G. (2000): Autoantigen-pulsed dendritic cells induce tolerance to experimental allergic encephalomyelitis (EAE) in Lewis rats. *Clin. Exp. Immunol.* 122: 437–444.

Huard, B., Arlettaz, L., Ambrose, C., Kindler, V., Mauri, D., Roosnek, E., Tschopp, J., Schneider, P. and French, L.E. (2004): BAFF production by antigen-presenting cells provides T cell co-stimulation. *Int. Immunol.* 16: 467–475.

Huard, B., Schneider, P., Mauri, D., Tschopp, J. and French, L.E. (2001): T cell costimulation by the TNF ligand BAFF. *J. Immunol.* 167: 6225–6231.

Hung, C.F., Hsu, K.F., Cheng, W.F., Chai, C.Y., He, L., Ling, M. and Wu, T.C. (2001): Enhancement of DNA vaccine potency by linkage of antigen gene to a gene encoding the extracellular domain of Fms-like tyrosine kinase 3-ligand. *Cancer Res.* 61: 1080–1088.

Hutloff, A., Dittrich, A.M., Beier, K.C., Eljaschewitsch, B., Kraft, R., Anagnostopoulos, L. and Kroczek, R.A. (1999): ICOS is an inducible T-cell co-stimulator structurally and functionally related to CD28. *Nature* 397: 263–266.

Illigens, B.M., Yamada, A., Fedoseyeva, E.V., Anosova, N., Boisgerault, F., Valujskikh, A., Heeger, P.S., Sayegh, M.H., Boehm, B. and Benichou, G. (2002): The relative contribution of direct and indirect antigen recognition pathways to the alloresponse and graft rejection depends upon the nature of the transplant. *Human Immunol.* 63: 912–925.

Iwai, H., Kozono, Y., Hirose, S., Akiba, H., Yagita, H., Okumura, K., Kohsaka, H., Miyasaka, N. and Azuma, M. (2002): Amelioration of collagen-induced arthritis by blockade of inducible costimulator-B7 homologous protein costimulation. *J. Immunol.* 169: 4332–4339.

Jahnsen, F.L., Moloney, E.D., Hogan, T., Upham, J.W., Burke, C.M. and Holt, P.G. (2001): Rapid dendritic cell recruitment to the bronchial mucosa of patients with atopic asthma in response to local allergen challenge. *Thorax* 56: 823–826.

Jonuleit, H., Schmitt, E., Schuler, G., Knop, J. and Enk, A.H. (2000): Induction of interleukin 10-producing, nonproliferating CD4+ T cells with regulatory properties by repetitive stimulation with allogeneic immature human dendritic cells. *J. Exp. Med*. 192: 1213–1222.

Jung, S., Unutmaz, D., Wong, P., Sano, G.-I., De los Santos, K., Sparwasser, T., Wu, S., Vuthoori, S., Ko, K., Zavala, F., Pamer Eric, G., Littman Dan, R. and Lang Richard, A. (2002): In vivo depletion of CD11c(+) dendritic cells abrogates priming of CD8(+) T cells by exogenous cell-associated antigens. *Immunity* 17: 211–220.

Jurk, M., Heil, F., Vollmer, J., Schetter, C., Krieg, A.M., Wagner, H., Lipford, G. and Bauer, S. (2002): Human TLR7 or TLR8 independently confer responsiveness to the antiviral compound R-848. *Nat. Immunol.* 3: 499.

Kalunian, K.C., Davis, J.C., Jr., Merrill, J.T., Totoritis, M.C. and Wofsy, D. (2002): Treatment of systemic lupus erythematosus by inhibition of T cell costimulation with anti-CD154: a randomized, double-blind, placebo-controlled trial. *Arthritis Rheum.* 46: 3251–3258.

Kaufmann, J., Wegener, W.A., Horak, I.D., Qidwai, M., Ding, C., Elmera, M., Kovacs, J., Goldenberg, D.M., Burmester, G.R., and Dörner, T. (2004): Pilot clinical trial of epratuzumab (humanized anti-CD-22 antibody) for immunotherapy in systemic lupus erythematosus (SLE). *Ann. Rheumatoid Diseases*: 63(Suppl. 1) Abst. THU0443.

Kavanaugh, A.F., Davis, L.S., Jain, R.I., Nichols, L.A., Norris, S.H. and Lipsky, P.E. (1996): A phase I/II open label study of the safety and efficacy of an anti-ICAM-1 (intercellular adhesion molecule-1; CD-54) monoclonal antibody in early rheumatoid arthritis. *J. Rheumatol*. 23: 1338–1344.

Kavanaugh, A.F., Davis, L.S., Nichols, L.A., Norris, S.H., Rothlein, R., Scharschmidt, L.A. and Lipsky, P.E. (1994): Treatment of refractory rheumatoid arthritis with a monoclonal antibody to intercellular adhesion molecule 1. *Arthritis Rheum.* 37: 992–999.

Kavanaugh, A.F., Schulze-Koops, H., Davis, L.S. and Lipsky, P.E. (1997): Repeat treatment of rheumatoid arthritis patients with a murine anti-intercellular adhesion molecule 1 monoclonal antibody. *Arthritis Rheum.* 40: 849–853.

Kayagaki, N., Yan, M., Seshasayee, D., Wang, H., Lee, W., French, D.M., Grewal, I.S., Cochran, A.G., Gordon, N.C., Yin, J., Starovasnik, M.A. and Dixit, V.M. (2002): BAFF/BLyS receptor 3 binds the B cell survival factor BAFF ligand through a discrete surface loop and promotes processing of NF-kB2. *Immunity* 17: 515–524.

Kazkaz, H. and Isenberg, D. (2004): Anti B cell therapy (rituximab) in the treatment of autoimmune diseases. *Curr. Opin. Pharmacol.* 4: 398–402.

Khan, W.N. (2001): Regulation of B lymphocyte development and activation by Bruton's tyrosine kinase. *Immunol. Res.* 23: 147–156.

Kiechl, S., Lorenz, E., Reindl, M., Wiedermann, C.J., Oberhollenzer, F., Bonora, E., Willeit, J. and Schwartz, D.A. (2002): TOLL-like receptor 4 polymorphisms and atherogenesis. *New England J. Med.* 347: 185–192.

Kim, K.D., Choe, Y.K., Choe, I.S. and Lim, J.S. (2001): Inhibition of glucocorticoid-mediated, caspase-independent dendritic cell death by CD40 activation. *J. Leukocyte Biol.* 69: 426–434.

Kim, S.H., Kim, S., Evans, C.H., Ghivizzani, S.C., Oligino, T. and Robbins, P.D. (2001): Effective treatment of established murine collagen-induced arthritis by systemic administration of dendritic cells genetically modified to express IL-4. *J. Immunol.* 166: 3499–3505.

Klinman, D.M. (2004): Immunotherapeutic uses of CpG oligodeoxynucleotides. *Nat. Rev. Immunol.* 4: 249–259.

Knoerzer, D.B., Karr, R.W., Schwartz, B.D. and Mengle-Gaw, L.J. (1995): Collagen-induced arthritis in the BB rat. Prevention of disease by treatment with CTLA-4-Ig. *J. Clin. Invest.* 96: 987–993.

Koopman, G., Dalgleish, A.G., Bhogal, B.S., Haaksma, A.G.M. and Heeney, J.L. (2001): Changes in dendritic cell subsets in the lymph nodes of rhesus macaques after application of glucocorticoids. *Human Immunol.* 62: 208–214.

Koyasu, S., Lawton, T., Novick, D., Recny, M.A., Siliciano, R.F., Wallner, B.P. and Reinherz, E.L. (1990): Role of interaction of CD2 molecules with lymphocyte function-associated antigen 3 in T-cell recognition of nominal antigen. *Proc. Natl. Acad. Sci. U.S.A.* 87: 2603–2607.

Kremer, J.M., Westhovens, R., Leon, M., Di Giorgio, E., Alten, R., Steinfeld, S., Russell, A., Dougados, M., Emery, P., Nuamah, I.F., Williams, G.R., Becker, J.-c., Hagerty, D.T. and Moreland, L.W. (2003): Treatment of rheumatoid arthritis by selective inhibition of T-cell activation with fusion protein CTLA4Ig. *New England J. Med.* 349: 1907–1915.

Krieg, A.M. (2004): Antitumor applications of stimulating toll-like receptor 9 with CpG oligodeoxynucleotides. *Curr. Oncol. Rep.* 6: 88–95.

Kugler, A., Stuhler, G., Walden, P., Zoller, G., Zobywalski, A., Brossart, P., Trefzer, U., Ullrich, S., Muller, C.A., Becker, V., Gross, A.J., Hemmerlein, B., Kanz, L., Muller, G.A. and Ringert, R.H. (2000): Regression of human metastatic renal cell carcinoma after vaccination with tumor cell-dendritic cell hybrids. *Nat. Med.* 6: 332–336.

Kumar, S., Boehm, J. and Lee, J.C. (2003): p38 MAP kinases: key signalling molecules as therapeutic targets for inflammatory diseases. *Nat. Rev. Drug Discovery* 2: 717–726.

Kurts, C., Cannarile, M., Klebba, I. and Brocker, T. (2001): Cutting edge: dendritic cells are sufficient to cross-present self-antigens to CD8 T cells in vivo. *J. Immunol.* 166: 1439–1442.

Kuwana, M. (2002): Induction of anergic and regulatory T cells by plasmacytoid dendritic cells and other dendritic cell subsets. *Human Immunol.* 63: 1156–1163.

Kuwana, M., Nomura, S., Fujimura, K., Nagasawa, T., Muto, Y., Kurata, Y., Tanaka, S. and Ikeda, Y. (2004): Effect of a single injection of humanized anti-CD154 monoclonal antibody on the platelet-specific autoimmune response in patients with immune thrombocytopenic purpura. *Blood* 103: 1229–1236.

Kwekkeboom, J., de Boer, M., Tager, J.M. and De Groot, C. (1993): CD40 plays an essential role in the activation of human B cells by murine EL4B5 cells. *Immunology* 79: 439–444.

Kwon, B.S., Hurtado, J.C., Lee, Z.H., Kwack, K.B., Seo, S.K., Choi, B.K., Koller, B.H., Wolisi, G., Broxmeyer, H.E. and Vinay, D.S. (2002): Immune responses in 4-1BB (CD137)-deficient mice. *J. Immunol.* 168: 5483–5490.

Laman, J.D., t Hart, B.A., Brok, H., van Meurs, M., Schellekens, M.M., Kasran, A., Boon, L., Bauer, J., de Boer, M. and Ceuppens, J. (2002): Protection of marmoset monkeys against EAE by treatment with a murine antibody blocking CD40 (mu5D12). *Eur. J. Immunol.* 32: 2218–2228.

Lanzavecchia, A. and Sallusto, F. (2001): Regulation of T cell immunity by dendritic cells. *Cell* 106: 263–266.

Lappin, M.B., Weiss, J.M., Delattre, V., Mai, B., Dittmar, H., Maier, C., Manke, K., Grabbe, S., Martin, S. and Simon, J.C. (1999): Analysis of mouse dendritic cell migration in vivo upon subcutaneous and intravenous injection. *Immunology* 98: 181–188.

Leandro, M.J., Edwards, J.C.W. and Cambridge, G. (2002): Clinical outcome in 22 patients with rheumatoid arthritis treated with B lymphocyte depletion. *Ann. Rheum. Diseases* 61: 883–888.

Lee, J., Chuang, T.-H., Redecke, V., She, L., Pitha Paula, M., Carson Dennis, A., Raz, E. and Cottam Howard, B. (2003): Molecular basis for the immunostimulatory activity of guanine nucleoside analogs: activation of Toll-like receptor 7. *Proc. Natl. Acad. Sci. U.S.A.* 100: 6646–6651.

Lee, S.-J., Myers, L., Muralimohan, G., Dai, J., Qiao, Y., Li, Z., Mittler, R.S. and Vella, A.T. (2004): 4-1BB and OX40 dual costimulation synergistically stimulate primary specific CD8 T cells for robust effector function. *J. Immunol.* 173: 3002–3012.

Legler, D.F., Loetscher, M., Roos, R.S., Clark-Lewis, I., Baggiolini, M. and Moser, B. (1998): B cell-attracting chemokine 1, a human CXC chemokine expressed in lymphoid tissues, selectively attracts B lymphocytes via BLR1/CXCR5. *J. Exp. Med.* 187: 655–660.

Levitskaya, J., Sharipo, A., Leonchiks, A., Ciechanover, A. and Masucci, M.G. (1997): Inhibition of ubiquitin/proteasome-dependent protein degradation by the Gly-Ala repeat domain of the Epstein-Barr virus nuclear antigen 1. *Proc. Natl. Acad. Sci. U.S.A.* 94: 12 616–12 621.

Lich, J.D., Jayne, J.A., Zhou, D., Elliott, J.F. and Blum, J.S. (2003): Editing of an immunodominant epitope of glutamate decarboxylase by HLA-DM. *J. Immunol.* 171: 853–859.

Liu, G. (2001): Small molecule antagonists of the LFA-1/ICAM-1 interaction as potential therapeutic agents. *Expert Opin. Therap. Pat.* 11: 1383–1393.

Loak, K., Li, D.N., Manoury, B., Billson, J., Morton, F., Hewitt, E. and Watts, C. (2003): Novel cell-permeable acyloxymethylketone inhibitors of asparaginyl endopeptidase. *Biol. Chem.* 384: 1239–1246.

Lopez, D. and Del Val, M. (1997): Cutting edge: selective involvement of proteasomes and cysteine proteases in MHC class I antigen presentation. *J. Immunol.* 159: 5769–5772.

Losy, J., Michalowska-Wender, G. and Wender, M. (2002): Interleukin 12 and interleukin 10 are affected differentially by treatment of multiple sclerosis with glatiramer acetate (Copaxone). *Folia Neuropathol.* 40: 173–175.

Lu, L., Lee, W.C., Takayama, T., Qian, S., Gambotto, A., Robbins, P.D. and Thomson, A.W. (1999): Genetic engineering of dendritic cells to express immunosuppressive molecules (viral IL-10, TGF-beta, and CTLA4Ig). *J. Leukocyte Biol.* 66: 293–296.

Lukacs, N.W., Prosser, D.M., Wiekowski, M., Lira, S.A. and Cook, D.N. (2001): Requirement for the chemokine receptor CCR6 in allergic pulmonary inflammation. *J. Exp. Med.* 194: 551–555.

Lynn, M., Wong, Y.N., Wheeler, J.L., Kao, R.J., Perdomo, C.A., Noveck, R., Vargas, R., D'Angelo, T., Gotzkowsky, S., McMahon, F.G., Wasan, K.M. and Rossignol, D.P. (2004): Extended in vivo pharmacodynamic activity of E5564 (a-D-glucopyranose) in normal volunteers with experimental endotoxemia. *J. Pharmacol. Exp. Therapeutics* 308: 175–181.

Maas, A. and Hendriks, R.W. (2001): Role of Bruton's tyrosine kinase in B cell development. *Develop. Immunol.* 8: 171–181.

MacDonald, K.P., Pettit, A.R., Quinn, C., Thomas, G.J. and Thomas, R. (1999): Resistance of rheumatoid synovial dendritic cells to the immunosuppressive effects of IL-10. *J. Immunol.* (Baltimore, Md.: 1950) 163: 5599–5607.

Mackay, F., Woodcock, S.A., Lawton, P., Ambrose, C., Baetscher, M., Schneider, P., Tschopp, J. and Browning, J.L. (1999): Mice transgenic for BAFF develop lymphocytic disorders along with autoimmune manifestations. *J. Exp. Med.* 190: 1697–1710.

Mahajan, S., Ghosh, S., Sudbeck, E.A., Zheng, Y., Downs, S., Hupke, M. and Uckun, F.M. (1999): Rational design and synthesis of a novel anti-leukemic agent targeting Bruton's tyrosine kinase (BTK), LFM-A13 [a-cyano-b-hydroxy-b-methyl-N-(2,5-dibromophenyl)propenamide]. *J. Biol. Chem.* 274: 9587–9599.

Maksymowych, W.P., Blackburn, W.D., Jr., Tami, J.A. and Shanahan, W.R., Jr. (2002): A randomized, placebo controlled trial of an antisense oligodeoxynucleotide to intercellular adhesion molecule-1 in the treatment of severe rheumatoid arthritis. *J. Rheumatol.* 29: 447–453.

Manoury, B., Hewitt, E.W., Morrice, N., Dando, P.M., Barrett, A.J. and Watts, C. (1998): An asparaginyl endopeptidase processes a microbial antigen for class II MHC presentation. *Nature* 396: 695–699.

Manoury, B., Mazzeo, D., Fugger, L., Viner, N., Ponsford, M., Streeter, H., Mazza, G., Wraith, D.C. and Watts, C. (2002): Destructive processing by asparagine endopeptidase limits presentation of a dominant T cell epitope in MBP. *Nat. Immunol.* 3: 169–174.

Manoury, B., Mazzeo, D., Li, D.N., Billson, J., Loak, K., Benaroch, P. and Watts, C. (2003): Asparagine endopeptidase can initiate the removal of the MHC class II invariant chain chaperone. *Immunity* 18: 489–498.

Marecki, S. and Kirkpatrick, P. (2004): Fresh from the pipeline: Efalizumab. *Nat. Rev. Drug Discovery* 3: 473–474.

Mathur, M., Herrmann, K., Qin, Y., Gulmen, F., Li, X., Krimins, R., Weinstock, J., Elliott, D., Bluestone, J.A. and Padrid, P. (1999): CD28 interactions with either CD80 or CD86 are sufficient to induce allergic airway inflammation in mice. *Am. J. Respiratory Cell Mol. Biol.* 21: 498–509.

Matsue, H., Matsue, K., Walters, M., Okumura, K., Yagita, H. and Takashima, A. (1999): Induction of antigen-specific immunosuppression by CD95L cDNA-transfected 'killer' dendritic cells. *Nat. Med.* 5: 930–937.

Matsue, H., Yang, C., Matsue, K., Edelbaum, D., Mummert, M. and Takashima, A. (2002): Contrasting impacts of immunosuppressive agents (rapamycin, FK506, cyclosporin A, and dexamethasone) on bidirectional dendritic cell-T cell interaction during antigen presentation. *J. Immunol.* 169: 3555–3564.

Melero, I., Bach, N., Hellstroem, K.E., Aruffo, A., Mittler, R.S. and Chen, L. (1998): Amplification of tumor immunity by gene transfer of the co-stimulatory 4-1BB ligand. Synergy with the CD28 co-stimulatory pathway. *Eur. J. Immunol.* 28: 1116–1121.

Meng, G., Rutz, M., Schiemann, M., Metzger, J., Grabiec, A., Schwandner, R., Luppa, P.B., Ebel, F., Busch, D.H., Bauer, S., Wagner, H. and Kirschning, C.J. (2004): Antagonistic antibody prevents toll-like receptor 2-driven lethal shock-like syndromes. *J. Clin. Invest.* 113: 1473–1481.

Menges, M., Rossner, S., Voigtlander, C., Schindler, H., Kukutsch, N.A., Bogdan, C., Erb, K., Schuler, G. and Lutz, M.B. (2002): Repetitive injections of dendritic cells matured with tumor necrosis factor a induce antigen-specific protection of mice from autoimmunity. *J. Exp. Med.* 195: 15–21.

Min, W.P., Gorczynski, R., Huang, X.Y., Kushida, M., Kim, P., Obataki, M., Lei, J., Suri, R.M. and Cattral, M.S. (2000): Dendritic cells genetically engineered to express Fas ligand induce donor-specific hyporesponsiveness and prolong allograft survival. *J. Immunol.* (Baltimore, Md.: 1950) 164: 161–167.

Mittler, R.S., Bailey, T.S., Klussman, K., Trailsmith, M.D. and Hoffmann, M.K. (1999): Anti-4-1BB monoclonal antibodies abrogate T cell-dependent humoral immune response in vivo through the induction of helper T cell anergy. *J. Exp. Med.* 190: 1535–1540.

Moingeon, P., Chang, H.C., Wallner, B.P., Stebbins, C., Frey, A.Z. and Reinherz, E.L. (1989): CD2-mediated adhesion facilitates T lymphocyte antigen recognition function. *Nature* 339: 312–314.

Moller, G.M., Overbeek, S.E., Van Helden-Meeuwsen, C.G., Van Haarst, J.M.W., Prens, E.P., Mulder, P.G., Postma, D.S. and Hoogsteden, H.C. (1996): Increased numbers of dendritic cells in the bronchial mucosa of atopic asthmatic patients: Downregulation by inhaled corticosteroids. *Clin. Exp. Allergy* 26: 517–524.

Momburg, F. and Hengel, H. (2002): Corking the bottleneck: The transporter associated with antigen processing as a target for immune subversion by viruses. *Curr. Top. Microbiol. Immunol.* 269: 57–74.

Monti, P., Mercalli, A., Leone, B.E., Valerio, D.C., Allavena, P. and Piemonti, L. (2003): Rapamycin impairs antigen uptake of human dendritic cells. *Transplantation* 75: 137–145.

Morel, P.A., Feili-Hariri, M., Coates, P.T. and Thomson, A.W. (2003): Dendritic cells, T cell tolerance and therapy of adverse immune reactions. *Clin. Exp. Immunol.* 133: 1–10.

Moreland, L.W., Alten, R., Van den Bosch, F., Appelboom, T., Leon, M., Emery, P., Cohen, S., Luggen, M., Shergy, W., Nuamah, I. and Becker, J.-C. (2002): Costimulatory blockade in patients with rheumatoid arthritis: a pilot, dose-finding, double-blind, placebo-controlled clinical trial evaluating CTLA-4Ig and LEA29Y eighty-five days after the first infusion. *Arthritis Rheum.* 46: 1470–1479.

Mori, S., Nakano, H., Aritomi, K., Wang, C.R., Gunn, M.D. and Kakiuchi, T. (2001): Mice lacking expression of the chemokines CCL21-ser and CCL19 (plt mice) demonstrate delayed but enhanced T cell immune responses. *J. Exp. Med.* 193: 207–218.

Moser, M., De Smedt, T., Sornasse, T., Tielemans, F., Chentoufi, A.A., Muraille, E., Van Mechelen, M., Urbain, J. and Leo, O. (1995): Glucocorticoids down-regulate dendritic cell function in vitro and in vivo. *Eur. J. Immunol.* 25: 2818–2824.

Mosyak, L., Zaller, D.M. and Wiley, D.C. (1998): The structure of HLA-DM, the peptide exchange catalyst that loads antigen onto class II MHC molecules during antigen presentation. *Immunity* 9: 377–383.

Moynagh, P.N. (2003): Toll-like receptor signalling pathways as key targets for mediating the anti-inflammatory and immunosuppressive effects of glucocorticoids. *J. Endocrinol.* 179: 139–144.

Muller, G., Muller, A., Tuting, T., Steinbrink, K., Saloga, J., Szalma, C., Knop, J. and Enk, A.H. (2002): Interleukin-10-treated dendritic cells modulate immune responses of naïve and sensitized T cells in vivo. *J. Invest. Dermatol.* 119: 836–841.

Nair, S. and Boczkowski, D. (2002): RNA-transfected dendritic cells. *Expert Rev. Vaccines* 1: 507–513.

Nakagawa, T.Y., Brissette, W.H., Lira, P.D., Griffiths, R.J., Petrushova, N., Stock, J., McNeish, J.D., Eastman, S.E., Howard, E.D., Clarke, S.R.M., Rosloniec, E.F., Elliott, E.A. and Rudensky, A.Y. (1999): Impaired invariant chain degradation and antigen presentation and diminished collagen-induced arthritis in cathepsin S null mice. *Immunity* 10: 207–217.

Nestle, F.O., Banchereau, J. and Hart, D. (2001): Dendritic cells: On the move from bench to bedside. *Nat. Med.* 7: 761–765.

Ng, L.G., Sutherland, A.P.R., Newton, R., Qian, F., Cachero, T.G., Scott, M.L., Thompson, J.S., Wheway, J., Chtanova, T., Groom, J., Sutton, I.J., Xin, C., Tangye, S.G., Kalled, S.L., Mackay, F. and Mackay, C.R. (2004): B cell-activating factor belonging to the TNF family (BAFF)-R is the principal BAFF receptor facilitating BAFF costimulation of circulating T and B cells. *J. Immunol.* 173: 807–817.

Ninomiya-Tsuji, J., Kajino, T., Ono, K., Ohtomo, T., Matsumoto, M., Shiina, M., Mihara, M., Tsuchiya, M. and Matsumoto, K. (2003): A resorcylic acid lactone, 5Z-7-oxozeaenol, prevents inflammation by inhibiting the catalytic activity of TAK1 MAPK kinase kinase. *J. Biol. Chem.* 278: 18485–18490.

Norbury, C.C., Malide, D., Gibbs, J.S., Bennink, J.R. and Yewdell, J.W. (2002): Visualizing priming of virus-specific CD8+ T cells by infected dendritic cells in vivo. *Nat. Immunol.* 3: 265–271.

Nozawa, K., Ohata, J., Sakurai, J., Hashimoto, H., Miyajima, H., Yagita, H., Okumura, K. and Azuma, M. (2001): Preferential blockade of CD8+ T cell responses by administration of anti-CD137 ligand monoclonal antibody results in differential effect on development of murine acute and chronic graft-versus-host diseases. *J. Immunol.* 167: 4981–4986.

Onuffer, J.J. and Horuk, R. (2002): Chemokines, chemokine receptors and small-molecule antagonists: recent developments. *Trends Pharmacol. Sci.* 23: 459–467.

O'Rourke M., G.E., Johnson, M., Lanagan, C., See, J., Yang, J., Bell John, R., Slater Greg, J., Kerr Beverley, M., Crowe, B., Purdie David, M., Elliott Suzanne, L., Ellem Kay, A.O. and Schmidt Christopher, W. (2003): Durable complete clinical responses in a phase I/II trial using an autologous melanoma cell/dendritic cell vaccine. *Cancer Immunol., Immunotherapy.* CII 52: 387–395.

O'Shea, J.J., Pesu, M., Borie, D.C. and Changelian, P.S. (2004): A new modality for immunosuppression: targeting the JAK/STAT pathway. *Nat. Rev. Drug Discovery* 3: 555–564.

Osugi, Y., Vuckovic, S. and Hart, D.N.J. (2002): Myeloid blood CD11c+ dendritic cells and monocyte-derived dendritic cells differ in their ability to stimulate T lymphocytes. *Blood* 100: 2858–2866.

Ozenci, V., Kouwenhoven, M., Huang, Y.M., Xiao, B.G., Kivisakk, P., Fredrikson, S. and Link, H. (1999): Multiple sclerosis: levels of interleukin-10-secreting blood mononuclear cells are low in untreated patients but augmented during interferon-b-1b treatment. *Scand. J. Immunol.* 49: 554–561.

Pashenkov, M., Teleshova, N., Kouwenhoven, M., Kostulas, V., Huang, Y.M., Soderstrom, M. and Link, H. (2002): Elevated expression of CCR5 by myeloid (CD11c+) blood dendritic cells in multiple sclerosis and acute optic neuritis. *Clin. Exp. Immunol.* 127: 519–526.

Pettit, A.R. and Thomas, R. (1999): Dendritic cells: the driving force behind autoimmunity in rheumatoid arthritis? *Immunol. Cell Biol.* 77: 420–427.

Piemonti, L., Monti, P., Allavena, P., Leone, B.E., Caputo, A. and Di Carlo, V. (1999a): Glucocorticoids increase the endocytic activity of human dendritic cells. *Int. Immunol.* 11: 1519–1526.

Piemonti, L., Monti, P., Allavena, P., Sironi, M., Soldini, L., Leone, B.E., Socci, C. and Di Carlo, V. (1999b): Glucocorticoids affect human dendritic cell differentiation and maturation. *J. Immunol.* 162: 6473–6481.

Piemonti, L., Monti, P., Sironi, M., Fraticelli, P., Leone, B.E., Dal Cin, E., Allavena, P. and Di Carlo, V. (2000): Vitamin D3 affects differentiation, maturation, and function of human monocyte-derived dendritic cells. *J. Immunol.* 164: 4443–4451.

Pilkington, K.R., Clark-Lewis, I. and McColl, S.R. (2004): Inhibition of generation of cytotoxic T lymphocyte activity by a CCL19/macrophage inflammatory protein (MIP)-3b antagonist. *J. Biol. Chem.* 279: 40 276–40 282.

Podolin, P.L., Capper, E.A., Bolognese, B.J., Chalupowicz, D.C., Dong, X., Fox, J.H., Gao, E.N., Johanson, R.A., Katchur, S., Marshall, L.A., Mayer, R.J., McQueney, M.S., Petrone, A., Tomaszek, T.A., Veber, D.F. and Yamashita, D.S. (2001): Inhibition of cathepsin S blocks invariant chain processing and antigen-induced proliferation in vitro, and reduces the severity of collagen-induced arthritis in vivo. *Inflamm. Res.* 50: S159.

Poe, J.C., Fujimoto, Y., Hasegawa, M., Haas, K.M., Miller, A.S., Sanford, I.G., Bock, C.B., Fujimoto, M. and Tedder, T.F. (2004): CD22 regulates B lymphocyte function in vivo through both ligand-dependent and ligand-independent mechanisms. *Nat. Immunol.* 5: 1078–1087.

Quesniaux, V.F.J. and Ryffel, B. (2004): Toll-like receptors: emerging targets of immunomodulation. *Expert Opin. Therapeutic Patents* 14: 85–100.

Ramanujam, M., Wang, X., Huang, W., Schiffer, L., Grimaldi, C., Akkerman, A., Diamond, B., Madaio, M.P. and Davidson, A. (2004): Mechanism of action of transmembrane activator and calcium modulator ligand interactor-Ig in murine systemic lupus erythematosus. *J. Immunol.* 173: 3524–3534.

Ramirez, M.C. and Sigal, L.J. (2004): The multiple routes of MHC-I cross-presentation. *Trends Microbiol.* 12: 204–207.

Rastetter, W., Molina, A. and White, C.A. (2004): Rituximab: expanding role in therapy for lymphomas and autoimmune diseases. *Annu. Rev. Med.* 55: 477–503.

Rawlings, D.J., Saffran, D.C., Tsukada, S., Largaespada, D.A., Grimaldi, J.C., Cohen, L., Mohr, R.N., Bazan, J.F., Howard, M. and et al. (1993): Mutation of unique region of Bruton's tyrosine kinase in immunodeficient XID mice. *Science* 261: 358–361.

Reits, E., Griekspoor, A. and Neefjes, J. (2002): Herpes viral proteins manipulating the peptide transporter TAP. *Curr. Top. Microbiol. Immunol.* 269: 75–83.

Reits, E., Neijssen, J., Herberts, C., Benckhuijsen, W., Janssen, L., Drijfhout, J.W. and Neefjes, J. (2004): A major role for TPPII in trimming proteasomal degradation products for MHC class I antigen presentation. *Immunity* 20: 495–506.

Rhodes, J. (2002): Discovery of immunopotentiatory drugs: current and future strategies. *Clin. Exp. Immunol.* 130: 363–369.

Riese, R.J., Mitchell, R.N., Villadangos, J.A., Shi, G.-P., Palmer, J.T., Karp, E.R., De Sanctis, G.T., Ploegh, H.L. and Chapman, H.A. (1998): Cathepsin S activity regulates antigen presentation and immunity. *J. Clin. Invest.* 101: 2351–2363.

Riese, R.J., Wolf, P.R., Brömme, D., Natkin, L.R., Villadangos, J.A., Ploegh, H.L. and Chapman, H.A. (1996): Essential role for cathepsin S in MHC class II-associated invariant chain processing and peptide loading. *Immunity* 4: 357–366.

Rock, K.L., Gramm, C., Rothstein, L., Clark, K., Stein, R., Dick, L., Hwang, D. and Goldberg, A.L. (1994): Inhibitors of the proteasome block the degradation of most cell proteins and the generation of peptides presented on MHC class I molecules. *Cell* 78: 761–771.

Roncarolo, M.-G., Levings, M.K. and Traversari, C. (2001): Differentiation of T regulatory cells by immature dendritic cells. *J. Exp. Med.* 193: F5–F9.

Roshak, A.K., Callahan, J.F. and Blake, S.M. (2002): Small-molecule inhibitors of NF-kB for the treatment of inflammatory joint disease. *Curr. Opin. Pharmacol.* 2: 316–321.

Rutz, M., Metzger, J., Gellert, T., Luppa, P., Lipford, G.B., Wagner, H. and Bauer, S. (2004): Toll-like receptor 9 binds single-stranded CpG-DNA in a sequence- and pH-dependent manner. *Eur. J. Immunol.* 34: 2541–2550.

Saegusa, K., Ishimaru, N., Yanagi, K., Arakaki, R., Ogawa, K., Saito, I., Katunuma, N. and Hayashi, Y. (2002): Cathepsin S inhibitor prevents autoantigen presentation and autoimmunity. *J. Clin. Invest.* 110: 361–369.

Saegusa, K., Ishimaru, N., Yanagi, K., Haneji, N., Nishino, M., Azuma, M., Saito, I. and Hayashi, Y. (2000): Treatment with anti-CD86 costimulatory molecule prevents the autoimmune lesions in murine Sjogren's syndrome (SS) through up-regulated Th2 response. *Clin. Exp. Immunol.* 119: 354–360.

Saeki, H., Moore, A.M., Brown, M.J. and Hwang, S.T. (1999): Cutting edge: secondary lymphoid-tissue chemokine (SLC) and CC chemokine receptor 7 (CCR7) participate in the emigration pathway of mature dendritic cells from the skin to regional lymph nodes. *J. Immunol.* (Baltimore, Md.: 1950) 162: 2472–2475.

Salmela, K., Wramner, L., Ekberg, H., Hauser, I., Bentdal, O., Lins, L.-E., Isoniemi, H., Backman, L., Persson, N., Neumayer, H.-H., Jorgensen, P.F., Spieker, C., Hendry, B., Nicholls, A., Kirste, G. and Hasche, G. (1999): A randomized multicenter trial of the anti-ICAM-1 monoclonal antibody (enlimomab) for the prevention of acute rejection and delayed onset of graft function in cadaveric renal transplantation: a report of the European Anti-ICAM-1 Renal Transplant Study group. *Transplantation* 67: 729–736.

Samady, L., Costigliola, E., MacCormac, L., McGrath, Y., Cleverley, S., Lilley, C.E., Smith, J., Latchman, D.S., Chain, B. and Coffin, R.S. (2003): Deletion of the virion host shutoff protein (vhs) from herpes simplex virus (HSV) relieves the viral block to dendritic cell activation: Potential of vhs- HSV vectors for dendritic cell-mediated immunotherapy. *J. Virol.* 77: 3768–3776.

Santiago-Schwarz, F., Anand, P., Liu, S. and Carsons, S.E. (2001): Dendritic cells (DCs) in rheumatoid arthritis (RA): progenitor cells and soluble factors contained in RA synovial fluid yield a subset of myeloid DCs that preferentially activate Th1 inflammatory-type responses. *J. Immunol.* (Baltimore, Md.: 1950) 167: 1758–1768.

Saric, T., Chang, S.-C., Hattori, A., York, I.A., Markant, S., Rock, K.L., Tsujimoto, M. and Goldberg, A.L. (2002): An IFN-g-induced aminopeptidase in the ER, ERAP1, trims precursors to MHC class I-presented peptides. *Nat. Immunol.* 3: 1169–1176.

Sasaki, M., Hasegawa, H., Kohno, M., Inoue, A., Ito, M.R. and Fujita, S. (2003): Antagonist of secondary lymphoid-tissue chemokine (CCR ligand 21) prevents the development of chronic graft-versus-host disease in mice. *J. Immunol.* 170: 588–596.

Sato, K., Yamashita, N., Baba, M. and Matsuyama, T. (2003a): Modified myeloid dendritic cells act as regulatory dendritic cells to induce anergic and regulatory T cells. *Blood* 101: 3581–3589.

Sato, K., Yamashita, N., Yamashita, N., Baba, M. and Matsuyama, T. (2003b): Regulatory dendritic cells protect mice from murine acute graft-versus-host disease and leukemia relapse. *Immunity* 18: 367–379.

Savary, C.A., Grazziutti, M.L., Melichar, B., Przepiorka, D., Freedman, R.S., Cowart, R.E., Cohen, D.M., Anaissie, E.J., Woodside, D.G., McIntyre, B.W., Pierson, D.L., Pellis, N.R. and Rex, J.H. (1998): Multidimensional flow-cytometric analysis of dendritic cells in peripheral blood of normal donors and cancer patients. *Cancer Immunol., Immunotherapy.* CII 45: 234–240.

Scheinecker, C., Zwolfer, B., Koller, M., Manner, G. and Smolen, J.S. (2001): Alterations of dendritic cells in systemic lupus erythematosus: phenotypic and functional deficiencies. *Arthritis Rheum.* 44: 856–865.

Schiemann, B., Gommerman, J.L., Vora, K., Cachero, T.G., Shulga-Morskaya, S., Dohbles, M., Frew, E. and Scott, M.L. (2001): An essential role for BAFF in the normal development of B cells through a BCMA-independent pathway. *Science* 293: 2111–2114.

Schmidtke, G., Holzhutter, H.-G., Bogyo, M., Kairies, N., Groll, M., De Giuli, R., Emch, S. and Groettrup, M. (1999): How an inhibitor of the HIV-I protease modulates proteasome activity. *J. Biol. Chem.* 274: 35 734–35 740.

Schuler, G., Schuler-Thurner, B. and Steinman, R.M. (2003): The use of dendritic cells in cancer immunotherapy. *Curr. Opin. Immunol.* 15: 138–147.

Schultz, K.R., Bader, S., Paquet, J. and Li, W. (1995): Chloroquine treatment affects T-cell priming to minor histocompatibility antigens and graft-versus-host disease. *Blood* 86: 4344–4352.

Schwarz, M.K. and Wells, T.N.C. (2002): New therapeutics that modulate chemokine networks. *Nat. Rev. Drug Discovery* 1: 347–358.

Seifert, U., Maranon, C., Shmueli, A., Desoutter, J.-F., Wesoloski, L., Janek, K., Henklein, P., Diescher, S., Andrieu, M., de la Salle, H., Weinschenk, T., Schild, H., Laderach, D., Galy, A., Haas, G., Kloetzel, P.-M., Reiss, Y. and Hosmalin, A. (2003): An essential role for tripeptidyl peptidase in the generation of an MHC class I epitope. *Nat. Immunol.* 4: 375–379.

Shen, L., Sigal, L.J., Boes, M. and Rock, K.L. (2004): Important role of cathepsin S in generating peptides for TAP-independent MHC class I crosspresentation in vivo. *Immunity* 21: 155–165.

Shi, G.-P., Villadangos, J.A., Dranoff, G., Small, C., Gu, L., Haley, K.J., Riese, R., Ploegh, H.L. and Chapman, H.A. (1999): Cathepsin S required for normal MHC class II peptide loading and germinal center development. *Immunity* 10: 197–206.

Shinomiya, M., Akbar, S.M.F., Shinomiya, H. and Onji, M. (1999): Transfer of dendritic cells (DC) ex vivo stimulated with interferon-gamma (IFN-g) down-modulates autoimmune diabetes in non-obese diabetic (NOD) mice. *Clin. Exp. Immunol.* 117: 38–43.

Shirahama-Noda, K., Yamamoto, A., Sugihara, K., Hashimoto, N., Asano, M., Nishimura, M. and Hara-Nishimura, I. (2003): Biosynthetic processing of cathepsins and lysosomal degradation are abolished in asparaginyl endopeptidase-deficient mice. *J. Biol. Chem.* 278: 33 194–33 199.

Shuford, W.W., Klussman, K., Tritchler, D.D., Loo, D.T., Chalupny, J., Siadak, A.W., Brown, T.J., Emswiler, J., Raecho, H., Larsen, C.P., Pearson, T.C., Ledbetter, J.A., Aruffo, A. and Mittler, R.S. (1997): 4-1BB Costimulatory signals preferentially induce CD8+ T cell proliferation and lead to the amplification in vivo of cytotoxic T cell responses. *J. Exp. Med.* 186: 47–55.

Sidiropoulos, P.I. and Boumpas, D.T. (2004): Lessons learned from anti-CD40L treatment in systemic lupus erythematosus patients. *Lupus* 13: 391–397.

Silverman, G.J. and Weisman, S. (2003): Rituximab therapy and autoimmune disorders: Prospects for anti-B cell therapy. *Arthritis Rheum.* 48: 1484–1492.

Simon, J.C., Tigelaar, R.E., Bergstresser, P.R., Edelbaum, D. and Cruz, P.D., Jr. (1991): Ultraviolet B radiation converts Langerhans cells from immunogenic to tolerogenic antigen-presenting cells. Induction of specific clonal anergy in

Sligh, J.E., Jr., Ballantyne, C.M., Rich, S.S., Hawkins, H.K., Smith, C.W., Bradley, A. and Beaudet, A.L. (1993): Inflammatory and immune responses are impaired in mice deficient in intercellular adhesion molecule 1. *Proc. Natl. Acad. Sci. U.S.A.* 90: 8529–8533.

Smits, H.H., De Jong, E.C., Schuitemaker, J.H.N., Geijtenbeek, T.B.H., Van Kooyk, Y., Kapsenberg, M.L. and Wierenga, E.A. (2002): Intercellular adhesion molecule-1/LFA-1 ligation favors human Th1 development. *J. Immunol.* 168: 1710–1716.

Sorbera, L.A., Leeson, P.A., Revel, L. and Bayes, M. (2002): Siplizumab: Antipsoriatic, treatment of transplant rejection. *Drugs Future* 27: 558–562.

Steinbrink, K., Graulich, E., Kubsch, S., Knop, J. and Enk, A. (2002): CD4+ and CD8+ anergic T cells induced by interleukin 10-treated human dendritic cells display antigen-specific suppressor activity. *Blood* 99: 2468–2476.

Steinbrink, K., Wolfl, M., Jonuleit, H., Knop, J. and Enk, A.H. (1997): Induction of tolerance by IL-10-treated dendritic cells. *J. Immunol.* 159: 4772–4780.

Steinman, L. (2004): Immune therapy for autoimmune diseases. *Science* 305: 212–216.

Streicher, H.Z., Berkower, I.J., Busch, M., Gurd, F.R. and Berzofsky, J.A. (1984): Antigen conformation determines processing requirements for T-cell activation. *Proc. Natl. Acad. Sci. U.S.A.* 81: 6831–6835.

Stohl, W. (2004): A therapeutic role for BLyS antagonists. *Lupus* 13: 317–322.

Stumbles, P.A., Strickland, D.H., Pimm, C.L., Proksch, S.F., Marsh, A.M., McWilliam, A.S., Bosco, A., Tobagus, I., Thomas, J.A., Napoli, S., Proudfoot, A.E.I., Wells, T.N.C. and Holt, P.G. (2001): Regulation of dendritic cell recruitment into resting and inflamed airway epithelium: use of alternative chemokine receptors as a function of inducing stimulus. *J. Immunol.* 167: 228–234.

Summers, K.L., Daniel, P.B., O'Donnell, J.L. and Hart, D.N. (1995): Dendritic cells in synovial fluid of chronic inflammatory arthritis lack CD80 surface expression. *Clin. Exp. Immunol.* 100: 81–89.

Sun, Y., Chen, H.M., Subudhi, S.K., Chen, J., Koka, R., Chen, L. and Fu, Y.-X. (2002): Costimulatory molecule-targeted antibody therapy of a spontaneous autoimmune disease. *Nat. Med.* 8: 1405–1413.

Sweeney, S.E. and Firestein, G.S. (2004): Signal transduction in rheumatoid arthritis. *Curr. Opin. Rheumatol.* 16: 231–237.

Tajima, K., Amakawa, R., Ito, T., Miyaji, M., Takebayashi, M. and Fukuhara, S. (2003): Immunomodulatory effects of cyclosporin A on human peripheral blood dendritic cell subsets. *Immunology* 108: 321–328.

Takahashi, C., Mittler, R.S. and Vella, A.T. (1999): Cutting edge: 4-1BB is a bona fide CD8 T cell survival signal. *J. Immunol.* 162: 5037–5040.

Takayama, T., Morelli, A.E., Robbins, P.D., Tahara, H. and Thomson, A.W. (2000): Feasibility of CTLA4Ig gene delivery and expression in vivo using retrovirally transduced myeloid dendritic cells that induce alloantigen-specific T cell anergy in vitro. *Gene Therapy* 7: 1265–1273.

Takayama, T., Nishioka, Y., Lu, L., Lotze, M.T., Tahara, H. and Thomson, A.W. (1998): Retroviral delivery of viral interleukin-10 into myeloid dendritic cells markedly inhibits their allostimulatory activity and promotes the induction of T-cell hyporesponsiveness. *Transplantation* 66: 1567–1574.

Takiguchi, M., Murakami, M., Nakagawa, I., Yamada, A., Chikuma, S., Kawaguchi, Y., Hashimoto, A. and Uede, T. (1999): Blockade of CD28/CTLA4-B7 pathway prevented autoantibody-related diseases but not lung disease in MRL/lpr mice. *Lab. Invest.* 79: 317–326.

Tan, J.T., Whitmire, J.K., Murali-Krishna, K., Ahmed, R., Altman, J.D., Mittler, R.S., Sette, A., Pearson, T.C. and Larsen, C.P. (2000): 4-1BB costimulation is required for protective anti-viral immunity after peptide vaccination. *J. Immunol.* 164: 2320–2325.

Tedder, T.F. and Engel, P. (1994): CD20: a regulator of cell-cycle progression of B lymphocytes. *Immunol. Today* 15: 450–454.

Tett, S., Cutler, D. and Day, R. (1990): Antimalarials in rheumatic diseases. *Bailliere's Clin. Rheumatol.* 4: 467–489.

Teunissen, M.B.M., Rongen, H.A.H. and Bos, J.D. (1994): Function of adhesion molecules lymphocyte function-associated antigen-3 and intercellular adhesion molecule-1 on human epidermal Langerhans cells in antigen-specific T cell activation. *J. Immunol.* 152: 3400–3409.

Thompson, J.S., Bixler, S.A., Qian, F., Vora, K., Scott, M.L., Cachero, T.G., Hession, C., Schneider, P., Sizing, I.D., Mullen, C., Strauch, K., Zafari, M., Benjamin, C.D., Tschopp, J., Browning, J.L. and Ambrose, C. (2001): BAFF-R, a newly identified TNF receptor that specifically interacts with BAFF. *Science* 293: 2108–2111.

Thurmond, R.L., Sun, S., Sehon, C.A., Baker, S.M., Cai, H., Gu, Y., Jiang, W., Riley, J.P., Williams, K.N., Edwards, J.P. and Karlsson, L. (2004): Identification of a potent and selective noncovalent cathepsin S inhibitor. *J. Pharmacol. Exp. Therap.* 308: 268–276.

Till, S.J., Jacobson, M.R., O'Brien, F., Durham, S.R., KleinJan, A., Fokkens, W.J., Juliusson, S. and Lowhagen, O. (2001): Recruitment of CD1a + langerhans cells to the nasal mucosa in seasonal allergic rhinitis and effects of topical corticosteroid therapy. *Allergy* (Copenhagen) 56: 126–151.

Trevor K.T., Cover, C., Ruiz Yvette, W., Akporiaye Emmanuel, T., Hersh Evan, M., Landais, D., Taylor Rachel, R., King Alan, D. and Walters Richard, W. (2004): Generation of dendritic cell-tumor cell hybrids by electrofusion for clinical vaccine application. *Cancer Immunol., Immunotherapy:* CII 53: 705–714.

Trevor, K.T., Hersh, E.M., Brailey, J., Balloul, J.-M. and Acres, B. (2001): Transduction of human dendritic cells with a recombinant modified vaccinia Ankara virus encoding MUC1 and IL-2. *Cancer Immunol. Immunotherapy* 50: 397–407.

Tunon-De-Lara, J.M., Redington, A.E., Bradding, P., Church, M.K., Hartley, J.A., Semper, A.E. and Holgate, S.T. (1996): Dendritic cells in normal and asthmatic airways: Expression of the a subunit of the high affinity immunoglobulin E receptor (FceRI-a). *Clin. Exp. Allergy* 26: 648–655.

Turtle, C.J. and Hart, D.N.J. (2004): Dendritic cells in tumor immunology and immunotherapy. *Curr. Drug Targets* 5: 17–39.

Uckun, F.M. and Mao, C. (2004): Tyrosine kinases as new molecular targets in treatment of inflammatory disorders and leukemia. *Curr. Pharmaceut. Design* 10: 1083–1091.

Ulevitch, R.J. (2004): Therapeutics targeting the innate immune system. *Nat. Rev. Immunol.* 4: 512–520.

Umezawa, K. and Chaicharoenpong, C. (2002): Molecular design and biological activities of NF-kB inhibitors. *Mol. Cells* 14: 163–167.

Van Tendeloo, V.F., Snoeck, H.W., Lardon, F., Vanham, G.L., Nijs, G., Lenjou, M., Hendriks, L., Van Broeckhoven, C., Moulijn, A., Rodrigus, I., Verdonk, P., Van Bockstaele, D.R. and Berneman, Z.N. (1998): Nonviral transfection of distinct types of human dendritic cells: high-efficiency gene transfer by electroporation into hematopoietic progenitor- but not monocyte-derived dendritic cells. *Gene Therapy* 5: 700–707.

Van Tendeloo, V.F., Van Broeckhoven, C. and Berneman, Z.N. (2001): Gene-based cancer vaccines: an ex vivo approach. *Leukemia* 15: 545–558.

Varona, R., Villares, R., Carramolino, L., Goya, I., Zaballos, A., Gutierrez, J., Torres, M., Martinez, A.C. and Marquez, G. (2001): CCR6-deficient mice have impaired leukocyte homeostasis and altered contact hypersensitivity and delayed-type hypersensitivity responses. *J. Clin. Investigation* 107: R37–45.

Vihinen, M., Mattsson, P.T. and Smith, C.I.E. (2000): Bruton tyrosine kinase (BTK) in X-linked agammaglobulinemia (XLA). *Frontiers Biosci.* [Electronic Publication] 5: D917–D928.

Villadangos, J.A., Riese, R.J., Peters, C., Chapman, H.A. and Ploegh, H.L. (1997): Degradation of mouse invariant chain: roles of cathepsins S and D and the influence of major histocompatibility complex polymorphism. *J. Exp. Med.* 186: 549–560.

Vinay, D.S., Choi, B.K., Bae, J.S., Kim, W.Y., Gebhardt, B.M. and Kwon, B.S. (2004):

CD137-Deficient mice have reduced NK/NKT cell numbers and function, are resistant to lipopolysaccharide-induced shock syndromes, and have lower IL-4 responses. *J. Immunol.* 173: 4218–4229.

Viney, J.L., Mowat, A.M., O'Malley, J.M., Williamson, E. and Fanger, N.A. (1998): Expanding dendritic cells in vivo enhances the induction of oral tolerance. *J. Immunol.* (Baltimore, Md.: 1950) 160: 5815–5825.

Vollmer, T., Key, L., Durkalski, V., Tyor, W., Corboy, J., Markovic-Plese, S., Preiningerova, J., Rizzo, M. and Singh, I. (2004): Oral simvastatin treatment in relapsing-remitting multiple sclerosis. *Lancet* 363: 1607–1608.

Vulcano, M., Struyf, S., Scapini, P., Cassatella, M., Bernasconi, S., Bonecchi, R., Calleri, A., Penna, G., Adorini, L., Luini, W., Mantovani, A., Van Damme, J. and Sozzani, S. (2003): Unique regulation of CCL18 production by maturing dendritic cells. *J. Immunol.* 170: 3843–3849.

Webb, L.M.C., Walmsley, M.J. and Feldmann, M. (1996): Prevention and amelioration of collagen-induced arthritis by blockade of the CD28 co-stimulatory pathway. Requirement for both B7-1 and B7-2. *Eur. J. Immunol.* 26: 2320–2328.

Weber, S.M., Chen, J.-M. and Levitz, S.M. (2002): Inhibition of mitogen-activated protein kinase signaling by chloroquine. *J. Immunol.* 168: 5303–5309.

Weber, S.M. and Levitz, S.M. (2000): Chloroquine interferes with lipopolysaccharide-induced TNF-alpha gene expression by a nonlysosomotropic mechanism. *J. Immunol.* (Baltimore, Md.: 1950) 165: 1534–1540.

Werther, W.A., Gonzalez, T.N., O'Connor, S.J., McCabe, S., Chan, B., Hotaling, T., Champe, M., Fox, J.A., Jardieu, P.M., Berman, P.W. and Presta, L.G. (1996): Humanization of an anti-lymphocyte function-associated antigen (LFA)-1 monoclonal antibody and reengineering of the humanized antibody for binding to rhesus LFA-1. *J. Immunol.* (Baltimore, Md.: 1950) 157: 4986–4995.

Wingren, A.G., Dahlenborg, K., Bjoerklund, M., Hedlund, G., Kalland, T., Sjoegren, H.O., Ljungdahl, A., Olsson, T., Ekre, H.P. and et al. (1993): Monocyte-regulated IFN-g production in human T cells involves CD2 signaling. *J. Immunol.* 151: 1328–1336.

Woltman, A.M., De Fijter, J.W., Kamerling, S.W.A., Paul, L.C., Daha, M.R. and Van Kooten, C. (2000): The effect of calcineurin inhibitors and corticosteroids on the differentiation of human dendritic cells. *Eur. J. Immunol.* 30: 1807–1812.

Woltman, A.M., De Filter, J.W., Kamerling, S.W.A., Van der Kooij, S.W., Paul, L.C., Daha, M.R. and Van Kooten, C. (2001): Rapamycin induces apoptosis in monocyte- and CD34-derived dendritic cells but not in monocytes and macrophages. *Blood* 98: 174–180.

Woltman, A.M., Massacrier, C., De Fijter, J.W., Caux, C. and Van Kooten, C. (2002): Corticosteroids prevent generation of CD34+-derived dermal dendritic cells but do not inhibit Langerhans cell development. *J. Immunol.* 168: 6181–6188.

Woltman, A.M., van der Kooij, S.W., Coffer, P.J., Offringa, R., Daha, M.R. and van Kooten, C. (2003): Rapamycin specifically interferes with GM-CSF signaling in human dendritic cells, leading to apoptosis via increased p27KIP1 expression. *Blood* 101: 1439–1445.

Xiao, B.G., Huang, Y.M., Yang, J.S., Xu, L.Y. and Link, H. (2001): Bone marrow-derived dendritic cells from experimental allergic encephalomyelitis induce immune tolerance to EAE in Lewis rats. *Clin. Exp. Immunol.* 125: 300–309.

Yacyshyn, B.R., Barish, C., Goff, J., Dalke, D., Gaspari, M., Yu, R., Tami, J., Dorr, F.A. and Sewell, K.L. (2002a): Dose ranging pharmacokinetic trial of high-dose alicaforsen (intercellular adhesion molecule-1 antisense oligodeoxynucleotide) (ISIS 2302) in active Crohn's disease. *Alimentary Pharmacol. Therap.* 16: 1761–1770.

Yacyshyn, B.R., Bowen-Yacyshyn, M.B., Jewell, L., Tami, J.A., Bennett, C.F., Kisner, D.L. and Shanahan, W.R., Jr. (1998): A placebo-controlled trial of ICAM-1 antisense oligonucleotide in the treatment of Crohn's disease. *Gastroenterology* 114: 1133–1142.

Yacyshyn, B.R., Chey, W.Y., Goff, J., Salzberg, B., Baerg, R., Buchman, A.L., Tami, J., Yu, R., Gibiansky, E., Shanahan, W.R., Anderson, F., Koval, G., Barish, C., Safdi, M., Taniguchi, D., Sutherland, L., Rutgeerts, P., Depew, W., Pruitt, R., Hanauer, S., Winston, B., Dolin, B., Koltun, W., McCabe, R., Scholmerich, J., Van Deventer, S., Wild, G., Breiter, J., Burakoff, R., Deren, J., Linne, J., Regueiro, M., Schwartz, H., Shivakumar, B., Binion, D., Cattano, C., Colombel, J., Galandiuk, S., Katz, J., Rustgi, V., Springgate, C., Varilek, G., Dalke, D., Herzog, L., Lamet, M., Pambianco, D., Singleton, J., Torres, E., Van Dullemen, H., Baldassano, R., Cortese, F., James, D., Moses, P., Raedler, A., Riff, D., Stanton, D. and Wilkofsky, S. (2002b): Double blind, placebo controlled trial of the remission inducing and steroid sparing properties of an ICAM-1 antisense oligodeoxynucleotide, alicaforsen (ISIS 2302), in active steroid dependent Crohn's disease. *Gut* 51: 30–36.

Yada, A., Ebihara, S., Matsumura, K., Endo, S., Maeda, T., Nakamura, A., Akiyama, K., Aiba, S. and Takai, T. (2003): Accelerated antigen presentation and elicitation of humoral response in vivo by FcgRIIB- and FcgRI/III-mediated immune complex uptake. *Cell. Immunol.* 225: 21–32.

Yamamoto, Y. and Gaynor, R.B. (2004): IkB kinases: key regulators of the NF-kB pathway. *Trends Biochem. Sci.* 29: 72–79.

Yarilin, D., Duan, R., Huang, Y.M. and Xiao, B.G. (2002): Dendritic cells exposed in vitro to TGF-b1 ameliorate experimental autoimmune myasthenia gravis. *Clin. Exp. Immunol.* 127: 214–219.

Yewdell, J.W. and Hill, A.B. (2002): Viral interference with antigen presentation. *Nat. Immunol.* 3: 1019–1025.

Yilmaz, A., Reiss, C., Tantawi, O., Weng, A., Stumpf, C., Raaz, D., Ludwig, J., Berger, T., Steinkasserer, A., Daniel, W.G. and Garlichs, C.D. (2004): HMG-CoA reductase inhibitors suppress maturation of human dendritic cells: new implications for atherosclerosis. *Atherosclerosis* (Amsterdam) 172: 85–93.

York, I.A., Chang, S.-C., Saric, T., Keys, J.A., Favreau, J.M., Goldberg, A.L. and Rock, K.L. (2002): The ER aminopeptidase ERAP1 enhances or limits antigen presentation by trimming epitopes to 8-9 residues. *Nat. Immunol.* 3: 1177–1184.

Youssef, S., Stueve, O., Patarroyo, J.C., Ruiz, P.J., Radosevich, J.L., Hur, E.M., Bravo, M., Mitchell, D.J., Sobel, R.A., Steinman, L. and Zamvil, S.S. (2002): The HMG-CoA reductase inhibitor, atorvastatin, promotes a Th2 bias and reverses paralysis in central nervous system autoimmune disease. *Nature* 420: 78–84.

Zhang, J., Roschke, V., Baker, K.P., Wang, Z., Alarcon, G.S., Fessler, B.J., Bastian, H., Kimberly, R.P. and Zhou, T. (2001): Cutting edge: a role for B lymphocyte stimulator in systemic lupus erythematosus. *J. Immunol.* 166: 6–10.

Zhang, M. and Coffino, P. (2004): Repeat sequence of Epstein-Barr virus-encoded nuclear antigen 1 protein interrupts proteasome substrate processing. *J. Biol. Chem.* 279: 8635–8641.

Zhang, Q.-H., Link, H. and Xiao, B.-G. (2004): Efficacy of peripheral tolerance induced by dendritic cells is dependent on route of delivery. *J. Autoimmun.* 23: 37–43.

Zhang, X., Gordon, J.R. and Xiang, J. (2002): Advances in dendritic cell-based vaccine of cancer. *Cancer Biotherapy Radiopharmaceut.* 17: 601–619.

Glossary

Activation-Induced Cell Death (Aicd) Apoptotic cell death that is triggered during lymphocyte activation. During a normal immune response, most antigen-specific lymphocytes undergo AICD. It ensures the rapid elimination of effector cells after their antigen-dependent clonal expansion.

Adaptive Immunity Immunity acquired through responses of antigen-specific B and T lymphocytes, resulting in immune memory.

ADCC Antibody-dependent cellular cytotoxicity.

Affinity A measure of the binding of a single antigen to a monovalent antigenic determinant. In contrast, "avidity" refers to combined affinities in multivalent binding, or the summation of multiple affinities, e.g., when a polyvalent antibody binds to a polyvalent antigen. Both terms represent the attraction between two molecules; the higher the affinity and avidity, the higher the probability they will bind and stay bound to one another.

AICD Activation-induced cell death.

AIRE Autoimmune regulator.

Allogeneic Tissues or cells originated from genetically different individuals of a same species. When transplanted, allogeneic material can elicit an immune response in the host, resulting in transplant rejection or graft-versus-host disease.

Alloreactive Responding to antigens that are distinct between members of the same species, such as MHC molecules or blood-group antigens.

Alzheimer's Disease Degenerative mental disease that is characterized by progressive brain deterioration and dementia, and by the presence of senile plaques, neurofibrillary tangles and neuropil threads. Disease onset can occur at any age, and women seem to be affected more frequently than men.

Amastigote Typical adult form in the life cycle of the genus *Leishmania*.

Anchor Motif Amino acid sequence pattern relevant for anchoring T cell epitopes in the peptide binding groove of MHC class I or II molecules.

Anergy   A state of unresponsiveness by T or B cells to their cognate antigens. T cell anergy is induced by stimulation through the T-cell receptor in the absence of ligation of CD28. Upon re-stimulation, these T cells are unable to produce interleukin-2 or proliferate, even in the presence of co-stimulatory signals.

Antibody-Dependent Cellular Cytotoxicity (ADCC)   A mechanism by which natural killer (NK) cells or neutrophils are targeted to IgG-coated cells, resulting in the lysis of the antibody-coated cells. A specific receptor for the constant region of IgG, known as FcγRIII (CD16), is expressed at the surface of NK cells and mediates ADCC.

Antigen Processing   Mechanism by which APCs generate linear peptides derived from antigenic proteins sampled from their external and internal environment to be presented to T cells in the context of MHC molecules.

Antigenic Variation   Changes in the composition, structure or amino acid sequence of antigenic components of pathogens recognized by T or B cells, which allow the microorganism to escape recognition by the adaptive immune system.

Apoptosis   Programmed cell death frequently used to delete unwanted, superfluous or potentially harmful cells, such as those undergoing transformation. Apoptosis involves cell shrinkage, chromatin condensation in the periphery of the nucleus, plasma-membrane blebbing and DNA fragmentation into segments of about 180 base pairs. Eventually, the cell breaks up into many membrane-bound 'apoptotic bodies', which are phagocytosed by neighboring cells.

Autoimmune Regulator (AIRE)   Protein contributing to immunological tolerance, and thus to the prevention of autoimmunity. AIRE is involved in the expression of ectopic proteins such as organ-specific self antigens by medullary thymic epithelial cells. This allows the establishment of central tolerance by negative selection and contributes to the prevention of organ-specific autoimmunity. AIRE-deficient mice and patients with genetic mutations in AIRE develop autoimmune polyendocrinopathy.

Biomarker   Gene or protein that is significantly altered (qualitatively or quantitatively) during the occurrence or the course of a disease so that it is used as an index of disease severity and/or progression of treatment efficacy or as a diagnostic tool.

Birbeck Granules   Membrane-bound, rod- or tennis racquet-shaped inclusions with a central linear, longitudinally striated nucleus, found in the cytoplasm of Langerhans cells. Their function is ill-defined; they appear to represent an antigen loading compartment for some CD1 molecules.

Blood–Brain Barrier   Selectively permeable cellular layer formed by brain microvascular endothelial cells, which are linked by tight junctions. It is crucial for the maintenance of homeostasis in the brain environment.

Bystander Stimulation   The activation and proliferation of cells after exposure to a pathogen in a manner that is independent of their antigenic specificity.

Calnexin Calcium-binding lectin-like protein in the endoplasmic reticulum. Together with calreticulin it acts as a chaperone for newly synthesized proteins and prevents ubiquitinglation and proteasomal degradation.

Calreticulin A calcium storage protein of the endoplasmic reticulum. Together with calnexin it acts as a chaperone for newly synthesized proteins by preventing them from randomly reacting with false polypeptides (e.g. for α-chains of class I HLA proteins).

CD Cluster of differentiation.

$CD4^+CD25^+$ Regulatory T Cells A specialized subset of $CD4^+$ T cells that can suppress the responses of other T cells. Subsets of these T cells are characterized by the expression of the α-chain of the interleukin-2 (IL-2) receptor, known as CD25. In some cases, suppression has been associated with the secretion of IL-10, transforming growth factor-β or both of these cytokines.

Central Tolerance Lack of self-responsiveness that is acquired as lymphoid cells develop. It is generated by the deletion of high-avidity autoreactive clones. For T cells, tolerance induction occurs in the thymus.

Chaperone Molecule that promotes the folding or prevents denaturation of another protein; see calnexin, calreticulin and HLA-DM.

Chemotaxis Movement of a cell along a concentration gradient of chemokines or anaphylatoxins. Chemotaxis is used to recruit immune cells to inflammatory sites or to promote their migration to lymphoid organs in response to chemokines.

Chimeric Antibody Antibody encoded by genes from more than one species, usually with antigen-binding regions from mouse genes and constant regions from human genes.

Choroid Plexus Site of production of cerebrospinal fluid in the adult brain. It is formed by invagination of ependymal cells into the ventricles, which become highly vascularized.

CLIP Class II MHC associated invariant chain (Ii) peptide that occupies the peptide-binding site of class II MHC molecules during folding in the endoplasmic reticulum and prevents binding of endogenous peptides. CLIP is removed in the MIIC vesicle by HLA-DM so that exogenous peptide can bind.

Cluster of Differentiation (CD) Set of surface markers on cells. They can be detected by means of the reaction with specific monoclonal antibodies. CD nomenclature is accepted by WHO (World Health Organization) and, at present, includes more than 300 proteins.

Congenic Genetically identical animal strains except for one allelic difference that does not correspond to an antigen able to elicit an immune response upon tissue transfer or transplantation from one strain to another.

Co-Stimulatory Signals Signals to a T cell provided by interaction with either a soluble or a membrane-bound molecule (such as CD80 and CD86) that have little or no effect alone, but either enhance or modify the physiological effect of the primary signal, which is mediated by engagement of the T-cell receptor.

CpG Sequences Oligodeoxynucleotide sequences that include a cytosine-guanosine sequence and certain flanking nucleotides. They induce innate immune responses through interaction with Toll-like receptor 9.

Cross Presentation Ability to present exogenous antigens on MHC class I molecules to CD8+ T lymphocytes instead of via the MHC II pathway. Dendritic cells and LSEC are very efficient cross-presenting cells.

Cross Talk Bidirectional exchange of information and signals between two cell types. It might involve signals that are mediated by cell–cell contact or by soluble factors, such as cytokines.

cSMAC Immunological synapse.

Danger Signals Agents that alert the innate immune system to danger associated with microbial invaders (exogenous danger signals) or with damaged cells (endogenous danger signals). Cell-wall components and other products of pathogens alert to the presence of potentially harmful invaders, usually by interacting with Toll-like receptors and other pattern-recognition receptors that are expressed by tissue cells or cells of the immune system, such as dendritic cells.

Darwinian Microenvironment Term used in the field of tumor immunology to describe local conditions in tissues where tumor cells proliferate, e.g., dominance of TH2 versus TH1 cytokine status.

Defensins A family of proteins exhibiting bactericidal properties. Defensins are small (2–6 kDa) cationic microbicidal peptides that participate in innate immunity. They are secreted by immune cells (particularly neutrophils), intestinal Paneth cells and epithelial cells. Their tertiary structures are stabilized by intradisulfide bridges and clusters of positively charged amino acids, which resemble those found in chemokines. Several defensins have chemotactic activity for leukocytes.

Demyelination Damage to the myelin sheath surrounding nerves in the brain and spinal cord, which affects the function of the nerves involved. It occurs in multiple sclerosis, a chronic disease of the nervous system affecting young and middle-aged adults, and in experimental autoimmune encephalomyelitis, which is a mouse model of multiple sclerosis.

Diapedesis The last step in the leukocyte–endothelial adhesion cascade. The cascade includes tethering, triggering, tight adhesion and transmigration. Diapedesis is the migration of leukocytes across the endothelium, which occurs by squeezing through the junctions between adjacent endothelial cells.

EAE Experimental allergic encephalomyelitis.

ECM   Extracellular matrix.

Editing   Quality control mechanism during antigenic peptide loading onto MHC molecules. Both editors, tapasin for MHC I and HLA-DM for MHC II molecules, favor binding of high-stability peptide ligands by inducing removal of low-stability ligands.

Endosomal Processing Pathway   Pathway for processing phagocytosed or endocytosed (exogenous) antigens for presentation on class II MHC molecules.

Exosomes   Small vesicles derived from endosomal multi-vesicular bodies that are secreted by APCs. They contain high concentrations of MHC molecules and tetraspanins. The physiological relevance of exosome secretion by APCs is still unknown.

Experimental Allergic (Autoimmune) Encephalomyelitis (EAE)   An experimental model of human multiple sclerosis that is induced in susceptible animals by immunization with myelin-derived antigens. EAE can be induced in various mammalian species, including mice. Several different mouse models have been established that develop either self-limiting (monophasic) or recurring (relapsing–remitting) disease. The animals develop an autoimmune paralytic disease with inflammation and demyelination in the brain and spinal cord.

Extracellular Matrix (ECM)   The complex, multi-molecular material that surrounds cells. The ECM comprises a scaffold on which tissues are organized, it provides cellular microenvironments and regulates various cellular functions.

Fc Fragment   The crystallizable fragment of an immunoglobulin molecule, formed by limited proteolysis. It represents the C-end moiety of both heavy chains connected by disulfide bonds and is responsible for binding of immunoglobulins to Fc receptors.

Fc Receptors (FcR)   Binding sites for Fc domains of immunoglobulins, expressed on the surface of leukocytes and a few other cell types. Each of the immunoglobulin classes has its specific FcR type. They are involved in phagocytosis and antibody-dependent cell-mediated cytotoxicity.

FDC   Follicular dendritic cell.

Fetal Thymic Organ Culture (FTOC)   Removal of fetal thymi between embryonic day 14 and 16 allows the analysis of several key processes in thymic development – including antigen-driven positive and negative selection events – using in vitro culture. Thymic lobes can also be used to allow the development of progenitor cells that are added to the cultures.

Fluorescence Resonance Energy Transfer (FRET)   Technology used to measure protein–protein interactions microscopically or by a FACS (fluorescence-activated cell sorter)-based method. Proteins fused to cyan, yellow or red, fluorescent dyes are assessed for interaction by measuring the energy transfer between fluorophores, which can only occur if proteins physically interact. FRET can also be used to examine the activation state of certain proteins if their activation results in specific protein–protein interactions.

Focal Adhesion Plaque The closest contact site of a cell with its environment, which is formed by integrin clustering. The integrins link the extracellular environment to the actin cytoskeleton by a complex assembly of adaptor proteins.

Follicular Dc (FDC) Cell with a dendritic morphology that is present in lymph nodes. These cells display on their surface intact antigens that are held in immune complexes. B cells present in the lymph node can interact with these antigens. FDCs are of non-haematopoietic origin and are functionally unrelated to dendritic cells.

FRET Fluorescence resonance energy transfer.

FTOC Fetal thymic organ culture.

α-Galcer–CD1d Tetramers Tetrameric complexes of CD1d molecules bound to α-galactosylceramide (α-GalCer), which have sufficient affinity for the T-cell receptor of invariant natural killer T cells to allow the detection of these cells by flow cytometry.

Gamma-IFN-Inducible Lysosomal Thiol Reductase (GILT) Enzyme responsible for the reduction of disulfide bonds in late endocytic compartments of APCs during antigen processing.

GILT γ-IFN-inducible lysosomal thiol reductase.

Graft-Versus-Host-Disease (GvHD) An immune response mounted against the recipient of an allograft by immunocompetent T cells that are derived from the graft. Typically, it is seen in the context of allogeneic bone-marrow transplantation. The extent of tissue damage and the severity of GvHD vary markedly, but it can be life threatening in severe cases and commonly affects the intestines, liver and skin.

GvHD Graft-versus-host-disease.

Hapten Small molecule that is immunogenic only when covalently linked to a carrier protein.

HEL Hen egg-white lysozyme.

Hen Egg-white Lysozyme (HEL) Protein classically used as foreign antigen for immunization studies in mice.

HEV High endothelial venules.

High Endothelial Venules (HEVs) Small veins that join capillaries to larger veins. They have a high-walled endothelium and are present in the paracortex of lymph nodes and tonsils, as well as in the interfollicular areas of Peyer's patches. HEVs are essential for entry of lymphocytes from the blood stream into secondary lymphoid organs.

HLA-DM Non-classical MHC class II molecule that facilitates removal of CLIP and binding of exogenous peptides to class II MHC molecules in the MIIC vesicle. Apart from that, it functions as a chaperone and peptide editor in the MHC II pathway.

HLA-DO Non-classical MHC class II molecule that is only expressed in B cells and thymic epithelial cells. It is bound to HLA-DM and facilitates antigenic peptide loading in deep endocytic compartments where immune complexes are processed.

IDO Indoleamin-2,3-dioxygenase.

Ii Invariant chain.

Immune Surveillance The constant sampling of external body surfaces by which the immune system detects antigen entry and conveys danger signals. This process ensures the body's integrity by discriminating between invasive pathogens and innocuous food antigens or commensal bacteria.

Immunodominant Epitopes Antigenic regions of a protein that are preferentially recognized by B or T cell receptors during an adaptive immune response.

Immunological Synapse A stable region of contact between a T cell and an antigen presenting cell that forms initially through cell–cell interaction of adhesion molecules. The mature immunological synapse contains two distinct membrane domains: a central cluster of T-cell receptors bound to MHC-peptide complexes, known as the central supramolecular activation cluster (cSMAC), and a surrounding ring of adhesion molecules known as the peripheral SMAC (pSMAC).

Immunoreceptor Tyrosine-based Activation Motif (ITAM) Region in cytoplasmic domains of cell-surface immune receptors, such as the T-cell receptor, the B-cell receptor, the receptor for IgE (FceR) and natural-killer-cell activating receptors. Following cell activation, the phosphorylated ITAM functions as docking site for SRC homology 2 (SH2)-domain-containing tyrosine kinases and adaptor molecules, thereby facilitating intracellular signaling cascades.

Indoleamin-2,3-Dioxygenase (IDO) Enzyme generating tryptophan metabolites that are thought to induce T cell deletion.

Innate Immunity Immunity present since birth and not dependent on prior antigen exposure. Innate immunity includes physical and chemical barriers to infection, phagocytes, complement, and Natural Killer cells.

Invariant Chain (Ii) Chaperone dedicated to MHC class II molecules. It trimerizes and forms nonameric complexes with MHC II in the ER. It directs MHC II to endosomal/lysosomal compartments and prevents premature antigenic peptide loading by occupying the MHC II antigen binding cleft.

ITAM Immunoreceptor tyrosine-based activation motif.

ITIM Immunoreceptor tyrosine-based inhibition motif.

Kupffer Cells Hepatic macrophages attached to liver sinusoids and specialized in phagocytosis of gut-derived IgA-coated bacteria circulating in portal venous blood.

Langerhans Cells Professional antigen-presenting dendritic cells that are localized in the skin epidermis.

Lipid Raft Membrane area, preferentially in the plasma membrane, that is rich in cholesterol, glycosphingolipids, several signaling proteins (such as SRC kinases, RAS, LAT and PAG) and glycosylphosphatidylinositol-anchored proteins. Also known as glycolipid-enriched membrane domains (GEMs) and detergent-insoluble glycolipid-enriched membranes (DIGs).

Lipoglycan Mainly microbial glycans that are acylated, such as lipoarabinomannan (LAM), recognized bt Toll-like receptors. LAM mediates entry of mycobacteria into macrophages.

Lipopolysaccharide (LPS) Cell envelope component of Gram-negative bacteria, denoted as "endotoxin". It consists of lipid A and a polysaccharide component. If released in high concentrations within the body, it may induce sepsis.

Liver Sinusoidal Endothelial Cells (LSEC) APC population lining the hepatic sinusoid and specialized in scavenging blood-borne macromolecular antigens. LSEC have been shown to promote immune tolerance by cross presentation.

LPS Lipopolysaccharide.

LSEC Liver sinusoidal epithelial cell.

Lysosome Endocytic organelle that is characterized by a low internal pH. It contains hydrolytic enzymes and is involved in the post-translational maturation of proteins, the degradation of receptors and the extracellular release of active enzymes. In APCs, lysosomes function as antigen processing compartments, denoted as MIICs.

Macrophage-DC Functional Unit (MDU) Concept of innate/adaptive immune surveillance generating tolerance in the steady state, with macrophages residing immobile in tissues and dendritic cells being highly mobile, connecting tissues with lymphoid organs.

Macropinocytosis Fluid-phase endocytosis for the uptake of large amounts of extracellular fluid; inducible in macrophages and constitutive in immature dendritic cells.

MALT (Mucosal-associated lymphoid tissue). Collections of lymphoid cells found along the mucus membranes of the respiratory (NALT), digestive (GALT), and urogenital tracts.

Matrix Metalloproteinases Family of tightly controlled peptide hydrolases that use a metal ion for their catalytic mechanism. These enzymes degrade the extracellular matrix in processes of cell migration that occur during embryogenesis, wound healing, inflammation and tumor dissemination.

MDU   Macrophage-DC functional unit.

Membrane Microdomaine   Array of locally organized membrane proteins and subsets of phospholipids and cholesterol. Lipid rafts and tetraspan microdomains are typical representatives that facilitate antigen processing and presentation in APCs.

Meninges   Surrounding membranes of the brain and spinal cord. There are three layers of meninges: the dura mater (outer), the arachnoid membrane (middle) and the pia mater (inner). Leptomeninges is the collective name for the arachnoid and the pia mater membranes.

MHC Tetramers   Reagents composed of four peptide–MHC complexes linked by biotinylation, which can be fluorescently labeled and used to detect antigen-specific T cells by flow cytometry.

Microfold (M) Cells   Specialized antigen-sampling cells that are located in the follicle-associated epithelium of the organized mucosa-associated lymphoid tissues. M cells deliver antigens by trans-epithelial vesicular transport from the aero-digestive lumen directly to the subepithelial lymphoid tissues of nasopharynx-associated lymphoid tissue and Peyer's patches.

Microglia   CNS-resident glial cells sharing phenotypical and lineage traits of bone marrow-derived monocytes/macrophages. Microglia promote neuronal and oligodendrocyte survival by producing growth factor and neurotrophins and under pathological conditions become capable of phagocytosis and antigen presentation.

MIIC   MHC class II compartment. Endocytic vesicle of antigen-presenting cells where class II MHC molecules bind processed exogenous antigen.

Mixed Lymphocyte Reaction (MLR).   Proliferative T cell response induced by T cell exposure to inactivated MHC-mismatched antigen presenting stimulator cells. The MLR is used for in vitro testing of histocompatibility.

MLR   Mixed lymphocyte reaction.

Mononuclear Phagocytic System   Group of bone-marrow-derived cells with different morphologies (monocytes, macrophages and dendritic cells), which are mainly responsible for phagocytosis, cytokine secretion and antigen presentation.

Multiple Sclerosis   Neurodegenerative disorder that is characterized by demyelination of bundles of nerve fibres in the central nervous system. Symptoms depend on the site of the lesion but include sensory loss, weakness in leg muscles, speech difficulties, loss of coordination and dizziness.

Multi-Vesicular Body (MVB)   Late endosome that contains several internal vesicles, denoted as "exosomes" after their extracellular secretion.

MVB   Multi-vesicular body.

Natural Killer T Cells (NKT Cells)   A heterogeneous subset of T cells, most of which express semi-invariant T-cell receptors and are restricted by the non-classical MHC class Ib molecule CD1d. These cells respond to the antigens α-galactosylceramide and glycerol-phosphatidylinositol in mice and have important functions in immunity against infections and malignancies. In mice, NKT cells were first identified by their expression of the cell-surface molecule NK1.1. The most abundant subset of NKT cells has a rearrangement of the T cell receptor (TCR) variable-gene segment Vα14 to the joining-region segment Jα18. The resulting TCR is known as Vα14 invariant (Vα14i). This TCR is autoreactive to CD1d, and Vα14i NKT cells respond strongly to α-galactosylceramide presented in the context of CD1d.

Negative Selection   The deletion of self-reactive thymocytes in the thymus. Thymocytes expressing T-cell receptors that strongly recognize self-peptide bound to self-MHC undergo apoptosis in response to the signaling generated by high-affinity binding.

NKT Cells   Natural killer T cells.

Non-obese Diabetic Mice (NOD mice)   Mice of the NOD strain spontaneously develop a form of autoimmunity that closely resembles human type 1 diabetes. Prevalence of disease is higher in female than male animals and in mice presenting a defect in NKT cells.

Non-Professional Antigen-Presenting Cells   Cells that can be induced to express molecules essential for antigen presentation, such as MHC class II molecules and the invariant chain. These cells often lack expression of co-stimulatory molecules, such as CD80 or CD86.

Opsonization   Literally means "preparation for eating". The coating of a bacterium with antibody and/or complement that leads to enhanced phagocytosis of the bacterium by phagocytic cells.

Oxidative Burst   Phenomenon observed with activated macrophages, neutrophilic granulocytes or microglial cells, characterized by the massive secretion of oxygen radicals in order to kill pathogens extracellularly.

PAMP   Pathogen-associated molecular pattern.

Pathogen-Associated Molecular Patterns (PAMPs)   Molecular patterns that are characteristic of prokaryotes and can, thereby, activate the mammalian innate immune system. Examples include terminally mannosylated and poly-mannosylated compounds, which bind the mannose receptor, and various microbial products, such as bacterial lipopolysaccharides, CpG-motif-containing DNA, flagellin and double-stranded RNA. PAMPs are thought to be recognized mainly by Toll-like receptors.

Pattern-recognition Receptor (PRR)   Host receptors, such as Toll-like receptors, that can sense pathogen-associated molecular patterns (PAMPs) and initiate signaling cascades (involving the activation of nuclear factor-κB) that lead to an innate immune response.

Peripheral Tolerance   Potentially autoreactive T cells that have escaped negative selection in the thymus (central tolerance) can be deleted or anergized by several mechanisms in the periphery. Deletion can be mediated by high-affinity T-cell receptor (TCR) cross linking or by CD95–CD95L-mediated apoptosis. Anergy can occur when incomplete activation signals are sent through the TCR (low-affinity interactions) or when there is a lack of co-stimulation during activation. Suppression can be mediated by cytokines produced by regulatory T cells.

Peyer's Patches   Organized lymphoid structures associated with the digestive tract in the small intestine, underlying M cells. Peyer's patches consist of a T-cell zone surrounding a B-cell zone, similar to germinal centres in lymph nodes.

Plasmacytoid DCs   Immature dendritic cells with a plasmacytoid morphology, resembling plasmablasts. They produce type I interferons in response to viral infection.

Programmed Cell Death   A common form of cell death, which is also known as apoptosis. Many physiological and developmental stimuli cause apoptosis, and this mechanism is frequently used to delete unwanted, superfluous or potentially harmful cells, such as those undergoing transformation. Apoptosis involves cell shrinkage, chromatin condensation in the periphery of the nucleus, plasma-membrane blebbing and DNA fragmentation. Eventually, the cell breaks up into many membrane-bound 'apoptotic bodies', which are phagocytosed by neighboring cells.

Proteasome   Cytoplasmic multisubunit proteases that process (cuts into peptides) most of the cytosolic and nuclear proteins in eukaryotic cells for transport and display on membrane class I MHC. Targeting of proteins to proteasomes most often occurs through the attachment of multiple ubiquitin tags.

PRR   Pattern-recognition receptor.

pSMAC   Immunological synapse.

RAG   (Recombination activating gene). Recombination-activating genes (Rag1 and Rag2) are expressed by developing lymphocytes and encode for the recombinase that is required for somatic recombination of both BCR and TCR genes. This process of rearrangement via cutting and splicing DNA segments ensures the production of functional BCR or TCR genes.

Receptor Editing   A molecular process that involves secondary rearrangements (mostly of the immunoglobulin light chains) that replace existing immunoglobulin molecules and generate a new antigen receptor with altered specificity.

SCID   Severe combined immunodeficiency. Disease where innate but no adaptive immune responses are possible.

Severe Combined Immunodeficiency (SCID)   Humans or mice with this rare genetic disorder lack functional T and B cells owing to a mutation in a gene that is involved in T cell and/or B cell development. Several forms of SCID have been described, including mutations in the common cytokine-receptor γ-chain of several interleukin receptors, Janus activated kinase 3 (JAK3) and adenosine deaminase.

SOCS Suppressor of cytokine signaling molecules.

SRC Family A group of structurally related cytoplasmic and/or membrane-associated enzymes that are named after the prototypical member, SRC. In haematopoietic cells, SRC kinases – such as LCK, FYN and LYN – are the first protein tyrosine kinases that are activated upon immunreceptor stimulation. They phosphorylate ITAMs that are present in the signal-transducing subunits of the immunoreceptors, thereby providing binding sites for SRC homology 2 (SH2)-domain-containing molecules, such as SYK.

Superantigen Molecule that circumvents the classical rules of MHC restriction in activation of T cells, thus activating many different clones of T cells, by bridging MHC class II molecules and TCR outside the normal binding site and irrespective of the groove occupancy.

Suppressor of Cytokine Signaling Molecules (SOCS) Intracellular proteins that are thought to block intracellular signal transduction from cytokine and hormone receptors.

Systemic Adjuvants Substances that help initiate a robust immune response. Typically, adjuvants contain a mixture of substances that mimic an active infection, such as bacterial cell-wall components, to simulate danger signals and emulsifiers to allow for the slow release of antigen.

TAP (Transporter associated with antigen processing). Heterodimeric transmembrane protein that shuttles peptides of variable lengths from the cytoplasm into the lumen of the endoplasmic reticulum, where they can bind class I MHC molecules. The transport is driven by ATP binding and hydrolysis.

Tapasin Chaperone and peptide editor in the endoplasmic reticulum that engages into complexes with calreticulin/calnexin, TAP, class I/β2m heterodimers and ERp57, thereby facilitating peptide loading onto class I MHC molecules.

Tetanus Toxin A neurotoxin released by *Clostridium tetani* spores. Epitopes derived thereof stand out for their high degree of promiscuity with regard to binding to human MHC II molecules.

Tetraspanins Superfamily of proteins characterized by four transmembrane domains. Tetraspanins cluster in membrane microdomains and associate with integrins and MHC molecules. They are expressed by a multitude of cells, including APCs and T cells, where they contribute to the stability of the immune synapse.

Tissue Tropism Homing of immune cells into particular tissues, controlled by the expression of specific adhesion molecules.

TLR Toll-like receptor.

Toll-like Receptor (TLR) A member of a family of receptors that recognize conserved motifs unique to microorganisms that are known as pathogen-associated molecular patterns (PAMPs). For example, TLR4 recognizes bacterial lipopolysaccharide (LPS), and TLR5 recognizes bacterial flagellin. TLR-mediated events signal to the host that a microbial pathogen has been encountered. Endogenous mammalian proteins – such as heat-shock proteins, DNA and extracellular-matrix components – which are characteristic of damaged tissues and typical of necrotic tumors, metastases, autoimmune diseases and infections, are also believed to activate TLRs.

Transcytosis   Transport of material across an epithelial layer by uptake on one side of the epithelial cell into a coated vesicle that might then be sorted through the trans-Golgi network and transported to the opposite side of the cell.

Transdifferentiation   Conversion of one type of APC into another, e.g., microglial cells or macrophages can become dendritic cells in the presence of GM-CSF and IL-4.

V(D)J Recombination   Somatic rearrangement of variable (V), diversity (D) and joining (J) regions of the genes that encode antigen receptors, leading to repertoire diversity of both T cell and B cell receptors.

Zymosan   Insoluble β-glucan carbohydrate derived from the cell wall of yeast.

Index

a

abatacept   554
ABC (ATP binding cassette)   60
actin cytoskeleton   93, 207, 214
activation induced apoptosis   275
activation induced cell death   345
active immunotherapy   182
acyl sulfotrehalose   134, 136
adaptive immunity   89, 255
adaptor protein (AP)   141 f
adenosine   299
adhesion molecule   113
adhesion receptor   200 ff
adjuvant   222 f, 235, 338, 545
AEP, see asparaginyl endopeptidase
agalactosyl IgG   95
agonistic peptide   111 ff
AIDS (acquired immune deficiency syndrome)   280
– opportunistic infection   279
AIDs dementia   280
AIRE   535
airway hyperresponsiveness (AHR)   359
alcohol induced liver disease   285
alicaforsen   557
allergic inflammation   360
allergic response   290
allergy   257, 543
allograft   442
– rejection   542, 554
– survival   555
allorecognition   542
alum (aluminium hydroxide)   223, 545
alveolar epithelium   286
alveolar macrophage   115, 261
Alzheimer's disease   355, 423, 455
aminopeptidase   52, 55, 71
amyloid β peptide   295
amyloid plaque   455

Anaplasma phagoctophilum   354
anchor   64
anchor motif   166, 169, 173
anchor residue   5, 53, 103, 106, 167
anergy   512
ankylosing spondylitis   173
anterior chamber associated immune deviation (ACAID)   352
anti-CD20 monoclonal   485, 487
anti-dsDNA antibody   465
anti-fungal resistance   430
anti-GpPI Ig   484
anti-inflammatory drug   544
anti-phospholipid antibody   368
antigen   4, 93
– capture   93
– unfolding   4
– binding groove   132
– depot   223
antigen processing   3 ff, 89, 99 ff
– major histocompatibility complex class I (MHC I)   53 ff
– major histocompatibility complex class II (MHC II)   89 ff
antigen processing machinery   53
antigen processing pathway   94
antigen uptake   93
antigenic peptide   64, 256
antiinflamatory TLR   294
antiinflammatory phenotype   261, 276
antiphospholipid antibody   370
AP 1 (activating protein-1)   203, 232
APC depletion   560
APC survival   558 f
apoptosis   34, 275, 301, 332, 377, 427, 533
– granulocyte   427
apoptotic cell   94, 256, 267
arachidonic acid   430
arginase   272, 291, 343 f

arthritis   104, 553, 555, 559
– adjuvant induced   104
– CIA   555, 559
– collagen induced   104, 553
arthritogenic antigen   461, 485
asparaginyl endopeptidase (AEP)   101, 421, 552
asthma   357 ff, 543
– pathogenesis   290 f
– IgE   360
atherosclerosis   284
atorvastatin   546
autoantibody   363 f, 368 ff, 461 ff
– anti-nuclear   373
autoantigen   362, 366
autoimmune disease   173, 182, 189, 542
autoimmune hepatitis   189
autoimmune susceptibility gene   373
autoimmune thyroid disease   374
autoimmunity   15, 257, 267, 334, 362, 376 f
– innate   376
– protective   377 f
autologous peptide   160
autoreactive hybridomas   467
auxiliary binding site   10

b

B7   202, 428, 514
B cell   75, 90 ff, 373, 463, 466, 473 f, 482 f, 560
– ablative therapy   488
– abnormality   481
– autoreactive   373, 466
– CD20   560
– checkpoint   474
– development   472
– elimination   487
– epitope   102
– homeostatis   475
– immune function   483
– longevity   474
– rheumatoid arthritis   482
– signal transduction   473
– therapeutic target   482
B cell receptor (BCR)   91, 108, 464, 467, 476
B cell repertoire   465
B lymphocyte stimulator   559
bacterial DNA   228
bacterial lipid   92
bafilomycin   98
Bap31   55, 210
Bap31 cargo receptor   66
bare lymphocyte syndrome (BLS)   425
BCG (Bacillus Calmette-Guérin)   235, 264

binding register   10
BiP   55, 65
Birbeck granule   91, 96, 141, 146
bleomycin hydrolase (BH)   55, 59
blood brain barrier (BBB)   442, 455
bone marrow chimera   534
bortezomib   551
breast carcinoma   561
brefeldin A   73 f, 76
Brugia malaya   343
Bruton's tyrosine kinase (BTK)   559
bystander activation   338 f

c

calnexin (CNX)   55, 63, 68, 103, 131
calreticulin (CRT)   55, 63, 65, 67, 131
cAMP   250, 263, 299, 346
cancer   543
cancer vaccine   190, 561
candida albicans   292
cardic allograft   505
cardiolipin   466
cartilage glycoprotein gp39   107
caspase 1   298
caspase 8   346
catalyst   105
β catenin   31
cathepsins   75, 76, 98 ff, 144, 250, 420 ff, 442 f
– Cat B   101
– Cat D   73, 101
– Cat E   101
– Cat F   104
– Cat L   104
– Cat S   77, 94, 98, 99, 101, 104, 421, 553
Cat S inhibitor   553
CCL20   30
CCR1   558
CCR2   35
CCR3   356
CCR5   280
CCR6   30, 32, 506, 558
CCR7   262 ff, 350, 359, 506, 558
CCR9   516
CD1   129 ff, 148 f, 267, 352
– antigen uptake   146
– biosynthesis   131
– trafficking   141 ff
– endosomal loading   148
– expression   131
– genes and classification   129
– ligand   134
– newly synthesized   131

- non endosomal loading 149
- pocket 135
- tissue distribution 140
- X-ray crystal structure 132
CD2 557
CD4$^+$ T cell counts 282
CD4$^+$8$^+$ thymocyte 534
CD8$^+$ T cell 51
CD9 109 ff
CD11b 33, 347
CD11c 450, 508
CD14 226, 342
CD18 34
CD19 476
CD20 560
CD21 477
CD22 479, 560
CD28 514, 555
CD34 291, 532
CD37 109 ff
CD38 264
CD40 42, 265, 554
CD45 446 ff
CD53 109 ff
CD54 42
CD63 109 ff
CD69 36
CD80 42, 428 f, 554
CD81 94, 109 ff
CD82 94, 109 ff
CD86 42, 94, 270, 428 ff
CD95 512
CD106 42
CD123 508
CD200 448
CD205 525
CD206 95
CD209 95
cdc2 cyclin-dependent protein kinase 203
CDR3 469
cell adhesion 555 ff
- anti-CD2 557
- anti-CD48 557
- anti-DC SIGN 557
- anti-ICAM 556
- anti-LFA 3 557
central nervous system (CNS) 441 f
ceramide 345
cervical lymph node 443
CG motif (CpG) 228
chaperone 55, 63, 64, 102 ff, 131, 183
chemokine 230, 339, 356, 431, 543
chemokine receptor 262, 557

chemotactic stimuli 415 f
chemotaxis 431, 557
Chlamydia pneumoniae 277
Chlamydia trachomatis 115
chloroquine 12, 76, 98, 148, 486, 546
cholera toxin 342
CIIV (class II MHC containing vesicles) 113
citrullinated peptide 461 f
citrullination 190
CLIP 94, 98, 104, 108, 111, 175, 183 ff, 188 f, 552
- antagonist of T$_H$1 cell 188
- HLA-DR3-CLIP crystal 185
- N terminal residues 185
- promiscuous binding 184
- self release 184
- self release mechanism 104
- sequence 175
class II associated invariant chain peptide (CLIP) 94, 98, 104, 108, 111, 175, 188 ff, 552
- antagonist of T$_H$1 cell 188
- HLA-DR3-CLIP crystal 185
- N terminal residues 185
- promiscuous binding 184
- self release 184
- self release mechanism 104
- sequence 175
class II MHC transactivator CIITA 101, 114
classical 508
clathrin coated pit 95
clathrin mediated endocytosis 141 f
claudin-1 31
CLIP, see class II associated invariant chain peptide
clonal deletion 335
clonal expansion 27, 507
CNS 91
- microenvironment 448
- pathology 456
co-chaperone 108
co-stimulation 201 f, 274, 553 f
- ICAM-1 201
co-stimulatory molecule 92 ff, 110, 112, 213, 427 f, 527, 554
- anti-CD40L 554
- CD40 / CD40L interaction 554
- CD80 554
- CD137 / CD137L interaction 555
- CD28 / B7 interaction 555
- CTLA4 Ig fusion protein 554
- dendritic cells 277

– negative co-stimulation 428
– positive co stimulation 428
– transport vesicle 210
– TLR signaling 230, 269
collagen induced arthritis (CIA) 478, 553
collectin 267
colorectal carcinoma 561
combinatorial peptide library 62
commensal bacteria 33
complement activation 349
complement cascade 221 f
complement receptor (CR) 419
complementarity determining region (CDRs) 463 f
complete Freund's adjuvant (CFA) 15, 223, 507
concanamycin A 148
concerted immune response 275
confocal microscopy 111
contact sensitization 36
COP-I vesicle 66
COP-II transport vesicle 66
20S core complex 56
corticosteroid 543 ff
COX 2 263, 281, 291, 299, 345 f, 366
COX 2 inhibitor 291
CpG 350
CpG DNA 92, 269
CpG mediated signal 486
CpG motif 235
CpG oligodeoxynucleotide (ODN) 549
CpG oligonucleotide 528
C1q 349
Crohn's disease 263, 270, 557
cross presentation 16, 73 ff, 161, 341, 423, 515 f, 528, 550
– TAP dependent 73
– TAP independent 74, 76, 553
– tolerance 34, 43, 510
cross priming 76, 515
cross tolerance 511, 515 f
cruzipain 344
cSMAC 201, 214
CTL epitope 53, 57
CTLA 4, see also cytotoxic T lymphocyte associated Ag 430, 514
CTLA4-Ig 514, 564
C type lectin 95 f, 115, 146, 267
CX_3CL1 31 f
CX_3CL1 (fractalkine) 31
CX_3CR1 31 ff
CXCR5 558
cyclosporin A (CysA) 545

cystatin C 98, 174
cytochalasin B 207
cytochalasin D 214, 423
cytochrome c 275
cytokine 339 f, 431
– anti-inflammatory 339
– suppresive 340
cytokine network 367
cytokine profile 338
cytoskeleton 214
cytosolic peptidase 59
cytotoxic lymphocyte 51
cytotoxic T lymphocyte associated antigen 4 (CTLA-4) 430, 514

d
danger 96
danger model 355
danger signal 354, 368
Darwinian microenvironment 291
DC, (dentritic cell) 28, 90 f, 333 ff, 504 ff, 510 ff, 523 ff
– adjuvant 507
– CD1d expressing 352
– $CD8^+$ dendritic cell 37 ff, 75, 511, 525
– $CD11^+c$ dendritic cell 37
– classical 426, 508
– conventional 525
– crosstalk 290
– dermal dendritic cell 36 f
– discovery 503
– dynamic of T/DC interaction 512
– hepatic dendritic cell 39
– homing 505
– immature 92
– interstitial DC 91
– maturation 263, 293 ff, 504
– mature 92
– monocyte derived 91, 111, 141, 357
– myeloid 338, 366
– neutrophil derived 426
– Peyer's patches 30, 32
– plasmacytoid (pDC) 39, 91, 338, 366, 508, 511, 525, 548
– semi mature 340, 505
– skin-derived dendritic cell 36
– tolerogenic 338, 362, 510, 563
– thymic 523 f
– trafficking 506
DC based cancer vaccine 561
DC based immunotherapy 517, 561
DC LAMP 529
DC precursor 289

DC SIGN (DC specific ICAM-3-Grabbing
    nonintegrin) (CD 209)   95 f, 115, 269, 273,
    557
– HIV   34 f, 281, 283
death by neglect   533
death receptor   348
DEC 205   29, 39, 95 f, 450, 510
decoy receptor   300
dectin 1   234, 267 f
defective ribosomal product (DRiPs)   52
defensin   29, 222
demyelination   423
dendritic cell marker   450
dendritic morphology   289
diabetes   16, 138, 542
dideoxymycobactin   136
dilysine motif   145
disease modifying antirheumatic drug
    (DMARD)   546
disulfide reduction   97
DM, see also HLA-DM   4, 13, 94 f
dome region   29
double positive thymocyte   16
draining lymph node   17, 289, 453, 504
Drosophila melanogaster   247, 253 ff
– acute phase response   253
– innate response   253
– macrophage function   255
– TNF/TNFR system   254
dsDNA   267
dsRNA   228, 269, 347, 350
dsRNA binding protein (DRBPs)   230
DTH response   558

e
E cadherin   31, 91
E selectin   203
EAE, see experimental autoimmune
    encephalomyelitis
early endosome   14
ectopic expression   179
Edman degradation   166, 172
efalizumab   556
empty MHC II molecule   105
endocytic compartment   140
endocytosis   73, 92, 95 ff, 552
endogenous alarm   355
endolysosomal hydrolase   73
endosomal compartment   108, 267
endosomal / lysosomal protease   100
endosomal tubulation   214
endothelial layer   259
endothelial like splenic stromal cell   511

endotoxin   42, 334
enlinmomab   556
entry receptor   35
eosinophil   339, 343, 356
eotaxin   356, 374
epithelial barrier   32
epithelial cell   93, 182
epithelial membrane protein 2 (EMP2)   213
epitope   177, 179, 182
– annexin II   182
– gp100   182
– HEL   5 ff
– HSA   5, 177
– β_2-m   75, 179
epratuzumab   560
Epstein-Barr virus (EBV)   63, 115, 190
ER (endoplasmic reticulum)   52 ff
ER like phagosome   516
ER, phagosome fusion   75
ER-resident aminopeptidase (ERAP1)   55, 71,
    550 f
ER signal sequence   63
ERp57   55, 63, 65 ff
ERp72   63
exogenous antigen   470
exosome   31, 515
experimental allergic encephalomyelitis   138
experimental autoimmune encephalomyelitis
    (EAE)   106, 449, 554
export of MHC II   109
extracellular fluid   93
extracellular proteolytic enzyme   422
ezrin   207

f
Fas mediated apoptosis   298
FasR   369
Fc receptor (FcR)   95 f, 444
Fcγ receptor   342, 369, 419
FcγRIIb   363, 369, 478
FcγRIII   369
fibrinogen   284
fibroblast   90
first line of defense   417
FK506 (tacrolimus)   545
flagellin   227
flanking residue   6, 185
Flt3 ligand (Flt3L)   562
fluid phase endocytosis   29, 73
fluorescent resonance energy transfer
    (FRET)   200, 208
follicle-derived CD4$^+$ T cell   485
follicular dendritic cell   29, 482

fractalkine   448
FRET experiment   209

g

α galactosylceramide (αGalCer)   139, 149, 351 ff, 362, 509
galiximab   554
ganglioside   134
gastrointestinal epithelial cell   28
gastrointestinal tract   28 ff
gelatinase A   423
gelatinase B   423
germinal center (GC)   108
germinal center reaction   471
germline gene   469
GILT (Gamma-interferon-inducible lysosomal thiol reductase)   94, 97
glatiramer acetate (GA)   547
glial cell   90
glomerulonephritis   365, 368
glucocorticoid   299, 346
glucose-6-phosphate isomerase (G6PI)   484
glutamate decarboxylase   107
glutathione   97
glycolipid antigen   129, 146, 351
glycosyl phosphatidylinositol (GPI)   139, 207
glycosylation   190
GM-CSF   256, 258, 278, 288, 346, 356, 358 f, 421 f
GM1 glycolipid   213
GM2 activator protein   148
gp96   63
gram negative bacteria   226
gram positive bacteria   227
granulocyte lineage   426
granzyme   369
Graves disease   189
green fluorescent protein (GFP)   445
gut-associated lymphoid tissue (GALT)   516
GVHD   555

h

H2-DM, see also HLA-DM   13, 422
haematopoietic stem cell   530
HCMV (Human Cytomegalovirus)   115, 551
heat shock protein   52, 96, 222
HEL (Hen egg-white lysosome)   4 ff, 106 ff, 165, 185
Helicobacter pylori   116
helminth infection   374
HepaDNA virus   44
hepatic fibrosis   285
hepatic sinusoid   39, 42, 257

hepatic stellate cell   40
Hepatitis B   44
Hepatitis B virus   72
Hepatitis C virus (HCV)   44
hepatocyte   284
Her2/neu   185
Hermansky-Pudlak syndrome type 2   142
Herpes simplex virus (HSV)   36, 115, 229
$\alpha\beta$ Ii heterotrimer   103
high endothelial venule   29, 263
hindering residue   5, 10
histone deacetylase   372
HIV (Human Immunodeficiency virus)   59, 63, 95, 190, 279, 557
HLA-DM   94, 98, 104 ff, 183 f, 422, 553
– catalyst   105
– chaperone   105, 183
– CLIP release   184
– conformational editor   106
– peptide editor   14, 72, 106, 110, 210
HLA-DO   107 ff
HLA-DR   281
Human cytomegalovirus (HCMV)   115, 551
Human immunodeficiency virus (HIV)   34 ff, 115
human X-linked agammaglobulinemia   559
humoral imprinting   484
hydrogen bond   64
hydrophobicity algorithm   60
hypergammaglobulinemia   481

i

IκB   232
ICAM 1 (Intercellular adhesion molecule 1)   200 ff, 206, 211, 556
– association with MHC I   211
– co-stimulatory molecule   201
– immunological synapse   201
– leukocyte adhesion   206
– lipid raft   206
– signaling   203
ICAM 2   95, 557
ICAM 3   95, 557
ICOS   555
ICP47   67
IDO, see also indoleamin-2,3-dioxygenase   281, 512
IFN-α 366 ff, 542
IFN-β 296, 362, 547
IFN-γ 42 f, 60, 66, 71, 114, 256, 296, 421 f
Igα (CD79a)   476
Igβ (CD79b)   476
IgA   29, 33, 41

IgE 359
IgG2a / IgG1 ratio 368 f
IgH gene 532
Ii (invariant chain) 420 f
IL-1β 92, 357 ff
IL-2 43, 274
IL-2 receptor 211
IL-3 259
IL-3Rα (CD123) 529
IL-4 39, 43, 339
IL-6 252, 259, 266, 343
IL-8 203, 270
IL-10 30, 39, 43, 258, 290, 298, 335, 371, 505
IL-12 30, 265, 332 f, 338
IL-12 / IL 10 ratio 364
IL-12p70 528 f
IL-13 344, 356
IL-15 260, 338, 339, 365, 375
IL-18 365, 375
IL-23 375, 454
imidazoquinoline 229, 235
immature DCs 258, 261, 264, 335
immune complexes (ICs) 363
immune escape 35
immune mediated inflammatory diesese (IMID) 541
immune privilege 348, 441 f
immune surveillance 27 ff, 138, 337
immunodominance 18, 57
immunoediting 291
immunogenicity 223
immunoglobulin 342
immunological synapse (IS) 11 f, 186, 199, 211, 556
immunoproteasome 52, 56, 60, 71
immunoreceptor tyrosine based activation motif (ITAM) 234 f, 476
immunoreceptor tyrosine based inhibitory motif (ITIM) 203, 477
immunosuppressive therapie 466, 544
incompletes Freund's adjuvant (IFA) 223
indoleamin-2,3-dioxygenase (IDO) 281, 512
infectious microorganism 27 ff
inflammation 334, 355, 430
inflammatory bowel disease (IBD) 374
inflammatory cytokine 504
inflammatory foci 276
inflammatory mediator 507
inflammatory reaction 481
inflammatory signal 90, 92
Influenza virus 166, 170, 229
inhibitory receptor 482
innate immunity 247, 253, 264, 332, 368

innate tolerance 332 f, 342
iNOS 260 ff, 271 ff, 343 ff, 433 f
– arginase ratio 344
insulin dependent diabetes mellitus (IDDM) 362, 478
insulin receptor 211
interfollicular T cell zone 30
interleukin 1 converting enzyme (ICE) 234
intestinal tract 28
intracellular pathogen 51
intracellular signaling 559 f
invariant chain (Ii) 98, 102 f, 144, 168
– sorting signal 103
IP-30 97
IRAK (IL-1R-associated kinase) 232, 267, 549
irradiated bone marrow chimeric rats and mice 444
islet antigen 374

j
JAK / STAT pathway 297
joint destruction 484
juvenile dermatomyositis 189

k
keratinocyte 90 f
kinetic stability 107
Kupffer cell 39, 41, 91, 257, 284

l
L-RAP aminopeptidase 73
L-SIGN 42 ff
lactacystin 53, 73 f, 76
lactoferrin 425 f
lamina propria 31
Langerhans cell (LC) 35, 91, 96, 141, 259, 504 f
langerin (CD207) 91, 96, 141, 146
laser scanning microscopy 513
Lassa virus 348
late endosome 14, 144
lectin pathway 349
legumain, *see also* asparaginyl endopeptidase (AEP) 99, 552
Leishmania 271, 350
Leishmania amazonensis 115
Leishmania donovani 342
Leishmania major 433, 506
length variant 165
leptin 262
leucine aminopeptidase (LAP) 55, 59
leucine rich motif 226
leucine rich repeat (LRRs) 234

leukocyte adhesion deficiency (LAD)   556
leukocyte trafficking   556
leupeptin   103
LFA-1 (lymphocyte function associated antigen-1)   200 ff
LFA-2   556
LFA-3   557
lipid A   226, 268
lipid antigen   129
lipid raft   112, 186, 204, 206, 208 ff, 267
– cholesterol extraction   209
– ICAM-1   206
– variation   209
lipid transport protein (LTPs)   148
lipoarabinomannan (ManLAM)   115, 136, 227, 276
lipopeptide   92
lipopeptide antigen   129
lipopeptide P$_3$CSK$_4$ 549
lipophosphoglycan (LPG)   271
lipopolysaccharide (LPS)   92, 113, 226, 269, 342
– tolerance   293
lipoprotein   268
liposome   4, 14, 236, 545
– editing function   14
Listeria monocytogenes   3, 52, 270
liver   38 ff
liver injury   285 f
liver sinusoidal endothelial cell (LSEC)   39 ff, 42 ff
LMP2   56, 60
LMP7   56, 60
local microenvironment   255
lung eosionophilia   288
lung infection   290
lung mucosa   286
lupus antigen   367
lupus nephritis   369
Ly6C   508, 527
lymph node   27, 32, 35
lymphoid follicle   29
lymphoid organ   91, 256
lymphoid precursor   531
lymphoid tissue   262, 337
lymphotoxin   482
lysosomal compartment   94, 144
lysosomal marker   109
lysosomal storage disease   552

m

M cell   29 ff, 32
M-CSF   258 f, 290
macropinocytosis   73, 75, 418
macrophage   73, 75, 90 ff, 289 f, 331 ff, 341 ff, 426 f
– antiinflammatory   333, 341
– CD36   248
– depletion   289
– differentiation   258 ff
– HIV-infected   280
– LPS tolerant   343
– microglia   257, 444
– nematode elicited   345
– phagocytosis   368
– phagosome   277
– phenotypic diversity   257, 332
– phosphatidylserine receptor   248
– proinflammatory   341
– proinflammatory macrophage   332
– receptor   276, 342
– scavenger receptor   248
– transcriptosome   277
– transdifferentiation   260
– tumor associated   290
– vitronectin   248
macrophage activating factor (MAF)   431
macrophage DC functional unit   335 f
macropinocytosis   29, 65, 75 f, 93
MAL / TIRAP (MyD88 adaptor like / TIR associated protein)   231
ManLAM   276
mannose receptor (CD206)   96
mannose receptor (MR)   42, 95, 146, 267
mannosyl phosphomycoketide   136
MAPPs (MHC associated peptide proteomics)   176 ff, 190 f
margin of safety   16
mass spectrometry   4, 51, 166, 170, 172
– electrospray ionization tandem   4, 172
– matrix assisted laser desorption   170
matrix metalloproteinase (MMP)   360, 422
MECL-1   56
medullary zone   533
melanoma   517, 561
melanoma antigen   182, 190
membrane microdomain   105, 109
memory cell   507
memory T cell   289
mesenteric lymph node   30 ff
methotrexate   560
MHC class I   52, 166, 211
– association with ICAM 1   211
– self peptide   166
MHC class II   256, 278, 424, 448
– anchor residue   169

– autoantibodies   461
– chaperones   183
– conformational change   105
– Edman microsequencing   168, 172
– empty molecule   105
– IFNγ induced   448
– peptide binding groove   102
– peptide loading in the ER   183
– self peptides   167 ff, 175
– X-ray crystal structure   102, 160 ff, 171
MHC class II compartments (MIICs)   75, 112 ff, 141 ff
MHC clustering   208
MHC guided processing   72, 100
MHC restriciton   103, 166
Michaelis-Menten kinetic   105
microbial challenge   339
microbial infection   265, 339
microbial product   504
microbial receptor   269
microbial signal   30
microdomain   210
microenvironment   331 ff, 373
microenvironmental instruction   373
microglia   91 ff, 257, 377, 441 ff, 454
– activation   377, 446
– antigen presenting cell   449
– CD45 expression   446
– cultured   449 f
– differentiation   447
– gene expression   446
– heterogeneity   454
– mesodermal origin   444
– parenchymal   444
– perivascular   444
– transdifferentiation   261
β_2 microglobulin (β_2m)   55, 64, 130, 145
microsomal triglyceride transfer protein (MTP)   149
migration inhibition factor (MIF)   343, 345 ff, 431
MIP 3α 506
MIP 3β 506
misfolded protein   75
missing self   349
mitogen activated protein kinase (MAPK)   203, 232
mixed lymphcyte reaction (MLR)   504, 515, 529
MOG   454
monensin   98
monoclonal antibody   89, 111, 168
monocyte   258, 332

monocyte derived DC   145
monophosphoryl lipid A (MPL)   236, 546
Mopeia virus   348
α MSH   448
mucins   29
mucosa associated lymphoid tissue (MALT)   29
multicatalytic protease complex   97
multidimensional protein indentification technology (MudPIT)   172, 176
multiple myeloma   561
multiple sclerosis (MS)   101, 362, 554
multivesicular body (MVBs)   112
multivesicular compartment   107
multivesicular endosome   109
Murine leukemia virus (MuLV)   168
myasthenia gravis (MG)   375
Mycobacteria   223, 276 f
mycobacterial infection   137
mycobacterial lipoarabinomannan (LAM)   146
Mycobacterium tuberculosis (MTB)   114 ff, 137, 265, 282, 339, 433
mycolate   134, 136
mycolylarabiongalactan peptidoglycan   114
mycoplasma lipopeptide   227
MyD88 (myeloid differential factor 88)   113, 231 f, 267 f, 549
myelin   450
myelin basic protein (MBP)   15, 101, 105
myeloid DC   96, 140, 260
myeloid hormone   531
myeloid lineage   456
myeloid precursor   531
myofibroblast   285

n

N-formylated methionine leucine phenylalanine peptide (fMLP)   420
N-glycanase   72
N-terminal trimming   71
natural killer T cell (NKT cell)   130 f, 138, 145, 351 ff
naturally processed self peptide   159 f
necrosis   265
Nef protein   59
negative selection   15, 168, 523, 533
neuron   448
neuropeptide   448
neurotoxin   377
neurotrophin   91, 448
neutrophil   339, 347, 432
neutrophil activation   427

neutrophil elastase   273
NFκB   252 ff, 260, 288, 345 f, 358, 549, 559
nitric oxide   430
NK cell   273, 332, 349 ff, 509
NKG2D   213
NKT cell, *see also* natural killer T cell   130 f, 138, 145, 351 ff
– CD4⁻   352
– CD 8⁻   352
– CD14⁺   352
– proinflammatory burst   353
Nod (nucleotide binding oligomerization domain)   233
NOD mice   373
non-Hodgkin's lymphoma (NHL)   560 f
non-small lung cell cancer   561

o

obesity   262
ornithine decarboxylase   273
osteoclast   260
osteopontin   339 f
ovalbumin   59, 76

p

p38 MAPK   93, 264, 272, 297, 549, 559
PA28 regulator   55 f
Pam3cys   268 f, 294
pancreatic beta cell   90
pathogen associated molecular pattern (PAMPs)   221 f, 224
pattern recognition receptor (PRRs)   42, 91, 252, 334, 504, 548 f
PDI (Protein disulfide isomerase)   63, 97
peptide binding motif   174
peptide library   167, 170, 173
peptide loading   69, 104
peptide loading complex (PLC)   63, 65, 66 f
– calnexin   64
– calreticulin   65
– ERp57   65
– tapasin   66
peptide MHC complex   11, 212
– conformational isomer   11
– type A and B   11
peptide processing   71
peptide regurgitation   76
peptide repertoire   72
peptide transporter   52
peptidoglycan (PGN)   227, 252, 268, 270
peripheral blood B cell   107
peripheral blood leukocytes   415
peripheral blood monocyte   428

peripheral deletion   512
peripheral tissue   27
peripheral tolerance   178, 503 ff, 512
– immature   505
peritoneal macrophage   261
Peyer's patch (PP)   29, 508, 558
PGE_2 (prostaglandin E2)   40, 263, 285 f, 290, 340 f, 346
phagocytic cell   3 ff, 261
phagocytic clearance   415 f
phagocytosis   29, 42, 73, 94, 221 f, 418
phagolysosome   34, 94
phagosome   75, 94
phosphatidylinositol   134
phosphatidylserine receptor   343
phosphoinositol phosphatase   477
pinocytosis   516
PKC-Θ   214
PKR (IFN inducible dsRNA dependent protein kinase)   234
plasma cell   466, 471
plasmacytoid DC   96, 260
plasmodium spp.   38, 41
plasticity   509
platelet activating factor (PAF)   430
PMA   214
PMN (*see also* polymorphonuclear neutrophilic)   417 f, 425 ff, 433
– antigen presenting cell   417
– antigen uptake   418
– cytokine secretion   433
– neutrophil activation   427
– neutrophil turnover   427
– precursor   425
podosome   93
polarization of T cell responses   91
Poliovirus   29
poly I : C   228
polyamine   344
polymorphism   66, 103
polymorphonuclear neutrophil (PMN)   415, 420, 424, 432 f
– antigen processing   420
– cell mediated immunity   432
– co-stimulatory molecule   424
– cytokine secretion   433
– expression of MHC class I/II   424
pool sequencing   167, 173
positive selection   168, 210, 533
post-translational modification   12, 51, 190
posttraumatic immunosuppression   344
precursor polypeptide   99
primary immunodeficiency   463

proinflammatory cytokine 284, 543
proinflammatory liver damage 301
proinflammatory stimuli 337
proinflammatory T cell 485
prostate carcinoma 561
prostate specific antigen (PSA) 420
protease 6, 250
– amino 6
– carboxypeptidase 6
– peptidase 6
proteasome 57 f, 75, 97, 515
– amino acid recycling 58
– cleavage specificity 57 f
– inhibitor 52 ff, 549 f
26S proteasome 56
protein disulfide isomerase (PDI) 97
protein tyrosine phosphatase 477, 480 ff
– SHIP 480 f
protein unfolding 97
proteolytic milieu 98
proteomics 51, 89
protozoon parasite 342
pseudodimer model 165
pSMAC 201
psoriasis 554, 556 f
puromycin sensitive aminopeptidase (PSA)
   55, 59

q
quality control 65
quinacrine 546

r
RANK 30
RANTES 203
rapamycin (sirolimus) 545
receptor clustering 208
receptor editing 467 f
receptor-mediated endocytosis 42, 65, 73,
   95 ff, 418
recycling endosome 141
redox state 97
19S regulatory complex 57
regulatory T cell 266, 513, 543, 563
– CD25$^+$ 513
– CD4$^+$ 513
– Th3 513
– Tr1 513
renal cell carcinoma 561
repetitive surfaces 222
respiratory burst 271
reticulo-endothelial system 42
retrograde transport 67, 113

reversed phase high performance liquid
   chromatography (RP-HPLC) 164 ff, 172
rheumatoid arthritis (RA) 189 f, 279, 432,
   542, 554
rheumatoid factor (RF) 469, 484
rheumatoid joint 292, 364
rheumatoid synovitis 485
ritonavir 551
rituximab 487, 560

s
Salmonella typhimurium 31, 34
saposin C 148
scavenger receptor (CD36) 42, 94, 276
Schistosoma mansoni (SM) 285
Sec61 71 ff
second line of defense 41
second signal 427
secondary lymphoid organ 230, 504
self peptide 161 ff, 170, 176 ff, 186
– acidic peptide elution 173
– length variant 165, 170
– medullary thymic epithelial cell 179
– monocyte derived DC 176
– myeloid dendritic cell 178
– origin 162, 170
– peripheral blood mononuclear cell 177
– sequence analysis 165
– splenocyte 181
– tetraspan microdomain 186
– tumor cell 182
self peptide epitope 175
self-reactive antibody 470
sentinels of the body 92, 441
sepsis 344
septic shock 293, 295
serine protease 273
serum amyloid A 252
shared epitope 462
Shigella 29
silent invader 354
Sjögrens syndrome 542, 553
SLE (*see also* systemic lupus erythematosus)
   363, 466 ff, 542, 554 f, 559
– anti-DNA antibody 470
– autoantibody 466
– IgV gene repertoire 466
– receptor editing 467
– V gene 468
SOCS 297 f
somatic hypermutation 464 f
sorting signal 103
specificity pocket 102, 167

spleen 38 ff
Src homology 2 (SH2) domain 203
ssRNA 229
Staphlococcus aureus 348
STAT1 (signal transducer and activator of transcription-1) 233
STAT3 298 f
STAT6 343 f, 372
statin 546
Streptococcus pneumoniae 99
substance P 448
sulfasalazine 486
sulfatide 134
superantigen 103
superdimers 111
supramolecular activation cluster 113
supramolecular assemblies 109
synovial fibroblast 341
synovial fluid macrophage 292
systemic lupus erythematosus (SLE) 363, 461 ff, 542, 554 f, 559

t
TACI 559
TAK1 (transforming growth factor β activated kinase) 232
TAP 67
tapasin 55, 61, 63, 66 ff
– chaperone 68
– TAP1 bridge 68
Tat protein 63
T/B cell aggregate 485
T cell 335, 338 f, 351, 369, 376
– alloreactive 351
– antigen non-specific 339
– antiinflammatory 335
– apoptosis 341, 369
– memory 338
– regulatory 335, 339, 376
T cell area 504
T cell epitope 102
T cell polarization 509
T cell proliferation 289
T cell receptor repertoire 464
T cell recirculation 516
T cell repertoire 260
T cell selection 523
tetanus toxoid (TT) 170, 421
tetraspan microdomain 107 ff, 110 ff, 185 f
tetraspan superfamily (TM4SF) 109, 209
tetraspan web 186
TGF-β 40, 248 ff, 263 ff, 290, 333 ff, 354, 505
Th1 43, 111, 189, 432, 449, 505, 510

Th1 polarization 339, 508
TH1/Th2 ratio 281, 509
Th2 30, 111 ff, 189, 432, 448
Theiler's virus 449
therapeutic MoAbs 489
therapeutic vaccine 223
thimet oligopeptidase (TOP) 55, 59
Thy 1 524
thymic cortex 111 ff
thymic cortical epithelium 533
thymic DC 523 ff, 530 ff
– cytokine production 528
– development of pDC 532
– differentiation 531
– lifespan 530
– maturation 527
– origin 530
– plasmacytoid 527
thymic epithelial cell 90, 101, 168, 175 ff
thymus 523, 529, 533
– cortex 529, 533
– cortico medullary junction 523, 529
– medulla 523, 529, 533
thyroid epithelial cell 90
TIR (Toll interleukin-1 receptor homologous region) 231
tissue distribution 140
tissue healing 344
tissue microenvironment 340
T lineage thymocyte 524, 530
TM4SF microdomain 212
TNF-α 42, 92, 259, 275
tolerance 32 ff, 41 ff, 93 ff, 101, 181, 331 ff, 371
– adaptive 334
– innate 334, 342, 371
– mucosally induced 334
– self peptide 181
– transplant 334
tolerogenicity 338
Toll family 224, 250
Toll / IL-1R (TIR) 226
Toll like receptor (TLR) 93 ff, 96, 214, 224 ff, 420
– adjuvant 545 f
– antagonist 546
– ligand 38, 225, 235 f, 252
– septic shock 549
– therapeutic potential 235
– TLR protein family 224
– TLR signaling 226, 230, 549
– TLR1 268
– TLR2 227, 276, 286, 549

- TLR3   228, 269
- TLR4   226, 254, 268, 284, 286, 548
- TLR5   227
- TLR6   268
- TLR7   229, 366
- TLR8   229 , 366
- TLR9   229, 267 ff, 286, 361 ff, 366, 486, 529, 546 ff
- TLR11   228
- TLR12   228
- TLR13   228

TPP II   55
TRAF6 (TNF receptor associated factor 6)   232
TRAM (Toll receptor associated molecule)   231
trans differentiation   426
transplantation   542
transporter associated with antigen presentation (TAP)   53 ff, 60 ff, 515, 550
- cytotoxic response   515
- transplant   515
- tumor cell   515
trauma   344
T regulatory (Tr) cell   564
TREM 2   454
TRIF (Toll receptor associated activator of interferon)   231
tripeptidyl peptidase II (TPP II)   55, 59, 550
Trojan peptide   63
trypanosoma cruzi   342, 351
tryptophan catabolism   512, 514
tryptophan depletion   341
tubular transport   112
tubule derived vesicle   113
tumor antigen   190, 561
tumor immunology   291

type A T cell   15
type B T cell   15
type 1 interferon   228, 508, 528 f
type 1 regulatory T cell (Tr1)   505
tyrosine based motif   141

u

ubiquitin   52, 56, 75
ulcerative colitis   361
UVB exposure   367

v

$V\beta$ gene usage   138
vaccine   222 f
van der Waal's interaction   134
vascular endothelial cell   90
Vesicular stomatitis virus (VSV)   165, 229
video microscopy   113
VIP   448
viral ds RNA   92
viral tropism   282
virosome   236
virus infection   51
virus-like particle   223, 236
vitronectin   280

w

web   110
Wegener's granulomatosis   425
wound healing   252

y

Ym1   345

z

zymosan   268
Zwitterionic polysaccharide (ZPS)   99